Lecture Notes in Artificial Intelligence 4632

Edited by J. G. Carbonell and J. Siekmann

Subseries of Lecture Notes in Computer Science

T0180367

Reda Alhajj   Hong Gao   Xue Li
Jianzhong Li   Osmar R. Zaïane (Eds.)

Advanced Data Mining and Applications

Third International Conference, ADMA 2007
Harbin, China, August 6-8, 2007
Proceedings

 Springer

Series Editors

Jaime G. Carbonell, Carnegie Mellon University, Pittsburgh, PA, USA
Jörg Siekmann, University of Saarland, Saarbrücken, Germany

Volume Editors

Reda Alhajj
University of Calgary, Computer Science Department
Calgary, AB, Canada
E-mail: alhajj@cpsc.ucalgary.ca

Hong Gao
Jianzhong Li
Harbin Institute of Technology, School of Computer Science and Technology
Harbin, China
E-mail: {honggao, lijzh}@hit.edu.cn

Xue Li
The University of Queensland
School of Information Technology and Electronic Engineering
Queensland, Australia
E-mail: xueli@itee.uq.edu.au

Osmar R. Zaïane
University of Alberta, Department of Computing Science
Edmonton, AB, Canada
E-mail: zaiane@cs.ualberta.ca

Library of Congress Control Number: 2007931453

CR Subject Classification (1998): I.2, H.2.8, H.3-4, K.4.4, J.3, I.4, J.1

LNCS Sublibrary: SL 7 – Artificial Intelligence

ISSN   0302-9743
ISBN-10   3-540-73870-3 Springer Berlin Heidelberg New York
ISBN-13   978-3-540-73870-1 Springer Berlin Heidelberg New York

Springer is a part of Springer Science+Business Media

springer.com

© Springer-Verlag Berlin Heidelberg 2007
Printed in Germany

Typesetting: Camera-ready by author, data conversion by Scientific Publishing Services, Chennai, India
Printed on acid-free paper   SPIN: 12098095   06/3180   5 4 3 2 1 0

Preface

The Third International Conference on Advanced Data Mining and Applications (ADMA) organized in Harbin, China continued the tradition already established by the first two ADMA conferences in Wuhan in 2005 and Xi'an in 2006. One major goal of ADMA is to create a respectable identity in the data mining research community. This feat has been partially achieved in a very short time despite the young age of the conference, thanks to the rigorous review process insisted upon, the outstanding list of internationally renowned keynote speakers and the excellent program each year. The impact of a conference is measured by the citations the conference papers receive. Some have used this measure to rank conferences. For example, the independent source cs-conference-ranking.org ranks ADMA (0.65) higher than PAKDD (0.64) and PKDD (0.62) as of June 2007, which are well established conferences in data mining. While the ranking itself is questionable because the exact procedure is not disclosed, it is nevertheless an encouraging indicator of recognition for a very young conference such as ADMA.

This year we had the pleasure and honour to host illustrious presenters. Our distinguished keynote speakers were Prof. Jaideep Srivastava from Minnesota, who had an enviable career in both industry and academia, and Prof. Geoff Web from Monash University, who is currently the editor-in-chief of the famous journal on Data Mining and Knowledge Discovery published by Springer Netherlands. Our invited speaker is Prof. Zhi-Hua Zhou, a rising star in the Chinese and international research community. Despite his young career, he has rightfully earned global recognition.

ADMA aims at bringing together researchers and practitioners to focus on advancements in data mining and peculiarities and challenges of real world applications using data mining. The major theme of the conference encompasses the innovative applications of data mining approaches to real-world problems that involve large data sets, incomplete and noisy data, or demand optimal solutions. While researchers are eager to submit their novel ideas and techniques covering the first part of the theme of the conference, *advanced data mining*, getting researchers and practitioners to share their experience with applications of data mining involving different knowledge of data in challenging application domains remains limited. This second part of the conference theme, *applications*, is important, as applications are very inspirational and educational, and the organizers continue to encourage the use of ADMA as a venue for sharing such valuable experiences with the data mining community.

This year ADMA received about 200 online submissions from 20 different countries, making it, yet again, a truly international conference. A rigorous process of pre-screening and review involved 92 well known international program committee members and 3 program co-chairs in addition to numerous external reviewers. This screening process yielded the remarkable papers organized in these proceedings in 44 regular papers and 15 short papers, bearing a total acceptance rate of 29%.

We witnessed this year a significant drop in the number of submissions and many authors chose to withdraw their papers after they were accepted. The main reason for this phenomenon is the fact that Thompson Scientific decided to take Springer's LNCS series out of the ISI journal citation index as of January 2007 arguing that LNCS publications are not journals. In fact, according to their recently published results for 2006, ISI evaluated a total of 25,576 articles published in all 373 journals they classify as computer science journals, whereas LNCS alone published roughly the same number of articles during the same period of time. However, LNCS together with its subseries LNAI and LNBI (like other proceedings from other publishers) will from 2007 onward be included in ISI's newly established Proceedings Index and, of course, will continue to be included in many other well known and frequently used bibliographic indices such as DBLP or CompuServe. Finally, despite the fact that LNCS is no longer covered by ISI/SCI, the standing and recognition of ADMA among the community remains unchanged.

June 2007

Reda Alhajj
Hong Gao
Xue Li
Jianzhong Li
Osmar R. Zaïane

Organization

ADMA 2007 was organized by Harbin Institute of Technology, China.

Organizers

General Co-chairs

Jianzhong Li   Harbin Institute of Technology, China
Osmar R. Zaïane   University of Alberta, Canada

Program Co-chairs

Reda Alhajj   University of Calgary, Canada
Hong Gao   Harbin Institute of Technology, China
Xue Li   Queensland University, Australia

Local Arrangement Co-chairs

Shengfei Shi   Harbin Institute of Technology, China
Lv Tianyang   Harbin Engineering University, China

Publicity Co-chairs

Haiwei Pan   Harbin Engineering University, China
Gang Li   Deakin Unversity, Australia

Finance Co-chairs

Jizhou Luo   Harbin Institute of Technology, China
Randy Goebel   Informatics Circle of Research Excellence, Canada

Registration Chair

Hongzhi Wang   Harbin Institute of Technology, China

Web Master

Osmar R. Zaïane   University of Alberta, Canada

Program Committee

Adam Krzyzak	Concordia University, Montreal, Canada
Ah-Hwee Tan	Nanyang Technological University, Singapore
Alfredo Cuzzocrea	University of Calabria, Italy
Alipio M. Jorge	University of Porto, Portugal
Andre Ponce Leao	University S.Paulo, Brazil
Andrew Kusiak	University of Iowa, USA
Ashkan Sami	Shiraz University, Iran
Arthur Tay	National University of Singapore, Singapore
Bart Goethals	University of Antwerp, Belgium
Carlos Soares	LIACC/Fac. of Economics, University of Porto, Portugal
Carlos Soares	University of Porto, Portugal
Cesar Rego	University of Mississippi, USA
Christophe Giraud-Carrier	Brigham Young University, USA
Daniel C. Neagu	University of Bradford, UK
Deepak S Padmanabhan	IBM India Research Lab, India
Desheng Dash Wu	University of Toronto, Canada
Dimitrios Katsaros	Aristotle University, Greece
Eamonn Keogh	University of California - Riverside, USA
Ee-Peng Lim	Nanyang Technological University, Singapore
Elena Baralis	Politecnico di Torino, Italy
Ezeife Christie	University of Windsor, Canada
Faruk Polat	Middle East Technical University, Turkey
Fernando Berzal	University of Granada, Spain
Francesco Bonchi	KDD Laboratory – ISTI CNR Pisa, Italy
Frans Coenen	The University of Liverpool, UK
Gang Li	Deakin University, Australia
Giovanni Semeraro	University of Bari, Italy
Giuseppe Manco	National Research Council of Italy, Italy
Grigorios Tsoumakas	Aristotle University, Greece
Guoren Wang	NorthEast University, China
Haiwei Pan	Harbin Engineering University, China
Hassan Abolhassani	Sharif University of Technology, Iran
Heng Tao Shen	University of Queensland, Australia
Jaideep Srivastava	University of Minnesota, USA
James Bailey	University of Melbourne, Australia
Jan Rauch	University of Economics, Prague, Czech Republic
Jean-Gabriel G Ganascia	LIP6 - University of Paris, France
Jeffrey Xu Yu	The Chinese University of Hong Kong, Hong Kong, China
Jeremy Besson	Insa-Lyon, France
Jimmy Huang	York University, Canada
Jing Liu	Xidian University, China
JingTao Yao	University of Regina, Canada
Joao Gama	University of Porto, Portugal
Junbin Gao	The University of New England, Australia

Kay Chen Tan	National University of Singapore, Singapore
Krzysztof Cios	University of Colorado at Denver, USA
Longin Jan Latecki	Temple University Philadelphia, USA
Lotfi A. Zadeh	Berkeley University of California, USA
Luis Torgo	University of Porto, Portugal
Mehmed Kantardzic	University of Louisville, USA
Mehmet Kaya	Firat University, Turkey
Michael Frank Goodchild	University of California, Santa Barbara, USA
Michael R. Berthold	University of Konstanz, Germany
Mohammad El-Hajj	University of Alberta, Canada
Mohammed Zaki	Rensselear Polytechnic Institute, USA
Naren Ramakrishnan	Virginia Tech, USA
Nasrullah Memon	Aalborg University, Denmark
Olfa Nasraoui	University of Louisville, USA
Ozgur Ulusoy	Bilkent University, Turkey
Paul Vitanyi	CWI and University of Amsterdam, The Netherlands
Peter Geczy	National Institute of Advanced Industrial Science and Technology, Japan
Philip S. Yu	IBM T.J. Watson Research Center, USA
Ping Jiang	Bradford University, UK
Raul Giraldez Rojo	University of Seville, Spain
Ricardo Vilalta	University of Houston, USA
Rui Camacho	University of Porto, Portugal
Sarah Zelikovitz	College of Staten Island, NY, USA
Shaobin Huang	Harbin Engineering University, China
Shashi Shekhar	University of Minnesota, USA
Shengfei Shi	Harbin Institute of Technology, China
Shichao Zhang	University of Technology, Sydney, Australia
Shuigeng Zhou	Fudan University, China
Shuliang Wang	Wuhan University, China
Shusaku Tsumoto	Shimane Medical University, Japan
Simsek Sule	University of Missouri-Rolla, USA
Sunil Choenni	University of Twente, Netherlands
Tan Kok Kiong	National University of Singapore, Singapore
Tansel Ozyer	TOBB University, Turkey
Vladimir Gorodetsky	Intelligent System Lab, The Russian Academy of Science, Russia
Wanquan Liu	Curtin University of Technology, Australia
Wei Wang	Fudan University, China
Xiangjun Dong	Shandong Institute of Light Industry, China
Xuemin Lin,	University of New South Wales, Australia
Yang ZHANG	Northwest A&F University, China
Yingshu Li	Georgia State University, USA
Yonghong Peng	University of Bradford, UK
Zbigniew W. Ras	University of North Carolina, USA
Zhanhuai Li	Northwest Polytechnical University, China

Zhaoyang Dong   The University of Queensland, Australia
Zhipeng Xie   Fudan University, China
Zijiang Yang   York University, Canada

Sponsoring Institutions

Harbin Institute of Technology
University of Alberta
Alberta Innovation and Science
Harbin Engineering University

Table of Contents

Invited Talk

Regular Papers

Short Papers

Mining Ambiguous Data with Multi-instance Multi-label Representation

Zhi-Hua Zhou

National Key Laboratory for Novel Software Technology
Nanjing University, Nanjing 210093, China
zhouzh@nju.edu.cn

Abstract. In traditional data mining and machine learning settings, an object is represented by an instance (or feature vector) which is associated with a class label. However, real-world data are usually ambiguous and an object may be associated with a number of instances and a number of class labels simultaneously. For example, an image usually contains multiple salient regions each can be represented by an instance, while in image classification such an image can belong to several classes such as *lion*, *grassland* and *tree* simultaneously. Another example is text categorization, where a document usually contains multiple sections each can be represented as an instance, and the document can be regarded as belonging to different categories such as *scientific novel*, *Jules Verne's writing* or even *books on travelling* simultaneously. Web mining is another example, where each of the links or linked pages can be regarded as an instance while the web page itself can be recognized as a *news page*, *sports page*, *soccer page*, etc. This talk will introduce a new learning framework, *multi-instance multi-label learning* (MIML), which is a choice in addressing such kind of problems.

R. Alhajj et al. (Eds.): ADMA 2007, LNAI 4632, p. 1, 2007.

DELAY: A Lazy Approach for Mining Frequent Patterns over High Speed Data Streams*

Hui Yang[1], Hongyan Liu[2], and Jun He[1]

[1] Information School, Renmin University of China, Beijing, 100872, China
{huiyang,hejun}@ruc.edu.cn
[2] School of Economics and Management, Tsinghua University, Beijing,
100084, China
hyliu@tsinghua.edu.cn

Abstract. Frequent pattern mining has emerged as an important mining task in data stream mining. A number of algorithms have been proposed. These algorithms usually use a method of two steps: one is calculating the frequency of itemsets while monitoring each arrival of the data stream, and the other is to output the frequent itemsets according to user's requirement. Due to the large number of item combinations for each transaction occurred in data stream, the first step costs lots of time. Therefore, for high speed long transaction data streams, there may be not enough time to process every transactions arrived in stream, which will reduce the mining accuracy. In this paper, we propose a new approach to deal with this issue. Our new approach is a kind of lazy approach, which delays calculation of the frequency of each itemset to the second step. So, the first step only stores necessary information for each transaction, which can avoid missing any transaction arrival in data stream. In order to improve accuracy, we propose monitoring items which are most likely to be frequent. By this method, many candidate itemsets can be pruned, which leads to the good performance of the algorithm, *DELAY*, designed based on this method. A comprehensive experimental study shows that our algorithm achieves some improvements over existing algorithms, *LossyCounting* and *FDPM*, especially for long transaction data streams.

1 Introduction

Data stream is a potentially uninterrupted flow of data that comes at a very high rate. Mining data stream aims to extract knowledge structure represented in models and patterns. A crucial issue in data stream mining that has attracted significant attention is to find frequent patterns, which is spurred by business applications, such as e-commerce, recommender systems, supply-chain management and group decision support systems. A number of algorithms [5-16] have been proposed in recent years to make this kind of searching fast and accurate.

* This work was supported in part by the National Natural Science Foundation of China under Grant No. 70471006,70621061, 60496325 and 60573092.

R. Alhajj et al. (Eds.): ADMA 2007, LNAI 4632, pp. 2–14, 2007.

But how fast does the task need to be done on earth (*challenge1*)? The high speed of streams answers the question: the algorithms should be as fast as the streams flow at least, that is to say, it should be so fast as to avoid missing data to guarantee the accuracy of mining results. Former algorithms [5-16] usually divide this process into two steps. One is calculating the frequency of itemsets while monitoring each arrival of the date stream (*step1*), and the other is to output the frequent itemsets according to user's requirement (*step2*). Due to the large number of item combinations for each transaction occurred in data stream, the first step costs lots of time. Therefore, for high speed long transaction data streams, there may be not enough time to process every transaction arrived in stream. As a result, some transactions may be missed, which will reduce the mining accuracy.

This problem can also lead to another challenge. If calculating the frequency of itemsets while monitoring each arrival of the date stream, the longer the transaction is, the more inefficiently the algorithm performs (*challenge2*). Unfortunately the transactions in data streams are often large, for example, sales transactions, IP packets, biological data from the fields of DNA and protein analysis.

Finding long pattern is also a challenge for mining of the static data set. Maxpattern [18] and closed-pattern [17] are two kinds of solutions proposed to solve this problem. These methods could avoid outputting some sub-patterns of frequent itemsets.

But for data stream what we concern more is how to reduce the processing time per element in the data stream. So the key to the efficiency of the algorithm is to reduce the number of candidates. Some papers [9,12,16] proposed several solutions of pruning method. But they all prune candidates by itemsets. In order to prune the candidate itemsets, frequency of itemsets must be calculated, which turns to the *challenge1*.

In this paper, we try to address the challenges discussed above. Our contributions are as follows.

(1) We propose a new approach, a kind of lazy approach to improve the processing speed per item arrival in data stream. This method delays calculation of the frequency of each itemset to step 2. So, the step 1 only stores necessary information for each transaction, which can avoid missing any transaction arrival in data stream. Furthermore, these two steps could be implemented in parallel, as they can be done independently.

(2) In order to improve the accuracy of mining result, we propose monitoring items which are most likely to be frequent. By this method, many candidate itemsets can be pruned. Based on this method, we develop an algorithm, *DELAY*, which can prune infrequent items and avoid generation of many subsets of transactions, especially for long transactions.

(3) We conducted a comprehensive set of experiments to evaluate the new algorithm. Experimental results show that our algorithm achieves some improvements over existing algorithms, *LossyCounting* and *FDPM*, especially for long transaction data streams.

The rest of the paper is organized as follows. In section 2, we review related work. In section 3, we formally formulate the problem of mining frequent itemsets over streams. Section 4 describes the proposed approach and algorithm, and section 5 gives the experimental results. Finally, Section 6 concludes our paper.

2   Background and Related Work

Throughout the last decade, a lot of people have implemented various kinds of algorithms to find frequent itemsets [2,4,17-20] from static data sets. In order to apply these algorithms to data stream, many papers [5-9] fall back on partition method such as sliding windows model proposed by Zhu and Shasha [5]. By this method, only part of the data streams within the sliding window are stored and processed when the data flows in. For example, a *lossyCounting* (a frequent item mining algorithm) based algorithm [9] divides a stream into batches, in which data is processed in a depth-first search style. For simplicity, we call this algorithm *LossyCounting* too. Time-fading model is a variation of sliding window model, which is suitable for applications where people are only interested in the most recent information of the data streams, such as stock monitoring systems. This model is implemented in [10,11,13], which gets more information and consumes more time and space in the meantime.

For the infinite of stream, seeking exact solution for mining frequent itemsets over data stream is usually impossible, which leads to approximate solution of this mining task [9,12,16]. Algorithms of this kind can be divided into two categories: false-positive oriented and false-negative oriented. The former outputs some infrequent patterns, whereas the latter misses some frequent patterns. *LossyCounting* [9] is a famous false-positive approximate algorithm in data stream. Given two parameters: support s and error ε, it returns a set of frequent patterns which are guaranteed by s and ε. Algorithm *FDPM* [16] is a false-negative approximate algorithm based on *Chernoff Bound* and has better performance than *LossyCounting*.

The above algorithms all perform well when transactions in data stream are not large or the stream does not flow at a high speed, that is to say, these algorithms could not meet the challenges mentioned in section 1. This could be explained by too much computation on data, so it could happen that next transaction has been here before last transaction is finished.

3   Problem Definition

Let $I = \{i_1, \ldots, i_m\}$ be a set of items. An itemset X is a subset of I. X is called k-itemset, if $|X| = k$, where k is the length of the itemset. A transaction T is a pair $(tid; X)$, where tid is a unique identifier of a transaction and X is an itemset. A data stream D is an open set of transactions. N represents the current length of the stream. There are two user-specified parameters: a support threshold $s \in (0, 1)$, and an error parameter $\varepsilon \in (0, 1)$. Frequent-patterns of a time interval mean the itemsets which appear more than sN times. When user submits a query for the frequent-patterns, our algorithm will produce the answers that following these guarantees:

(1) All itemsets whose true frequency exceed $(s + \varepsilon)N$ are output.
(2) No itemset with true frequency less than sN is output.There are no false positive.
(3) Estimated frequencies are less than the true frequencies by at most εN.

4   A New Approach

The strongpoint of our approach is the quick reaction to data stream, so the algorithm could achieve good performance for data streams flowing at high speed. Our approach is also a two-step method. In the first step (step 1), we just store necessary information of stream in a data structure. Frequent itemsets are found until the second step (step 2), the query for them comimg. In the first step, some itemsets which do not exceed a threshold are pruned so as to save space. The criterion of pruning is whether the count of every item in the itemset exceeds a threshold. The second step is a pattern fragment growth step which is the same as the second step of *FP-growth*[19]. So in this paper we focus on the first step.

4.1   Data Structures

There are two main data structures in this algorithm: *List* and *Trie*. The *List* is used to store possible frequent items; and the *Trie* is used to store possible frequent itemsets.

***List*:** a list of counters, each of which is a triple of $(itemid, F, E)$, where *itemid* is a unique identifier of an item, F is the estimation of the item's frequency; and E is the maximum error between F and the item's true frequency.

***Trie*:** a lexicographic tree, every node is a pair of (P, F), where P is a pointer that points one counter of *List*. In this way the association between itemsets and items is constructed and that is why we could prune itemsets by items. F is the estimated frequency of itemset that consists of the items from the root of *Trie* down to this node.

***List.update*:** A frequent itemset consists of frequent items. So, if any item of an itemset is infrequent, then the itemset can not be frequent. Since data stream flows rapidly, frequent itemsets are changing as well. Some frequent itemsets may become non-frequent and some non-frequent itemsets may become frequent. Therefore, technique to handle *concept-drifting*[3] needs to be considered. In this paper, we dynamically maintain a *List*, in which every item's estimated frequency and estimated error is maintained by a frequency ascending order. The method used to update the *List* is based on *space-saving*[1].

***Trie.update*:** While updating *List*, the nodes which point the items deleted for becoming infrequent will also be deleted. This is just the method by which we implement itemsets' pruning by items. For each transaction arrived in the data stream, its subset consisting of items maintained in the *List* will be inserted into the *Trie* .

A example *List* and *Trie* for a stream with two transactions is shown in Fig 1 (a).

4.2   Algorithm *DELAY*

Based on the new approach, we develop an algorithm, *DELAY*, which is shown as followed.

Algorithm: DELAY (data stream S, support s, error ε)
Begin
1. List.length = $\lceil m/\varepsilon \rceil$;
2. For each transaction t in S
3.   List.update (t);
4.   Delete the items from t which are not in the list;
5.   Insert t into trie;
6.   If user submits a query for frequent-patterns
7.   FP-growth(trie, s);
8.   end if
9. end for
End.

Procedure List.update (transaction t, tree trie)
Begin
1. for each item, e, in t
2.   If e is monitored, then increment the F of e;
3.   else
4.   let e_m be the element with least frequency, min
5.   delete all node from trie which point to e_m;
6.   Replace em with e;
7.   Increment F;
8.   Assign E_i the value min;
9.   end if
10. end for
End.

The main steps of *DELAY* are as follows. First, we define the length of *List*, *l*, to be $\lceil m/\varepsilon \rceil$ (line 1),m is the average length of transactions in data stream. Then, for every transaction *t* of data stream *S*, we update the *List* with the items of the transaction by procedure *List.update* (line 3). For those items which are not monitored in the *List*, delete them from *t* (line 4). Then insert transaction *t* into the *Trie* (line 5). Whenever a user submits a query for frequent itemsets, a procedure *FP-growth*[19] will be used to find and output answers using the information stored in *Trie* (line 6-8).

The procedure of *List.update* is similar to *space-saving*[1]. If we observe an item, *e*, that is monitored, we just increase its *F* (line 2). If *e* is not monitored, give it the benefit of doubt, and find from the *List* item e_m, the item that currently has the least estimated hits, *min* (line 4). Nodes of the *Trie* which point to item e_m are deleted (line 5). Then, item e_m is replaced by e (line 6). Assign F_m the value $(min+1)$ (line 7). For each monitored item e_i, we keep track of its over-estimation error, E_i, resulting from the initialization of its counter when it was inserted into the *List*. That is, when starting to monitor e_i, set $E_i = min$, the estimated frequency of the evicted item e_m.

An example of this algorithm is shown in Fig 1. Fig 1 (a) and (b) show the change of *List* and *Trie* without any item of *List* being replaced. In (c), with the coming of transaction with items CF, the number of unique items has exceeded the length of *List*, so the last item E in the *List* is replaced by F with frequency 2, and the nodes in *Trie* which point to E are deleted (the red one). In (d), transaction CF is inserted into *Trie*.

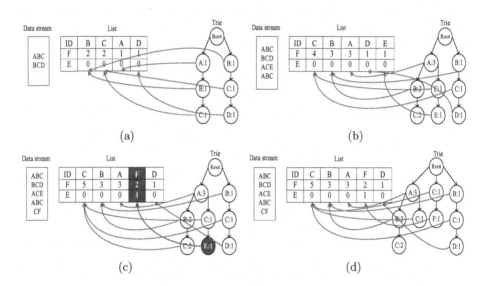

Fig. 1. An example

4.3 Properties of *DELAY*

Lemma 1. *For any item(e_i, F_i, E_i) monitored in* List, $E_i \leq \varepsilon N$

This is proved in [1].

Lemma 2. *Given a support threshold s, for a stream of length N, the error of frequent itemset E_p is bounded by εN, that is, $f_p - F_p \leq \varepsilon N$.*

Proof. If itemset p is frequent, then $F_p \geq sN$;

Assume $e_1(e_1, F_1)$ is the item with the least estimated frequency in p, and its real frequency is f_1.

Assume $e_2(e_2, F_2)$ is the item with the least real frequency in p, and its real frequency is f_2.

$DELAY$ is a false-negative algorithm, so $f_i \geq F_i$.

So, $E_p = f_p - F_p = f_2 - F_1 \leq f_1 - F_1 \leq \varepsilon N$.

Theorem 1. *An itemset p with $f_p \geq (s + \varepsilon)N$, must be found by* DELAY.

Proof. For itemset p, according to lemma 2, $f_p \leq F_p + \varepsilon N$. If $f_p \geq (s + \varepsilon)N$ then $F_p \geq sN$, so our algorithm $DELAY$ will output itemset p.

Theorem 2. *Time spent in step 1 is bounded by $O(Nm + Nt)$, where N is the length of data stream, m is the average length of transactions in data stream, and t is the time of inserting one itemset into* Trie.

Proof. In step 1 each transaction needs to be processed by *List.update* first. This part consumes time $O(Nm)$. Then the transaction is inserted into *Trie*. The time complexity of this part is difficult to estimate. Here we assume the time of inserting one itemset into *Trie* is t ignoring the difference of length between itemsets. For $DELAY$, one transaction one itemset needs to be inserted into the *Trie*. so the time complexity of this part is $O(Nt)$. So time spent in step 1 is bounded by $O(Nm + Nt)$. In the following we compare the times of inserting between $LossyCounting$ [9] and $DELAY$.

4.4   Comparison with Existing Work

4.4.1   Comparison with *LossyCounting*

Theorem 2 proves that time spent in the step 1 by $DELAY$ is about $O(Nm+Nt)$. $LossyCounting$ [9] has a bound of $O((2^m)N + \frac{(2^m)N}{k\varepsilon}t)$, where m is the average length of transactions in data stream, k is the buffer size, N and t with the same definition of Theorem 2. Let's compare the time bound of $DELAY$ and $LossyCounting$.

The time of inserting itemsets into *Trie* is difficult to estimate, so we just compare the times of inserting. The times of inserting in $DELAY$ is N, that is one transaction one insertion into *Trie*, whereas the times of inserting itemsets in $LossyCounting$ is $\frac{2^m N}{k\varepsilon}$. In $LossyCounting$ itemsets whose frequencies exceed εk will be inserted into *Trie*, and every subsets of them will be inserted respectively too. $Nm + Nt < 2^m N + \frac{2^m N}{k\varepsilon}t$, when $k < \frac{2^m}{\varepsilon}$, that is, given $m = 30$, $\varepsilon = 0.001$, only when $k > 10^{12}$, $DELAY$ will consume more time than $LossyCounting$ in step 1, but 10^{12} is a huge number for memory space. So $DELAY$ usually consumes less time. So we can see that the average length of transaction, m, is the determinant of which one performs well, which is also demonstrated in the following experiments.

4.4.2   Comparison with *FDPM*

$FDPM$ proposed in [16] finds frequent patterns through two steps. First calculate the frequency of itemsets, and then prune them based on *Chernoff Bound*. The merit of it is in the space bound. As the method of calculating the frequency of itemsets is not given in the paper, we could not estimate the time bound. But as long as the algorithms need to calculate frequency of itemsets, they will consume more time than the step 1 of $DELAY$ for the same data set, which will be proved in the following section.

5   Experimental Results

In this section we conducted a comprehensive set of experiments to evaluate the performance of *DELAY*. We focus on three aspects: time, space and sensitivity to parameter. Algorithm *DELAY* is implemented in C++ and run in a Pentium IV 2.4GHz PC with 1.0 G RAM and 40G Hard disk. In the experiments, we use the synthetic datasets generated by IBM data generator[21].

5.1   *DELAY*

In this section, we design two sets of experiments to test the performance of DELAY.

5.1.1   Time and Space
We fix $s = 1\%$, $\epsilon = s/10$, the average length of transaction $L = 30$, and vary the length of data stream from 100k to 1000k. Fig.2(a) shows the time used in step 1 (*step1 time*) and the time used in step 2(*step2 time*), and Fig.2(b) shows the memory consumption.

As shown in Fig.2(a) *step1 time* is linear with the length of data stream, accounting for only a small part of the total runtime, which means that this algorithm could deal with streams flowing at a high speed. Fig.2(b) shows that the increase of the memory consumed slows down with the increase of the length of stream.

This experiment proves that *DELAY* could potentially handle large-scale data stream of high speed, consuming only limited memory space.

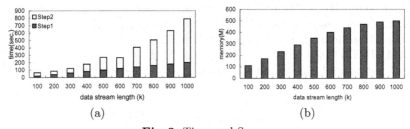

Fig. 2. Time and Space

5.1.2   Sensitivity to the Support Threshold
In this set of experiments, We generate dataset T30.I10.D1000k, set $\epsilon = s/10$, and vary s from 0.1% to 1%. Fig.3(a), (b), (c) and (d) show the total runtime, *step1* time per pattern, *step2* time per pattern, memory and memory per pattern respectively as the support varies.

As shown in Fig.3(a), *step1 time* remains almost stable as support varies. This is because *step1 time* is only relative to the length of stream, and *DELAY* do nothing different for different support value. Fig.3(b) show that the average run time per frequent pattern (itemset) decreases as support decreases though the total time is increased. Fig.3(c) and Fig.3(d) indicate that more memory is needed as support decreases, but the memory consumed per frequent pattern keeps steady on the whole.

The behavior of *DELAY* with the variation of error level, ϵ, is similar to the results of this set of experiments. Due to space limitation, we do not give the results here.

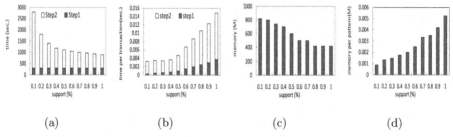

(a)   (b)   (c)   (d)

Fig. 3. Sensitivity to the support threshold

5.2   Comparison with *LossyCounting* and *FDPM*

In this section, we compare *DELAY* with *LossyCounting* and *FDPM* in the following aspects: reaction time, total run time, memory requirements under different dataset size and support levels.

5.2.1   Average Length of Transactions

We fix $s = 1\%$, $\varepsilon = s/10$, the length of data stream = 100k, and vary average length of transaction from 10 to 60.

Fig.4 shows the change of run time of *DELAY*, *LossyCounting* and *FDPM* as the length of transactions increases. As shown in Fig.4, *DELAY* significantly outperforms *LossyCounting* and *FDPM* on the running time, and the excellence of *DELAY* is more evident when transaction becomes longer. The reason has been explained in section 4.3.

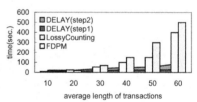

Fig. 4. Transaction length

5.2.2   Length of Stream

In this section we test the algorithms with two data sets: T30.I10.D?k and T15.I6.D?k. For this set of experiments, we fix $s = 1\%$, $\varepsilon = s/10$, and report the time and memory usage as the length of data stream increases from 100k to 1000k.

For T30.I10.D?k, Fig.5 shows the results. As shown in Fig.5(a), *DELAY* significantly outperforms *LossyCounting* and *FDPM* on the running time for the stream with relatively longer transactions. Fig.5(b) shows that the increasing speed of memory by *DELAY* as the increase of stream size is faster than *LossyCounting* and FDPM, but the ability of processing streams of higher speed is worthy of the sacrifice of memory.

For T15.I10.D?k, Fig.5(c) shows the results. As shown in Fig.5(c), though *DELAY* does not perform better than *LossyCounting* and *FDPM* do when transactions are short, but the *step1* time of *DELAY* shows that *DELAY* could work on streams flowing at high speed.

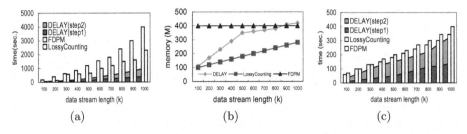

Fig. 5. Length of Stream

5.2.3   Support Threshold

In this section we test the algorithms with two data sets: T30.I10.D1000k and T15.I6.D1000k. We fix $\varepsilon = s/10$, and vary s from 0.1% to 1%.

For T30.I10.D1000k, Fig.6(a), (b), (c) and (d) show the results. As shown in Fig.6(a) and (b), *DELAY* outperforms *LossyCounting* and *FDPM* for streams of long transactions as support increases. Fig.6(c) and (d) show that *DELAY* consumes a bit more memory than *LossyCounting* and *FDPM*.

For T15.I10.D1000k, Fig.6(e) and (f) show the results. As shown in these Figures, the *step1 time* of *DELAY* is almost unchanged with the varying of support level. Though *DELAY* does not significantly outperform *LossyCounting* and *FDPM* for short transactions overall, *DELAY* could perform better when support level is relatively low.

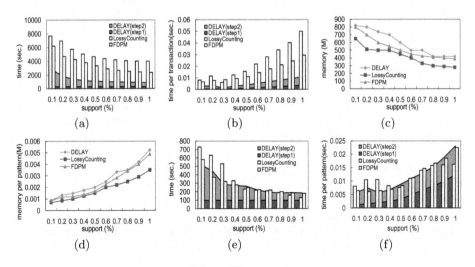

Fig. 6. Varying s

5.2.4　Recall and Precision

In order to evaluate the accuracy of these algorithms, we test the recall and precision for streams with different average length of transactions, different support threshold and different error rate. Recall and precision are defined as follows. Given a set A of true frequent itemsets and and a set B of frequent itemsets output by the algorithms. The recall is $\frac{|A \cap B|}{|A|}$ and the precision is $\frac{|A \cap B|}{|B|}$.

Fixing $s = 1\%$, $\varepsilon = s/10$, the length of data stream 1000k, we vary the average length of transactions from 10 to 50. Table 1 shows the recall and precision of *DELAY*, *LossyCounting* and *FDPM*. *DELAY* is a false-negative algorithm, which ensures its precisions to be 100%, *DELAY* achieves a little higher recall than *FDPM* on the average.

For dataset T30.I10.D1000k, setting $\varepsilon = s/10$, Table 2 lists the recall and precision of *DELAY*, *LossyCounting* and *FDPM* with different support thresholds. It shows that recall increase as support thresholds decrease. That is because the error of items is bound by εN, and the error distributes to lots of itemsets. When the support decreases, there will be more frequent itemsets, so the error per itemset becomes less.

For dataset T30.I10.D1000k, Table 3 shows the recall and precision of *LossyCounting* and *DELAY* as ε increases from 0.01% to 0.1% and support is fixed to be 1%. It tells us that *DELAY* can get high recall under the condition of maintaining precision to be 1.

Table 1. Varying transaction length

L	DELAY		FDPM		LC	
	R	P	R	P	R	P
10	1	1	1	1	1	0.89
20	1	1	1	1	1	0.6
30	0.96	1	0.93	1	1	0.54
40	0.92	1	0.93	1	1	0.51
50	0.9	1	0.9	1	1	0.42

Table 2. Varying support s

S %	DELAY		FDPM		LC	
	R	P	R	P	R	P
0.1	0.97	1	1	1	1	0.89
0.2	0.96	1	0.99	1	1	0.97
0.4	0.93	1	0.99	1	1	0.97
0.6	0.85	1	0.98	1	1	0.87
0.8	0.89	1	0.98	1	1	0.73
1	0.78	1	0.96	1	1	0.79

Table 3. Varying error ε

error%	DELAY		FDPM		LC	
	R	P	R	P	R	P
0.01	1	1	1	1	1	0.99
0.02	1	1	0.99	1	1	0.99
0.04	1	1	0.99	1	1	0.98
0.06	1	1	0.98	1	1	0.93
0.08	0.96	1	0.98	1	1	0.9
0.1	0.97	1	0.96	1	1	0.85

6　Conclusions

In this paper, we propose a lazy approach for mining frequent patterns over a high speed data stream. Based on this approach, we develop an algorithm, *DELAY*, which does not calculate the frequency of itemsets as soon as the data arrive in data stream like other algorithms, but only stores necessary information. The frequency is not calculated until the query for frequent itemsets comes, which can avoid missing any transaction arrival in data stream. This kind of delay also helps this method to perform well for long transaction data streams. In order to reduce the information needed to store, we propose monitoring items which are more likely to be frequent. In the meantime, this algorithm can also guarantee a predefined error rate.

References

1. Metwally, A., Agrawal, D., Abbadi, A.E.: Efficient Computation of Frequent and Top-k Elements in Data Streams. In: Eiter, T., Libkin, L. (eds.) ICDT 2005. LNCS, vol. 3363, Springer, Heidelberg (2004)
2. Bayardo Jr., R.J.: Efficiently Mining Long Patterns from Databases. In: Proceedings of the ACM SIGKDD Conference (1998)
3. Wang, H., Fan, W., Yu, P.S., Han, J.: Mining Concept-Drifting Data Streamsusing Ensemble Classifiers. In: ACM SIGKDD Int'l Conf. on Knowledge Discovery and Data Mining (August 2003)
4. Agrawal, R., Imielinski, T., Swami, A.: Mining Association Rules between Sets of Items in Massive Databases. In: Int'l Conf. on Management of Data (May 1993)
5. Zhu, Y., Shasha, D.: StatStream: Statistical Monitoring of Thousands of Data Streams in Real Time. In: Int'l Conf. on Very Large Data Bases (2002)
6. Chi, Y., Wang, H., Yu, P.S., Richard, R.: Moment: Maintaining Closed Frequent Itemsets over a Stream Sliding Window. In: IEEE Int'l Conf. on Data Mining (November 2004)
7. Chang, J.H., Lee, W.S.: A Sliding Window Method for Finding Recently Frequent Itemsets over Online Data Streams. Journal of Information Science and Engineering (2004)
8. Cheng, J., Ke, Y., Ng, W.: Maintaining Frequent Itemsets over High-Speed Data Streams. In: Ng, W.-K., Kitsuregawa, M., Li, J., Chang, K. (eds.) PAKDD 2006. LNCS (LNAI), vol. 3918, Springer, Heidelberg (2006)
9. Manku, G.S., Motwani, R.: Approximate Frequency Counts over Data Streams. In: Int'l Conf. on Very Large Databases (2002)
10. Chang, J.H., Lee, W.S., Zhou, A.: Finding Recent Frequent Itemsets Adaptively over Online Data Streams. In: ACM SIGKDD Int'l Conf. on Knowledge Discovery and Data Mining (August 2003)
11. Giannella, C., Han, J., Pei, J., Yan, X., Yu, P.S.: Mining Frequent Patterns in Data Streams at Multiple Time Granularities. In: Data Mining: Next Generation Challenges and Future Directions, AAAI/MIT Press, Cambridge (2003)
12. Li, H.-F., Lee, S.-Y., Shan, M.-K.: An Efficient Algorithm for Mining Frequent Itemsets over the Entire History of Data Streams. In: Int'l Workshop on Knowledge Discovery in Data Streams (September 2004)
13. Chang, J.H., Lee, W.S.: A Sliding Window Method for Finding Recently Frequent Itemsets over Online Data Streams. Journal of Information Science and Engineering (2004)
14. Charikar, M., Chen, K., Farach-Colton, M.: Finding Frequent Items in Data Streams. Theoretical Computer Science (2004)
15. Lin, C.-H., Chiu, D.-Y., Wu, Y.-H., Chen, A.L.P.: Mining Frequent Itemsets from Data Streams with a Time-Sensitive Sliding Window. In: SIAM Int'l Conf. on Data Mining (April 2005)
16. Yu, J.X., Chong, Z.H., Lu, H.J., Zhou, A.Y.: False positive or false negative: Mining frequent Itemsets from high speed transactional data streams. In: Nascimento, M.A., Kossmann, D. (eds.) VLDB 2004. Proc. of the 30th Int'l Conf. on Very Large Data Bases, pp. 204–215. Morgan Kaufmann Publishers, Toronto (2004)
17. Pasquier, N., Bastide, Y., Taouil, R., Lakhal, L.: Discovering frequent closed itemsets for association rules. In: Beeri, C., Bruneman, P. (eds.) ICDT 1999. LNCS, vol. 1540, pp. 398–416. Springer, Heidelberg (1998)

18. Bayardo Jr., R.J.: Efficiently mining long patterns from databases. In: Haas, L.M., Tiwary, A. (eds.) Proceedings of the 1998 ACM SIGMOD International Conference on Management of Data. SIGMOD Record, vol. 27(2), pp. 85–93. ACM Press, New York (1998)

19. Han, J., Pei, J., Yin, Y., Mao, R.: Mining frequent patterns without candidate generation: A frequent-pattern tree approach. Data Mining and Knowledge Discovery (2003)

20. Zaki, M.J.: Scalable algorithms for association mining. IEEE Transactions on Knowledge and Data Engineering 12(3), 372–390 (2000)

21. Agrawal, R., Srikant, R.: Fast algorithms for mining association rules. In: Proc. of 20th Intl. Conf. on Very Large Data Bases, pp. 487–499 (1994)

Exploring Content and Linkage Structures for Searching Relevant Web Pages

Darren Davis and Eric Jiang

University of San Diego
5998 Alcala Park, San Diego, CA 92110, USA
{ddavis-08,jiang}@sandiego.edu

Abstract. This work addresses the problem of Web searching for pages relevant to a query URL. Based on an approach that uses a deep linkage analysis among vicinity pages, we investigate the Web page content structures and propose two new algorithms that integrate content and linkage analysis for more effective page relationship discovery and relevance ranking. A prototypical Web searching system has recently been implemented and experiments on the system have shown that the new content and linkage based searching methods deliver improved performance and are effective in identifying semantically relevant Web pages.

Keywords: Web mining, hyperlink analysis, information retrieval, singular value decomposition.

1 Introduction

In general, the search problem spans a spectrum of activities ranging from a precise query for a specific subject to a non-specific desire for learning what information is available. Keyword search systems are commonly used to address these needs and have worked well in some cases. However, they are limited by a vague conception of the user's intent as represented by the few words typical of a keyword search [1]. These limitations mean that keyword search may not be a panacea for information retrieval on the Web and different information search methodologies need to be developed. In this work, we investigate approaches of search where the input is a Web page URL, and the search system returns a list of pages that are relevant to the query URL.

The query URL presumably contains useful content, but the user may desire additional pages that elaborate upon the subject matter found in the query page. Such pages could provide additional information or another perspective on the topic. Thus, Web pages that are relevant to the query address the same topic as the original page, in either a broad or a narrow sense [2]. If the category in which the pages are related is too broad, however, the returned pages may not be very useful. In this work, we aim to find pages with the same specific topic. For example, if the input page describes a security vulnerability, a page that addresses the same topic may discuss procedures for removing the vulnerability.

R. Alhajj et al. (Eds.): ADMA 2007, LNAI 4632, pp. 15–22, 2007.

Relevant page search could have several important benefits. A web site URL often provides copious information that can be used to infer the user's information need. Thus, without significantly increasing the demands made on the user, the user can specify much more information about their intended results. Such a mechanism could be invaluable for document-specific needs such as cross-referencing or locating additional information. It could also be useful in a general case where a topic cannot be readily simplified to a few words or a phrase, but a site is available to provide an example of the topic. In addition to direct use in information retrieval, relevant page search can assist other content organization processes such as clustering, topic separation or disambiguation, or a Web page recommendation system based on pages in a user's search history.

The paper is organized as follows. Some related work is presented in Section 2. In Section 3, we discuss our approaches that integrate content and linkage analysis, thereby enhancing available information about page relationships and improving relevant page searching performance. A prototypical content and linkage based search system has been implemented and is described in Section 4. The experiments on the system and the evaluations of the ranking strategies are presented in Section 5. Some concluding remarks are provided in Section 6.

2   Background

Web searching differs from traditional information retrieval because of the Web's massive scale, diversified content and in particular, its unique link structure. The hyperlinks on the Web not only serve as navigational paths between pages, but also define the context in which a Web page appears and reflect the semantic associations of Web pages [7]. The page information embedded in hyperlinks is useful and can be used for ranking relevant Web pages.

Due to a growing interest in hyperlink analysis, several hyperlink-oriented methodologies have been developed and successfully applied to Web-related fields such as search engines, Web mining, clustering and visualization. The hyperlink analysis has also been used in finding Web pages that are relevant to a topic defined by a user-specified keyword set or Web page. One well-known and representative work in this area is the HITS (Hyperlink Induced Topic Search) algorithm [6] that applies an iteration process to identify pages with topic-related links (hubs) and topic-related content (authorities) within a page neighborhood. The HITS algorithm has offered an interesting perspective on page hyperlink analysis. However, it suffers from the problem of *topic drift* when the majority of neighborhood pages are on a different topic from the query. Several HITS extensions have been proposed that address the problem by either adding linkage weights or expanding the page neighborhood ([1], [2]).

More recently, [5] describes a more direct algorithm named LLI (Latent Linkage Information) for finding relevant pages through page similarity. For a query page, it constructs a page neighborhood in two steps. First, it builds one *reference page set* (Pu) by selecting a group of its parent (inlink) pages and the other reference page set (Cu) by a group of its child (outlink) pages. In the second step, it builds one *candidate page set* (BPS) by adding a group of child pages from each of the Pu pages. Similarly, it builds the other candidate set (FPS) by adding a group of parent pages from each of the Cu pages. Both candidate sets BPS and FPS are presumably rich in relevant pages, and their pages are ranked and returned by the algorithm. The neighborhood page construction is finalized by merging some of the reference pages and their outlink sets.

For page ranking in LLI, two matrices (Pu-BPS, Cu-FPS) are constructed to represent the hyperlink relations among the pages. In both Pu-BPS and Cu-FPS matrices, each column represents a reference page and each row represents a candidate page. The binary matrix entries indicate if there are page links between the corresponding reference and candidate pages. LLI then applies the singular value decomposition (SVD) on the matrices to reveal deeper relationships among the pages in the rank-reduced SVD spaces.

3   New Approaches (MDP and QCS)

In this section, we present two approaches (MDP, QCS) for finding pages relevant to a given query page. They build on the LLI algorithm [5] and utilize both page linkage and content information for accurate page similarity assessment.

To gather Web page content, terms are read from the page title, meta-description and keywords, and body text, but we limit the term extraction to the first 1000 words, as in [1]. Next, common words and word suffixes are removed to improve term matching [8]. Finally, we use the vector space model [9] to represent page content and determine content similarity between pages. Each page has an associated content vector that is represented by the *log(tf)-idf* term weighting scheme [3], and the dot product between two content vectors produces a normalized similarity score for the two pages.

These content scores can be incorporated into LLI in various ways. During the neighborhood page construction, we use content scoring to eliminate candidate pages that have scores below the median score value. This decreases the computational load and potentially reduces the influence of irrelevant pages.

Our other content integration efforts concern page ranking. With content similarity available, dichotomous link/no-link matrix entries are replaced with content similarity scores between the two pages. There are three primary areas where we integrated content information into the page ranking algorithm. First is the representation of the query vector. Not all pages are equal in the number of relevant pages that they bring into the page neighborhood, and their influence should vary accordingly. The SVD techniques explored in LLI address this by assigning higher influences to reference pages that have more links to candidate pages, and we refine this by using content similarity scores. Reference pages deemed more relevant by content scoring should correspond to a larger value in the query vector. Secondly, the nonzero matrix entries are modified by content relevance. By keeping the zero-nonzero status of the entries, we can retain the linkage information between pages while allowing a continuous range of values to incorporate content information. Third, content is also used to influence the integration between page sets and the overall ranking of the pages. This can be especially useful to reconcile scores from two sets that may differ widely in their scales. We describe two approaches for addressing the first two issues, followed by several variations in page score integration.

In the *multilevel dot product* (MDP) approach, content-vector dot products between two pages determine each nonzero entry. In the query vector representation, the ith entry is the dot product between the query and the ith parent or child page. In the matrices that represent the page sets, a nonzero entry is given by the dot product between the corresponding candidate (BPS or FPS) page and reference (Pu or Cu) page. This approach thus captures content relationship information between all page levels: the query, the first link level, and the candidate pages. However, some dot products could be zero, which would discard the hyperlink information between the two pages. Thus, the dot products, originally in the range $[0, 1]$ inclusive, are rescaled to $[\mu, 1]$, $0 < \mu \leq 1$. We use the empirically determined value $\mu = 0.01$ in our experiments. This allows the content similarity scores to influence the results while preserving the benefits obtained from the linkage structure. After these modifications, the SVD can be applied as before.

The second approach is *query-candidate scoring* (QCS). This method measures content similarity directly between the query page and the candidate pages, thus eliminating the need to store content information for the reference-level pages. This can reduce the memory requirement and accelerate the page ranking process. The query representation is determined as follows: for each entry, there exist pages in the candidate set that exhibit the appropriate linkage relationship with the corresponding reference page. Dot products are computed between the query content vector and all of these linked pages, and the mean score becomes the value of the entry. A nonzero matrix score is set as a weighted average of a content score and a link score. More specifically, a page's content score is the dot product between the query content vector and the associated candidate page, normalized by division with the maximum score in the set. A page's link score is the number of in- or out-links that the page has with pages in the corresponding reference set. This score is also divided by the maximum link score in the set. These normalization procedures ensure that the range of possible scores is filled, allowing score influences to be properly balanced. The final score for a page is given by a linear combination of both content and link scores. We use a content weight of 0.7 and a link weight of 0.3 in our experiments. As with MDP, the algorithm can then proceed with the SVD.

Once the SVD has been computed and the cosine similarity scores between the query vector and the projected candidate pages in SVD spaces are obtained, various options exist to produce the final page ranking. One option, used in [5], is to use the scores for page ranking without further processing. We call this the *direct* method. It should be noted that a list of scores is generated for each of the page sets (BPS and FPS) in an SVD space. Therefore, there is no guarantee that both lists are comparable. This is particularly the case when certain queries create large size differences between the two page sets, producing two SVD spaces with widely different dimensions. There are several possible ways to mitigate this compatibility problem. One approach is to balance the scores between the two sets. In our experiments with MDP and QCS, we accomplished this by dividing each score in each candidate set by the mean of the top ten scores in the set.

We consider two additional options that allow further and fine-grained integration of the page scores. The first is the *weighted ordinal* method, where pages in each set are sorted by their cosine score and the unreturned pages with the highest score in each set are iteratively compared. During the comparison, each page is assigned a new composite score equal to the weighted sum of cosine, content, and link scores. The content and link scores are obtained in the same way as the QCS matrix entry weighting scheme, but they are not projected through a SVD space. In our experiments we used cosine, content, and link weights of 0.6, 0.3, and 0.1, respectively. This approach preserves within-set ordering but allows changes in between-set ordering. The second score integration option works similarly by computing the composite scores for all pages, but it does not take the orders of original cosine scores into consideration. In this approach, pages are returned according to their composite scores. This allows changes in both between- and within-set ordering but still benefits from the cosine information. It is referred to as the *weighted non-ordinal* method.

The MDP and QCS approaches are summarized as follows:

Algorithm (MDP, QCS)

1. Construct the reference page sets Pu and Cu (MDP, QCS)
2. Extract reference page content (MDP)
3. Construct the candidate page sets BPS and FPS (MDP, QCS)
4. Extract candidate page content (MDP, QCS)
5. Filter candidate pages by content (MDP, QCS)
6. Set the query vector by query-reference dot products (MDP) or by linked-page content scores (QCS)
7. Set the matrix entries of Pu-BPS and Cu-FPS by reference-candidate dot products (MDP) or by combinations of query-candidate dot products and linkage frequency (QCS)
8. Perform SVD on Pu-BPS and Cu-FPS (MDP, QCS)
9. Compute similarity scores between the query and candidate pages in the rank-reduced SVD spaces (MDP, QCS)
10. Balance the scores from the two SVD spaces (MDP, QCS)
11. Rank candidate pages by one of the score integration options (direct, weighted ordinal, weighted non-ordinal) (MDP, QCS)

---

4   A Prototypical System for Searching Relevant Pages

In order to evaluate and compare the performance of our algorithms, a prototypical relevant page searching system has been implemented. The system takes a query URL and identifies its reference and candidate pages. Next, the system accepts a list of ranking strategies to be performed along with their corresponding parameters. The ranking strategies share the same set of initial pages, but each ranking trial is presented with the same set of initial pages regardless of its serial position. This independence includes the possible elimination of candidate pages from the page neighborhood, as discussed in Section 3. By making the initial page read common to all ranking approaches but keeping the rankings otherwise independent, we can ensure higher consistency in our evaluation of the ranking algorithms.

The ranking strategies implemented in the system are LLI, MDP, and QCS, along with several variations. LLI is coded as a replicated model in [5] that ranks pages by direct score comparison. Score balancing is applied to MDP and QCS, and one can also choose cosine, content, and link weights for the weighted ordinal and weighted non-ordinal rankings of these two approaches. The user is permitted to choose an arbitrary number of ranking configurations before the program exits. For MDP and QCS, all three page integration strategies (direct, weighted ordinal, weighted non-ordinal) are automatically performed. All page information is available throughout program execution, but no cache of page content is maintained between program runs.

5 Experiments

In this section, we compare and evaluate the performance of LLI and the two approaches, MDP and QCS, which are presented in this work. Note that, given the volatile nature of the Internet, we could not exactly replicate the results of LLI reported in [5]. The MDP and QCS algorithms are configured with the score integration techniques (direct, weighted-ordinal, and weighted-non-ordinal).

For performance comparison, we investigated the top ten results returned by the algorithms for various input URLs. Then, results were subjectively classified into four categories: pages whose content has no relevance to the query, pages whose content is relevant in a broad sense, pages that contain some narrow-topic related information, and pages that are exclusively devoted to the same topic as defined in a narrow sense. Fig. 1 summarizes the top returned pages from LLI and the variations of MDP and QCS for three quite different query pages.

The first query page[1] concerns deprecated thread methods in the Java API. The results in Fig. 1 have indicated an improved performance of MDP and QCS over LLI for the query. A further analysis illustrates a weakness in LLI that was addressed in our approaches. In this case, some pages contain very few in- or out-links, and after merging there may only be one or two reference pages left. The resulting matrix has only one or two columns, which inflates the scoring of all the corresponding candidate pages. This makes it difficult to integrate the rankings of the candidate pages from the different sets. Our methods address this issue by allowing a continuous range of scores within the linkage matrices, and by providing post-SVD adjustments to the ranking scores. Thus, while the benefits from the SVD-based rankings are preserved, further content and

---

[1] http://java.sun.com/j2se/1.3/docs/guide/misc/threadPrimitiveDeprecation.html

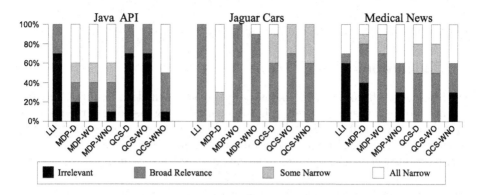

Fig. 1. Result relevance comparison for three query pages

link-based analysis, especially with the weighted non-ordinal method, can provide further differentiation between pages.

The second query presented in Fig. 1 is the Jaguar home page[2] and was also used in [5]. This is a good example to demonstrate the algorithm behavior when very little content (but much misleading) information is available on query pages. From Fig. 1, it can be seen that even with little useful content information in the query page, the performance of the MDP and QCS algorithms is still equal or superior to LLI.

The last query we present in Fig. 1 is a medical news query page[3] that discusses a potential risk factor for early-onset Parkinson's disease. It reflects a popular Web search activity that people turn to for finding health related information. Fig. 1 shows that in this case all MDP and QCS approaches perform competitively to LLI and in particular, both MDP and QCS with weighted-non-ordinal deliver the best results.

Across both techniques and all results, we found that the weighted-ordinal method produces results that are similar or identical to the direct method. An examination of program output indicates that this may be because pages with a high original ranking but lower composite ranking must be listed before all later pages from the same set, even if the later pages were assigned a higher composite relevance score. This blocking of later results is addressed by the weighted non-ordinal method, which generally delivers improved results as compared to the weighted-ordinal method. For both MDP and QCS, the weighted non-ordinal method also consistently performs better that LLI.

We have also observed another interesting property of our approaches. Although not indicated by the categorization of the results reported here, we found that our results generally favor pages with textual content about the topic in question, while LLI often favors link indices. For example, in the Jaguar case, many LLI results contain lists of links to official Jaguar sites for different countries. Our algorithms return some of these pages, but also include pages

---

[2] http://www.jaguar.com (likely the page has been updated since [5])
[3] http://www.medicalnewstoday.com/medicalnews.php?newsid=51960

such as a Jaguar fan club, which arguably offers content beyond pointers to other pages.

6   Conclusions

The Internet has become an enormous and popular information resource. Searching this information space, however, is often a frustrating task due to the massive scale and diversity of the digital data. It has been an active research field for developing efficient and reliable searching techniques. In this work, we investigate the approaches for searching relevant Web pages to a query page. Based on the algorithm LLI [5] that applies SVD to reveal a deep linkage association among pages, we propose two approaches (MDP and QCS) that integrate content and linkage analysis for more effective page relationship discovery and relevance ranking. Both approaches incorporate content and link information in building a neighborhood of pages and their rank-reduced SVD spaces, as well as in scoring page relevance. We have also developed a prototypical search system that implements LLI, MDP and QCS algorithms. The experiments of the system have indicated that the proposed content and linkage based searching methods (MDP and QCS) deliver improved performance and are effective in identifying more semantically relevant Web pages. As future work, we plan to conduct extensive testing of our approaches, and explore methodologies to associate and rank all neighborhood pages in a unified information space.

References

1. Bharat, K., Henzinger, M.: Improved Algorithms for Topic Distillation in a Hyperlinked Environment. In: Proceedings of 21st International ACM Conference on Research and Development in Information Retrieval, pp. 104–111 (1998)
2. Dean, J., Henzinger, M.: Finding Related Pages in the World Wide Web. In: Proceedings of 8th International World Wide Web Conference, pp. 389–401 (1999)
3. Dumais, S.: Improving the retrieval of Information from External Sources. Behavior Research Methods, Instruments, and Computers 23(2), 229–232 (1991)
4. Golub, G., Van loan, C.: Matrix Computations, 3rd edn. John-Hopkins, Baltimore (1996)
5. Hou, J., Zhang, Y.: Effectively Finding Relevant Web Pages from Linkage Information. IEEE Transactions on Knowledge and Data Engineering 15(4), 940–951 (2003)
6. Kleinberg, J.: Authoritative Sources in a Hyperlinked Environment. In: Proceedings of 9th ACM-SIAM Symposium on Discrete Algorithms, ACM Press, New York (1998)
7. Kleinberg, J., Kumar, R., Raghaven, P., Rajagopalan, S., Tomkins, A.: The Web as a graph: measurements, models, and methods. In: Asano, T., Imai, H., Lee, D.T., Nakano, S.-i., Tokuyama, T. (eds.) COCOON 1999. LNCS, vol. 1627, pp. 1–17. Springer, Heidelberg (1999)
8. Porter, M.F.: An Algorithm for Suffix Stripping. Program (14), 130–137 (1980)
9. Salton, G., Wong, A., Yang, C.S.: A Vector Space Model for Automatic Indexing. Communications of the ACM 18(11), 613–620 (1975)

CLBCRA-Approach for Combination of Content-Based and Link-Based Ranking in Web Search

Hao-ming Wang and Ye Guo

Department of Computer Science, Xi'an University of Finance & Economics,
Xi'an, Shaanxi 710061, P.R. China
{hmwang,yeguo}@mail.xaufe.edu.cn

Abstract. There were two kinds of methods in information retrieval, based on content and based on hyper-link. The quantity of computation in systems based on content was very large and the precision in systems based on hyper-link only was not ideal. It was necessary to develop a technique combining the advantages of two systems. In this paper, we drew up a framework by using the two methods. We set up the transition probability matrix, which composed of link information and the relevant value of pages with the given query. The relevant value was denoted by TFIDF. We got the CLBCRA by solving the equation with the coefficient of transition probability matrix. Experimental results showed that more pages, which were important both in content and hyper-link, were selected.

1 Introduction

With information proliferation on the web as well as popularity of Internet, how to locate related information as well as providing accordingly information interpretation has created big challenges for research in the fields of data engineering, IR as well as data mining due to features of Web (huge volume, heterogeneous, dynamic and semi-structured etc.). [1]

While web search engine can retrieve information on the Web for a specific topic, users have to step a long ordered list in order to locate the valuable information, which is often tedious and less efficient due to various reasons like huge volume of information.

The search engines are based on one of the two methods, the content of the pages and the link structure. The first kind of search engineers works well for traditional documents, but the performance drops significant when applied to the web pages. The main reason is that there is too much irrelevant information contained in a web page. The second one takes the hyperlink structures of web pages into account in order to improve the performance. The examples are Pagerank and HITS. They are applied to Google and the CLEVER project respectively.

R. Alhajj et al. (Eds.): ADMA 2007, LNAI 4632, pp. 23–34, 2007.

However, these algorithms have shortcomings in that (1) the weight for a web page is merely defined; and (2) the relativity of contents among hyperlinks of web pages are not considered. [2]

In this paper, we combine the contents and the links among the pages in order to refine the retrieval results. Firstly, we compute the the similarity of pages to the query, the TFIDF is often used. And then, we compute the new pagerank by the similarity and the out-link information of each page. As the page set, which is computed in this algorithm, includes all the pages we can find. The new pagerank is called `Content and Link Based Complete Ranking Algorithm(CLBCRA)`.

This paper is organized as follows: Section 2 introduces the concept of `Pagerank` and `TFIDF`. Section 3 describes the algorithm of `CLBCRA`. Section 4 presents the experimental results for evaluating our proposed methods. Finally, we conclude the paper with a summary and directions for future work in Section 5.

2   Related Works

2.1   Pagerank

PageRank was developed at Stanford University by Larry Page and Sergey Brin as part of a research project about a new kind of search engine. The project started in 1995 and led to a functional prototype, named Google, in 1998. [3,4]. The algorithm can be described as:

Let u be the web page. Then let F_u be the set of pages u points to and B_u be the set of pages that point to u. Let $N_u = |F_u|$ be the number of links from u and let c be a factor used for normalization (so that the total rank of all web pages is constant).

$$R(u) = c \sum_{v \in B_u} \frac{R(v)}{N_v}.$$

PageRank is a probability distribution used to represent the likelihood that a person randomly clicking on links will arrive at any particular page. PageRank can be calculated for any-size collection of documents.

The formula uses a model of a random surfer who gets bored after several clicks and switches to a random page. The PageRank value of a page reflects the chance that the random surfer will land on that page by clicking on a link. It can be understood as a `Markov chain` in which the states are pages, and the transitions are all equally probable and edges are the links between pages.

The PageRank values are the entries of the `dominant eigenvector` of the `modified adjacency matrix`. As a result of Markov theory, it can be shown that the PageRank of a page is the probability of being at that page after lots of clicks.

2.2   HITS

The HITS is another algorithm for rating, ranking web pages. [5] HITS uses two values for each page, the `authority` value and the `hub` value. Authority and hub values are defined in terms of one another in a mutual recursion. An authority value is computed as the sum of the scaled hub values that point to that page. A hub value is the sum of the scaled authority values of the pages it points to. Relevance of the linked pages is also considered in some implementations.

The algorithm is similar to PageRank, in that it is an iterative algorithm based purely on the linkage of the pages on the web. However it does have some major differences from Pagerank:

- It is executed at query time, and not at indexing time, with the associated hit on performance that accompanies query-time processing.
- It is not commonly used by search engines.
- It computes two scores per page (hub and authority) as opposed to a single score.
- It is processed on a small subset of relevant pages, not all pages as was the case with PageRank.

2.3   TFIDF

TFIDF is the most common weighting method used to describe documents in the Vector Space Model (VSM), particularly in IR problems. Regarding text categorization, this weighting function has been particularly related to two important machine learning methods: kNN (k-nearest neighbor) and SVM(Support Vector Machine). The TFIDF function weights each vector component (each of them relating to a word of the vocabulary) of each document on the following basis.

Assuming the document d is represented by vector $\tilde{d} = (\widetilde{w^{(1)}}, \widetilde{w^{(2)}}, ..., \widetilde{w^{(N)}})$ in a vector space. Each dimension $\widetilde{w^{(i)}}$ ($i \in [1, N]$) of $\tilde{d}$ represents the weight of the feature w_i, which is the word selected from the document d. [6,7,8] The N means number of all features.

The values of the vector elements $w^{(i)}$ ($i \in [1, N]$) are calculated as a combination of the statistics $TF(w_i, d)$(Term Frequency) and $DF(w_i)$ (Document Frequency).

$$w^{(i)} = TF(w_i, d) \times IDF(w_i).$$

Where $TF(w_i, d)$ is the number of word w_i occurred in document d. $IDF(w_i)$ can be computed by the number of documents N_{all} and the number of documents $DF(w_i)$, in which the word w_i occurred at least once time.

$$IDF(w_i) = log\frac{N_{all}}{DF(w_i)}.$$

We can construct the vector $\tilde{q}$ of a query q by using the similar way just as we do for the documents. The cosine distance between the vector $\tilde{d}$ and the $\tilde{q}$, which means the similarity of them, can be computed. The bigger the value, the more similar they are.

2.4   Computing the Pagerank

According to the Markov theory, the Pagerank is the `dominant eigenvector` of the `adjacency matrix`. In order to get the real, positive, and the biggest eigenvalue, the adjacency matrix should be changed in 2 steps:

- guarantees the matrix is row-stochastic;
- guarantees the matrix is irreducible by adding the link pair to each page.

The second step means add the links between all pages. In Ref. [9, 10], it was pointed out that the modification of matrix might change the eigenvector order of the matrix or change the importance of the pages. The author then went on to explain a new algorithm with the same complexity of the original PageRank algorithm that solves this problem.

2.5   Precision and Recall

For a retrieval system, there are 2 sides should be considered, the *precision* and the *recall*. Just as the illustrator in Fig.2.5.

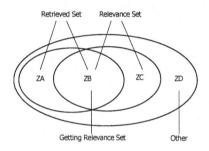

Fig. 1. Concept of Information Retrieval

For a given query Q, we can define,

- $ZA \cup ZB$: all retrieval pages set;
- $ZB \cup ZC$: all relevance pages set;
- ZA: pages set which retrieved but not relevance to the query;
- ZB: pages set which retrieved and relevance to the query indeed;
- ZC: pages set which relevance but could be retrieved;
- ZD: all the other pages set;
- Precision: $\dfrac{ZB}{ZA + ZB}$;
- Recall: $\dfrac{ZB}{ZB + ZC}$.

3   New Model of CLBCRA

We donate the query from the user with Q, all of the pages the retrieval system can select form the $Set\_1 = \{S_i, i \in [1, m]\}$. The pages in Set_1 link to other pages, which part of them belong to the Set_1, and the others form the $Set\_2 = \{S_i, i \in [1, n]\}$. All other pages in the system form the Set_3. [11,12,13]They are shown in Fig.2.

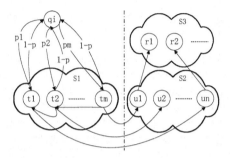

Fig. 2. New Model

In this model, there are 4 parts,

(1) Set S_0: all of the queries,which the user submits to the retrieval system, form the query set. In our illustrator, the $q_i \in S_0$;
(2) Set S_1: the pages are retrieved by the system according to the query q_i; just as the $t_1, t_2, \cdots, t_m$;
(3) Set S_2: the pages can be reached in 1 step from the pages in set S_1, just as the $u_1, u_2, \cdots, u_n$;
(4) Set S_3: all other pages out of the S_0, S_1 and S_2.

3.1   Transition Probability

When user submits a query to the retrieval system, he gets feedback results, S_1, from the system. Some of pages in S_1 are indeed relevant to the query and others are not. Meanwhile, some of pages in S_2 and S_3 are relevant to the query, but they are not retrieved by the system.

When user arrive the page in S_1, he will check the page carefully. If the page is relevant to the query, he will look through the page and move to other pages following the links. If the page is not relevant to the query, he has two choice: going back to the query or moving to other page randomly. In our experiment, we assume the user go back to the query.

(1) $q \to t_i, t_i \in Set\_1$

Assuming he moves from the q to the page $t_i, t_i \in S_1$ with the probability p_i, it is always true that $\sum_i p_i = 1$.

$$p_i = t_{0i} = \frac{\delta_{0i}}{\sum_i \delta_{0i}}. \tag{1}$$

where δ_{0i} is the relevant value of query q and page t_i. The value can be computed by many ways, such as the TFIDF. The value should be nonnegative.

(2) Page $i \to j$

No matter page i is in S_1 or not, the user should check it in order to determine whether the page relevant to the query q or not. We get,

$$\Sigma t_{ij} = \begin{cases} \mu & p_i(q) > 0 \quad \wedge \quad relevance(i, q) > 0; \\ 1 - \mu & p_i(q) > 0 \quad \wedge \quad relevance(i, q) = 0; \\ \nu & p_i(q) = 0 \quad \wedge \quad relevance(i, q) > 0; \\ 1 - \nu & p_i(q) = 0 \quad \wedge \quad relevance(i, q) = 0. \end{cases} \tag{2}$$

Where μ is the sum of probability the user jumps to other pages following the hyper-link from page i when page i is in S_1 and indeed relevant to the query q. Otherwise, the user jumps back to the q with the probability $1 - \mu$.

The ν is the sum of probability the user jumps to other pages following the hyper-link from page i when page i is not in S_1 but it is indeed relevant to the query q. Otherwise, the user jumps back to the q with the probability $1 - \nu$.

According to the Fig.2, we can define the transition probability T as,

(1) $(t_{0i}, \forall i)$

It means the probability from query q to page $i, i \in S_1$.

$$t_{0i} = \frac{\delta_{0i}}{\sum\limits_i \delta_{0i}}. \tag{3}$$

(2) $(t_{ij}, \forall i, j)$

It means the probability from page i, which is relevant to the query, move to page j. We assume that the out-links of page i have the same probability to be selected.

$$t_{ij} = \begin{cases} \mu * m_{ij} & p_i(q) > 0; \\ \nu * m_{ij} & Otherwise. \end{cases} \tag{4}$$

where $m_{ij} = \dfrac{1}{\sum\limits_j linknum(i \to j)}$.

(3) $(t_{i0}, \forall i)$

It means the probability of returning to query q when the user finds the page i is not relevant to the query q.

$$t_{i0} = \begin{cases} 1 - \mu & p_i(q) > 0; \\ 1 - \nu & Otherwise. \end{cases} \tag{5}$$

Gathering the q, S_1, S_2 and S_3, we get the transition probability matrix T shown in (6) .

$$\mathbf{T} = \begin{pmatrix} 0 & P'(q) & 0 & 0 \\ (1-\mu)U_1 & \mu * M_{11} & \mu * M_{12} & 0 \\ (1-\nu)U_2 & \nu * M_{21} & \nu * M_{22} & \nu * M_{23} \\ (1-\nu)U_3 & \nu * M_{31} & \nu * M_{32} & \nu * M_{33} \end{pmatrix} \tag{6}$$

where $P'(q) = (p_1, p_2, \cdots, p_n)$ is probability vector of query q links to all pages. $U_i, (i = 1, 2, 3)$ is the $n_i \times 1$ vector. And $n_i, (i = 1, 2, 3)$ is the page number of set $S_i, (i = 1, 2, 3)$. $M_{ij}, (i, j = 1, 2, 3)$ is the adjacency matrix of set $S_i, (i = 1, 2, 3)$.
In the formula (6), if the $\nu \ll 1$ is true, it can be changed to

$$\mathbf{T} = \begin{pmatrix} 0 & P'(q) & 0 & 0 \\ (1-\mu)U_1 & \mu * M_{11} & \mu * M_{12} & 0 \\ U_2 & 0 & 0 & 0 \\ U_3 & 0 & 0 & 0 \end{pmatrix}$$

and

$$\mathbf{T'} = \begin{pmatrix} 0 & (1-\mu)U_1 & U_2 & U_3 \\ P(q) & \mu * M'_{11} & 0 & 0 \\ 0 & \mu * M'_{12} & 0 & 0 \\ 0 & 0 & 0 & 0 \end{pmatrix} \tag{7}$$

3.2   Computing the Eigenvalue

Assuming $QQ = (x_0, X'_1, X'_2, X'_3)'$, we get $T' * QQ = QQ$.

$$\begin{pmatrix} 0 & (1-\mu)U_1 & U_2 & U_3 \\ P(q) & \mu * M'_{11} & 0 & 0 \\ 0 & \mu * M'_{12} & 0 & 0 \\ 0 & 0 & 0 & 0 \end{pmatrix} \begin{pmatrix} x_0 \\ X_1 \\ X_2 \\ X_3 \end{pmatrix} = \begin{pmatrix} x_0 \\ X_1 \\ X_2 \\ X_3 \end{pmatrix}$$

$$(1-\mu) * |X_1| + |X_2| + |X_3| = x_0; \tag{8}$$
$$x_0 * P(q) + \mu * M'_{11} * X_1 = X_1; \tag{9}$$
$$\mu * M'_{12} * X_1 = X_2; \tag{10}$$
$$X_3 = 0; \tag{11}$$

From (9),

$$(I - \mu * M'_{11}) * X_1 = x_0 * P(q)$$
$$\Rightarrow X_1 = x_0 * (I - \mu * M'_{11})^{-1} * P(q)$$
$$\Rightarrow X_1 = x_0 * V.$$

where $V = (I - \mu * M'_{11})^{-1} * P(q)$.
From (10),

$$X_2 = x_0 * \mu * M'_{12} * V.$$

From (8),

$$(1-\mu) * |x_0 * V| + |x_0 * \mu * M'_{12} * V| + |0| = x_0$$
$$\Rightarrow (1-\mu) * |V| + \mu * |M'_{12} * V| = 1$$
$$\Rightarrow (1-\mu) * |V| = 1 - \mu * |M'_{12} * V|$$
$$\Rightarrow \mu * |M'_{12} * V| = 1 - (1-\mu) * |V|. \tag{12}$$

As $(x_0 + |X_1| + |X_2| + |X_3| = 1)$ is always true, we get

$$x_0 + |x_0 * V| + |x_0 * \mu * M'_{12} * V| + |0| = 1$$
$$\Rightarrow x_0 * (1 + |V| + |\mu * M'_{12} * V|) = 1$$

Changing (12),

$$x_0 * (1 + |V| + 1 - (1 - \mu) * |V|) = 1$$
$$\Rightarrow x_0 * (2 + |V| - |V| + \mu * |V|) = 1$$
$$\Rightarrow x_0 * (2 + \mu * |V|) = 1$$
$$\Rightarrow x_0 = (2 + \mu * |V|)^{-1}.$$

Above all, we get

$$\begin{cases} x_0 = (2 + \mu * |V|)^{-1} \\ X_1 = x_0 * V \\ X_2 = x_0 * \mu * M'_{12} * V \\ X_3 = 0. \end{cases} \tag{13}$$

Where $V = (I - \mu * M'_{11})^{-1} * P(q)$.

From the formula (13), we can conclude that the final feedback pages for a given query q belong to the set S_1 and S_2 only when $\nu \ll 1$ is true.

So, we can set $S = S_1 \cup S_2$.

4 Experimental

4.1 Experimental Setup

We construct experiment data set in order to verify the retrieval method of our approach described in Section 3.

The experiment data set is constructed by using the **TREC WT10g** test collection, which contains about 1.69 million Web pages (http://trec.nist.gov/),100 queries and the links among all pages. Stop words have been eliminated from all web pages in the collection based on the stop-word list and stemming has been performed using Porter Stemmer. [14]

4.2 Constructing the Test Data-Set

We Select all 100 queries $q_i, (i \in [1, 100])$ respectively. For each of the query q, we do the steps just as fellows:

(1) Computing the relevant value of the query q to all pages in WT10g; Selecting the $Top - N, (N = 500, 1000, 5000, 10^4, 1.5 * 10^4, 3 * 10^4)$ pages to construct

the data set $S_{1i}, (i = [1,6])$ respectively. In order to illustrate the method, we assume the relevant value is TFIDF. For example,

$$S_{16} = \{t_i | tfidf(q, d_i) > 0 \wedge tfidf_1 \geq tfidf_2 \geq \cdots \geq tfidf_n > 0, i \in [1, 30000]\};$$

(2) Drawing up all links, which the source page is belonged to S_{1i}, to construct the link set $L_{1i}, (i = [1,6])$.

$$L_{1m} = \{l_{ij} | \exists link(t_i \rightarrow t_j), t_i \in S_{1m}\};$$

(3) Constructing the data set $S_{2i}, (i = [1,6])$ by collecting the target pages of links appeared in L_{1i}. If the page appeared in the S_{1i} and S_{2i} , it will be deleted from S_{2i}.

$$S_{2m} = \{t_j | link(t_i \rightarrow t_j) \in L_{1m} \wedge t_i \in S_{1m} \wedge t_j \notin S_{1m}\}.$$

(4) Combining S_{1i} and S_{2i}.

$$S_i = S_{1i} \cup S_{2i}, i \in [1,6].$$

4.3   Irreducible and Aperiodic

The transition probability matrix has the property just as,

(1) Irreducible
According to the consist of the transfer matrix T, we can get $T = q \cup S_1 \cup S_2$. The links just as fellow are always exist.
   − Link $q_i \rightarrow t_i, t_i \in S_1$;
   − Link $t_i \rightarrow q_i, t_i \in S_1$;
   − Link $t_i \rightarrow u_j, t_i \in S_1 \wedge u_j \in S_2$;
   − Link $u_j \rightarrow q_i, u_j \in S_2$;
That means we can reach each other pages from one of the pages in set S. So, the matrix T is irreducible.
(2) Aperiodic
In the matrix T, all elements in the diagonal are positive except the $t_{00} = 0$. $T_{ii} > 0, (i \in (0, N])$ is always true. So, the matrix T is aperiodic.

So, the transition probability matrix T has a real, positive, and the biggest eigenvalue.

4.4   Experiment Results

In the follow description, we show the results when we deal with the top $3 * 10^4$ pages of TFIDF.

(1) ν
We use the formula (13) to compute the new pagerank only when $\nu \ll 1$ is true.
   We define the ν as the sum of probability of page i when it is not belong to S_1, but it is indeed relevant to query q. The average ν is shown in Table 1.The result shows that $\nu = 3.025 * 10^{-5} \ll 1$ is true. So, we can compute the new pagerank by using formula (13).

Table 1. Result of ν for 30000 pages

	S1	S2	Precision	Recall	\nu
Q1	25767	113119	0.000776	0.909091	0.000012
Q2	30000	55940	0.006933	0.773234	0.00013
Q3	4390	20952	0.022096	0.941748	0.000061
Q4	30000	50399	0.0043	0.921429	0.000081
Q5	30000	52262	0.000733	0.916667	0.000014
Q6	30000	52026	0.0003	0.642857	0.000006
Q7	21126	78488	0.002887	0.884058	0.000038
Q8	30000	68911	0.000633	0.542857	0.000012
...	...	...	...	...	...
Q100	30000	48004	0.001533	0.779661	0.000029
Avg			0.0052235	0.8266822	3.025E-05

(2) Number(Relevance page)/ Number(All page)

We compute the new pagerank according to the formula (13) and sort them decreasingly. The ratio of number of relevance pages in top-N pages to the all the top-N pages can be computed.

The result is showed in Fig.3. We can find that the value of TFIDF is the biggest. The CLBCRA value increases with the increase of the number of pages in the data-set.

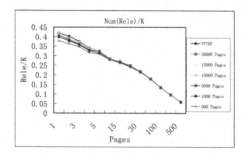

Fig. 3. Relevance pages / All pages

(3) Discussing the result

Considering the formula (13), we can find the solution of x_0, X_1, and X_2 are depended on V, where $V = (I - \mu * M'_{11})^{-1} * P(q)$.

If the $(\mu * M'_{11})$ is small, the V and the $P(q)$ will be very similar. As the M'_{11} is the truth, the μ decides the value of V. With the changing of μ, we can get different values of V.

In our experiment, because the μ is small, the final solution is similar to the $P(q)$, the TFIDF value.

5   Conclusion

This paper introduces two kinds of methods of information retrieval on the web, based on the hyper-link and based on the content. Both of them have shortages, such as the quantity of computation and the precision of retrieval, etc.

This paper draws a new framework by combining the TFIDF and Pagerank in order to support the precise results to users. We set up the computation model and get the final solution. We test the framework by using TREC WT10g test collection. The result shows that the precision of new method approaches the TFIDF's. But the new framework has less quantity of computation than TFIDF.

However, in order to satisfy the users' actual information need, it is more important to find relevant web page from the enormous web space. Therefore, we plan to address the technique to provide users with personalized information.

Acknowledgements

This work was supported by project 06JK300 and 2005F08 of Shaanxi Province, P.R.China.

References

1. Raghavan, S., Garcia-Molina, H.: Complex queries over web repositories. In: VLDB 2003. Proceedings of 29th International Conference on Very Large Data Bases, pp. 33–44. Morgan Kaufmann, Berlin, Germany (2004)
2. Delort, J.-Y., Bouchon-Meunier, B., Rifqi, M.: Enhanced web document summarization using hyperlinks. In: HYPERTEXT 2003. Proceedings of the 14th ACM conference on Hypertext and hypermedia, pp. 208–215. ACM Press, New York (2003)
3. Brin, S., Page, L.: The anatomy of a large-scale hypertextual web search engine. In: Proceedings of the 7th international conference on World Wide Web, pp. 107–117 (1998)
4. Page, L., Brin, S., Motwani, R., Winograd, T.: The pagerank citation ranking: Bringing order to the web. Technical report, Stanford Digital Library Technologies Project (1998)
5. Kleinberg, J.M.: Authoritative sources in a hyperlinked environment. In: SODA '98. Proceedings of the ninth annual ACM-SIAM symposium on Discrete algorithms, Philadelphia, PA. Society for Industrial and Applied Mathematics, pp. 668–677 (1998)
6. Steinberger, R., Pouliquen, B., Hagman, J.: Cross-lingual document similarity calculation using the multilingual thesaurus eurovoc. In: Gelbukh, A. (ed.) CICLing 2002. LNCS, vol. 2276, pp. 415–424. Springer, Heidelberg (2002)
7. Guo, G., Wang, H., Bell, D.A., Bi, Y., Greer, K.: An knn model-based approach and its application in text categorization. In: Gelbukh, A. (ed.) CICLing 2004. LNCS, vol. 2945, pp. 559–570. Springer, Heidelberg (2004)
8. Soucy, P., Mineau, G.W.: Beyond tfidf weighting for text categorization in the vector space model. In: IJCAI-05. Proceedings of the Nineteenth International Joint Conference on Artificial Intelligence, Edinburgh, Scotland, July 30-August 5, 2005, pp. 1130–1135 (2005)

9. Tal-Ezer, H.: Faults of pagerank / something is wrong with google mathematical model (2005)
10. Gyöngyi, Z., Garcia-Molina, H., Pedersen, J.: Combating web spam with trustrank. In: VLDB 2004. Proceedings of the Thirtieth International Conference on Very Large Data Bases, Toronto, Canada, August 31 - September 3, pp. 576–587 (2004)
11. Podnar, I., Luu, T., Rajman, M., Klemm, F., Aberer, K.: A peer-to-peer architecture for information retrieval across digital library collections. In: Gonzalo, J., Thanos, C., Verdejo, M.F., Carrasco, R.C. (eds.) ECDL 2006. LNCS, vol. 4172, pp. 14–25. Springer, Heidelberg (2006)
12. Buntine, W.L., Aberer, K., Podnar, I., Rajman, M.: Opportunities from open source search. In: Skowron, A., Agrawal, R., Luck, M., Yamaguchi, T., Morizet-Mahoudeaux, P., Liu, J., Zhong, N. (eds.) Web Intelligence, pp. 2–8. IEEE Computer Society Press, Los Alamitos (2005)
13. Aberer, K., Klemm, F., Rajman, M., Wu, J.: An architecture for peer-to-peer information retrieval. In: Callan, J., Fuhr, N., Nejdl, W. (eds.) Workshop on Peer-to-Peer Information Retrieval (2004)
14. Sugiyama, K., Hatano, K., Yoshikawa, M., Uemura, S.: Improvement in tf-idf scheme for web peges based on the contents of their hyperlinked neighboring pages. Syst. Comput. Japan 36(14), 56–68 (2005)

Rough Sets in Hybrid Soft Computing Systems

Renpu Li, Yongsheng Zhao, Fuzeng Zhang, and Lihua Song

School of Computer Science and Technology, Ludong University,
Yantai 264025, China
lip0109@sina.com

Abstract. Soft computing is considered as a good candidate to deal with imprecise and uncertain problems in data mining. In the last decades research on hybrid soft computing systems concentrates on the combination of fuzzy logic, neural networks and genetic algorithms. In this paper a survey of hybrid soft computing systems based on rough sets is provided in the field of data mining. These hybrid systems are summarized according to three different functions of rough sets: preprocessing data, measuring uncertainty and mining knowledge. General observations about rough sets based hybrid systems are presented. Some challenges of existing hybrid systems and directions for future research are also indicated.

1 Introduction

Data mining is a process of nontrivial extraction of implicit, previously unknown and potentially useful information (such as knowledge rules, constraints, regularities) from data in databases [1]. Soft computing [2], a consortium of methodologies in which fuzzy sets, neural networks, genetic algorithms and rough sets are principle members, has been widely applied to deal with various problems that contain uncertainty or imprecision in many fields, especially in data mining [3].

Each soft computing methodology has its own powerful properties and offer different advantages. For example, Fuzzy sets are famous at modeling human reasoning and provide a natural mechanism for dealing with uncertainty. Neural networks are robust to noise and have a good ability to model highly non-linear relationships. Genetic algorithm is particularly useful for optimal search. Rough sets are very efficient in attribute reduction and rule extraction.

On the other hand, these soft computing techniques also have some restrictions that do not allow their individual application in some cases. Fuzzy sets are dependent on expert knowledge. The training times of neural networks are excessive and tedious when the input data are large and most neural network systems lack explanation facilities. The theoretical basis of genetic algorithm is weak, especially on algorithm convergence. Rough sets are sensitive to noise and have the NP problems on the choice of optimal attribute reduct and optimal rules.

In order to cope with the drawbacks of individual approaches and leverage performance of data mining system, it is natural to develop hybrid systems by integrating two or more soft computing technologies. In the last decades research on hybrid soft computing systems concentrates on the combination of neural networks, fuzzy sets

R. Alhajj et al. (Eds.): ADMA 2007, LNAI 4632, pp. 35–44, 2007.

and genetic algorithms [4], and the most notable achievement is neuro-fuzzy computation [5]. Comparatively, hybrid systems based on rough sets, which has been categorized as a soft computing technology only in the recent years, is scarce.

The rest of this paper is organized as follows: In section 2 rough sets and soft computing are briefly introduced. In section 3 hybrid soft computing systems based on rough sets are summarized according to three different functions of rough sets in these systems: preprocessing data, measuring uncertainty and mining knowledge. Some generalized observations are provided in section 4. Finally challenges of existing hybrid systems based on rough sets and directions for future research are indicated in section 5.

2 Rough Sets and Soft Computing

Rough set theory, which was introduced by Pawlak [6] in the early 1980s, is a mathematical approach that can be employed to handle imprecision, vagueness and uncertainty.

Rough sets have many important advantages for data mining [7], such as providing efficient algorithms for finding hidden patterns in data, finding minimal sets of data, generating sets of decision rules from data, and offering straightforward interpretation of obtained results.

In the last two decades, rough sets have widely been applied to data mining and rapidly established themselves in many real-life applications such as medical diagnosis, control algorithm acquisition and process control and structural engineering.

Soft computing, a term coined by Zadeh [2], is a consortium of methodologies that works synergetically and provides in one form or another flexible information processing capability for handling real life ambiguous situations. Its aim is to exploit the tolerance for imprecision, uncertainty, approximate reasoning, and partial truth in order to achieve tractability, robustness, and low-cost solutions. The guiding principle is to devise methods of computation that lead to an acceptable solution at low cost by seeking for an approximate solution to a problem, possibly formulated in an imprecisely way.

Fuzzy logic, neural networks and Probabilistic reasoning are the initial components of soft computing. Later other methodologies like genetic algorithm, chaotic system, and rough sets became members of soft computing one by one. Traditional research of soft computing concentrates on the combination of neural networks, fuzzy sets and genetic algorithm [4], among them neuro-fuzzy computation is the most notable achievement [5].

It is only in recent years that rough sets are studies as a soft computing technology along with other soft computing technologies. As Pawlak said, "...the theory (rough set theory) is not competitive but complementary to other methods and can also be often used jointly with other approaches (e.g. fuzzy sets, genetic algorithms, statistical methods, neural networks etc.)" [8]. The combination of rough sets and other soft computing technologies such as fuzzy sets, genetic algorithms and neural networks has attracted much attention [9], [10] and is growing into an important issue of soft computing.

3 Hybrid Soft Computing Systems Based on Rough Sets

A data mining process can be viewed as consisting of three phases, shown in Fig. 1. Data preprocessing converts raw data into input data of mining algorithm. Many tasks, such as attribute reduction and data discretization, can be included in this phase. An efficient data preprocessing phase not only provides data that can be handled by mining algorithm, but improves efficiency and performance of data mining system by removing redundant or irrelevant data. Mining algorithm is the most important part in the whole process, which extracting patterns from preprocessed data. Pattern is comprehensible for experts of the data mining system, but not for common users. So the phase of interpretation and visualization are needed. By this phase pattern is transformed into knowledge that can be easily understood and applied by users.

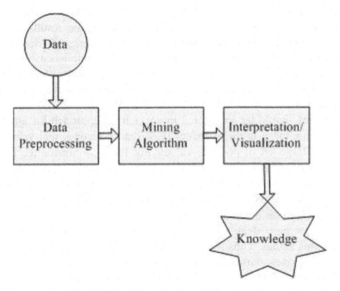

Fig. 1. Overview of a data mining process

Based on the available literatures on rough sets based hybrid soft computing systems, the functions of rough sets in a data mining process can be classified into three categories: preprocessing data, measuring uncertainty, mining knowledge. Each category is described below along with a survey of the existing hybrid systems belong to the category.

3.1 Attribute Reduction by Rough Sets

In this type of hybridization, rough sets are usually used to reduce the size of a dataset by removing some redundant or irrelevant attributes while preserving the classification performance of dataset.

Attribute reduction plays an important role in data mining. Irrelevant and redundant attributes generally affect the performance of mining algorithms. A good choice of a useful attribute subset helps in devising a compact and efficient learning algorithm as well as results in better understanding and interpretation of data in data mining.

Rough sets can be a useful tool for pre-processing data for neural network by attribute reduction to reduce the network's input vector. In [11], a hybrid model integrating rough sets and an artificial neural network is presented to forecast an air-conditioning load. Rough sets are applied to find relevant factors to the load, which are used as inputs of an artificial neural-network to predict the cooling load. Experimental results from a real air-conditioning system show that the hybrid forecasting model is better than comparable models and its relative error is within 4%.

Similarly, in this context rough sets can be integrated with genetic algorithm [12], [13], [14]. In [12], a hybrid approach is proposed to bankruptcy prediction. It firstly uses a rough sets model to identify subsets of potentially important explanatory variables, and then a genetic programming algorithm is applied to construct a bankruptcy prediction model based on these variables. The experimental results show that this hybrid system is more efficient and more effective than an original rough sets model.

Sometimes the process of attribute reduction was accomplished by two steps for addressing noisy data [15], [16]. In the first step rough sets are used to delete irrelevant and redundant attributes, and in the second step neural networks are used to delete noisy attributes.

Two constraints should be considered in a rough set approach for attribute reduction [10]. First, finding a minimal reduct is NP-hard. Second, real-valued attributes can have a very large set of possible values. For the first constraints, Li et al. [17] used genetic algorithm to get the minimal reduct of attributes and Slezak et al. [18] also provided a method of order based genetic algorithm to find optimal approximate entropy reduct. The second constraint was solved in [19] through reducing attribute value field by a rough set based algorithm.

Rough sets can only deal with discrete values. However, real-world data are usually continuous values. Thus it needs a step to transform the continuous valued attributes to discrete ones, namely it should partition the domains of every continuous valued attribute into intervals. In [9] Dependency factor of rough sets is used as the fitness function of GA to get the optimal cutpoints of intervals in order to ensure the maximum consistency of the discrete data.

3.2 Measuring Uncertainty in Data by Rough Sets

In this context rough sets are not the main technique to directly extract knowledge from data, but are used as an assistant tool to estimate some system parameters by its analysis capability for uncertainty in data.

The choice of architecture is a difficult task for neural networks. Rules generated from data by rough sets can be used to determine the architecture of a neural network, including the number of hidden layers, the number of hidden layer nodes and the initial weights of networks [20], [21], [22], [23]. This hybridization helps neural networks possess the characteristics of short learning time, good understandability and strong generalization ability [24], [25].

Qin and Mao [20] proposed an algorithm based on rough sets for fixing the optimal number of hidden layer units. The results on Chinese medicine practical projects show that this algorithm is of wide applicability for the study of neural network architecture.

Yasdi [21] used rough sets for the design of knowledge–based networks in the rough-neuro framework. This methodology consists of generating rules from training examples by using rough sets concepts and mapping them into a single layer of connection weights of a four-layered neural network. In this rough-neuro framework, rough sets were used to speed up or simplify the process of using neural networks. Pal et al. [22] designed a rough-fuzzy MLP for voice recognition. The rules obtained by rough sets are encoded into a neural network, while the confidences of the rules are used to initial the weights of the neural network. The results show that the performance of rough-fuzzy MLP is superior to that of fuzzy MLP and that of MLP without prior knowledge.

On the other hand, rough sets also can be used to adjust the structure of a trained neural network [26], [27]. In [26], the structure of a trained neural network is optimized by rough sets. Dependency of rough set rule is used in an iterative algorithm until a minimal number of rules of the model are selected. Hassan and Tazaki [27] present a hybrid decision support system for medical diagnosis in which rough sets are used to delete the nodes of a neural network.

Fuzzy sets are dependent on expert knowledge, while rough sets can deal with data without any preliminary or additional information. So introduction of rough sets can enlarge the application area of fuzzy sets. Base on rough sets [28] proposes a parameter-free roughness measure for fuzzy sets, which does not depend on parameters that are designed as thresholds of definiteness and possibility in membership of the objects to a fuzzy set.

3.3 Mining Knowledge by Rough Sets

Rule is the most common form of knowledge generated by rough sets based method. Many rule generation algorithms based on rough sets have been proposed to obtain the final knowledge in hybrid soft computing systems [16].

When rough sets is used as the main technique of mining rules from data, other soft computing technologies is also used as the assistant tools to deal with some problems that rough sets can not accomplish.

Some rule extraction algorithms based on rough sets may produce large amounts of rules that are not all useful according to some measures such as accuracy and coverage, and the work picking out the interesting ones from these rules is time consuming. In order to address this problem, Huang and Dai [29] presented a hybrid system based on rough sets and genetic algorithms, rough set is used to extract rules and genetic algorithm is exploited to find the optimal probabilistic rules that have the highest accuracy and coverage, and shortest length. Experimental results show that it run more efficiently than those traditional methods.

In [30], a hybrid scheme based on fuzzy sets and rough sets are proposed for breast cancer detection. Fuzzy sets are firstly used to pre-processing breast cancer images for enhancing the contrast of images. Rough sets-based approaches are applied for attribute reduction and rule extraction. Experimental results show that the hybrid scheme performs well reaching over 98% in overall accuracy. In [31] fuzzy sets are introduced to a rough sets-based information measure for getting the reducts with a better performance.

In order to generating from data with quantitative values which are common in real-world application, Hong et al. [32] proposed an algorithm to produce a set of maximally general fuzzy rules for an approximate coverage of training examples based on the variable rough set model form noisy quantitative training data. Each quantitative value is firstly transformed into a fuzzy set of linguistic terms using membership fuctions. Then fuzzy rules are generated by a rough set-based method.

Rule generation from artificial neural networks has received a great deal of research due to its capability of providing some insight to the user about the symbolic knowledge embedded within the network [33].

In order to improve the classification accuracy of rules, many schemas were proposed. Ahn et al. [34] proposed a hybrid intelligent system by combining rough set approach with neural network for predicting the failure of firms based on the past financial performance data. When a new object is predicted by rule set, it is fed into the neural network if it does not match any of the rules. Thus they use rough sets as a preprocessing tool for neural networks. Though this approach can get high classification accuracy, some knowledge in neural networks is still hidden and not comprehensible for user.

It seems that using fuzzy rules instead of classical decision rules may improve classification results. In [35], a hybrid system combing rough sets and fuzzy sets is provided to improve classification process. Fuzzy sets support approximate reasoning and rough sets are responsible for data analyzing and process of automatic fuzzy rules generation. Drwal and Sikora [36] proposed a rough-fuzzy hybridization scheme for case generation. In this method, fuzzy sets are firstly used to represent a pattern in terms of its membership to linguistic variables. This gives rise to efficient fuzzy granulation of the feature space. Then rough sets are used to obtain fuzzy rules that contain informative and irreducible information both in terms of features and patterns. These fuzzy rules represent different clusters in the granular feature space, which can be mapped to different cases by fuzzy membership functions. Experimental results obtained on three UCI data sets show that this method is superior to traditional methods in terms of classification accuracy, case generation, and retrieval times.

4 A General Observation

Generally, in the two former hybridization types rough sets are used as an assistant tool for mining knowledge from data, and in the latter type as a major element of hybrid systems to directly extracting rule knowledge from data.

Combination of rough sets and neural networks is popular for complementary features of the two technologies. Attribute reduction of rough sets can shorten training time of neural networks, and rules extracted from a trained neural network by rough sets will greatly improve the interpretation of knowledge embedded in trained neural network. On the other hand, neural network is also a good candidate helps rough set deal with noise in the data.

When integrated with rough sets, genetic algorithms are involved in various optimization and search processes, especially for the NP problem of rough set-based algorithms, like optimal reduct and optimal rule set. While Fuzzy sets are suitable for handling the problems related to fuzzy cases.

It is no doubt that hybridization of soft computing technologies can improve performance of a data mining system in such areas as tractablility, low cost, and robustness. According to the specific problems to be solved, hybrid systems can be summarized in Table 1 and Table 2. Table 1 shows typical problems of rough sets that are relievable by other soft computing technologies, while obvious problems of other soft computing technologies with corresponding solution of rough sets are listed in Table 2.

Table 1. Problems of rough sets and corresponding solutions provided by other soft computing techologies

Problem	Solution
Sensitivity to noise	Filtrating noise in data by neural networks [15, 16]
NP-hard in attribute reduct and rule extraction	GA-based optimation method [17-19, 29]
Low generalization of rule prediction	Predicted with neural networks classifier or integrated with fuzzy rules [34-36]
Difficulty to deal with fuzzy features of data	Fuzzy sets [30-32]

Table 2. Problems of other soft computing tecnnologies and corresponding solutions provided by rough sets

Problem	Solution
Time-consuming of neural network training	Reducing data by rough set-based pre-processing [11]
Lack of explanation of trained neural networks	Rule extraction by rough sets from data or from trained neural networks [33]
Uncertainty on initial architecture and parameters of neural networks	Measuring uncertainty by rough sets [20][21][22][23]
Dependence for domain information of fuzzy sets	Rough sets can analyze data without any preliminary or additional information about data [28]
Time-consuming of convergence of genetic algorithm search	Reducing attributes by rough sets[12][13][14]

5 Conclusions

In this paper we presented an overview of hybrid soft computing systems from a rough set perspective. These systems are discussed and summarized from three categories according to different functions of rough sets in a data mining process: preprocessing data, measuring uncertainty, and mining knowledge. From these systems it can be immediate concluded that rough sets are complementary with other soft computing technologies including neural networks, genetic algorithms and fuzzy sets. Hybridization of rough sets and other soft computing technologies is a promising approach for developing robust, high-efficient, and low-cost data mining systems.

Although hybrid soft computing systems based on rough sets have been successfully applied in many areas, some of the challenges still exist. For example:

- Lack of general theoretical framework and design principle for integration of rough sets and other technologies.
- Efficient mechanism to handle complex data which is dynamic, incomplete, sequential, etc.
- Integration of extensive rough set theory and other soft computing methodologies.
- Quantitative evaluation of system performance

Some aspects of hybrid soft computing systems based on rough sets are as follows:

1) At present most systems are based on classical rough set theory. However, many limitations of classical rough set theory have been widely realized and corresponding extensive theory for rough sets have been proposed. How to integrate efficiently extensive rough set theory with neural networks and genetic algorithms needs to be developed in near future. Some attempts in this direction are described in [37].

2) Facing complex data, efficient hybridization mechanism including management for missing values and update of knowledge is a potential issue [38].

3) Interaction with user is of great significance for successful application of a data mining system. So the related researches such as utilization of expert knowledge and visualization of mined knowledge need to be studied more deeply.

4) Novel combinations are still needed for better utilizing the advantages of rough sets and other technologies and developing high-performance, low-cost and robust data mining systems.

Acknowledgement

This work is supported by the Youth Natural Science Foundation of Ludong University under contract 22480301.

References

1. Chen, M., Han, J., Yu, P.S.: Data Mining: An Overview Form a Database Perspective. IEEE Trans. Knowledge and Data Engineering 8, 866–883 (1996)
2. Zadeh, L.A.: Fuzzy Logic, Neural Networks, and Soft Computing. Commun. ACM 37, 77–84 (1994)
3. Mitra, S., Pal, S.K., Mitra, P.: Data Mining in Soft Computing Framework: A Survey. IEEE Trans. Neural Networks 11, 3–14 (2002)

4. Bonissone, P.P., Chen, Y.-T., Goebel, K., et al.: Hybrid Soft Computing Systems - Industrial and Commercial Applications. Proceedings of the IEEE 9, 1641–1667 (1999)
5. Mitra, S., Hayashi, Y.: Neuro-fuzzy Rule Generation: Survey in Soft Computing Framework. IEEE Trans. Neural Networks 11, 748–768 (2000)
6. Pawlak, Z.: Rough Sets. Inter. J. of Computer and Information Sciences 5, 341–356 (1982)
7. Pawlak, Z.: Rough Sets and Intelligent Data Analysis. Information Sciences 147, 1–12 (2002)
8. Pawlak, Z.: Data Mining – A Rough Set Perspective. In: Zhong, N., Zhou, L. (eds.) Methodologies for Knowledge Discovery and Data Mining. LNCS (LNAI), vol. 1574, pp. 3–12. Springer, Heidelberg (1999)
9. Li, Y., Jiang, J.-P.: The Integrated Methodology of Rough Sets Theory, Fuzzy Logic and Genetic Algorithms for Multisensor Fusion. In: Proceedings of the American Control Conference. IEEE Arlington VA, pp. 25–27. IEEE Computer Society Press, Los Alamitos (2001)
10. James, F.P., Marcin, S.S.: Rough Neurocomputing: A Survey of Basic Models of Neurocomputation. In: Alpigini, J.J., Peters, J.F., Skowron, A., Zhong, N. (eds.) RSCTC 2002. LNCS (LNAI), vol. 2475, pp. 308–315. Springer, Heidelberg (2002)
11. Hou, Z., Lian, Z., et al.: Cooling-load prediction by the combination of rough set theory and an artificial neural-network based on data-fusion technique. Applied Energy 83, 1033–1046 (2006)
12. Thomas, E.M., Terje, L.: Genetic Programming and Rough Sets: A Hybrid Approach to Bankruptcy Classification. European Journal of Operational Research 138, 436–451 (2002)
13. Zhai, L.-Y., Khoo, L.-P., Fok, S.-C.: Feature Extraction Using Rough Sets and Genetic Algorithms – An application for the Simplification of Product Quality Evaluation. Computers & Industrial Engineering 43, 661–676 (2002)
14. Khoo, L.-P., Zhai, L.-Y.: A Prototype Genetic Algorithm-enhanced Rough Set-based Rule Induction System. Computers in Industry 46, 95–106 (2001)
15. Chakraborty, B.: Feature Subset Selection by Neuro-rough Hybridization. In: Ziarko, W., Yao, Y. (eds.) RSCTC 2000. LNCS (LNAI), vol. 2005, pp. 519–526. Springer, Heidelberg (2001)
16. Li, R., Wang, Z.: Mining Classification Rules Using Rough Sets and Neural Networks. European Journal of Operational Research 157, 439–448 (2004)
17. Li, Q.D., Chi, Z.X., Shi, W.B.: Application of Rough Set Theory and Artificial Neural Network for load forecasting. In: Proceedings of the first Int. Conf. on Machine Learning and Cybernetics, pp. 1148–1152. IEEE Computer Society Press, Beijing (2002)
18. Slezak, D., Wroblewski, J.: Order Based Genetic Algorithms for The Search of Approximate Entropy Reduct. In: Wang, G., Liu, Q., Yao, Y., Skowron, A. (eds.) RSFDGrC 2003. LNCS (LNAI), vol. 2639, pp. 308–311. Springer, Heidelberg (2003)
19. Jelonek, J., Krawiec, K., Slowinski, R.: Rough Set Reduction of Attributes and Their Domains for Neural Networks. Computational Intelligence 11, 339–347 (1995)
20. Qin, Z., Mao, Z.: A New Algorithm for Neural Network Architecture Study. In: Proceedings of the 3rd World Congress on Intelligent Control and Automation, pp. 795–799. IEEE Computer Society Press, Hefei (2000)
21. Yasdi, R.: Combining Rough Sets Learning and Neural Learning Method to Deal with Uncertain and Imprecise Information. Neurocomputation 7, 61–84 (1995)
22. Pal, S.K., Mitra, S., Mitra, P.: Rough-fuzzy MLP: Modular Evolution, Rule Generation, and Evaluation. IEEE Trans. on Knowledge and Data Engineering 15, 14–25 (2003)

23. Huang, X.M., Yi, J.K., Zhang, Y.H.: A method of constructing fuzzy neural network based on rough set theory. In: International Conference on Machine Learning and Cybernetics, pp. 1723–1728. IEEE Computer Society Press, Xi'an (2003)
24. Tian, K., Wang, J., et al.: Design of a novel neural networks based on rough sets. In: Chinese Control Conference, pp. 834–838. IEEE Computer Society Press, Harbin (2006)
25. Wu, Z.C.: Research on Remote Sensing Image Classification Using Neural Network Based on Rough Sets. In: International Conference on Info-tech and Info-net, pp. 279–284. IEEE Press & PPTPH, Beijing (2001)
26. Yon, J.H., Yang, S.M., Jeon, H.T.: Structure Optimization of Fuzzy-Neural Network Using Rough Set Theory. In: Proceedings of 1999 IEEE International Conference on Fuzzy Systems, pp. 1666–1670. IEEE Computer Society Press, Seoul (1999)
27. Hassan, Y., Tazaki, E.: Decision Making Using Hybrid Rough Sets and Neural Networks. Int. J. of Neural System 12, 435–446 (2002)
28. Huynh, V.N., Nakamori, Y.: An Approach to Roughness of Fuzzy Sets. In: Proceedings of the 13th IEEE International Conference on Fuzzy Systems, pp. 115–120. IEEE Computer Society Press, Budapest (2004)
29. Hang, X., Dai, H.: An Optimal Strategy for Extracting Probabilistic Rules by Combining Rough Sets and Genetic Algorithm. In: Grieser, G., Tanaka, Y., Yamamoto, A. (eds.) DS 2003. LNCS (LNAI), vol. 2843, pp. 153–165. Springer, Heidelberg (2003)
30. Hassanien, A.: Fuzzy rough sets hybrid scheme for breast cancer detection. Image and Vision Computing 25, 172–183 (2007)
31. Hu, Q., Yu, D., Xie, Z.: Information-preserving hybrid data reduction based on fuzzy-rough techniques. Pattern Recognition Letters 27, 414–423 (2006)
32. Hong, T.-P., Wang, T.-T., Chien, B.-C.: Learning Approximate Fuzzy Rules from Training Examples. In: Proceedings of the 2001 IEEE International Conference on Fuzzy Systems, pp. 256–259. IEEE Computer Society Press, Melbourne (2001)
33. Fu, L.: Rule Generation from Neural Networks. IEEE Trans. on Systems, Man and Cybernetics 24, 1114–1124 (1994)
34. Ahn, B.S., Cho, S.S., Kim, C.Y.: The Integrated Methodology of Rough Set Theory and Artificial Neural Network for Business Failure Prediction. Expert Systems with Application 18, 65–74 (2000)
35. Pal, S.K., Mitra, P.: Case Generation Using Rough Sets with Fuzzy Representation. IEEE Trans. on Knowledge and Data Engineering 16, 292–300 (2004)
36. Drwal, G., Sikora, M.: Induction of Fuzzy Decision Rules Based Upon Rough Sets Theory. In: IEEE Conf. on Fuzzy Systems, pp. 1391–1395. IEEE Computer Society Press, Gliwice (2004)
37. Su, C.-T., Hsu, J.-H.: Precision parameter in the variable precision rough sets model: an application. Omega 34, 149–157 (2006)
38. Sikder, I., Gangopadhyay, A.: Managing uncertainty in location services using rough set and evidence theory. Expert Systems with Applications 32, 386–396 (2007)

Discovering Novel Multistage Attack Strategies

Zhitang Li, Aifang Zhang, Dong Li, and Li Wang

Computer Science and Technology Department, Huazhong University of Science and
Technology, Wuhan, Hubei, 430074, China
{leeying, frost}@mail.hust.edu.cn

Abstract. In monitoring anomalous network activities, intrusion detection
systems tend to generate a large amount of alerts, which greatly increase the
workload of post-detection analysis and decision-making. A system to detect
the ongoing attacks and predict the upcoming next step of a multistage attack in
alert streams by using known attack patterns can effectively solve this problem.
The complete, correct and up to date pattern rule of various network attack
activities plays an important role in such a system. An approach based on
sequential pattern mining technique to discover multistage attack activity
patterns is efficient to reduce the labor to construct pattern rules. But in a
dynamic network environment where novel attack strategies appear
continuously, the novel approach that we propose to use incremental mining
algorithm shows better capability to detect recently appeared attack. In order
to improve the correctness of results and shorten the running time of the mining
algorithms, the directed graph is presented to restrict the scope of data queried
in mining phase, which is especially useful in incremental mining. Finally, we
remove the unexpected results from mining by computing probabilistic score
between successive steps in a multistage attack pattern. A series of experiments
show the validity of the methods in this paper.

Keywords: alert correlation, sequential pattern, multistage attack, incremental
mining.

1 Introduction

In order to discover potential security threats in protected network, more and more
security devices such as firewall, IDS and vulnerability scanner are deployed. These
various devices in different locations can reduce false positives and false negatives
resulting from dependence on single device. They also can support and supplement
each other so that complicated multistage can be disclosed.

However, with the growing deployment of these devices, the large volume of alerts
gathered from these devices often overwhelm the administrator, and make it almost
impossible to discover complicated multistage attacks in time. It is necessary to
develop a real-time system to detect the ongoing attacks and predict the upcoming
next step of a multistage attack, using known attack patterns. Although such a
detection system effectively reduces the need for operation staff, it requires
constructing and maintaining rule base. So it is a key mission to make sure that the
pattern rule is correct, complete and up to date.

R. Alhajj et al. (Eds.): ADMA 2007, LNAI 4632, pp. 45–56, 2007.
© Springer-Verlag Berlin Heidelberg 2007

In this paper, a classical data mining algorithm is used to help us discover attack activity pattern, construct and maintain composite attack activity pattern base. It can overcome the highly dependent on knowledge of experts, time-consuming and error-prone drawbacks in previous approaches using manual analysis. Unfortunately, for a dynamic network environment where novel attack strategies appear continuously, the method shows a limited capability to detect the newly-appeared attack patterns. We can address the problem by presenting a novel approach using incremental mining algorithm to discover new composite attack activity patterns that appear recently.

To trim away any data that we knew to be irrelevant before starting the mining activities, the directed graph is used to restrict the scope of data queried. We remove the unexpected results by computing relativity score. The two methods can improve the correctness of results and shorten the running time of the algorithm.

The remainder of this paper is organized as follows. Section 2 discusses the related work. We propose to use the sequential pattern mining algorithm to discovery multistage attack patterns in Section 3. In Section 4, we briefly discuss how to update patterns using incremental mining. We present the directed graph and probabilistic correlation approaches to improve the running speed of mining algorithms and the correctness of results in Section 5. Finally, the paper is summarized and future work is pointed out in Section 6.

2   Related Work

IDS can report single-step attack activities, but can not discover complicated multistage attacks. In order to discover composite multistage attacks, approaches for alert correlation are proposed in many papers. They dedicate to disclose the logical association between network attack activities by analyzing their corresponding alerts. They are classified as fellows: pre-/post-condition method, aggregation method, intention-based method and model-based method.

Ning et al.[1][2][3], Cuppens and Mi`ege [4][5][6] build alert correlation systems based on matching the pre-/post-conditions of individual alerts. The idea of this approach is that prior attack steps prepare for later ones. The correlation engine searches alert pairs that have a consequence and prerequisite matching. One challenge to this approach is that a new attack cannot be paired with any other attacks because its prerequisites and consequences are not defined. This approach is highly dependent on the priori knowledge and expert experience. It is also difficult to enumerate all possible matching conditions between attacks.

Valdes et al. [7] propose a probabilistic-based approach to correlate security alerts by measuring and evaluating the similarities of alert attributes. They use a similarity metric to fuse alerts into meta-alerts to provide a higher-level view of the security state of the system. Alert aggregation is conducted by enhancing or relaxing the similarity requirements in some attribute fields. This method relies too much on the expert knowledge.

Ming Yuh Huang [8] proposes the method based on intrusion intention and intrusion strategy to detect and predict the composite attacks by analyzing the potential intention instead of the inherent logic of attack activities.

Sheyner and J. W. Wing et al. [9] propose a model checking based technique to automatically construct attack graphs. Although it helps facilitate the task of defining attack graphs, it has the limitation of scalability especially for larger network and systems.

In addition, In [10][11],Wenke Lee and Xinzhou Qin proposed a GCT-based and Bayesian-based correlation approach without the dependence of the prior knowledge of attack transition patterns. However, these methods are only suitable for discovering causal or statistical relationship between alerts.

James J. and Ramakrishna[12] used Association rule mining algorithm Apriori to discover the casual relationships between an attacker and the combination of alerts generated in IDS as a result of their behavior in a network.

All these approaches overcome the drawback of high dependence on knowledge of experts, time-consuming and error-prone in previous approaches with manual definition.

Algorithm to discover frequent episodes in a single long sequence and its generalization can be found in [13]. Several algorithms were developed for the mining of sequential patterns [14][15]. The incremental update of sequential pattern is presented in [16][17].

So in this paper, as in [18], we also use a classical sequential pattern mining algorithm, to discover attack behavior patterns, and provide the foundation for definition of patterns.

The approach show degraded detection capability in a dynamic network environment where novel attack strategies appear continuously. To keep patterns up to date, we presented an incremental mining algorithm based on the GSP that can discover novel sequential patterns by only focus on the recent data sequence. The directed graph and Bayesian network are used to improve running speed of the algorithm and correctness of the results.

3   Generating Pattern Rules

Our motivating application are rule construction in the multistage composite attack detection system used for monitoring the protected network, where thousands of alerts accumulate daily; there can be hundreds of different alert types. Rule construction requires experts to identify problem patterns, a process that is time-consuming and error-prone. In this paper, we use GSP, an algorithm for discovering sequential patterns in data mining field, to discover multistage attack behavior patterns. The idea of this method comes from the observation that among all these intrusion threat, most are intended attacks and are of multiple steps taken by attackers with specific attack intents. Multistage attack strategy taken by the attacker usually has relatively fixed attack pattern and occurs in a confined time span.

3.1 Problem Statement

We collect alerts from the different devices deployed in the protected network for a long time. All the alerts stored in database can be viewed as a global sequence of alerts sorted by ascending DetectTime timestamp. Sequences of alerts describe the behavior and actions of attackers. We can find out multistage attack strategies by analyzing this alert sequence. A sequential pattern is a collection of alerts that occur relatively close to each other in a given order frequently. Once such patterns are known, the rules can be produced for describing or predicting the behavior of the sequence of network attack.

In order to discover correct and interesting patterns, the alerts need to be reduced by filtering background noise, aggregation, clustering, verifying and so on. In this way, a data set comprised of rich multistage attack alerts but few false alerts are gained. Therefore, the problem of mining attack activity patterns becomes to find out the frequent alert subsequences among the global alert sequence that are not less than the user-specified minimum frequency and complete in user-specified time span.

3.2 Preparation for GSP

The GSP algorithm is initially used for mining frequent sequential patterns in transaction database to discovery the customer purchase patterns. All the transactions of a customer that are sorted in time order is a customer sequence. The problem of mining sequential patterns is to find the maximal sequences among all customer sequences that have a certain user-specified minimum support. Each such maximal sequence represents a sequential pattern.

For our database, there is only a single global alert sequence. So we have to modify the algorithm slightly to meet our requirement. Here the concept of time window is introduced to represent the time span between the first step and the last step of a multistage attack. It is based on our observation that most attackers complete their attacks in a certain time span. In this paper, we postulate that all multistage attacks finish in the time window, which is not perfect but can account for most situations.

We divide the global alert sequence according to time window size. It leads to multiple candidate subsequences generated, and a subsequence is a series of alerts in the same time window. Since attack behavior occurrence time span is stochastic, the number of alerts in subsequence is variable. Our algorithm only focuses on the attack class and timestamp attributes in the alerts. Here we use respective attack classes instead of the alerts in order to omit the other concrete attributes and discovery generalized patterns. Note that all the attack classes are normalized before the algorithm executes. In order to run quickly, we map the attack classes to integer number.

Table 1 shows a part of the global alert sequence with its DetectTime timestamp ascending. Table 2 illustrates the multiple subsequences generated by dividing the global alert sequence according to sliding time window size. To simplify the problem, we only list four subsequences.

Table 1. Part of global alert sequence

Attack Class	DetectTime	Other attributes
...	...	...
7	06-08-09-11:12:14	...
4	06-08-09-11:13:18	...
5	06-08-09-11:34:46	...
2	06-08-09-11:34:46	...
3	06-08-10-00:06:53	...
5	06-08-10-00:12:47	...
...	...	...

Table 2. Alert subsequences

Subsequence id	Attack class sequence
1	<(7,5),2,6,4>
2	<7,6,4,(6,5)>
3	<7,2,6,4>
4	<7,6,5>

3.3 Using GSP to Discover Sequential Patterns

The problem of mining attack behavior patterns transforms to mine the frequent attack class sequences from these candidate alert subsequences in database. These subsequences are taken as customer sequences in GSP. Consequently, GSP can execute in the same way as it does on transaction database.

Table 3 illustrates the complete course of generating maximal sequence with support 40%. The results of GSP is {<7,2,6,4>, <7,6,5>, <4,5>}. Other sequences are deleted because they are subsequences included in <7,2,6,4>.

Table 3. Generation of Maximal Sequence with support 40%

Subsequence id	1-sequence	2-sequence	3-sequence	4-sequence
<(7,5),2,6,4>	<7> <2>	<7,2>,<7,6>,<7,4>,	<7,2,6 >,<7,2,4>,	<7,2,6,4>
<7,6,4,(6,5)>	<6> <4>	<7,5>,<2,6>,<2,4>,	<7,6,4>,<7,6,5>,	
<7,2,6,4>	<5>	<6,4>,<6,5>,<4,5>	<2,6,4>	
<7,6,5>				

In fact, not all maximal sequences found out here are correct and interesting. We have to analyze the results carefully and remove the illogical patterns. We discuss these in a later section in this paper. The results after removed can be transformed into rules automatically. In contrast with other approaches, the mining algorithm provides a means of reducing the amount of labor required to discover relationships between attack behaviors and define rules.

4 Incremental Mining

Unfortunately, the real-time detection system using known attack patterns often shows degraded detection capability in a dynamic network environment where novel attack strategies often appear. The reason is that the rules can not be updated in time. The aim of this section is to address the challenge of how to update correlation rules in time, overcoming the limitation in other methods that can not discover novel attack strategies.

4.1 Problem Statement

The delay of pattern update results from the fact that GSP can only discover sequential patterns in historical data. Whereas, update involving in new attack type or new attack sequences is made to the alert database continuously and frequent re-executions of the algorithm on the updated database are infeasible when the data sequence increases to a huge order of magnitude. On the other hand, any delay in updating the rule base could result in potentially undetected attacks. So we presented a novel method using an incremental mining algorithm that can discover novel sequential patterns by only focus on the recent data sequence.

4.2 Incremental Update

The algorithm is based on the sequential nature of data, and only focus on the updated data sequence. Let |DB| be the number of data sequences in the original database DB and min_sup be the minimum support. After some update of the database, new alerts are appended to DB. These alerts are also sorted by timestamp in the increment database db. UD is the updated database combining the whole data sequence from DB and db, UD = DB $\cup$ db. Let S_k^{DB} be the set of all frequent k-sequences in DB, S_k^{UD} be the set of all frequent k-sequences in UD, and the set of all sequential patterns in DB and UD be S^{DB} and S^{UD} respectively.

Assume that for each sequential pattern s in *DB*, its support count, denoted by S_{count}^{DB}, is available. The support count can get easily by record it into another database or file during the running process of mining sequential pattern algorithm. For db, we also divide the data into multiple subsequences by time window size used in the pattern discovery section. Let |db| be the number of data sequences in the increment database db. Without loss of generality, with respect to the same minimum support *min_sup*, a sequence s is a frequent sequence in the updated database *UD*, if its support count S_{count}^{UD} is greater than *min_sup* (total subsequences in *UD*).

We can observe the changes that a sequential pattern s in S^{DB} might not be in S^{UD} because of database update. On the other hand, a sequence s' that is not in S^{DB} might turn out to be in S^{UD}. Consequently, the problem of incremental update on sequential patterns is to find the new set S^{UD} of frequent sequences in UD.

Different from customer transaction database, merge data sequence from DB and db is not needed, because there are no new alert appended to the old alert sequences in DB thanks to the sequential feature of alert data division. This makes the problem simple, and we can only use the second part of FASTUP algorithm [17] to discover new attack patterns.

4.3 Discussion

In contrast to re-mining, the incremental update algorithm which effectively utilizes discovered knowledge is the key to improve mining performance. By counting over appended data sequences instead of entire updated database in most cases, fast filtering patterns found in last mining and successive candidate sequence reductions together make efficient update on sequential patterns possible.

In fact, in order to update rule base in time, we can also focus on the min_support on the db rather than UD. A sequence whose support count exceeds min_support×|db| is taken as a frequent pattern, then is considered to be added into rule base. Our method provides a means of reducing the amount of labor required to keep the rules current, at the same time significantly reducing the amount of time which elapses from the appearance of a new attack pattern to transformation to the corresponding rule in the real-time detection system. If the alerts arriving at the detection system match a predefined rule, an alarm is triggered and displayed on the console for inspection. Accordingly, network administrator can be aware of the threat as soon as possible and take deliberate action to prevent the target host of an attack from further compromise.

5 Improvement

Using the mining algorithm directly is often less efficient and correct, owing to the large volumes of data that required analysis. Sometimes, certain alerts in the overall sequence even are not related to one another. It was beneficial to trim away any data that we knew to be irrelevant before starting the mining activities. So in this paper we present directed graph to restrict the scope of query in mining algorithm.

Because the mining algorithm only focus on the sequence and frequency of attack class, the result transforming to rules directly often come into being redundancy even errors, so we have to analyze the result further using probabilistic correlation.

5.1 Directed Graph

In order to facilitate a novel technique for filtering the alerts which must be analyze during the mining process, we generated a directed graph which modeled the alerts to be analyzed. Each entry in the alert database included both the source IP address and destination IP address. We deduced the direction of each potential attack from this information. We then generated a directed graph $G = (V, E)$ such that each IP address was represented as a vertex in the graph, and each edge was represented by a detected alarm. The edge was drawn from the source IP address toward the destination IP address, corresponding to the direction of the alert.

With the directed graph drawn, we can easily identify the connected components in the graph. Only the alerts involving in the same connected component are likely to have the same intent. So we limit our each mining activity solely to alerts in the same connected component, exploring only relationships between alerts which could legitimately exist.

The method is useful to improve the running speed of mining algorithm and the correctness of the result by removing the irrelevant alerts from consideration. It is especially efficient to restrict the scope of data queried by incremental update algorithm because the number of newly appended alerts is limited and they occur in a relatively shorter time span. We also observe that most of alerts in a short time span are always involved in limited hosts.

If we were unable to safely remove significant numbers of rows from consideration by filtering on connected component, the time required for the mining algorithm to generate results grew rapidly. A side effect produced by this complication was the generation of a very large number of rules by the algorithm. This number of rules on its own is of limited value, as it does not solve the problem of limiting the amount of data which must be examined manually by operational staff. However, the vast majority of the time, the count of rules for a single network on a single day was below 100.

5.2 Probabilistic Correlation

The patterns discovered by mining algorithm are not absolutely correct and interesting because the algorithm focuses only on the temporal and attack class constraints. And the frequent occurrences of certain alerts that may not be completely removed by pre-analysis can also lead to generating improper patterns, which will result in the false alarms of real-time detection system. So we have to analyze the results carefully and remove the illogical patterns. For example, it is more rational that an exploit follows a probe than the other way round. Therefore, we can remove the obviously illogical sequence. However, we can not remove all improper sequences so easily because no more precise information can be used in the results of sequential patterns mining algorithm.

For a pattern which is not obviously false, we need to search the candidate subsequences in the alert database which support it for detailed information. Then the pattern probability score is computed. Because the alerts have the same format with multiple attributes, we can define the appropriate similarity function for each attribute. Then the alert similarity is defined as the weighted sum of the attribute similarity. The weight values can be set empirically and tuned in practice. The alert similarity reveals the logical relationship between the two successive attack steps in a multistage attack sequence. The higher the score is, the more likely that they belong to the same attack sequence. The pattern probability score is the average of alert similarity in each candidate subsequence supporting the pattern.

Only those patterns whose pattern probability score between each two successive steps exceeds a certain threshold value can be considered as rule and added into rule base. At the same time, the score is also recorded as a reference for match in real-time detection system.

6 Experiment

To evaluate the effectiveness and validity of our method, we performed a series of experiments using DARPA 2000 benchmark repository, scenario-specific datasets [19] and live data collected from our network center. We describe and report our experiment results in this section.

6.1 DARPA 2000 Experiment

DARPA 2000 datasets consist of two intrusion scenarios, LLDDOS1.0 and LLDDOS2.0.2. Each dataset includes the network traffic collected from both the DMZ and the inside part of the evaluation network. LLDOS 1.0 contains a series of attacks in which an attacker scans for victim target, exploits vulnerability, illegally breaks in, sets up Trojan Mstream DDos software and finally launches DDOS attack. LLDOS 2.0.2 includes a similar sequence of attacks run by an attacker which is a little more sophisticated than the former one. The DARPA 1999 dataset contains 20 days' different attack traffics. And the two dataset share the same network topology.

We used playback technique [19] to replay each of the LLDDOS1.0 and LLDDOS2.0.2 dataset 10 times separately and each time we replayed one-day attack traffic of DARPA1999 dataset simultaneously as the background attack traffics. We chose Snort V2.4.5, which is configured according to the network settings of DARPA2000 Evaluation. Among many features of the alerts generated by snort, we only preserve several essential ones: DetectTime time stamp, source and destination address, source and destination port, alert type.

Then we gathered the alerts and stored them in the database sorting by ascending time order, then preprocessed by the alert aggregation and verification. We give the parameters with the minimum support 0.05 and attack sequence time window size 2 hours.

Using the alerts in database, we first executed the sequential pattern mining algorithm on the earliest 10% alerts. Then we executed incremental update algorithm on the following 20%. We repeated the process 9 times. Finally we find out that the incremental update algorithm gained the same results with sequential pattern algorithm, both of them discovered 41 patterns from the total 40 attack scenarios. 38 of them are correct and 3 are false. This shows that the two methods can obtain identical results from the same alert dataset. However, the further analysis shows that the incremental update shows better capability in rapid rule generation. The table shows the number of the attack activity patterns each time the incremental update algorithm gained. We can find that most patterns can be found in the earliest 10% of data, and less and less can be added in the following parts. This is feasible because the main multistage attack traffics is mostly similar. Only the patterns whose support are lower than min_support were found in the following re-execution of the incremental update we they satisfied the min_support.

Table 4. Attack Activity Patterns Gained by Incremental Update Algorithm

Alert data	10%	10%	10%	10%	10%	10%	10%	10%	10%	10%
Patterns	32	35	37	38	38	40	40	41	41	41

After the directed graph is introduced, the running time of sequential pattern mining algorithm decreases to 20% while the number of attack patterns discovered decreases to 39. This is rational because the number of alerts queried is reduced and the irrelative alerts are removed. Then we use probabilistic correlation to remove another one improper pattern.

6.2 Live Network Experiment

In order to valid the validity of the methods in real network environment, we deploy a honey-net to attract various attacks. We deliberately make some hosts in our lab network run a considerable amount of services with known vulnerabilities. We also deploy many security devices, such as Snort, Cisco Router, iptables, Cisco firewall etc. The topology of experiment network is shown in Figure 1.

We gathered the alerts generated by security devices in the honey-net continuously and restarted incremental mining algorithm every 8 hours. During two-week experiment, 49 multistage attacks were launched randomly and 45 frequent patterns were found. The first day we gained 17 sequential pattern, and 9 in the second day. The number of new patterns decreased with time going. The reason is that the discovered patterns become more and the number of newly-appeared patterns might reduce. This is in accord with our observation that for a given network, most attack actions are similar unless a new type of attack occurs. As time goes on, the database become larger, running time of mining algorithm in section 3 become longer and frequent re-executions become more difficult. However, the performance of incremental update did not decrease obviously because it mainly explores the newly added data. In fact in a network which generates large number of alerts, months of data will make GSP run for quite a long time. The incremental mining algorithm is significantly valuable in the dynamic network, because it can efficiently shorten the time gap between the appearance of a novel attack pattern and generation of corresponding rule.

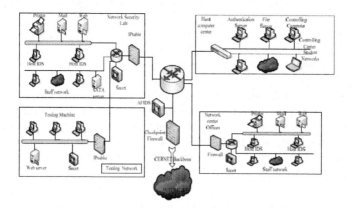

Fig. 1. The network topology of experiment

7 Conclusion and Future Work

The real-time detection system using known multistage attack patterns can help network administrator realize the state of protected network correctly and quickly. However, it is critical to insure the patterns are complete, correct, and up to date. Sequential pattern mining is a classical algorithm in data mining field, which is initially used to discover customer purchase patterns. In this paper, we introduce time window to divide the single long alert sequence into multiple subsequences and then use sequential pattern mining algorithm to search for attack patterns in alert stream. To adapt the dynamic network environment, at the same time to avoid wasting time in re-execution of mining algorithm, incremental mining is presented to find out new patterns. Both of the above methods reduce the time for rule generation. In order to get the correct and interesting patterns, directed graph and probabilistic correlation are presented, the former can restrict the scope of query by sequential mining algorithm while the latter can remove the illogical results. We evaluate our approaches using DARPA 2000 benchmark repository datasets and real data collected from our test network. The experiments show that our approach can effectively discover new and interesting attack patterns in reality.

We will continue our research on multistage attack pattern discovery, with focus on complementary approaches that can be incorporated with the methods in this paper. We can also make progress to improve the correctness of result from pattern discovery.

Acknowledgments. We would like to thank the anonymous reviewers for their valuable comments and suggestion.

References

1. Ning, P., Cui, Y., Reeves, D.S.: Constructing attack scenarios through correlation of intrusion alerts. In: Proceedings of the 9th ACM Conference on Computer and Communications Security, Washington, DC, November 18-22, 2002, pp. 18–22 (2002)
2. Ning, P., Cui, Y., Reeves, D.S., Xu, D.: Techniques and tools for analyzing intrusion alerts. ACM Transactions on Information and System Security 7, 274 (2004)
3. Ning, P., Xu, D.: Alert correlation through triggering events and common resources, Tucson, AZ (2004)
4. Cuppens, F.: Managing alerts in multi-intrusion detection environment. In: Proceedings 17th annual computer security applications conference, New Orleans (2001)
5. Cuppens, F., Miège, A.: Alert correlation in a cooperative intrusion detection framework. In: Proceedings of the 2002 IEEE symposium on security and privacy, IEEE Computer Society Press, Los Alamitos (2002)
6. Cuppens, F., Autrel, F., Mie'ge, A., Benferhat, S.: Correlation in an intrusion detection process. In: SECI02. Proceedings SÉcurité des communications sur internet (2002)
7. Valdes, A., Skinner, K.: Probabilistic alert correlation. In: Lee, W., Mé, L., Wespi, A. (eds.) RAID 2001. LNCS, vol. 2212, Springer, Heidelberg (2001)
8. Hang, M.Y, Wicks, T.M.: A large-scale distributed intrusion detection framework based on attack strategy analysis. Computer network, 2465–2475 (1999)

9. Sheyner, O., Haines, J., Jha, S., Lippmann, R., Wing, J.M.: Automated generation and analysis of attack graphs. In: Proceedings of the 2002 IEEE Symposium on Security and Privacy, Oakland, CA, May 2002 (2002)
10. Lee, W., Qin, X.: Statistical Causality Analysis of INFOSEC Alert Data. In: Vigna, G., Krügel, C., Jonsson, E. (eds.) RAID 2003. LNCS, vol. 2820, Springer, Heidelberg (2003)
11. Qin, X., Lee, W.: Discovering novel attack strategies from INFOSEC alerts. In: Sophia Antipolis, France (2004)
12. Treinen, J.J., Thurimella, R.: A Framework for the Application of Association Rule Mining in Large Intrusion Detection Infrastructures. In: Zamboni, D., Kruegel, C. (eds.) RAID 2006. LNCS, vol. 4219, pp. 1–18. Springer, Heidelberg (2006)
13. Mannila, H., Toivonen, H., Verkamo, A.I.: Discovering Frequent Episodes in Sequences. In: KDD'95. Proceedings of the First International Conference on Knowledge Discovery and Data Mining, Montreal, Canada, pp. 210–215 (1995)
14. Srikant, R., Agrawal, R.: Mining Sequential Patterns: Generalizations and Performance Improvements. In: KDD'95. Advances in Database Technology –5th International Conference on Knowledge Discovery and Data Mining, Montreal, Canada, pp. 269–274 (1995)
15. Agrawal, R., Srikant, R.: Mining sequential patterns. In: Research Report RJ 9910, IBM Almaden Research Center, San Jose, California (October 1994)
16. Masseglia, F., Poncelet, P., Teisseire, M.: Incremental mining of sequential patterns in large databases. Data Knowledge 46(1), 97–121 (2003)
17. Lin, M.-Y., Lee, S.-Y.: Incremental Update on Sequential Patterns in Large Databases. In: ICTAI98. Proceedings of the 10th IEEE International Conference on Tools with Artificial Intelligence, Taipei, Taiwan, R.O.C, pp. 24–31 (1998)
18. Wang, L., Li, Z.-t., Fan, J.: Learning attack strategies through attack sequence mining method. International Conference on Communication Technology (2006)
19. MIT Lincoln Lab: DARPA Intrusion Detection Scenario Specific Data Sets (2000), http://www.ll.mit.edu/IST/ideval/data/2000/2000_data_index.html
20. MIT Lincoln Lab: Tcpdump File Replay Utility. http://ideval.ll.mit.edu/IST/ideval/tools/tools_index.html

Privacy Preserving DBSCAN Algorithm for Clustering

K. Anil Kumar and C. Pandu Rangan

Department of Computer Science and Engineering
Indian Institute of Technology - Madras
Chennai - 600036, India
kumark@cse.iitm.ernet.in,
rangan@iitm.ernet.in

Abstract. In this paper we address the issue of privacy preserving clustering. Specially, we consider a scenario in which two parties owning confidential databases wish to run a clustering algorithm on the union of their databases, without revealing any unnecessary information. This problem is a specific example of secure multi-party computation and as such, can be solved using known generic protocols. However there are several clustering algorithms are available. They are applicable to specific type of data, but DBSCAN [4] is applicable for all types of data and the clusters obtained by DBSCAN are similar to natural clusters. However, DBSCAN [4] algorithm is basically designed as an algorithm working on a single database. In this paper we proposed a protocols for how the distances are measured between data points, when the data is distributed across two parties. By using these protocols we propose the first novel method for running DBSCAN algorithm operating over vertically and horizontally partitioned data sets, distributed in two different databases in a privacy preserving manner.

1 Introduction

A key problem that arises with large collection of information is that of *confidentiality*. The need for privacy is due to law or can be motivated by business interests. However, there are situations where *sharing of data* that can lead to mutual gain. Recently, more emphasis has been placed on preserving the privacy of user-data aggregations, ex. databases of personal information. Access to these collection is, however, enormously useful. It is from this balance between privacy and utility that the area of *privacy preserving data mining* [1,7] emerged. The main objective in privacy preserving data mining is to develop algorithms for modifying the original data in some way, so that the private data and private knowledge remain private after the mining process. In this paper we are concentrating particularly on privacy preserving clustering algorithms.

Consider the following scenario, *Let parties X and Y own private databases D_X and D_Y. The parties wish to get clusters on their joint data, apply a clustering algorithm (ex: DBSCAN [4]) to the joint database $D_X \cup D_Y$ without revealing any unnecessary information about their individual databases. i.e. the only information learned by X about D_Y is that which can be learned from the output*

R. Alhajj et al. (Eds.): ADMA 2007, LNAI 4632, pp. 57–68, 2007.
© Springer-Verlag Berlin Heidelberg 2007

of the clustering algorithm, and vice versa. We are giving two kinds of solutions for the above problem, one with the help of trusted third party (TTP) and another TTP not available.

There are several applications of clustering. Any application of clustering where there are privacy concerns is a possible candidate for out privacy-preserving clustering algorithm. For example, suppose network traffic is collected at two ISPs, and the two ISPs want to cluster their joint network traffic without revealing their individual traffic data. Out DBSCAN for horizontally partitioned algorithm can be used to obtain joint clusters while respecting the privacy of the network traffic at the two ISPs. Another example, two organizations, an internet marketing company and an on-line retail company, have datasets with different attributes for a common set of individuals. These organizations decides to share their data for clustering to find the optimal customer targets so as to maximize using each other's data without learning anything about the attribute value of each other? For this problem to solve our DBSCAN for vertically partitioned algorithm is used.

2   Related Work

Recently privacy preserving data mining has been a very active area of research. Privacy preserving data mining introduced by Rakesh Agrawal and Ramakrishnan Srikanth [1]. Lindall and Pinkas [1,7] arouse as a solution to this problem by allowing parties to cooperate in the extraction of knowledge without any of the cooperating parties having to reveal their individual data to any other party. Initial focus in this area was on construction of decision trees from distributed datasets. There is also a significant body of research on privacy preserving mining of association rules [10]. We will focus only on *privacy preserving clustering.* Privacy preserving clustering has been previously addressed Jha et al [6] gave a distributed privacy preserving algorithm for k-means clustering when the data is horizontally partitioned. Oliveira and Zaiane [8] gave a privacy preserving solution for clustering based on perturbation. In their work they used geometric data transformation methods for perturbing the data and they give solution for hierarchical and partition based clustering on perturbed data. Vaidya and Clifton [11] introduced the problem of privacy preserving clustering when the data is vertically partitioned. Prasad and Rangan [9] proposed a solution, privacy preserving BIRCH. Their algorithm works for vertically partitioned, large databases. Jagannathan and Wright [5] introduced the notion of arbitrarily partitioned databases and gave a privacy preserving solution of k-means clustering. In this paper we present a privacy preserving protocol for DBSCAN clustering which works for large databases.

3   Preliminaries

In this section we briefly describe the primitives that we used in our work for clustering the distributed data securely.

Millionaire's protocol: The purpose of this primitive is to compare two private numbers and decide which one is larger. This problem was first proposed by Yao in 1982 and is widely known as Yao's Millionaire problem [12]. Essentially the problem is Alice and Bob are two millionaires who want to find out who is richer without revealing the precise amount of their wealth to each other. Among several solutions, we adopt the Cachin's [3] solution based on the ϕ-hiding assumption. The communication complexity of Cachin's scheme is $O(l)$, where l is the number of bits of each input number and the computation complexity is also linear on the number of bits used to represent both the numbers.

Secure scalar product protocol: The purpose of this primitive is to find securely $\vec{X} \cdot \vec{Y}$ where Alice has the private vector $\vec{X} = (x_1, x_2, \cdots, x_n)$ and Bob has the private vector $\vec{Y} = (y_1, y_2, \cdots, y_n)$ and $\vec{X} \cdot \vec{Y} = \sum_{i=1}^{n} x_i y_i$. We use the private homomorphic dot product protocol given by Goethals et al. [2] for secure computation of scalar product.

4   DBSCAN Algorithm

DBSCAN (Density Based Spatial Clustering of Applications with Noise) [4] is a density-based algorithm which can detect arbitrary shaped clusters. Density is defined as a minimum number of points within a certain distance of each other. It takes input parameters as $MinPts$, indicates minimum number of points in any cluster and for each point in a cluster there must be another point in the cluster whose distance from it is less than a threshold input value Eps. DBSCAN starts from an arbitrary object and if neighborhood around it within a given radius (Eps) satisfies at least the minimum number of tuples ($MinPts$) this object is a *core object* and the search recursively continues with its neighborhoods and stops at the border objects were all the points within the cluster must be in the neighborhood of one of its core objects. Another arbitrary ungrouped object is selected and the process is repeated until all data points in the data set have been placed in the clusters. All the non-core objects which are not in the neighborhood of any of the core objects are labeled as noise. DBSCAN doesn't need the number of final clusters to be given in advance.

5   Secure DBSCAN Algorithm

5.1   Distributed Data Mining

The distributed data mining model assumes that the data sources are distributed across multiple sites. The data could be partitioned into many ways as horizontally, vertically or arbitrarily. **Vertically partitioning** of data implies that though different sites gather information about the same set of entities, but they collect different feature sets. **Horizontal partitioning** of data implies that though different sites collect the same set of information, but about different entities.

5.2   Problem Statement

Assume that the database $DB = \{t_1, t_2, \ldots, t_n\}$, consists of n records. Each record t_i is described by the values of m attributed of object t_i by $(A_{i_1}, A_{i_2}, \ldots, A_{i_m})$. For the simplification purposes let us assume that data is distributed between two parties (A and B). The $A's$ share is represented by DB_A and the $B's$ share is represented by DB_B, such that $DB = DB_A \# DB_B$. The partitioned data clustering attempts to form a cluster on DB_A and DB_B where A has no knowledge of DB_B and B has no knowledge of DB_A. DB_A and DB_B on vertically and horizontally partitioned databases are defined as follows.

Definition 1. *For* **Vertically partitioned database,** *$A's$ share is represented by $DB_A = \{t_{A_1}, t_{A_2}, \ldots, t_{A_n}\}$, and the $B's$ share is represented by $DB_B = \{t_{B_1}, t_{B_2}, \ldots, t_{B_n}\}$. For each record t_i A has values of k attributed record, $t_{A_i} = (A_{i_1}, A_{i_2}, \ldots, A_{i_k})$ and B has $(m - k)$ attributed record, $t_{B_i} = (A_{i_{k+1}}, A_{i_{k+2}}, \ldots, A_{i_n})$. The actual record $\{t_i\} = \{t_{A_i} \# t_{B_i}\}$.*

Definition 2. *For* **Horizontally partitioned database,** *$A's$ share is represented by $DB_A = \{t_{A_1}, t_{A_2}, \ldots, t_{A_k}\}$ and $B's$ share is represented by $DB_B = \{t_{B_1}, t_{B_2}, \ldots, t_{B_l}\}$, such that $l = n - k$. Each record t_{A_i} or t_{B_i} are described by the values of m attributed record as $(A_{i_1}, A_{i_2}, \ldots, A_{i_m})$. The complete database is described by $\{t_1, t_2, \ldots, t_k, t_{k+1}, t_{k+2}, \ldots, t_n\} = \{t_{A_1}, t_{A_2}, \ldots, t_{A_k}, t_{B_1}, t_{B_2}, \ldots, t_{B_l}\}$.*

5.3   Protocols

In this section we will describe some of the protocols which are used in our actual algorithm. These protocols are used to know whether two given data points are neighbors or not in a secured fashion. In general there are two approaches for designing privacy preserving algorithms. The first approach with the help of Trusted Third Party (TTP), i.e. send the whole data to the TTP, and apply the clustering algorithm, finally distribute the result to all the parties. A second approach is designing privacy preserving algorithms by using the protocols from the secure-multi-party computation literature. i.e. without having TTP.

Solutions using TTP. If TTP is available then these protocols are secure, so we are able to compute the neighbors of any other data point by using these protocols in a secured manner. The advantage of these protocols is that computation complexity is less but communication complexity is more.

Protocol 1. *Party A has private input $\{X_{A_1}, X_{A_2}, \ldots, X_{A_n}\}$ and B has private input $\{X_{B_1}, X_{B_2}, \ldots, X_{B_n}\}$. A and B need to find whether $\{\{(X_{A_1} + X_{B_1}) \leq Eps^2\}, \{(X_{A_2} + X_{B_2}) \leq Eps^2\}, \ldots, \{(X_{A_n} + X_{B_n}) \leq Eps^2\}\}$ or not. The condition here is party A should not know the values $\{X_{B_1}, X_{B_2}, \ldots, X_{B_n}\}$ and party B should not know the values $\{X_{A_1}, X_{A_2}, \ldots, X_{A_n}\}$ at the end of the protocol. Moreover, TTP is available.*

1. Party A does

1.1 *Generate a shared key $\{SK_{A_1,TP}, SK_{A_2,TP}, \ldots, SK_{A_n,TP}\}$ with TTP.*

1.2 *Compute* $\{Y_{A_1}, Y_{A_2}, \ldots, Y_{A_n}\} = \{(X_{A_1} + SK_{A_1,TP}), (X_{A_2} + SK_{A_2,TP}),$
$\ldots, (X_{A_n} + SK_{A_n,TP})\}$ *and sends* $\{Y_{A_1}, Y_{A_2}, \ldots, Y_{A_n}\}$ *to party B.*

2. Party B does

2.1 *Generate a shared key* $\{SK_{B_1,TP}, SK_{B_2,TP}, \ldots, SK_{B_n,TP}\}$ *with TTP.*

2.2 *Compute* $\{Y_{B_1}, Y_{B_2}, \ldots, Y_{B_n}\} = \{(Y_{A_1} + X_{B_1} + SK_{B_1,TP}), (Y_{A_2} + X_{B_2} +$
$SK_{B_2,TP}), \ldots, (Y_{A_n} + X_{B_n} + SK_{B_n,TP})\}$ *and sends* $\{Y_{B_1}, Y_{B_2}, \ldots, Y_{B_n}\}$
to TTP.

3. Party TTP does

3.1 *Compute* $\{res_1, res_2, \ldots, res_n\} = \{(Y_{B_1} - SK_{A_1,TP} - SK_{B_1,TP}), (Y_{B_2} -$
$SK_{A_2,TP} - SK_{B_2,TP}), \ldots, (Y_{B_n} - SK_{A_n,TP} - SK_{B_1,TP})\}$

3.2 *For i from 1 to n do*
IF $\{res_i \leq Eps^2\}$ *THEN* $finalres_i = 1$
ELSE $finalres_i = 0$

3.3 *Return finalres array to both parties A and B.*

Theorem 1. *The communication complexity of protocol 1 is six messages and the computational complexity is negligible.*

Protocol 2. *Party A has private input* $\{X_{A_1}, X_{A_2}, \ldots, X_{A_n}\}$ *and B has private input* $\{Y_{B_1}, Y_{B_2}, \ldots, Y_{B_n}\}$. *A and B need to find whether* $\{(X_{A_1} - Y_{B_1})^2 + (X_{A_2} - Y_{B_2})^2 + \ldots + (X_{A_n} - Y_{B_n})^2\} \leq Eps^2$ *or not. The condition here is party A should not know the values* $\{Y_{B_1}, Y_{B_2}, \ldots, Y_{B_n}\}$ *and B should not know the values* $\{X_{A_1}, X_{A_2}, \ldots, X_{A_n}\}$ *at the end of the protocol. Moreover, TTP is available.*

1. Party A does

1.1 *Generate a shared key* $\{SK_{A_1,TP}, SK_{A_2,TP}, \ldots, SK_{A_n,TP}\}$ *with TTP*

1.2 *Compute* $\{T_{A_1}, T_{A_2}, \ldots, T_{A_n}\} = \{(X_{A_1} + SK_{A_1,TP}), (X_{A_2} + SK_{A_2,TP}),$
$\ldots, (X_{A_n} + SK_{A_n,TP})\}$ *and sends* $\{T_{A_1}, T_{A_2}, \ldots, T_{A_n}\}$ *to party B.*

2. Party B does

2.1 *Generate a shared key* $\{SK_{B_1,TP}, SK_{B_2,TP}, \ldots, SK_{B_n,TP}\}$ *with TTP.*

2.2 *Compute* $\{T_{B_1}, T_{B_2}, \ldots, T_{B_n}\} = \{(T_{A_1} - Y_{B_1} + SK_{B_1,TP}), (T_{A_2} - Y_{B_2} +$
$SK_{B_2,TP}), \ldots, (T_{A_n} - Y_{B_n} + SK_{B_n,TP})\}$ *and sends* $\{T_{B_1}, T_{B_2}, \ldots, T_{B_n}\}$
to TTP.

3. Party TTP does

3.1 *Compute* $res = \{(T_{B_1} - SK_{A_1,TP} - SK_{B_1,TP})^2 + (T_{B_2} - SK_{A_2,TP} -$
$SK_{B_2,TP})^2 + \ldots + (T_{B_n} - SK_{A_n,TP} - SK_{B_1,TP})^2\}$

3.2 *IF* $\{res \leq Eps^2\}$ *THEN*
Return Y to both parties A and B
ELSE Return N to both parties A and B.

Theorem 2. *The communication complexity of protocol 2 is six messages and the computational complexity is negligible.*

There is no complex operations like encryption or decryption functions. So if TTP is available then the above protocols are secure and simple.

Without using TTP. In some situations where TTP is not available, then we may use the following protocols. In order to solve this we are using secure scalar product protocol and the millionaire's protocols. In these protocols computation and communication complexities are more when compared with the protocols with TTP.

Protocol 3. *Party A has private input $\{X_{A_1}, X_{A_2}, \ldots, X_{A_n}\}$ and B has private input $\{X_{B_1}, X_{B_2}, \ldots, X_{B_n}\}$. A and B need to find whether $\{\{(X_{A_1} + X_{B_1}) \leq Eps^2\}, \{(X_{A_2} + X_{B_2}) \leq Eps^2\}, \ldots, \{(X_{A_n} + X_{B_n}) \leq Eps^2\}\}$ or not. The condition here is party A should not know the values $\{X_{B_1}, X_{B_2}, \ldots, X_{B_n}\}$ and party B should not know the values $\{X_{A_1}, X_{A_2}, \ldots, X_{A_n}\}$ at the end of the protocol. Moreover TTP is not available.*

1. *B generates a vector of random values $\{R_{B_1}, R_{B_2}, \ldots, R_{B_n}\}$*
2. *A constructs its vector as $\{(\frac{X_{A_1}}{Eps^2}, 1), (\frac{X_{A_2}}{Eps^2}, 1), \ldots, (\frac{X_{A_n}}{Eps^2}, 1)\}$ and B constructs its vector as $\{(R_{B_1}, R_{B_1}(\frac{X_{B_1}}{Eps^2})), (R_{B_2}, R_{B_2}(\frac{X_{B_2}}{Eps^2})), \ldots, (R_{B_n}, R_{B_n}(\frac{X_{B_n}}{Eps^2}))\}$. Now parties A and B invokes the **secure scalar product protocol** [2] and gets the result as $\{(R_{B_1}(\frac{X_{A_1}+X_{B_1}}{Eps^2})), (R_{B_2}(\frac{X_{A_2}+X_{B_2}}{Eps^2})), \ldots, (R_{B_n}(\frac{X_{A_n}+X_{B_n}}{Eps^2}))\}$ and without B knowing the result.*
3. *Now A has $\{(R_{B_1}(\frac{X_{A_1}+X_{B_1}}{Eps^2})), (R_{B_2}(\frac{X_{A_2}+X_{B_2}}{Eps^2})), \ldots, (R_{B_n}(\frac{X_{A_n}+X_{B_n}}{Eps^2}))\}$ and B has the value $\{R_{B_1}, R_{B_2}, \ldots, R_{B_n}\}$*
4. *Now both parties A and B run the millionaire protocol [3] to decide whether $(R_{B_i}(\frac{X_{A_i}+X_{B_i}}{Eps^2})) > R_{B_i}$, for each $i = 1$ to n.*
5. *If $(R_{B_i}(\frac{X_{A_i}+X_{B_i}}{Eps^2})) > R_{B_i}$ then decide that $(X_{A_i} + X_{B_i}) > Eps^2$ else they decide $(X_{A_i} + X_{B_i}) \leq Eps^2$*

Theorem 3. *The communication complexity of protocol 3 is $(6n + 1)$ messages and the computational complexity of protocol 3 is $2n$ encryptions, $2n$ decryptions.*

Protocol 4. *Party A has private input $\{X_{A_1}, X_{A_2}, \ldots, X_{A_n}\}$ and B has private input $\{Y_{B_1}, Y_{B_2}, \ldots, Y_{B_n}\}$. A and B need to find whether $\{(X_{A_1} - Y_{B_1})^2 + (X_{A_2} - Y_{B_2})^2 + \ldots + (X_{A_n} - Y_{B_n})^2\} \leq Eps^2$ or not. The condition here is party A should not know the values $\{Y_{B_1}, Y_{B_2}, \ldots, Y_{B_n}\}$ and B should not know the values $\{X_{A_1}, X_{A_2}, \ldots, X_{A_n}\}$ at the end of the protocol. Moreover, TTP is not available.*

1. *Party B genereates a random number R.*
2. *A constructs $\{\frac{X_{A_1}^2}{Eps^2}, \frac{1}{Eps^2}, \frac{-2X_{A_1}}{Eps^2}, \frac{X_{A_2}^2}{Eps^2}, \frac{1}{Eps^2}, \frac{-2X_{A_2}}{Eps^2}, \ldots, \frac{X_{A_n}^2}{Eps^2}, \frac{1}{Eps^2}, \frac{-2X_{A_n}}{Eps^2}\}$ as it's vector and B constructs $\{R, RY_{B_1}^2, RY_{B_1}, R, RY_{B_2}^2, RY_{B_2}, \ldots, R, RY_{B_n}^2, RY_{B_n}\}$ as it's vector. Now A and B invokes the **secure scalar product protocol** [2] and gets $R(\frac{(X_{A_1}-Y_{B_1})^2 + (X_{A_2}-Y_{B_2})^2 + \ldots + (X_{A_n}-Y_{B_n})^2}{Eps^2})$ without B knowing the result.*
3. *Now party A has $R(\frac{(X_{A_1}-Y_{B_1})^2 + (X_{A_2}-Y_{B_2})^2 + \ldots + (X_{A_n}-Y_{B_n})^2}{Eps^2})$ and party B has the value R.*
4. *Party A and party B run the cachin's millionaire protocol [3] to decide whether $R(\frac{(X_{A_1}-Y_{B_1})^2 + (X_{A_2}-Y_{B_2})^2 + \ldots + (X_{A_n}-Y_{B_n})^2}{Eps^2}) > R$ or not.*

5. If $R(\frac{(X_{A_1}-Y_{B_1})^2+(X_{A_2}-Y_{B_2})^2+...+(X_{A_n}-Y_{B_n})^2)}{Eps^2}) > R$ then parties A and B decide $\{(X_{A_1} - Y_{B_1})^2 + (X_{A_2} - Y_{B_2})^2 + \ldots + (X_{A_n} - Y_{B_n})^2\} > Eps^2$ else they decide $\{(X_{A_1} - Y_{B_1})^2 + (X_{A_2} - Y_{B_2})^2 + \ldots + (X_{A_n} - Y_{B_n})^2\} \le Eps^2$.

Theorem 4. *The communication complexity of protocol 4 is $(8n + 1)$ messages and the computational complexity of protocol 4 is $3n$ encryptions, $3n$ decryptions.*

5.4 Proposed Algorithms

DBSCAN for Vertically Partitioned data: The DBSCAN for vertically partitioned data starts by choosing an arbitrary point. In step 3 of algorithm 1 we are calling the ComputeDistance function which takes arguments as the selected data point, number of attributes, and number of records. This function simply computes the distance between the chosen point and the remaining points at both A and B. Party A has $dist_A$ and party B has $dist_B$. We want to find $dist_A + dist_B \le Eps^2$, protocol 1 or protocol 3 provides solution for this. Now we are able to say the chosen point is core point or not. If it is a core point the cluster is formed in steps 9 to 32. Otherwise chosen point is marked as noise. We repeat the process until all points are classified.

DBSCAN for Horizontally Partitioned Data: The DBSCAN for horizontally partitioned data starts by choosing an arbitrary point at party A. Party A finds all the neighbors of chosen point. From steps 3 to 6, party B knows the neighbors of chosen point. Step 10 checking the core point condition for chosen point. If the chosen point is a core point then cluster is formed with all its density reachable points at both parties A and B from steps 13 to 30. The above procedure is repeated until all points are classified at party A. If there are any unclassified points at party B, then B performs local DBSCAN algorithm.

Proof of Correctness: In this section we show that the secure DBSCAN algorithms (Algorithm 2 and Algorithm 3) computes the same set of clusters over horizontal or vertical partitioned databases as its non secure variant DBSCAN computes on $DB = DB_X \# DB_Y$.

Theorem 5. *DBSCAN for vertically partitioned data computes the same set of clusters as its non secure variant DBSCAN does with the input $DB = DB_A \# DB_B$*

Proof. The DBSCAN algorithm for vertically partitioned data starts with an unclassified point is same as actual DBSCAN algorithm. To know the chosen point p is core point or not, we will find the neighbors of p using the protocol 1 or protocol 2. In normal DBSCAN algorithm to know a datapoint q is neighbor of p or not, we will use $(q_{A_1} - p_{A_1})^2 + (q_{A_2} - p_{A_2})^2 + \ldots + (q_{A_k} - p_{A_k})^2 + (q_{A_{k+1}} - p_{A_{k+1}})^2 + \ldots + (q_{A_m} - p_{A_m})^2 \le Eps^2$. But in vertically partitioned DBSCAN algorithm party A computes $dist_A = (q_{A_1} - p_{A_1})^2 + (q_{A_2} - p_{A_2})^2 + \ldots + (q_{A_k} - p_{A_k})^2$ and party B computes $dist_B = (q_{A_{k+1}} - p_{A_{k+1}})^2 + \ldots + (q_{A_m} - p_{A_m})^2$. We need $dist_A + dist_B \le Eps^2$, this is done using the protocol 1 or protocol 3. In

---

Algorithm 1. DBSCAN algorithm for Vertically Partitioned Data

---

    input : $p, neighbors, clsId, DB$
    output: DB (with some classified data points)
1 /* First Party A contains k attributes, second party B contains $(m - k)$ attributes for the n tuple database. */
2 Select a point $p = p_A \# p_B$ which is not classified.
3 Party X does: $dist_A \leftarrow ComputeDistance(p_A, k, n)$
4 Party Y does: $dist_B \leftarrow ComputeDistance(p_B, m - k, n)$
5 Invoke protocol 1 or protocol 3 with A contains $dist_A$, and B contains $dist_B$ by both parties A and B. According to the obtained result decide whether p is core point or not.
6 /* DB.changeClId(p.clID) - Assigns $clsID$ (cluster Id) to data point p in the DB (data base).*/
7 /* seeds.append(x) - Adds x into the list $seeds$ at the end.*/
8 /* seeds.first() - Getting the first item from the list $seeds$ */
9 /* result.get(i) - Returns the ith item from the list $result$ */
10 **if** p *is core point* **then**
11   seeds $\leftarrow$ neighbors obtained from protocol
12   Assign $CLSID$ to p.
13   **while** $seeds \neq empty$ **do**
14   currentp $\leftarrow$ seeds.first()
15   party A does : $result_A \leftarrow$ computeDistance($currentp_A, k, n$)
16   party B does : $result_B \leftarrow$ computeDistance($currentp_B, m - k, n$)
17   Invoke protocol 1 or protocol 2 with A contains $result_A$ and B contains $result_B$ by both parties A and B. and store the output of protocol 1 or protocol 2 in $result$
18   **if** $|result| \geq MinPts$ **then**
19   **for** $i \leftarrow 1$ to $|result|$ **do**
20   resultp $\leftarrow$ result.get(i)
21   **if** $resultp.clsID == Unclassified$ **then**
22   seeds.append(resultp)
23   DB.changeclId(resultp,clsId)
24   **end**
25   **else if** $resultp.clsID == Noise$ **then**
26   DB.changeclId(resultp,clsId)
27   **end**
28   **end**
29   **end**
30   **end**
31 **end**
32 **else** Mark p as noise.
33 Repeat the above procedure until all points are clustered.

---

protocol 1 simply pass $dist_A$ and $dist_B$ to TTP and get the required result to both the parties A and B. In protocol 3 we have used the secure scalar product and millionaire's protocol solution. These protocols already proven to be correct *in [2,3]. The remaining steps are same as the actual DBSCAN algorithm. So the clusters formed by the DBSCAN algorithm is same as the clusters formed by the DBSCAN algorithm for vertically partitioned data.*

Theorem 6. *DBSCAN for horizontally partitioned data computes the same set of clusters as its non secure variant DBSCAN does with the input $DB = DB_A \# DB_B$*

Proof.   The DBSCAN algorithm for horizontally partitioned data starts with an unclassified point is same as actual DBSCAN algorithm. To know the chosen point p is core point or not, we will find the neighbors of p using normal DB-SCAN algorithm which are at the same party and use the protocol 2 or protocol 4 to know the neighbors which are at the other party. In normal DBSCAN algorithm to know a datapoint q is neighbor of p or not, we will use $(q_{A_1} - p_{A_1})^2 + (q_{A_2} - p_{A_2})^2 + \ldots + (q_{A_k} - p_{A_k})^2 + (q_{A_{k+1}} - p_{A_{k+1}})^2 + \ldots + (q_{A_m} - p_{A_m})^2 \leq Eps^2$. If both p and q are at the same party then that party easily says whether p is neighbor of q or not. Suppose if p is at party A and q is at party B then to compute the above distance we will use the protocol 2 or protocol 4. In protocol 2 simply pass p and q to TTP and get the required result to both the parties A and B. In protocol 4 we have used the secure scalar product and millionaire's protocol solution [2,3]. The remaining steps are same as the actual DBSCAN algorithm. So the clusters formed by the DBSCAN algorithm is same as the clusters formed by the DBSCAN algorithm for horizontally partitioned data.

Security Analysis: In this section we have to prove that the secure DBSCAN algorithm for vertically and horizontally partitioned data is secure against semi honest adversaries.

Theorem 7. *DBSCAN for vertically partitioned data is secure against semi honest adversaries.*

Proof.   In order to prove DBSCAN for vertically partitioned data is secure against semi honest adversaries, it is sufficient to prove protocol 1 or protocol 3 are secure. In these protocols only the data is transfered between from one party

---

Algorithm 2. computeDistance

    input   : point,attrCnt,recCnt,DB
    output: dist

1 **for** $i \leftarrow 1$ *to recCnt* **do**
2   **if** $i \neq p$ **then**
3      $dist_i \leftarrow \text{dist}(i,p)$
4      $/ * \{dist(i,p) \leftarrow (i_{A_1} - p_{A_1})^2 + \ldots + (i_{A_{attrCnt}} - p_{A_{attrCnt}})^2\} * /$
5   **end**
6 **end**
7 dist $\leftarrow \{dist_1, dist_2, \ldots, dist_{attrcnt}\}$

Algorithm 3. DBSCAN algorithm for Horizontally Partitioned Data

---

    input : $p, neighbors, clsId, DB$
    output: Formed Clusters
1 /* First party A contains k tuples of data, and second party B contains $(k + 1)$ to n tuples.*/
2 Select a point p which is not classified at party A.
3 $neighbor_A \leftarrow$ the neighbors at party A.
4 **for** $i = (k+1)$ to n **do**
5   Invoke protocol 2 or protocol 4 to know whether i is neighbor of p or not as p at party A and i at party B. Update $neighbor_B$ if i is neighbor of p
6 **end**
7 {/*Condition to know p is a core point or not, computed using protocol 1 A contains $|neighbor_A|$, Y contains $|neighbor_B|$ and Eps as $MinPts$*/} /* seeds.append(x) - Adds x into the list $seeds$ at the end.*/
8 /* seeds.first() - Getting the first item from the list $seeds$ */
9 /* result.get(i) - Returns the ith item from the list $result$ */
10 **if** $(|neighbor_A| + |neighbor_B|) \leq Minpts$ **then**
11   Party A does: $seeds_A \leftarrow neighbors_A$, Assign CLSID to p
12   Party B does: $seeds_B \leftarrow neighbors_B$.
13   Party A does:
14   **while** $seeds_A \neq empty$ **do**
15   currentp $\leftarrow seeds_A$.first()
16          $result_A \leftarrow$ neighbors of currentp at party A
17   **for** $i \leftarrow (k+1)$ to n **do**
18   Invoke protocol 2 or protocol 4 to know whether i is neighbor of currentp or not as p at party A and i is at party B. Update $neighbor_B$ if i is the neighbor of p.
19   **end**
20   **if** $|neighbor_A| + |neighbor_B| \geq MinPts$ **then**
21   **for** $i \leftarrow 1$ to $|result_A|$ **do**
22                 $resultp_A \leftarrow result_A$.get(i)
23   **if** $resultp_A.clID ==$ Unclassified or $resultp.clID ==$ Noise **then**
24   **if** $resultp_A.clID ==$ Unclassified **then** $seeds_A$.append($resultp_A$)
25   Assign CLSID to $resultp_A$
26   **end**
27   **end**
28   **end**
29   **end**
30   Party B does: Repeat steps 14 to 29 if $seeds_B$ not empty, by replacing A by B and B by A. Repeat steps 14 to 29 if $seeds_A$ not empty by A, $seeds_B$ not empty by B.
31 **end**
32 **else** Mark p as noise.
33 Repeat the above procedure until all points are clustered at party A.
34 Perform normal DBSCAN algorithm if party B contains any unclassified data.

---

to another. In protocol 1 all the data is passed to TTP, if the TTP is honest then we say that protocol 1 is secure. In protocol 3 we are using the secure scalar product protocol and millionaire's protocol. These protocols are already proven to be secure against semi honest adversaries. So DBSCAN for vertically partitioned data is also secure.

Theorem 8. *DBSCAN for horizontally partitioned data is secure against semi honest adversaries.*

Proof.   In order to prove DBSCAN for horizontally partitioned data is secure against semi honest adversaries, it is sufficient to prove protocol 2 or protocol 4 are secure. In these protocols only the data is transfered between from one party to another. In protocol 2 all the data is passed to TTP, if the TTP is honest then we say that protocol 2 is secure. In protocol 4 we are using the secure scalar product protocol and millionaire's protocol. These protocols are already proven to be secure against semi honest adversaries. So DBSCAN for horizontally partitioned data is also secure against semi honest adversaries.

Complexity Analysis: DBSCAN algorithm for vertically partitioned databases with TTP, invokes n times protocol 1. In protocol 1 the communication complexity is 6 messages, and computational complexity is additions or subtractions. So overall complexity is $(6n^2)$ messages and $O(n)$ additions. Similarly for horizontally partitioned databases with TTP the communication complexity is $(6n^2)$ messages, $O(n)$ additions.

DBSCAN algorithm for vertically partitioned databases without TTP, invokes n times protocol 3. In protocol 3 we are invoking n times secure scalar product protocol with vector size 2, so it takes $6n$ communication complexity, $2n$ encryptions, $2n$ decryptions as computational complexity. The millionaire's protocol is also invoked n times, so the communication complexity is (nl) and computational complexity is negligible. Overall communication complexity for DBSCAN over vertically partitioned databases is $(6n^2) + (N^2l)$ and computational complexity is $2n^2$ encryptions and $(2n^2)$ decryptions. Similarly the communication complexity for DBSCAN over horizontally partitioned databases is $(8n^2) + (N^2l)$ and computational complexity is $3n^2$ encryptions and $(3n^2)$ decryptions.

6   Conclusion and Future Work

While most existing privacy preserving algorithms work only for specific data, DBSCAN is applicable for all types of data. The clusters formed by DBSCAN are similar to natural clusters. In this paper we presented a protocols for secure distance metrics for both vertical and horizontal partitioned databases. By using those protocols we present a DBSCAN algorithm for distributed data. As a direction for future work, we wish to look into how can this protocol be extended to arbitrary partitioned data. Also we wish to look into any privacy preserving algorithm for categorical and mixed types of data.

References

1. Agrawal, R., Srikant, R.: Privacy preserving data mining. In: Proceedings of the 2000 ACM SIGMOD Conference on Management of Data, Dallas, TX, May 2000, pp. 439–450. ACM Press, New York (2000)
2. Goethals, B., Laur, S., Lipmaa, H., Mielikainen, T.: On private scalar product computation for privacy-preserving data mining. In: Park, C.-s., Chee, S. (eds.) ICISC 2004. LNCS, vol. 3506, pp. 104–120. Springer, Heidelberg (2005)
3. Cachin, C.: Efficient private bidding and auctions with an oblivious third party. In: SIGSAC. Proceedings of 6th ACM Computer and communications security, pp. 120–127. ACM Press, New York (1999)
4. Ester, M., Kriegel, H.-P., Sander, J., Xu, X.: A density-based algorithm for discovering clusters in large spatial databases with noise. In: SIGKDD96. Proceedings of 2nd International Conference on Knowledge discovery and data mining, Portland, Oregon, pp. 226–231 (1996)
5. Jagannathan, G., Wright, R.N.: Privacy-preserving distributed k-means clustering over arbitrarily partitioned data. In: Proceedings of the 11th ACM SIGKDD International Conference on Knowledge Discovery and Data Mining, Chicago, Illinois, August 2005, pp. 593–599. ACM Press, New York (2005)
6. Jha, S., Kruger, L., McDaniel, P.: Privacy Preserving Clustering. In: di Vimercati, S.d.C., Syverson, P.F., Gollmann, D. (eds.) ESORICS 2005. LNCS, vol. 3679, pp. 397–417. Springer, Heidelberg (2005)
7. Lindell, Y., Pinkas, B.: Privacy preserving data mining. In: Bellare, M. (ed.) CRYPTO 2000. LNCS, vol. 1880, pp. 36–54. Springer, Heidelberg (2000)
8. Oliveira, S., Zaiane, O.R.: Privacy preserving clustering by data transformation. In: Proceedings of the 18th Brazilian Symposium on Databases, Marnaus, pp. 304–318 (2003)
9. Krishna Prasad, P., Pandu Rangan, C.: Privacy preserving BIRCH algorithm for clustering over vertically partitioned databases. In: Jonker, W., Petković, M. (eds.) SDM 2006. LNCS, vol. 4165, pp. 84–99. Springer, Heidelberg (2006)
10. Rizvi, S., Haritsa, J.R.: Maintaining data privacy in association rule mining. In: VLDB 2002. Proceedings of the 28th International Conference on Very Large Data Bases, Washington, DC, August 2003, pp. 206-215 (2003)
11. Vaidya, J., Clifton, C.: Privacy-preserving k-means clustering over vertically partitioned data. In: Proceedings of the 9th ACM SIGKDD International Conference on knowledge Discovery and Data Mining, Washington, DC, August 2003, ACM Press, New York (2003)
12. Yao, A.C.: Protocols for secure computation. In: Proceedings of 23rd IEEE Symposium on Foundations of Computer Science, pp. 160–164. IEEE Computer Society Press, Los Alamitos (1982)

A New Multi-level Algorithm Based on Particle Swarm Optimization for Bisecting Graph

Lingyu Sun[1], Ming Leng[2], and Songnian Yu[2]

[1] Department of Computer Science,
Jinggangshan College, Ji'an, PR China 343009
sunlingyu@jgsu.edu.cn
[2] School of Computer Engineering and Science,
Shanghai University, Shanghai, PR China 200072
lengming@shu.edu.cn,
snyu@staff.shu.edu.cn

Abstract. An important application of graph partitioning is data clustering using a graph model — the pairwise similarities between all data objects form a weighted graph adjacency matrix that contains all necessary information for clustering. The min-cut bipartitioning problem is a fundamental graph partitioning problem and is NP-Complete. In this paper, we present a new multi-level algorithm based on particle swarm optimization (PSO) for bisecting graph. The success of our algorithm relies on exploiting both the PSO method and the concept of the graph core. Our experimental evaluations on 18 different graphs show that our algorithm produces encouraging solutions compared with those produced by MeTiS that is a state-of-the-art partitioner in the literature.

1 Introduction

Partitioning is a fundamental problem with extensive applications to many areas using a graph model, including VLSI design [1], task scheduling [2], knowledge discovery [3], data clustering [4] and parallel processing [5]. Given the attributes of the data points in a dataset and the similarity or affinity metric between any two points, the symmetric matrix containing similarities between all pairs of points forms a weighted adjacency matrix of an undirected graph. Thus the data clustering problem becomes a graph partitioning problem [4]. The *min-cut bipartitioning problem* is a fundamental partitioning problem and is NP-Complete [6]. It is also NP-Hard to find good approximate solutions for this problem [7]. The survey by Alpert and Kahng [1] provides a detailed description and comparison of various such schemes which can be classified as *move-based* approaches, *geometric representations*, *combinatorial* formulations, and *clustering* approaches.

Most existing partitioning algorithms are heuristics in nature and they seek to obtain reasonably good solutions in a reasonable amount of time. Kernighan and Lin (KL) [8] proposed a heuristic algorithm for partitioning graphs. The KL algorithm is an iterative improvement algorithm that consists of making several improvement passes. It starts with an initial bipartitioning and tries to improve it by every pass. A pass consists of the identification of two subsets of vertices,

R. Alhajj et al. (Eds.): ADMA 2007, LNAI 4632, pp. 69–80, 2007.
© Springer-Verlag Berlin Heidelberg 2007

one from each part such that can lead to an improved partitioning if the vertices in the two subsets switch sides. Fiduccia and Mattheyses (FM) [9] proposed a fast heuristic algorithm for bisecting a weighted graph by introducing the concept of cell *gain* into the KL algorithm. These algorithms belong to the class of *move-based* approaches in which the solution is built iteratively from an initial solution by applying a move to the current solution. Move-based approaches are the most frequently combined with stochastic hill-descending algorithms such as those based on Tabu Search[10], Genetic Algorithms [11], Neural Networks [12], Ant Colony Optimization[13], etc., which allow movements towards solutions worse than the current one in order to escape from local minima.

As the problem sizes reach new levels of complexity recently, a new class of partitioning algorithms have been developed that are based on the multi-level paradigm. The multi-level graph partitioning schemes consist of three phases [14],[15],[16]. The *coarsening phase* is to reduce the size of the graph by collapsing vertex and edge until its size is smaller than a given threshold. The *initial partitioning phase* is to compute initial partition of the coarsest graph. The *refinement phase* is to project successively the partition of the smaller graph back to the next level finer graph while applying an iterative refinement algorithm.

In this paper, we present a multi-level algorithm which integrates a new PSO-based refinement approach and an effective matching-based coarsening scheme. Our work is motivated by the discrete particle swarm optimization algorithm of Kennedy [17] who limits the trajectories of a population of *"particles"* in the probability that a coordinate will take on a zero or one value and Karypis who introduces the concept of the graph *core* for coarsening the graph in [15] and supplies **MeTiS** [14], distributed as open source software package for partitioning unstructured graphs. We test our algorithm on 18 graphs that are converted from the hypergraphs of the ISPD98 benchmark suite [18]. Our comparative experiments show that the proposed algorithm significantly produces partitions that are better than those produced by **MeTiS** in a reasonable time.

The rest of the paper is organized as follows. Section 2 provides some definitions and describes the notation that is used throughout the paper. Section 3 describes the motivation behind our algorithm. Section 4 presents a new multi-level PSO refinement algorithm. Section 5 experimentally evaluates our algorithm and compares it with **MeTiS**. Finally, Section 6 provides some concluding remarks and indicates the directions for further research.

2   Mathematical Description

A graph $G=(V,E)$ consists of a set of vertices V and a set of edges E such that each edge is a subset of two vertices in V. Throughout this paper, n and m denote the number of vertices and edges respectively. The vertices are numbered from 1 to n and each vertex $v \in V$ has an integer weight $S(v)$. The edges are numbered from 1 to m and each edge $e \in E$ has an integer weight $W(e)$. A decomposition of a graph V into two disjoint subsets V^1 and V^2, such that $V^1 \cup V^2 = V$ and $V^1 \cap V^2 = \varnothing$, is called a *bipartitioning* of V. Let $S(A) = \sum_{v \in A} S(v)$ denotes the size of a subset $A \subseteq V$. Let ID_v be denoted as v's *internal degree* and is equal to the sum of the edge-weights of the adjacent vertices of v that are in the same side of the partition as v, and v's *external degree* denoted by ED_v is equal to the sum of edge-weights of the adjacent vertices of v that are in different sides. The *cut* of a *bipartitioning* $P=\{V^1,V^2\}$ is the sum of weights of edges which contain two vertices in V^1 and V^2 respectively. Naturally, vertex v belongs at the boundary if and only if $ED_v > 0$ and the *cut* of P is also equal to $0.5 \sum_{v \in V} ED_v$.

Given a balance constraint r, the *min-cut bipartitioning problem* seeks a solution $P=\{V^1,V^2\}$ that minimizes $cut(P)$ subject to $(1-r)S(V)/2 \leq S(V^1), S(V^2) \leq (1+r)S(V)/2$. A *bipartitioning* is *bisection* if r is as small as possible. The task of minimizing $cut(P)$ can be considered as the *objective* and the requirement that solution P will be of the same size can be considered as the *constraint*.

3 Motivation

PSO is a population-based optimization method first proposed by Kennedy [19]. This method is inspired by the choreography of bird flock and has been shown to be effective in optimizing difficult multidimensional problems. In PSO, each single solution, called *particle*, is a bird in the search space. All of the *particles* have *fitness values*, which are evaluated by a *fitness function* to be optimized. Each *particle* has an *adaptable velocity* and flies through the multidimensional search space in search for the optimum solution. Moreover, each *particle* has a *memory*, remembering the best *position* of the search space it has ever visited. Thus, its movement is an aggregated acceleration towards its best previous *position* and towards the best previous *position* of any *particle* in the neighborhood. The PSO algorithms reported in the literatures are classified into two groups: continuous PSO [19] and discrete PSO [17]. In continuous PSO, the *particles* operate in continuous search space, where the trajectories are defined as changes in position on some number of dimensions. But in discrete PSO, the *particles* operate on discrete search space, and the trajectories are defined as changes in the probability that a coordinate will take on a value from feasible discrete values.

In this paper, we address a discrete PSO algorithm to search for good approximate solutions of the *min-cut bipartitioning problem*. In the problem, each *particle* of the *swarm* can be considered as a *bipartitioning* X and the solution space is $|V|$-dimensional. Naturally, the *fitness function* to be minimized is defined as $cut(X)$. Suppose that the scale of *swarm* is N, then the *position* of the ith (i =1,2,...,N) *particle* can be represented by an n-dimensional vector $(|V|=n)$, $\overrightarrow{X}_i = (x_{i1}, x_{i2}, \ldots, x_{in})^T$, where x_{id} (d =1,2,...,n) denotes the dth dimension of the *position* vector $\overrightarrow{X}_i$ and is restricted to zero and one value. The *velocity* of this *particle* can be represented by another n-dimensional vector, $\overrightarrow{Y}_i = (y_{i1}, y_{i2}, \ldots, y_{in})^T$, where each y_{id} represents the probability of the bit x_{id} taking the value one. The best previously visited *position* of the ith *particle* is denoted by $\overrightarrow{Q}_i = (q_{i1}, q_{i2}, \ldots, q_{in})^T$. Defining κ as the index of the best *particle* in the *swarm* and the best previously visited *position* of the *swarm* is denoted by $\overrightarrow{Q}_\kappa = (q_{\kappa 1}, q_{\kappa 2}, \ldots, q_{\kappa n})^T$. Let the superscripts t denotes the iteration number, then the standard discrete PSO algorithm is manipulated according to the following two equations:

$$\overrightarrow{Y}_i^{t+1} = \alpha \cdot \overrightarrow{Y}_i^t + \overrightarrow{U}[0, \varphi_1] \otimes (\overrightarrow{Q}_i^t - \overrightarrow{X}_i^t) + \overrightarrow{U}[0, \varphi_2] \otimes (\overrightarrow{Q}_\kappa^t - \overrightarrow{X}_i^t) \qquad (1)$$

$$x_{id}^{t+1} = \begin{cases} 0 & \text{if random}(0,1) \geq \text{sigmoid}(\ y_{id}^{t+1}\ ) \\ 1 & \text{if random}(0,1) < \text{sigmoid}(\ y_{id}^{t+1}\ ) \end{cases} \quad \text{for all} \quad \begin{matrix} i \in \{1, 2, \ldots, N\} \\ d \in \{1, 2, \ldots, n\} \end{matrix} \qquad (2)$$

where $\otimes$ denotes point-wise vector multiplication; $\overrightarrow{U}[min, max]$ is a function that returns a vector whose positions are randomly generated following the uniform distribution between min and max; the function random$(0,1)$ generates random number selected from a uniform distribution in $[0.0, 1.0]$ and sigmoid(y_{id}) is a sigmoid limiting transformation; α is called the *inertia weight*; φ_1, φ_2 are two positive constants, called *cognitive* and *social* parameter respectively.

However, the space for feasible solutions of the *min-cut bipartitioning problem* is prohibitively large. The standard discrete PSO algorithm is not usually seen as an effective approach for the problem because the computing times are quite large, especially when $|V|$ is more than several thousands. We are forced to recourse to the multi-level paradigm and propose an effective multi-level PSO refinement algorithm (MPSOR) that combines the PSO method with the multi-level paradigm. The MPSOR algorithm is a refinement algorithm which improves the quality of the level l graph $G_l(V_l, E_l)$ partitioning P_{G_l} with a boundary refinement policy, instead of trying to run the standard discrete PSO algorithm directly on the original graph. Furthermore, the MPSOR algorithm combines PSO with local optimization heuristics to form PSO-*local search hybrids* that applies the FM algorithm to every *particle* to achieve local optimum and maintain the balance constraint of the partitioning $\overrightarrow{X}_i$. The use of both heuristics defines a hybrid strategy that allows the search process to escape from local minima, while simultaneously strengthens effectiveness and achieves significant speedups for high quality partitionings with well-designed heuristics.

Algorithm 1 (algorithm for determining the cores hierarchy[21])

```
    INPUT:graph G(V,E)
  OUTPUT:table core with core number for each vertex
  Begin
    Compute the degrees of vertices;
    Order the set of vertices V in increasing order of their degrees;
    For Each v ∈ G in the order Do Begin
      core[v]=degree[v];
      For Each u ∈ Neighbors(v) Do Begin
        If degree[u] >degree[v] Then Begin
          degree[u]=degree[u]-1;
          Reorder V accordingly;
```

End If
  End For
 End For
End
Return table *core* with core number

In [14], Karypis presents the sorted heavy-edge matching (SHEM) algorithm that identifies and collapses together groups of vertices that are highly connected. Firstly, SHEM sorts the vertices of the graph ascendingly based on the *degree* of the vertices. Next, the vertices are visited in this order and SHEM matches the vertex v with unmatched vertex u such that the weight of the edge $W(v,u)$ is maximum over all incident edges. In [20], Sediman introduces the concept of the graph *core* firstly that the *core* number of a vertex v is the maximum order of a *core* that contains that vertex. Vladimir gives an $O(m)$-time algorithm for cores decomposition of networks shown in Algorithm 1 and $O(m \cdot \log(n))$-time algorithm to compute the *core* numbering in the context of sum-of-the-edge-weights in [22] respectively. In [15], Amine and Karypis introduce the concept of the graph *core* for coarsening the *power-law* graphs. In [10], Leng present the core-sorted heavy-edge matching (CSHEM) algorithm that combines the concept of the graph *core* with the SHEM scheme. Firstly, CSHEM sorts the vertices of the graph descendingly based on the *core* number of the vertices by the algorithm in [22]. Next, the vertices are visited in this order and CSHEM matches the vertex v with its unmatched neighboring vertex whose edge-weight is maximum. In case of a tie according to edge-weights, we will prefer the vertex that has the highest *core* number.

In our multi-level algorithm, we adopt the MPSOR algorithm during the *refinement phase*, the greedy graph growing partition (GGGP) algorithm [14] during the *initial partitioning phase*, an effective matching-based coarsening scheme during the *coarsening phase* that uses the CSHEM algorithm on the original graph and the SHEM algorithm on the coarser graphs. The pseudocode of our multi-level algorithm is shown in Algorithm 2. Each *particle i* of the population maintains *position* vector, *velocity* vector, *personal best position* vector on the graph $G_l(V_l,E_l)$ which be denoted as $\vec{X}_{li}$, $\vec{Y}_{li}$ and $\vec{Q}_{li}$ respectively. In the *initial partitioning phase*, we should initial the population on the smallest graph for the first call of the MPSOR algorithm. During the *refinement phase*, we also should project successively all the *particles's* $\vec{X}_{li}$, $\vec{Y}_{li}$ and $\vec{Q}_{li}$ back to the next level finer graph for the next call of the MPSOR algorithm.

Algorithm 2 (our multi-level algorithm)

  INPUT:original graph $G(V,E)$
  OUTPUT:the partitioning P_G of graph G
  /*coarsening phase*/
  $l = 0$
  $G_l(V_l,E_l) = G(V,E)$
  $G_{l+1}(V_{l+1},E_{l+1}) = \text{CSHEM}(G_l(V_l,E_l))$

While ($|V_{l+1}| > 20$) do
   $l = l + 1$
   $G_{l+1}(V_{l+1},E_{l+1})$=SHEM($G_l(V_l,E_l)$)
End While
/*initial partitioning phase*/
P_{G_l}=GGGP(G_l)
For i = 1 to N do
   $\overrightarrow{X}_{li} = P_{G_l}$
   $\overrightarrow{Q}_{li} = \overrightarrow{X}_{li}$
   $\overrightarrow{Y}_{li} = \overrightarrow{U}[-y_{max}, +y_{max}]$
End For
/*refinement phase*/
While ($l \geq 1$) do
   $\overrightarrow{Q}_{l\kappa}$=MPSOR($G_l,P_{G_l}$)
   /*Prepare for the next call of MPSOR*/
   For i = 1 to N do
    Project the ith particle's velocity $\overrightarrow{Y}_{li}$ to $\overrightarrow{Y}_{(l-1)i}$;
    Project the ith particle's position $\overrightarrow{X}_{li}$ to $\overrightarrow{X}_{(l-1)i}$;
    Project the ith particle's best position $\overrightarrow{Q}_{li}$ to $\overrightarrow{Q}_{(l-1)i}$;
   End For
   Project the swarm's best position $\overrightarrow{Q}_{l\kappa}$ to $P_{G_{l-1}}$;
   $l = l - 1$
End While
P_G=MPSOR(G_l,P_{G_l})
Return P_G

4   A New Multi-level Particle Swarm Optimization Refinement Algorithm

The pseudocode of the MPSOR algorithm is shown in Algorithm 3. Each *particle i* stores the *internal* and *external degrees* and boundary vertices which be denoted as ID^i, ED^i and *boundary hash-table^i* respectively. Informally, the MP-SOR algorithm works as follows: At time zero, an initialization phase takes place during which all the *particles's ID^i*, ED^i and *boundary hash-table^i* are computed. The main structure of MPSOR consists of a nested loop. The outer loop detects the stopping condition whether MPSOR is run for a fixed number of iterations NC_{max}. In the inner loop of the MPSOR algorithm, $\overrightarrow{Y}_{li}$ of each *particle* is adjusted with a boundary refinement policy. It is necessary to impose a maximum value y_{max} on *velocity* y_{id}. If the y_{id} exceeded this threshold, it will be set equal to y_{max}. Each *particle's position* $\overrightarrow{X}_{li}$ also needs to update by moving probabilistically vertices and maybe violates the balancing constraint b. Finally, we choose repetitively the highest-gain vertex to move from the larger side and satisfy the balance constraint of $\overrightarrow{X}_{li}$ by using the early-exit FM (FM-EE) algorithm [16].

It may be possible to improve the quality of $\overrightarrow{X}_{li}$ by using the local heuristic algorithm until the local optimum is reached.

Algorithm 3 (MPSOR)

INPUT:the graph G_l, the *inertia weight* α,the scale of *swarm* N
the *cognitive* parameter φ_1, the *social* parameter φ_2
the maximum *velocity* y_{max}, the maximum cycles NC_{max}
OUTPUT:the *swarm*'s *best position* $\overrightarrow{Q}_{l\kappa}$
/*Initial heuristic information*/
$t = 0$
For $i = 1$ to N do
  For every vertex v in V_l do
$$ID_v^i = \sum_{(v,u)\in E_l \wedge x_{iv}^t = x_{iu}^t} W(v, u)$$
$$ED_v^i = \sum_{(v,u)\in E_l \wedge x_{iv}^t \neq x_{iu}^t} W(v, u)$$
    Store v in *boundary hash-table^i* if and only if $ED_v^i > 0$;
  End For
End For
For $t = 1$ to NC_{max} do /*the outer loop*/
  For $i = 1$ to N do /*the inner loop*/
    For $d = 1$ to $|V_l|$ do
      If the dth vertex in *boundary hash-table^i* then
$$y_{id}^{t+1} = \alpha \cdot y_{id}^t + random(0,1) \cdot \varphi_1 \cdot (q_{id}^t - x_{id}^t) + random(0,1) \cdot \varphi_2 \cdot (q_{i\kappa}^t - x_{id}^t)$$
$$y_{id}^{t+1} = sign(y_{id}^{t+1}) \cdot min(abs(y_{id}^{t+1}), y_{max})$$
      If $(random(0,1) < \frac{1.0}{1.0+exp(-1.0\times y_{id}^{t+1})})$ then
        $x_{id}^{t+1} = 1$
      Else
        $x_{id}^{t+1} = 0$
      End If
      If $x_{id}^{t+1} \neq x_{id}^t$ then
        Move the dth vertex to the other side;
        Update ID^i, ED^i, *boundary hash-table^i*;
      End If
      End If
    End For
    Apply the FM-EE algorithm to refine $\overrightarrow{X}_{li}$ byID^i, ED^i,*boundary hash-table^i*;
    Update the ith *particle*'s *best position* $\overrightarrow{Q}_{li}$;
  End For
  Update the *swarm*'s *best position* $\overrightarrow{Q}_{l\kappa}$;
End For /* $t \leq NC_{max}$*/
Return the *swarm*'s *best position* $\overrightarrow{Q}_{l\kappa}$

Because the MPSOR algorithm aggressively moves the best admissible vertex to refine $\overrightarrow{X}_{li}$ by the FM-EE algorithm, it must examine and compare a number of boundary vertices by the *bucket* that allows to storage, retrieval and update the gains of vertices very quickly. It is important to obtain the efficiency of MPSOR by using the *bucket* with the last-in first-out scheme. The *internal* and *external degrees* of all vertices help MPSOR to facilitate computation of vertex *gain* and judgement of boundary vertex. We also use a *boundary hash-table* to store the boundary vertices whose *external degree* is greater than zero. During each iteration of MPSOR, ID^i, ED^i of each *particle* are kept consistent with respect to the partitioning $\overrightarrow{X}_{li}$. This can be done by updating the *degrees* of the moved vertex's neighboring vertices. Of course, the *boundary hash-table^i* might change as the partitioning $\overrightarrow{X}_{li}$ changes. For example, due to a move in an other boundary vertex, a boundary vertex would no longer be such a boundary vertex and should be removed from the *boundary hash-table*. Furthermore, a no-boundary vertex can become such a vertex if it is connected to a boundary vertex which is moved to the other side and should be inserted in the *boundary hash-table*.

5   Experimental Results

We use the 18 graphs in our experiments that are converted from the hypergraphs of the ISPD98 benchmark suite [18] and range from 12,752 to 210,613 vertices. Each hyperedge is a subset of two or more vertices in hypergraph. We convert hyperedges into edges by the rule that every subset of two vertices in hyperedge can be seemed as edge [10]. We create the edge with unit weight if the edge that connects two vertices doesn't exist, else add unit weight to the weight of the edge. Next, we get the weights of vertices from the benchmark suite. Finally, we store 18 edge-weighted and vertex-weighted graphs in format of **MeTiS** [14]. The characteristics of these graphs are shown in Table 2.

We implement our multi-level algorithm in ANSI C and integrate with the leading edge partitioner **MeTiS**. In the evaluation of our algorithm, we must make sure that the results produced by our algorithm can be easily compared against those produced by **MeTiS**. We use the same balance constraint b and random seed in every comparison. In the scheme choices of three phases offered by **MeTiS**, we use the SHEM algorithm during the *coarsening phase*, the GGGP algorithm during the *initial partitioning phase* that consistently finds smaller edge-cuts than other algorithms, the boundary KL (BKL) refinement algorithm during the *refinement phase* because BKL can produce smaller edge-cuts when coupled with the SHEM algorithm. These measures are sufficient to guarantee that our experimental evaluations are not biased in any way.

The quality of partitions is evaluated by looking at two different quality measures, which are the minimum *cut* (MinCut) and the average *cut* (AveCut). To ensure the statistical significance of our experimental results, two measures are obtained in twenty runs whose random seed is different to each other. For all experiments, we use a 49-51 *bipartitioning* balance constraint by setting b to 0.02. Furthermore, we adopt the experimentally determined optimal set of parameters values for MPSOR, $\alpha=1$, $\varphi_1=0.5$, $\varphi_2=0.5$, $N=20$, $y_{max}=4$, $NC_{max}=15$.

Table 1. The characteristics of 18 graphs to evaluate our algorithm

benchmark	vertices	hyperedges	edges
ibm01	12752	14111	109183
ibm02	19601	19584	343409
ibm03	23136	27401	206069
ibm04	27507	31970	220423
ibm05	29347	28446	349676
ibm06	32498	34826	321308
ibm07	45926	48117	373328
ibm08	51309	50513	732550
ibm09	53395	60902	478777
ibm10	69429	75196	707969
ibm11	70558	81454	508442
ibm12	71076	77240	748371
ibm13	84199	99666	744500
ibm14	147605	152772	1125147
ibm15	161570	186608	1751474
ibm16	183484	190048	1923995
ibm17	185495	189581	2235716
ibm18	210613	201920	2221860

Table 2. Min-cut bipartitioning results with up to 2% deviation from exact bisection

benchmark	Metis(α)		our algorithm(β)		ratio(β:α)		improvement	
	MinCut	AveCut	MinCut	AveCut	MinCut	AveCut	MinCut	AveCut
ibm01	517	1091	259	542	0.501	0.497	49.9%	50.3%
ibm02	4268	11076	3319	5795	0.778	0.523	22.2%	47.7%
ibm03	10190	12353	5049	7484	0.495	0.606	50.5%	39.4%
ibm04	2273	5716	2007	3457	0.883	0.605	11.7%	39.5%
ibm05	12093	15058	8043	11445	0.665	0.760	33.5%	24.0%
ibm06	7408	13586	1864	8754	**0.252**	0.644	**74.8%**	35.6%
ibm07	3219	4140	2202	2908	0.684	0.702	31.6%	29.8%
ibm08	11980	38180	12168	16444	**1.016**	**0.431**	**-1.6%**	**56.9%**
ibm09	2888	4772	2835	3138	0.982	0.658	1.8%	34.2%
ibm10	10066	17747	5598	7999	0.556	0.451	44.4%	54.9%
ibm11	2452	5095	2439	3179	0.995	0.624	0.5%	37.6%
ibm12	12911	27691	10514	14466	0.814	0.522	18.6%	47.8%
ibm13	6395	13469	4206	6983	0.658	0.518	34.2%	48.2%
ibm14	8142	12903	8025	11441	0.986	0.887	1.4%	11.3%
ibm15	22525	46187	14015	32102	0.622	0.695	37.8%	30.5%
ibm16	11534	22156	9862	15818	0.855	0.714	14.5%	28.6%
ibm17	16146	26202	15014	21330	0.930	0.814	7.0%	18.6%
ibm18	15470	20018	15388	20496	0.995	**1.024**	0.5%	**-2.4%**
average					**0.759**	**0.649**	**24.1%**	**35.1%**

Table 2 presents *min-cut bipartitioning* results allowing up to 2% deviation from exact bisection and Fig. 1 illustrates the MinCut and AveCut comparisons of two algorithms on 18 graphs. As expected, our algorithm reduces the AveCut by -2.4% to 56.9% and reaches 35.1% average AveCut improvement. We also obtain 24.1% average MinCut improvement and between -1.6% and 74.8% improvement in MinCut. All evaluations that twenty runs of two algorithms on 18 graphs are run on an 1800MHz AMD Athlon2200 with 512M memory and can be done in four hours.

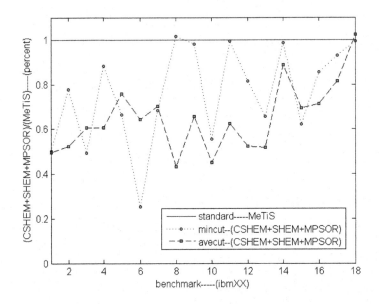

Fig. 1. The MinCut and AveCut comparisons of two algorithms on 18 graphs

6   Conclusions

In this paper, we have presented an effective multi-level algorithm based on PSO. The success of our algorithm relies on exploiting both the PSO method and the concept of the graph core. We obtain excellent *bipartitioning* results compared with those produced by **MeTiS**. Although it has the ability to find cuts that are lower than the result of **MeTiS** in a reasonable time, there are several ways in which this algorithm can be improved. For example, we note that adopting the CSHEM algorithm alone leads to poorer experimental results than the combination of CSHEM with SHEM. We need to find the reason behind it and develop a better matching-based coarsening scheme coupled with MPSOR. In the MinCut evaluation of benchmark ibm08, our algorithm is 1.6% worse than **MeTiS**. Therefore, the second question is to guarantee find good approximate solutions by setting optimal set of parameters values for MPSOR.

Acknowledgments

This work was supported by the international cooperation project of Ministry of Science and Technology of PR China, grant No. CB 7-2-01, and by "SEC E-Institute: Shanghai High Institutions Grid" project. Meanwhile, the authors would like to thank professor Karypis of university of Minnesota for supplying source code of **MeTiS**. The authors also would like to thank Alpert of IBM Austin research laboratory for supplying the ISPD98 benchmark suite.

References

1. Alpert, C.J., Kahng, A.B.: Recent directions in netlist partitioning. Integration, the VLSI Journal 19, 1–81 (1995)
2. Khannat, G., Vydyanathant, N.: A hypergraph partitioning based approach for scheduling of tasks with batch-shared I/O. In: IEEE International Symposium on Cluster Computing and the Grid, pp. 792–799 (2005)
3. Hsu, W.H., Anvil, L.S.: Self-organizing systems for knowledge discovery in large databases. In: International Joint Conference on Neural Networks, pp. 2480–2485 (1999)
4. Ding, C., He, X., Zha, H., Gu, M., Simon, H.: A Min-Max cut algorithm for graph partitioning and data clustering. In: Proc. IEEE Conf Data Mining, pp. 107–114 (2001)
5. Hendrickson, B., Leland, R.: An improved spectral graph partitioning algorithm for mapping parallel computations. SIAM Journal on Scientific Computing 16, 452–469 (1995)
6. Garey, M.R., Johnson, D.S.: Computers and intractability: A guide to the theory of NP-completeness. WH Freeman, New York (1979)
7. Bui, T., Leland, C.: Finding good approximate vertex and edge partitions is NP-hard. Information Processing Letters 42, 153–159 (1992)
8. Kernighan, B.W., Lin, S.: An efficient heuristic procedure for partitioning graphs. Bell System Technical Journal 49, 291–307 (1970)
9. Fiduccia, C., Mattheyses, R.: A linear-time heuristics for improving network partitions. In: Proc. 19th Design Automation Conf. pp. 175–181 (1982)
10. Leng, M., Yu, S.: An effective multi-level algorithm for bisecting graph. In: Li, X., Zaïane, O.R., Li, Z. (eds.) ADMA 2006. LNCS (LNAI), vol. 4093, pp. 493–500. Springer, Heidelberg (2006)
11. Żola, J., Wyrzykowski, R.: Application of genetic algorithm for mesh partitioning. In: Proc. Workshop on Parallel Numerics, pp. 209–217 (2000)
12. Bahreininejad, A., Topping, B.H.V., Khan, A.I.: Finite element mesh partitioning using neural networks. Advances in Engineering Software, 103–115 (1996)
13. Leng, M., Yu, S.: An effective multi-level algorithm based on ant colony optimization for bisecting graph. In: Ng, W.-K., Kitsuregawa, M., Li, J., Chang, K. (eds.) PAKDD 2006. LNCS (LNAI), vol. 3918, pp. 138–149. Springer, Heidelberg (2006)
14. Karypis, G., Kumar, V.: MeTiS 4.0: Unstructured graphs partitioning and sparse matrix ordering system. Technical Report, Department of Computer Science, University of Minnesota (1998), Available on the WWW at URL http://www.cs.umn.edu/~metis

15. Amine, A.B., Karypis, G.: Multi-level algorithms for partitioning power-law graphs. Technical Report, Department of Computer Science, University of Minnesota (2005), Available on the WWW at URL http://www.cs.umn.edu/~metis
16. Karypis, G., Aggarwal, R., Kumar, V., Shekhar, S.: Multilevel hypergraph partitioning: Application in VLSI domain. In: Proc. Design Automation Conf, pp. 526–529 (1997)
17. Kennedy, J., Eberhart, R.C.: A discrete binary version of the particle swarm algorithm. In: IEEE International Conference on Systems, Man, and Cybernetics, pp. 4104–4108 (1997)
18. Alpert, C.J.: The ISPD98 circuit benchmark suite. In: Proc. Intel Symposium of Physical Design, pp. 80–85 (1998)
19. Kennedy, J., Eberhart, R.: Particle swarm optimization. Proc. IEEE Conf. Neural Networks IV, pp. 1942–1948 (1995)
20. Seidman, S.B.: Network structure and minimum degree. Social Networks, 269–287 (1983)
21. Batagelj, V., Zaveršsnik, M.: An O(m) Algorithm for cores decomposition of networks. Journal of the ACM, 799–804 (2001)
22. Batagelj, V., Zaveršnik, M.: Generalized cores. Journal of the ACM, 1–8 (2002)

A Supervised Subspace Learning Algorithm: Supervised Neighborhood Preserving Embedding

Xianhua Zeng[1,2] and Siwei Luo[1]

[1] School of Computer and Information Technology, Beijing Jiaotong University,
Beijing 100044, China
[2] School of Computer Science, China West Normal University, Sichuan 637002, China
xianhuazeng2005@163.com

Abstract. Neighborhood Preserving Embedding (NPE) is an unsupervised manifold learning algorithm with subspace learning characteristic. In fact, NPE is a linear approximation to Locally Linear Embedding (LLE). So it can provide an unsupervised subspace learning technique. In this paper, we proposed a new Supervised Neighborhood Preserving Embedding (SNPE) algorithm which can use the label or category information of training samples to better describe the intrinsic structure of original data in low-dimensional space. Furthermore, when a new unknown data needs to be processed, SNPE, as a supervised subspace learning technique, may be conducted in the original high-dimensional space. Several experiments on USPS digit database demonstrate the effectiveness of our algorithm.

1 Introduction

In many pattern recognition tasks, the recognition results are only simple class labels, but the inputs of recognition tasks contain rich structure information. The high-dimensional inputs are commonly controlled by several intrinsic variables. It is reasonable that low-dimensional codes, determined by intrinsic variables, are first found by using unsupervised learning and discriminative learning can then be used to model the dependence of the class label on the low-dimensional codes [1]. So the dimensionality reduction of data is very useful for pattern recognition. ISOMAP [2] and LLE [3], [8], [9] are two very important nonlinear methods to find the low-dimensional embeddings of data set sampled from high-dimensional space. A linear dimensionality reduction algorithm, called Neighborhood Preserving Embedding (NPE), had been proposed by He X.F. [4]. NPE is a linear approximation to the LLE algorithm. Moreover, NPE is different from PCA [5] which aims at preserving the global Euclidean structure. NPE aims at preserving the local manifold structure. So NPE is a different subspace learning algorithm from PCA.

   In some visualization tasks, high dimensional data are sampled from multiple classes and the class labels are known. In classification tasks, the class labels of all training data must be known. The information provided by these class labels may be used to guide the procedure of dimensionality reduction [6]. In this paper, we propose a new Supervised Neighborhood Preserving Embedding (SNPE) algorithm which can use the label information of training samples to better describe the intrinsic structure

R. Alhajj et al. (Eds.): ADMA 2007, LNAI 4632, pp. 81–88, 2007.

of original data in low-dimensional space. And SNPE is capable of controlling the degree of label or category information by using supervised factor. Experiments on the benchmark USPS digit data set demonstrate that SNPE is a powerful feature extraction and visualization method, and that SNPE can yield very promising recognition results.

The rest of this paper is organized as follows: Section 2 gives a brief description of Neighborhood Preserving Embedding. In section 3, SNPE is proposed and discussed in detail. The experimental results are shown in Section 4. Finally, we give some concluding remarks and future works in Section 5.

2 Neighborhood Preserving Embedding (NPE)

In this Section, we briefly introduce Neighborhood Preserving Embedding [4]. NPE is a linear approximation to the LLE algorithm [3], [7], [8], [9]. The algorithmic procedure is stated as follows:

Step I, Constructing a neighborhood graph G
Let G denote a graph with n nodes. The i-th node corresponds to the data point X_i. There are two ways to construct the neighborhood graph G: a) connecting nodes i and j if X_j is among the K nearest neighbors of X_i; b) connecting nodes i and j if $\|X_j\text{-}X_i\| < \varepsilon$.

Step II, Computing the reconstruction weights
In this step, we compute the weights on the edges. According to the adjacency graph G, every data point in training set is reconstructed by the linear combination of its weighed K nearest data points. Let W denote the weight matrix. The weight W_{ij}, on the edge from node i to node j, can be computed by minimizing the following objective function:

$$\begin{cases} Min\sum_i \| X_i - \sum_{j=1}^{K} W_{ij} X_j \|^2 \\ Subject \quad to: \sum_{j=1}^{K} W_{ij} = 1 \end{cases} \tag{1}$$

where K is the number of the neighbors of node X_i, and X_j is the j-th neighbor of node X_i.

Step III, Finding the basis vectors of the subspace
In the step, we fix the weight matrix W while optimizing the coordinates Y_i. Suppose that there exists a linear transformation $Y^T = a^T X$, where a is the basis vector. The basis vector a is computed by solving the minimization problem:

$$\begin{cases} Min \quad a^T XMX^T a \\ Subject \quad to: a^T XX^T a = 1 \end{cases} \tag{2}$$

where $M = (I - W)^T (I - W)$ and $X = [X_1, X_2, ..., X_n]$.

Using **Lagrange Multiplication**, the basis vector a is given by the minimum eigenvalue solution to the following generalized eigenvector problem:

$$(XX^T)^+(XMX^T)a = \lambda a \tag{3}$$

where $(XX^T)^+$ is pseudoinverse matrix of XX^T. It is easy to analyze that XX^T and XMX^T are symmetric and positive semi-definite. So the basis vector a corresponds to the eigenvector of the minimum eigenvalue.

Finally, the basis vectors $a_1, a_2, ..., a_d$ of the d-dimensional subspace correspond to the d smallest eigenvectors of the matrix $(XX^T)^+(XMX^T)$.

Step IV, Computing the embedding coordinate

Let $A = [a_1, a_2, ..., a_d]$. If X_i denotes the i-th observed data point in the high-dimensional space, then the d-dimensional embedding Y_i is computed by the formulation: $Y_i = A^T X_i$.

3 Supervised NPE

In some visualization tasks, high dimensional data are sampled from multiple classes and the class labels are known. In classification tasks, the class labels of all training data must be known. The information provided by these class labels may be used to guide the procedure of dimensionality reduction [6]. The multiple manifolds of different classes are properly separated and the manifold structure of per class can be preserved. This can be achieved by artificially increasing the distances between samples belonging to different classes, but leaving them unchanged if samples are from the same class. Due to seeking the supervised subspace on all classes in this paper, the changes of distances between samples should be controlled within certain bound. This paper proposes a new supervised subspace learning algorithm based on Neighborhood Preserving Embedding, called as supervised NPE. SNPE has the same steps with NPE except the first step. So we only analyze how to construct the neighborhood graph in SNPE. The easiest method is to select the neighbors Xj of data point Xi from the class that Xi itself belongs to. To make the algorithm more robust, a more sophisticated method is proposed in this section.

Let $\nabla = [\delta_{ij}]_{n \times n}$ denote the dissimilar matrix between different classes, where $\delta_{ij} = 1$ if Xi and Xj are sampled from different classes and $\delta_{ij} = 0$ otherwise. A new distance matrix can be calculated by adding distance between samples in different classes. That is,

$$D_{new} = D + \alpha \max(D)\nabla \tag{4}$$

where D is Euclidean distance matrix, $max(D)$ denotes the maximum Euclidean distance between two points in the whole data set, and $\alpha \in [0,1]$ controls the amount which label information may be incorporated.

Notice that SNPE introduces an additional parameter α which controls the amount of supervision. When $\alpha=0$, SNPE is the same with NPE; when $\alpha=1$, SNPE uses the fully supervised information and this means that neighbors of a sample in per class will always be picked from that same class; when $0<\alpha<1$, SNPE gives a partially supervised mapping from high dimension to low dimension and this may lead to better generalization than $\alpha=1$ on previously unseen samples. In practice, it is possible that the new distance matrix D_{new} leads to K–nearest neighbors of a sample in per class, only picked from that same class after α is bigger than the certain value, and the structure information can be preserved, as shown in section 4.

Finally, a supervised neighborhood graph G is constructed by using the new distance matrix D_{new}. The neighborhood can be defined as the K-nearest points or the points whose dissimilarity is less than a certain positive value ε. In this paper, the neighborhood is defined as the K-nearest points.

4   Experimental Results on the USPS Database

In this section, the application of SNPE has been discussed in feature extraction, data visualization and classification experiments. The data set in experiments is the well known US Postal Service (USPS) handwritten digits recognition corpus. It contains 11000 normalized grey images of size 16×16, with 1100 images from each of the ten class digits: 0,1,2,3,4,5,6,7,8,9. In Fig. 1, we have displayed 30 sample images of all digits of the USPS digit data set. In all experiments, every image corresponds to a 256-dimensional vector.

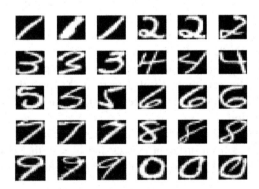

Fig. 1. 30 sample images from USPS digit data set. For each digit, there are 1100 images with different handwritten style.

4.1   Feature Extraction Analysis and Data Visualization

In the experiments, SNPE was compared to NPE and PCA which are the two widely used subspace learning techniques. The subspace is spanned by the basis vectors obtained from subspace learning algorithm. Therefore, any image in the subspace can be represented as a linear combination of the basis vectors. The basis vectors can be displayed as a sort of feature images. The nearest neighbor parameter K=25 in SNPE experiments. For simplicity, the visualization experimental data set consists of 1500 handwritten digit images which are respectively selected 500 samples from three class digits: "0", "1" and "2". In section 2, 3, we have discussed how to learn a supervised subspace by SNPE. Fig. 2 shows the several basis vectors obtained by PCA, NPE and SNPE. Fig. 2(b)(c)(d) show the first five basis images in low subspace obtained by SNPE when the supervised factor α=0, 0.1 and 0.5$\leq\alpha\leq$1. The different value of supervised factor α retains the information amount of between-class in supervised subspace. The partially SNPE (0<α<1) is a trade-off between the unsupervised NPE (α=0) and the fully SNPE (α=1). Fig. 3 shows the 2D visualization of the 1500 digit images in experimental data set. Fig. 3(b)(c)(d) show the change between the class structure and the neighborhood structure, and the neighborhood structure information

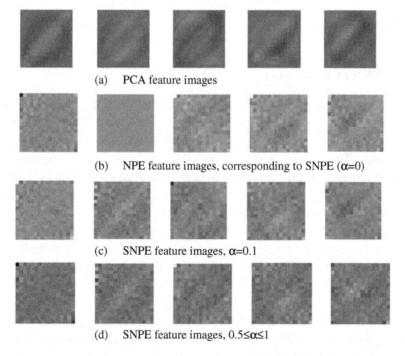

(a)   PCA feature images

(b)   NPE feature images, corresponding to SNPE (α=0)

(c)   SNPE feature images, α=0.1

(d)   SNPE feature images, 0.5$\leq\alpha\leq$1

Fig. 2. The first 5 basis vectors by using PCA and SNPE (α=0,0.1,0.5 and1.0)

is preserved while the between-class structure information is considered by using the supervised factor. Comparing SNPE to PCA, PCA pursues the global Euclidean structure while SNPE incorporates the neighborhood structure information and the class structure information. When the value of supervised factor α varies from 1 to 0, SNPE gradually decays into NPE. When $\alpha>0$, SNPE produces the better 2-dimensional visualization of USPS digit data than do PCA and NPE, as shown in Fig. 3.

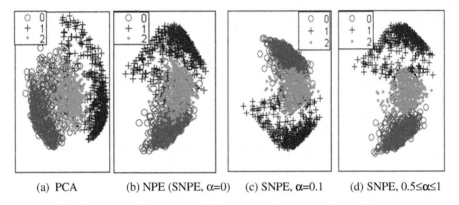

(a) PCA   (b) NPE (SNPE, $\alpha=0$)   (c) SNPE, $\alpha=0.1$ (d) SNPE, $0.5\leq\alpha\leq1$

Fig. 3. The 2-D visualization of the 1500 digit images sampled from the 3 class digits: "0", "1" and "2" by using PCA, LDA and SNPE ($\alpha=0$, 0.1 and $0.5\leq\alpha\leq1$)

4.2 SNPE for Classification

In classification experiments, SNPE is used to the full USPS digit data set mentioned previously. From this data set, 500 samples per class are randomly selected as a training set (containing 5000 images), and the rest (600 samples) per class are used for testing. The nearest neighbor parameter $K=25$ in SNPE experiments. In recognition step, we apply the Nearest Neighbor classification (NN) to PCA subspace, LDA subspace (the most $c-1$ basis vectors where c is the number of sample classes) and supervised subspaces (obtained by using SNPE at the different values of supervised factor α). In general, the recognition error varies with the dimension of the subspace. Fig. 4 shows the plots of test error versus dimension by using PCA+NN, LDA+NN, NPE+NN and SNPE+NN. Note that the test error=$1-P/N$, where N denotes the number of test samples and P denotes the number of correctly recognizing test samples. The results confirm that SNPE ($a>0$) generally leads to better classification performance than PCA and NPE when the reduction dimension from 24 to 40, but in others it is not significantly better. And the classification performance of SNPE+NN can't be further improved after $a\geq0.5$.

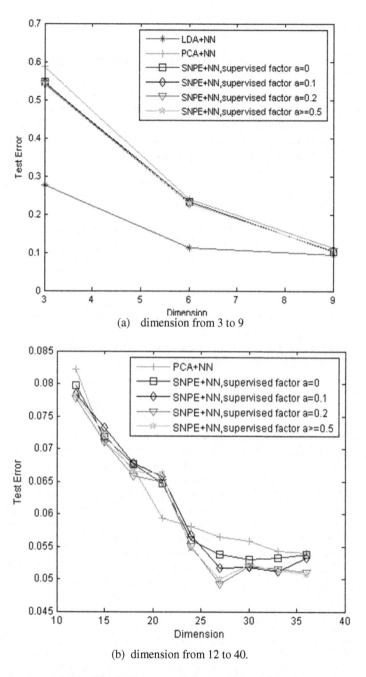

(a)   dimension from 3 to 9

(b) dimension from 12 to 40.

Fig. 4. Test errors of LDA, PCA,NPE(corresponding to SNPE, supervised factor α=0) and SNPE (supervised factor $0<\alpha\leq1$) on USPS digit data set; Notice that the dimension of LDA subspace is less than 10 on handwritten digit set with 10 classes. (a-b) shows the plots of test error versus dimension from 3 to 40.

5 Conclusions

This paper proposes a Supervised Neighborhood Preserving Embedding algorithm (SNPE) which can control the degree of using the label or category information to better describe the intrinsic structure of original data in low-dimensional space. Experiments on the benchmark USPS digit data set demonstrate that SNPE is a powerful feature extraction and visualization method, and that SNPE can yield very promising recognition results. Further research will analyze how to estimate the supervised factor α and the nearest neighbors K in SNPE.

Acknowledgements

The research is supported by: National Natural Science Foundations of China (No. 60373029) and Doctoral Foundations of China (No. 20050004001).

References

1. Hinton, G.E., Salakhutdinov, R.R.: Reducing the Dimensionality of Data with Neural Networks. Science 313, 504–507 (2006)
2. Tenenbaum, J., De Silva, V., Langford, J.: A Global Geometric Framework for Nonlinear Dimensionality Reduction. Science 290, 2319–2323 (2000)
3. Roweis, S., Saul, L.: Nonlinear Dimensionality Reduction by Locally Linear Embedding. Science 290, 2323–2326 (2000)
4. He, X.F., Deng, C., Yan, S.C., Zhang, H.J.: Neighborhood Preserving Embedding 2, 1208–1213 (2005)
5. Jolliffe, I.T.: Principal Component Analysis. Springer, New York (1989)
6. Geng, X., Zhan, D.C., Zhou, Z.H.: Supervised Nonlinear Dimensionality Reduction for Visualization and Classification. IEEE Transactions on Systems, Man, and Cybernetics - Part B: Cybernetics 6, 1098–1107 (2005)
7. Saul, L., Roweis, S.: Think Globally, Fit locally: Unsupervised Learning of Low Dimensional Manifolds. Journal of Machine Learning Research 4, 119–155 (2002)
8. de Ridder, D., Duin, R.P.W.: Locally Linear Embedding for Classification. Technical Report PH-2002-01,Pattern Recognition Group, Dept. of Imaging Science & Technology, Delft University of Technology, Delft, The Netherlands (2002)
9. Ridder, D., Vojtech, F.: Robust Manifold Learning. Research Report CTU-CMP-2003-08, Dept. of Imaging Science & Technology, Delft University of Technology, Delft, The Netherlands (2003)
10. Ye, J.P., Qi, L.: A Two-stage Linear Discriminant Analysis via QR-decomposition. IEEE Transactions on Pattern Analysis and Machine Intelligence 6, 929–941 (2005)
11. Seung, H.S., Lee, D.D.: The Manifold Ways of Perception. Science 290, 2268–2269 (2000)
12. Luo, S.W., Zhao, L.W.: Manifold Learning Algorithm Based on Spectral Graph Theory. Journal of Computer Research and Development, China 7, 1173–1179 (2006)
13. He, X.F., Niyogi, P.: Locality Preserving Projections. Advances in Neural Information Processing Systems 16, pp. 153–160. MIT Press, Cambridge (2004)

A *k*-Anonymity Clustering Method for Effective Data Privacy Preservation

Chuang-Cheng Chiu and Chieh-Yuan Tsai

Department of Industrial Engineering and Management, Yuan Ze University, Taiwan
cytsai@saturn.yzu.edu.tw

Abstract. Data privacy preservation has drawn considerable interests in data mining research recently. The *k*-anonymity model is a simple and practical approach for data privacy preservation. This paper proposes a novel clustering method for conducting the *k*-anonymity model effectively. In the proposed clustering method, feature weights are automatically adjusted so that the information distortion can be reduced. A set of experiments show that the proposed method keeps the benefit of scalability and computational efficiency when comparing to other popular clustering algorithms.

Keywords: Data privacy preservation, *k*-Anonymity, Clustering, C-means clustering algorithm, Feature weighting.

1 Introduction

Rapid advances in database technologies enabled organizations to accumulate vast amounts of data in recent years. Data mining has been a common methodology to retrieve and discover useful knowledge from these growing data [1]. In many industrial applications, many personal details and sensitive information are contained in these data such as financial transactions, telephone communication traffic, health care records, and so on. The knowledge extracted from these data may unwittingly uncover personal sensitive information. Therefore, before conducting data mining, these data must be protected through some privacy-preserving techniques. This makes privacy-preserving becomes an important issue in data mining fields in recent years [2, 3].

The *k*-anonymity model, proposed by Sweeney [4], is a simple and practical privacy-preserving approach and is extensively studied recently [5, 6, 7]. The *k*-anonymity model ensures that each record in the table is identical to at least (*k*-1) other records with respect to the privacy-related features. Therefore, no privacy-related information can be inferred from the *k*-anonymity protected table during a data mining process. For example, patient diagnosis records without conducting the *k*-anonymity model is shown in Fig. 1(a) [8]. It is clear that a diagnosis classifier can be developed using these data to predict patient's illness based on features of Zip, Gender, and Age. If the hospital simply publishes the table to other organizations for classifier development, the organizations might extract patients' disease history by joining this table with other tables. Conversely, if the *k*-anonymity model is conducted for these data, data values in features Zip, Gender, and Age might been

R. Alhajj et al. (Eds.): ADMA 2007, LNAI 4632, pp. 89–99, 2007.

generalized as capsule values shown in Fig 1(b). For each patient in the table, we can find that at least two patients have the same Zip, Gender, and Age feature values with him/her. Therefore, when the hospital publishes such a *k*-anonymity protected table to other organizations, the organizations still develops an illness-diagnosing classifier from this table similarly. Importantly, the organizations can not uncover additional information from each patient's generalized feature values. The purpose of data privacy preservation is then achieved.

Zip	Gender	Age	Diagnosis
47918	Male	35	Cancer
47906	Male	33	HIV+
47918	Male	36	Flu
47916	Female	39	Obesity
47907	Male	33	Cancer
47906	Female	33	Flu

Zip	Gender	Age	Diagnosis
4791*	Person	[35-39]	Cancer
4790*	Person	[30-34]	HIV+
4791*	Person	[35-39]	Flu
4791*	Person	[35-39]	Obesity
4790*	Person	[30-34]	Cancer
4790*	Person	[30-34]	Flu

(a)   Patient diagnosis records in a hospital

(b) The *k*-anonymity protected table of (a) when *k* = 3

Fig. 1. An example of data privacy preservation using the *k*-anonymity model

In the *k*-anonymity model, the *quasi-identifier feature set* consists of features in a table that potentially reveals private information, possibly by joining with other tables. In addition, the *sensitive feature* is a feature serves as the class label of each record. As shown in Fig. 1(b), the set of three features {Zip, Gender, Age} is the quasi-identifier feature set, while the feature {Diagnosis} is the sensitive feature. For each record in this table, its feature values in the quasi-identifier feature set are generalized as capsule feature values, while its value of sensitive feature are not generalized. Through generalization, an *equivalence class* is the set composed of records in the table which has the same values on all features in the quasi-identifier feature set. The 1st, 3rd and 4th records in Fig. 1(b) are assembled to form one equivalence class, while the 2nd, 5th and 6th records are assembled to form another equivalence class. The number of records in each equivalence class must be not less than *k*, which is called as the *k*-anonymity requirement. The value of *k* is specified by users according to the purpose of their applications. The records in Fig. 1(b) satisfy 3-anonymity requirement since the numbers of records in its two equivalence classes are both equal to three.

To ensure data mining performance, usability should be taken into account when constructing the *k*-anonymity protected table [8]. The less the information distortion in the *k*-anonymity protected table makes, the larger the table usability is. Therefore, a *k*-anonymity model must minimize the information distortion from its original table. Unfortunately, the computational complexity of finding an optimal solution for *k*-anonymity model has been shown to be NP-hard [9]. In recent years, many clustering techniques based on heuristic scheme have been developed to conduct the *k*-anonymity protected table [5, 7, 8, 9, 10, 11]. Clustering [12] aims at grouping a set of objects into clusters so that objects in a cluster are similar to each other and are different from objects in other clusters. In the *k*-anonymity protected table, if the records that will be assembled as an equivalence class are more similar to each other, it retrenches the more information distortion for generalizing the equivalence class. That is the reason why the *k*-anonymity model can be addressed from the viewpoint of clustering. Among various types of clustering methods, hierarchical clustering methods are frequently used to conduct the *k*-anonymity protected table [5, 8, 9, 10]. Although their efforts are admirable in the issue, their computational efficiency may degenerate when the amount of records increases. Furthermore, how to define a proper similarity/dissimilarity measure between two equivalence classes is another challenge when using hierarchical clustering methods.

A novel clustering method to construct the *k*-anonymity protected table is proposed in this paper. In the proposed method, a Weighted Feature C-means clustering algorithm (WF-C-means) is proposed to partition all records into equivalence classes. For enhancing clustering quality, WF-C-means adaptively adjusts the weight of each quasi-identifier feature based on the importance of the feature to clustering quality. The operational procedure in WF-C-means is similar to the C-means algorithm [13] which has good scalability for large data, so that the computational efficiency of WF-C-means is practicable in practice. After completing the clustering, a class-merging mechanism merges equivalence classes to make sure that all equivalence classes satisfy the *k*-anonymity requirement. All records in each equivalence class are generalized to be the same with the class center in the class. Through our experiments, the proposed clustering method outperforms existing methods in terms of information distortion measure and computational efficiency.

2 The Proposed *k*-Anonymity Clustering Method

The core of the proposed clustering method for constructing the *k*-anonymity protected table consists of a Weighted Feature C-means clustering algorithm (WF-C-means) and a class-merging mechanism, and is introduced respectively in detail as follows.

2.1 A Weighted Feature C-Means Clustering Algorithm

Let a table $\mathbf{T}=\{\mathbf{r}_1,\ldots,\mathbf{r}_m,\ldots,\mathbf{r}_M\}$ include M records and a quasi-identifier feature set $\mathbf{F}=\{\mathbf{f}_1,\ldots,\mathbf{f}_n,\ldots,\mathbf{f}_N\}$ comprise N features. A record $\mathbf{r}_m=(r_{m1},\ldots,r_{mn},\ldots,r_{mN})$ is composed of N quasi-identifier feature values where r_{mn} is the value of the nth quasi-identifier feature $\mathbf{f}_n$ in the mth record $\mathbf{r}_m$. Noted that the sensitive feature values can not be generalized as capsule values so that the sensitive feature is not involved in the record $\mathbf{r}_m$.

The development of the proposed WF-C-means is derived from the C-means clustering algorithm [13]. WF-C-means aims at partitioning all M records in the table $\mathbf{T}$ into C equivalence classes. The number of equivalence classes, C, depends on the value of k specified in the *k*-anonymity model, which is shown as Equation (1):

$$C = M \setminus k \tag{1}$$

where "\" is the integer division operator. For example, in the 3-anonymity model a table of 100 records will be divided into 33 equivalence classes (100\3=33). Let $\mathbf{C}=\{\mathbf{C}_1,\ldots,\mathbf{C}_i,\ldots,\mathbf{C}_C\}$ be the set of the C equivalence classes and $\mathbf{A}=\{\mathbf{a}_1,\ldots,\mathbf{a}_i,\ldots,\mathbf{a}_C\}$ be the set of the C class centers in $\mathbf{C}$ where $\mathbf{a}_i=(a_{i1},\ldots,a_{in},\ldots,a_{iN})$ is the class center of the ith equivalence class $\mathbf{C}_i$ and a_{in} is the value of the nth quasi-identifier feature $\mathbf{f}_n$ in the ith class center $\mathbf{a}_i$.

Accordingly, the dissimilarity between a record $\mathbf{r}_m$ and a class center $\mathbf{a}_i$, termed as $\mathrm{diss}(\mathbf{r}_m, \mathbf{a}_i)$, can be defined as:

$$\mathrm{diss}(\mathbf{r}_m, \mathbf{a}_i) = \sum_{n=1}^{N} w_n \times \mathrm{diss}(r_{mn}, a_{in}) \tag{2}$$

where $w_n \in \mathbf{w}$ is the weight of the quasi-identifier feature $\mathbf{f}_n$ and $\mathbf{w} = \{w_1, \ldots, w_n, \ldots, w_N\}$ is the set of the N weights associated with N quasi-identifier features in $\mathbf{F}$, $\sum_{n=1}^{N} w_n = 1$, $0 \le w_n \le 1$. Furthermore, $\mathrm{diss}(r_{mn}, a_{in})$ is the dissimilarity between $\mathbf{r}_m$ and $\mathbf{a}_i$ in terms of the nth quasi-identifier feature. Numerical features and categorical features have their respective formulas to evaluate the value of $\mathrm{diss}(r_{mn}, a_{in})$. In this paper we assume all quasi-identifier features in $\mathbf{F}$ are numerical since we emphasize the introduction of this clustering algorithm. Therefore, the evaluation formula of $\mathrm{diss}(r_{mn}, a_{in})$ can be defined as Equation (3). Noted that the details about the dissimilarity evaluation for categorical features can be referred to [10].

$$\mathrm{diss}(r_{mn}, a_{in}) = (r_{mn} - a_{in})^2 \tag{3}$$

The objective of WF-C-means, equivalent to C-means, is to minimize the sum of the dissimilarities between all M records to their corresponding class centers, which can be expressed as follows:

$$\textit{Minimize } S(\mathbf{U}, \mathbf{A}, \mathbf{w}) = \sum_{m=1}^{M} \sum_{i=1}^{C} u_{mi} \times \mathrm{diss}(\mathbf{r}_m, \mathbf{a}_i) = \sum_{m=1}^{M} \sum_{i=1}^{C} \sum_{n=1}^{N} u_{mi} \times w_n \times \mathrm{diss}(r_{mn}, a_{in}) \tag{4}$$

subject to

$$\begin{cases} \sum_{i=1}^{C} u_{mi} = 1 \\ u_{mi} \in \{1, 0\} \quad \text{for } m = 1, 2, \cdots, M; \ i = 1, 2, \cdots, C; \ n = 1, 2, \cdots, N \\ \sum_{n=1}^{N} w_n = 1 \\ \quad w_n \ge 0 \end{cases} \tag{5}$$

where $\mathbf{U}$ is a matrix of size $M \times C$ that stores the record-class memberships and $u_{mi} \in \{1, 0\}$ is an element in $\mathbf{U}$ that represents the membership of the record $\mathbf{r}_m$ with the ith cluster $\mathbf{C}_i$. If $u_{mi} = 1$, $\mathbf{r}_m$ belongs to $\mathbf{C}_i$. If $u_{mi} = 0$, by contrast, $\mathbf{r}_m$ does not belong to $\mathbf{C}_i$.

The WF-C-means algorithm solves the described optimization problem by iteratively solving the following three reduced problems until all elements in the record-class membership matrix $\mathbf{U}$ remain the same without being changed.

1. Problem P_1: Fix $\mathbf{A} = \hat{\mathbf{A}}$ and $\mathbf{w} = \hat{\mathbf{w}}$ to solve the reduced problem $S(\mathbf{U}, \hat{\mathbf{A}}, \hat{\mathbf{w}})$.

2. Problem P_2: Fix $\mathbf{U} = \hat{\mathbf{U}}$ and $\mathbf{w} = \hat{\mathbf{w}}$ to solve the reduced problem $S(\hat{\mathbf{U}}, \mathbf{A}, \hat{\mathbf{w}})$.

3. Problem P_3: Fix $\mathbf{A} = \hat{\mathbf{A}}$ and $\mathbf{U} = \hat{\mathbf{U}}$ to solve the reduced problem $S(\hat{\mathbf{U}}, \hat{\mathbf{A}}, \mathbf{w})$.

The purpose of solving P_1 is to assign a record to an equivalence class in which the class center is closest to the record. Therefore, the procedure for solving P_1 is called a record-assignment procedure and is expressed in Equation (6):

$$\begin{cases} u_{mi} = 1, \text{ if } \sum_{n=1}^{N} w_n \times \text{diss}(r_{mn}, a_{in}) \le \sum_{n=1}^{N} w_n \times \text{diss}(r_{mn}, a_{jn}) \\ u_{mi} = 0, \text{Otherwise} \end{cases} \text{ for } 1 \le i, j \le C, \ j \ne i \quad (6)$$

Accordingly, the purpose of solving the problem P_2 is to update all K class centers in the C classes respectively. Therefore, the procedure for solving P_2 is called as a center-updating procedure and is expressed in Equation (7):

$$a_{in} = \sum_{m=1}^{M} u_{mi} \times r_{mn} \Big/ \sum_{m=1}^{M} u_{mi} \quad \text{for } i = 1,2,\cdots,C; \ n = 1,2,\cdots,N \quad (7)$$

The difference between WF-C-Means and C-means is that WF-C-Means further solves the weight-adjusting problem P_3 but C-means does not. The weight of a quasi-identifier feature should reflect the importance of the feature to the clustering quality, measured by how the feature can achieve the clustering objective function of minimizing the separations within clusters and maximizing the separations between clusters simultaneously. If a feature is important, increasing its feature weight should make the clustering objective function be easily achieved. Therefore, the goal of sub-problem P_3 is to

$$\textit{Maximize } V(\hat{\mathbf{U}}, \hat{\mathbf{A}}, \mathbf{w}, \hat{g}) = \frac{S'(\hat{\mathbf{A}}, \mathbf{w}, \hat{g})}{S(\hat{\mathbf{U}}, \hat{\mathbf{A}}, \mathbf{w})} = \frac{\sum_{n=1}^{N} \left[w_n \times \left(\sum_{i=1}^{C} \| \mathbf{C}_i \| \times \text{diss}(a_{in}, g_n) \right) \right]}{\sum_{n=1}^{N} \left[w_n \times \left(\sum_{m=1}^{M} \sum_{i=1}^{C} u_{mi} \times \text{diss}(r_{mn}, a_{in}) \right) \right]} \quad (8)$$

subject to

$$\begin{cases} \sum_{n=1}^{N} w_n = 1 \\ w_n \ge 0 \end{cases} \text{ for } n = 1,2,\cdots,N \quad (9)$$

where $S(\hat{\mathbf{U}}, \hat{\mathbf{A}}, \mathbf{w})$ is the sum of all separations within clusters and $S'(\hat{\mathbf{A}}, \mathbf{w}, \hat{g})$ is the sum of all separations between clusters. Noted that $g=(g_1,...,g_n,...,g_N)$ is the global center of all M records in the table $\mathbf{T}$, and its nth feature value, g_n, can be evaluated by $g_n = \sum_{m=1}^{M} r_{mn} \big/ M$. In addition, $\| \mathbf{C}_i \|$ represents the number of records in the ith cluster $\mathbf{C}_i$ such that $\sum_{i=1}^{C} \| \mathbf{C}_i \| = M$.

Let $e_n = \sum_{m=1}^{M} \sum_{i=1}^{C} u_{mi} \times \text{diss}(r_{mn}, a_{in})$ be the sum of separations within clusters in terms of $\mathbf{f}_n$ and $f_n = \sum_{i=1}^{C} \| \mathbf{C}_i \| \times d(a_{in}, g_n)$ be the sum of separations between clusters in terms of $\mathbf{f}_n$. Accordingly, Equation (8) can be rewritten as:

$$\textit{Maximize } V(\hat{\mathbf{U}}, \hat{\mathbf{A}}, \mathbf{w}, \hat{g}) = \frac{\sum_{n=1}^{N} w_n \times f_n}{\sum_{n=1}^{N} w_n \times e_n} \quad (10)$$

subject to

$$\begin{cases} \sum_{n=1}^{N} w_n = 1 \\ w_n \ge 0 \end{cases} \quad \text{for } n = 1, 2, \cdots, N \tag{11}$$

This research proposes an adaptive weight-adjusting principle to derive **w** from Equation (10). Let $\{w_1^{(s)}, \cdots, w_n^{(s)}, \cdots, w_N^{(s)}\}$ be the set of the N feature weights at the sth iteration (i.e. current iteration) in WF-C-Means. Each feature weight $w_n^{(s+1)}$ for $n=1$, 2, …,N at the $(s+1)$th iteration (i.e. next iteration) in WF-C-Means will be adjusted by adding an adjustment margin Δw_n, which is shown as Equation (12).

$$w_n^{(s+1)} = w_n^{(s)} + \Delta w_n \quad \text{for } n = 1, 2, \cdots, N \tag{12}$$

The adjustment margin Δw_n for feature $\mathbf{f}_n$ is evaluated based on how important the feature contributes to clustering quality. From Equation (10), we know that feature $\mathbf{f}_n$ possessing a high (f_n/e_n) value should have a high weight value. Therefore, adjustment margin Δw_n can be derived according to its (f_n/e_n) value using the following equation:

$$\Delta w_n = \frac{f_n/e_n}{\sum_{n=1}^{N}(f_n/e_n)} \quad \text{for } n = 1, 2, \cdots, N \tag{13}$$

Accordingly, an adjusted feature weight $w_n^{(s+1)}$ can be rewritten as:

$$w_n^{(s+1)} = w_n^{(s)} + \frac{f_n/e_n}{\sum_{n=1}^{N}(f_n/e_n)} \quad \text{for } n = 1, 2, \cdots, N \tag{14}$$

In addition, the adjusted weight in Equation (14) need to be normalized as the value between 0 to 1 to satisfy the constraint of $\sum_{n=1}^{N} w_n^{(s+1)} = 1$. Therefore, a simple normalization function $f(t_n)$ defined in Equation (15) is used in this paper.

$$f(t_n) = \frac{t_n}{\sum_{n=1}^{N} t_n} \quad \text{for } n = 1, 2, \cdots, N \tag{15}$$

Through the normalization function, each adjusted feature weight $w_n^{(s+1)}$ can be derived as:

$$w_n^{(s+1)} = f(w_n^{(s+1)}) = \frac{w_n^{(s)} + \dfrac{f_n/e_n}{\sum_{n=1}^{N}(f_n/e_n)}}{\sum_{n=1}^{N} w_n^{(s)} + \sum_{n=1}^{N}\left(\dfrac{f_n/e_n}{\sum_{n=1}^{N}(f_n/e_n)}\right)} \quad \text{for } n = 1, 2, \cdots, N \tag{16}$$

With Equation (16), the adjusted feature weights can be derived in the weight-adjusting procedure and feed back to the beginning of the record-assignment procedure in the WF-C-means algorithm for the next iteration. The pseudo-code of the WF-C-means algorithm is summarized in Fig. 2.

Input: a table **T** contains M records in which each record has N quasi-identifier features;
the value of k in the k-anonymity model.

1: Calculate the number of equivalence classes C using Equation (1).
2: Randomly select C records from **T** as the class centers of the C equivalence classes.
3: Let the weight of each quasi-identifier feature be (1/N).
4: **Repeat**
5:      Form C equivalence classes by assigning each record to its closest class center using Equation (6).
6:      Update the class center in each equivalence class using Equation (7).
7:      Adjust the feature weight of each quasi-identifier feature using Equation (16).
8: **Until** all elements in the record-class membership matrix do not change

---

Fig. 2. The pseudo-code of the WF-C-means algorithm

2.2 A Class-Merging Mechanism

After executing the proposed WF-C-means algorithm, a few equivalence classes may violate the k-anonymity requirement because they are possibly located at the purlieus of data distribution or even they are outliers. Assume there are P illegal equivalence classes violating the k-anonymity requirement among all C equivalence classes, so that other $(C-P)$ equivalence classes are legal. In the proposed method, a class-merging mechanism is developed to eliminate the illegal equivalence classes by means of merging them with legal equivalence classes.

Let the distance between two equivalence classes $\mathbf{C}_i$ and $\mathbf{C}_j$ be defined as the dissimilarity between their class centers $\mathbf{a}_i$ and $\mathbf{a}_j$, which is expressed as:

$$\text{distance}(\mathbf{C}_i, \mathbf{C}_j) = \text{diss}(\mathbf{a}_i, \mathbf{a}_j) = \sum_{n=1}^{N} w_n \times \text{diss}(a_{in}, a_{jn}) \tag{17}$$

where the two class centers $\mathbf{a}_i$ and $\mathbf{a}_j$, and w_n for $n=1,2,\ldots,N$ are known after executing the WF-C-means algorithm. When merging x equivalence classes, the class center of a new equivalence class, termed as $\mathbf{a}^{new} = (a_1^{new},\ldots,a_n^{new},\ldots,a_N^{new})$, can be defined as Equation (18):

$$a_n^{new} = \frac{\sum_{i=1}^{x}(a_{in} \times \|\mathbf{C}_i\|)}{\|\mathbf{C}_i\|} \quad \text{for } n = 1,\ldots,N \tag{18}$$

where $\|\mathbf{C}_i\|$ is the number of records in the equivalence class $\mathbf{C}_i$. Noted that the number of records in the new equivalence class equals to $\sum_{i=1}^{x}\|\mathbf{C}_i\|$.

For an illegal equivalence class, its merging target is the legal equivalence class with closest distance evaluated by Equation (17). For a legal equivalence class, on the other hand, it may receive the merging requests from several illegal equivalence classes so that it will be merged with these illegal equivalence classes simultaneously. The class center of the new equivalence class can be found easily by Equation (18). The pseudo-code of the class-merging procedure in the proposed mechanism is summarized in Fig. 3. After performing the class-merging mechanism, all records in each equivalence class $\mathbf{C}_i$ are generalized to be the same with the class center $\mathbf{a}_i$ of $\mathbf{C}_i$.

Input: P illegal equivalence classes and $(C - P)$ legal equivalence classes which are generated from WF-C-means
1:    **For** each illegal equivalence class
2:        Calculate the distances with the $(C - P)$ legal classes respectively using Equation (17).
3:        Select and mark the nearest legal class with it.
4:    **For** each legal equivalence class
5:        **If** the class has been marked by any illegal equivalence class
6:            Collect the illegal equivalence classes which have done a mark on it.
7:            Merge these collected illegal equivalence classes with it as a new equivalence class.

Fig. 3. The pseudo-code of the class-merging procedure in the proposed mechanism

3   Experiments

To show performance of the proposed k-anonymity clustering method, a series of experiments using Iris, Wine, and Zoo datasets from UCI machine learning repository [15] are conducted. For each dataset, its original predictive features are all in the quasi-identifier feature set while its class-label feature is the sensitive feature. Since hierarchical clustering methods have been adopted most frequently in previous studies, the experiment result of the proposed method is compared with the results of three common hierarchical clustering methods. They are single-link, complete-link, and average-link clustering methods [16]. In a hierarchical clustering method, all records are initially considered as independent equivalence classes and are merged progressively until the number of records in each equivalence class is not less than k.

3.1   Information Distortion

The amount of information distortion can be evaluated from the difference between the original table and the k-anonymity protected table. For each record in the k-anonymity protected table, its feature values in the quasi-identifier feature set are generalized as the feature values of the equivalence class center which is closet to it. Therefore, the amount of information distortion of a k-anonymity protected table can be calculated using Equation (4). The less the amount of information distortion, the larger the usability of the k-anonymity protected table is.

When using the hierarchical clustering methods, all the feature weights in Equation (4) are set as $1/N$ uniformly and the class center of an equivalence class is considered as the mean of all records in the equivalence class. In addition, the parameter, k, is tested using 2, 4, 8, 16, respectively, for each dataset. The plots of the amounts of information distortion with respect to k using the three datasets for the four clustering methods are illustrated in Fig. 4.

From Fig. 4, it is clear that the amount of information distortion increases as k grows no matter which clustering method is used to develop the k-anonymity model. Among these four clustering methods, the proposed k-anonymity clustering method is the best one to restrain information distortion for all k values and datasets.

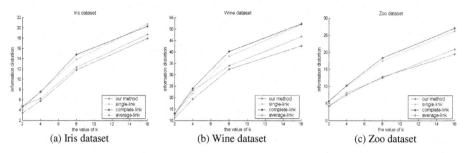

Fig. 4. The plots of the information distortion with respect to *k* using the three datasets for the four clustering methods

3.2 Classification Error Rate

In the study, we assume that the one nearest neighbor (1NN) classification technique is used to classify unknown data based on the *k*-anonymity protected table in following data mining tasks. In the classification task, each record in the original table serves as a testing sample to measure the classification error rate. Therefore, the less the classification error rate, the larger the usability of the *k*-anonymity protected table is. Table 1 shows classification error rates for the four clustering methods using the three datasets when *k* = 3. It is clear that the performance of the proposed method is superior to other three clustering methods in Wine and Zoo datasets, while only slightly inferior to the average-link hierarchical clustering method in the Iris dataset. For our method, in addition, the classification error rate increases lightly using the *k*-anonymity protected table when comparing to the one using the original table. However, the data privacy in the *k*-anonymity protected table can be strongly preserved.

Table 1. The classification error rates using the three datasets for the four clustering methods

Dataset	Original table	*k*-anonymity Protected table			
		Our method	Hierarchical clustering method		
			single-link	complete-link	average-link
Iris	4.47%	6.67%	8.67%	9.33%	**6.00%**
Wine	15.73%	**17.42%**	22.47%	21.91%	19.66%
Zoo	56.25%	**59.38%**	65.63%	62.50%	62.50%

(Noted that the value of *k* is set as 3 in the *k*-anonymity model.)

3.3 Computational Efficiency

In this section, the computational efficiency of constructing the *k*-anonymity protected table using the four clustering methods is evaluated by measuring their execution time. The experiment environment is set identically with the one in Section 3.1. All experiments are implemented with Excel VBA programming language, and run on an AMD K7 2.5G personal computer with 512 MB memory. The plots of the execution time with respect to *k* for the three datasets using the four clustering algorithms are illustrated in Fig. 5.

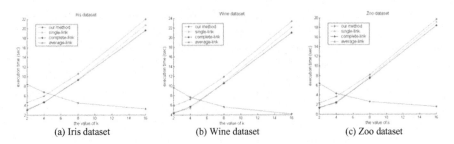

| (a) Iris dataset | (b) Wine dataset | (c) Zoo dataset |

Fig. 5. The plots of the execution time with respect to k for the three datasets using the four clustering algorithms

The execution time using the proposed clustering method decreases as k increases, which is totally different to the trends using the three hierarchal clustering methods. Moreover, the execution time using proposed clustering method is less than 10 seconds for all k value settings and datasets. The computational complexity of the proposed clustering method is $O(M \times C) \cong O(M \times (M / k)) = O(M^2 / k)$, while the computational complexity of a hierarchical clustering method equals to $O(M^2 \log M)$. It is obvious that the proposed method is superior to the hierarchical clustering method in terms of computational efficiency.

4   Conclusion

In this paper we propose a novel C-means type clustering method for the k-anonymity model, which is distinct from the typical hierarchical clustering methods. For restraining information distortion in the k-anonymity protected table, the proposed method adaptively adjusts the weight of each quasi-identifier feature based on the importance of the feature to clustering quality. The experiment results in Section 3.1 and Section 3.2 also confirms that the proposed method enables the k-anonymity protected table restrain its information distortion. In addition, the experiment result in Section 3.3 indicates that the computational efficiency of the proposed clustering method is superior to the hierarchical clustering method for the k-anonymity model.

In this paper only quasi-identifier features with numerical values are considered. However, quasi-identifier features with categorical values are also common in practice. In the future, we will focus on developing a dissimilarity-evaluating approach which takes different types of feature values into account simultaneously.

References

1. Tan, P.N., Steinbach, M., Kumar, V.: Introduction to Data Mining, pp. 487–559. Addison-Wesley, Boston (2005)
2. Agrawal, R., Srikant, R.: Privacy-Preserving Data Mining. SIGMOD Record 29, 439–450 (2000)
3. Lindell, Y., Pinkas, B.: Privacy Preserving Data Mining. Journal of Cryptology 15, 177–206 (2003)

4. Sweeney, L.: k-Anonymity: A Model for Protecting Privacy. International Journal of Uncertainty, Fuzziness and Knowlege-Based Systems 10, 557–570 (2002)
5. Domingo-Ferrer, J., Torra, V.: Ordinal, Continuous and Heterogeneous k-Anonymity through Microaggregation. Data Mining and Knowledge Discovery 11, 195–212 (2005)
6. LeFevre, K., DeWitt, D.J., Ramakrishnan, R.: Incognito: Efficient Full-Domain k-Anonymity. In: Proceedings of the ACM SIGMOD International Conference on Management of Data, pp. 49–60 (2005)
7. Li, X.-B., Sarkar, S.: A Tree-Based Data Perturbation Approach for Privacy-Preserving Data Mining. IEEE Transactions on Knowledge and Data Engineering 18, 1278–1283 (2006)
8. Byun, J.-W., Kamra, A., Bertino, E., Li, N.: Efficient k-Anonymization Using Clustering Techniques. To appear in the International Conference on Database Systems for Advanced Applications (2007)
9. Meyerson, A., Williams, R.: On the Complexity of Optimal k-Anonymity. In: Proceedings of the 18th ACM SIGACT-SIGMOD-SIGART Symposium on Principles of Database Systems, pp. 223–228 (2004)
10. Jiuyong, L., Wong, R.C.-W., Fu, A.W.-C., Jian, P.: Achieving k-Anonymity by Clustering in Attribute Hierarchical Structures. In: Tjoa, A.M., Trujillo, J. (eds.) DaWaK 2006. LNCS, vol. 4081, pp. 405–416. Springer, Heidelberg (2006)
11. Aggarwal, C.C.: On k-Anonymity and the Curse of Dimensionality. In: Proceedings of the 31st International Conference on Very Large Data Bases, pp. 901–909 (2005)
12. Jain, A., Dube, R.: Algorithms for Clustering Data. Prentice Hall, New Jersey (1988)
13. McQueen, J.: Some Methods for Classification and Analysis of Multivariate Observations. In: Proceedings of the 5th Berkeley Symposium on Mathematical Statistics and Probability, pp. 281–297 (1967)
14. Hillier, F.S., Lieberman, G.J.: Introduction to Operation Research. McGraw-Hill, New York (2001)
15. Newman, D.J., Hettich, S., Blake, C.L., Merz, C.J.: UCI Repository of Machine Learning Databases (1998), available at http://www.ics.uci.edu/~mlearn/MLSummary.html
16. Jain, A.K., Murty, M.N., Flynn, P.J.: Data Clustering: A Review. ACM Computer Survey 31, 264–323 (1999)

LSSVM with Fuzzy Pre-processing Model Based Aero Engine Data Mining Technology*

Xuhui Wang[1,**], Shengguo Huang[1], Li Cao[1], Dinghao Shi[2], and Ping Shu[2]

[1] College of civil aviation, Nanjing University of Aeronautics and Astronautics,
Nanjing, 210016
Tel.: +8625-84892273,
{wxhui, huangsg, caoli}@nuaa.edu.cn
[2] General Civil Aviation Administration of China, Center of Aviation Safety Technology,
Aviation Safety Institute Technology Lab, Beijing, 100028
{shidh, shup}@mail.castc.org.cn

Abstract. The operations of aircraft fleets typically result in large volumes of data collected during the execution of various operational and support processes.This paper reports on an Airlines-sponsored study conducted to research the applicability of data mining for processing engine data for fault diagnostics. The study focused on three aspects: (1) understanding the engine fault maintenance environment, and data collection system; (2) investigating engine fault diagnosis approaches with the purpose of identifying promising methods pertinent to aircraft engine management; and (3) defining a Support Vector Machines model with Fuzzy clustering to support the data mining work in aero engine fault detection. Results of analyses of maintenance data and flight data sets are presented. Architecture for mining engine data is also presented.

Keywords: Data analysis, support vector machines, fuzzy clustering, engine diagnostics.

1 Introduction

The development of FDAMS (Flight Data Acquisition and Management System) on civil aircraft results in large volumes of data storage such as engine condition data, inspection data and crew operation data. This data is collected with the purpose of recording important events and activity for recall during future analysis [1]. However, review and analysis of engine data is typically complicated, and requires significant human involvement. As a result, the data accumulates much faster than it can be processed. Motivated by the possibility that such large quantities of flight data could contain valuable information that could help in a better recognition of undesirable events for engine system, and, thus, in foreseeing, preventing, or more efficiently handling their future occurrences, the airlines focus their interest on the knowledge

---

* This work is supported by National 863 Program (2006AA12A108) and NSFC (79870032).
** Corresponding author.

R. Alhajj et al. (Eds.): ADMA 2007, LNAI 4632, pp. 100–109, 2007.

discovery of massive data. The objective was to investigate the implementation issues to be overcome in order to apply the emerging data mining techniques for improving future Civil Aviation maintenance and logistics management [2, 3].

In this paper, we construct a SVM (Support Vector Machine) model to accomplish aero engine data mining task, and Fuzzy algorithm is used to pre-process the input data. This approach has successfully applied to a series of commercial aircraft engines, getting high accuracy and consistently. This paper is organized as follows: section 2 provides a brief overview of Civil Aviation maintenance system, and KDD algorithm for engine diagnostic and prognostic. Section 3 describes the model combine SVM with fuzzy cluster to realize the vector sparsity. Section 4 proposes fuzzy SVM based methodology for engine data mining model. Section 5 presents mining result of the running-data collecting from certain air craft engine within airlines. Section 6 draws a conclusion for aero engine data analysis with SVM data mining algorithm.

2 Model Based Engine Data Mining

Aircraft engine fault diagnostic is required for all commercial airlines. Typically, on-board data collection sub-systems recorder key parameters of engine performance during different flight phases including takeoff, climb and cruise. Due to the large-capacity and high dimension characters of engine data, the diagnostic based on flight become a hard work. Direct trend analysis for most of the sensor collected raw data is of little value for anomaly diction since trends may vary under various operation conditions. Therefore, an engine fault diagnostic is typically developed to derive residuals from sensor data for anomaly detection purposes. Models include manufacturer-provided baseline model, system identification approach using Markov chain model and etc. By comparing the collected sensor data with engine model data and analyzing the deviation trend over a continuous period, one can detect the performance shift and anomaly of on-wing engine. Diagnostic of aero engine can be regard as data mining within large scale engine performance database. Data mining algorithms such as decision trees, neural network and genetic algorithm are applied to aero engine fault diagnostic and anomaly diction. Here, we propose a SVM with Fuzzy cluster method to mining the knowledge from the engine gas path parameter database.

3 Method

In this section we briefly sketch the ideas behind SVM for classification [4, 5, 6]. Given the training data$\{(x_i, y_i)\}_{i=1}^{N}$, $x_i \in \mathbf{R}^m$, $y_i \in \{\pm1\}$, for the case of two-class pattern recognition, SVM first maps x from input data x into a high dimensional feature space by using a nonlinear mapping φ, $z = \varphi(x)$, like figure 1. In case of linearly separable data, SVM then searches for a hyperplane $\mathbf{W}^T\mathbf{Z}+b$ in the feature space for which the separation between the positive and negative examples is maximized like figure 2. The W for this optimal hyperplane can be written as $W = \sum_{i=1}^{N} \alpha_i y_i z_i$ where $\alpha = (\alpha_1 \ldots \alpha_N)$ can be found by solving the following quadratic programming (QP) problem:

$$max: \qquad \alpha^T - \frac{1}{2}\alpha^T Q\alpha,$$

$$subject \quad to: \quad \begin{cases} a \geq 0. \\ a^T Y = 0. \end{cases} \qquad (1)$$

Where $Y^T = (y_1. \ldots y_N)$ and Q is a symmetric $N \times N$ matrix with elements $Q_{ij} = y_i y_j z_i^T z_j$. Notice that Q is always positive semi-infinite and so there is no local optimum for the QP problem. For those α_i that are nonzero, the corresponding training examples must lie closest to the margins of decision boundary (by the Kuhn-Tucker theorem [4]), and these examples are called the support vectors (SVs).

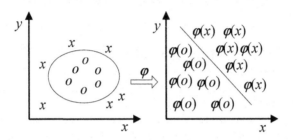

Fig. 1. Nonlinear kernel function

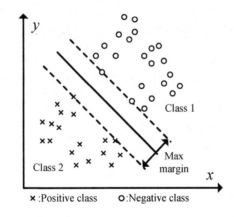

Fig. 2. The structure of a simple SVM

To obtain Q_{ij}, one does not need to use the mapping φ to explicitly get z_i and z_j. These expensive calculations can be reduced significantly by using a suitable kernel function K such that $K(x_i, x_j) = z_i^T z_j$, Q_{ij} is then computed as $Q_{ij} = y_i y_j K(x_i, x_j)$. By using different kernel functions, the SVM can construct a variety of classifiers, some of which as special cases coincide with classical architectures:

Polynomial classifiers of degree p: $K(x_i, x_j) = (x_i^T x_j + 1)^p$,

Radial basis function (RBF) classifier: $K(x_i, x_j) = e^{-\|x_i - x_j\|^{2/\sigma}}$,

Neural networks (NN): $K(x_i, x_j) = \tanh(kx_i x_j + \theta)$.

In RBF case, SVM automatically determines the number (the SVs) and locations (the SVs) of RBF centers and gives excellent result compared to classical RBF [6]. During testing, for a test vector $x \in R^m$, we first compute:

$$a(x, W) = W^T Z + b$$
$$= \sum_i a_i y_i K(x, x_i) + b. \tag{2}$$

and then its class label $o(x, W)$ is 1 if $a(x, W) > 0$, otherwise, it is -1.

The above algorithm for separable data can be generalized to the nonseparable data by introducing nonnegative slack variables ξ_i, $i = 1 \ldots N$ [12]. In case of LSSVM (Least Squares Support Vector Machines), the resultant problem becomes:

$$min: \qquad \frac{1}{2}\|W\|^2 + \frac{C}{2}\sum_{i=1}^{N}\xi_i^2,$$

$$subject \quad to: \quad \begin{cases} \xi \geq 0, \\ y_i a(x, W) \geq 1 - \xi_i, \end{cases} \quad \forall i \in \{1...N\}. \tag{3}$$

Thus, once an error occurs, the corresponding ξ_i which measures the (absolute) difference between $a(x, W)$ and y_i must exceed unity, so $\Sigma_i \xi_i$ is an upper bound on the number of training error. C is a constant controlling the tradeoff between training error and model complexity. Again, minimization of the above equation can be transformed to a QP problem: maximize (1) subject to the constraints $0 \leq \alpha \leq C$ and $\alpha^T Y = 0$. Its convergence behavior is nothing but solving a linearly constrained convex quadratic programming problem.

We evaluated the performance of the classifier by computing the percentages of correct classification (CC), sensitivity (SE) and specificity (SP), which are defined as follows:

$$CC = 100*(TP + TN)/N. \tag{3}$$

$$SE = 100*TP/(TP + FN). \tag{4}$$

$$SP = 100*TN/(TN + FP). \tag{5}$$

where N was the total number of the patients studied, TP was the number of true positives, TN was the number of the true negatives, FN was the number of false negatives, and FP was the number of false positives [7].

Although SVM has good performance for classification, regression, its limitation is obviously due to the algorithm complexity. In case of large scale data process task, new optimization has been introduced to enhance SVM efficiency [8]. In this paper, we applied Fuzzy clustering method to data pre-processing in order to reduce the data scale [9]. Fuzzy c-means clustering minimizes the sum of squared errors:

$$min: \quad \sum_{k=1}^{n} \sum_{i=1}^{c} \left(u_{ik} \right)^{m} \left(\|x_k - v_i\|^2 \right),$$

$$subject \quad to: \quad \begin{cases} 0 \leq u_{ik} \leq 1, \quad \forall i,k. \\ \sum_{i-1}^{c} u_{ik} = 1, \quad \forall k. \end{cases} \tag{6}$$

where n=number of individuals to be clustered, c=number of clusters, u_{ik} =degree of membership of individual k in cluster i, x_k =a vector of h characteristics for individual k, v_i =a vector of the cluster means of the h characteristics for cluster i, and m=the weighting exponent. Eq. (7) represents the sum of squared errors and is a goal function that the fuzzy c-means algorithm tries to minimize. The values of c and m are empirically determined and are constant once they are selected. The cluster means are given by:

$$v_i = \frac{\sum_{k=1}^{n} \left(u_{ik} \right)^{m} x_k}{\sum_{k=1}^{n} \left(u_{ik} \right)^{m}}, \qquad \forall i. \tag{7}$$

and the degrees of membership are given by:

$$u_{ik} = \frac{1}{\sum_{j=1}^{c} \left(\dfrac{\|x_k - v_i\|^{2/(m-1)}}{\|x_k - v_j\|^{2/(m-1)}} \right)}, \tag{8}$$

$$for \quad x_i \neq v_j; \forall i,k; \quad and \quad m > 1.$$

Solution is obtained by iteration through these conditions. An iterative algorithm, also called "alternating optimization," is used to solve these equations and to identify clusters and associated cluster memberships. It starts with an initial solution for U_0 (Eq. (9)) and loops through a cycle of estimates for U_{t-1} (Eq. (9)) $\rightarrow V_t$ (Eq. (8)) $\rightarrow U_t$ (Eq. (9)). The iteration stops when the difference between U_t and U_{t-1} is very small [9]. The algorithm was initialized by generating a random matrix of cluster memberships (U_0), as suggested by the literature [10]. Similar to many studies in the literature [11, 12], the algorithm was run multiple times (10 times) with different random starting values. The runs generated similar results all within 150 iterations. There was no limit for the number of iterations. However, the algorithm stopped when the difference between U_t and U_{t-1} was very small. The literature defines this difference as epsilon (ε) and was set in the program at 0.0001 [13].

Application of Fuzzy clustering for SVM is shown in figure 3. The margin band of SVM classifier is expanded clearly. Fuzzy Pre-processing discards the non-support vectors. This method will be helpful if the number of support vectors is very large. Furthermore, the generalization ability of SVM model is enhanced.

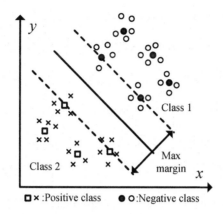

Fig. 3. The structure of a SVM with Fuzzy Pre-processing

From figure 3, we can see, training data points are obviously reduced, and this will help the sparseness of input vectors. Also, time complexity of training the LSSVM model is decreased.

4 Model Based Aero Engine Fault Diagnostic

The main gas path components of the gas turbines (GT), which are compressor, combustor and turbines, are usually very reliable. But could result in low availability of whose unit if an unexpected outage is encountered, and it would take some considerable time to effect repairs on them. Improving availability and reducing life-cycle costs of the GT require maintenance schemes such as CBM (Condition Based Maintenance) that advocates maintenance only when it is necessary at the appropriate time. For the health of the engine to be regularly monitored for gas path faults, such measurable parameters as aero engine gas temperature, gas-generator's burner fuel flow and gas-generator's relative shaft-speed are required.

Traditional gas path fault diagnosis techniques such as visual inspection, fault trees, fault matrixes and gas path analysis (GPA) [14], have their limitations. Current researches have been focus on the application of such advanced techniques as artificial neural networks (ANN) [15]. The main method used in this paper is SVM with Fuzzy cluster approach. SVM has shown its good generalization for model-based classification. The input of SVM model is the vector of engine parameters from sensors, and output is the fault classes of engine gas path. Historical diagnosis information is used for training SVM model, and Fuzzy clustering algorithm is adopted for pre-processing. Figure 4 shows the general diagnostics strategy.

To facilitate this model-based engine data mining, the Airlines made available engine data sets of sample data from actual aviation operations. The data set consists of data collected during ground maintenance activities. Parameters values are decoded from QAR (Quick Access Recorder).

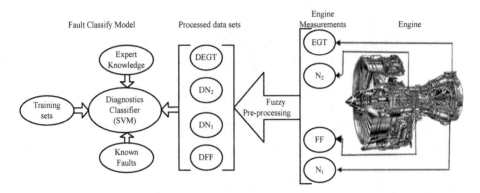

Fig. 4. Schematic flow diagram of fault-diagnosis algorithm

The data set is the aircraft health inspection and maintenance activity data containing logs of maintenance activity recorded on all aircraft. The logs include types of maintenance actions, faults, parts consumed, removed, etc. This data is collected as part of the Aircraft Logbook Forms, Maintenance Forms, and Flying Time entry process. A proper and exhaustive analysis of this data would require interaction with the domain-expert to identify the cause-effect relationships among attributes and to identify the "decision" variables of interest. In the absence of a description accompanying the data, our initial task is first to understand the data collection environment, and subsequently to identify pre-processing requirements. Sample analyses are performed. Attribute of data is shown below:

Table 1. Sample recorders of gas path fault

A/C SER_NO	FAULT_NO	ENGINE	DEGT (deg)	DFF (lbs/hr)	DN_1 (%rpm)	DN_2 (%rpm)
B-2056	1	PW4000	-2	-0.4	0.4	-0.1
B-2056	5	PW4000	8	1.3	0.3	0.4
B-2471	12	PW4000	-1	-0.2	0.3	0
B-2471	11	PW4000	86	16.2	0.8	5.3
B-2471	7	PW4000	33	-0.4	0.4	-0.3
B-2471	4	PW4000	11	11.7	0.1	2.3

DEGT denotes the deviation of current engine gas temperature with standard engine gas temperature, DFF denotes the deviation of current fuel flow with standard fuel flow, and DN_x denotes the deviation of current engine shaft speed with standard engine shaft speed. Standard values are set by aircraft manufacturer.

In case of experience of previous engine data mining, we knew that overlapping points, noise points and error recorder would decrease the knowledge acquisition ability. So some data pre-processing method were introduced to discard the useless or overlapping points, here, we introduce Fuzzy c-means clustering algorithm to accomplish the data pre-processing.

In Fuzzy pre-processing, each feature takes new feature value according to its old value. In the end of algorithm, a list of the estimated degree of membership of object to each of the c cluster is printed. The object data can be assigned to the cluster for which the degree of membership is higher. Contrary to other clustering method the Fuzzy c-means is more flexible because it shows those objects that have some interface with more than one cluster in the partition.

5   Case Study

Standard LSSVM is a binary classifier, therefore, we use ECOC（Error Correcting Output Coding）method as multi-classifier expanding algorithm. The size of original train data set is 450, each recorder includes four dimensions, and output class (engine fault model) is 12. LSSVM modeling parameter is optimized by experiment [16], optimization process is shown as figure 5.

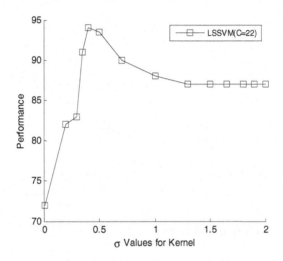

Fig. 5. Performance of LSSVM against σ values

In our experimental, engine data set is firstly pre-processed by Fuzzy c-means Pre-processing and then classified by LSSVM classifier. This data set includes 450 samples with 4 features and 12 output labels. 5-fold cross validation is adopted. Classification ability and computer complexity are presented in Table 2.

Table 2. Comparing between Standard LSSVM and LSSVM with Fuzzy Pre-processing

method	Parameter	Support vectors	Training time	Testing accuracy	Sensitivity (SE)	Specificity (SP)
Standard LSSVM	σ=0.4 C=22	49	6.924(s)	79.6%	100%	7.31%
Fuzzy LSSVM	σ=0.4 C=22	27	3.817(s)	92.3%	95%	93.3%

The saving of training time is attributable to reduce of support vectors. In additional, to compare the classification performance of standard LSSVM and Fuzzy LSSVM, receiver operator characteristic (ROC) curves method is preferred. ROC curves of both classifiers are shown in figure 6.

ROC curves is a statistical comparing method which uses the rates of true positive and false positive. Areas under ROC curves are represented by Az value. This value is related to the accuracies of classifiers. Higher values represent higher classification accuracies, while lower values represent lower classification accuracies. ROC curves show that there is a significant difference between computed areas for two classifiers (Az = 0.95 for LSSVM with fuzzy but Az = 0.328 for LSSVM without fuzzy).

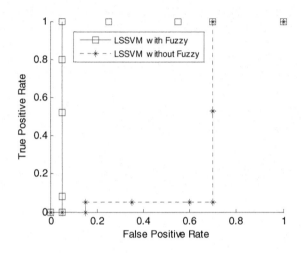

Fig. 6. ROC curves for LSSVM with fuzzy and LSSVM without fuzzy

6   Conclusion and Future Work

In this paper, a data pre-processing method have been developed using Fuzzy c-means clustering and a engine data mining system is built by associating this fuzzy method with LSSVM classifier.

In application phase of this study, developed LSSVM with Fuzzy Pre-processing method is applied to PW4000 engine dataset and 92.3% classification rate is obtained. This rate is the highest classification rate in literature. In addition to, with standard LSSVM (without Fuzzy Weighting Pre-processing) 79.6% classification rate is obtained. This result shows that Fuzzy Pre-processing extremely increases the classification rate of LSSVM for current data set.

According to the application results, LSSVM with Fuzzy Pre-processing showed a considerably high performance with regard to the classification accuracy especially for engine dataset. Also, ROC curves are used to compare the accuracy of proposed system statistically. As shown in ROC curves while under the area of ROC curves for standard LSSVM is 0.328, this area for LSSVM with Fuzzy Pre-processing is 0.95. According to these results, proposed system is effective and reliable.

Although developed method is built as an offline diagnosing system, it can be rebuilt as an online diagnosing system in the future.

Acknowledgments. This paper is supported by the National High Technology Research and Development Program of China 863 program (No. 2006AA12A108) and National Natural Science Foundation of China (No.79870032).

References

1. Mathur, A.: Data Mining within an Advanced Diagnostic and Prognostic System for Rotorcraft Maintenance, SBIR Phase I Final Report, U.S. Army Aviation and Missile Command, Research, Development and Engineering Center, Report no. USAAMCOM TR 01-D-17 (2001)
2. Kusiak, A.: Results of a Rule Induction Technique Applied to a Sample of Fault, Actions and Aircraft Data, Correspondence under QSI, sub-contract no. QSI-SC-01-003 to University of Iowa (2001)
3. Chidester, T.R.: Understanding Normal and Atypical Operations through Analysis of Flight Data. In: Proceedings of the 12th International Symposium on Aviation Psychology, Dayton, Ohio (2003)
4. Vapnik, V.: Statistical Learning Theory. Addison-Wesley, Reading (1998)
5. Suykens, J.T, Van, G.l.: Least Squares Support Vector Machines. Singapore World Scientific (2002)
6. Zhang, L., Zhou, W.D., Jiao, L.C.: Hidden space support vector machines. IEEE Transactions on Neural Networks 15, 1424–1434 (2004)
7. Eberhart, R.C, Dobbins, R.W.: Neural Network PC Tools. Academic Press, San Diego (1990)
8. Lin, C.F., Wang, S.D.: Fuzzy Support Vector Machines. IEEE Trans. Neural Networks 13(2), 464–471 (2002)
9. Yu, H., Yang, J.: Classifying large data sets using SVMs with hierarchical clusters. In: Proceedings of the 9th ACM SIGKDD International Conference on Knowledge Discovery and Data Mining, pp. 306–315. ACM Press, Washington, DC (2003)
10. Oleg, S., Piany, k.h.: Analytically tractable case of fuzzy c-means clustering. Pattern Recognition 39, 35–46 (2006)
11. Eschrich, S., Ke, J. (eds.): Fast accurate fuzzy clustering through data reduction. IEEE Trans. Fuzzy Syst. 11, 262–270 (2003)
12. Łęski, J.: ε-Insensitive fuzzy c-medians clustering. Bull. Polish Acad.: Tech. 50, 361–374 (2002)
13. Łęski, J.: ε-insensitive fuzzy c-regression models: introduction to ε-insensitive fuzzy modeling. IEEE Trans. Systems Man Cybernet.-Part B: Cybernet 34, 4–15 (2004)
14. Urban, L.: A Gas Path Analysis Applied to Turbine Engine Condition Monitoring, AIAA 1082 (1972)
15. Joly, R.B, Ogaji, S.R.: Gas-turbine diagnostics using artificial neural-networks for a high bypass ratio military turbofan engine. Applied Energy 78, 397–418 (2004)
16. Chapelle, O., Vapnik, V.N.: Choosing Multiple Parameters for SVM. Mach Learn 46, 131–143 (2002)

A Coding Hierarchy Computing Based Clustering Algorithm

Jing Peng[1], Chang-jie Tang[2], Dong-qing Yang[1], An-long Chen[2], and Lei Duan[2]

[1] School of Electronics Engineering and Computer Science; Peking University; Beijing 100871, China
[2] School of computer Science and Engineering, Sichuan University, Chengdu 610065, China
pj@pku.edu.cn

Abstract. In actual databases, there are a lot of hierarchy coding data, existing clustering algorithms don't consider the special treatment of these data structure, so lead nonideal performance and clustering result. This paper proposes a new clustering algorithm to deal with the hierarchy coding data structure (HCDS) that exists in many applications. The main contributions include: (1) proposes a new concept for HCDS and corresponding definitions. (2) Proposes and implements a new clustering algorithm–CHCC (Coding Hierarchy Computing Based Clustering Algorithm) based on HCDS. (3) Proposes a fast algorithm for hierarchy coding structure processing. (4) Applies the algorithm into the clustering analysis of transient population for public security, and through extensive experiments, proves the validity and efficiency of the algorithm.

Keywords: Coding hierarchy computing; k-medoids; Clustering; Data mining.

1 Introduction

Data clustering [1] is an important branch of data mining. It does clusters process without domain knowledge. The clustering results are naturally computed from the actual distribution of data sets.

Clustering algorithms have been studied extensively for more than 40 years and a lot of algorithms have been proposed, including: (1) methods based on division, such as *k-means* and *k-medoids* [2], (2) methods based on hierarchy, such as CURE and BIRCH, (3) methods based on density, such as DBSCAN and OPTICS, (4) methods based on network, such as STING, CLIQUE and WaveCluster, (5)methods based on models, such as COBWEB; (6) methods based on mixture methods[3]. Recent studies have focused on clustering stream data [4,5], constraint-based clustering [6], Outlier detection [7] and high-dimensional clustering [8].

The main data structures of variants that exist in clustering algorithms processing include interval-scaled variants, binary variants, nominal, ordinal variants, and ratio variants [1]. For the other data structures are often converted into interval-scaled variants or nominal structure variants, i.e. distance between two objects are defined as 1 if two objects are equal or 0 otherwise.

R. Alhajj et al. (Eds.): ADMA 2007, LNAI 4632, pp. 110–121, 2007.

In actual application, a lot of special information data are coded in hierarchy, such as regionalism, domain, with following features:

(1) Variants can be divided into several parts, and the relationships of these parts are hierarchy;

(2) High degree of similarity lies among those variants that have more same prefixes; otherwise there is low degree of similarity.

If these hierarchy data sets are simply treated as interval-scaled variants, the similar between different data is hardly described. For example, in the practice of regionalism, in an area, the actual distance between two cities, though they have neighboring regionalism codes, would not be shorter than those that don't have neighboring codes.

Secondly, it will come out some meaningless results when interval-scaled variants are used in clustering computing.

Example 1. The median of region codes "510204" and "100101" is "305152.5", which is an inexistent regionalism.

These results of clustering computing are hard to be interpreted in reality, and are meaningless to users. If these hierarchy data sets are treated as nominal variants, it will lose a lot of information, and it will deduct some unreasonable information, such as the distance between two different cities in different provinces equals to the distance between those in same area.

To deal with those problems above, we propose a new data clustering algorithm-CHCC. The main contributions of this paper are:

(1) Propose a new distance computing method for variants of HCDS (hierarchy coding data structure). (2) Modify k-medoids clustering algorithm, and propose a fast computing clustering median algorithm for hierarchy coded data sets. (3) Successfully apply CHCC algorithm into the clustering analysis of transient population for public security, and through extensive experiments, prove the validity and efficiency of the algorithm.

2 Preliminaries

In real applications and databases, a lot of data, such as postcodes, regionalism and vehicle types, are variants of HCDS. These variants can be divided into smaller disjoint parts, denoted as $P=(V_1,V_2,...,V_n)$, every $V_i(i=1,..n)$ is a nominal variant, while any $V_i(i=2,...,n)$ has a direct ancestor node V_{i-1}, and the data range of V_i is decided by V_{i-1}.

Example 2. Region code '510104'(means Qinyang district in Chengdu city, Sichuan province) can be denoted as $P_1=(51,01,04)$, where n=3, $V_1=51$(Sichuan Province), $V_2=01$(Chengdu City), and $V_3=04$(Qinyang district).

Thus, a hierarchy coding can be transformed into a Balance Tree (Figure. 1). Any hierarchy coding variant P can be considered as a set of nodes, and this set contains all the nodes, which are on the path from root to leaf. Besides, the relation between each leaf on the tree and its relative path is one-to-one, thus each hierarchy coding variant P can be represented by the leaf, denoted as T(P). The weights of each path between two layers of the Balance Tree are denoted as $W_1, W_2, \ldots W_{n-1}$, ($W_i > 0, i = 1, \ldots, n-1$). W_i is the weight of the path between a node in layer i and a node in layer i+1, and the weights represent the degree of difference between two nodes. Smaller the value of weights are, more similar the two nodes are. The weight set is usually replaced by weight sum: $W_i' = \sum_{k=i}^{n} W_k, i = 1, \ldots, n$. The balance tree is called hierarchy coding tree, denoted as T. Let's look at a hierarchy coding instance, a regionalism, and the regionalism of District Qingyang, City Chengdu(510104) can be expressed as the bold path in the Figure 1.

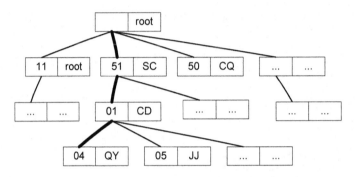

Fig. 1. Example of hierarchy coding balance tree

To formalize the above discussion, we give the following definitions.

Definition 1 (Coding distance). Let P_1 and P_2 be variants of HCDS under the same value domain. The weight sum of shortest path from $T(P_1)$ to $T(P_2)$ in Tree T is called the coding's distance, denoted as $d(P_1, P_2)$.

Definition 2 (Median of hierarchy coding variants set). Let p be a variant of HCDS, and $p = ArgMin\{E_i \mid E_i = \sum_{p_j \in C} |d(p_i, p_j)|, p_i \in C\}$, where C is variants set of HCDS, then p is said to be the Median of set C, denoted as C_{Centre}.

Actually, in Definition 2, median of the set is the variant that has the smallest sum of distance to the other variants. Based on Definition 2, a fast method to compute median of hierarchy coding variants in clustering is proposed in the following sections.

3   CHCC Algorithm

The steps of CHCC algorithm are as following: (1) Organize n data tuples into k ($k \leq n$) partitions. (2) Each partition represents a cluster. (3) Tuples in same cluster are similar; tuples in different cluster are dissimilar.

The basic strategy of this algorithm is: (1) For every cluster, randomly select a tuple to represent it; (2) For the rest tuples, they are assigned to the nearest cluster according to their distance to these represent tuples. (3) Improve the quality of clustering by replacing the represent tuple with non-represent tuple and repeating the processes of Step 1 and Step 2 until the distance between every tuple and its cluster does not change anymore or reaches the maximum iterative times.

The main tasks of CHCC algorithm contain the following parts.

- Proposed distance computing method for hierarchy coding variants.
- Proposed hierarchy coding variants' median computing algorithm in cluster.
- Amended k-medoids algorithm, and reduces the iterative times.
  The details will be introduced in the following sections.

3.1 Distance Computing Algorithm

When computing the distance between two data tuples, it will calculate: (a) the distance on every dimensionality of tuple based on the data structure; (b) the total distance between tuples by using weighted Euclidean distance or weighted Manhattan distance. For the data structure of hierarchy coding variant, we have algorithm as following:

> **Algorithm 1.** Hierarchy coding distance computing (hd-Distance)
> **Input**: Pi, Pj *(hierarchy coding variants)*
> **Output**: d(Pi,Pj)
> **Begin**
> 1 $k \leftarrow 1$;
> 2 **While** $(k <= n)$ **And** $(P_{ik} = P_{jk})$
> 3 $k \leftarrow k+1$;
> 4 $rc \leftarrow 0$;
> 5 **For Each** m **In** $[k .. n-1]$ **Do**
> 6 $rc \leftarrow rc + 2 * w_m$
> 7 **Return** rc
> **End.**

The n in hd-Distance shows that hierarchy coding variants have *n* non-intersections. $W_m(m=1,...,n-1)$ represents a path weight from layer *m* to layer *m+1*.

3.2 Algorithm for Computing Median of Hierarchy Coding Variants

Many clustering methods compute the median or represent point of a cluster by average method to deal with interval-scaled variants. However, it cannot deal with hierarchy coding variants, because this kind of variants cannot be simply mapped into one dimension space. To deal with the problem, the main idea of the algorithm is:

1) Create a Balance Tree structure to store different coding; let tree represent the hierarchy of the coding.

2) Visit every hierarchy coding variants in turn, and assign them to related nodes of the tree, then revise the related node parameters.

3) Bring out the median according to the statistical result of all the parameters in the Balance Tree.

The structure of node in Balance Tree is as following:

TCodingNodeStructure = RECORD

Parent:	**POINTER**;	{Parent pointer}
FirstChild:	**POINTER**;	{First Child pointer}
MinWeightChild:	**POINTER**;	{Min Weight Child pointer}
Value:	**DATA**;	{data of this node}
Count:	**INTEGER**;	{total number of leaf nodes}
Weight:	**FLOAT**;	{Weight}
NextSilbing:	**POINTER**;	{Next Brother pointer}

END.

An instance of this node structure is shown as below.

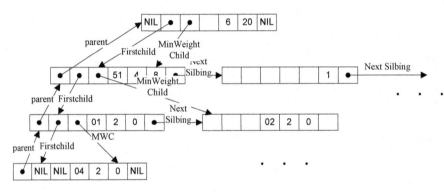

Fig. 2. An instance of node structure

The "Min Weight Child" represents the node which has the minimum weight among all of child nodes. "Weight" stands for the weight value of this node, and the definition and computing method will be described later in this paper. The formal description of dispose hierarchy coding variants is as following:

Algorithm 2. CHCC_CountTree
Input: T *(current clustering tree)*, P*(hierarchy coding variant)*
Output: T *(current clustering tree)*
Begin
  1 Root←GetRootNode(T)
   2 CurrentNode←SearchTree(Root);
   3 RegressTree(CurrentNode)
   4 Return T
End.

Function SearchTree(NodeParent)
Begin
   1   **For Each** k **In [1..n] Do**
   2   CurrentNode←NodeParent.FirstChild;
   3   **If** CurrentNode=**NULL Then**
   4   CurrentNode=NewNode(NodeParent,NULL,P_k)
   5   **Else Begin**
   6   Found←**False;**
   7   **While Do**
   8   **If** CurrentNode.Value = P_k **Then** Found←**True; break;**
   9   **If** CurrentNode. NextSilbing=**NULL Then break;**
   10   CurrentNode←CurrentNode.NextSilbing
   11 **End While**
   12   **If** Found **Then** CurrentNode.Count←CurrentNode.Count+1
   13   **Else** CurrentNode←NewNode(NodeParent, CurrentNode, P_k)
   14 **End**
   15 NodeParent←CurrentNode
   16 End For
End.

Function RegressTree(CurrentNode)
Begin
   1 For k = n Downto 1 Do
   2   NodeParent←CurrentNode.parent
   3   tWeight←CurrentNode.Weight+(NodeParent.Count-CurrentNode.Count)*2*W'_{k-1}
   4   NodeParent.Weight←NodeParent.Weight+ 2*W'_{k-1}
   5   **If** tWeight< NodeParent.Weight **Then**
   6   NodeParent.Weight ←tWeight
   7   NodeParent.MinWeightChild ←CurrentNode **End if**
   8   CurrentNode←NodeParent
   9 End For
End.

The main processes in Algorithm 2 are:

(1) From top to bottom, looking for an appropriate place for the hierarchy coding variant, if finds it, the corresponding statistical numbers add 1, otherwise, create a new node.

(2) From bottom to top, estimate the value of weight in turn, if the value has been changed, then amend the property of "MinWeightChild". P_k stands for the k-th part value of hierarchy coding variant P. W'_k stands the weight sum of path from layer k to the nodes in layer n.

Example 6. Figure 2 gives an instance of tree structure. Consider process of P=("51","01","05") .

1) Estimate whether or not P has a corresponding node in every layer of the tree, and processes it. The result is as shown in Figure 3.
2) From leaf node, trace back to the root step by step. In each step, estimate the corresponding weight, and adjust the "weight" and "MinWeightChild". The final result is shown in Figure 4.

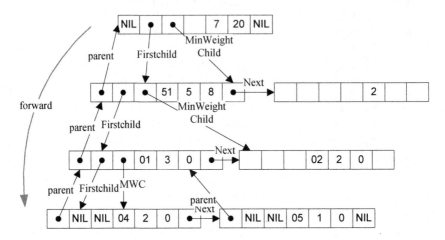

Fig. 3. An instance process of hierarchy coding variant

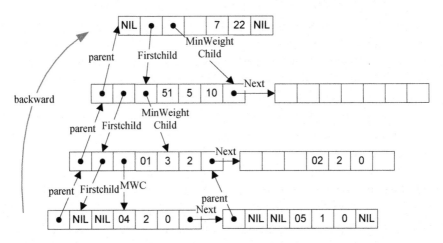

Fig. 4. Process result of hierarchy coding variant

Actually, algorithm CHCC_CountTree finishes disposing single hierarchy coding variant. In the main algorithm processing, it visits all the tuple in turn, and when it encounters hierarchy coding variant, it calls CHCC_CountTree to dispose the variant. When it finishes all the data, it can get the code of median by searching the Balance Tree. The formalized description of getting median value in hierarchy coding variants is as the following:

Algorithm 3. hd_Final
Input: T *(hierarchy coding tree of cluster)*
Output: P*(median code of cluster)*
Begin
  1 CurrentNode←GetRootNode()
  2 **For Each** k **In [1..n] Do**
3   CurrentNode.Node←CurrentNode.MinWeightChild
4   Pk←CurrentNode.Value
  5 **End For**
  6 **Return** P
  End.

P_k stands for the *k*-th part value of hierarchy coding variant P.

3.3 Algorithm CHCC_MAIN

Algorithm CHCC_MAIN is the main procedure of clustering. It completes the whole clustering process and returns the result, and it will call the median algorithm and distance algorithm depending on the data structure in its processes. According to a practice application, user will choose CHCC_MAIN to call the median algorithm of hierarchy coding variants.

CHCC algorithm differs from the traditional k-medoids algorithm. CHCC algorithm computes the nearest point (or tuple) to the median of every cluster in each iterative computing, while the traditional k-medoids algorithm randomly selects a point from the cluster, and then chooses the better of new point and old point. The traditional k-medoids algorithm consumes less time in every iterative computing. However, it will take much more times of iterative computing than CHCC algorithm. Through the real data comparison experiment in Section 5, it will show that as a whole, the method of this paper consumes much less time than traditional k-medoids algorithm.

4 Algorithm Analyses

Theorem 1. The median of hierarchy coding variants which computed by CHCC algorithm measures up the criterion in Definition 2.

Proof: The median of hierarchy coding variants is computed by Algorithm 2 and Algorithm 3 of CHCC. In Algorithm 3, the process of computing median starts from the root of the clustering tree, and then searches "MinWeightChild", i.e., the minimum weight node, on every node in turn and the final search result is the median value. Thus to prove the theorem, it is enough to prove each node's MinWeightChild points the median in its subtree.

By the hierarchy coding variant inserting process in Algorithm 2 that the MinWeightChild of current node's parent does or does not to be amended is determined by the weights comparison between parent's and current node's. In the algorithm, the actual comparison is between NodeParent.Weight+2*W'_{k-1} and tWeight

(tWeight= CurrentNode.Weight + (NodeParent.Count- CurrentNode.Count)*2*W'_{k-1}). If the latter is bigger, amend the parent's MinWeightChild pointer and pointing to current node. If the current node is a leaf node, the current node's weight formula, shown in bracket behind tWeight above, represents the sum of distances from current node to the brother nodes, and these distances are computed in one subtree, whose root is the parent of current node. At this time, MinWeightChild of the current node's parent consequentially points to the sub node where the median locates; meanwhile, the weight of the parent node is the sum of distances from median to other sub nodes. The rest may be deduced by analogy, every node's weight computed by Algorithm 2 represents the sum of distances from the median to other nodes in the subtree, whose root is right this node.

The reason to compare NodeParent.Weight+2*W'_{k-1} and tWeight in algorithm is that when it is inserted a new node, the total number (Count) of a parent's leaf nodes increases 1, and Weight (sum of distance from median to other nodes) accordingly increases 2*W'_{k-1} (under the condition of parent's old MinWeightChild differing current node)

Based on the discussion above, when the algorithm insert a hierarchy coding variant, and after it processes in turn, the node's pointer of MinWeightChild nodes in every layer surely points to the median node in the subtree, whose root is right this node(Although the algorithm would not guarantee each node's Weight is right in the tree, that MinWeightChild points right position is guaranteed.). So, based on algorithm 3, the final result surely is the median of the hierarchy coding variants set.   □

5   Experiments and Capability Analysis

CHCC algorithm is successfully applied in the analysis of transient population for public security. The purpose of population information analysis is to study different crowd transient moving patterns and moving trends of public population, and find out abnormal moving information and patterns. Thus, it can provide decision-making for public security preventing and striking crimes, and clues to detect criminal cases.

Through CHCC algorithm, automatic clustering different transient moving population information and relative analyses are achieved in practices. Through the analysis of the tuples that are far away from clustering median, the isolated data are obtained. In this way, it can provide useful information for detecting criminal cases.

The hardware for the experiments environment is Athlon 64 3000+, 1G memory, 160G hard drive, and the developing tool is DELPHI7.0, database is Oracle 9i.

Sample data for the experiments are transient population information logged in hotels in resent 2 years. The total number of data records is 3,700,000. The information these data contain are Name, ID, Origin Regionalism, Hotel Regionalism, Check-in Time, Check out Time. Before experiments, these data were read into memory, and then relative preprocess work was done. So, the experiment time does not contain data loading and preprocess time.

Capability Comparison between CHCC algorithm and naive algorithm.
According to the Definition 2, the naive algorithm computing hierarchy coding
variant median is implemented in another way here, and the algorithm computes
distances from each point to others and sum of these distances, and then selects the
point which has smallest sum of distances as median of the set. It is obviously that the
time complexity of naive algorithm is $O(n^2)$. If the hierarchy coding variant median is
computed by naive algorithm, the time complexity of clustering algorithm will
increase to $O(n^2 k)$.

In the experiment, it was randomly selected 500~5,000 data records, and chose
Origin Regionalism to be processed by CHCC algorithm median computing method
and naive algorithm respectively, and then compared the time consumption between
these two algorithms. The result of this experiment is as shown in the left figure
below. In the Figure 5, x-coordinate stands for data set size, y-coordinate stands for
CHCC's elapsed time in millisecond.

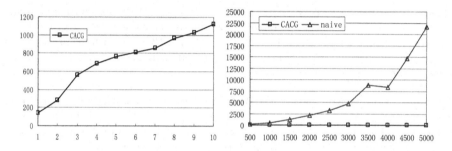

Fig. 5. Median computing comparing of CHCC and naive algorithm

The experiment result shows that when data records are less than 5000; the time
consumption of CHCC algorithm can be ignored, while it grows fast in naive
algorithm. This shows that the capability of CHCC algorithm is much better than
naive algorithm. To test the capability of this algorithm, data records is increased to
50,000~500,000, and the CHCC algorithm time consumption result is as shown in
right Figure 5. In the right of Figure 5, the X-coordinate stands for dataset, and unit is
dataset/50,000, Y-coordinate stands for time consumption, and unit is milliseconds. It
can be seen that the time consumption of the algorithm increases linearly with data
records, but even to 500,000 records, the time consumption is till less than 1.2
seconds.

Relationship between CHCC algorithm and data size. The purpose of this
experiment is to test the relationship of CHCC algorithm time consumption and data
size in actual dataset. In the clustering, 4 key properties have been used in the
experiment, they are Origin Regionalism, Hotel Regionalism, Hotel ID number,
Check-in Time (the first 3 dimensions are hierarchy coding structure), and k is set
to 8. The experiment result is as shown in the figure below:

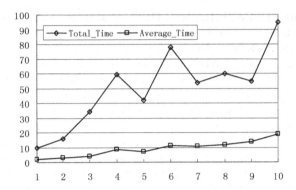

Fig. 6. CHCC 's elapsed time and data size

X-coordinate stands for data set size/50,000, Y-coordinate stands for the algorithm elapsed time in second. Total_Time stands for time consumption, and it represents CHCC's elapsed time in each iterative computing under current data size. From the experiment result, it is clear to see that when CHCC algorithm processes 4 dimensions clustering, the average time consumption in every iterative computing increases linearly by data size.

Capability Comparison between CHCC Algorithm and k-Medoids Algorithm. The purpose of this experiment is to compare the difference between CHCC algorithm and k-medoids algorithm, and k is set to 10. Data set of this experiment is same as Experiment 3. The result of the experiment is shown in Figure 7:

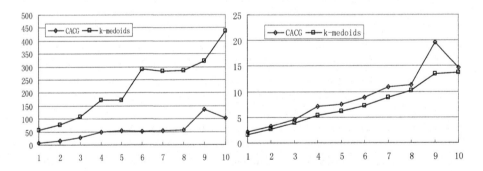

Fig. 7. Comparing of CHCC and k-medoids

In the left figure, there is total time consumption comparison of the two algorithms, and right is the comparison in single iterative computing. X-coordinate stands for data size/50,000, and Y-coordinate stands for the algorithm's elapsed time in second.

Figure 7 shows that at the single iterative computing time consumption aspect, k-medoids algorithm is less than CHCC algorithm, this is because that to get clustering median, CHCC algorithm visits data one time more than k-medoids does in the iterative process. However, CHCC algorithm obviously has advantage in time consumption, and its average elapsed time is about 25% as that of k-medoids algorithm. This is because CHCC algorithm can greatly decrease the iterative times than k-medoids algorithm does.

6 Conclusions

To deal with the hierarchy coding data, this paper proposes a novel data clustering method – CHCC. The method includes relative distance algorithm and median algorithm of hierarchy coding variants. Through amend k-medoids algorithm, CHCC greatly increase the performance of clustering task. Extensive experiments show that CHCC algorithm is validity and high efficiency in actual dataset.

Acknowledgments. This research was supported by the National Natural Science Foundation of China under Grant No. 60473051, 60473071 and 60503037, the China Postdoctoral Science Foundation under Grant No. 20060400002, the Sichuan Youth Science and Technology Foundation of China under Grant No. 07ZQ026-055, the National High-tech Research and Development of China under Grant No. 2006AA01Z230 and the Natural Science Foundation of Beijing Natural Science Foundation under Grant No. 4062018.

References

1. Han, J., Kamber, M.: Data Mining: Concepts and Techniques, 2nd edn. pp. 383–466. Morgan Kaufmann Publishers, Seattle (2006)
2. Chaturvedi, A., Green, P., Carroll, J.: K-modes clustering. J. Classification 18, 35–55 (2001)
3. Mei, Q., Zhai, C.: A mixture model for contextual text mining.In KDD, pp. 649–655 (2006)
4. Aggarwal, C.C., Han, J., Wang, J., Yu, P.S.: A framework for clustering evolving data streams. In: Aberer, K., Koubarakis, M., Kalogeraki, V. (eds.) Databases, Information Systems, and Peer-to-Peer Computing. LNCS, vol. 2944, pp. 81–92. Springer, Heidelberg (2004)
5. Aggarwal, C., Han, J., Wang, J., Yu, P.S.: A framework for projected clustering of high dimensional data streams. In: Proc. 2004 Int. Conf. Very Large Data Bases (VLDB'04), Toronto, Canada, pp. 852–863 (August 2004)
6. Basu, S., Bilenko, M., Mooney, R.J.: A probabilistic framework for semi-supervised clustering. In: Proc. 2004 ACM SIGKDD Int. Conf. Knowledge Discovery in Databases (KDD'04), Seattle, WA, pp. 59–68 (August 2004)
7. Aggarwal, C.C., Yu, P.S.: Outlier detection for high dimensional data. In: Proc. 2001 ACM-SIGMOD Int. Conf. Management of Data (SIGMOD'01), Santa Barbara, CA, pp. 37–46 (May 2001)
8. Wang, H., Wang, W., Yang, J., Yu, P.S.: Clustering by pattern similarity in large data sets. In: Proc. 2002 ACM-SIGMOD Int. Conf. Management of Data (SIGMOD'02), Madison, WI, pp. 418–427 (June 2002)

Mining Both Positive and Negative Association Rules from Frequent and Infrequent Itemsets

Xiangjun Dong[1,2], Zhendong Niu[3], Xuelin Shi[4], Xiaodan Zhang[3],
and Donghua Zhu[1]

[1] School of Management and Economics, Beijing Institute of Technology,
Beijing 100081, P.R. China
dxj@sdili.edu.cn, zhudh111@bit.edu.cn
[2] School of Information Science and Technology, Shandong Institute of Light Industry,
Jinan 250353, P.R. China
[3] School of Computer Science and Technology, Beijing Institute of Technology,
Beijing 100081, P.R. China
{zniu, zxd75}@bit.edu.cn
[4] School of Information Science and Technology, Beijing University of Chemical
Technology, Beijing, 100029, P.R. China
shixl@mail.buct.edu.cn

Abstract. A lot of new problems may occur when we simultaneously study *positive and negative association rules* (PNARs), i.e., the forms $A \Rightarrow B$, $A \Rightarrow \neg B$, $\neg A \Rightarrow B$ and $\neg A \Rightarrow \neg B$. These problems include how to discover infrequent itemsets, how to generate PNARs correctly, how to solve the problem caused by a single minimum support and so on. Infrequent itemsets become very important because there are many valued negative association rules (NARs) in them. In our previous work, a MLMS model was proposed to discover simultaneously both frequent and infrequent itemsets by using multiple level minimum supports (MLMS) model. In this paper, a new measure *VARCC* which combines correlation coefficient and minimum confidence is proposed and a corresponding algorithm *PNAR_MLMS* is also proposed to generate PNARs correctly from the frequent and infrequent itemsets discovered by the MLMS model. The experimental results show that the measure and the algorithm are effective.

1 Introduction

Mining association rules in database has received much attention recently. Many efforts have been devoted to design algorithms for efficiently discovering association rules of the form $A \Rightarrow B$, whose support (s) and confidence (c) meet some user specified minimum support (ms) and minimum confidence (mc) thresholds respectively. This is the famous traditional support-confidence framework [1]. The rules of the form $A \Rightarrow B$ have been studied widely since then, while the rules of the form $A \Rightarrow \neg B$ have attracted a little attention until recently some papers [2,3,4,5,6,7,8,9,10,11,12] proposed that the rules of the form $A \Rightarrow \neg B$ could play the same important roles as rules of the form $A \Rightarrow B$. The rules of the form $A \Rightarrow \neg B$, which suggest that a customer

R. Alhajj et al. (Eds.): ADMA 2007, LNAI 4632, pp. 122–133, 2007.

would not buy item B after buying item A, together with the rules of the forms $\neg A \Rightarrow B$ and $\neg A \Rightarrow \neg B$ are called **negative association rules** (**NARs**) and the rules of the form $A \Rightarrow B$ **positive association rules**(**PARs**) [4]. Some new problems would occur when we consider both **positive and negative association rules** (**PNARs**) simultaneously. Let's look at an example first.

Example 1 (Adopted from [4]). Suppose we have market basket data from a grocery store, consisting of n baskets. Let us focus on the purchase of tea (t) and coffee (c). Suppose $s(c) = 0.6$, $s(t) = 0.4$, $s(t \cup c) = 0.05$, $ms=0.2$ and $mc= 0.52$.

Because $s(t \cup c) = 0.05 < ms$, $t \cup c$ is an *infrequent itemset* (*inFIS*), $t \Rightarrow c$ cannot be extracted as a rule in the traditional support-confidence framework. But $s(t \cup \neg c) = s(t) - s(t \cup c) = 0.4 - 0.05 = 0.35 > ms$, and $c(t \Rightarrow \neg c)=s(t \cup \neg c)/s(t) = 0.35/0.4 = 0.875 > mc$. Therefore, $t \Rightarrow \neg c$ should be extracted as a negative rule.

From example 1, we can see that infrequent itemsets become very important because there are many valued NARs in them. NARs are discussed in some papers. Negative relationships between two itemsets were first mentioned in [2]. [3] proposed an approach to find strong negative association rules. [7] proposed an approach based on taxonomy to find negative association rules. The approach employs a hierarchical graph-structured taxonomy of domain terms. [8] proposed an approach for mining confined positive and negative association rules. In our previous work, we proposed an approach that also can find all the rules of the four forms $A \Rightarrow B$, $A \Rightarrow \neg B$, $\neg A \Rightarrow B$ and $\neg A \Rightarrow \neg B$ by using chi-squared test and multiple minimum confidences [9]. An approach of multilevel positive and negative association rule mining for spatial databases was discussed in [10]. An extended form for negative association rules and the corresponding mining algorithm were discussed in [11]. In [12] the authors discussed the elicitation of fuzzy association rules from positive and negative examples.

However, these papers did not concentrate on discovering NARs from infrequent itemsets. There are many mature techniques to discover frequent itemsets, while few papers study the approach of how to discover infrequent itemsets and how to discover NARs from them. [5,6] proposed an approach to find positive association rules from frequent itemsets and find negative association rules from infrequent itemsets. However, there are a bit more constraints to the defined infrequent itemsets, which means that only a few infrequent itemsets can be discovered and in turn only a few negative association rules can be obtained. And the negative association rules in frequent itemsets have not been discovered.

In order to get more valued NARs, more infrequent itemsets should be discovered. Unfortunately, infrequent itemsets are the complement of frequent itemsets theoretically and they are too large to be discovered completely in real application. In fact, how to discover infrequent itemsets is a challenging problem. The difficulty is how to ensure the number of infrequent itemsets is moderate, i.e., neither too few to get valued NARs nor too many to be easily processed. So some constraints must be given to infrequent itemsets. In our previous work [13], a multiple level minimum support (MLMS) model is proposed to solve this problem in a degree. That is, different minimum supports are assigned to itemsets with different lengths for the reason that using a single minimum support to all the k-itemsets ($k=1$, $2...n$) is unfair. The MLMS model can discover both infrequent itemsets and frequent itemsets simultaneously. It is different from [14] which gives every item a different minimum support to discover only frequent itemsets. In this paper we try to find an approach to discover PNARs from the FIS and the inFIS discovered by the MLMS model.

The main contributions of this paper are as follows:

1. A new measure *VARCC* which is combined correlation coefficient and minimum confidence is proposed.
2. An algorithm *PNAR_MLMS* based on the measure *VARCC* is proposed to discover PARs only from the FIS and NARs from both the FIS and the inFIS discovered by the MLMS model.

The rest of the paper is organized as follows: Section 2 is an introduction of the MLMS model. The application of correlation coefficient in PNARs and the measure *VARCC* are discussed in Section 3. Section 4 is algorithm design. Section 5 is experimental results and Section 6 is conclusions and future work.

2 MLMS Model

Let $I=\{i_1, i_2,..., i_n\}$ be a set of n distinct literals called items, and TD a transaction database of variable-length transactions over I, and the number of transactions in TD is denoted as $|TD|$. Each transaction contains a set of item $i_1, i_2, ..., i_m \in I$ and each transaction is associated with a unique identifier TID. A set of distinct items from I is called an itemset. The number of items in an itemset is the length of the itemset. An itemset of length k are referred to as k-itemset. Each itemset A $(A \subseteq I)$ has an associated statistical measure called support, denoted by $s(A)$, $s(A)=A.count/|TD|$, where $A.count$ is the number of transactions in TD containing itemset A. The support of a rule $A \Rightarrow B$ is denoted as $s(A \cup B)$ or $s(A \Rightarrow B)$, where $A, B \subseteq I$,and $A \cap B = \Phi$. The confidence of the rule $A \Rightarrow B$ is defined as the ratio of $s(A \cup B)$ to $s(A)$, i.e., $c(A \Rightarrow B) = s(A \cup B)/s(A)$.

The MLMS model is to assign different minimum supports to itemsets with different length [13]. Let $ms(k)$ be the minimum support of k-itemsets $(k=1,2,...,n)$, $ms(1) \geq ms(2) \geq ,..., \geq ms(n) \geq ms> 0$, ms be a threshold for infrequent itemsets. For any k-itemset A,

if $s(A) \geq ms(k)$, then A is a frequent itemset; and
if $s(A) < ms(k)$ and $s(A) \geq ms$, then A is an infrequent itemset.

The MLMS model allows users to control the number of frequent and infrequent itemsets conveniently by setting different $ms(k)$ and ms. The value of $ms(k)$ and ms are given by users or experts.

Algorithm. *Apriori_MLMS*
Input: *TD*: Transaction Database; *ms(k)*, *ms*: minimum support threshold;
Output: *FIS*: frequent itemsets; *inFIS*: infrequent itemsets;
(1) $FIS=\Phi$; $inFIS=\Phi$;
(2) $temp_1 =\{A \mid A \in 1\text{-itemsets}, s(A) \geq ms\}$;
  $FIS_1 = \{A \mid A \in temp_1 \text{ and } s(A) \geq ms(1)\}$;
  $inFIS_1 = temp_1 - FIS_1$;

```
(3) for (k=2;temp_{k-1}≠Φ;k++) do begin
    (3.1) C_k = apriori_gen(temp_{k-1}, ms);
    (3.2) for each transaction t∈TD do begin
          /*scan transaction database TD*/
          C_t=subset(C_k,t);
          for each candidate c ∈ C_t
              c.count++;
      end;
    (3.3) temp_k = {c|c∈C_k and (c.count/|TD|)≥ms};
          FIS_k = {A|A∈temp_k and A.count/|TD|≥ms(k)};
          inFIS_k = temp_k - FIS_k;
  end;
(4) FIS = ∪_kFIS_k; inFIS = ∪_kinFIS_k;
(5) return FIS and inFIS;
```

The algorithm **Apriori_MLMS** can discover both frequent and infrequent itemsets in a given transaction database *TD*. There are four kinds of sets: FIS_k, $inFIS_k$, $temp_k$ and C_k. $Temp_k$ contains the itemsets whose support meets the constraint *ms*. C_k contains the itemsets generated by the procedure *apriori_gen* which is the same as the procedure in traditional *Apriori* and is omitted here. More details about the explanation of the algorithm can be found in [13].

Here is an example. The transaction database is shown in Table 1 (Adopted from [4]). Let $ms(1)=0.5$, $ms(2)=0.4$, $ms(3)=0.3$, $ms(4)=0.2$ and $ms=0.15$, the results generated by algorithm **Apriori_MLMS** are shown in Table 2, where the itemsets with non background are frequent itemsets and the itemsets with gray background are infrequent itemsets. From Table 2, we can see that the itemsets with the same support but different length is either frequent itemsets or infrequent itemsets. For example, $s(AB)=s(ABD)=0.3$, AB is an infrequent itemset, while ABD is a frequent itemset.

The support and confidence of non-existing items are difficult to be calculated directly. But according to our previous work [9], we can calculate them in the following way:

Table 1. Transaction Database

TID	itemsets
T_1	A, B, D
T_2	A, B, C, D
T_3	B, D
T_4	B, C, D, E
T_5	A, C, E
T_6	B, D, F
T_7	A, E, F
T_8	C, F
T_9	B, C, F
T_{10}	A, B, C, D, F

(1) $s(\neg A) = 1-s(A)$;

(2) $s(A \cup \neg B) = s(A)-s(A \cup B)$; $s(\neg A \cup B) = s(B)- s(A \cup B)$;

(3) $s(\neg A \cup \neg B) = 1- s(A)- s(B) + s(A \cup B)$;

(4) $c(A \Rightarrow \neg B) = 1 - c(A \Rightarrow B)$;

(5) $c(\neg A \Rightarrow B) = \dfrac{s(B)- s(A \cup B)}{1-s(A)}$;

(6) $c(\neg A \Rightarrow \neg B) = \dfrac{1- s(A)- s(B) + s(A \cup B)}{1-s(A)} = 1- c(\neg A \Rightarrow B)$.

Table 2. FIS_k and $inFIS_k$ in the transasction database in table 1

1-itemsets	s(*)	2-itemsets	s(*)	2-itemsets	s(*)	3-itemsets	s(*)
A	0.5	BD	0.6	CD	0.3	ABD	0.3
B	0.7	BC	0.4	CF	0.3	BCD	0.3
C	0.6	AB	0.3	AE	0.2	ABC	0.2
D	0.6	AC	0.3	AF	0.2	ACD	0.2
E	0.3	AD	0.3	CE	0.2	BCF	0.2
F	0.5	BF	0.3	DF	0.2	BDF	0.2
						4- itemsets	
						ABCD	0.2

3 Application of Correlation Coefficient in PNARs and the Measure VARCC

Correlation coefficient measures the degree of linear dependency between a pair of random variables. Theoretically, it is defined as the covariance between two variables, divided by their standard deviations (σ):

$$\rho_{AB} = \frac{Cov(A,B)}{\sigma_A \sigma_B},$$

(1)

where $Cov(A,B) = E(AB) - E(A)E(B)$ and $E(*)$ is the expected value.

ρ_{AB} has the following three possible cases.

(1) If $\rho_{AB} > 0$, then A and B are positively correlated. The more events A occur, the more events B do.

(2) If $\rho_{AB} = 0$, then A and B are independent (for binary variables). The occurrence of event A has nothing to do with the occurrence of event B.

(3) If $\rho_{AB} < 0$, then A and B are negatively correlated. The more events A occur, the less events B do.

If the two variables are independent, then $\rho_{AB} = 0$. The converse, however, is not necessarily true. It is possible that $\rho_{AB} = 0$ when the variables have strong non-linear dependencies. Fortunately, such a problem does not exist for binary variables.

The range of ρ_{AB} is between -1 and $+1$. The correlation coefficient and its strength are discussed in [15]. According to this book, a variable α ($0 \leq \alpha \leq 1$) is used to express the correlation strength. $\alpha = 0.5$, the strength is large, 0.3, moderate, and 0.1, small. This means that a rule whose correlation of less than 0.1 is unvalued. So we use $\rho_{AB} \geq \alpha$ ($0 \leq \alpha \leq 1$) as a constraint to prune less valued rules. In real application, users can sort the rules by their correlations to get the rules with high correlations.

The contingency table of itemsets A and B is shown in table 3, where n denotes the total number of transactions in a database. The purpose of the data expressed in the form of support here is to associate the correlation coefficient with the concepts of association rules easily.

Therefore, the correlation coefficient between A and B can be written as

$$\rho_{AB} = \frac{s(A \cup B) - s(A)s(B)}{\sqrt{s(A)(1 - s(A))s(B)(1 - s(B))}}, \tag{2}$$

where $s(*) \neq 0, 1$.

The equation 2 is obtained assuming that the contingency table is constructed using the entire data. For finite samples, the above equation is equivalent to Pearson's Φ-coefficient.

Now we consider the relationship among the correlations between the itemsets A and B, A and $\neg B$, $\neg A$ and B and $\neg A$ and $\neg B$.

Theorem 1. If $\rho_{AB} \geq \alpha$ ($0 \leq \alpha \leq 1$), then
(1) $\rho_{\neg AB} \leq -\alpha$; (2) $\rho_{A \neg B} \leq -\alpha$; (3) $\rho_{\neg A \neg B} \geq \alpha$; and vice versa.

Proof: We only prove (1) here, (2), (3) can be proved similarly.

$$\rho_{\neg AB} = \frac{s(\neg A \cup B) - s(\neg A)s(B)}{\sqrt{s(\neg A)(1 - s(\neg A))s(B)(1 - s(B))}} = \frac{-(s(A \cup B) - s(A)s(B))}{\sqrt{s(A)(1 - s(A))s(B)(1 - s(B))}} = -\rho_{AB} \leq -\alpha.$$

Table 3. Contingency table of itemsets A and B

	A	$\neg A$	Σ
B	$s(A \cup B)*n$	$s(\neg A \cup B)*n$	$s(B)*n$
$\neg B$	$s(A \cup \neg B)*n$	$s(\neg A \cup \neg B)*n$	$s(\neg B)*n$
Σ	$s(A)*n$	$s(\neg A)*n$	n

The Theorem 1 shows that the correlation strength of the four form rules $A \Rightarrow B$, $A \Rightarrow \neg B$, $\neg A \Rightarrow B$ and $\neg A \Rightarrow \neg B$ is the same. So we can only calculate ρ_{AB} in the algorithm in section 4. The theorem 1 also shows that correlation coefficient can ensure the correctness of mining positive and negative association rules because at most two rules can be extracted among the four rules between itemsets A and B. They are $A \Rightarrow B$ and $\neg A \Rightarrow \neg B$ (or $\neg A \Rightarrow B$ and $A \Rightarrow \neg B$).

According to above discussion and the minimum confidence constraint, we propose a new measure *VARCC* (*Valid Association Rule based on Correlation coefficient and Confidence*) which combines correlation coefficient ρ and the minimum confidence mc. That is, a rule $A \Rightarrow B$ can be extracted as a valid association rule if it meets $VARCC(A, B, \alpha, mc) = 1$, where

$$VARCC(A, B, \alpha, mc) = \frac{\rho_{AB} - \alpha + c(A \Rightarrow B) - mc + 1}{|\rho_{AB} - \alpha| + |c(A \Rightarrow B) - mc| + 1}. \tag{3}$$

According to equation (3), we can easily get:

$$VARCC(A, \neg B, \alpha, mc) = \frac{\rho_{A \neg B} - \alpha + c(A \Rightarrow \neg B) - mc + 1}{|\rho_{A \neg B} - \alpha| + |c(A \Rightarrow \neg B) - mc| + 1}. \tag{4}$$

$$VARCC(\neg A, B, \alpha, mc) = \frac{\rho_{\neg AB} - \alpha + c(\neg A \Rightarrow B) - mc + 1}{|\rho_{\neg AB} - \alpha| + |c(\neg A \Rightarrow B) - mc| + 1}. \tag{5}$$

$$VARCC\,(\neg A,\,\neg B,\,\alpha,\,mc) = \frac{\rho_{\neg A\neg B} - \alpha + c(\neg A \Rightarrow \neg B) - mc + 1}{|\,\rho_{\neg A\neg B} - \alpha\,| + |\,c(\neg A \Rightarrow \neg B) - mc\,| + 1}\,. \qquad (6)$$

Now we give the following definition.

Definition 1. Let I be a set of items, TD a database, $A,\,B \subseteq I$ and $A \cap B = \Phi$. $s(A)$, $s(B) \neq 0,1$. FIS and $inFIS$ are generated by $Apriori\_MLMS$. α $(0 \leq \alpha \leq 1)$ is the correlation strength, and mc is the minimum confidence threshold. If $|\rho_{AB}| < \alpha$, we don't generate rules from $A \cup B$, else

(1) if $A \cup B \subseteq FIS$ and $VARCC(A,\,B,\,\alpha,\,mc) = 1$, then $A \Rightarrow B$ is a PAR;
(2) if $A \cup B \subseteq FIS$ and $VARCC(A,\,\neg B,\,\alpha,\,mc) = 1$, then $A \Rightarrow \neg B$ is a NAR;
(3) if $A \cup B \subseteq FIS$ and $VARCC(\neg A,\,B,\,\alpha,\,mc) = 1$, then $\neg A \Rightarrow B$ is a NAR;
(4) if $A \cup B \subseteq FIS$ and $VARCC(\neg A,\,\neg B,\,\alpha,\,mc) = 1$, then $\neg A \Rightarrow \neg B$ is a NAR;
(5) if $A \cup B \subseteq inFIS$ and $VARCC(A,\,\neg B,\,\alpha,\,mc) = 1$, then $A \Rightarrow \neg B$ is a NAR;
(6) if $A \cup B \subseteq inFIS$ and $VARCC(\neg A,\,B,\,\alpha,\,mc) = 1$, then $\neg A \Rightarrow B$ is a NAR; and
(7) if $A \cup B \subseteq inFIS$ and $VARCC(\neg A,\,\neg B,\,\alpha,\,mc) = 1$, then $\neg A \Rightarrow \neg B$ is a NAR.

Definition 1 shows that PARs are only generated from frequent itemsets, while NARs are generated from both frequent itemsets and infrequent itemsets.

4 Algorithm Design

Algorithm. *PNAR_MLMS*
Input: *mc*: minimum confidence; α $(0 \leq \alpha \leq 1)$: correlation strength;
Output: *PAR*: set of PARs; *NAR*: set of NARs;
(0) **CALL** *Apriori_MLMS*;
(1) *PAR* $= \Phi$; *NAR* $= \Phi$;
(2) /*generating all positive & negative association rules from *FIS*.*/
 for any itemset X in *FIS* **do** **begin**
  for any itemset $A \cup B = X$ **and** $A \cap B = \Phi$ **do begin**
   calculate ρ_{AB} with Equation 2;
   if $\rho_{AB} \geq \alpha$ **then** **begin**
    (2.1)/*generate rules of the forms $A \Rightarrow B$ and $\neg A \Rightarrow \neg B$. */
    if $VARCC(A,\,B,\,\alpha,\,mc) = 1$ **then** *PAR* = *PAR* $\cup \{A \Rightarrow B\}$;
    if $VARCC(\neg A,\,\neg B,\,\alpha,\,mc) = 1$ **then** *NAR* = *NAR* $\cup \{\neg A \Rightarrow \neg B\}$;
   end;
   if $\rho_{AB} \leq \alpha$ **then begin**
    (2.2)/*generate rules of the forms $A \Rightarrow \neg B$ and $\neg A \Rightarrow B$. */

```
        if VARCC(A, ¬B, α, mc)=1 then    NAR = NAR
        ∪{A⇒¬B};
        if VARCC(¬A, B, α, mc)=1   then   NAR = NAR
        ∪{¬A⇒B};
     end;
   end;
 end;
```
(3) /*generating all negative association rules from in-
FIS.*/
```
   for any itemset X in inFIS do  begin
     for any itemset A∪B = X and A∩B =Φ do begin
       calculate ρ_AB with Equation 2;
       if ρ_AB ≥ α then
          (3.1)/*generate rules of the form ¬A⇒¬B. */
          if VARCC(¬A, ¬B, α, mc)=1   then NAR = NAR
          ∪{¬A⇒¬B};
       if ρ_AB ≤ α then begin
          (3.2)/*generate rules of the forms A⇒¬B and
¬A⇒B.  */
          if VARCC(A, ¬B, α, mc)=1 then    NAR = NAR
          ∪{A⇒¬B};
          if VARCC(¬A, B, α, mc)=1   then   NAR = NAR
          ∪{¬A⇒B};
       end;
     end;
   end;
(4) return PAR and NAR;
```

Step (0) calls the algorithm *Apriori_MLMS* to generate the sets *FIS* and *inFIS*. Step (1) initializes *PAR* and *NAR* with empty set. Step (2) generates all the four form rules from frequent itemsets. Step (3) generates all NARs from infrequent itemsets. Step (4) returns the result and finishes the whole algorithm.

Example 2. Take (1) *BD* and (2) *DF* in table 2 for example. Suppose $\alpha=0.3$ and $mc=0.7$.

(1) Because $BD \in FIS$, and $\rho_{BD}=(0.6-0.7*0.6)/(0.7*0.6*0.3*0.4)\approx0.8>\alpha$, step(2.1) runs. $c(B\Rightarrow D)=0.6/0.7\approx0.857$, $c(\neg B\Rightarrow\neg D)=1$,

$$VARCC(B, D, \alpha, mc)= \frac{0.8-0.3+0.857-0.7+1}{|0.8-0.3|+|0.857-0.7|+1}=1,$$ thus, $B\Rightarrow D$ is a valid PAR.

$$VARCC(\neg B, \neg D, \alpha, mc)=\frac{0.8-0.3+1-0.7+1}{|0.8-0.3|+|1-0.7|+1}=1,$$ thus, $\neg B\Rightarrow\neg D$ is a valid NAR.

(2) Because $DF \in inFIS$, and $\rho_{DF} =(0.2-0.5*0.6)/(0.5*0.6*0.5*0.4)\approx-0.41<-\alpha$, step(3.2) runs. $c(D\Rightarrow\neg F)\approx0.67$, $c(\neg D\Rightarrow F)=0.75$,

$$VARCC(D, \neg F, \alpha, mc) = \frac{0.41 - 0.3 + 0.67 - 0.7 + 1}{|0.41 - 0.3| + |0.67 - 0.7| + 1} < 1, \text{ thus, } D \Rightarrow \neg F \text{ is not a valid}$$

NAR.

$$VARCC(\neg D, F, \alpha, mc) = \frac{0.41 - 0.3 + 0.75 - 0.7 + 1}{|0.41 - 0.3| + |0.75 - 0.7| + 1} = 1, \text{ thus, } \neg D \Rightarrow F \text{ is a valid}$$

NAR.

5 Experimental Results

The experimental dataset records areas of www.microsoft.com that each user visited in a one-week timeframe in February 1998. Summary statistical information of the dataset is: 32711 training instances, 5000 testing instances, 294 attributes and the mean area visits per case is 3.0 (http://www.cse.ohio-state.edu/~yanghu/CIS788_dm_proj.htm#datasets). Now we demonstrate that our methods are effective from the following aspects.

1. The measure *VARCC* can generate PARs correctly and the algorithm *PNAR_MLMS* can discover PARs and NARs from itemsets discovered by the MLMS model.

Fig. 1 shows the comparison of the numbers of PARs generated by traditional *Apriori* and *PNAR_MLMS* when $\alpha=0$ and $ms(k)=ms$. The number of PARs generated by *PNAR_MLMS* is less than that generated by traditional *Apriori*. This means that the measure *VARCC* can delete those misleading rules generated by traditional *Apriori* and can generate PARs correctly, just as the measure chi-squared test can do [9].

Table 4 shows the numbers of rules at different single supports and multiple level supports when $\alpha=0$ and $mc=0.2$. Let's take the rules of the form $A \Rightarrow B$ for example, the number is 44, 47, 52, 62, 83 when $k=2$ and $ms(*)$ =0.025, 0.020, 0.017, 0.013, 0.01 respectively. For the rules of the four forms in frequent itemsets, when $ms(1)=0.025$, $ms(2)=0.02$, $ms(3)=0.017$, $ms(4)=0.013$ and $ms=0.01$, the numbers (47,14,9,60) at $k=2$ are equal to that when $ms(*)=0.02$; the numbers (83,6,3,138) at $k=3$ are equal to that when $ms(*)=0.017$; the numbers (32,0,0,36) at $k=4$ are equal to that when $ms(*)=0.013$. These data are emphasized with gray background in Table 4. The number of total PARs when $ms(1)=0.025$, $ms(2)=0.02$, $ms(3)=0.017$, $ms(4)=0.013$ and $ms=0.01$ is 162, which is neither so small that will lose many PARs (for example, the total number of PARs is 70 when $ms(*)=0.025$, and the number of PARs is 0 at $k=4$), nor is so big that will give user extra difficulties to choose right rules (the total number of PARs is 342 when $ms(*)=0.01$). These data show that the algorithm *PNAR_MLMS* can discover PARs from itemsets with different lengths and can control the number of PARs in an appropriate degree by setting appropriate $ms(k)$. These data further show that the MLMS model is effective. Moreover, the algorithm *PNAR_MLMS* also generates many NARs from infrequent itemsets.

2. The correlation strength α can be used to prune less valued rules effectively.

Table 4. The numbers of rules in itemsets with different lengths generated by *PNAR_MLMS* (*α*=0 and *mc*=0.2)

Supports		itemsets	$A{\Rightarrow}B$	$A{\Rightarrow}\neg B$	$\neg A{\Rightarrow}B$	$\neg A{\Rightarrow}\neg B$	Total
	k=2	*FIS*	44	12	8	56	120
ms(*)=0.025	*k* =3	*FIS*	26	2	1	40	69
	Total		70	14	9	96	189
	k =2	*FIS*	47	14	9	60	130
ms(*)=0.02	*k* =3	*FIS*	54	2	1	88	145
	k =4	*FIS*	11	0	0	14	25
	Total		112	16	10	162	300
	k =2	*FIS*	52	16	10	70	148
	k =3	*FIS*	83	6	3	138	230
ms(*)=0.017	*k* =4	*FIS*	11	0	0	14	25
	Total		146	22	13	222	403
	k =2	*FIS*	62	24	13	90	189
	k =3	*FIS*	133	12	5	240	390
ms(*)=0.013	*k* =4	*FIS*	32	0	0	36	68
	Total		227	36	18	366	647
	k =2	*FIS*	83	38	19	122	262
	k =3	*FIS*	202	22	9	362	595
ms(*)=0.01	*k* =4	*FIS*	57	0	0	77	134
	Total		342	60	28	561	991
	k =2	*FIS*	47	14	9	60	130
ms(1)=0.025		*inFIS*	-	24	10	62	96
ms(2)=0.02	*k* =3	*FIS*	83	6	3	138	230
ms(3)=0.017		*inFIS*	-	16	6	224	246
ms(4)=0.013	*k* =4	*FIS*	32	0	0	36	68
ms=0.01		*inFIS*	-	0	0	41	41
	Total		162	60	28	561	811

Note: *ms*(*) denotes *ms*(1), *ms*(2), *ms*(3), *ms*(4) and *ms*.

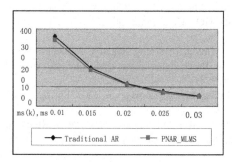
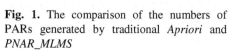
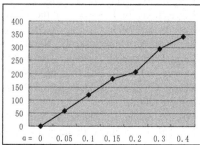

Fig. 1. The comparison of the numbers of PARs generated by traditional *Apriori* and *PNAR_MLMS*

Fig. 2. The changes of the pruned rules with the different *α*

Fig. 2 shows the changes of the pruned rules with the different α when $ms(1)=0.025$, $ms(2)=0.02$, $ms(3)=0.017$, $ms(4)=0.013$, $ms=0.01$ and $mc=0.2$. With α increasing, the number of pruned rules also increases. This means that the correlation strength α can be used to prune less valued rules effectively.

6   Conclusions and Future Work

Infrequent itemsets become very important because there are many valued negative association rules in them. In this paper, a new measure *VARCC* is proposed and an algorithm *PNAR_MLMS* based on the measure is designed to generate PARs from frequent itemsets and to generate NARs from both frequent itemsets and infrequent itemsets discovered by the MLMS model. The correlation strength α is used to prune less valued rules. The experimental results show that the measure *VARCC* and the algorithm *PNAR_MLMS* can work effectively. For future work, we will look for a new approach to further decrease the total number of rules because the rules are still too many to be chosen easily.

Acknowledgements

This work was partially supported by Program for New Century Excellent Talents in University, China; Huo Ying-dong Education foundation, China (91101); Excellent Young Scientist Foundation of Shandong Province, China (2006BS01017); and the Scientific Research Development Project of Shandong Provincial Education Department, China (J06N06).

References

1. Agrawal, R., Imielinski, T., Swami, A.: Mining Association Rules between Sets of Items in Large Database. In: Proceedings of the 1993 ACM SIGMOD International Conference on Management of Data, pp. 207–216. ACM Press, New York (1993)
2. Brin, S., Motwani, R., Silverstein, C.: Beyond Market: Generalizing Association Rules to Correlations. In: Processing of the ACM SIGMOD Conference, pp. 265–276. ACM Press, New York (1997)
3. Savasere, A., Omiecinski, E., Navathe, S.: Mining for Strong Negative Associations in a Large Database of Customer Transaction. In: Proceedings of the 1998 International Conference on Data Engineering, pp. 494–502 (1998)
4. Zhang, C., Zhang, S.: Association Rule Mining. LNCS (LNAI), vol. 2307. Springer, Heidelberg (2002)
5. Wu, X., Zhang, C., Zhang, S.: Mining both Positive and Negative Association Rules. In: Proceedings of the 19th International Conference on Machine Learning, pp. 658–665 (2002)
6. Wu, X., Zhang, C., Zhang, S.: Efficient Mining of Both Positive and Negative Association Rules. ACM Transactions on Information Systems 22, 381–405 (2004)

7. Yuan, X., Buckles, B.P., Yuan, Z., Zhang, J.: Mining Negative Association Rules. In: Proceedings of The Seventh IEEE Symposium on Computers and Communications, pp. 623–629. IEEE Computer Society Press, Los Alamitos (2002)
8. Antonie, M-L., Zaiane, O.: Mining Positive and Negative Association Rules: An Approach for Confined Rules. In: Boulicaut, J.-F., Esposito, F., Giannotti, F., Pedreschi, D. (eds.) PKDD 2004. LNCS (LNAI), vol. 3202, pp. 27–38. Springer, Heidelberg (2004)
9. Dong, X., Sun, F., Han, X., Hou, R.: Study of Positive and Negative Association Rules Based on Multi-confidence and Chi-Squared Test. In: Li, X., Zaïane, O.R., Li, Z. (eds.) ADMA 2006. LNCS (LNAI), vol. 4093, pp. 100–109. Springer, Heidelberg (2006)
10. Sharma, L.K., Vyas, O.P., Tiwary, U.S., Vyas, R.: A Novel Approach of Multilevel Positive and Negative Association Rule Mining for Spatial Databases. In: Perner, P., Imiya, A. (eds.) MLDM 2005. LNCS (LNAI), vol. 3587, pp. 620–629. Springer, Heidelberg (2005)
11. Gan, M., Zhang, M., Wang, S.: One Extended Form for Negative Association Rules and the Corresponding Mining Algorithm. In: Yeung, D.S., Liu, Z.-Q., Wang, X.-Z., Yan, H. (eds.) ICMLC 2005. LNCS (LNAI), vol. 3930, pp. 1716–1721. Springer, Heidelberg (2006)
12. De Cock, M., Cornelis, C., Kerre, E.E.: Elicitation of Fuzzy Association Rules from Positive and Negative Examples. Fuzzy Sets and Systems. Fuzzy Sets in Knowledge Discovery, vol. 149(1), pp. 73–85. Elsevier, Amsterdam (2005)
13. Dong, X., Zheng, Z., Niu, Z., Jia, Q.: Mining Infrequent Itemsets based on Multiple Level Minimum Supports. In: ICICIC07. Proceedings of the Second International Conference on Innovative Computing, Information and Control, Kumamoto, Japan, September 2007 (to appear, 2007)
14. Liu, B., Hsu, W., Ma, Y.: Mining Association Rules with Multiple Minimum Supports. In: KDD-99. Proceedings of the ACM SIGKDD International Conference on Knowledge Discovery & Data Mining, San Diego, CA, August 15-18, 1999, pp. 337–341. ACM Press, New York (1999)
15. Cohen, J.: Statistical Power Analysis for the Behavioral Sciences, 2nd edn. Lawrence Erlbaum, New Jersey (1988)

Survey of Improving Naive Bayes for Classification

Liangxiao Jiang[1], Dianhong Wang[2], Zhihua Cai[1], and Xuesong Yan[1]

[1] Faculty of Computer Science, China University of Geosciences
Wuhan, Hubei, P.R. China, 430074
ljiang@cug.edu.cn
[2] Faculty of Electronic Engineering, China University of Geosciences
Wuhan, Hubei, P.R. China, 430074
wangdh@cug.edu.cn

Abstract. The attribute conditional independence assumption of naive Bayes essentially ignores attribute dependencies and is often violated. On the other hand, although a Bayesian network can represent arbitrary attribute dependencies, learning an optimal Bayesian network classifier from data is intractable. Thus, learning improved naive Bayes has attracted much attention from researchers and presented many effective and efficient improved algorithms. In this paper, we review some of these improved algorithms and single out four main improved approaches: 1) Feature selection; 2) Structure extension; 3) Local learning; 4) Data expansion. We experimentally tested these approaches using the whole 36 UCI data sets selected by Weka, and compared them to naive Bayes. The experimental results show that all these approaches are effective. In the end, we discuss some main directions for future research on Bayesian network classifiers.

Keywords: Bayesian network classifiers, naive Bayes, feature selection, local learning, structure extension, data expansion, classification.

1 Introduction

A Bayesian network consists of a structural model and a set of conditional probabilities. The structural model is a directed graph in which nodes represent attributes and arcs represent attribute dependencies. Attribute dependencies are quantified by conditional probabilities for each node given its parents. Bayesian networks are often used for classification problems, in which a learner attempts to construct a classifier from a given set of training examples with class labels. Assume that A_1, $A_2, \cdots$, A_n are n attributes (corresponding to attribute nodes in a Bayesian network). An example E is represented by a vector $(a_1, a_2, \cdots, a_n)$, where a_i is the value of A_i. Let C represent the class variable (corresponding to the class node in a Bayesian network). We use c to represent the value that C

R. Alhajj et al. (Eds.): ADMA 2007, LNAI 4632, pp. 134–145, 2007.

takes and $c(E)$ to denote the class of E. The Bayesian classifier represented by a Bayesian network is defined in Equation 1.

$$c(E) = \arg\max_{c \in C} P(c)P(a_1, a_2, \cdots, a_n|c). \tag{1}$$

Assume all attributes are independent given the class. This is called the attribute conditional independence assumption:

$$P(E|c) = P(a_1, a_2, \cdots, a_n|c) = \prod_{i=1}^{n} P(a_i|c). \tag{2}$$

Then, the resulting classifier is called a naive Bayesian classifier [1], or simply naive Bayes (NB):

$$c(E) = \arg\max_{c \in C} P(c) \prod_{i=1}^{n} P(a_i|c). \tag{3}$$

In NB, each attribute node has the class node as its parent, but does not have any parent from attribute nodes. Although NB is easy to construct, because the values of $P(a_i|c)$ can be easily estimated from training examples, the attribute conditional independence assumption made by the naive approach harms the classification performance of naive Bayes when it is violated.

In order to relax this assumption effectively, an appropriate language and efficient machinery to represent and manipulate independence assertions are needed [8]. Both are provided by Bayesian networks [2]. Unfortunately, however, it has been proved that learning an optimal Bayesian networks is NP-hard [3]. In order to avoid the intractable complexity for learning Bayesian networks, learning improved naive Bayes has attracted much attention from researchers. Related work can be broadly divided into four approaches:

1. Feature selection: Selecting attributes subsets in which attributes satisfy the attribute independence assumption.
2. Structure extension: Extending the structure of naive Bayes to represent the dependencies among attributes.
3. Local learning: Employing the principle of local learning to find a local training data set and use it to build a naive Bayes.
4. Data expansion: Expanding the training data and build a naive Bayes on the expanded training data.

In this paper, a survey of improving naive Bayes for classification is provided. We organize the rest of the paper as follows. In section 2, we review some improved algorithms of naive Bayes keeping to four main improved approaches we summarized. In section 3, we describe the experimental setup and results in detail. In section 4, we make conclusions and outline our main directions for future research.

2   Improved Naive Bayesian Classifiers

In this section, we try our best to review some improved algorithms of naive Bayes keeping to four main improved approaches we summarized.

2.1   Improve Naive Bayes Using Feature Selection

The feature selection approach improves the classification performance of naive Bayes by removing redundant and/or irrelevant attributes from training data sets, and only selecting those that are most informative in classification task. This approach works well with the hypotheses that it can improve naive Bayes' classification accuracy in domains that include redundant and/or irrelevant attributes without reducing naive Bayes' classification accuracy in domains that don't. In fact, any of this kind improved algorithms is a variant of naive Bayes that only uses a subset of the given attributes in making predictions. In other words, the improved naive Bayes with feature selection classifies the test instance E using Equation 4 to replace Equation 3.

$$c(E) = \arg \max_{c \in C} P(c) \prod_{i=1}^{k} P(a_i|c). \tag{4}$$

where $a_i(i = 1, 2, \ldots, k)$ respectively is the value of the selective attribute $A_i(i = 1, 2, \ldots, k)$, k is the number of selective attributes.

Obviously, the basic idea of these improved algorithms is how to efficiently select relevant attributes subsets from training data sets. To achieve this, many feature selection algorithms have been presented and demonstrates significant improvement over naive Bayes.

For example, Langley and Sage [4] presented an algorithm called selective Bayesian classifiers (simply SBC). It uses a forward greedy search method to select an attribute subset through the whole space of attributes. In detail, it uses naive Bayes' classification accuracy to evaluate alternative subsets of attributes and consider adding each unselected attribute which can improve the classifier's accuracy at most on each iteration.

For another example, Jiang et al. [5] presented an algorithm called evolutional naive Bayes (simply ENB). It selects an attribute subset through the whole space of attributes by carrying out a evolutional (genetic) search process. ENB uses naive Bayes' classification accuracy as the fitness function to evaluate alternative subsets of attributes and selects the individual (hypotheses) with maximum classification accuracy after an appointed number of generations.

Moreover, Kohavi and George [6] explored the relation between optimal feature subset selection and relevance, and proposed a wrapper method for feature subset selection. Ratanamahatana and Gunopulos [7] proposed another feature selection approach by building decision trees.

2.2   Improve Naive Bayes Using Structure Extension

Just as we discussed before, The main problem confronting naive Bayes is its attribute conditional independence assumption. In real-world data mining applications, this assumption is unrealistic. Thus, extending the structure of naive Bayes and using directed arcs to explicitly represent attribute dependencies is a direct way to relax this unrealistic assumption. The resulting model is essentially a Bayesian network. Thus, learning the structure of Bayesian networks is unavoidable. However, learning the optimal structure is a NP-hard problem [3]. In practice, imposing restrictions on the structures of Bayesian networks is necessary. For example, learning tree augmented naive Bayes (simply TAN) [8] leads to acceptable computational complexity and a considerable improvement over naive Bayes. TAN assumes that the structure among all attributes nodes is tree-like structure.

A number of TAN learning algorithms have been proposed, among which TAN based on the ChowLiu algorithm [9], simply CL-TAN [8] demonstrates significant improvement over naive Bayes in terms of classification accuracy. In CL-TAN, the class node directly points to all attributes nodes and each attribute node except for the root node of the tree only has one parent from another attribute node.

The SuperParent algorithm is another TAN learning algorithm, simply SP-TAN [10]. SP-TAN is a greedy heuristic search algorithm in which an arc of achieving the highest accuracy improvement is selected in each step. According to their experimental results, SP-TAN significantly outperforms CL-TAN. But, one disadvantage of SP-TAN is its relative higher time complexity.

After this, Zhang and Ling [11] observed that the dependence among attributes tends to cluster into groups in many real-world domains with a large number of attributes. Based on their observation, they presented an improved algorithm simply called StumpNetwork. Their motivation is to enhance the efficiency of SP-TAN while maintaining very similar predictive accuracy.

One unavoidable issue in learning TAN is structure learning. Thus, a model that avoids structure learning, and is still able to represent attribute dependencies to some extent, is desirable. To achieve this, Jiang et al. [12] presented an algorithm called one dependence augmented naive Bayes, simply ODANB. For each one attribute node, ODANB selects the other attribute node with maximum conditional mutual information as its attribute parent.

One main problem confronting ODANB is that at most one attribute parent is allowed for each attribute, ignoring the influences from other attributes. Aiming at overcoming this limitation, Webb et al. presented an algorithm called averaged one-dependence estimators, simply AODE [13]. In AODE, an aggregate of one-dependence classifiers are learned and the prediction is produced by directly averaging the predictions of all these qualified one-dependence classifiers. WAODE [14] is its improved algorithm by assigning different weights to different one-dependence classifiers. HNB [15] presented by Zhang et al. is another example. In HNB, a hidden parent node is created for each attribute node to integrate the influences from all other attributes nodes.

2.3   Improve Naive Bayes Using Local Learning

The basic idea of the local learning approach is that building a naive Bayes on a local training data set not on the whole training data set. Although the attribute conditional independence assumption of naive Bayes is always violated on the whole training data, it is expected that the dependencies within the local training data is weaker than that on the whole training data. Thus, the naive Bayes built on the local training data performance better. In addition, it has been observed that the accuracy of naive Bayes doesn't scale up in some larger databases [16]. This feature makes it especially fit to be a local model embedded into another model, such as a decision tree, a k-nearest neighbor.

For decision tree, naive Bayes tree (simply NBTree) presented by Kohavi [16] is a typical example. It is a hybrid algorithm by combining decision tree with naive Bayes. Learning an NBTree is similar to C4.5 [17] except for its score function of evaluating split attributes. After a tree is grown, a naive Bayes is constructed for each leaf using the data associated with that leaf. NBTree classifies a test instance by sorting it to a leaf and applying the naive Bayes in that leaf to assign a class label to it. Their experimental results proved that NBTree significantly outperforms naive Bayes and C4.5.

Talking of local learning, k-nearest neighbor (simply KNN) is absolutely necessary. The idea of combining KNN with naive Bayes is quite straightforward. Like all lazy learning methods, the training data is simply stored, and learning is deferred until classification time. Whenever a test instance is classified, a local naive Bayes is trained using the k nearest neighbors of the test instance, with which the test instance is classified.

In recently years, researchers have done considerable work on combining KNN with naive Bayes. For example, Frank et al. [18] present an algorithm called locally weighted naive Bayes, simply LWNB. In LWNB, k nearest neighbors of the test instance are firstly found and each of them is weighted in terms of its distance to the test instance. Then a local naive Bayes is built from the locally weighted training instances. Although it is a k-related algorithm, its classification performance is not particularly sensitive to the size of k as long as it is not too small. In ICLNB [19], Jiang et al. presented another method called instance cloning to weight the local training instances.

In fact, two main work before this, LBR [20] and SNNB [21], demonstrate significant improvement over naive Bayes. LBR (lazy Bayesian rule) does not directly use the k nearest neighbors of the test instance as the training data for the local naive Bayes. Instead, before classifying a test instance, LBR generates a rule most appropriate to the test instance. The training instances that satisfy the antecedent of the rule are chosen as the training data for the local naive Bayes, and this local naive Bayes only uses those attributes that do not appear in the antecedent of the rule. In SNNB (selective neighborhood naive Bayes), multiple naive Bayes are learned by using different k values and a local naive Bayes is trained for each k value. The most accurate one is used to classify the test instance. Compared to LWNB, SNNB isn't a k-related algorithm, But it has relative higher time complexity in lazily searching a best k value for each test instance at test time. Responding to its shortcoming, DKNAW is presented by Jiang et al. [22]. In DKNAW, a best k value is eagerly learned for each training data at training time.

2.4   Improve Naive Bayes Using Data Expansion

One practical problem in learning a Bayesian Network classifier is the high variance due to the lack of training data. More precisely, the probability estimates are prone to being unreliable. Intuitively, when the training data are not enough, the underlying distribution would not be clearly reflected. If there are more training data even with the same pattern, the underlying distribution would be reinforced in the training data, and thus could be more easily learned. Thus, adding more instances to the training set, called data expansion, would help the learning algorithm. To draw more training instances, however, the underlying distribution should be known, which is not the case in reality. One way to overcome this issue is to replicate the existing instances.

The basic idea for data expansion is adding some clones of the existing training instances to the training data set. Now, let us firstly define an important concept called *similarity* between two instances with nominal attributes used in data expansion. Assume that an instance x is represented by a vector of attribute-values as follows.

$$< a_1(x), a_2(x), \ldots, a_n(x) >,$$

where $a_i(x)$ is the value of attribute A_i of instance x. Then the *similarity* between two instances x and y is defined as:

$$s(x, y) = \sum_{i=1}^{n} \delta(a_i(x), a_i(y)), \tag{5}$$

where

$$\delta(a_i(x), a_i(y)) = \begin{cases} 1 \ a_i(x) = a_i(y) \\ 0 \ a_i(x) \neq a_i(y). \end{cases} \tag{6}$$

$s(x, y)$ is a simple function that counts the number of identical attributes of x and y and roughly reflects the extent of similarity between two instances.

We proposed two approaches to expand training data. The first one is LNB [23]. Given a training data set $\mathbf{T}$ and a test instance x, LNB firstly uses Equation 5 to calculate the similarity $s(x, y)$ between the test instance x and each training instance y in $\mathbf{T}$. Then, $s(x, y)$ clones of y are added into $\mathbf{T}$ to expand the training data set $\mathbf{T}$. At last, a naive Bayesian classifier is built on this expanded $\mathbf{T}$ to classifies this test instance x.

The second one is IGCNB [24]. Different from LNB, IGCNB conducts a greedy search process to expand the current training data set $\mathbf{T}$ by cloning some instances, on which a naive Bayes is built. Although they all achieve significant improvement over NB, they are lazy learning algorithms. For each test instance, they must acquire an expanded training data set and build a naive Bayes. This leads to a relatively high time complexity with them. Although LNB and IGCNB are originally designed for improving the ranking performance of NB, their classification performance significantly outperforms NB in the same way.

3   Experimental Methodology and Results

We conducted experiments under the framework of Weka[25] to study the effect of four improved approaches on the performance of naive Bayesian classifiers. We ran our experiments on 36 UCI data sets [26] selected by Weka, which represent a wide range of domains and data characteristics listed in Table 1. In our experiments, we adopted the following three preprocessing steps.

1. Replacing missing attribute values: We don't handle missing attribute values. Thus, we used the unsupervised filter named *ReplaceMissingValues* in Weka to replace all missing attribute values in each data set.
2. Discretizing numeric attribute values: We don't handle numeric attribute values. Thus, we used the unsupervised filter named *Discretize* in Weka to discretize all numeric attribute values in each data set.
3. Removing useless attributes: Apparently, if the number of values of an attribute is almost equal to the number of instances in a data set, it is a useless attribute. Thus, we used the unsupervised filter named *Remove* in Weka to remove this type of attributes. In these 36 data sets, there are only three such attributes: the attribute "Hospital Number" in the data set "colic.ORIG", the attribute "instance name" in the data set "splice" and the attribute "animal" in the data set "zoo".

We conduct extensive empirical comparison for naive Bayes and its related improved algorithms ENB, SP-TAN, LWNB (k=50), and LNB in terms of classification accuracy. We use the implementation of NB (NaiveBayes) and LWNB (LWL with NaiveBayes as the basic classifier) in Weka system. We use the implementation of SP-TAN kindly provided by Geoffrey I. Webb. Besides, we implement ENB and LNB in Weka system.

The classification accuracy of each classifier on each data set was obtained via 1 run of 10-fold cross validation. Run with the various algorithms were carried out on the same training sets and evaluated on the same test sets. In particular, the cross-validation folds are the same for all the experiments on each data set. Finally, we compare naive Byes and its each improved algorithm via two-tailed t-test with significantly different probability of 95% [27].

Table 2 shows the classification accuracy and standard deviation of naive Bayes and its each improved algorithm on each data set, and their average values are summarized at the bottom of the tables. The symbols v and * in the tables respectively denotes statistically significant improvement and degradation over naive Bayes with a 95% confidence level. Our experiments show that all algorithms using these four improved approaches significantly outperform naive Bayes. Now, we summarize the highlights briefly as follows:

1. ENB significantly outperforms NB. In the 36 data sets we test, ENB wins in 9 data sets, surprisingly loses in 0 data sets, and almost ties all the other data sets. ENB's average accuracy is 83.22%, much higher than that of NB (82.41%). This fact proves that feature selection is an effective approach to improve NB.

2. SP-TAN significantly outperforms NB. In the 36 data sets we test, SP-TAN wins in 11 data sets, surprisingly loses in 0 data sets, and almost ties all the other data sets. SP-TAN's average accuracy is 84.76%, much higher than that of NB (82.41%). This fact proves that structure extension is an effective approach to improve NB.

Table 1. Description of data sets used in the experiments. All these data sets are the whole 36 UCI data sets selected by Weka. We downloaded these data sets in format of arff from main web of Weka.

No.	Dataset	Instances	Attributes	Classes	Missing	Numeric
1	anneal	898	39	6	Y	Y
2	anneal.ORIG	898	39	6	Y	Y
3	audiology	226	70	24	Y	N
4	autos	205	26	7	Y	Y
5	balance-scale	625	5	3	N	Y
6	breast-cancer	286	10	2	Y	N
7	breast-w	699	10	2	Y	N
8	colic	368	23	2	Y	Y
9	colic.ORIG	368	28	2	Y	Y
10	credit-a	690	16	2	Y	Y
11	credit-g	1000	21	2	N	Y
12	diabetes	768	9	2	N	Y
13	Glass	214	10	7	N	Y
14	heart-c	303	14	5	Y	Y
15	heart-h	294	14	5	Y	Y
16	heart-statlog	270	14	2	N	Y
17	hepatitis	155	20	2	Y	Y
18	hypothyroid	3772	30	4	Y	Y
19	ionosphere	351	35	2	N	Y
20	iris	150	5	3	N	Y
21	kr-vs-kp	3196	37	2	N	N
22	labor	57	17	2	Y	Y
23	letter	20000	17	26	N	Y
24	lymph	148	19	4	N	Y
25	mushroom	8124	23	2	Y	N
26	primary-tumor	339	18	21	Y	N
27	segment	2310	20	7	N	Y
28	sick	3772	30	2	Y	Y
29	sonar	208	61	2	N	Y
30	soybean	683	36	19	Y	N
31	splice	3190	62	3	N	N
32	vehicle	846	19	4	N	Y
33	vote	435	17	2	Y	N
34	vowel	990	14	11	N	Y
35	waveform-5000	5000	41	3	N	Y
36	zoo	101	18	7	N	Y

Table 2. The detailed experimental results on classification accuracy and standard deviation. NB: Naive Bayes; ENB: Evolutional Naive Bayes; SP-TAN: TAN based on the SuperParent algorithm; LWNB: Locally Weighted Naive Bayes (k=50); LNB: Lazy Naive Bayes. v, * : statistically significant improvement or degradation over naive Bayes with a 95% confidence level.

Datasets	NB	ENB	SP-TAN	LWNB	LNB
anneal	94.32±2.38	97.44±2.29 v	97.44±1.82 v	98.89±0.74 v	97.44±1.58 v
anneal.ORIG	87.53±4.69	89.2±3.44	89.08±3.72	92.43±3.35	88.53±3.05
audiology	71.23±7.03	72.92±7.42	71.66±7.34	77.83±6.39 v	78.32±7.12
autos	64.83±11.18	74.57±8.73 v	75.50±8.99 v	79.90±5.22 v	74.57±10.6 v
balance-scale	91.36±1.38	91.36±1.38	91.36±1.38	85.44±3.66 *	91.04±1.55
breast-cancer	72.06±7.97	70.67±8.77	71.01±9.97	73.18±9.42	71.74±8.74
breast-w	97.28±1.84	96.99±1.85	97.42±1.76	96.71±2.04	97.42±1.89
colic	78.81±5.05	80.69±8.15	79.63±4.58	80.15±2.67	80.99±5.8
colic.ORIG	75.26±5.26	75.81±6.45	75.8±6	75.26±4.55	75.81±5.19
credit-a	84.78±4.28	86.09±3.94	84.78±4.06	85.51±3.13	84.93±3.94
credit-g	76.3±4.76	76±3.46	75.3±4.24	73.6±3.63	76.8±4.18
diabetes	75.4±5.85	77.62±4.92	75.79±5.71	72.53±5.12	75.78±5.94
glass	60.32±9.69	61.26±7.33	62.19±10.5	64.05±7.75	59.85±7.39
heart-c	84.14±4.16	83.46±6.71	82.45±6.38	79.16±6.01 *	81.48±6.21
heart-h	84.05±6.69	79.67±8.13	84.05±5.85	83.38±6.36	84.05±4.4
heart-statlog	83.7±5	80.74±5.18	82.96±4.35	81.85±5.08	82.22±5.47
hepatitis	83.79±8.79	81.71±9.47	83.79±8.21	85.08±6.2	85.67±7.57
hypothyroid	92.79±1.02	93.43±0.62	93.35±1.03 v	93.29±0.7	92.84±0.88
ionosphere	90.89±3.49	89.45±4.07	90.32±3.34	92.32±2.98	91.44±3.82
iris	94.67±8.2	96.67±4.71	93.33±7.7	94.67±5.26	96.67±4.71
kr-vs-kp	87.89±1.81	94.65±1.38 v	95.50±1.37 v	97.87±0.73 v	88.67±1.64 v
labor	93.33±11.65	91±12.38	93.33±11.7	91.67±14.2	90±14.05
letter	70±0.81	70.59±1.00 v	83.62±1.05 v	92.74±0.48 v	78.22±0.96 v
lymph	85.67±9.55	83.05±10.7	86.38±7.99	83.67±10.8	86.33±8.8
mushroom	95.57±0.45	99.57±0.37 v	99.91±0.08 v	100.0±0.00 v	99.37±0.24 v
primary-tumor	46.89±4.32	43.32±7.91	47.5±3.25	44.49±6.62	47.5±4.9
segment	88.92±1.95	91.00±2.06 v	94.37±1.21 v	95.19±1.05 v	91.26±1.67 v
sick	96.74±0.53	97.51±0.72 v	97.61±0.84 v	98.17±0.38 v	97.08±0.54
sonar	77.5±11.99	78.43±9.48	76.07±11.1	83.26±8.39	79.4±8.88
soybean	92.08±2.34	92.37±2.31	93.11±2.43	93.7±3.11	94.43±2.29
splice	95.36±1	96.11±1.36	95.45±1.18	94.36±1.4	95.86±0.87
vehicle	61.82±3.54	61.47±4.95	70.10±2.75 v	71.99±4.61 v	67.86±4.73 v
vote	90.14±4.17	95.63±3.33 v	92.43±3.58	96.10±3.05 v	90.82±3.72
vowel	67.07±4.21	68.18±3.82	92.42±3.27 v	94.75±1.49 v	87.68±2.22 v
waveform-5000	79.96±1.92	81.06±1.78 v	82.18±1.78 v	78.5±1.89	81.52±1.74 v
zoo	94.18±6.6	96.18±6.54	94.18±6.6	96.18±6.54	96.18±6.54
Mean	82.41±4.88	83.22±4.92	84.76±4.64	85.50±4.30	84.44±4.55

3. LWNB significantly outperforms NB. In the 36 data sets we test, LWNB wins in 11 data sets, surprisingly loses in 2 data sets, and almost ties all the other data sets. LWNB's average accuracy is 85.50%, much higher than that

of NB (82.41%). This fact proves that local learning is an effective approach to improve NB.

4. LNB significantly outperforms NB. In the 36 data sets we test, LNB wins in 9 data sets, surprisingly loses in 0 data sets, and almost ties all the other data sets. LNB's average accuracy is 84.44%, much higher than that of NB (82.41%). This fact proves that data expansion is an effective approach to improve NB.

4   Conclusions and Future Work

In this paper, we single out four main approaches for improving naive Bayes' classification accuracy: 1) Feature selection; 2) Structure extension; 3) Local learning; 4) Data expansion. Keeping to these approaches, we review some of its improved algorithms and experimentally tested these algorithms using the whole 36 UCI data sets selected by Weka, and compared them to naive Bayes. The experimental results show that all these improved algorithms significantly outperform naive Bayes. Thus, we can conclude that all these improved approaches are effective.

Bayesian network has been widely used in various data mining applications as an effective classification model. Learning (or improving) it for accurate classification is one of the main research directions, but not all [28,29], because an accurate class probability estimation [30,31,32] and ranking [33,34,35] also is desirable in many real-world data mining applications. Thus, investigating and improving its performance of class probability estimation and ranking is one of the main direction for our future work.

Acknowledgements

Many thanks to Geoffrey I. Webb for kindly providing us with the implementation of SP-TAN.

References

1. Langley, P., Iba, W., Thomas, K.: An analysis of Bayesian classifiers. In: Proceedings of the Tenth National Conference of Artificial Intelligence, pp. 223–228. AAAI Press, Stanford (1992)
2. Pearl, J.: Probabilistic Reasoning in Intelligent Systems. Morgan Kaufmann, San Francisco, CA (1988)
3. Chickering, D.M.: Learning Bayesian networks is NP-Complete. In: Fisher, D., Lenz, H. (eds.) Learning from Data: Artificial Intelligence and Statistics V, pp. 121–130. Springer, Heidelberg (1996)
4. Langley, P., Sage, S.: Induction of selective Bayesian classifiers. In: Proceedings of the Tenth Conference on Uncertainty in Artificial Intelligence, pp. 339-406 (1994)
5. Jiang, L., Zhang, H., Cai, Z., Su, J.: Evolutional Naive Bayes. In: Proceedings of the 1st International Symposium on Intelligent Computation and its Applications, ISICA, China University of Geosciences Press, pp.344–350 (2005)

6. Kohavi, R., John, G.: Wrappers for Feature Subset Selection. Artificial Intelligence journal, special issue on relevance 97(1-2), 273–324 (1997)
7. Ratanamahatana, C.A., Gunopulos, D.: Scaling up the Naive Bayesian Classifier: Using Decision Trees for Feature Selection. In: proceedings of Workshop on Data Cleaning and Preprocessing (DCAP 2002), at IEEE International Conference on Data Mining (ICDM 2002), Maebashi, Japan (2002)
8. Friedman, Geiger, Goldszmidt.: Bayesian Network Classifiers. Machine Learning 29, 131–163 (1997)
9. Chow, C.K., Liu, C.N.: Approximating discrete probability distributions with dependence trees. IEEE Trans. on Information Theory 14, 462C–467 (1968)
10. Keogh, E., Pazzani, M.: Learning augmented Bayesian classifiers: A comparison of distribution-based and classification-based approaches. In: Proceedings of the International Workshop on Artificial Intelligence and Statistics, pp. 225C–230 (1999)
11. Zhang, H., Ling, C.X.: An improved learning algorithm for augmented naive Bayes. In: Cheung, D., Williams, G.J., Li, Q. (eds.) PAKDD 2001. LNCS (LNAI), vol. 2035, pp. 581–586. Springer, Heidelberg (2001)
12. Jiang, L., Zhang, H., Cai, Z., Su, J.: One Dependence Augmented Naive Bayes. In: Li, X., Wang, S., Dong, Z.Y. (eds.) ADMA 2005. LNCS (LNAI), vol. 3584, pp. 186–194. Springer, Heidelberg (2005)
13. Webb, G.I., Boughton, J., Wang, Z.: Not so naive bayes: Aggregating one-dependence estimators. Machine Learning 58, 5–24 (2005)
14. Jiang, L., Zhang, H.: Weightily Averaged One-Dependence Estimators. In: Yang, Q., Webb, G. (eds.) PRICAI 2006. LNCS (LNAI), vol. 4099, pp. 970–974. Springer, Heidelberg (2006)
15. Zhang, H., Jiang, L., Su, J.: Hidden Naive Bayes. In: Proceedings of the 20th National Conference on Artificial Intelligence, AAAI 2005, pp. 919–924. AAAI Press, Stanford (2005)
16. Kohavi, R.: Scaling Up the Accuracy of Naive-Bayes Classifiers: A Decision-Tree Hybrid. In: Proceedings of the Second International Conference on Knowledge Discovery and Data Mining (KDD-96), pp. 202–207. AAAI Press, Stanford (1996)
17. Quinlan, J.R.: C4.5: Programs for Machine Learning. Morgan Kaufmann, San Mateo, CA (1993)
18. Frank, E., Hall, M., Pfahringer, B.: Locally Weighted Naive Bayes. In: Proceedings of the Conference on Uncertainty in Artificial Intelligence (2003), pp. 249–256. Morgan Kaufmann, Seattle (2003)
19. Jiang, L., Zhang, H., Su, J.: Instance Cloning Local Naive Bayes. In: Kégl, B., Lapalme, G. (eds.) Canadian AI 2005. LNCS (LNAI), vol. 3501, pp. 280–291. Springer, Heidelberg (2005)
20. Zheng, Z., Webb, G.I.: Lazy Learning of Bayesian Rules. Machine Learning 41(1), 53–84 (2000)
21. Xie, Z., Hsu, W., Liu, Z., Lee, M.: A Selective Neighborhood Based Naive Bayes for Lazy Learning. In: Chen, M.-S., Yu, P.S., Liu, B. (eds.) PAKDD 2002. LNCS (LNAI), vol. 2336, pp. 104–114. Springer, Heidelberg (2002)
22. Jiang, L., Zhang, H., Cai, Z.: Dynamic K-Nearest-Neighbor Naive Bayes with Attribute Weighted. In: Wang, L., Jiao, L., Shi, G., Li, X., Liu, J. (eds.) FSKD 2006. LNCS (LNAI), vol. 4223, pp. 365–368. Springer, Heidelberg (2006)
23. Jiang, L., Guo, Y.: Learning Lazy Naive Bayesian Classifiers for Ranking. In: Proceedings of the 17th IEEE International Conference on Tools with Artificial Intelligence, ICTAI 2005, pp. 412–416. IEEE Computer Society Press, Los Alamitos (2005)

24. Jiang, L., Zhang, H.: Learning Instance Greedily Cloning Naive Bayes for Ranking. In: Proceedings of the 5th IEEE International Conference on Data Mining, ICDM 2005, pp. 202–209. IEEE Computer Society Press, Los Alamitos (2005)

25. Witten, I.H., Frank, E.: Data Mining: Practical machine learning tools and techniques, 2nd edn. Morgan Kaufmann, San Francisco (2005), http://prdownloads.sourceforge.net/weka/datasets-UCI.jar

26. Merz, C., Murphy, P., Aha, D.: UCI repository of machine learning databases. In: Dept of ICS, University of California, Irvine (1997), http://www.ics.uci.edu/mlearn/MLRepository.html

27. Nadeau, C., Bengio, Y.: Inference for the generalization error. In: Advances in Neural Information Processing Systems 12, pp. 307–313. MIT Press, Cambridge (1999)

28. Hand, D.J., Till, R.J.: A simple generalisation of the area under the ROC curve for multiple class classification problems. Machine Learning 45, 171–186 (2001)

29. Ling, C.X., Huang, J., Zhang, H.: AUC: a statistically consistent and more discriminating measure than accuracy. In: Proceedings of the International Joint Conference on Artificial Intelligence IJCAI03, Morgan Kaufmann, San Francisco (2003)

30. Lowd, D., Domingos, P.: Naive Bayes Models for Probability Estimation. In: Proceedings of the Twenty-Second International Conference on Machine Learning, pp. 529–536. ACM Press, New York (2005)

31. Jiang, L., Zhang, H.: Learning Naive Bayes for Probability Estimation by Feature Selection. In: Lamontagne, L., Marchand, M. (eds.) Canadian AI 2006. LNCS (LNAI), vol. 4013, pp. 503–514. Springer, Heidelberg (2006)

32. Grossman, D., Domingos, P.: Learning Bayesian Network Classifiers by Maximizing Conditional Likelihood. In: Proceedings of the Twenty-First International Conference on Machine Learning, pp. 361–368. ACM Press, Banff, Canada (2004)

33. Zhang, H., Su, J.: Naive Bayesian classifiers for ranking. In: Boulicaut, J.-F., Esposito, F., Giannotti, F., Pedreschi, D. (eds.) ECML 2004. LNCS (LNAI), vol. 3201, pp. 501–512. Springer, Heidelberg (2004)

34. Zhang, H., Jiang, L., Su, J.: Augmenting Naive Bayes for Ranking. In: Proceedings of the 22nd International Conference on Machine Learning, ICML 2005, pp. 1025–1032. ACM, New York (2005)

35. Jiang, L., Zhang, H., Cai, Z.: Discriminatively Improving Naive Bayes by Evolutionary Feature Selection. Romanian Journal of Information Science and Technology 9(3), 163–174 (2006)

Privacy Preserving BIRCH Algorithm for Clustering over Arbitrarily Partitioned Databases

P. Krishna Prasad and C. Pandu Rangan

Department of Computer Science and Engineering
Indian Institute of Technology - Madras
Chennai - 600036, India
pkp@cse.iitm.ernet.in,
rangan@iitm.ernet.in

Abstract. BIRCH algorithm [22] is a well known algorithm for clustering for effectively computing clusters in a large data set. As the data is typically distributed over several sites, clustering over distributed data is an important problem. The data can be distributed in horizontal, vertical or arbitrarily partitioned databases. But, because of privacy issues no party may share its data to other parties. The problem is how the parties can cluster the distributed data without breaching privacy of others data. The solutions in arbitrarily partitioned database setting generally work for both horizontal and vertically partitioned databases. In our work we give a procedure for securely running BIRCH algorithm over arbitrarily partitioned database. We introduce secure protocols for distance metrics and give a procedure for using these metrics in securely computing clusters over arbitrarily partitioned database.

1 Introduction

Sharing of data and carrying out collaborative data mining has emerged as a powerful tool for analysis for mutual profitability among several related business houses. However, the sharing of data in collaboration has raised number of ethical issues like privacy, data security, and intellectual property rights [6,17]. From a philosophical point of view, Schoeman [18] and Walters [19] gave three possible definitions for privacy: 1) Privacy as the right of a person to determine which personal information about himself/herself may be communicated to others 2) Privacy as the control over access to information about oneself 3) Privacy as limited access to a person and to all the features related to the person. There are privacy rules and regulations like HIPAA, GLBA and SOX [12] that restrict the companies to share their data in raw form to any other parties.

Privacy preserving data mining introduced by Agarwal and Srikant [1] and Lindell and Pinkas [10], arose as a solution to this problem by allowing parties to cooperate in the extraction of knowledge without any of the cooperating parties having to reveal their individual data items to any other party.

R. Alhajj et al. (Eds.): ADMA 2007, LNAI 4632, pp. 146–157, 2007.
© Springer-Verlag Berlin Heidelberg 2007

Data that is of interest is considered to be partitioned to the following three ways: 1) Horizontally partitioned 2) vertically partitioned 3) arbitrarily partitioned [7]

We provide a privacy preserving solution to an important data mining problem- namely, clustering of data and this problem was addressed by Oliveira and Zaiane [13] and Vaidya and Clifton [15]. We consider arbitrarily partitioned database in our work, as this is the generic partition where the solutions in this setting are applicable to both horizontally as well as vertically partitioned setting.

2   Related Work

Oliveira and Zaiane [13] gave a privacy preserving solution for clustering based on perturbation. In their work they used geometric data transformation methods for perturbing the data and they give solution for hierarchical and partition based clustering on perturbed data.

Based on secure multi party computation approach, we have solutions 1) solutions for horizontally partitioned databases 2) solutions for vertically partitioned databases and 3) solutions for arbitrarily partitioned databases.

In horizontally partitioned databases: Jha et al., [9] proposed a solution,called privacy preserving $k - means$ algorithm. The crucial step in their algorithm is computing mean vectors. They used oblivious transfer [11] and homomorphic encryption scheme [5,14] for computing these vectors. The claim is that, by knowing only mean vectors neither party can recompute individual vectors. Jagannathan et al., gave a solution known as *recluster*, their solution uses divide, conquer and merge strategy along with some cryptographic primitives like homomorphic encryption scheme [5,14], secure scalar product protocol [2] and Yao's circuit evaluation protocol [21] to compute cluster centers securely.

In vertically partitioned databases: Vaidya and Clifton [15] proposed a solution, called *privacy preserving $k - means$ clustering*. Their approach is based on secure multi party computation approach and they used cryptographic primitives like secure scalar product protocol [21] and Yao circuit evaluation protocol for readjusting mean vectors securely. Prasad and Rangan [16] proposed a solution, called *privacy preserving BIRCH*. Their algorithm works for vertically partitioned, large databases. They used cryptographic primitives like secure scalar product protocol [2], protocol for Yao's millionaire problem [4,20], secure min index in vector sum protocol [16] and threshold sum protocol [16] for computing clusters securely.

In arbitrarily partitioned databases: Jagannathan and Wright [7] introduced the notion of arbitrarily partitioned databases and gave a privacy preserving solution of $k - means$ clustering. They used secure scalar product protocol [2] and Yao's circuit evaluation protocol [21] for computing mean vectors of the clusters securely. Their solution suffers from the facts of limitation of $k - means$ algorithm and computational overhead of Yao's circuit evaluation protocol.

3   Secure BIRCH Algorithm

3.1   Arbitrarily Partitioned Databases

In the two party distributed data setting, two parties (Alice and Bob) hold data forming a (virtual) database consisting of their joint data. More specifically, the virtual database $DB = t_1, t_2, \cdots, t_n$ consists of n objects, or records. Each object t_i is described by the values of m attributes of object t_i by $(x_{i_1}, x_{i_2}, \cdots, x_{i_m})$. The share of Alice database for the virtual database is represented by DB_A and Bob's share for the virtual database is DB_B

[7] introduced the concept of arbitrarily partitioned data. In arbitrarily partitioned data, there may not exist a simple pattern of how data is shared between the parties. For each t_i, Alice knows the values for a subset of attributes, and Bob knows the values for the remaining attributes. That is, each t_i is partitioned into disjoint subsets t_{i_A} and t_{i_B} such that Alice knows t_{i_A} and Bob knows t_{i_B}.

3.2   Problem Statement

The problem of clustering the arbitrarily partitioned database is defined as follows:

Definition 1. *Given an arbitrarily partitioned database $DB = DB_A \# DB_B = \{t_{1_A} \# t_{1_B}, t_{2_A} \# t_{2_B}, \cdots, t_{n_A} \# t_{n_B}\}$ of tuples and an integer value k, the clustering problem in arbitrarily partitioned database is to define a mapping $f : DB_A \# DB_B \rightarrow \{1, 2, \cdots k\}$ where each $t_{i_A} \# t_{i_B}$ is assigned to one cluster $K_j, 1 \leq j \leq k$. This mapping should be done by both the parties, without A knowing DB_B and without B knowing DB_A. Cluster K_j, contains precisely those tuples mapped to it; that is, $K_j = \{t_{i_A} \# t_{i_B} | f(t_{i_A} \# t_{i_B}) = K_j, 1 \leq i \leq n$ and $t_{i_A} \# t_{i_B} \in DB_A \# DB_B$.*

3.3   Cluster Features

The concept of cluster feature (CF) is at the core of the BIRCH's algorithm for computing clusters [22]. The CF is a triple, $\langle N, \overrightarrow{LS}, SS \rangle$, where N is the number of data points in a cluster, $\overrightarrow{LS}$ is the linear sum of N data points and SS is the square sum of the N data points. CF summarizes the information that we maintain in a cluster. The clustering feature (CF) vector for arbitrarily partitioned database is defined as follows:

Definition 2. *Given N $m-$ dimensional arbitrarily partitioned data points of a cluster: $\{\overrightarrow{X}_{i_A} \# \overrightarrow{X}_{i_B}\}$ where $1 \leq i \leq N$, and $\overrightarrow{X}_{i_A} \# \overrightarrow{X}_{i_B} = \overrightarrow{X}_i$, the arbitrarily partitioned clustering feature is defined as a triple: $CF_A \# CF_B = (N, \overrightarrow{LS}_A + \overrightarrow{LS}_B, SS_A + SS_B)$, where $\overrightarrow{LS}_A$ is the linear sum and SS_A is the square sum of*

N data points at party A, similarly $\overrightarrow{LS}_B$ is the linear sum and SS_B is the square sum of the corresponding N data points at party B.

Based on similarity measure two clusters are merged by adding their corresponding cluster features. Theorem for adding two arbitrarily partitioned clusters is stated as follows:

Theorem 1. *(CF additivity theorem for arbitrarily partitioned data): Assume that $CF_{1_A} \# CF_{1_B} = (N_1, \overrightarrow{LS}_{1_A} + \overrightarrow{LS}_{1_B}, SS_{1_A} + SS_{1_B})$ and $CF_{2_A} \# CF_{2_B} = (N_2, \overrightarrow{LS}_{2_A} + \overrightarrow{LS}_{2_B}, SS_{2_A} + SS_{2_B})$ are the CF vectors of two clusters. Then the CF vector of the two disjoint clusters is: $CF_1 + CF_2 = (CF_{1_A} \# CF_{1_B}) + (CF_{2_A} \# CF_{2_B}) = (CF_{1_A} + CF_{2_A}) \# (CF_{1_B} + CF_{2_B})$.*

3.4 Secure Distance Metrics

By using CFs of the clusters one can compute distance metrics like radius R of a cluster, diameter D of a cluster and average inter cluster distance D_2 [22]. Given below are the distance metrics for arbitrarily partitioned database case with partitioned cluster features.

$$R = \left(\frac{SS_A - \frac{1}{N}(LS_A LS_A^T)}{N} - \frac{\frac{2}{N}(LS_A LS_B^T)}{N} + \frac{SS_B - \frac{1}{N}(LS_B LS_B^T)}{N} \right)^{\frac{1}{2}}$$

$$D = \left(\frac{2NSS_A - 2(LS_A LS_A^T)}{N(N-1)} - \frac{4(LS_A LS_B^T)}{N(N-1)} + \frac{2NSS_B - 2(LS_B LS_B^T)}{N(N-1)} \right)^{\frac{1}{2}}$$

$$D_2 = \left(\frac{N_2 SS_{1_A} + N_1 SS_{2_A} - 2LS_{1_A} LS_{2_A}^T}{N_1 N_2} - \frac{2(LS_{1_A} LS_{2_B}^T + LS_{1_B} LS_{2_A}^T)}{N_1 N_2} + \right.$$
$$\left. \frac{N_2 SS_{1_B} + N_1 SS_{2_B} - 2LS_{1_B} LS_{2_B}^T}{N_1 N_2} \right)^{\frac{1}{2}}$$

Given below are the protocols for securely computing the squares of the above mentioned distance metrics for arbitrarily partitioned databases. At the end of the execution of these protocols each party will have random shares for those distance metrics. The combination of those random shares is the exact square of the distance metrics.

Protocol for computing the radius R of a cluster

1. Alice computes $a = \frac{SS_A - \frac{1}{N}(LS_A LS_A^T)}{N}$
2. Bob computes $b = \frac{SS_B - \frac{1}{N}(LS_B LS_B^T)}{N}$
3. They both invoke the secure scalar product protocol[2] with private inputs LS_A from Alice and LS_B from Bob to get the random shares for $s = \frac{\frac{2}{N}(LS_A LS_B^T)}{N}$. Let the random shares be s_a and s_b, where $s_a + s_b = s$.
4. Now Alice has the share $a - s_a$ and Bob has the share $b - s_b$, where $a - s_a + b - s_b = R^2$

Protocol for computing diameter D of a cluster
1. Alice computes $a = \frac{2NSS_A - 2(LS_A LS_A^T)}{N(N-1)}$.
2. Bob computes $b = \frac{2NSS_B - 2(LS_B LS_B^T)}{N(N-1)}$.
3. They both invoke the secure scalar product protocol [2] with private inputs LS_A from Alice and LS_B from Bob to get the random shares for $s = \frac{4(LS_A LS_B^T)}{N(N-1)}$. Let the random shares be s_a and s_b, where $s_a + s_b = s$.
4. Now Alice has the share $a - s_a$ and Bob has the share $b - s_b$ where,
$a - s_a + b - s_b = D^2$

Protocol for computing D_2
1. Alice computes $a = \frac{N_2 SS_{1_A} + N_1 SS_{2_A} - 2LS_{1_A} LS_{2_A}^T}{N_1 N_2}$
2. Bob computes $b = \frac{N_2 SS_{1_B} + N_1 SS_{2_B} - 2LS_{1_B} LS_{2_B}^T}{N_1 N_2}$
3. They both invoke secure scalar product protocol [2] with private inputs LS_{1_A}, LS_{2_A} from Alice and LS_{1_B}, LS_{2_B} from Bob to get the random shares for $s = \frac{2(LS_{1_A} LS_{2_B}^T + LS_{1_B} LS_{2_A}^T)}{N_1 N_2}$. Let the random shares be s_a and s_b, where $s_a + s_b = s$.
4. Now Alice has the share $a - s_a$ and Bob has the share $b - s_b$, where
$a - s_a + b - s_b = D_2^2$

The above protocols for secure distance metrics computes the squares of radius R, diameter D and inter cluster distance D_2 correctly and securely. The main step in the above protocols is the invocation of secure scalar product protocol [2]. As this protocol is proved to be correct, the above secure distance metrics gives the same distance metrics as in the centralized database case. As secure scalar product protocol [2] is proved to be secure against semi honest adversaries, by composition theorem, these protocols for secure distance metrics are secure against semi honest adversaries. In our discussion we will use the notation $distance^2(CF_{i_A}\#CF_{i_B}, CF_{j_A}\#CF_{j_B})$ for invoking the protocol for D_2 to compute inter cluster distance between the clusters $CF_{i_A}\#CF_{i_B}$ and $CF_{j_A}\#CF_{j_B}$.

3.5   Secure Construction of CF Tree

In arbitrarily partitioned database case, each party learns the structure of the CF tree and the entries of the tree are the corresponding shares of CFs of the virtual database. That is each party in its memory will have its share of CF based on its input database in the node. As in centralized database case a node in CF tree has to be split for several times. Given the shares of entries of a node in a CF tree, Algorithm 1 demonstrates how to split a node securely in arbitrarily partitioned database.

Secure CF Tree insertion
We now give the procedure for how to insert a new Ent securely in the CF tree in the partitioned case. The secure CF Tree insertion procedure is similar to

the insertion procedure in single data base case but with invocation of secure protocols. Here the new entry Ent is partitioned into Ent_A and Ent_B where $Ent = Ent_A \# Ent_B$ where A has Ent_A and B has Ent_B.

Both the parties learn the tree structure but the entries in the tree will be different for each party. Alice inserts its partitioned share of CF in its tree and at the same time party Bob inserts its partitioned share of CF in its tree at the corresponding node in its tree. Given below is the algorithm for inserting a new partitioned entry in to the CF tree securely.

1. *Identifying the appropriate leaf:* (1) apply $dist^2()$ for $Ent_A \# Ent_B$ to the entries of the current node to get the shares of distances for Alice as $X = \{a_1, a_2, \cdots, a_p\}$ and for Bob as $Y = \{b_1, b_2, \cdots, b_p\}$ (2) invoke *secure min index in vector sum protocol* with the input X by Alice and with the input Y by Bob to get the index **j** and (3) both the parties share the index **j** and they descend through the corresponding branch from the current node at their respective trees.

2. *Modifying the leaf:* When the new entry $Ent_A \# Ent_B$ reaches the leaf node, it finds the closest leaf entry say $L_{i_A} \# L_{i_B}$ by using the same procedure described in the Step 1. The threshold violating condition is checked by the following procedure:

   (a) Let $L_{inew_A} \# L_{inew_B} = L_{i_A} \# L_{i_B} + Ent_A \# Ent_B$. A has L_{inew_A} and B has L_{inew_B}.
   (b) Invoke secure protocol for diameter for computing D^2 and let the shares of D^2 for Alice and Bob be **a** and **b** respectively.
   (c) Invoke *secure threshold sum protocol [16]* with private inputs a and b and the threshold T^2 to check securely if $a + b$ is greater than T^2.
   (d) If $D \leq T$, then $L_{i_A} \# L_{i_B} \leftarrow L_{inew_A} \# L_{inew_B}$. That is Alice updates L_{i_A} with L_{inew_A} and Bob updates L_{i_B} with L_{inew_B}. If $D > T$, then a new entry for $Ent_A \# Ent_B$ is added to the leaf. If there is a space for this new entry, we are done, otherwise we have to split the leaf node. Splitting is done by invoking the algorithm 1 with the shares of leaf entries as private inputs.

3. *Modifying the path to the leaf:* After inserting $"Ent_A \# Ent_B''$ into a leaf, each party must update the CF information for each non leaf entry on the path to the leaf. In absence of split it simply involves adding partitioned CF vector at each party in its tree to reflect the addition of $Ent_A \# Ent_B$. If there is a leaf split, each party inserts a new non leaf entry in its parent node. If there is space in the parent node for this newly created leaf, at all higher levels each party only need to update the shared CF vectors to reflect the addition of $Ent_A \# Ent_B$. That is Alice updates its share of CF vectors in its tree and Bob updates its share of CF vectors in its tree. If there is no space for the new leaf in its parent, each party should split its parent as it was done to split the leaf node and this change should be propagated up to root. If root is split, the tree height increases by one.

4. *A Merging Refinement:* Merging refinement has three main steps,
    (a) Identify node N_t such that the propagation of split of leaf stops.
    (b) Find the two closest entries in N_t as follows:

    Each party calculates its shares of distances for all the pairs of CFs in the node N_t: $\{dist^2(CF_{i_A}\#CF_{i_B}, CF_{j_A}\#CF_{j_B})\}$ for $i \neq j$. If there are p such pairs, set of shares from Alice is $X = \{a_1, a_2, \cdots, a_p\}$ and the corresponding set of shares from Bob is $Y = \{b_1, b_2, \cdots, b_p\}$ where $a_i + b_i = dist^2()$ for the ith pair. Invoke secure min index in vector sum protocol [16] to compute the minimum distance index and let the corresponding pair be $(CF_{m_A}\#CF_{m_B}, CF_{n_A}\#CF_{n_B})$.
    (c) If the pair $(CF_{m_A}\#CF_{m_B}, CF_{n_A}\#CF_{n_B})$ is not the pair corresponding to the split, merge them and corresponding child nodes (Each party can merge its share of CF entries in its corresponding tree). If there are more entries in the two child nodes than one page can hold, split the merging result again by using the Algorithm 1.

Algorithm 1. SecureNodeSplit

1: {/*Input : Two private vectors of partitioned CFs of size p, $X = \{CF_{1_A}, CF_{2_A}, \cdots, CF_{p_A}\}$ and $Y = \{CF_{1_B}, CF_{2_B}, \cdots, CF_{p_B}\}$ by Alice and Bob respectively, where $CF_{i_A}\#CF_{i_B} = CF_i*/\}$

2: {/*Output: Two sets $node_1$ and $node_2$ of CFs based on minimum distance criteria*/}

3: Calculate   distance   for   each   pair   of CFs   Get   shares   of $dist^2(CF_{i_A}\#CF_{i_B}, CF_{j_A}\#CF_{j_B})$ for $i = 1, 2, \cdots, p$, for $j = 1, 2, \cdots, p$ and $i \neq j$. The shares for all the distance pairs for Alice is $S_X = \{a_1, a_2, \cdots, a_{((p-1)p)/2}\}$ and the corresponding shares of distances for Bob is $S_Y = \{b_1, b_2, \cdots, b_{((p-1)p)/2}\}$.

4: Invoke *secure min index in vector sum protocol [16]* with maximum index as the criteria instead of minimum index for the private input S_X from Alice and for the private input S_Y from Bob to get the index. Let $(CF_{m_A}\#CF_{m_B}, CF_{n_A}\#CF_{n_B})$ be the pair of partitioned CFs corresponding to the index. Therefore, $CF_{m_A}\#CF_{m_B}$ and $CF_{n_A}\#CF_{n_B}$ are the two farthest clusters in the input.

5: **for** $i = 1, 2, \cdots, p$, **do**

6:    Get the shares for Alice and Bob for $dist^2(CF_{i_A}\#CF_{i_B}, CF_{m_A}\#CF_{m_B})$. Let the share of Alice be a_1 and the share of Bob be b_1. (ie., $a_1 + b_1 = dist^2(CF_{i_A}\#CF_{i_B}, CF_{m_A}\#CF_{m_B})$).

7:    Get the shares for party A and party B for $dist^2(CF_{i_A}\#CF_{i_B}, CF_{n_A}\#CF_{n_B})$. Let the share for party A be a_2 and the share for party B be b_2. (ie., $a_2 + b_2 = dist^2(CF_{i_A}\#CF_{i_B}, CF_{n_A}\#CF_{n_B})$).

8:    Now Alice has $S = \{a_1, a_2\}$ and Bob has $T = \{b_1, b_2\}$. Now invoke *protocol for secure min index in vector sum [16]* with private inputs S and T. If the index is 1 put $CF_{i_A}\#CF_{i_B}$ in the set $node1$ (ie., put CF_{i_A} in A's $node1$ and CF_{i_B} in B's $node1$) else put $CF_{i_A}\#CF_{i_B}$ in $node2$ (ie., put CF_{i_A} in A's $node2$ and CF_{i_B} in B's $node2$).

9: **end for**

Algorithm 2 describes the computation of clusters securely in arbitrarily partitioned database.

Algorithm 2. SecureBIRCH

1: *Phase 0:* Alice and Bob agree on the parameters $(\mathcal{B}, T, L)$ where $\mathcal{B}$ is the branching factor, T is the threshold factor and L is the maximum number of leaf entries in a leaf.

2: *Phase 1:* Load into each party's memory the CF tree of shared CF's that are formed by using the procedure **secure CF tree insertion** given in Section 3.5.

3: *Phase 2:* Condense the CF tree. This phase is same as the phase 1. Each party scans the leaf entries in the initial CF tree to rebuild a smaller CF tree by using the procedure **secure CF tree insertion** given in Section 3.5.

4: *Phase 3:* Use hierarchical clustering such as the secure single link clustering given in [16].

5: *Phase 4:* For each point in the partitioned database calculate the $dist^2()$ to the clusters formed in phase 3. Use **Protocol 1** to find out the minimum distance cluster and label the point with that cluster.

3.6   Proof of Correctness

In this section we show that the algorithm secure BIRCH (Algorithm 2) computes the same set of clusters over arbitrarily partitioned databases as its non secure variant BIRCH computes on $DB = DB_A \# DB_B$.

Lemma 1. *The CF tree computed in phase 1 of secure BIRCH is same as the one computed by BIRCH on input $DB = DB_A \# DB_B$.*

Proof. Given the fact that the data point $d_i = d_{i_A} \# d_{i_B} \in DB$ is arbitrarily partitioned and the comparison of distances based on square of their distances is same as the actual distances(ie given two distance metrics D_1 and D_2, if $D_1 < D_2$ then $D_1^2 < D_2^2$) and as the secure distance metrics are correct, the step 1 of secureCFTree insertion identify the same leaf node in the CF tree for a new entry as if it were done in centralized database.

Algorithm SecureNodeSplit uses secure min index in vector sum protocol [16] for identifying two farthest CFs in the node and also it uses secure min index in vector sum protocol [16] for distributing the remaining CFs in the node two form two separate nodes. As secure min index in vector sum protocol [16] is proved to be correct, the algorithm SecureNodeSplit (Algorithm 1) splits the node exactly same as it would be done in centralized database case.

Step 2 of SecureCFTreeInsertion uses secure threshold sum protocol [16] and algorithm SecureNodeSplit (Algorithm 1) for modifying the leaf. As secure threshold sum protocol and algorithm SecureNodeSplit is proved to be correct, Step 2 modifies the leaf same way as it would modify in the centralized database case.

Step3 of the algorithm SecureCFTreeInsertion for modifying the path to the leaf is same as in centralized database since, each party observes same type of changes in its version of the CF tree . Hence step 3 is same as in centralized database.

Step 4 in SecureCFTree insertion for merging refinement uses secure min index in vector sum protocol [16] for merging two CFs and algorithm SecureNodeSplit for splitting a node. Since secure min index in vector sum protocol [16] and the algorithm *SecureNodeSplit* are proved to be correct, step 4 is same as the step merging refinement in centralised database.

As steps 1 to 4 of algorithm *SecureCFTreeInsertion* are proved to be same as the steps 1 to 4 of algorithm *CFTreeInsertion* in centralized database, phase1 of *SecureBIRCH* constructs the same *CFTree* as in its centralized database case.

Theorem 2. *SecureBIRCH computes the same set of clusters as its non secure variant BIRCH does with the input $DB = DB_A \# DB_B$.*

Proof. The algorithm SecureCFTreeInsertion is invoked in phase1 and phase 2 where the algorithm CFTreeInsertion is invoked in its centralized database variant. As the algorithm SecureCFTreeInsertion is proved to be correct [16], phases 1 and 2 of the algorithm SecureBIRCH is same as the phases 1 and 2 of the algorithm BIRCH. Phase 3 of the algorithm SecureBIRCH uses the algorithm SecureSingleLinkClustering for computing global clusters, where SingleLinkClustering is used in phase 3 of BIRCH algorithm. As the algorithm SecureSingleLinkClustering is proved to be correct [16], phase 3 of SecureBIRCH is same as phase 3 of BIRCH. Phase 4 of SecureBIRCH uses secure min index in vector sum protocol [16] for labeling the clusters. As secure min index in vector sum protocol is proved to be correct, phase 4 labels the clusters correctly. Hence SecureBIRCH computes the same set of clusters as BIRCH computes on centralized database.

3.7   Security Analysis

In this section we prove that the algorithm SecureBIRCH (Algorithm 2) is secure against semi honest adversaries.

Lemma 2. *Algorithm SecureNodeSplit is secure against semi honest adversaries*

Proof. In algorithm SecureNodeSplit, the data is shared by using secure min index in vector sum protocol [16]. secure min index in vector sum protocol is proved be secure against semi honest adversaries [16]. Hence by composition theorem, the algorithm SecureNodeSplit (Algorithm 1) is secure against semi honest adversaries.

Lemma 3. *Algorithm SecureCFTreeInsertion is secure against semi honest adversaries*

Proof. The algorithm SecureCFTreeInsertion uses secure min index in vector sum protocol, secure threshold sum protocol [16] and algorithm SecureNodeSplit for sharing data. secure min index in vector sum protocol [16], secure threshold sum protocol [16] the algorithm SecureNodeSplit (by lemma 2) are proved to be secure against semi honest adversaries. Hence by composition theorem, the algorithm SecureCFTreeInsertion is secure against semi honest adversaries.

Theorem 3. *Algorithm Secure BIRCH is secure against semi honest adversaries*

Proof. Phases 1 and 2 of SecureBIRCH are the invocation of the algorithm SecureCFTreeInsertion. Phase 3 is the invocation of the algorithm SecureSingleLinkClustering. Phase 4 is the invocation of secure min index in vector sum protocol [16]. SecureCFTreeInsertion is proved to be secure against semi honest adversaries by lemma 3, the algorithm SecureSingleLinkClustering is proved to be secure against semi honest adversaries [16] and secure min index in vector sum protocol is secure against semi honest adversaries. Hence, by composition theorem, the algorithm SecureBIRCH is secure against semi honest adversaries.

3.8   Complexity Analysis

The communication and computation complexities of the algorithm Secure-BIRCH (Algorithm 2) is given in Table 1. In the table, B is the branching factor of CF tree, n is the number of records in the combined data base, m is the total number of attributes, K is the number of clusters formed in CF tree construction (Phases 1 and 2) and g is the final global clusters formed.

Table 1. Complexity analysis of SecureBIRCH

	Communication Complexity	Computation Complexity
Phase1	$O(Bnm \log_B K) + O(KL^2m)$ values	$O(Bnm \log_B K) + O(KL^2m)$ enc $O(Bnm \log_B K) + O(KL^2m)$ dec
Phase2	$O(Bnm \log_B K) + O(KL^2m)$ values	$O(Bnm \log_B K) + O(KL^2m)$ enc $O(Bnm \log_B K) + O(KL^2m)$ dec
Phase3	$O(K^2m)$ values	$O(K^2m)$ enc and $O(K^2m)$ dec
Phase4	$O(nmg)$ values	$O(nmg)$ enc and $O(nmg)$ dec
Total	$O(Bnm \log_B K) + O(KL^2m) + O(K^2m) + O(nmg)$ values	$O(Bnm \log_B K) + O(KL^2m) + O(K^2m) + O(nmg)$ enc $O(Bnm \log_B K) + O(KL^2m) + O(K^2m) + O(nmg)$ dec

4   Special Cases

Both horizontally partitioned and vertically partitioned data can be viewed as specific cases of arbitrarily partitioned data.

For Horizontally Partitioned Databases: In horizontally partitioned databases Alice has u number of records and Bob has v number of records, where $u + v = n$ and n is the total number of records. Consider the records owned by the Alice as first u records and Bob's records as the remaining records as in the centralized database case. For the firs u records, Bob will put u records with the entry values as zero in each record entry. Alice appends its database with v records with each entry as zero in the records with its set of records. Now both invoke the secure BIRCH algorithm with their modified private inputs. Here no party needs to know the other party's records for clustering data. At the end each party clusters its data as if it is done on centralized database.

Communication complexity and computation are same as for the arbitrarily partitioned databases with m records from each party as private input. Here $m = u + v$.

For Vertically Partitioned Databases: In this setting Alice has m records with p attributes and Bob has m records with q attributes. Alice appends with each of its record with q zeros and Bob prepends each of its records with p number of zeros. Now Both will have m records with each record having $p + q = n$ values. Now they invoke secure BIRCH algorithm for clustering over arbitrarily partitioned databases with each of their modified database as private inputs.

The communication and computation complexities are same as for the secure BIRCH for arbitrarily partitioned databases with m records and n attributes from each party. The complexities are slightly higher compared to the specific solution the secure BIRCH for vertically partitioned database [16].

5   Conclusion and Future Work

In this paper we presented protocols for secure distance metrics for arbitrarily partitioned databases. These protocols for secure distance metrics are used for computing clusters in arbitrarily partitioned database. We also presented secure algorithms for building CF tree, splitting a node and single link clustering. These are in turn used in secure BIRCH algorithm. We also proved the correctness of the algorithm and the security of the algorithm. As a direction for future work, one may attempt privacy preserving clustering over arbitrarily partitioned database for categorical attributes and mixed attributes.

References

1. Agrawal, R., Srikant, R.: Privacy preserving data mining. In: Proceedings of the 2000 ACM SIGMOD, Dallas, TX (May 14-19, 2000)
2. Goethals, B., Laur, S., Lipmaa, H., Mielikainen,: On private scalar product computation for privacy-preserving data mining. In: Park, C.-s., Chee, S. (eds.) ICISC 2004. LNCS, vol. 3506, pp. 2–3. Springer, Heidelberg (2005)
3. Brankovic, L., Estivill-Castro, V.: Privacy issues in knowledge discovery and data mining. In: AICE99. Proceedings of Australian Institute of Computer Ethics Conference, Melbourne, Victoria, Australia (1999)
4. Cachin, C.: Efficient private bidding and auctions with an oblivious third party. In: SIGSAC. Proceedings of 6th ACM Computer and communications security, pp. 120–127. ACM Press, New York (1999)
5. Damgard, I., Jurik, M.: A Generalisation, a Simplification and Some Applications of Paillier's Probabilistic Public-Key System. In: Kim, K.-c. (ed.) PKC 2001. LNCS, vol. 1992, pp. 13–15. Springer, Heidelberg (2001)
6. Froomkin, M.: Death of Privacy? Stanford Law Review 52, 1461–1543 (2000)
7. Jagannathan, G., Wright, R.N.: Privacy-preserving distributed k-means clustering over arbitrarily partitioned data. In: Proceedings of the 11th ACM SIGKDD, Chicago, Illinois, USA, August 21-24 2005, pp. 21–24. ACM, New York (2005)
8. Jain, A.K., Dubes, R.C.: Algorithms for Clustering Data, Ch. 3. Prentice-Hall Inc, Englewood Cliffs (1988)

9. Jha, S., Kruger, L., McDaniel, P.: Privacy Preserving Clustering. In: di Vimercati, S.d.C., Syverson, P.F., Gollmann, D. (eds.) ESORICS 2005. LNCS, vol. 3679, pp. 397–417. Springer, Heidelberg (2005)
10. Lindell, Y., Pinkas, B.: Privacy preserving data mining. In: Bellare, M. (ed.) CRYPTO 2000. LNCS, vol. 1880, pp. 36–54. Springer, Heidelberg (2000)
11. Naor, M., Pinkas, B.: Oblivious transfer and polynomial evaluation(extended abstract). In: Proceedings of the 31st ACM Symposium on Theory of Computing, Atanta, GA, USA, pp. 245–254 (May 1-4, 1999)
12. Natan, R.B.: Implementing Database Security and Auditing, ch. 11. Elsevier, North-Holland (2005)
13. Oliveira, S., Zaiane, O.R.: Privacy preserving clustering by data transformation. In: Proceedings of the 18th Brazilian Symposium on Databases, pp. 304–318 (2003)
14. Paillier, P.: Public-key Cryptosystems Based on Composite Degree Residuosity Classes. In: Stern, J. (ed.) EUROCRYPT 1999. LNCS, vol. 1592, pp. 2–6. Springer, Heidelberg (1999)
15. Vaidya, J., Clifton, C.: Privacy-preserving k-means clustering over vertically partitioned data. In: Proceedings of the 9th ACM SIGKDD, Washington, DC, USA (August 24-27, 2003)
16. Krishna Prasad, P., Pandu Rangan, C.: Privacy preserving BIRCH algorithm for clustering over vertically partitioned databases. In: Jonker, W., Petković, M. (eds.) SDM 2006. LNCS, vol. 4165, pp. 84–99. Springer, Heidelberg (2006)
17. Quittner, J.: The Death of Privac. In Time Magazine (August 1997)
18. Schoeman, F.: Philosophical Dimensions of Privacy: An Anthology. Cambridge University Press, Cambridge (1984)
19. Walters, G.J.: Privacy and security: an ethical analysis. ACM SIGCAS Computers and Society 31(2), 8–23 (2001)
20. Yao, A.C.: Protocols for secure computation. In: Proceedings of 23rd IEEE Symposium on Foundations of Computer Science, pp. 160–164. IEEE Computer Society Press, Los Alamitos (1982)
21. Yao, A.C.: How to generate and exchange secrets. In: Proceedings of the 27th IEEE Symp. on Foundations of Computer Science, Toronto, Ontario, Canada, October 27 - 29, 1986, pp. 27–29. IEEE Computer Society Press, Los Alamitos (1986)
22. Zhang, T., Ramakrishnan, R., Livny, M.: BIRCH: An efficient Data Clustering Method of Very Large Databases. In: Proceedings of the ACM SIGMOD, Montreal, Canada, June 1996, pp. 103–114. ACM Press, New York (1996)

Unsupervised Outlier Detection in Sensor Networks Using Aggregation Tree*

Kejia Zhang, Shengfei Shi, Hong Gao, and Jianzhong Li

Database Research Center
School of Computer Science & Technology
Harbin Institute of Technology, Harbin, China
veron1967@163.com

Abstract. In the applications of sensor networks, outlier detection has attracted more and more attention. The identification of outliers can be used to filter false data, find faulty nodes and discover interesting events. A few papers have been published for this issue. However some of them consume too much communication, some of them need user to pre-set correct thresholds, some of them generate approximate results rather than exact ones. In this paper, a new unsupervised approach is proposed to detect global top n outliers in the network. This approach can be used to answer both snapshot queries and continuous queries. Two novel concepts, *modifier set* and *candidate set* for the global outliers, are defined in the paper. Also a *commit-disseminate-verify* mechanism for outlier detection in aggregation tree is provided. Using this mechanism and the these two concepts, the global top n outliers can be detected through exchanging short messages in the whole tree. Theoretically, we prove that the results generated by our approach are exact. The experimental results show that our approach is the most communication-efficient one compared with other existing methods. Moreover, our approach does not need any pre-specified threshold. It can be easily extended to multi-dimensional data, and is suitable for detecting outliers of various definitions.

1 Introduction

Wireless Sensor Network (WSN) consists of hundreds or thousands of small, resource limited sensor nodes except one or more powerful nodes gathering the information of others [3,4]. We call the powerful node the sink or the base station (BS). Since sensor nodes (except the sink) are battery-powered and it is usually impractical to install new batteries, the WSN designers must find ways

* Supported by the Key National Natural Science Foundation of China under Grant No. 60533110; National Grand Fundamental Research 973 Program of China under Grant No. 2006CB303000; the Key National Natural Science Foundation of Heilongjiang Province; the National Natural Science Foundation of China under Grant No. 60473075; Program for New Century Excellent Talents in University "NCET" under Grant No. NCET-05-0333.

R. Alhajj et al. (Eds.): ADMA 2007, LNAI 4632, pp. 158–169, 2007.

to save energy and prolong the living time of WSN. The researches [1,2] reflect that for a sensor node, the communication cost is often several orders of magnitude higher than the computation cost. Therefore, minimizing communication overhead becomes the most important designing goal.

Generally, given a dataset S, an outlier is a data deviating markedly from other members of S. In a dataset, we can calculate *deviation rank* for each data to indicate its rank of deviating from others. Using different measurement criteria, data's deviation rank has different meanings. For example, for data x, we can use the *distance to kth nearest neighbors*[1] to measure its deviation rank [6], and we can also use the inverse of the number of neighbors within distance α, namely the inverse of $N(x, \alpha)$, to measure x's deviation rank [7]. Now the definition of outlier is: given a dataset, an outlier is a data whose deviation rank is larger than a given threshold. We can see that to define an outlier, a given deviation rank measurement criterion and a pre-specified threshold are needed.

In sensor networks, it is more useful to detect global outliers. Assume there are m sensor nodes in the network, and datasets $S_1, \ldots, S_m$ are the sample sets of these nodes. A global outlier is an outlier in the global set $\bigcup_{1 \leq i \leq m} S_i$. Compared with the data produced by sensors in the entire history, users are more interested in the data produced in a sliding window W of size $|W|$, especially in the current sliding window. Assuming the current time is t, the current sliding window can be seen a time interval from $t - |W|$ to t. Therefore, in a sensor network, $S_1, \ldots, S_m$ can be seen the datasets sampled by sensor nodes in current sliding window. Besides, since it is too difficult to set a correct threshold to define outliers while we don't know the distribution of data, our approach detects global top n outliers with largest deviation rank in global range, and just needs users to give the criterion to measure data's deviation rank.

Several characteristics of WSNs make the identification of outliers useful for analysis. The limited resource and capability make the data generated by sensor nodes undependable and inaccurate. Especially with power exhausting, the probability of generating erroneous data will grow rapidly. Therefore, we need a method to filter the data much different from others, i.e., outliers. Besides, we can also identify the faulty nodes by finding the nodes always generating outliers. Even if we are certain of the quality of the readings reported by sensor nodes, detection of outliers provides an efficient way to focus on the interesting events. For example, to supervise the process of manufacture, a machine is fitted with sensors monitoring its operation. When we find that the readings of temperature or vibration amplitude generated by a sensor are always much different from the data sampled by other nodes, there must be something wrong in the part this sensor monitoring.

To this day, we have already found a few papers [8,9,10,11,14,15,16] studying the problem of outlier detection in sensor networks, but works are still far from enough. [9,10] detect global outliers whose deviation rank in global range is larger than a pre-specified threshold. They need user to pre-set a correct threshold to find outliers. They use *kernel density estimator* to estimate the distribution of

---

[1] Here "neighbors" means the data around a given data point.

sensor readings, and use the probability density function to estimate $N(x, \alpha)$ for each data point x. Their approaches generate approximate results, which may filter many interesting events with small probability of occurrence. Even if we can tolerate the approximate results, we can not give a correct threshold to define outliers while we don't know the distribution of all the original sensor readings. [11] intends to clean outliers for better analysis. It uses *wavelet approximation* to clean local sudden-burst outliers in every node, and uses *dynamic time wrapping* to detect and remove outliers of small range (2 hops) which last for a long period. If we use this approach to detect outliers, it can only give an approximate result of outliers in small area, but not the exact global result in the whole network. [8] is similar to our approach. By using the definitions of outlier's *support set* and exchanging messages among all the nodes, it can exactly detect global top n outliers. It has two main disadvantages. One is that since it does not adopt any structure and every node uses broadcast to communicate, there is too much communication wasted while exchanging global information among all the nodes. The other is that its algorithm will end when all the nodes stop transmitting message, so it can not give a definite ending condition which can be judged by the sink. In other articles, [14] revises outliers using EM algorithm; [15] detects spatio-temporal outliers in stable distribution; [16] constructs a Bayesian Belief Network over WSN, and detects outliers by computing the likelihood of tuples.

In this paper, we propose a communication efficient approach to detect global top n outliers. We do not need any pre-specified threshold except the criterion (function) to measure outlier's deviation rank. Our approach is suitable to answer both snapshot and continuous queries. The basic idea is to build all the nodes in a structure of aggregation tree. Then let every node in the tree transmit some useful data to its parent after gathering all the messages from its children. The sink uses the information from its children to calculate global top n outliers, and floods these outliers for verification. If any node finds that it has two kinds of data which may modify the global result, it will send them to its parent in an appropriate time interval. All these processes are done level-by-level in the tree. Through a few such *commit-disseminate-verify* processes, all the nodes will agree on the global result calculated by the sink. The contributions of this paper are:

- We propose a communication efficient unsupervised approach for outlier detection in sensor networks. In our approach, we adopt the structure of aggregation tree. And we design a *commit-disseminate-verify* mechanism for outlier detection using aggregation tree. This mechanism can exactly detect outliers through exchanging short messages in the whole tree.
- We develop the definition of outliers support set and define two kinds of data which may affect the result of global outliers: *modifier set* and *candidate set* for the global outliers. By only transmitting these two kinds of data to upper node, every node can save a lot of communication. And the messages in the network are very short.
- We prove the correctness of our approach theoretically, i.e., the outliers reported by our algorithm are exact global top n outliers.

2   Preliminary

In this section, we introduce some necessary background definitions and notations.

The definition of data's deviation rank can be seen in section 1. For a data x in dataset S, we use $R(x, S)$ to denote the deviation rank of x in S. Here R is a deviation rank function. For different measurement criteria, $R(x, S)$ equals different values. We suppose in all cases, R satisfies the following theorem [8]:

Theorem 1. *For given $S_1 \subseteq S_2, R(x, S_1) \geq R(x, S_2)$.*

This theorem implies that the data belonging to a larger dataset has fewer opportunities to become an outlier.

We use the expression $\mathscr{A}[S]$ to denote the top n outliers in dataset S detected with algorithm $\mathscr{A}$. In $\mathscr{A}$, the deviation rank function is R. So we have:

$$\mathscr{A}[S] = \{x_1, \ldots, x_n \in S \mid \forall 1 \leq i \leq n \text{ and } y \in S \setminus \mathscr{A}[S], \ R(y, S) \leq R(x_i, S)\}$$

Definition 1 (Support set). *Given a set $S_0 \subseteq S$ and a deviation rank function R, S_0 is called the* support set *of $x \in S$ over S if $R(x, S_0) = R(x, S)$ [8]. In all the support sets of x over S, we choose one with smallest size as the* smallest support set *of x over S, and use the expression $[S|x]$ to denote it. For given $S' \subseteq S$, we write $[S|S']$ to denote $\bigcup_{x \in S'}[S|x]$.*

In a distributed system, there are sets $S_1, \ldots, S_m$. We want to detect global top n outliers, i.e., $\mathscr{A}[\bigcup_{1 \leq i \leq m} S_i]$. Assume we have found global top n outliers and their deviation rank using a subset of $\bigcup_{1 \leq i \leq m} S_i$. Both the outliers and their deviation rank may be not right for not using the whole set. $A_g = \{x_1, \ldots, x_n\}$ denotes current global outliers. $R_g = \{R(x_1), \ldots, R(x_n)\}$ denotes their current deviation rank in global range.

Definition 2 (Modifier set). *For set S_i, its* modifier set *for global outliers consists of the data whose distance to a global outlier is less than the global outlier's deviation rank. We use $M(S_i)$ to denote it, and*

$$M(S_i) = \{x | x \in S_i, \exists current\ global\ outlier\ x_g \in A_g, \ |x - x_g| < R(x_g)\}$$

Here we assume R is a function of distance. If using the inverse of $N(x, \alpha)$ to measure data's deviation rank, α should be written instead of $R(x_g)$ in the formula.

Definition 3 (Candidate set). *For set S_i, its* candidate set *for global outliers consists of the data which may be candidates of correct global outliers, i.e., the data whose local deviation rank is larger than a global outlier's. We use $C(S_i)$ to denote it:*

$$C(S_i) = \{x | x \in S_i, \exists current\ global\ outlier\ x_g \in A_g, \ R(x, S_i) > R(x_g)\}$$

3   Unsupervised Outlier Detection Using Aggregation Tree

In this section, we describe our approach for efficiently detecting global top n outliers. In our algorithm, we assume there is a topological tree rooted at the sink. There are various methods [5] for constructing the aggregation tree according to different application requirements. And we do not rely on a specific tree construction algorithm as long as there is one.

For clearness, we first describe a simple aggregation tree construction algorithm. At the beginning, the sink broadcasts a beaconing message including its own ID and its depth 0. When a node, say x, receives a broadcast message at its first time from a node y, x assigns its depth to be the depth of y plus one, and its parent to be y. Then, x broadcasts the message recursively. This process continues until all the nodes have received beaconing message. For energy efficiency, we can join the outlier detection query into the beaconing message. When we disseminate a new query, the aggregation tree is constructed.

3.1   Algorithm for Snapshot Query

Snapshot query is a kind of queries which collect data about now or some other time point. In our model, snapshot query queries once about the global top n global outliers in current sliding window. Through the dissemination of query, an aggregation tree is constructed. To answer this kind of queries, our algorithm uses the *commit-disseminate-verify* mechanism. For a node N_i in the tree, we use S_i to denote the sample set of N_i in current sliding window. We define the *knowledge* of N_i as $\overline{S_i} = S_i \cup \{\text{data received from } N_i\text{'s children}\}$. And we use dataset HS_i to record the data N_i has sent to its parent.

Initially, every node commits for the first time. Through the first committing, we hope that every node N_i in the tree has a coarse view about the data generated in the sub-tree rooted at N_i. And we also hope not to consume too much communication. Therefore, every node must transmit the most useful dataset for outlier detection to its parent. For node N_i, it sends its local outliers $A_i = \mathscr{A}[\overline{S_i}]$ and their smallest support set $SA_i = [\overline{S_i}|A_i]$ to its parent after incorporating all the data from its children into its $\overline{S_i}$. Meanwhile, N_i sets $HS_i = A_i \cup SA_i$. At the end of this process, the sink calculates the global top n outliers for the first time which may be not right. We do this because though the outliers of parent (the top n outliers in the dataset of the sub-tree rooted at the parent) may not be outliers of its children, they come from the union of these children's outliers with high probability, especially when the data from different nodes is nearly the same.

To disseminate, the sink floods the unverified global outliers and their deviation rank. After receiving that, every node verifies and commits again. For node N_i, it calculates its $M(\overline{S_i})$, sets $Info_i^1 = M(\overline{S_i}) \setminus HS_i$, and waits for enough time to receive messages from all its children. After incorporating all the data from its children into its $\overline{S_i}$, N_i calculates its $C(\overline{S_i})$, and sets

$Info_i^2 = C(\overline{S_i}) \cup [\overline{S_i}|C(\overline{S_i})] \setminus HS_i$. If there is remaining data in $Info_i^1$ and $Info_i^2$, N_i sends $(Info_i^1, Info_i^2)$ to parent node and adds them into HS_i. Why we append the smallest support set of $C(\overline{S_i})$ to $Info_i^2$? The reason is that upper node can re-calculate deviation rank of the data in $C(\overline{S_i})$ exactly after incorporating $Info_i^2$ into a larger dataset.

After gathering all the committed information from its children, the sink re-calculates global top n outliers. If the result is different from last one (either the outliers or their deviation rank), the sink will disseminate again until the result does not change.

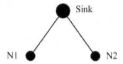

Fig. 1. A simple example. $S_1 = \{0.7, 1.3, 1.6, 2.0, 2.4, 2.8, 3.2, 8.6\}$. $S_2 = \{4.5, 8.2, 9.1, 9.3, 9.7, 10.2, 11.6, 12.1\}$. $S(sink) = \{3.0, 5.9, 6.4, 7.1, 7.9, 10.8, 11.3, 13.0\}$.

Next, we give a simplest example to demonstrate our algorithm. As shown in figure 1, the sink only has two children N_1 and N_2. We use *distance to the nearest neighbor* to measure outliers' deviation rank. The number of outliers we want to detect is $n = 1$. At the beginning, the sample set of N_1 N_2 and the sink are S_1 S_2 and $S(sink)$, as shown in the caption of the figure. We can see that the global outlier is 4.5.

To commit for the first time, N_1 calculates its local outlier $A_1 = \{8.6\}$ and A_1's smallest support set $SA_1 = \{3.2\}$, sends $A_1 \cup SA_1$ to the sink, and sets $HS_1 = \{3.2, 8.6\}$. Similarly, N_2 sends $\{4.5, 8.2\}$ to the sink and records them. After receiving, the knowledge of the sink becomes $\{3.0, 3.2, 4.5, 5.9, 6.4, 7.1, 7.9, 8.2, 8.6, 10.8, 11.3, 13.0\}$. Then, the sink finds that the global outlier is 13.0 and its deviation rank is 1.7 (distance to 11.3), which are not verified.

To disseminate, the sink broadcasts the binary group (13.0, 1.7). After receiving that, N_1 and N_2 verifies and commits again. N_2 checks if it has some data whose distance to 13.0 is less than 1.7, and sets its $M(\overline{S_2}) = \{11.6, 12.1\}$, $Info_2^1 = M(\overline{S_2}) \setminus HS_2 = M(\overline{S_2})$. Then N_2 checks if it has some data whose local deviation rank is larger than 1.7, and sets its $C(\overline{S_2}) = \{4.5\}$, $Info_2^2 = C(\overline{S_2}) \cup [\overline{S_2}|C(\overline{S_2})] \setminus HS_2 = \{4.5, 8.2\} \setminus \{4.5, 8.2\} = \emptyset$. Finally, N_2 sends $(\{11.6, 12.1\}, \emptyset)$ to the sink. Meanwhile, N_1 finds that its $Info_1^1 = \emptyset$ and $Info_1^2 = \emptyset$, and sends nothing to the sink. After receiving messages from its children, the knowledge of the sink becomes $\{3.0, 3.2, 4.5, 5.9, 6.4, 7.1, 7.9, 8.2, 8.6, 10.8, 11.3, 11.6, 12.1, 13.0\}$. Then the sink finds that the global outlier has become 4.5 with deviation rank 1.3, so it disseminates again and floods (4.5, 1.3).

Receiving the information about the new outlier, both N_1 and N_2 find that they have nothing to send back to the sink. So the sink will affirm that the global outlier is really 4.5 with deviation rank 1.3. N_1 and N_2 can know that too.

3.2   Update of Outliers for Continuous Query

Continuous query is a kind of queries which collect data for a long period. In our model, continuous query queries continuously about the global top n outliers in current sliding window for a long period. As time advances and the current sliding window moves forward, the sample set of every node changes continuously, so the result of global outliers must change in time. It is easy to modify the above algorithm to answer continuous queries.

Algorithm 1. Find_Outliers (N_i)

Input of N_i: outlier detection algorithm $\mathscr{A}$ and window size τ
Output of sink node : global top n outliers
Initially:
1. Every leaf node N_i calculate its $A_i \leftarrow \mathscr{A}[\overline{S_i}]$ and $SA_i \leftarrow [\overline{S_i}|A_i]$, then sends $A_i \cup SA_i$ to parent.
2. After incorporating all the data from its children into its $\overline{S_i}$, the intermediate node calculates its own A_i and SA_i, then sends them to parent node.
3. After receiving from all its children, the sink calculates global outliers $A_g = \{x_1, \ldots, x_n\}$ and their deviation rank $R_g = \{R(x_1), \ldots, R(x_n)\}$, then floods (A_g, R_g).
Case receive new global outliers' information or there is any change in $\overline{S_i}$:
4. Update the information about A_g and R_g
5. Retire the data older than τ in $\overline{S_i}$
6. $Info_i^1 \leftarrow M(\overline{S_i}) \setminus HS_i$
7. IF N_i is leaf node:
8. $Info_i^2 \leftarrow C(\overline{S_i}) \cup [\overline{S_i}|C(\overline{S_i})] \setminus HS_i$
9. $Message \leftarrow (Info_i^1, Info_i^2)$ and send $Message$ to parent
Case receive $Message = (Info_j^1, Info_j^2)$ from child N_j:
10. $\overline{S_i} \leftarrow \overline{S_i} \cup Info_j^1 \cup Info_j^2$
11. $Info_i^1 \leftarrow Info_i^1 \cup Info_j^1$
Wait for enough time to receive data from every child, then:
12. IF N_i is not the sink:
13. $Info_i^2 \leftarrow C(\overline{S_i}) \cup [\overline{S_i}|C(\overline{S_i})] \setminus HS_i$
14. $Message \leftarrow (Info_i^1, Info_i^2)$ and send $Message$ to parent
15. ELSE $//N_i$ is the sink node
16. $A_g \leftarrow \mathscr{A}[\overline{S_i}]$ $R_g \leftarrow \{R(x, \overline{S_i}) | x \in A_g\}$
17.   IF there is any change in A_g and R_g:
18.   Flood (A_g, R_g) to every node.
19.   ELSE
20.   Report A_g as global top n outliers

The algorithm designed for both snapshot query and continuous query is given in algorithm 1. The inputs of every node are the deviation rank function R, the number of reported outliers n and the sliding window size τ. R and n can be contained in algorithm $\mathscr{A}$, which we use to detect local outliers in every node. In algorithm 1, A_g denotes global top n outliers detected by the sink which may be not right, and R_g denotes these outliers' deviation rank calculated by the sink. Node N_i has child N_j.

Initially, every node commits for the first time (lines 1-3). If the sink finds new result of global outliers, it disseminates (lines 16-18). Otherwise, the sink reports global top n outliers (lines 19, 20). Receiving the information of new global outliers, every node verifies and assigns its $Info_i^1$ (lines 4, 6). After waiting for enough time to receive data from its children, every node assigns its $Info_i^2$ and commits (lines 7, 8, 12-14). As time advances, if there is any change in $\overline{S}_i$, N_i verifies by checking whether there is new data added into its $Info_i^1$ and $Info_i^2$ (lines 6, 13, 14). If there is useful data to send, N_i will transmit them in an appropriate time interval.

For further optimizing, as time advances, we could let every node delay for a fixed time to accumulate more changes before verifying and committing. Of course, this will cause the result of global outliers to defer to a certain extent. However, the exactness of the result can not be suppressed much in case that the sample set of every node changes smoothly. Even if we can not tolerate the delay of results and we need to detect outliers exactly in time, our experiments show that our approach performs much better than other existing methods.

4   Proof of Correctness

Our algorithm intends to find the exact global top n outliers. As described in the following two theorems, we can prove the correctness of our algorithm in two steps. Below, we use US to denote the union of all the original data in current sliding window, i.e., $US = \bigcup_i S_i$. And $\overline{S}$ denotes the final knowledge of the sink.

Theorem 2. *If the network setting and every node's sample set do not change, the algorithm will definitely terminate. When the algorithm terminates, for each global outlier x in A_g, its deviation rank calculated by the sink, i.e., $R(x)$, is the correct deviation rank in global range.*

Proof. Since network setting and every node's sample do not change, line 20 in algorithm 1 can make sure that the algorithm will definitely terminate. Lines 6 and 11 assure that for each outlier x in A_g, any data point closer to x than its current deviation rank $R(x)$ must be added into node's $Info_i^1$ and sent up until to the sink. When the algorithm terminates, at least one support set of x will be received by the sink (Otherwise, there must be some data closer to x than $R(x)$ not received by the sink). Therefore, when the algorithm terminates, $R(x)$ equals the deviation rank calculated using all the original sensor readings, i.e., $R(x) = R(x, US)$.

Theorem 3. *If the network setting and every node's sample set do not change, when the algorithm terminates, the outliers reported by the sink, i.e., A_g, must be equal to the correct global top n outliers $\mathscr{A}[US]$.*

Proof. Suppose to the contrary that there is a data x belonging to $\mathscr{A}[US]$ but not in A_g. Therefore, for at least one outlier y in A_g, there must be $R(x, US) > R(y, US)$. According to theorem 2, $R(y, US) = R(y)$, so we have $R(x, US) > R(y)$.

a) If x is in $\overline{S}$, according to theorem 1 and $\overline{S} \subseteq US$, we know that $R(x) = R(x, \overline{S}) \geq R(x, US)$. Then we can conclude that $R(x) > R(y)$, which means x should be in instead of y in A_g.   **b)** If x is in the knowledge of node N_i(i.e., $\overline{S_i}$) but not in $\overline{S}$. Since $\overline{S_i} \subseteq US$, we have $R(x, \overline{S_i}) \geq R(x, US) > R(y)$ according to theorem 1. Lines 8, 9, 13 and 14 in algorithm 1 assure that any such data x must be added into N_i's $Info_i^2$, and sent to upper node until to the sink. Basing on the proof **a** and **b**, we can conclude that the existing of such data x is impossible by all means, and $A_g = \mathscr{A}[US]$.

The proof of correctness is under such a hypothesis: for every node in the network, its sample set does not change. That does not imply the algorithm is incorrect while answering the continuous query. As long as we can make sure that there is no change in every node's sample set when the algorithm is executing, i.e., the calculation of outliers is faster than the updating of data, the algorithm's results are certainly correct at any moment.

5   Experimental Evaluation

5.1   Experimentation Setup

We build a simulator implemented with 1000 lines of C++ code to evaluate the performance of our method. And we collect two kinds of data to measure the efficiency of approaches (y-coordinate), they are: (a) Average total number of packets received per node. (b) Average total number of data points/bytes received per node. Because the receiving energy consumption is the dominate term in energy use [13], we believe that these two kinds of data can give a good estimation of energy consumption.

We compared our approach with two approaches: centralized approach and the approach proposed by [8]. In centralized approach, all the nodes in the network periodically send their original data to the sink, and the sink detects outliers using its local algorithm.

The dataset we use comes from [12], which is a real-world dataset collected by Berkeley research lab. Because our algorithm has good scalability of data dimensionality (generally, data dimensionality does not affect the exactness and efficiency of our algorithm), we only use one attribute (temperature) to compose data point and measure the performance. One data points include following features: (1) ID of the node producing the data point. (2) The time stamp when the data point is generated. (3) The data value (temperature). Furthermore, for the reason that the environment in Berkeley's lab is very steady and the data from [12] is too smooth to reflect some abnormal events happening, we randomly insert some synthetic data with abnormal values into the dataset. Our algorithm detects outliers measured by *distance to k-nearest neighbor*. In our experiments, we set $k = 4$.

In the simulated experimentation, 50 nodes are deployed in a $250m \times 250m$ area. The node's transmission range is assigned to be 40m and each node has 4~6 neighbors. Unlike in actual environment, we assume that the communication in the network is reliable and the transmission failure rate is zero. All the experiments run 1000 seconds of simulation time. As shown in the following graph, we evaluate the performance for different parameters (x-coordinate) of a) the size of sliding window. b) the number of reported outliers n.

In these figures, *Centralized* denotes the naive centralized approach gathering sensor readings from all the nodes and calculating the outliers in the sink. *IN-Broadcast* denotes the in-network outlier detection approach proposed by [8]. *NS-Tree* denotes the unsupervised approach using our algorithm. *NS-Delay-x* denotes the improved version of our approach. In this approach, each leaf node delays x seconds to accumulate more changes before reporting to upper node rather than updates the knowledge and reports to upper node immediately. The idea of this approach can be seen in section 3.2.

5.2   Impact of Sliding Window Size

As shown in figure 2, NS-Delay-5 is the most communication-efficient approach in the four, no matter how big the sliding window is. In NS-Delay-5, every leaf node delay 5 seconds to accumulate more changes before reporting to upper node. Even if we can not tolerate the 5 seconds delay of the result, the approach NS-Tree without any delay performs much better than the other two. With the size of sliding window growing, except the centralized approach, the communication overhead of the other three approaches decrease gradually. The reason for this phenomenon is that in larger dataset, more benefit can be achieved through analysis in-network and there are more chances to reduce communication.

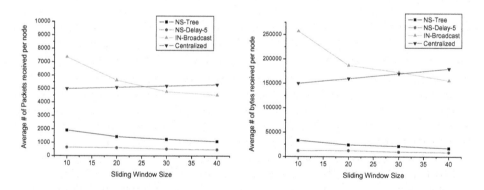

Fig. 2. Impact of sliding window size ($k = 4, n = 4$)

In figure 2, we find a problem that while the window size is smaller that 30, the IN-Broadcast approach performs even worse than the centralized method. Since each node in IN-Broadcast uses broadcast to communicate and the direction of transmission is not controlled by the algorithm, there is too much communication wasted in this approach. But IN-Broadcast has less average number of packets

sent per node than the centralized method, because broadcast means that one
node sends while many nodes receive.

5.3   Impact of Number of Reported Outliers

As shown in figure 3, NS-Tree and NS-Delay-5 still perform much better than
the other two. As the number of reported outliers increases, the communication overhead of centralized approach does not change, because no matter how
many outliers to detect, the sink still gathers data from all the nodes. On the
other side, the communication overhead of the other three approaches increases
with n. That is because more outliers we want to detect, more communication
we consume. Like in figure 2, when n is larger than 4, IN-Broadcast performs
even worse than the naive centralized method. Since the in-network analysis in
IN-Broadcast is complex and not efficient, this approach by in-network analysis is not as good as naive centralized method while reporting a number of
outliers.

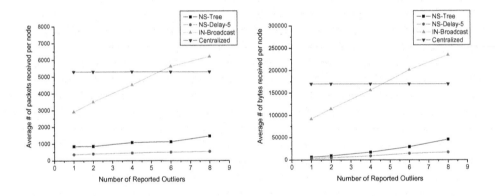

Fig. 3. Impact of number of reported outliers ($WindowSize = 40$, $k = 4$)

6   Summary

In this paper, we study the problem of outlier detection in sensor networks and
propose a unsupervised approach to detect global top n outliers in a sliding window. Our approach is base on the model of aggregation tree and we design a
commit-disseminate-verify mechanism for outlier detection in aggregation tree.
Moreover, we develop the definition of outlier's support set and define two kinds
of data which may affect the result of global outliers. By exchanging short messages in the whole tree, every node will get the information of correct global
outliers. We proved the correctness of our algorithm strictly. Compared with
existing approaches, our method is more communication-efficient and has advantages of unsupervised, result exactness, latency decrease and definite ending
condition. The experiments confirm the efficiency of our method.

In the future, our works may be applying the idea of our algorithm in some other applications, designing an approximation model to decrease communication overhead further, or making node find some other useful knowledge through analysis in-network.

References

1. Shnayder, V., Hempstead, M., Chen, B.R., Allen, G.W., Welsh, M.: Simulating the Power Consumption of Large-scale Sensor Network Applications. In: SenSys (2004)
2. Gupta, P., Kumar, P.R.: The Capacity of Wireless Networks. IEEETrans. Information Theory 46(2), 388–404 (2000)
3. Warneke, B., Last, M., Liebowitz, B., Pister, K.: Smart Dust: Communicating with A Cubic-millimeter Computer. IEEE Computer Magazine, pp. 44–51 (January 2001)
4. Gunopulos, D., Kollios, G., Tsotras, J., Domeniconi, C.: Approximating Multi-Dimensional Aggregate Range Queries over Real Attributes. In: SIGMOD (2000)
5. Madden, S., Franklin, M.J., Hellerstein, J.M., Hong, W.: The Design of An Acquisitional Query Processor for Sensor Networks. In: SIGMOD (2003)
6. Ramaswamy, S., Rastogi, R., Shim, K.: Efficient Algorithms for Mining Outliers from Large Datasets. In: SIGMOD (2000)
7. Knorr, E., Ng, R.: Algorithms for Mining Distance-Based Outliers in Large Datasets. In: VLDB, 24–27 (1998)
8. Branch, J., Szymanski, B., Giannella, C., Wolff, R.: In-Network Outlier Detection in Wireless Sensor Networks. In: ICDCS (2006)
9. Subramaniam, S., Palpanas, T., Papadopoulos, D., Kalogeraki, V., Gunopulos, D.: Online Outlier Detection in Sensor Data Using Non-Parametric Models. In: VLDB (2006)
10. Palpanas, T., Papadopoulos, D., Kalogeraki, V., Gunopulos, D.: Distributed Deviation Detection in Sensor Networks. ACM SIGMOD 32(4), 77–82 (2003)
11. Zhuang, Y., Chen, L.: In-network Outlier Cleaning for Data Collection in Sensor Networks. In: CleanDB (2006)
12. Intel Berkeley Research Lab. http://db.lcs.mit.edu/labdata/labdata.html
13. Crossbow Technology Inc. http://www.xbow.com/
14. Ash, J.N., Moses, R.L.: Outlier Compensation in Sensor Network Self-localization via the EM Algorithm. In: ICASSP (2005)
15. Jun, M.C., Jeong, H., Kuo, C.J.: Distributed Spatio-temporal Outlier Detection in Sensor Networks. In: SPIE (2005)
16. Janakiram, D., Reddy, A.M., Kumar, A.P.: Outlier Detection in Wireless Sensor Networks using Bayesian Belief Networks. In: Comsware (2006)

Separator: Sifting Hierarchical Heavy Hitters Accurately from Data Streams

Yuan Lin and Hongyan Liu

Dept. of Management Science and Engineering,
Tsinghua University, Beijing 100084, China
{liny.05, liuhy}@sem.tsinghua.edu.cn

Abstract. In this paper, we present a new algorithm, *Separator*, for accurate and efficient Hierarchical Heavy Hitter (HHH) detection, an emerging research area of data stream mining. Existing algorithms exploit either bottom-up or top-down processing strategy to solve this problem, whereas we propose a novel combination of these two strategies. Based on this strategy and a devised compact data structure, we implement our algorithm. It is theoretically proved to have tight error bound and small space usage. Comprehensive experiments conducted also verify its accuracy and efficiency.

Keywords: data stream mining, hierarchical heavy hitters, algorithms.

1 Introduction

Along with the development of information technology and various networks, there have been more and more applications related with stream data recently. Decisions in these applications often have to be made depending on stream data analysis and executed in real-time style. Data mining tools have been widely used to support decision-making because of their ability to discover interesting patterns from vast datasets efficiently. Likewise, data stream mining techniques could be capable assistants to the real-time decision-making based on stream data [1].

Among various data stream mining topics, Hierarchical Heavy Hitter (HHH) [2] detection is an emerging research area, which intends to find frequent items as well as their hierarchical clusters. It is a generalization of Heavy Hitter (HH) detection— finding "flat" frequent items. In many applications of data streams, the data are hierarchically organized, and HH algorithms [3]–[9] may not identify all meaningful patterns, e.g., those frequent clusters constituted by infrequent items. Besides, the hierarchical aggregation enables a choice among levels of detail, which, analogous to the OLAP operations in data warehouses, can better the understanding of data.

In this paper, we propose a compact data structure and an efficient algorithm, *Separator*, to detect HHHs from one-dimensional data streams with high accuracy. Our algorithm allows users to predetermine the memory usage by defining a proper error bound. To the best of our knowledge, no previous algorithms have provided similar functions. We conduct comprehensive experiments to study the performances of existing HHH algorithms and ours, using different evaluation metrics such as

R. Alhajj et al. (Eds.): ADMA 2007, LNAI 4632, pp. 170–182, 2007.
© Springer-Verlag Berlin Heidelberg 2007

recall, precision and their variants. Results show that our algorithm achieves a satisfying balance between result correctness and resource requirements.

The rest of the paper is organized as follows. Section 2 presents the HHH problem as well as the notions used in this paper. Section 3 reviews related work. We expound our algorithm *Separator* in Section 4 and experimentally evaluate it in Section 5. Section 6 concludes this paper.

2 Problem Specification

The inputs of the HHH detection problem we discuss include a data stream S and its one-dimensional domain hierarchy T. T naturally forms a tree structure with a height of $h + 1$; the levels from bottom to top are named as $L_0, L_1, \ldots,$ and L_h. Every element in S amounts to a leaf node of T. Let N be the current length of S. All leaf nodes constitute a set A_0 of size b_0, where $b_0 \ll N$. All the nodes at L_k constitute a set A_k of size b_k. $p(c)$ and $a(c)$ represent the parent and the ancestor node of node c respectively. If c is an internal node, then *subtree*(c) rooted at it contains all c's descendants.

Example 1. *Consider a data Stream S = {111, 111, 112, 112, 121, 123, 122, 121, 111, 131, 111, 112}. N = 12, a user-defined support φ=0.3. Fig. 1 illustrates the domain hierarchy T with h = 2. The set of leaf nodes A_0 = {111, 112, 121, 122, 123, 131}, b_0 = 6. The set of L_1 nodes, A_1 = {110, 120, 130}. A_2 = {100}. The parent of "111" is "110"; both "110" and "100" are its ancestors. f^* and df^* are defined below.*

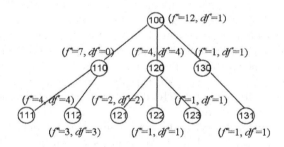

Fig. 1. An example of domain hierarchy T

Definition 1 (Total Frequency). If c is a leaf node, its exact total frequency $f^*(c)$ is its total occurrences in the stream S. If c is an internal node, $f^*(c) = \sum f^*(q)$, $\forall q \in A_0 \wedge q \in \{subtree(c) - c\}$. The estimation of $f^*(c)$ contains the lower bound $f_{min}(c)$ and the upper bound $f_{max}(c)$, i.e., $f_{min}(c) \leq f^*(c) \leq f_{max}(c)$.

Simply reporting items with high total frequency will produce a query answer infested with identified HHHs as well as all their ancestors, which provide little new information and can be easily inferred. Hence, we bring in the concept of *discounted frequency* introduced in [2] and [10].

Definition 2 (Discounted Frequency). Given a support threshold φ ($0 < \varphi < 1$). If c is a leaf node, its exact discounted frequency $df^*(c) = f^*(c)$; its estimated discounted frequency $df(c) = f_{max}(c)$. If c represents an internal node at L_k $(1 \leq k \leq h)$, $df^*(c) = f^*(c) - \sum f^*(q), \forall q \in \{subtree(c) - c\} \wedge df^*(q) \geq \lfloor \varphi N \rfloor$; $df(c) = f_{max}(c) - \sum f_{min}(q), \forall q \in \{subtree(c) - c\} \wedge df(q) \geq \lfloor \varphi N \rfloor$.

During the process of HHH detection, a synopsis structure [11] T' is maintained in the memory to summarize data stream and provide approximate answers. Since we focus on insert-only data streams (or the Cash Register Model [12]) and counter-based solutions, T' in this paper is comprised of counters, storing current frequent items. Each entry of T' is a HHH candidate, corresponding to a certain node in T. Based on T' and a user-defined support threshold φ, HHH detection algorithms report estimated frequent nodes as hierarchical heavy hitters.

Definition 3 (Hierarchical Heavy Hitter). The set of HHHs at L_k $(0 \leq k \leq h)$, $HHH_k = \{c \mid c \in A_k \wedge df(c) \geq \lfloor \varphi N \rfloor\}$.

3 Related Work

HHH detection is firstly discussed in [10], without practical online solution to the problem. After that, some *hierarchy-aware* online algorithms were proposed, such as the hierarchical adaptations of *LossyCounting* in [2] and the top-down approach proposed in [13]. *LossyCounting* [6] is famous for its efficiency and resources economization, but its frequent deletion and reallocation imply a possible accuracy defect when it is extended to HHH detection. Four strategies were presented in [2] to deal with hierarchical information; strategy 2 and 4 are respectively more accurate than Strategy 1 and 3. The upper bound on their space usage is $O(h \cdot \varepsilon^{-1} \cdot \log(\varepsilon N))$. Different from the bottom-up processing style used in [2], i.e., using specific HHHs to infer general ones [14], the top-down style algorithm in [13] starts from the root of T'. Element counts are intercepted by their ancestors, and only those real HHHs can stretch down to the levels they belong to. This algorithm uses a space of $O(h^2/\varepsilon)$.

Out of the concern for applying HHH detection to real-time decision-making, we are especially interested in correctness. Thus our algorithm uses for reference a "flat" algorithm called *SpaceSaving* [4], which is highly accurate and efficient, and can answer frequent-item queries with tight error bound and small space usage.

Existing HHH algorithms have not achieved a satisfying balance between accuracy and resource requirements for real-time style decision-making. Some is efficient yet not accurate enough, whereas others improve their accuracy by storing verbose information. Besides, there is little comparison among existing algorithms. In addition, accuracy has not been sufficiently studied in previous work. Some widely accepted accuracy metrics, such as recall and precision, are seldom used as evaluation criteria.

4 Our Approach

4.1 Data Structure

SpaceSaving employs a counter-based data structure "*Stream-Summary*" [4]. Each counter monitors a stream element, as an entry. All counters with the same counts are linked together and attached to a common bucket. All buckets are kept in a link list, sorted by the counts of their counters. This structure helps *SpaceSaving* to rapidly locate entries to be deleted, as well to update entry values without order violation.

The main body of our data structure (indicated as D, shown in Fig. 2) consists of r ($1 \leq r \leq h$) *Stream-Summary* structures, each corresponding to a level of the domain tree T. We denote them $D[0]$, $D[1]$,..., $D[r]$. $D[k]$ contains at most $M_k = \min$ (b_k, $\left\lceil h \cdot \varepsilon^{-1} \cdot (h-k)^{-1} \right\rceil$) ($0 \leq k \leq r$, $k \neq h$) counters, where b_k is the number of different internal nodes at L_k. With this formula, users can make a tradeoff between the memory needed and the error bound. We discuss the role of M_k in overestimation error control in Section 4.3. In our experiments, a tighter bound is used and we set M_0 to $\min$ (b_0, $\left\lceil (1/\varepsilon)^{1/\alpha} \right\rceil$) (when $\alpha > 1.0$) or $\min$ (b_0, $\left\lceil 1/\varepsilon \right\rceil$) (when $\alpha \leq 1.0$) for the Zipfian data, where α is the skew factor.

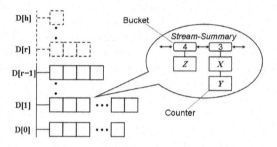

Fig. 2. The data structure D of *Separator*

At lower levels, M_k is often smaller than b_k, but b_k decreases quickly from bottom to top. L_h has $b_h = 1$. Thus, a *Stream-Summary* will be the first to have $M_k = b_k$. Assume it is $D[r]$. Then we do not allocate any counter to L_{r+1} and higher levels, i.e., $D[k]$ ($r+1 \leq k \leq h$) are null. If $D[h]$ is not null, it has only one counter.

In $D[k]$ ($0 \leq k \leq r$), every bucket records a value F, which represents the counts of all counters attached to it. Each occupied counter monitors a node at L_k, recording its maximal possible overestimation Δ. We call $(c, F(c), \Delta(c))$ an entry corresponding to node c. A node at L_k of the domain tree T is monitored *iff.* it has a corresponding entry in $D[k]$. Assume node c at L_k is monitored. In terms of the notations given in Section 2, we have: (1) If $k = 0$, $f_{min}(c) = F(c) - \Delta(c)$. (2) If $k > 0$, $f_{min}(c) = F(c) - \Delta(c)$ $+ \sum(F(q) - \Delta(q))$, $\forall q \in subtree(c) \wedge q$ is monitored; or $f_{min}(c) = F(c) - \Delta(c) +$ $\sum f_{min}(q)$, $\forall q \in A_{k-1} \cap subtree(c) \wedge q$ is monitored. (3) $f_{max}(c) = f_{min}(c) + \Delta(c)$. The minimal bucket value of $D[k]$ is denoted $D[k].minf$. Our adapted *Stream-Summary* structures are initially empty, with r variables set to control later counter allocation in case of the sparsity of upper-level counters when the data are highly skewed.

4.2 Algorithm *Separator*

On the arrival of an element e with addition v (=1), an update operation on $D[0]$ is initiated, denoted $D[0].update(e, 1, 0, true)$ as shown in Fig. 3. If e has already been monitored in $D[0]$, we simply update the counter. Otherwise, if there are still empty counters in $D[0]$, one of them will be allocated to monitor e. In the situation that neither a corresponding counter nor an empty one exists in $D[0]$, $D[1].update(p(e), 1,$ $0,$ true) will be next triggered to find a counter for e's parent $p(e)$ in $D[1]$, following the above rules. If necessary, more update operations will be executed on even higher levels to find counters for e's ancestors. Except for some conditions discussed later, such a counter can always be found. In the worst case, it locates in the top *Stream-Summary* $D[r]$, where every node at L_r is monitored by a counter. If the picked counter is an empty one, we set its F to v and Δ to 0; otherwise we just increase its F by v and leave its Δ unchanged. Since the above description is similar to a "floating" process, we refer to this kind of updates as the "floating" updates.

Assume the finally picked counter q belongs to $D[k]$ $(1 \leq k \leq h)$ and the node it monitors is c. If $F(c) + v > D[k-1].minf$, v will be instead added to the F value of e's ancestor at L_{k-1}, which is also c's child, indicated as $d(c)$. However, $D[k-1]$ does not have a counter monitoring $d(c)$ or any empty counters; otherwise, not q but a counter in $D[k-1]$ would be picked earlier. Under this circumstance, the counter of entry t, which has $D[k-1].minf$, will be taken to monitor $d(c)$. We reset it by assigning $\Delta(d(c))$ $= F(c)$ and $F(d(c)) = v + \Delta(d(c))$. Since this kind of updates operates in the opposite direction of the "floating" updates, we call them the "sinking" updates. $D[k-1].minf$ is the "sinking" threshold of L_k $(1 \leq k \leq h)$. It guarantees the error bound of *Separator*, as proved in Section 4.3. As for the deleted entry t, $F(t) - \Delta(t)$ will be added to its closest ancestor's F via a series of "floating" updates starting from $D[k]$. The whole process looks like separating a mixture: merged components gradually separate due to different density, and finally settle on different layers where they belong. Below we sketch the update operation; the following example shows how the updates are done.

```
Bool D[k].update(c, v, error, sign)
    if node c is monitored in this structure
        if k > 0 and F(c) + v + error > D[k-1].minf
            return D[k-1].UPDATE(d(c), v + error, F(c), false);
        else
            F(c) := F(c) + v + error; return true;
    else if sign = true
        if there are vacant counters
            allocate one to monitor c; return true;
        else if k < h
            return D[k+1].UPDATE(p(c), v, error, sign);
        else if there is not any counter
            return false;
        else
            locate a counter with D[k].minf;
            {Assume this counter is for node t};
            if D[k+1].UPDATE(t, D[k].minf - Δ(t), 0, true)
                c replaces the counter of t;
                F(c) := v + error; Δ(c) := error;
            else return false;
```

Fig. 3. The update operation. Assume it is executed in $D[k]$ $(0 \leq k \leq h)$. c represents a node at L_k. "*sign* = true" indicates a "floating" update; "*sign* = false" indicates a "sinking" update.

Example 2. *Consider a data Stream S = <111, 111, 112, 112, 121, 122, 111, 121, 113, 123, 131, 113>. In its domain tree T, A_0 = {111, 112, 113, 121, 122, 123, 131}, b_0 = 7; A_1 = {110, 120, 130}, b_1 = 3; A_2 = {100}, b_2 = 1. h = 2, N = 12. If ε = 0.25, we have M_0 = min (7, 4) = 4, M_1 = min (3, 6) = 3 = b_1, and thus M_2 = 0. The stream fraction S_1 yields Fig. 4(a). Through some D[0].update, "111" "112" "121" and "122" take all the counters of D[0]. The subsequent elements "111" and "121" are both monitored, so their arrivals just change the position of their counters. Then "113" arrives. Although D[0].update("113", 1, 0, true) cannot find a counter for it, the next D[1].update ("110", 1, 0, true) finds an empty counter to monitor its parent "110", as shown in Fig. 4(b). When "113" reappears (Fig. 4(c)), D[1].update("110", 1, 0, true) finds the corresponding counter (Step 1 in Fig. 4(c)). Then, since F(110) + v = 2 > D[0].minf (= 1), D[0].update("113", 1, F(110), false) is triggered, taking the counter for "122", which has D[0].minf, to monitor "113" (Step 2). We transfer the information of "122" to its parent "120" by D[1].update ("120", F(122) − Δ(122), 0, true) (Step 3). Since F(122) − Δ(122) + F(120) = present D[0].minf (= 2), the whole procedure ends up. Fig. 4 (d) shows the final structure D, where "113" has replaced "122" and the counter for "120" has been moved to a bucket with bigger value.*

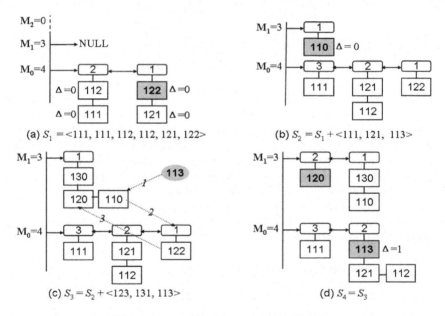

Fig. 4. An example of the updates in *Separator*

When the data are weakly skewed or uniformly distributed, the update operation sometimes cannot find any counter for a new element, as illustrated below. Fortunately, practical data streams typically exhibit significant skewness [15].

Example 3. *Consider a data Stream S = <117, 131, 127, 149, 112, 168, 155, 151, 163>. ε = 0.5, M_0 = 2, M_1 = 4, M_2 = 1. All counters of D are taken after processing the fraction S_1, as shown in Fig. 5(a). Then the arrival of "151" (v=1) starts some* *floating updates, among which D[2].update("100", 1, 0, true) locates "100", an ancestor of "151" (Step 1 in Fig. 5(b)). Since F(100) + v = 2 > D[1].minf, we have to reset the counter for "160" to monitor "150", the ancestor of "151" at L₁ (Step 2). The resulting D[2].update should have transferred the information of "160" to the counter for its parent "100", however, F(160)− Δ(160) + F(100) = 2 >D[1].minf. So the counter for "110" is taken to monitor the re-sinking "160" (Step 3). These "floating" and "re-sinking" keep going on. In the end, we cannot find any counter in D[1] to monitor the re-sinking "120" (Step 10).*

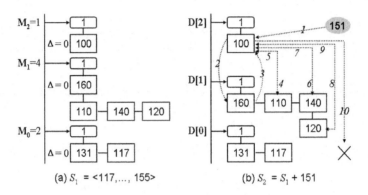

Fig. 5. An example of counter shortage

In order to solve this problem, an extra structure E is created as soon as counter shortage occurs for the first time. Like D, it is comprised of $(h + 1)$ *Stream-Summaries*. Every time counter shortage happens, we firstly compress D from bottom to up. Each entry in $D[k]$ $(0 \le k <h)$ whose $F \le \varepsilon N$ are removed, with its $(F − \Delta)$ added to its parent entry in $D[k+1]$. If the parent entry does not exist, we will insert one, setting its Δ to the F value of its closest ancestor existing in D. Then we transfer the entries left at each level of D to the same level of E, making D vacant for new elements. Within E, later transferred entries are integrated with earlier ones, and those becoming infrequent will be removed from E through a compression similar to the previous one executed on D.

4.3 Space Complexity and Error Bound

Lemma 1. *Consider a stream fraction S_p of length N_p, which proceeds between two successive counter shortages. For any node c in the domain tree T, its overestimation during this period of time, $\Delta_p(c)$, is no larger than εN_p.*

Proof. During this period of time, *Separator* adds every count to one and only one counter in D, so the aggregate counts of every $D[k]$ $(0 \le k \le h)$ are no larger than N_p. Any node c monitored in $D[k]$ at the end of this period has its Δ assigned when taking its counter. If this counter was empty, $\Delta_p(c) = 0$. Otherwise, $\Delta_p(c) = F(p(c))$, where $p(c)$ is c's parent node. Assume it is in $D[k].update(0 \le k \le h)$ that c took its counter. $M_k \cdot D[k].minf \le \sum F(t)$ ($\forall$ entry t in $D[k]$) $\le N_p$, where $M_k \le h \cdot \varepsilon^{-1} \cdot (h−k)^{-1}$ is the number of counters in $D[k]$. Hence $D[k].minf \le N_p / M_k \le \varepsilon N_p \cdot (h−k) \cdot h^{-1} \le \varepsilon N_p$. We indicate the minimal bucket value of $D[k]$ before $D[k].update$ as $D[k].pre\_minf$. $D[k].update$ is started once $F(p(c)) + v > D[k].pre\_minf$. Before that, $F(p(c)) \leq D[k].pre\_minf \leq D[k].minf \leq \varepsilon N_p$. Thus $\Delta_p(c) \leq \varepsilon N_p$ no matter $\Delta_p(c) = 0$ or $F(p(c))$. If c is not monitored in D at the end of this period, then c did not appear in S_p or its entry in D was deleted. In the first case, $\Delta_p(c) = 0$. In the second case, $\Delta_p(c)$ amounts to the $F(c)$ when c was ultimately deleted. We assume c was monitored in $D[k]$ ($0 \leq k < h$) at that deletion. Since entry deletion in $Separator$ appears in the form of entry replacement, c was deleted with a value of $D[k].minf$. As proved before, $D[k].minf \leq \varepsilon N_p$. So $\Delta_p(c) \leq \varepsilon N_p$.   ∎

Theorem 1. *For any node c of the domain tree T, the absolute value of its overestimation $|\Delta(c)| \leq \varepsilon N$, where N is the current stream length.*

Proof. Consider a data Stream S of length N. When $N = N_1$, counter shortage happens for the first time, and it reoccurs when $N = N_2$ (the current stream length) $> N_1$. Assume a node $c \in A_k$ ($0 \leq k \leq h$), whose maximal possible overestimation at these two times are respectively Δ_1 and Δ_2. From Lemma 1, $\Delta_1 \leq \varepsilon N_1$, $\Delta_2 \leq \varepsilon (N_2 - N_1)$. Therefore, $|\Delta(c)| \leq |\Delta_1 \pm \Delta_2| \leq \varepsilon N_2$.

Lemma 2. *For a given ε and h, data structure D needs no larger than $O(h^2 / \varepsilon)$ space.*

$$Proof. \sum_{i=0}^{h-1} M_i = \sum_{i=0}^{h-1} \frac{h}{(h-i)\varepsilon} = \frac{h}{\varepsilon} \sum_{i=0}^{h-1} \frac{1}{h-i} \leq \frac{h^2}{\varepsilon} .$$ ∎

Lemma 3. *The current stream length, N, is the upper bound of the aggregate counts of any level in E.*

Theorem 2. *Given ε and h, Separator finds HHHs using no larger than $O(h^2/\varepsilon)$ space.*

Proof. From Lemma 3, $\sum F(t) \leq N$, $\forall$ entry t in $E[k]$ ($0 \leq k \leq h$). Hence the entries at each level of E, whose $F > \varepsilon N$, are less than $\sum F(t) / \varepsilon N \leq N /\varepsilon N = 1/\varepsilon$. On compressing E, the total number of counters will keep to be h/ε, though the distribution of counters among different levels may vary. During merging D and E, the space usage at most doubles. Thus the space complexity of E is $O(h/\varepsilon)$. From Lemma 2, D needs no more than h^2/ε space. So the space requirements of $Separator$, which mostly come from structure D and E, are no larger than $O(h^2/\varepsilon)$.   ∎

5 Experimental Evaluation

A number of experiments are conducted to compare $Separator$ with some other HHH detection algorithms, which are briefly referred to as follows: (1) *Top-T*: the top-down approach proposed in [13]; (2) *LC-T-S2*: implementation of Strategy 2 in [2]; (3) *LC-T-S4*: implementation of Strategy 4 in [2], using a tree structure; (4) *LC-A-S4*: implementation of Strategy 4, using a set of *Stream-Summary* structures. We expect *LC-A-S4* to outperform *LC-T-S4* in accuracy.

Different HHH detection algorithms were evaluated based on their abilities to deal with the best and the worst kinds of data streams (or refer to as the "best" and "worst" cases). By saying "the best case", we mean highly skewed data streams with a shallow/bushy domain hierarchy. According to our preliminary experiments, all algorithms perform better when dealing with this kind of streams. Conversely, a weakly skewed or uniformly distributed data stream with a deep/thin hierarchy is "the worst case", challenging the algorithms by more time/space usage and lower accuracy.

Four accuracy metrics are used: (1) recall based on total frequency (*Recall*), (2) precision based on total frequency (*Precision*), (3) recall based on discounted frequency (*D-Recall*), and (4) precision based on discounted frequency (*D-Precision*). We define (1) and (3) as a percentage of the number of detected correct entries to the number of all correct entries, (2) and (4) as a percentage of the number of detected correct entries to the entire output. To (1) and (2), an entry is correct *iff.* its exact total frequency $f^* \geq \lfloor \varphi N \rfloor$. To (3) and (4), a correct entry is a HHH with $df^* \geq \lfloor \varphi N \rfloor$. The resource metrics include runtime and space usage. For the sake of fairness, the space used by auxiliary structures, such as hash tables, was also counted in.

All algorithms were implemented in C++ and compiled by VC++ 6.0. We ran our experiments on an Intel Pentium D 3.40GHz PC with 1.00GB RAM, using synthetic datasets generated by our developed generator. It produces data streams with domain hierarchy information embedded in each element. The dataset is controlled by several parameters such as number of leaf nodes (i.e., b_0), number of levels (i.e., $h+1$), the average number of child nodes (i.e., avg) and *Zipf* [16] skew factor. The best kind of data stream we used was generated with $\alpha = 1.5$, $h = 3$ and $avg = 9$; the worst kind of stream was generated with $\alpha = 0.5$, $h = 6$ and $avg = 3$.

5.1 Accuracy and Resource Requirements

The following experiments are conducted with the stream length $N = 10^6$, the number of leaf nodes $b_0 = 700$. $\varphi = 0.02$, $\varepsilon = 0.1\varphi$. The results are shown in Fig. 6, 7 and 8.

Our results verify the space efficiency of Strategy 4 in [2], but the algorithms show a weakness in accuracy. Comparing with *LC-T-S4*, *LC-A-S4* returns more accurate answers at the expense of more runtime and memory. Both their strengths and weakness stem from the fact that Strategy 4 allows to delete intermediate nodes. Unlike them, *LC-T-S2* and *Top-T* only prune "fringe nodes" with no children. The preservation of upper-level information results in their estimation correctness, but maintaining complete synopses entails more space, especially in the worst case. Due to its higher compression frequency, *LC-T-S2* is more space efficient than *Top-T*.

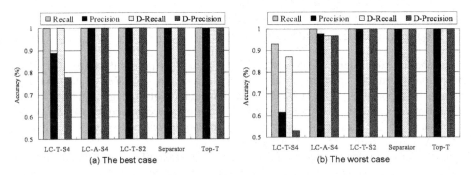

Fig. 6. The accuracy comparison

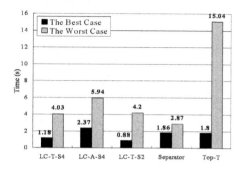

Fig. 7. The runtime comparison

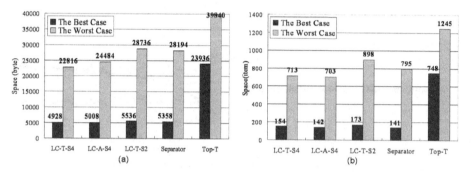

Fig. 8. The comparison of space usage. For each algorithm, we recorded its space-usage values (in forms of both byte and item number) after every stream item, and chose the biggest one for comparison.

Separator combines the error-control strategies of top-down [13] and bottom-up [2] methods, which lead to its correctness. It also benefits from the space-control strategy of *SpaceSaving*. As shown in Fig. 8, *Separator* has space usage lower than *LC-T-S2* and *Top-T*. From Fig. 7, *Separator* is the fastest of all algorithms in the worst case.

5.2 Scalability

Here we compare the scalability of different algorithms as (1) the data stream grows, and (2) the error rate varies.

In the first set of experiments, N varied from 10^4 to 10^6, $b_0 = 700$, $\varphi = 0.02$, $\varepsilon = 0.1\varphi$. The results are shown in Fig. 9. The space usage of *LossyCounting*-adapted algorithms slightly fluctuates around certain values, with *LC-T-S2* requiring the most. The serrated curve in Fig. 9(c) indicates the weak, long-interval space compression of *Top-T*, which is the least space efficient of all. The runtime of *Separator* increases most slowly in the worst case. Its space usage grows at the beginning, but soon settles down at a level close to that of *LC-T-S2*.

In the second set of experiments, $N = 10^6$, $b_0 = 700$, $\varphi = 0.02$, ε varied from 2×10^{-6} to 0.02. The results are shown in Fig. 10. Bottom-up algorithms run slower as ε rises, whereas the top-down *Top-T* becomes faster. As a combination of bottom-up and

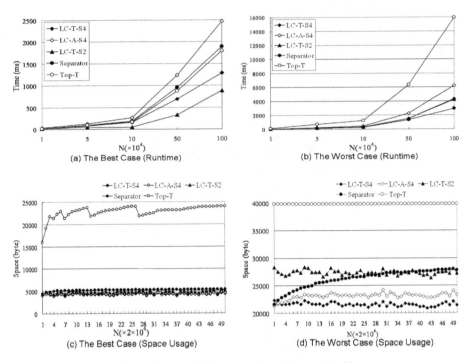

Fig. 9. The scalability comparison (stream length)

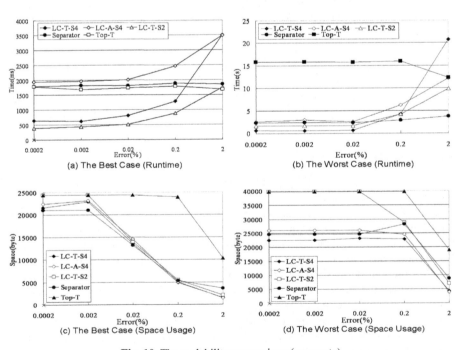

Fig. 10. The scalability comparison (error rate)

top-down strategies, *Separator* inherently counterbalances these two tendencies. The space of all algorithms extends as ε decreases. Among them, *Top-T* and *LC-T-S2* gradually exceeds the other three. *Separator* has relatively slow space expansion, actually the slowest in the best case.

6 Conclusions

Concluding the above, we may state the following:

--*Separator* detects hierarchical heavy hitters from one-dimensional insert-only data streams, utilizing a *Stream-Summary*-based synopsis as well as a combination of top-down and bottom-up processing strategies.

--Whereas existing algorithms either trade accuracy for time/space efficiency, or vice versa, *Separator* achieves a satisfying balance between accuracy and resource requirements. It has better scalability as the stream length or the error rate changes.

--*Separator* facilitates users' choice between accuracy and space usage.

Acknowledgments. This work was supported in part by the National Natural Science Foundation of China under Grant No. 70471006 and 70621061.

References

1. Gaber, M.M., Zaslavsky, A., Krishnaswamy, S.: Mining Data Streams: A Review. SIGMOD Record. 2, 18–26 (2005)
2. Cormode, G., Korn, F., Muthukrishnan, S., Srivastava, D.: Finding Hierarchical Heavy Hitters in Data Streams. In: Proc. 29th ACM VLDB, pp. 464–475 (2003)
3. Estan, C., Varghese, G.: New Directions in Traffic Measurement and Accounting. In: Proc. 1st ACM SIGCOMM Workshop on Internet Measurement, pp. 75–80 (2001)
4. Metwally, A., Agrawal, D., Abbadi, A.E.: Efficient Computation of Frequent and Top-k Elements in Data Streams. In: Eiter, T., Libkin, L. (eds.) ICDT 2005. LNCS, vol. 3363, pp. 398–412. Springer, Heidelberg (2004)
5. Charikar, M., Chen, K., Farach-Colton, M.: Finding Frequent Items in Data Streams. In: Widmayer, P., Triguero, F., Morales, R., Hennessy, M., Eidenbenz, S., Conejo, R. (eds.) ICALP 2002. LNCS, vol. 2380, pp. 693–703. Springer, Heidelberg (2002)
6. Manku, G., Motwani, R.: Approximate Frequency Counts over Data Streams. In: Proc. 28th ACM VLDB, pp. 346–357 (2002)
7. Demaine, E.D., López-Ortiz, A., Munro, J.I.: Frequency Estimation of Internet Packet Streams with Limited Space. In: Möhring, R.H., Raman, R. (eds.) ESA 2002. LNCS, vol. 2461, pp. 348–360. Springer, Heidelberg (2002)
8. Cormode, G., Muthukrishnan, S.: Whats Hot and What's Not: Tracking Most Frequent Items Dynamically. In: Proc. 22nd ACM PODS, pp. 296–306 (2003)
9. Cormode, G., Muthukrishnan, S.: An Improved Data Stream Summary: The Count-Min Sketch and Its Applications. In: Farach-Colton, M. (ed.) LATIN 2004. LNCS, vol. 2976, pp. 29–38. Springer, Heidelberg (2004)
10. Estan, C., Savage, S., Varghese, G.: Automatically Inferring Patterns of Resource Consumption in Network Traffic. Computer Communication Review. 4, 137–150 (2003)

11. Gibbons, P.B., Matias, Y.: Synopsis Data Structures for Massive Data Set. DIMACS Series in Discrete Mathematics and Theoretical Computer Science, 39–70 (1999)
12. Muthukrishnan, S.: Data Streams: Algorithms and Applications. In: ACM-SIAM Symp. Discrete Algorithms (2003), Available http://athos.rutgers.edu/m uthu/stream-1-1.ps
13. Zhang, Y., Singh, S., Sen, S., Duffield, N., Lund, C.: Online Identification of Hierarchical Heavy Hitters: Algorithms, Evaluation, and Applications. In: Proc. IMC, pp. 101–114 (2004)
14. Hershberger, J., Shrivastava, N., Suri, S., Toth, C.D.: Space Complexity of Hierarchical Heavy Hitters in Multi-Dimensional Data Streams. In: Proc. 24th PODS, pp. 338–347 (2005)
15. Cormode, G., Muthukrishnan, S.: Summarizing and Mining Skewed Data Streams. In: Proc. 5th SDM (2005)
16. Knuth, D.E.: The Art of Programming. Addison-Wesley, London (1973)

Spatial Fuzzy Clustering Using Varying Coefficients

Huaqiang Yuan, Yaxun Wang, Jie Zhang, Wei Tan, Chao Qu, and Wenbin He

Department of Computer Science, DongGuan University of Technology
DongGuan, GuangDong 523808, China
hyuan66@163.com

Abstract. To consider spatial information in spatial clustering, the Neighborhood Expectation-Maximization (NEM) algorithm incorporates a spatial penalty term in the objective function. Such an addition leads to multiple iterations in the E-step. Besides, the clustering result depends mainly on the choice of the spatial coefficient, which is used to weigh the penalty term but is hard to determine a priori. Furthermore, it may not be appropriate to assign a fixed coefficient to every site, regardless of whether it is in the class interior or on the class border. In estimating class posterior probabilities, sites in the class interior should receive stronger influence from their neighbors than those on the border. To that end, this paper presents a variant of NEM using varying coefficients, which are determined by the correlation of explanatory attributes inside the neighborhood. Our experimental results on real data sets show that it only needs one iteration in the E-step and consequently converges faster than NEM. The final clustering quality is also better than NEM.

1 Introduction

Compared to conventional data, the attributes under consideration for spatial data include not only non-spatial normal attributes, but also spatial attributes that describe the object's spatial information such as location and shape. The assumption of independent and identical distribution is no longer valid for spatial data. In practice, almost every site is related to its neighbors. To that end, Ambroise et al. proposed the Neighborhood Expectation-Maximization (NEM) algorithm [1], which incorporates a spatial penalty term in the objective function to encourage neighboring sites with similar class posterior probabilities. In contrast to the standard EM algorithm [2] that maximizes likelihood alone, such an addition involves multiple iterations in the E-step. Besides, the clustering results rely heavily on the spatial coefficient, which specifies the degree of spatial smoothness in the clustering solution but is hard to determine a priori in practice. Furthermore, it may not be appropriate to assign a fixed coefficient to every site, regardless of whether it is in the class interior or on the class border.

Based on the observation above, this paper presents a Neighborhood EM algorithm using Varying coefficients (NEMV). Rather than set empirically, the

R. Alhajj et al. (Eds.): ADMA 2007, LNAI 4632, pp. 183–190, 2007.

coefficient is determined by the correlation of true explanatory attributes inside the neighborhood. Our experimental results on real data sets show that it only needs one iteration in the E-step and consequently converges faster than NEM. The final clustering quality is also consistently better than NEM.

The rest of the paper is organized as follows. Section 2 reviews the problem background and related work. In Section 3, we first outline NEM and then present our NEMV algorithm. Experimental evaluation is reported in Section 4. Finally Section 5 concludes this paper with a summary and discussion of future work.

2   Background and Related Work

In this section, we first introduce the background by formulating the problem. Then we briefly review related work.

2.1   Problem Formulation

The goal of spatial clustering is to partition data into groups or clusters so that pairwise dissimilarity, in both attribute space and spatial space, between those assigned to the same cluster tend to be smaller than those in different clusters. Let S denote the set of locations, e.g., the set of triple (index, latitude, longitude). Spatial clustering can be formulated as an unsupervised classification problem. We are given a spatial framework of n sites, $S = \{s_i\}_{i=1}^n$ with a neighbor relation $N \subseteq S \times S$. Sites s_i and s_j are neighbors iff $(s_i, s_j) \in N, i \neq j$. Let $N(s_i) \equiv \{s_j : (s_i, s_j) \in N\}$ denote the neighborhood of s_i. We assume N is given by a contiguity matrix W whose $W(i,j) = 1$ iff $(s_i, s_j) \in N$ and $W(i,j) = 0$ otherwise. Associated with each s_i, there is a d-dimensional feature vector of normal attributes $\mathbf{x}_i \equiv \mathbf{x}(s_i) \in \Re^d$. We need to find a many-to-one mapping $f : \{\mathbf{x}_i\}_{i=1}^n \to \{1, ..., K\}$. If each object $\mathbf{x}_i$ has a true class label $y_i \in \{1, ..., K\}$, naturally the ultimate goal is to maximize similarity between obtained clustering and true classification. However, since the class information is unavailable during learning, the objective in practice is to optimize some criterion function such as likelihood. Besides, spatial clustering imposes the following constraint of spatial autocorrelation. y_i is not only affected by $\mathbf{x}_i$, but also by $(\mathbf{x}_j, y_j)$ of its neighbors $N(s_i)$. Hence it is more appropriate to model the distribution of y_i with $P(y_i \mid \mathbf{x}_i, \{(\mathbf{x}_j, y_j) : s_j \in N(s_i)\})$ instead of $P(y_i | \mathbf{x}_i)$.

2.2   Related Work

Many methods have been proposed to incorporate spatial information in the clustering process. The simplest one is adding spatial information, e.g., spatial coordinates, directly into datasets [3]. Others achieve this goal by modifying existing algorithms, e.g., allowing an object assigned to a class if and only if this class already contains its neighbor [4]. Another class, where our algorithm falls, selects a model that encompasses spatial information [1]. This can be achieved by modifying a criterion function that includes spatial constraints [5], which mainly comes from image analysis where Markov random field and EM style algorithms were intensively used [6,7].

Clustering using mixture models with EM can be regarded as a soft K-means algorithm in that the output is posterior probability rather than hard classification. It does not account for spatial information and usually cannot give satisfactory performance on spatial data. NEM extends EM by adding a spatial penalty term in the criterion, but this makes it need more iterations in each E-step. If further information about structure is available, the structural EM algorithm may be used to learn Bayesian networks for clustering [8]. In our case, we assume that soft constraints can be derived with locations of sites. Another relevant problem is semi-supervised clustering, where some pairs of instances are known belonging to same or different clusters [9]. In their case, the goal is to fit the mixture model to the data while minimizing the violation of hard constraints.

3   The NEMV Algorithm

In this section, we first outline the basics of NEM. Then we present the NEMV algorithm.

3.1   NEM for Spatial Clustering

We assume the data $X = \{\mathbf{x}\}_{i=1}^n$ come from a mixture model of K components $f(\mathbf{x}|\Phi) = \sum_{k=1}^K \pi_k f_k(\mathbf{x}|\theta_k)$, where π_k is k-th component's prior probability, missing data (cluster label) $y \in \{1, ..., K\}$ indicate which component $\mathbf{x}$ comes from, i.e., $p(\mathbf{x}|y = k) = f_k(\mathbf{x}|\theta_k)$, and Φ denotes the set of all parameters. Because it is hard to directly maximize the sample likelihood $L(\Phi) = \sum_{i=1}^n \ln[f(\mathbf{x}_i|\Phi)]$, EM tries to iteratively maximize L in the context of missing data y. Let $\overline{P}$ denote a set of fuzzy classifications representing the grade of membership of $\mathbf{x}_i$ to class (component) k, i.e., $\{\overline{P}_{ik} \equiv \overline{P}(y_i = k)\}$. As highlighted in [10], the new objective function U of NEM that incorporates a spatial penalty term can be written as

$$U(\overline{P}, \Phi) = F(\overline{P}, \Phi) + \beta G(\overline{P})$$

where

$$F(\overline{P}, \Phi) = E_{\overline{P}}[\ln(P(\{\mathbf{x}, y\}|\Phi))] + H(\overline{P})$$
$$= \sum_{i=1}^n \sum_{k=1}^K \overline{P}_{ik} \ln(\pi_k f_k(\mathbf{x}|\theta_k)) - \sum_{i=1}^n \sum_{k=1}^K \overline{P}_{ik} \ln \overline{P}_{ik}$$

$$G(\overline{P}) = \frac{1}{2} \sum_{i=1}^n \sum_{j=1}^n W_{ij} \sum_{k=1}^K \overline{P}_{ik} \overline{P}_{jk}$$

Compared to standard EM, in addition to maximizing $L(\Phi)$ which is achieved by maximizing $F(\overline{P}, \Phi)$, NEM also tries to increase $G(\overline{P})$, the spatial penalty that encourages neighboring sites with similar class posterior probabilities. The spatial penalty is weighted by $\beta > 0$, a fixed coefficient that determines the degree of smoothness in the solution clustering. U can be maximized by alternately estimating its two parameters $\overline{P}$ and Φ. Starting from an initial $\overline{P}^0$, NEM iterates the following two steps:

1. M-step: With $\overline{P}^t$ fixed, set $\Phi^t = \text{argmax}_\Phi U(\overline{P}^t, \Phi)$, which is exactly the same as the M-step in EM, for G dose not depend on Φ.
2. E-step: With Φ^t fixed, set $\overline{P}^{t+1} = \text{argmax}_{\overline{P}} U(\overline{P}, \Phi^t)$ by applying Eq. (1) repeatedly until convergence.

$$\overline{P}^*_{ik} = \frac{\pi_k f_k(\mathbf{x}_i|\theta_k)\exp\left(\beta \sum_{j=1}^n W_{ij}\overline{P}^*_{jk}\right)}{\sum_{l=1}^K \pi_l f_l(\mathbf{x}_i|\theta_l)\exp\left(\beta \sum_{j=1}^n W_{ij}\overline{P}^*_{jl}\right)} \tag{1}$$

3.2   NEM with Varying Coefficients

EM is not appropriate for spatial clustering since it does not account for spatial information. In contrast, although NEM incorporates spatial information, it requires multiple iterations in E-step and the spatial coefficient is hard to determine a priori. To overcome these difficulties, we propose NEMV, which is based on the observation that it may not be appropriate to assign a constant coefficient to every site. For those in the class interior, the whole neighborhood is from the same class and hence the site should receive more influence from its neighbors, especially when their posterior estimates are accurate. For those on the class border, because their neighbors are from different classes, its own class membership should be determined mostly by its own explanatory attributes.

Along this line, NEMV employs a site-sensitive spatial coefficient, the local Moran's I measure, which is determined by the correlation of explanatory attributes inside the neighborhood [11]. Let z_{ip} denote the normalized p-th attribute of site s_i, i.e., $z_{ip} = x_{ip} - \overline{x}_p$, where $\overline{x}_p$ is the global mean of the p-th attribute. Let σ_p denote the global standard deviation of the p-th attribute. Then, for the p-th attribute at site s_i, the local I measure is defined as $I_{ip} = \frac{z_{ip}}{\sigma_p^2} \sum_j W_{ij} z_{jp}$, where W is a row-normalized (sum to 1) version of the original binary W. A high I (e.g., $I > 1$) implies a high local spatial autocorrelation at site s_i, which is likely to occur in the class interior. In NEMV, β_i is obtained by first averaging I_{ip} over all attributes and then normalizing to $[0, 1]$, i.e., $I_i = \text{mean}_p(I_{ip})$, $\beta_i = \frac{I_i - \min_i\{I_i\}}{\max_i\{I_i\} - \min_i\{I_i\}}$. Then the new penalty and criterion become

$$G = \frac{1}{2} \sum_{i=1}^n \beta_i \sum_{j=1}^n W_{ij} \sum_{k=1}^K \overline{P}_{ik}\overline{P}_{jk}$$
$$U(\overline{P}, \Phi) = F(\overline{P}, \Phi) + G(\overline{P})$$

Besides, we find that one iteration is usually enough for the E-step in NEMV. Therefore, NEMV proceeds as follows.

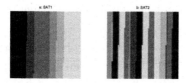

Fig. 1. Satimage data with site's location synthesized: (a) SAT1 (contiguity ratio 0.96) and (b) SAT2 (contiguity ratio 0.89)

1. E-step: Set $\overline{P}^t = \mathrm{argmax}_{\overline{P}} U(\overline{P}, \varPhi^{t-1})$ by applying Eq. (1) once, where β has been replaced by β_i in Eq. (1).
2. M-step: Set $\varPhi^t = \mathrm{argmax}_{\varPhi} U(\overline{P}^t, \varPhi)$, which is exactly the same as the M-step in EM.

4 Experimental Evaluation

In this section, we first introduce the clustering validation measures used in our experiments. Then we report comparative results on two real datasets.

4.1 Performance Criteria

If every site has a true class label, they can be used to evaluate the final clustering quality via external validation measures. Let $C, Y \in \{1, ..., K\}$ denote the true class label and the cluster label, respectively. Then clustering quality can be measured with conditional entropy $H(C|Y)$ defined in Eq. (2), which equals zero if their distributions are the same. We also use a more intuitive measure, error rate $E(C|Y)$, which computes the misclassified fraction of data in each cluster after assuming the true class label should be the major class in the cluster. The above two measures are only for the discrete target value. When the target variable C is continuous, we calculate the standard deviation defined in Eq. (3), where $\mathrm{std}(\cdot)$ denotes the standard deviation operator and $(C|Y = k)$ denotes the C's values in cluster $Y = k$.

$$H(C|Y) = \sum_{k=1}^{K} P(Y = k) \times H(C|Y = k) \qquad (2)$$

$$S(C|Y) = \sum_{k=1}^{K} P(Y = k) \times \mathrm{std}(C|Y = k) \qquad (3)$$

4.2 Experimental Results

Satimage Dataset. We first evaluate NEMV on a real landcover dataset, Satimage, which is available at the UCI repository. It consists of the four multi-spectral values of pixels in a satellite image together with the class label from a

Table 1. Clustering performance on the Satimage dataset: [+]SAT1 and [*]SAT2

			SAT1		SAT2	
	supervised	EM	NEM	NEMV	NEM	NEMV
entropy	0.5121	0.6320	0.5391	0.5094	0.5635	0.5340
error	0.1508	0.2315	0.2039	0.1816	0.2142	0.2004
$-U(10^4)$	5.1884[+]	5.1406[+]	5.1029	5.0926	5.1416	5.1102
	5.2274[*]	5.1717[*]				
$-L(10^4)$	5.8128	5.7711	5.8207	5.7842	5.8141	5.7804

Table 2. Clustering performance on the Satimage dataset by NEMV with varying number of iterations of E-step

	SAT1				SAT2		
#E-step	1	10	20	30	1	5	10
entropy	0.5094	0.5092	0.5088	0.5086	0.5340	0.5332	0.5330
error	0.1816	0.1813	0.1810	0.1809	0.2004	0.2001	0.2000
$-U(10^4)$	5.0926	5.0916	5.0913	5.0912	5.1102	5.1101	5.1099
$-L(10^4)$	5.7842	5.7836	5.7834	5.7833	5.7804	5.7802	5.7801

six soil type set. Because the dataset is given in random order and there is no spatial location, we synthesize their spatial coordinates and allocate them in a 64×69 grid. 4-neighborhood is used in construction of W and contiguity ratio is computed as the fraction of edges shared by the pixels from the same class. To emphasize spatial autocorrelation, we generate two images SAT1 and SAT2 in Fig. 1(a) and (b) with high contiguity ratios 0.96 and 0.89, respectively.

We test NEM and set $\beta = 1$ empirically to maximize U. With random initialization, Table 1 gives the average results of 10 runs recorded at maximum L for EM, and maximum U for NEM and NEMV. The U values for EM are computed using the definition in NEM. For clarity, we report $-L$ and $-U$ so that all criteria in the tables are to be minimized. Note that due to different β used in NEM and NEMV, it is meaningless to compare U for them. For comparison, we also list the results under supervised mode where each component's parameters are estimated with all data from a single true class. We can see that the entropy and error generally decrease as $-U$, rather than $-L$, decreases. NEMV gives better results than NEM. As expected, both of them perform better on SAT1 than on SAT2, for the former's contiguity ratio is higher and hence fits our assumption more.

To see if one iteration of E-step is really enough in NEMV, we perform a series of experiments by varying the number of iterations of E-step. The average results of 10 runs are shown in Table 2. Note that 30/10 is the number of iterations of E-step we used in the standard NEM. Although the computational cost has been increased by an order of magnitude, we can see that the improvement is not significant, especially in error rate and U.

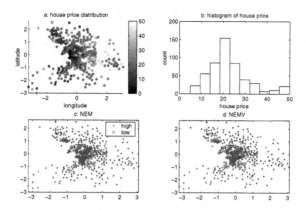

Fig. 2. (a) shows house price distribution in 506 towns in Boston area. The corresponding histogram is plotted in (b). Two sample clustering results are shown in (c) and (d) for NEM and NEMV, respectively.

Table 3. Clustering performance on the house dataset

	EM	NEM	NEMV
std	8.3377	8.3486	8.3088
$-U(10^4)$	1.2580	1.2675	1.2557
$-L(10^4)$	1.3942	1.4014	1.3966

House Dataset. We also evaluate NEMV on a real house price dataset with 12 explanatory variables, such as nitric oxides concentration and crime rate. The clustering performance is evaluated with the target variable, median values of owner-occupied homes, which is expected to has a small spread in each cluster. Fig. 2(a) shows the true house values of 506 towns in Boston area. Their histogram is plotted in Fig. 2(b). Using a Gaussian mixture of two components, we set $\beta = 1$ for NEM and about 20 iterations are needed for convergence in its E-step. Table 3 gives the average results of 10 runs. One can see that NEM performance is slightly worse than EM in terms of either standard deviation or U. But NEMV still gives the best results. Two sample clustering results are shown in Fig. 2(c) and (d) for NEM and NEMV, respectively. We can see that NEM yields a clustering with even stronger spatial continuity than that of NEMV, which may not be appropriate for such a mixed dataset with many borders sites on the class boundary.

5 Conclusion

Compared to EM, the incorporation of a weighted spatial penalty term into the criterion function makes NEM need multiple iterations in each E-step. Besides, it is difficult to determine the spatial coefficient, on which the clustering results depend heavily. This paper presented a variant of NEM algorithm using Variable coefficients (NEMV). The site-sensitive coefficients are determined by the local Moran's I measure using correlation of explanatory attributes inside the neighborhood. Empirical results on real data sets indicated that it not only led to better results than NEM, but also converged faster with only one iteration needed in the E-step. For future work, we plan to investigate online or stochastic versions of EM to reduce dependence on initialization. Other optimization techniques, such as genetic algorithms [12], are worth trying to improve convergence rate and final clustering quality. Finally, theoretical analysis and justification are also needed for NEMV.

References

1. Ambroise, C., Govaert, G.: Convergence of an EM-type algorithm for spatial clustering. Pattern Recognition Letters 19(10), 919–927 (1998)
2. Dempster, A.P., Laird, N.M., Rubin, D.B.: Maximum likelihood from incomplete data via the EM algorithm. Journal of Royal Statistical Society B39, 1–38 (1977)
3. Guo, D., Peuquet, D., Gahegan, M.: Opening the black box: Interactive hierarchical clustering for multivariate spatial patterns. In: Proceedings of the 10th ACM International Symposium on Advances in Geographic Information Systems, pp. 131–136 (2002)
4. Legendre, P.: Constrained clustering. In: Legendre, P., Legendre, L. (eds.) Developments in Numerical Ecology, NATO ASI Series G 14, pp. 289–307 (1987)
5. Rasson, J.P., Granville, V.: Multivariate discriminant analysis and maximum penalized likelihood density estimation. Journal of the Royal Statistical Society B57, 501–517 (1995)
6. Geman, S., Geman, D.: Stochastic relaxation, Gibbs distributions and the Bayesian restoration of images. IEEE Transactions on Pattern Analysis and Machine Intelligence 6, 721–741 (1984)
7. Solberg, A.H., Taxt, T., Jain, A.K.: A markov random field model for classification of multisource satellite imagery. IEEE Transactions on Geoscience and Remote Sensing 34(1), 100–113 (1996)
8. Pena, J.M., Lozano, J.A., Larranaga, P.: An improved Bayesian structural EM algorithm for learning Bayesian networks for clustering. Pattern Recognition Letters 21(8), 779–786 (2000)
9. Basu, S., Bilenko, M., Mooney, R.J.: A probabilistic framework for semi-supervised clustering. In: Proceedings of the 10th ACM SIGKDD International Conference on Knowledge Discovery and Data Mining, pp. 59–68 (2004)
10. Hathaway, R.J.: Another interpretation of the EM algorithm for mixture distributions. Statistics and Probability Letters 4, 53–56 (1986)
11. Shekhar, S., Chawla, S.: Spatial Databases: A Tour. Prentice-Hall, Englewood Cliffs (2002)
12. Pernkopf, F., Bouchaffra, D.: Genetic-based EM algorithm for learning gaussian mixture models. IEEE Transactions on Pattern Analysis and Machine Intelligence 27(8), 1344–1348 (2005)

Collaborative Target Classification for Image Recognition in Wireless Sensor Networks

Xue Wang, Sheng Wang, and Junjie Ma

State Key Laboratory of Precision Measurement Technology and Instruments, Department
of Precision Instruments, Tsinghua University, Beijing 100084, P.R. China
wangxue@mail.tsinghua.edu.cn,
wang_sheng00@mails.tsinghua.edu.cn,
mjj@mails.tsinghua.edu.cn

Abstract. Target classification, especially visual target classification, in complex situations is challenging for image recognition in wireless sensor networks (WSNs). The distributed and online learning for target classification is significant for highly-constrained WSNs. This paper presents a collaborative target classification algorithm for image recognition in WSNs, taking advantages of the collaboration for the data mining between multi-sensor nodes. The proposed algorithm consists of three steps, target detection and feature extraction are based on single-sensor node processing, whereas target classification is implemented by collaboration between multi-sensor nodes using collaborative support vector machines (SVMs). For conquering the disadvantages of inevitable missing rate and false rate in target detection, the proposed collaborative SVM adopts a robust mechanism for adaptive sample selection, which improves the incremental learning of SVM by just fusing the information from a selected set of wireless sensor nodes. Furthermore, a progressive distributed framework for collaborative SVM is also introduced for enhancing the collaboration between multi-sensor nodes. Experimental results demonstrate that the proposed collaborative target classification algorithm for image recognition can accomplish target classification quickly and accurately with little congestion, energy consumption and execution time.

1 Introduction

Wireless sensor networks (WSNs) are the key technology for the future world [1]. Because of the advantages on powerful sensing ability and intelligent information processing, WSNs are suitable for many new applications [2]. In the near future, the development of visual sensor networking technology employing content-rich vision-based sensors will require efficient distributed processing for target detection and classification. Some effective techniques for target classification have been presented in previous literatures [3, 4, 5]. But all those techniques just aim to achieve one kind of specific target classification task and they are all not competent for visual target classification. Moreover, the intrinsic characteristics of WSNs determine that the classifier should be trained and updated dynamically according to different tasks.

R. Alhajj et al. (Eds.): ADMA 2007, LNAI 4632, pp. 191–202, 2007.

Unfortunately, the previously presented techniques can not achieve this goal. It is well known that support vector machines (SVMs) have been successfully used as classification tools in WSNs [6, 7]. But training a SVM calls for solving a quadratic programming (QP) problem in a number of coefficients equal to the number of training examples [8], so it becomes infeasible in WSNs. Recently, various incremental learning methods of SVMs are proposed [8, 9, 10, 11, 12]. The key idea is to preserve only the current estimation of the decision boundary at each incremental step along with the next batch of data. However, because the missing rate and false rate in target detection is inevitable, it's very important to select correct set of samples for training SVMs. Moreover, for balancing the performance and consumption, collaborative signal processing (CSP) algorithms are also desired to aggregate the distributed data and make decisions in a reliable and efficient manner.

In this paper, we direct our research efforts to the collaborative target classification for image recognition in WSNs with a focus on collaborative SVM algorithm which is based on SV-L-incremental SVM [9] and collaborative sample selection mechanism. And we also propose a distributed visual target classification system for WSNs, where target detection and feature extraction are based on single-sensor node processing and target classification requires multi-sensor nodes collaboration.

The structure of this paper is as follows. Section II introduces the principle of incremental SVMs. Section III presents the collaborative visual target classification system in WSNs. Section IV analyzes the performance and energy consumption of target classification and compares the proposed collaborative SVM with classical SVM and SV-L-incremental SVM. And finally, section VI presents the conclusions.

2   The Principle of Incremental SVM

2.1   Basis SVM

Given a training set $S = \{(x_i, y_i)\}_{i=1}^n$, basis SVM classifier is used to find an optimal hyperplane to separate the feature vectors that belong to two classes denoted by $y_i = \{-1, +1\}$, while the distance from the hyperplane to either class is maximized. The hyperplane is determined by a vector w with minimal norm and an offset vector b. To find such a hyperplane, one must solve the following quadratic problem [13]

$$\min_{w, \xi} \Phi(w, \xi) = \frac{1}{2}\|w\|^2 + C\sum_{i=1}^n \xi_i \tag{1}$$

subject to

$$y_i(w \cdot x_i + b) \geq 1 - \xi_i \quad \text{and} \quad \xi_i \geq 0 \quad \text{for } i = 1, 2, \ldots n \tag{2}$$

where b determines the offset of the plane from the origin, the set of variables $\{\xi_i\}_{i=1}^n$ measures the amount of violation of the constraints, and C is a parameter that defines the cost of constraint violation.

A SVM is essentially a linear classifier operating in a higher dimensional space. In SVMs, a transformation $\phi(x)$ is used to convert the data from an N-dimensional input space to a Q-dimensional feature space which is relatively computation-intensive. An equivalent kernel function can be used to perform this transformation. The kernel function $K(x \cdot y)$ is defined as follows:

$$K(x \cdot y) = \phi(x) \cdot \phi(y). \tag{3}$$

With the kernel functions, the basic form of SVM is as follows:

$$f(x) = \text{sign}\left(\sum_{i=1}^{l} v_i \left(\phi(x) \cdot \phi(x_i) \right) + b \right). \tag{4}$$

The parameter v_i is used as weighting factors to determine which input vectors are actually support vectors (SVs) $(0 < v_i < \infty)$. In the numerical examples presented in this paper, we use a third degree polynomial kernel:

$$k(x_i, x_j) = \left(x_i^T x_j + 1 \right)^3 \tag{5}$$

The corresponding decision rule can be expressed as follows: a new test vector **x** belongs to class 1 when $f(x) > 0$ while x belongs to class -1 when $f(x) < 0$.

In SVMs, the location of hyperplane is specified via the weights on training samples. Only training samples that lie close to decision boundary between two classes receive non-zero weights. In fact, since their design allows the number of SVs to be small compared to the total number of training samples, they provide a compact representation of data, to which new examples can be added as they become available.

2.2 Incremental Training of SVM

Because of compact representation, SVMs promise to be an effective tool for incremental learning. Training an SVM on the SVs alone results in the same decision function as training on the whole data set. Then, for each new batch of data, a SVM can be trained on the new data and the SVs from the previous learning step [10]. Thus, instead of transmitting all the measurements, only the current estimation of the hyperplane is transmitted, which reduces very importantly the energy spent.

But the incremental SVM is based on the assumption that the batch of data is appropriate. If the batch of data is not appropriate or the concept of interest is time-varying, the statistical properties of new batch and the old batches are different. Because SVM is robust against outliers in the data, the old SVs which are acquired from the old batches in previous learning step will have little influence in the results when an SVM is trained on these SVs together with the second batch of data. The reason is that the old SVs are almost considered as outliers since it is a desired property of the SVM algorithm. The statement of target and environment may be time-varying and place-varying, so this effect is even more accentuated for adaptively updating the rules of classification in distributed sequential training of a SVM in WSNs. Then Ruping [9] proposed a SV-L-incremental SVM which makes the error on old SVs more costly than the error on new samples. This can be easily achieved by training the SVM with respect to a new loss function. Let $(x_i, y_i)_{i \in S}$ be old SVs and $(x_i, y_i)_{i \in I}$ be new examples. The improved SVM cost function is defined as follow.

$$\Phi\left(w,\xi\right)=\frac{1}{2}\|w\|^{2}+C\left(\sum_{i\in I}\xi_{i}+L\sum_{j\in S}\xi_{j}\right) \tag{6}$$

where the L is the extra factor which is used to modify the punishment of the error on the old support vector. A natural choice for L is to let L be the number of examples in the previous batch divided by the number of SVs. The key motivation behind this method is the idea for approximating the average error of an arbitrary decision function over all examples by the average error over just the SVs.

3   Collaborative SVM Based Target Classification

Although SV-L-incremental algorithm emphasizes the influence of old SVs, imprecise samples still act on the training process. Collaboration of sensor nodes can reduce their impact and decrease energy consumption and time delay by purposefully integrating the samples from specific sensor nodes. Generally, target classification involves following steps: target detection, feature extraction and target classification.

3.1   The Basis of Target Detection and Feature Extraction

Background subtraction method which learns and models background scenes statistically to detect foreground objects is adopted for target detection, because it is demonstrated as a low-cost simple but efficient method for target detection [14]. The output result is minimum boundary rectangle (MBR) of target.

The initial background model for a stationary pixel location $x=\left[m(x),n(x),d(x)\right]$ is

$$\begin{bmatrix} m(x) \\ n(x) \\ d(x) \end{bmatrix} = \begin{bmatrix} \min_{i}\{V^{i}(x)\} \\ \max_{i}\{V^{i}(x)\} \\ \max_{i}\{|V^{i}(x)-V^{i-1}(x)|\} \end{bmatrix}, \text{where } |V^{i}(x)-\lambda(x)|<2*\sigma(x). \tag{7}$$

where $V^{i}(x)$ is intensity of location in ith image of N consecutive images array V. $\sigma(x)$ and $\lambda(x)$ are standard deviation and median of intensities at location x.

During tracking, a change map is dynamically constructed by three components:

- A detection support map (gS):

$$gS\left(x,t\right)=\begin{cases} gS\left(x,t-1\right)+1 & \text{if } x \text{ is background pixel} \\ gS\left(x,t-1\right) & \text{if } x \text{ is foreground pixel} \end{cases} \tag{8}$$

- A motion support map (mS):

$$mS\left(x,t\right)=\begin{cases} mS\left(x,t-1\right)+1 & \text{if } M\left(x,t\right)=1 \\ mS\left(x,t-1\right) & \text{if } M\left(x,t\right)=0 \end{cases} \tag{9}$$

where

$$M\left(x,t\right)=\begin{cases} 1 & \text{if } \left(|I\left(x,t\right)-I\left(x,t+1\right)|>2*\sigma\right)\wedge\left(|I\left(x,t-1\right)-I\left(x,t\right)|>2*\sigma\right) \\ 0 & \text{otherwise} \end{cases} \tag{10}$$

- A change history map (hS):

$$hS(x,t) = \begin{cases} 255 & \text{if } x \text{ is foreground pixel} \\ hS(x,t-1) - \dfrac{255}{N} & \text{otherwise} \end{cases} \tag{11}$$

Let $m^c(x)$, $n^c(x)$, $d^c(x)$ be background model parameters currently being used; new background model parameters $m(x)$, $n(x)$, $d(x)$ are determined as follows:

$$\left[m(x), n(x), d(x) \right] =$$

$$\begin{cases} \left[m^b(x), n^b(x), d^b(x) \right] & \text{if } \left(gS(x) > k*N \right) \\ \left[m^f(x), n^f(x), d^f(x) \right] & \text{if } \left(gS(x) < k*N \wedge mS(x) < r*N \right) \\ \left[m^c(x), n^c(x), d^c(x) \right] & \text{otherwise,} \end{cases} \tag{12}$$

where k and r are typically 0.8 and 0.1, respectively.

Pixel x from image I^t is a foreground pixel if:

$$B(x) = \begin{cases} 0 \text{ background} & \left(\left(I^t(x) - m(x) \right) < kd_\mu \right) \vee \left(\left(n(x) - I^t(x) \right) < kd_\mu \right) \\ 1 \text{ foreground} & \text{otherwise.} \end{cases} \tag{13}$$

Here, we consistently use $k = 2$ in our system as presented in [14]. With the presented method, one example of target detection is shown in Fig.1.

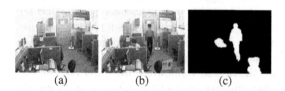

(a)   (b)   (c)

Fig. 1. Example of target detection: (a) background, (b) foreground, and (c) detection result

Features extraction refers to a process whereby a data space is transformed into a feature space. Here, principle component analysis (PCA) is used, because it can maximize the rate of decrease of variance and is therefore the right choice. It's verified that the closer the eigenvalues are to zero, the more effective the dimensionality reduction will be in preserving the information content of the original input data. So the eigen-image can be extracted as follows [15]:

First, calculate the difference between average image of M images and each image:

$$\Phi_i = \Gamma_i - \frac{1}{M} \sum_{i=1}^{M} \Gamma_i, \tag{14}$$

Then, form difference matrix $A = \left[\Phi_1 \dots \Phi_M \right]$ and compute covariance matrix C:

$$C = AA^T = \frac{1}{M} \sum_{i=1}^{M} \Phi_i \Phi_i^T \tag{15}$$

Next, calculate the eigenvectors of the covariance matrix, and construct the eigen-subspace by selecting a subset of K eigenvectors with the largest eigenvalues. Here, K is set to 50 for decreasing computational complexity in the polynomial computation in each node and ensuring the eigen-subspace can retain enough intrinsic information of the original space. For each new image, the eigen-image is calculated by projecting its difference image Φ to the eigen-subspace for providing the compact representation.

3.2 The Principle of Collaborative SVM

As mentioned before, target classification is performed by the collaboration of multi-sensor nodes using collaborative SVM which is based on the SV-L-incremental SVM with collaborative samples (nodes) selection in the progressive distributed framework for balancing the performance and consumption.

3.2.1 Progressive Distributed Framework for Collaborative SVM

Notice that under client/server framework, centralized SVM approach should send the measurements to sink node, where data processing takes place [16]. However, direct communication of raw data between sensors and sink node is infeasible. Compared with client/server framework, progressive distributed framework performs local processing in selected node for integrating new samples with old SVs, and then current node searches a new optimal node and transmits current SVs. The incremental learning is looped one by one. Finally, last node returns the result to sink node. Because the progressive distributed framework dynamically chooses a proper set of sensor nodes according to the current predictions of contributions and transmits the information in order, the congestion, energy consumption and execution time can be decreased. While the number of sensor nodes becomes greater, sensors are organized into local spatial clusters [17]. Each cluster has a cluster head for receiving the data from all other sensors in the cluster and performing data fusion. The progressive distributed framework can also be adopted between cluster head nodes.

In the progressive distributed framework, the order and number of sensors on the route traversed by the information package have a significant impact on the overall performance of incremental SVM. In practice, the routing algorithm involves a trade off between the cost and performance. The routing objective is to find a path for the incremental training of SVM that satisfies the desired classification accuracy while minimizing the energy consumption.

3.2.2 Energy-Efficient Routing

Because of the energy constraint in wireless sensor networks, energy consumption metric φ_{Cost} is most important in routing method, which normally contains three basic types: sensing cost φ_s, communication cost φ_t and information fusion cost φ_f, where communication cost is most important in WSNs. The energy $P_{j,r}$ received by sensor S_j has following relation with the energy $P_{i,t}$ transmitted by sensor S_i according to the Friis free space propagation model [18]:

$$\frac{P_{i,t}}{P_{j,r}\left(d_{i,j}\right)} = \frac{\left(4\pi\right)^2 d_{i,j}^2 \beta}{G_{i,t}G_{j,r}\lambda^2} \tag{16}$$

where $G_{i,t}$ is the gain of sensor S_i as a transmitter, $G_{j,r}$ is the gain of sensor S_j as a receiver, λ is the wavelength, β is the system loss factor. The physical distance $d_{i,j}$ between S_i and S_j is computed from their spatial locations. It means that the communication cost is proportional to the square of Euclidean distance between source and destination, so we can use $d_{i,j}^2 / d_0^2$ as a crude measure of the consumed energy, where d_0 is the predefined standard distance.

Furthermore, for prolonging the lifetime, we should consume the energy evenly among all nodes. Here, entropy theory is adopted which can measure the randomness of a given random variable to score the equality of reserved energy in each node.

$$H\left(S_{it}\right) = -\sum_{it \in S_{it}} p\left(E_{it}\right) \log p\left(E_{it}\right) \tag{17}$$

where E_{it} is the amount of energy reserved in ith sensor node at instant t, $H\left(S_{it}\right)$ is the entropy of energy reserve in WSNs. The bigger the entropy is, the more evenly the reserved energy is, so we use $1/H\left(S_{it}\right)$ to score the impact of energy consumption. The combined energy consumption metric is as follow.

$$\varphi_{Cost}\left(S_i\right) = \frac{d_{i,j}^2}{H\left(S_{it}\right)} \tag{18}$$

Then, routing of incremental SVM is determined with energy consumption metric.

3.2.3 Feedback Sensor Selection

For ensuring the performance of incremental SVM, effective samples must be taken into consideration as much as possible. However, the effectiveness is evaluated by many factors, such as precision of sensors, environment, noise and so on. So, it is difficult to directly evaluate the contribution of samples before transmission [19].

Because the empirical error on the new batch drastically outweighs the error on the old SVs, if the samples in the new batch are imprecise, the old SVs will be considered as outliers. That is, if the statistical properties of old SVs are coherent with target, the new batch of data can be considered to imprecise while the old SVs are almost considered as outliers in new SVM. Given the total number of old SVs N_s and the number of old SVs considered as outliers N_o, the ratio of N_o to N_s is considered as the metric of the effectiveness of new data, which is called effectiveness metric φ_{effect}. If $\varphi_{effect} > \gamma$, new data is considered to imprecise, where γ is the predefined threshold. Then the old SVs will be transferred to the new sensor node instead of the new SVs for reserving the appropriate statistical properties. However, this method strongly depends on the assumption that the statistical properties of old SVs are coherent with target. If old SVs are inferred from imprecise data, the evaluated φ_{effect} is meaningless. For solving this problem, the old SVs are discarded if n batches of data are considered to imprecise continuously. Then new SVs will be inferred from current batch of data.

Besides, the energy-efficient routing method doesn't consider the contribution of each sensor node. With the consideration of energy saving, the incremental SVM need not to access all sensor nodes. An access probability parameter $\varphi_{access(i)}$ is used to evaluate the contribution of the data in sensor node i. The more contribution the data have, the more probability the relative sensor node should be accessed.

During the incremental training of SVM, the order of sensors on the route traversed by the information package is determined by the energy-efficient routing method and the number of accessed sensors is determined by the access probability. The access probability equals to 1 in initialization. After a full access of all sensor nodes, the access probability $\varphi_{access(i)}$ is equal to the effectiveness metric φ_{effect} of corresponding sensor i. In next iteration, the sensor i is selected in the probability of $\varphi_{access(i)}$. Under this framework, the number of accessed sensor nodes can be decreased, which can accordingly reduce the energy consumption. This is the key motivation of sensor selection.

With the guidance of energy-efficient routing and feedback sensor selection strategy, the collaborative SVM can select the proper set of samples for incremental training, which can largely reduce the time delay and energy consumption of WSNs. The outstanding performance of proposed collaborative SVM is discussed below.

4   Experimental Results

In this section, a multi-target classification experiment is described for surveillance of people, the SVM classifier is train for human and non-human classification. For other applications, different classifiers can be trained and combined for more complex and particular classification. Because of the space limitation, just a small number of nodes are deployed and tested. But this scenario can also be considered as a part of WSN.

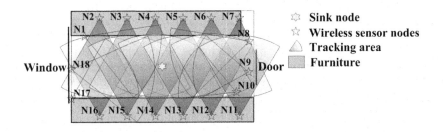

Fig. 2. The setup scenario of the distributed wireless sensor network

4.1   Deployment of Experiment

A WSN which consists of 18 wireless sensor nodes and 1 sink node is deployed in a room (see Fig.2 as a reference of the node coverage). The low deployed position (1.5m~2.5m) and complex situations increase the difficulties of target detection, classification and tracking. Each sensor node contains one image/pyroelectric-infrared sensor pair which has 60°visual angle and 3.6mm camera lens. The target detection and feature extraction is processed in each node at a frame rate of 10Hz with video down-sampled to 160×120 pixels. As illustrated in Fig.3, the raw sample binary images acquired by each node are scaled to have a resolution of 32×32 pixels, which are used for feature extraction using PCA. These were produced in different dates with different groups of people from different sensor nodes. During training of collaborative SVM, each information package carries old SVs, historical effectiveness metric information and a list of passed itinerary. Moreover, each node can locally estimate the rough energy consumption and share the reserved energy information per minute. Table.1. illustrates the communication parameters of WSN.

(a)   (b)

Fig. 3. Reduced resolution training images of (a) human and (b) non-human

Table 1. Communication parameters of WSN

Parameters	Value
Data rate	19.2kbps
Basic frequency	900MHz
MAC protocol	CSMA/CA
Bandwidth	7.2MHz

Fig. 4. The results of multi-taget detecting and classifying, where the solid rectangle represents human target and the dashed rectangle represents non-human target

4.2 Multi-target Classification Results and Performance Comparison

As shown in Fig.4, with extensive tracking experiments in individual sensor nodes, we can make a conclusion that the proposed collaborative target classification system which adopts collaborative SVM for image recognition performs well on target detection and classification. The classification accuracy of collaborative SVM for human type and non-human type are 91.3% and 95.1% respectively in a large set of over 1000 images.

The misclassifications are caused by the irregular poses of human-being, such as stoop, squat and grovel. Then the performance of the proposed collaborative SVM is compared with classical SVM and SV-L-incremental SVM. The results are illustrated in Table 2, the collaborative SVM performs better than other two algorithms, while SV-L-incremental SVM has the same accuracy with the classical SVM.

Table 2. Comparison of cross-validated accuracy of the classifier trained on classical SVM, SV-L-incremental SVM, collaborative SVM

Category	Classical SVM	SV-L-incremental SVM	Collaborative SVM
Human	89.8%	89.9%	91.3%
Non-human	93.9%	93.3%	95.1%

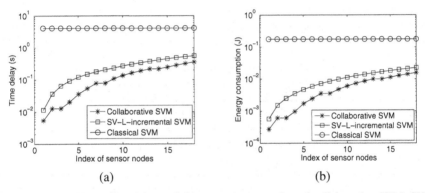

(a)   (b)

Fig. 5. Compare the (a) time delay and (b) energy consumption of collaborative SVM, SV-L-incremental SVM and Classical SVM during one pass

Furthermore, the benefits in terms of energy consumption and time delay in WSNs using these three algorithms are investigated. The total energy consumed in three algorithms can be approximately expressed as the sum of the energy consumed for the transmission of data in all sensor nodes. The time delay in classical SVM can be expressed as the maximum time delay of each sensor node after a whole access of all sensor nodes, while the time delay in two incremental SVM algorithms are expressed as the sum of time delay during one pass across all sensor nodes. Fig.5 illustrates the details of energy consumption and time delay of three algorithms during one pass. The results show the time delay and energy consumption of two incremental SVM algorithms as a function of the consecutive incremental steps, while the time delay and energy consumption of classical SVM just have a stationary value because the data transmission is carried out confusedly by each sensor node at the same time. Moreover, the time delay and energy consumption of two incremental SVM algorithms are both much less than classical SVM algorithm. It's because that the incremental SVM algorithms transmits the data in order, while the classical SVM transmits the data confusedly which will increase the time delay and energy consumption. Furthermore, the proposed collaborative SVM performs better than the SV-L-incremental SVM, because the collaborative SVM collect the information

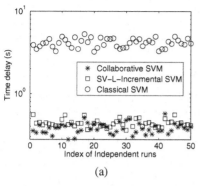 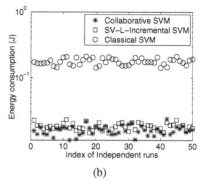

(a)   (b)

Fig. 6. Compare the (a) time delay and (b) energy consumption of collaborative SVM, SV-L-incremental SVM and Classical SVM in 50 independent runs

from selected set of sensor nodes instead of all sensor nodes. As illustrated in Fig. 5, compared to classical SVM and SV-L-incremental SVM, the collaborative SVM can reduces the energy consumption and time delay by more than 90% and 30%.

Fig. 6 shows the details of total energy consumption and time delay in 50 independent runs, which verify that the outstanding performance of collaborative SVM is robust in both time delay and energy consumption. The time delay and energy consumption of collaborative SVM is always the least between three algorithms, which implies that the collaborative SVM is suitable for highly-constrained WSNs.

The comparison results of classification accuracy, time delay and energy consumption present the outstanding performance of collaborative SVM. The results also illustrate that the collaborative SVM is an effective target classification algorithm for image recognition in WSNs.

5   Conclusions

In this paper, we focused on target classification in WSNs and introduced a collaborative SVM algorithm which is based on SV-L-incremental SVM and collaborative sample (sensor) selection mechanism. The proposed collaborative SVM algorithm can complete the classification task progressively with the information from selected set of sensor nodes. The experimental results verify that the proposed collaborative SVM algorithm with the progressive distributed framework can effectively determine the order and number of nodes on the routes traversed by the SV-L-incremental SVM with the consideration of energy consumption and contribution of each sensor node. The collaborative SVM performs better than classical SVM and SV-L-incremental SVM in classification, and it is also much more efficient in terms of energy consumption and time delay than the classical SVM and SV-L-incremental SVM, since it reduces the energy consumption and time delay by more than 90% and 30% respectively.

Acknowledgments. This paper is supported by the National Basic Research Program of China (973 Program) under Grant No. 2006CB303000 and National Natural Science Foundation of China under Grant No. 60673176, 60373014 and 50175056.

References

1. Li, D., Wong, K.D., Hu, Y.H., Sayeed, A.M.: Detection, Classification and Tracking of Targets. IEEE Signal Processing Magazine 19(2), 17–29 (2002)
2. Chong, D., Kumar, S.P.: Sensor Networks: Evolution, Opportunities and Challenges. Proceedings of the IEEE 91(8), 1247–1256 (2003)
3. Wang, T.-Y. (ed.): A Combined Decision Fusion and Channel Coding Scheme for Distributed Fault-Tolerant Classification in Wireless Sensor Networks. IEEE Transactions on Wireless Communications 5(7), 1695–1705 (2006)
4. D'Costa, A., Ramachandran, V., Sayeed, A.M.: Distributed Classification of Gaussian Space-Time Sources in Wireless Sensor Networks. IEEE Journal on Selected Areas in Communications 22(6), 1026–1036 (2004)
5. Kotecha, J.H., Ramachandran, V., Sayeed, A.M.: Distributed Multitarget Classification in Wireless Sensor Networks. IEEE Journal on Selected Areas in Communications 23(4), 703–713 (2005)
6. Duarte, M.F., Hu, Y.H.: Vehicle Classification in Distributed Sensor Networks. Journal of Parallel and Distributed Computing 64(7), 826–838 (2004)
7. Flouri, K., Beferull-Lozano, B., Tsakalides, P.: Training A SVM-based Classifier in Distributed Sensor Networks. In: Proc. 14nd European Signal Processing Conference, Florence, Italy, pp. 1–5 (2006)
8. Shilton, A., Palaniswami, M., Ralph, D., Tsoi, A.C.: Incremental Training of Support Vector Machines. IEEE Trans on Neural Networks 16(1), 114–131 (2005)
9. Ruping, S.: Incremental Learning with Support Vector Machines. In: Proc. IEEE Int. Conf. on Data Mining, San Jose, CA, USA, pp. 641–642 (2001)
10. Syed, N., Liu, H., Sung, K.: Incremental Learning with Support Vector Machines. In: Proc. of the 16th Int. Joint Conf. on Artificial Intelligence, Stockholm, pp. 352–356. Morgan Kaufmann, San Francisco (1999)
11. Domeniconi, C., Gunopoulos, D.: Incremental Support Vector Machine Construction. In: IEEE Int. Conf. on Data Mining, San Jose, CA, USA, pp. 589–592 (2001)
12. Diehl, C.P., Cauwenberghs, G.: SVM Incremental Learning, Adaptation and Optimization. In: Proc. Int. Joint Conf. on Neural Networks, Portland, OR, pp. 2685–2690 (2003)
13. Bennett, K., Campbell, C.: Support Vector Machines: Hype or Hallelujah. SIGKDD Explorations, 1–13 (2000)
14. Haritaoglu, I., Harwood, D., Davis, L.S.: W4: Real-Time Surveillance of People and Their Activities. IEEE Trans. on Pattern Analysis and Machine Intelligence 22(8), 809–830 (2000)
15. Lu, S., Zhang, J., Feng, D.: Classification of Moving Humans Using Eigen-Features and Support Vector Machines. In: Gagalowicz, A., Philips, W. (eds.) CAIP 2005. LNCS, vol. 3691, pp. 522–529. Springer, Heidelberg (2005)
16. Wang, X., Wang, S., Ma, J.-J.: Dynamic Deployment Optimization in Wireless Sensor Networks. Lecture Notes in Control and Information Sciences, vol. 344, pp. 182–187 (2006)
17. Wang, X., Jiang, A.-G., Wang, S.: Mobile Agent Based Wireless Sensor Network for Intelligent Maintenance. In: Huang, D.-S., Zhang, X.-P., Huang, G.-B. (eds.) ICIC 2005. LNCS, vol. 3645, pp. 316–325. Springer, Heidelberg (2005)
18. Rappaport, T.S.: Wireless Communications: Principles and Practice, 2nd edn. Prentice-Hall, Englewood Cliffs, NJ (2002)
19. Wang, X., Wang, S., Ma, J.-J.: An Improved Particle Filter for Target Tracking in Sensor System. Sensors 7(1), 144–156 (2007)

Dimensionality Reduction for Mass Spectrometry Data

Yihui Liu

School of Computer Science and Information Technology,
Shandong Institute of Light Industry,
Jinan, Shandong, China, 250353
Yihui_liu_2005@yahoo.co.uk

Abstract. In this paper multilevel wavelet analysis is performed for high dimensional mass spectrometry data. A set of wavelet approximation coefficients at different scale is used to characterize the features of mass spectrometry data. Approximation coefficients compress mass spectrometry data and act as "fingerprint" of mass spectrometry data. Support vector machine is used to classify the different tissue based on these wavelet features. 2 and 3 fold cross validation experiments are performed on 2 datasets based on approximation coefficients at 1st, 2nd and 3rd level decomposition respectively. A highly competitive accuracy in comparison to the best performance of other kinds of classification models is achieved.

1 Introduction

Mass spectrometry is being used to generate protein profiles from human serum, and proteomic data obtained from mass spectrometry have attracted great interest for the detection of early-stage cancer. Surface enhanced laser desorption/ionization time-of-flight mass spectrometry (SELDI-TOF-MS) in combination with advanced data mining algorithms, is used to detect protein patterns associated with diseases [1,2,3,4,5]. As a kind of MS-based protein Chip technology, SELDI-TOF-MS has been successfully used to detect several disease-associated proteins in complex biological specimens such as serum [6,7,8].

The researchers [9] employ principle component analysis (PCA) for dimensionality reduction and linear discriminant analysis (LDA) coupled with a nearest centroid classifier [10] for classification. In [11], the researchers compare two feature extraction algorithms together with several classification approaches on a MALDI TOF acquired data. The T-statistic was used to rank features in terms of their relevance. Support vector machines (SVM), random forests, linear/quadratic discriminant analysis (LDA/QDA), knearest neighbors, and bagged/boosted decision trees were subsequently used to classify the data. More recently, in [12], both the GA approach and the nearest shrunken centroid approach have been found inferior to the boosting based feature selection approach. The researcher [13] examines the performance of the nearest centroid classifier coupled with the following feature selection algorithms. Student-t test, Kolmogorov-Smirnov test, and the P-test are univariate statistics used for filter-based feature ranking. Sequential forward selection and a modified version

R. Alhajj et al. (Eds.): ADMA 2007, LNAI 4632, pp. 203–213, 2007.
© Springer-Verlag Berlin Heidelberg 2007

of sequential backward selection are also tested. Embedded approaches included shrunken nearest centroid and a novel version of boosting based feature selection. In addition, several dimensionality reduction approaches are also tested. Yu et al. [14] develop a hybrid method for dimensionality reduction and test on a published ovarian high-resolution SELDI-TOF dataset. They use a four-step strategy for data preprocessing based on: (1) binning, (2) Kolmogorov–Smirnov test, (3) restriction of coefficient of variation and (4) wavelet analysis. They use 1st level wavelet analysis.

In this study we perform multilevel wavelet decomposition on mass spectrometry data. Firstly we filter mass spectrometry data with small profile variance, which means that data with a variance less than the threshold are removed from mass spectrometry data. Then multi-level wavelet analysis is performed on the filtered mass spectrometry data. A vector of approximation coefficients at different levels is extracted to compress the mass spectrometry data and characterize the "fingerprint" of mass spectrometry data. Finally the extracted wavelet features are input into SVM classifier to distinguish the diagnostic classes.

2 Wavelet Analysis

For one dimensional wavelet analysis, a signal can be represented as a sum of wavelets at different time shifts and scales (frequencies) using discrete wavelet analysis (DWT). The DWT is capable of extracting the features of transient signals by separating signal components in both time and frequency. According to DWT, a time-varying function (signal) $f(t) \in L^2(R)$ can be expressed in terms of $\phi(t)$ and $\psi(t)$ as follows:

$$f(t) = \sum_k c_0(k)\phi(t-k) + \sum_k \sum_{j=1} d_j(k) 2^{\frac{-j}{2}} \psi(2^{-j}t - k)$$

$$= \sum_k c_{j0}(k) 2^{\frac{-j0}{2}} \phi(2^{-j0}t - k) + \sum_k \sum_{j=j0} d_j(k) 2^{\frac{-j}{2}} \psi(2^{-j}t - k)$$

where $\phi(t), \psi(t), c_0$, and d_j represent the scaling function, wavelet function, scaling coefficients at scale 0, and wavelet detailed coefficient at scale j, respectively. The variable k is the translation coefficient for the localization of a signal for time. The scales denote the different (high to low) frequency bands. The variable symbol j_0 is scale number selected.

The wavelet filter-banks approach was developed by Mallat [15]. We use Daubechies wavelet of order 7 (db7) [16] for wavelet analysis of mass spectrometry data and the boundary values are symmetrically padded. Multilevel discrete wavelet transform (DWT) is performed on mass spectrometry data. Given a mass spectrometry data ms of length N, the DWT consists of $Log_2 N$ levels at most. The first level produces, starting from ms, two sets of coefficients: approximation coefficients and detail coefficients. These vectors are obtained by convolving ms with the low-pass filter for approximation, and with the high-pass filter for detail, followed by dyadic decimation (downsampling). The next level splits the approximation coefficients in two parts using the same scheme, and so on.

Multilevel wavelet analysis makes it possible to compress the mass spectra and find "trend" of mass spectra. Approximation coefficients act as the "fingerprint" of mass spectra. If the first levels of the decomposition can be used to eliminate a large part of the noise, the successive approximations appear less and less noisy; however, they also lose progressively more high-frequency information. In our study we use approximation coefficients at 1st, 2nd and 3rd level to represent the mass spectra, remove noise, and reduce dimensionality of mass spectra.

3 Support Vector Machine

The SVMs originate from the idea of the structural risk minimization developed by Vapnik [17]. SVMs are an effective algorithm to find the maximal margin hyperplane to separate two classes of patterns. A transform to map nonlinearly the data into a higher dimensional space allows a linear separation of classes, which could not be linearly separated in the original space. In this study radial basis functions (RBF) $K(x_i, x_j) = e^{-\|x_i - x_j\|^2 / r1}$, where $r1$ is a strictly positive constant, is used. Apparently the linear kernel is less complex than the polynomial and the RBF kernels. The RBF kernel usually has better boundary response as it allows for extrapolation, and most high dimensional data sets can be approximated by Gaussian-like distributions similar to those used by RBF networks [18].

4 Results

In this study we use Correct Rate, Sensitivity, Specificity, Positive Predictive Value (PPV), Negative Predictive Value (NPV) and Balanced Correct Rate (BACC) to evaluate the performance of the proposed method. Let TP, TN, FP and FN stand for the number of true positive (cancer), true negative (control), false positive and false negative samples. Sensitivity is defined as $\dfrac{TP}{TP + FN}$; Specificity is defined as $\dfrac{TN}{TN + FP}$; PPV (Positive Predictive Value) is defined as $\dfrac{TP}{TP + FP}$; NPV (Negative Predictive Value) is defined as $\dfrac{TN}{TN + FN}$; Correct rate is defined as $\dfrac{TP + TN}{TP + TN + FP + FN}$. BACC (Balanced Correct Rate) is defined as $\dfrac{1}{2}(\dfrac{TP}{TP + FN} + \dfrac{TN}{TN + FP})$, which is the average of sensitivity and specificity.

Low resolution ovarian dataset (8-7-02)
The sample set includes 91 controls and 162 ovarian cancers with 15154 features. They are provided by National Cancer Institute (http://home.ccr.cancer.gov/ncifdaproteomics/ppatterns.asp).

● **Filter mass spectra**

The mass spectrometry data with 15154 features is quite large and a lot of the features do not show key changes. To make it easier to find the key features, the first thing to do is to reduce the size of the dataset by removing the features that do not show large change. We filter out mass spectrometry data with small variance over time [19]. We select 20% threshold, which means that data with a variance less than the threshold are removed from mass spectrometry data. Figure 1 shows the original ovarian mass spectra and features with variance greater than 20th percentile. The dimensionality is reduced to 12123 after filtering.

● **Extract wavelet features**

We use Daubechies wavelet of order 7 (db7) for wavelet analysis of mass spectrometry data and the boundary values are symmetrically padded. Multilevel discrete wavelet transform (DWT) is performed on mass spectrometry data. Figure 2 shows the approximation coefficients at 1st level, 2nd level and 3rd level wavelet decomposition. The dimensionality of approximation coefficients at different level is shown in Table1. Approximation coefficients at 3rd level, which remove the detail changes based on first, second and third derivative, only have 1526 dimensionality.

● **The performance of SVM Classifier**

In our study we use K fold cross validation experiments. K fold cross validation randomly generates indices, which contain equal (or approximately equal) proportions of the integers 1 through K that define a partition of the N observations into K disjoint subsets. In K fold cross validation, K-1 folds are used for training and the last fold is used for evaluation. This process is repeated K times, leaving one different fold for evaluation each time. We run 20 times for each K fold cross validation experiments, where K=2 and 3. Table 2 shows the performance of approximation coefficients at 1st level, 2nd level and 3rd level decomposition.

The researcher [13] used a three fold cross validation procedure to test this mass spectrometry data using different methods. For all experiments dataset was split into three subsets of equal size. Each test fold used one of the three subsets with the remaining two subsets used for training. Their results are shown in Table 3. For PCA/LDA and Boosted FE methods, 100% correct rate is achieved. Our method also achieves 100% correct rate when we perform 3 fold cross validation experiment using 1526 approximation coefficients at 3rd level decomposition.

Figure 3 shows the details at 3 levels and approximation at 3rd level. As illustrated in Figure 3, a large part of the noise can be eliminated from mass spectra by removing the details at the first levels of decomposition. The successive approximations appear less and less noisy. So the best performance 100% is achieved at 3rd level decomposition.

Table 1. Feature number for low resolution ovarian dataset

	Original	Filtered	1st level	2nd level	3rd level
Feature number	15154	12123	6068	3040	1526

This Table shows the feature number at 1st level, 2nd level and 3rd level decomposition.

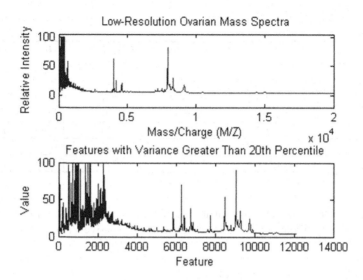

Fig. 1. Features for low resolution ovarian dataset. This Figure shows that the original low resolution ovarian mass spectra and features with variance greater than 20th percentile.

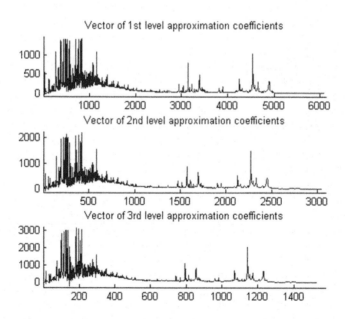

Fig. 2. Approximation coefficients for low resolution ovarian mass spectra. This Figure shows approximation coefficients at 1st level, 2nd level and 3rd level decomposition.

Table 2. Performance of wavelet features at 1st level, 2nd level and 3rd level decomposition

Wavelet features	K fold	Correct rate	Sensitivity	Specificity	PPV	NPV
1st Level	2	0.9781	0.9560	0.9905	0.9825	0.9757
	3	0.9791	0.9608	0.9894	0.9808	0.9782
2nd Level	2	0.9860	0.9630	0.9989	0.9979	0.9796
	3	0.9977	0.9937	1.0000	1.0000	0.9965
3rd Level	2	0.9993	0.9980	1.0000	1.0000	0.9989
	3	1.0000	1.0000	1.0000	1.0000	1.0000

This Table shows performance of low resolution ovarian dataset. PPV stands for Positive Predictive Value; NPV stands for Negative Predictive Value.

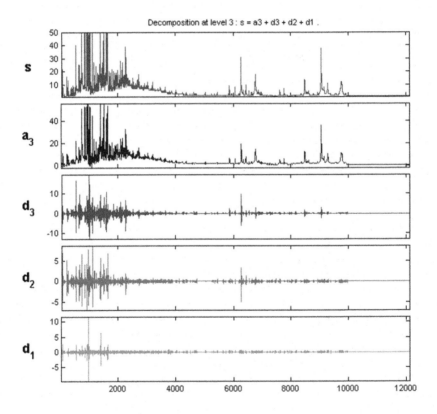

Fig. 3. Approximation at 3rd level and details at 3 levels for low resolution ovarian dataset

Table 3. Performance of different methods [13]

Methods	Correct rate	Sensitivity	Specificity	PPV
No FE	0.837	0.822	0.846	0.891
PCA	0.901	0.867	0.920	0.926
PCA/LDA	1.000	1.000	1.000	1.000
SFS	0.992	0.989	0.994	0.994
SBS	0.901	0.911	0.895	0.942
P-test	0.980	0.956	0.994	0.976
T-test	0.837	0.822	0.846	0.897
KS-test	0.984	0.978	0.988	0.988
NSC(20)	0.972	0.978	0.969	0.988
Boosted	0.980	0.989	0.975	0.994
Boosted FE	1.000	1.000	1.000	1.000

This Table shows performance of low resolution ovarian dataset of 3 fold cross validation experiments. PPV stands for Positive Predictive Value; NPV stands for Negative Predictive Value.

Prostate cancer dataset
This data was collected using the H4 protein chip (JNCI Dataset 7-3-02) [20]. There are 322 samples: 190 samples with benign prostate hyperplasia with PSA levels greater than 4, 63 samples with no evidence of disease and PSA level less than 1, 26 samples with prostate cancer with PSA levels 4 through 10, and 43 samples with prostate cancer with PSA levels greater than 10. Each sample is composed of 15154 features. We combined benign prostate hyperplasia samples and those with no evidence of disease to form the normal class. The rest of the samples are classed into cancer category. We have 69 cancer samples and 253 normal samples.

- **Filter mass spectra**
The original prostate mass spectrometry data has 15154 features. We filter out mass spectrometry data with 20% variance and feature number with variance greater than 20th percentile reduces to 12123. Figure 4 shows the original prostate mass spectrometry data and features with variance greater than 20th percentile.

- **Extract wavelet features**
We use Daubechies wavelet of order 7 (db7) for wavelet decomposition of mass spectrometry data with symmetrically padded boundary values. Multilevel discrete wavelet transform (DWT) is performed on mass spectrometry data. Figure 5 shows the approximation coefficients at 1st level, 2nd level and 3rd level decomposition. The dimensionality of approximation coefficients is shown in Table 4.

- **The performance of SVM Classifier**
In this study we test wavelet features at 1st level, 2nd level and 3rd level decomposition. For wavelet features of each level, we run K fold cross validation where K=2 and 3. We run 20 times for each K fold cross validation experiments. Table 5 show the performance of approximation coefficients at 1st level, 2nd level and 3rd level decomposition respectively.

The researcher [13] performed 3 fold cross validation using different methods and their performance is shown in Table 6. Their best BACC result is 90.6% using Boosted FE. Our method achieves 90.92% BACC result for 3 fold cross validation experiments at 2nd level decomposition, which performs better than T-test, P-test, PCA/LDA, SFS, boosted method, etc. When the decomposition level is getting higher, approximations of mass spectra appear more smooth less noisy gradually. However with the higher frequencies removed, some discriminating information hidden in mass spectra is lost. So the performance of approximation coefficients at 3rd level is a little bit lower than one at 2nd level. Normally first 3 levels of decomposition is acceptable, removing noise and keeping most discriminating information contained in mass spectra.

Table 4. Feature number for prostate cancer dataset

	Original	filtered	1st level	2nd level	3rd level
Feature number	15154	12123	6068	3040	1526

This Table shows the feature number at 1st level, 2nd level and 3rd level decomposition.

Table 5. Performance of approximation coefficients at different levels

Wavelet features	K fold	Correct rate	Sensitivity	Specificity	PPV	NPV	BACC
1st level	2	0.9345	0.9684	0.8103	0.9493	0.8748	0.8893
	3	0.9419	0.9689	0.8427	0.9576	0.8810	0.9058
2nd level	2	0.9286	0.9558	0.8287	0.9534	0.8364	0.8923
	3	**0.9401**	**0.9633**	**0.8551**	**0.9606**	**0.8640**	**0.9092**
3rd level	2	0.9238	0.9515	0.8221	0.9515	0.8221	0.8868
	3	0.9268	0.9548	0.8240	0.9521	0.8326	0.8894

This Table shows performance of low resolution prostate cancer dataset. PPV stands for Positive Predictive Value; NPV stands for Negative Predictive Value.

Table 6. Performance of different methods [13]

	BACC	Specificity	Sensitivity	PPV
No FE	0.777	0.698	0.855	0.439
PCA	0.516	0.540	0.493	0.248
PCA/LDA	0.667	0.710	0.623	0.431
SFS	0.827	0.929	0.725	0.728
SBS	0.729	0.806	0.652	0.498
P-test	0.728	0.877	0.580	0.572
T-test	0.709	0.897	0.522	0.575
KS-test	0.784	0.857	0.710	0.579
NSC(20)	0.736	0.833	0.638	0.529
Boosted	0.810	0.881	0.739	0.627
Boosted FE	0.906	1.000	0.812	1.000

This Table shows the performance of prostate cancer dataset for 3 fold cross validation experiments. PPV stands for Positive Predictive Value; BACC stands for Balanced Correct Rate.

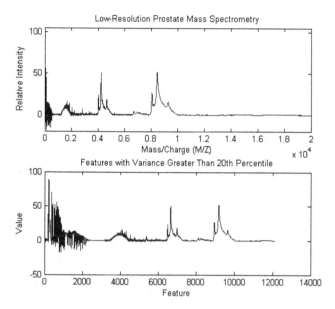

Fig. 4. Features for low resolution prostate mass spectra. This Figure shows that the original prostate mass spectra and features with variance greater than 20th percentile.

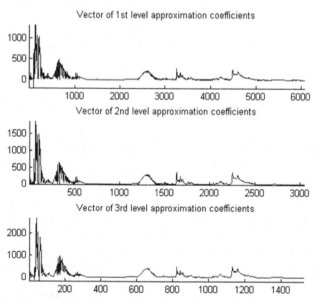

Fig. 5. Approximation coefficients of prostate mass spectra. This Figure shows approximation coefficients at 1st level, 2nd level and 3rd level decomposition.

5 Conclusions

In this paper we use multilevel wavelet analysis to extract the features of high dimensional mass spectrometry data and reduce dimensionality. A set of wavelet approximation coefficients at different level decomposition acts as "fingerprint" of mass spectrometry data. The approximation coefficients compress high dimensional mass spectrometry data and extract the essential information of mass spectrometry data. Experimental results suggest that approximation coefficients are efficient way to reduce the dimensionality of high dimensional mass spectrometry data and characterize the features of mass spectrometry data.

Acknowledgements

This work was supported by Shandong Institute of Light Industry under Grant 12041653.

References

1. Petricoin, E.F., Ardekani, A.M., Hitt, B.A., Levine, P.J., Fusaro, V.A., Steinberg, S.M., Mills, G.B., Simone, C., Fishman, D.A., Kohn, E.C., Liotta, L.A.: Use of proteomic patterns in serum to identify ovarian cancer. The Lancet 359, 572–577 (2002)
2. Sorace, J.M., Zhan, M.: A data review and re-assessment of ovarian cancer serum proteomic profiling. BMC Bioinform 4 (2003)
3. Michener, C.M., Ardekani, A.M., Petricoin, E.F., Liotta 3rd, L.A., Kohn, E.C.: Genomics and proteomics: application of novel technology to early detection and prevention of cancer. Cancer Detect Prev. 26, 249–255 (2002)
4. Petricoin, E.F., Zoon, K.C., Kohn, E.C., Barrett, J.C., Liotta, L.A.: Clinical proteomics: translating benchside promise into bedside reality. Nat. Rev. Drug. Discov. 1, 683–695 (2002)
5. Srinivas, P.R., Verma, M., Zhao, Y., Srivastava, S.: Proteomics for cancer biomarker discovery. Clin. Chem. 48, 1160–1169 (2002)
6. Herrmann, P.C., Liotta, L.A., Petricoin 3rd, E.F.: Cancer proteomics: the state of the art. Dis. Markers 17, 49–57 (2001)
7. Jr, G.W., Cazares, L.H., Leung, S.M., Nasim, S., Adam, B.L., Yip, T.T., Schellhammer, P.F., Gong, L., Vlahou, A.: Proteinchip surface enhanced laser desorption/ionization (SELDI) mass spectrometry: a novel protein biochip technology for detection of prostate cancer biomarkers in complex protein mixtures. Prostate Cancer Prostatic Dis. 2, 264–276 (1999)
8. Vlahou, A., Schellhammer, P.F., Mendrinos, S., Patel, K., Kondylis, F.I., Gong, L., Nasim, S., Wright Jr.: Development of a novel proteomic approach for the detection of transitional cell carcinoma of the bladder in urine. Am. J. Pathol. 158, 1491–1520 (2001)
9. Lilien, R.H., Farid, H., Donald, B.R.: Probabilistic disease classification of expression-dependent proteomic data from mass spectrometry of human serum. Computational Biology 10 (2003)
10. Park, H., Jeon, M., Rosen, J.B.: Lower dimensional representation of text data based on centroids and least squares. BIT 43, 1–22 (2003)

11. Wu, B., Abbott, T., Fishman, D., McMurray, W., Mor, G., Stone, K., Ward, D., Williams, K., Zhao, H.: Comparison of statistical methods for classifcation of ovarian cancer using mass spectrometry data. BioInformatics 19 (2003)
12. Jeffries, N.O.: Performance of a genetic algorithm for mass spectrometry proteomics. BMC Bioinformatics 5 (2004)
13. Levner, I.: Feature selection and nearest centroid classification for protein mass spectrometry. BMC Bioinformatics 6 (2005)
14. Yu, J.S., Ongarello, S., Fiedler, R., Chen, X.W., Toffolo, G., Cobelli, C., Trajanoski, Z.: Ovarian cancer identification based on dimensionality reduction for high-throughput mass spectrometry data. Bioinformatics 21, 2200–2209 (2005)
15. Mallat, S.: A theory for multiresolution signal decomposition: The wavelet representation. IEEE Transactions on Pattern Analysis and Machine Intelligence 11, 674–693 (1989)
16. Daubechies, I.: Orthonormal bases of compactly supported wavelets. Communications on Pure and Applied Mathematics 41, 909–996 (1988)
17. Vapnik, V.N.: Statistical learning theory. Wiley, New York (1998)
18. Burges, C.: A Tutorial on Support Vector Machines for Pattern Recognition. Kluwer Academic Publishers, Dordrecht (1998)
19. Kohane, I.S., Kho, A.T., Butte, A.J.: Microarrays for an Integrative Genomics. MIT Press, Cambridge (2003)
20. Petricoin, E.F., Ornstein 3rd, D.K., Paweletz, C.P., Ardekani, A., Hackett, P.S., Hitt, B.A., Velassco, A., Trucco, C., Wiegand, L., Wood, K., Simone, C.B., Levine, P.J., Linehan, W.M., Emmert-Buck, M.R., Steinberg, S.M., Kohn, E.C., Liotta, L.A.: Serum proteomic patterns for detection of prostate cancer. J. Natl. Cancer Inst. 94, 1576–1578 (2002)

The Study of Dynamic Aggregation of Relational Attributes on Relational Data Mining

Rayner Alfred

Universiti Malaysia Sabah,
School of Engineering and Information Technology,
88999, Kota Kinabalu, Sabah, Malaysia
ralfred@ums.edu.my

Abstract. Most aggregation functions are limited to either categorical or numerical values but not both values. In this paper, we define three concepts of aggregation function and introduce a novel method to aggregate multiple instances that consists of both the categorical and numerical values. We show how these concepts can be implemented using clustering techniques. In our experiment, we discretize continuous values before applying the aggregation function on relational datasets. With the empirical results obtained, we demonstrate that our transformation approach using clustering techniques, as a means of aggregating multiple instances of attribute's values, can compete with existing multi-relational techniques, such as Progol and Tilde. In addition, the effect of the number of interval for discretization on the classification performance is also evaluated.

Keywords: Data Summarization, Multiple Instance Aggregation, Clustering, Relational data Mining.

1 Introduction

Most databases employ the relational model for data storage. To use this data in a traditional propositional learner, a data transformation process or a propositionalization step has to take place. *Propositionalization* methods can be classified along two dimensions: *complete* versus *partial* and *selection-based* versus *aggregation-based*. In a *complete propositionalization* approach, no information is lost in the transformation. However, this approach may result in an exponential number of derived attributes [13]. On the other hand, *partial propositionalization* approaches, where the goal is to automatically generate a small but relevant set of derived attributes, are much more common in practice.

In a *selection-based propositionalization*, the derived attributes are binary and each attribute represents a particular selection of the instances. Many systems have been used for *selection-based propositionalization* [6]. For instance, the LINUS system [14,15] and its descendants are probably the most well known selection-based propositionalization. In an *aggregation-based propositionalization* (known as database-oriented propositionalization), the derived attributes are typically numeric and each attribute is defined by an aggregation function that aggregates for each

R. Alhajj et al. (Eds.): ADMA 2007, LNAI 4632, pp. 214–226, 2007.

instance the values of a particular attribute in the related tuples, depending on the type of attribute (e.g. *average, sum, min, max* for numeric values and *count, mode* for nominal values). Approaches for propositionalization that are based on aggregate functions are widely used in the database area [11,12] and most of the learning tasks concentrate on the classifications problems. The application of aggregate functions, as a means of data transformation or propositionalization, ensures that data are appropriately summarized such that only one row per example in the target table. RELAGGS system [16], RSD [11] and the POLKA system [17] are some of the examples of *aggregation-based propositionalization* systems. Krogel *et al.* [20] has outlined a detailed comparison of both approaches and has been shown that an undesirable bias [19] exists on both kinds of propositionalizations, as the *selection-based* does not consider *multiple instance* problem [3]. On the other hand, the aggregation-*based* propositionalization functions handle *multiple instance* problem but they are limited to categorical or numerical values only.

In this paper, we try to overcome this bias by introducing a data transformation algorithm called Dynamic Aggregation of Relational Attributes, hence called DARA [1], that aggregates selected *multiple instance* datasets stored in relational database. Our approach is based on clustering technique that combines concepts from *selection-based* and aggregation-*based* propositionalization. We also evaluate the effect of discretization methods, implemented in the DARA algorithm, on learning from three of the most commonly used datasets used in relational data mining, namely the Musk [3], Financial [4] and Mutagenesis [2] datasets.

The paper is structured as follows. In Section 2, we define three kinds of aggregation and give an overview of Dynamic Aggregation of Relational Attributes (DARA) algorithm and the framework in which DARA works. Section 3 discusses three different methods of discretization of continuous attributes in relational database. Section 4 describes the experimental setup and evaluation. We draw conclusions in Section 5.

2 Aggregation and Dynamic Aggregation of Relational Attributes

Aggregation is normally associated with data reduction in relational database. In fact, aggregation is one of the feature construction techniques in a relational domain, and an essential component of a relational model induction which has a significant impact on generalization performance for domains with important *one-to-many* relationships. In this section, we describe three types of aggregation in a relational data model and introduce the technique of dynamic aggregation of relational attributes (DARA). Aggregation can be defined as a summarization of the underlying pattern or distribution from which the related objects were sampled. For example, *count, mode* aggregation functions for nominal values and *avg, sum, min* and *max* for the numerical values are some of the aggregation functions in relational database system. Therefore, the design and selection of appropriate aggregations depend on the characteristics of the value distributions. Generally, there are three types of aggregation: *Vertical, Horizontal* and *Cross Aggregations*. Before proceeding, we need to define some concepts used in designing the DARA algorithm.

Definition 2.1. A *Relationship* represents an association between two or more tables. (e.g. Each employee in table *Employee* is assigned to projects in table *Project*.)

Definition 2.2. A *one-to-one* (1:1) relationship is when at most one instance of a table, Table 1, is associated with one instance of another table, Table 2. (For example, employees in the company are each assigned their own office.)

Definition 2.3. A *one-to-many* (1:N) relationships is when for one instance of Table 1, there are zero or many instances of Table 2, but for one instance of Table 2, there is only one instance of Table 1. (For example, a department has many employees.)

Definition 2.4. A *many-to-many* (M:N) relationship, sometimes called non-determinate, is when for one instance of Table 1, there are zero or many instances of Table 2 and for one instance of Table 2, there are zero or many instances of Table 1. (For example, a single employee can be assigned to many projects; conversely, a single project can have assigned to it many employees.)

2.1 Degrees of Aggregation

2.1.1 Vertical Aggregation

Let Φ denote the aggregation of multiset values to a categorical or numerical value. Given a set of multiple instances or rows $\{R_1,..,R_n\}$ that correspond to the same value of primary key (ID) and a multiset values $M_v = \{A1_1,...,A1_n\}$ of attribute A1, (see Fig. 1), a *vertical aggregation*, $\Phi^V(\text{ID}, M_v)$, is a mapping from multiset values $\{A1_1,...,A1_n\}$ of attribute A1 to a categorical or numerical value, λ_V. It is also known as a basic aggregation. As a result, we are left with a single row, with a value of (ID_1, λ_V), after the vertical aggregation operation is performed. Most relational database systems have these aggregation functions for multiset values, e.g. *mode* and *count* aggregation functions for nominal values and *sum, avg, min, max, count* functions can be used to aggregate numerical values. However, the aggregation functions for nominal values can capture only limited discriminative information.

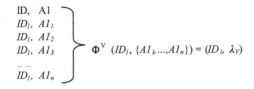

Fig. 1. Vertical Aggregation (ID is the primary key)

A vertical aggregation can be computed by projecting the column and aggregating the column's multiset values based on a predefined mapping function (*sum, min, max, count, avg, mode*), grouped by the key attribute. Generally, this type of aggregation is done for records with multiple instances in a single table.

Definition 2.5. Since the join operation is restricted to tables with common column only, two tables T1 and T2, can be joined to become a single table, if and only if they have a common attribute, whose domains are the same. This is to ensure the integrity of relationship between records in multiple tables is maintained.

Definition 2.6. Let *attr(T)* be a list of attributes in Table T. Given two tables, *T1* and *T2*, with attributes $\{A1,...,An\} \subset attr(T1)$ and $\{B1,...,Bn\} \subset attr(T2)$ and *T1* and *T2* have *one-to-many* (1:n) relationship for individual record *I* in *T1*, the *vertical aggregation* Φ^V of attribute *Bm* would be $\Phi^V (I, \Pi_{T2.Bm} (T1 _{T1.Ai = T2.Bj} \otimes_{1 : n} T2)) \rightarrow \lambda_V$, where

1. $\{Bm, Bj\} \subset attr(T2)$ and $Ai \subset attr(T1)$
2. $(T1 _{T1.Ai = T2.Bj} \otimes_{1 : n} T2)$ is the equijoin operation of *T1* and *T2* based on *T1.Ai = T2.Bj*, with *one-to-many* (1:n) relationship between *T1* and *T2*
3. $\Pi_{T2.Bm}$ is the projection of attribute *Bm* from table *T2*
4. λ_V denotes a categorical or numerical value resulted from the vertical aggregation function Φ^V.

2.1.2 Horizontal Aggregation

Given a set of distinct instances identified by a unique key ID and a list of attributes $L_A = (A1, ..., An)$ describing each row, a *horizontal aggregation* $\Phi^H(ID, L_A)$ is a mapping from a set of attributes' values into a categorical or numerical value, λ_H, where the number of attributes is greater than one.

ID,	A1,	A2,	...,	An		Horizontal Aggregation
ID_1,	$A1_1$,	$A2_1$,	...,	An_3	$\rightarrow$	$\Phi^H(ID_1, \{A1_1, A2_1,...,An_3\}) = (ID_1, \lambda_{H1})$
ID_2,	$A1_2$,	$A2_1$,	...,	An_1	$\rightarrow$	$\Phi^H(ID_2, \{A1_2, A2_1,...,An_1\}) = (ID_2, \lambda_{H2})$
ID_3,	$A1_1$,	$A2_1$,	...,	An_3	$\rightarrow$	$\Phi^H(ID_3, \{A1_1, A2_1,...,An_3\}) = (ID_3, \lambda_{H1})$
...	...	...	...	...	...	...
ID_n,	$A1_n$,	$A2_1$,	...,	An_n	$\rightarrow$	$\Phi^H(ID_n, \{A1_n, A2_1,...,An_n\}) = (ID_n, \lambda_{Hn})$

Fig. 2. Horizontal Aggregation

In Fig. 2, with the assumption that all attributes from A3 through An-1 have the same values for all rows, ID_1 and ID_3 will have the same mapping values since they have the same values for all their attributes. Horizontal aggregation normally describes the association of two or more attributes in a table. Given *n* attributes in the form of a feature vector $(A_1, ..., A_n)$, one can map this feature vector to a sequential categorical or a predefined symbolic value (e.g. increasing, decreasing, stable, associate) or to a numerical value (e.g. differences or summation).

Definition 2.7. Let $\{A1,...,An\} \subset attr(T)$, the *horizontal aggregation* Φ^H of attributes *A1*, ..., *An* (for individual record *I*) would be $\Phi^H (I, \Pi_{A1...An} (T)) \rightarrow \lambda_H$, where

1. $\Pi_{A1...An}$ is the projection or selection of columns *A1* through *An*
2. λ_H denotes a categorical or numerical value from the horizontal aggregation function Φ^H.

The mapping function for horizontal aggregation is more complicated compared to the vertical aggregation since the aggregation's values are reflected by the attributes' independencies or dependencies (e.g. time-series data). Logic-based propositionalization [7] proposes a horizontal aggregation (known as selection-based) that expresses a feature vector which implements a Boolean conditioning.

2.1.3 Cross Aggregation

A *cross aggregation* is a mapping of a specific record, with a set of n attributes that contains multiset values to a categorical or numerical value, λ_{VH}. A *cross aggregation* Φ^C is a set of two operations that starts with horizontal Φ^H aggregation operation followed by the vertical Φ^V aggregation operation. For example, suppose we have a table that contains records with two attributes ($A1$ and $A2$) as shown in Fig. 3. The first step is to perform a horizontal aggregation on these attributes followed by the vertical aggregation of the multiple instances of an object.

ID, A1, A2	Horizontal Aggregation	Vertical Aggregation
$ID_1, A1_1, A2_1 \rightarrow \Phi^H(ID_1,\{A1_1, A2_1\}) = (ID_1, \lambda_{H1})$		
$ID_1, A1_2, A2_1 \rightarrow \Phi^H(ID_1,\{A1_2, A2_1\}) = (ID_1, \lambda_{H2}) \rightarrow \Phi^V(ID_1,\{\lambda_{H1},\lambda_{H2}\})=(ID_1, \lambda_{V1})$		
$ID_2, A1_2, A2_1 \rightarrow \Phi^H(ID_2,\{A1_1, A2_1\}) = (ID_2, \lambda_{H1})$		
$ID_2, A1_2, A2_2 \rightarrow \Phi^H(ID_2,\{A1_2, A2_1\}) = (ID_2, \lambda_{H2}) \rightarrow \Phi^V(ID_2,\{\lambda_{H1},\lambda_{H2}\})=(ID_2, \lambda_{V1})$		
$ID_3, A1_1, A2_1 \rightarrow \Phi^H(ID_3,\{A1_1, A2_1\}) = (ID_3, \lambda_{H1}) \rightarrow \Phi^V(ID_3,\{\lambda_{H1}\})=(ID_3, \lambda_{V3})$		
... ..., ...		
$ID_n, A1_1, A2_1 \rightarrow \Phi^H(ID_n,\{A1_1, A2_1\}) = (ID_n, \lambda_{H1}) \rightarrow \Phi^V (ID_n,\{\lambda_{H1}\})=(ID_n, \lambda_{V3})$		

Fig. 3. Cross Aggregation

Definition 2.8. Let $attr(T)$ be a list of attributes in Table T. Given two tables, $T1$ and $T2$, with attributes $\{A1,...,An\} \subset attr(T1)$ and $\{B1,...,Bn\} \subset attr(T2)$ and $T1$ has *many-to-many* (M:N) relationship with $T2$ for an individual record I, the *cross aggregation* Φ^C of attribute $Bi,...,Bj$ would be $\Phi^C(I, \Pi_{Bi...Bj}(T1_{T1.Ai\,=\,T2.Bk} \otimes_{m\,:\,n} T2)) = \Phi^V(I, \Phi^H(I, \Pi_{Bi...Bj}(R1_{R1.Ai\,=\,R2.Aj} \otimes_{m\,:\,n} R2))) \rightarrow \lambda_C$, where

1. $\{Bi,...,Bj\} \subset attr(T2)$, $Bk \subset attr(T2)$ and $Ai \subset attr(T1)$
2. $(T1_{\,T1.Ai\,=\,T2.Bk} \otimes_{m\,:\,n} T2)$ is the equijoin of $T1$ and $T2$ based on $T1.Ai = T2.Bk$
3. $\Pi_{Bi...Bj}$ is the projection or selection of columns Bi through Bj
4. λ_C denotes the categorical or numerical value resulted from the cross aggregation function Φ^C
5. $\Phi^V (I, \Phi^H(I))$ denotes the operation of horizontal aggregation is performed followed by the vertical aggregation operation, for an individual record I.

For example, consider finding a frequent pattern in employee's account balances for 12 months. This aggregation requires the employees' salary at the beginning of every month and the balance of the account at the end of each month. In particular, cross aggregation has received little treatment due to the fact that this technique requires domain knowledge of the problem at hand to properly aggregate the dataset. However, with Dynamic Aggregation of Relational Attributes (DARA), one requires no background knowledge to perform the cross aggregation and data summarization can be obtained dynamically by mapping these records into a predefined symbol using clustering techniques.

2.2 Dynamic Aggregation of Relational Attributes

In our implementation, the Dynamic Aggregation of Relational Attributes (DARA) system performs all three degrees of aggregation, particularly the cross aggregation. The process of cross aggregation in DARA is done through the data generalization. The generalization task is divided into three stages; Data-Patterns Transformation, Computation of Component Magnitudes, and Pattern-Based Clustering.

2.2.1 Data-Patterns Transformation

In relational database, records are stored separately in different tables and they are associated through the matching of primary and foreign keys. With a high degree of *one-to-many* association, a single record, R, stored in a main table is associated with a large volume of records stored in another table. Let R denotes a set of m records stored in the target table and let S denotes a set of n records $(T_1, T_2, T_3, \ldots, T_n)$, stored in the non-target table. Let S_i is in the subset of S, $S_i \in S$, and is associated with a single record R_a stored in the target table, $R_a \in R$. Thus, the association of these records can be described as $R_a \rightarrow S_i$. Since a record can be characterized based on the bag of term/records that are associated with it, we use the vector space model to cluster these records, as described in the work of Salton *et al.* [7]. In vector space model, a record is represented as a vector or 'bag of terms', i.e., by the terms it contains and their frequency, regardless of their order. These terms are encoded and represent instances stored in the non-target table referred by a record stored in the target table. The non-target table may have a single attribute or multiple attributes and the process of encoding the terms is as follows;

Case I: *Non-target table with a single attribute*

- Step 1) Compute the cardinality of the attribute's domain in the non-target table. For continuous values, discretizes them and take the number of bins as the cardinality of the attribute's domain
- Step 2) To encode values, find the appropriate number of bits, n, that can represent different values for the attribute's domain, where $2^{n-1} < |\text{Attribute's Domain}| \leq 2^n$. For example, if a city's attribute has 5 different values (London, New York, Chicago, Paris, Kuala Lumpur), then we just need 3 ($5 < 2^3$) bits to represent each of those value (001, 010, 011, 100, 101).
- Step 3) Each encoded term will be added to the bag of terms that describes the characteristics of the record associated with them.

Case II: *Non-target table with multiple attributes*

- Step 1) Repeat step 1) and step 2) in Case I, for all attributes
- Step 2) For each instance of a record stored in the non-target table, concatenate p-number of columns' values, where p is less than or equal to the total number of attributes. For example, let $F = (F_1, F_2, F_3, \ldots, F_k)$ denotes k field columns or attributes in the non-target table. Let $F_1 = (F_{1,1}, F_{1,2}, F_{1,3}, \ldots, F_{1,n})$ denotes n values that are allowed to be used by field/column F_1. So, we can have instances of

record in the non-target table with these values $(F_{1,a}, F_{2,b}, F_{3,c}, F_{4,d}..., F_{k-1,b}, F_{k,n})$, where $a \leq |F_1|, b \leq |F_2|, c \leq |F_3|, d \leq |F_4|,..., b \leq |F_{k-1}|, n \leq |F_k|$. If $p = 1$, we have $F_{1a}, F_{2,b}, F_{3,c}, F_{4,d}..., F_{k-1,b}, F_{k,n}$ as the produced terms. If $p = 2$, then we have $F_{1,a}F_{2,b}, F_{3,c}F_{4,d}..., F_{k-1,b}F_{k,n}$ (provided we have even number of fields). Finally, if we have $p = k$, then we have $F_{1,a}F_{2,b}F_{3,c}F_{4,d}...F_{k-1,b}F_{k,n}$ as a single term produced.

- Step 3) For each encoded term, add this term to the bag of terms and these terms describe the characteristics of a record associated with them.

The encoding process to transform relational datasets into data represented in a vector-space model has been implemented in DARA. Given this data representation, we can use clustering techniques [20,21,22] to cluster them, as a means of aggregating them. DARA algorithm simply assigns each record in the target table with the cluster number. Each cluster then can generate more information by looking at the most frequent patterns that describe it.

2.2.2 Computation of Component Magnitudes

The second stage of data generalization is computing the component magnitudes of the vector in order to cluster these records and map these records into group in which each member of the group has the same characteristics. At this stage, a vector represents a record with series of patterns that characterize itself. In other words, we have a series of patterns (bag of patterns) that characterize a single record. The component magnitudes of the vector (record) is computed, by computing the *pattern-frequency* and *inverse-record frequency* (1). Each record is viewed as a vector whose dimensions correspond to the bag of patterns. The component magnitudes of the vector are the *pf-irf* weights of the patterns adapted from [7].

$$pf\text{-}irf = pf(p, r) \cdot irf(p) \tag{1}$$

$$irf(p) = \log \frac{|R|}{rf(p)} \tag{2}$$

$$sim(r_i, r_j) = \frac{r_i \cdot r_j}{||r_i|| \cdot ||r_j||} \tag{3}$$

pf-ipf, as described in (1), is the product of pattern frequency $pf(p, r)$ and the inverse record frequency (2). Pattern frequency (1) refers to the number of times pattern p occurs in the corresponding record r. In inverse record frequency (2), $|R|$ is the number of records in the table and $rf(p)$ is the number of records in which pattern p occurs at least once. The similarity between two records is then (3) where r_i and r_j are vectors with *pf-irf* coordinates as described above.

2.2.3 Weighted-Based Clustering

Once we have computed the *pf-irf* weights for each patterns for every record, then we can compute the distance between each record and cluster them based on their weights. In our experiment, we use the partitioning clustering (k-means) algorithm to cluster these records. Partitioning clustering algorithms [9] divide a data set into a number of clusters, typically by trying to minimize some criterion or error function

[8]. In short, DARA algorithm treats records in relational database as a bag of patterns and clusters these records based on the created patterns that they have. In the next section, we describe three methods of discretization. Continuous numbers in relational database need to be discretized first, before the dynamic aggregation procedure described in this section can be executed.

3   Types of Discretization

The motivation for discretization continuous features is to obtain higher accuracy rates, although this operation may affect the efficiency of the induction procedures. In this experiment, we used three methods of discretization; *Equal Width*, *Equal Height* and *Equal Weight* implemented in the DARA algorithm.

Equal width interval binning is the simplest method to discretize data and has often been applied as a means for producing nominal values from continuous ones. It involves sorting the observed values of a continuous feature and find the minimum, V_{min}, and maximum, V_{max}, value, the interval can be find by dividing the range of observed values for the variable into k equally sized bins, where k is a parameter supplied by the user; $interval = (V_{max} - V_{min})/k$. The boundaries then can be constructed at $V_{min}, + i(interval)$ where $i = 1, ..., k-1$. This method is applied to each continuous feature independently regardless of the structure of the whole database and the instance class information.

Equal Height interval binning discretize data so that each bin will have approximately the same number of samples. It involves sorting the observed values together with the record's ID. If $|R|$ refers to the size of the records and the $V[|R|]$ refers to the size of the array that stores the sorted values, then the boundaries can be constructed at $V[(|R|/k) \times i]$ where $i = 1, ..., k-1$. The result is a collection of k bins of roughly equal size and this algorithm is class-blind and does not take into consideration the structure of the database.

Finally, *equal weight* interval binning considers not only the distribution of numeric values present, but also the groups they appear in. It is observed that larger groups have a bigger influence on the choice of boundaries because they have more contributing numeric values. In *equal weight* interval binning, numeric values are weighted $w(v) = 1/|class_v|$, where w is the weight function and v is the value being considered and $|class_v|$ is the size of the class that v belongs to. Instead of producing bins of equal size, we compute boundaries to obtain bins of equal weight. The algorithm starts by computing the size of each class, then it moves through the sorted arrays of values, keeping a running sum of weights wt. Whenever wt reaches a target boundary ($|$number of classes$|$/bins), the current numeric value is added as one of the boundaries, and the process is repeated until k times (k is the number of bins).

4   Experimental Evaluations

In this experiment, we implement the aggregation and discretization methods, described in Section 2 and 3, in the DARA algorithm, in conjunction with C4.5 classifier (J48 in WEKA) [10], as an induction algorithm that is run on the

DARA's discretized and transformed data representation. We then evaluate the effectiveness of each discretization method with respect to C4.5. The C4.5 induction algorithm [5] is a state-of-the-art top-down method for inducing decision trees. All experiments with DARA and C4.5 were performed using 10-fold cross validation estimation and were carried out with five different values of the number of bins, b = 2, 4, 6, 8, 10. We chose three well-known datasets; Mutagenesis [2], Musk [3] and Financial [4].

4.1 Mutagenesis Dataset

The data in mutagenesis domain [2] describes 188 molecules falling in two classes, *mutagenic* (active) and *non-mutagenic* (inactive) and 125 of these molecules are mutagenic. The description consists of the atoms and bonds that make up the compound. Thus, a molecule is described by listing its atoms *atom(AtomID, Element, Type, Charge)* and the bonds *bond(Atom1, Atom2, BondType)* between atoms. In this experiment, we use three different sets of background knowledge; B1, B2 and B3.

- B1: The atoms in the molecule are given, as well as the bonds between them; the type of each bond is given as well as the element and type of each atom.
- B2: Data in B1 and continuous values about the charge of atoms are added
- B3: Data in B2 and two continuous values describing each molecule are added, which are the log of compound's octanol/water partition coefficient (*logP*) and energy of the compounds lowest unoccupied molecular orbital ($^{\mathcal{E}}LUMO$).

Table 1. Performance of C4.5 on pre-processed Mutagenesis dataset B1, B2, B3 by DARA and C4.5

Algorithm \Bin	b = 10	b = 8	b = 6	b = 4	b = 2	b = 0	Ave
Mutagenesis B1							
Width	78.0	79.1	79.4	79.3	79.3	69.5	79
Height	71.9	72.4	73.0	69.9	74.4	69.5	72
Weight	70.5	72.4	73.6	69.9	74.4	69.5	72
Mutagenesis B2							
Width	75.6	77.4	80.8	80.5	80.2	73.0	79
Height	77.1	77.8	72.1	76.2	73.0	73.0	75
Weight	77.1	77.1	76.5	74.8	73.0	73.0	76
Mutagenesis B3							
Width	81.5	82.4	79.9	80.9	79.6	79.9	81
Height	83.2	83.1	81.0	83.9	82.3	79.9	83
Weight	81.6	82.9	82.0	79.8	82.3	79.9	82

Table 1 gives a detailed overview of the accuracy estimates from *10-fold cross validation* result for different setting of bins, *b*, tested for B1, B2 and B3. The performance accuracies are high when the number of bins, *b*, is 6 or 8 and the *equal width* interval binning produces higher percentage of performance accuracy compared to the *equal height* and *equal weight*.

4.2 Musk

This database describes molecules occurring in different conformations [3]. Each molecule is either *musk* or *non-musk*, and one of the conformations determines this property. The molecules will be modeled by two relations called molecule and conformation, associated by a *one-to-many* association. The *conformation* relation consists of 166 continuous features and the *molecule* relation consists of the molecules' identifier and class. The database comes in two versions, large and small musk datasets, and we used the small datasets. In this experiment, we run the DARA process with $c = 1$ and $c = 64$ (c is the length of attribute merged to become a single pattern) and Table 2 gives a detailed overview of the accuracy estimates from *10-fold cross validation* result for different setting of bins, b, with $c = 1$ and $c = 64$, tested. The performance accuracy is higher when we concatenate more attributes and cluster them into groups. It also reveals that the accuracy performance is higher when the number of bins is 8-10.

Table 2. Performance of C4.5 on pre-processed Musk dataset with $c = 1$ by DARA

bin	$b = 10$	$b = 8$	$b = 6$	$b = 4$	$b = 2$	$b = 0$	Ave
$C = 1$							
Width	67.5	68.0	67.5	69.8	72.4	60.0	69.0
Height	64.6	61.9	59.0	58.7	65.0	60.0	61.8
Weight	63.0	59.1	64.0	56.3	61.2	60.0	60.7
$C = 64$							
Width	88.5	82.9	78.0	70.5	54.8	88.5	74.9
Height	88.5	88.5	82.7	82.7	58.8	88.5	80.2
Weight	82.9	88.5	82.7	82.9	61.7	88.5	79.8

4.3 Financial Dataset

The financial dataset is taken from the Discovery Challenge organized at PKDD 1999 and PKDD 2000 [4]. The database is based on the data from a Czech bank. It describes the operations of 5369 clients holding 4500 accounts. The data is stored in seven relations. In this experiment, we perform a loan prediction, *successful* and *unsuccessful* loan. We have chosen the *account* relation as the target table and the *transaction* relation as a table that is joined by a *one-to-many* association with the target relation. We consider that the key parameter for this analysis should be derived from "behaviour" of the individual accounts by summarizing the *transaction* relation. The description of loan classes is as follows. There a four types of loan status: *A* if contract has been finished without problems, *B* if contract finished but loan not paid, *C* for running contract, OK so far, and *D* for running contract, where client is in debt.

The measurements of accuracy performance for the three methods (Equal Width, Equal Height and Equal Weight) are 69%, 67% and 71% respectively. The accuracy performance with the Equal Weight discretization method in DARA process is measured the highest, 71%, compared to the DARA process with Equal Width or Equal Height discretization method, which produces 69% and 67% of performance accuracies. We are unable to compare with previously published results due to scalability problems.

The data summarization approach, done by DARA, proved particularly successful on large datasets. Table 3 shows the results of DARA-based performance accuracy compared to other previously published results. Data summarization can be performed separately from the target relation, and it makes DARA more scalable and flexible in characterizing a specific item stored in the target relation that has *one-to-many* association to other non-target relations.

Table 3. Result on Musk, Mutagenesis and Financial Dataset

Algorithm	Small Musk	Mutagenesis B1	B2	B3	Financial
PROGOL	-	76%	81%	83%	-
FOIL	-	61%	61%	83%	-
Tilde	87%	75%	79%	85%	-
DARA + C4.5	88%	79%	79%	83%	71%
RollUp	89%	86%	85%	89%	100%

5 Conclusions

The multiple instances problem is an important problem that arises in real-world tasks where the training examples are ambiguous: a single record may have many associations with feature vectors that describe it, and yet only a few of those vectors may be relevant for the observed classification of the record. In this paper, we have presented an algorithm to aggregate values in a multiple instances problem, in which a novel method for aggregating multiset instances of a single object from multiple tables, with a high degree of *one-to-many* association, is introduced. The basic idea is to treat a series of records, associated with a single record in the target table, as a bag of terms, and take an unsupervised clustering method to aggregate them. Our approach combines the *selection-based* and *aggregation-based* propositionalization in learning data from multiple tables. In addition to that, our experiments reveal that all discretization methods in DARA algorithm help us to get higher percentage of accuracy, particularly when the number of bins is 6 or 8. The experimental results demonstrate that our approach of dynamic aggregation of relational attributes using DARA is at least competitive with existing multi-relational techniques, such as Progol and Tilde. Our approach has one major difference with these techniques, which may be the source of the good performance, namely the use of the aggregates to summarize a block of data, containing multiple instances of a single object, without the requirement of any domain knowledge.

The ability of describing the group as a whole is the main difference between our aggregation approach and first order logic (FOL) system (e.g. Progol, Tilde). In FOL, the characterization is based on the occurrence of one or more records in the group that maximizes certain properties, and it does not consider the multiple instance problem. On the other hand, DARA algorithm takes all records into consideration as each record has some influence on the value of the aggregation. As a result, FOL and DARA produce two different sets of feature-spaces, though there is still some overlapping space. We have presented a method called dynamic aggregation of relational attributes (DARA) with discretization to propositionalise a multi-relational database, such that the resulting view can be analysed by existing propositional methods. The DARA method has shown a good performance on three well-known datasets in term of performance accuracy. Further work should be done to improve this method by developing more specific rules for finding the right number of clusters for mapping purposes.

References

1. Alfred, R., Kazakov, D.: Pattern-Based Transformation Approach to Relational Domain Learning Using DARA. In: Crone, S.F., Lessmann, S., Stahlbock, R. (eds.) The Proceedings of the International Conference on Data Mining, Las Vegas, Nevada, June 25-29, 2006, pp. 296–302. CSREA Press (2006)
2. Srinivasan, A., Muggleton, S.H., Sternberg, M.J.E., King, R.D.: Theories for mutagenicity: A Study in first-order and feature-based induction. Artificial Intelligence, 85 (1996)
3. Dietterich, T.G., Lathrop, R.H., Lozano-Perez, T.: Solving the multiple-instance problem with axis-parallel rectangles. Artificial Intelligence 89(1-2), 31–71 (1997)
4. Workshop notes on Discovery Challenge In: Żytkow, J.M., Rauch, J. (eds.): Principles of Data Mining and Knowledge Discovery. LNCS (LNAI), vol. 1704. Springer, Heidelberg (1999)
5. Quinlan, J.R.: C4.5: Programs for Machine Learning, Morgan Kaufmann, Los Altos, Cal
6. Kramer, S., Lavrač., N., Flach, P.: Propositionalization approaches to relational data mining. In: Dzeroski, S., Lavrač, N. (eds.) Relational Data mining, Springer, Heidelberg (2001)
7. Salton, G., Michael, J.: Introduction to Modern Information Retrieval. McGraw-Hill, Inc, New York (1986)
8. Bezdek, J.C.: Some new indexes of cluster validity. IEEE Transaction System, Man, Cybern. B 28, 301–315 (1998)
9. Boley, D.: Principal direction divisive partitioning. Data Mining and Knowledge Discovery 2(4), 325–344 (1998)
10. Witten, I., Frank, E.: Data Mining: Practical Machine Learning Tools and Techniques with Java Implementations. Morgan Kaufman, Seattle (1999)
11. Knobbe, A.J., de Haas, M., Siebes, A.: Propositionalisation and Aggregates. In: Siebes, A., De Raedt, L. (eds.) PKDD 2001. LNCS (LNAI), vol. 2168, pp. 277–288. Springer, Heidelberg (2001)
12. Perlich, C., Provost, F.: Aggregation-Based Feature Invention and Relational Concept Classes. In: KDD 2003. Proceedings of the Ninth ACM SIGKDD International Conference on Knowledge Discovery and Data Mining, ACM Press, New York (2003)
13. De Raedt, L.: Attribute-value learning versus inductive logic programming: The missing links (extended abstract). In: Page, D.L. (ed.) Inductive Logic Programming. LNCS, vol. 1446, pp. 1–8. Springer, Heidelberg (1998)
14. Lavrač, N., Džeroski, S.: Inductive Logic Programming: Techniques and Applications. Ellis Horwood (1994)
15. Lavrač, N., Džeroski, S., Grobelnik, M.: Learning nonrecursive definitions of relations with LINUS. In: Kodratoff, Y. (ed.) Machine Learning - EWSL-91. LNCS, vol. 482, pp. 265–281. Springer, Heidelberg (1991)
16. Krogel, M.A., Wrobel, S.: Transformation-Based Learning Using Multirelational Aggregation. In: Rouveirol, C., Sebag, M. (eds.) ILP 2001. LNCS (LNAI), vol. 2157, Springer, Heidelberg (2001)

17. Lavrač, N., Železny, F., Flach, P.A.: RSD: Relational subgroup discovery through first-order feature construction. In: Matwin, S., Sammut, C. (eds.) ILP 2002. LNCS (LNAI), vol. 2583, Springer, Heidelberg (2003)
18. Krogel, M.A., Rawles, S., Železny, F., Flach, P.A., Lavrač, N., Wrobel, S.: Comparative evaluation of approaches to propositionalization. In: Horváth, T., Yamamoto, A. (eds.) ILP 2003. LNCS (LNAI), vol. 2835, pp. 197–214. Springer, Heidelberg (2003)
19. Blokceel, H., Bruynooghe, M.: Aggregation versus Selection Bias, and relational neural networks. In: Kurumatani, K., Chen, S.-H., Ohuchi, A. (eds.) IJCAI-WS 2003 and MAMUS 2003. LNCS (LNAI), vol. 3012, Springer, Heidelberg (2004)
20. Agrawal, R., Gehrke, J., Gunopulos, D., Raghavan, P.: Automatic subspace clustering of high dimensional data for data mining applications. In: Proceedings of the 1998 ACM SIGMOD International Conference on Management of Data, pp. 94–105. ACM Press, New York (1998)
21. Hofmann, T., Buhnmann, J.M.: Active data clustering. In: Advance in Neural Information Processing System (1998)
22. Hartigan, J.A.: Clustering Algorithms. Wiley, New York (1975)

Learning Optimal Kernel from Distance Metric in Twin Kernel Embedding for Dimensionality Reduction and Visualization of Fingerprints

Yi Guo[1], Paul W. Kwan[1], and Junbin Gao[2],[*]

[1] School of Math, Stat. & Computer Science,
University of New England, Armidale, NSW 2351, Australia
{yguo4,kwan}@turing.une.edu.au
[2] School of Computer Science,
Charles Sturt University, Bathurst, NSW 2795, Australia
jbgao@csu.edu.au

Abstract. Biometric data like fingerprints are often highly structured and of high dimension. The "curse of dimensionality" poses great challenge to subsequent pattern recognition algorithms including neural networks due to high computational complexity. A common approach is to apply dimensionality reduction (DR) to project the original data onto a lower dimensional space that preserves most of the useful information. Recently, we proposed Twin Kernel Embedding (TKE) that processes structured or non-vectorial data directly without vectorization. Here, we apply this method to clustering and visualizing fingerprints in a 2-dimensional space. It works by learning an optimal kernel in the latent space from a distance metric defined on the input fingerprints instead of a kernel. The outputs are the embeddings of the fingerprints and a kernel Gram matrix in the latent space that can be used in subsequent learning procedures like Support Vector Machine (SVM) for classification or recognition. Experimental results confirmed the usefulness of the proposed method.

1 Introduction

Biometric information based on a person's physiological and behavioural traits is gaining acceptance as a method for uniquely verifying his/her real identity. Among these traits, fingerprint, face, speech, iris and hand geometry are most commonly used. Recognition systems utilizing biometric data have been applied in many real world scenarios such as national security, medical services, etc. An early application of this technology was Automatic Fingerprint Identification System (AFIS) found in law enforcement. In recent years, a number of non-forensic applications have been proliferating very quickly [19,11].

Biometric data like fingerprints are often highly structured and high dimensional. For example, a fingerprint is characterized by numerous ending points

---

[*] Corresponding author.

R. Alhajj et al. (Eds.): ADMA 2007, LNAI 4632, pp. 227–238, 2007.
© Springer-Verlag Berlin Heidelberg 2007

and ridge bifurcations that are called minutiae. Figure 1 shows an image of fingerprint indicating the minutiae. In order to analyse a fingerprint, a vector containing the coordinates, type information and tangent angles (if necessary) for all feature points is normally created which is clearly residing in a very high dimensional space. Similar problem applies to face, iris and palm biometrics. The original images sampled from human subjects are either converted to key features' descriptors or represented directly by vectors of image pixels. Other types of biometric data such as voices, gaits and keystrokes fall into the time series category which are also high dimensional.

All of these biometric data pose great challenges to subsequent pattern recognition algorithms such as neural networks due to high computational complexity. The "curse of dimensionality" causes the number of hidden layers and neurons to increase dramatically, thereby worsening the performance of the network by having too many free parameters that are redundant. In addition, high dimensionality demands large storage requirement in practical recognition systems. Not only can this affect the performance such as response time, but it may also limit the usefulness to some extent. To address the problem of high dimensionality, some algorithms adopt a multi-step strategy [8] to reduce computation.

Another commonly adopted approach to address the dimensionality problem is to apply dimensionality reduction (DR) methods to project the original input onto a much lower dimensional "latent space" that preserves most of the useful information. The idea is to reduce the complexity of the problem while minimizing the information loss. Furthermore, if the data were represented by vectors of dimension less than 3, they could be visualized in a Euclidean space to facilitate understanding. It could also benefit researchers in interpreting the relationships in the input data and design data-oriented algorithms.

Recent years have witnessed significant advances in DR methods that are widely applied in robotics, bioinformatics, information retrieval, etc. DR methods can be categorized into linear methods like PCA (Principal Component Analysis) [7], ICA (Independent Component Analysis) [3] and nonlinear methods such as ISOMAP [18], Laplacian Eigenmaps (LE) [1], Locally Linear Embedding (LLE) [15], etc. The latter category is attracting more attention since the linear assumption on which linear methods are based is violated in most cases. Some of these DR methods have been successfully applied in biometric information processing on face and iris data [10,11] in order to simplify the problems at hand.

However, DR methods applying to fingerprints occurred to a much lesser extent. One reason might be the rather specific form of their representation. Algorithms based only on the raw fingerprint images often failed due to very little difference in appearance between the digitalized images. Therefore, minutiae that reveal unique personal identity have to be extracted from the raw images and used. It is hard to define an accurate distance metric between fingerprints by the Euclidean norm which are often used in most DR methods. However, more accurate distances can be defined in other ways such as the Enhanced Shape Context (ESC) distance reported in [9]. By using the ESC, we can construct a pairwise distance matrix for the input fingerprints and input it into the appropriate DR methods. In this paper, we choose TKE [6] as the candidate DR method because it only depends on the pairwise similarity information. However, in our earlier paper on TKE, a kernel defined on the input data is required but the ESC can only provide the distance information.

In this paper, we will explain how to convert the ESC distance matrix to a kernel Gram matrix that can be applied in TKE. A big advantage of TKE which is absent in most DR methods is that it can learn a valid kernel on the embeddings (the lower dimensional representations of the original data) which can then be utilized in subsequent learning tasks such as SVM or other kernel based algorithms.

The rest of this paper is organized as follow. We will review briefly DR methods applied to biometric data in Section 2. In Section 3, we will describe TKE to some details. In Section 4, we will explain how to convert the ESC distance metric on input fingerprints to a quasi kernel, followed by the procedure to learn an optimal kernel in the latent space for fingerprint clustering and visualization. In Section 5, experimental results will be presented, and in Section 6 we conclude and discuss our future work.

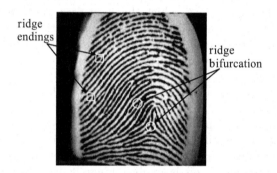

Fig. 1. Fingerprint with minutiae marked

2   Related Work

Applications of DR methods on biometric data can be traced back to a lower dimensional reconstruction technique proposed by Sirovich and Kirby [17] in 1987 for face images that maximizes the variance by assuming that the distribution of the data was Gaussian. In the following years, linear methods such as PCA and ICA appeared to dominate the research in this field due to their simplicity and low cost of implementation.

In recent years, Discrete Cosine Transform (DCT) has been used to reduce the dimensionality of the space of face images by truncating the high frequency components since only a small number of these components will be sufficient in preserving most of the important facial features [13]. The results of the DR process were input to a Multi-Layer Perceptron (MLP) for subsequent classification.

Having recognized the limitations of linear methods, [2] proposed the Curvilinear Component Analysis (CCA) and applied it on face images. The result was used in two learning algorithms, MLP and Support Vector Machine (SVM) for gender classification. CCA is composed by a two-layer structure. The first layer performs vector quantization (vectorization) on the input data and the second was called the projection layer which acts as a topographic mapping of the structure obtained from the first layer. Then, a linear kernel was used in subsequent SVM learning. However a linear kernel might not be most suitable for the task that involves revealing the non-linear information, and in fact SVM does not restrict the choice to linear kernels.

In [10], a revised Locally Linear Embedding (LLE) called weighted LLE was applied in face recognition. The spatial and temporal weights reflecting the possibility of appearance variance were incorporated in LLE by influencing the neighborhood searching and adjacency graph construction.

A new DR method was presented in [20] for face and palm biometrics named Unsupervised Discriminant Projection (UDP) which is quite similar to the Fisher Linear Discriminant Analysis (LDA) and can be considered an unsupervised version of LDA. Initially, it constructs a neighborhood graph on the input data by selecting k nearest neighbors and thereby creates several neighborhoods. Then, it finds an optimal hyperplane onto which the input data will be projected according to the LDA-like criterion, that is to maximize the ratio of inter neighborhood scatter to inner neighborhood scatter. Because the scatter of the projected data on the optimal hyperplane is included in the criterion, the authors claim that UDP preserves not only the locality but also the non-locality. As such, it will facilitate classification because the result of UDP would have compact and far apart clusters.

All the DR approaches mentioned above are mainly for face or palm images because these data are readily converted into vectorial form. Much less attention were paid to other structured data such as fingerprints due to their intrinsic complexities. Furthermore, there is an obvious gap between the result of DR and the subsequent learning algorithms because it purely finds the low-dimensional embeddings for the input data and nothing additional. In other words, the target of DR is quite simple, that is to reduce the dimensionality of the input data and apply some classifiers to the simplified data with much less computational cost. To the subsequent learning algorithms that actually operate on the much lower dimensional space output by DR, they have no idea how different the input and output spaces is and how much information might have been lost due to the DR process.

In our opinion, DR methods should provide us with more by preserving as much information on the original data as possible, thus allowing them to be used with confidence in subsequent processing. By analyzing the structure of TKE, we found that its potential can be further explored to fill that gap as described in the previous paragraph. The key lies in the kernel in the latent or output space that can be learned by TKE. To provide the background necessary for such understanding, we will summarize TKE in the next section.

3   Twin Kernel Embedding

With no loss of generality, the following notations will be adopted. The data in the input space are denoted by $\mathbf{y}_i$ $(i = 1, \ldots, N)$ while $\mathbf{x}_i$ $(i = 1, \ldots, N)$ their embeddings in a low-dimensional space or the so-called latent space[1]. Notice that $\mathbf{y}_i$ does not have to be a vector while it can be any object, for example, a fingerprint. In addition, $\mathbf{Y}$ and $\mathbf{X}$ will be used to denote respectively the set of input objects and the set of embedded objects. If the objects were vectorial, $\mathbf{Y}$ (and $\mathbf{X}$) would denote a matrix consisting of rows of vectors. Furthermore, $a \cdot b$ denotes the inner product of a and b.

Twin Kernel Embedding (TKE) preserves the similarity structure of input data in the latent space by matching the similarity relations represented by two kernel Gram matrices, that is one for input data and the other for their embeddings. It simply minimizes the following objective function

$$-\mathrm{Vec}\mathbf{K_y} \cdot \mathrm{Vec}\mathbf{K_x}, \tag{1}$$

where Vec denotes the vec operator on matrix and $\mathbf{K_y}$ and $\mathbf{K_x}$ the kernel Gram matrices derived from the kernel functions $k_y(\cdot, \cdot)$ and $k_x(\cdot, \cdot)$ which are defined on the input data and the embeddings respectively. The idea is to preserve the similarities among input data and reproduce them in a lower dimensional latent space expressed also in similarities among embeddings. To explain this clearer, we can just regard $\mathrm{Vec}\mathbf{K_y} \cdot \mathrm{Vec}\mathbf{K_x}$ as a linear kernel which is a measure of similarity between the variables involved in the kernel function. The larger the value of the kernel, the more similar those two variables are. Therefore, we minimize (1) in order to make $\mathbf{K_x}$ and $\mathbf{K_y}$ as similar as possible.

To avoid trivial solutions, two regularization terms on the kernel and the embeddings are introduced and the objective function becomes

$$L = -\mathrm{tr}(\mathbf{K_y}\mathbf{K_x}) + \lambda_k \mathrm{tr}(\mathbf{K_x}\mathbf{K_x}) + \lambda_x \mathrm{tr}(\mathbf{XX}^\top), \tag{2}$$

where we use the fact that $\mathrm{Vec}\mathbf{K_y} \cdot \mathrm{Vec}\mathbf{K_x} = \mathrm{tr}(\mathbf{K_y}\mathbf{K_x})$. The second term is a ridge regularizer on the kernel to make sure that the norm of the kernel is controlled. The third term is a penalty for too large a norm for the embeddings in order to ensure that their coordinates are relatively small. λ_k and λ_x are tunable parameters for controlling the strength of the regularization and are supposed positive.

In order to capture the nonlinear structure, $k_x(\cdot, \cdot)$ should be chosen to be nonlinear. If we use the common RBF kernel

$$k_x(\mathbf{x}_i, \mathbf{x}_j) = \gamma \exp\left(-\sigma \frac{\|\mathbf{x}_i - \mathbf{x}_j\|^2}{2}\right). \tag{3}$$

in (2), there would be no closed form solution for $\mathbf{X}$ and hence a gradient-based algorithm for optimization has to be employed based on the requirement that $k_x(\cdot, \cdot)$

---

[1] We prefer to use the term latent space because it is frequently used in related literature.

is differentiable. The initialization of $\mathbf{X}$ is also required to start the optimization. KLE, KPCA and other methods which can work with kernels can be applied here. The dimension of the embeddings is assigned according to the requirement of the application, which is normally 2. A by-product of this optimization process is that we can get the optimal hyper-parameters for the kernel function $k_x(\cdot, \cdot)$ as well. It ensures that the kernel we pick is well adjusted.

Similar to UDP, TKE is also designed to preserve both locality and non-locality simultaneously. This is achieved by filtering the entries in $\mathbf{K_y}$. In other words, not all entries will remain in the optimization process but only those that convey most of the similarity information in the input data. This filtering is fulfilled by the k-nearest neighboring process. Given an object $\mathbf{y}_i$ in the input space, only those objects whose similarities (in the sense of kernel values) to $\mathbf{y}_i$ are among the k nearest neighbors of $\mathbf{y}_i$ are selected to retain their original values while all others are set to 0. The variable $k(> 1)$ in k-nearest neighboring controls the locality that the algorithm will preserve. It can be interpreted as constructing a weighted adjacency graph in the feature space while the weights of edges are evaluated by a kernel as in KLE [5]. However, it also works without filtering in which case TKE becomes a global approach. Because TKE attempts to match $\mathbf{K_x}$ with $\mathbf{K_y}$ and that RBF kernel (3) cannot return a value of 0 other than two points being very far apart in the latent space, TKE seeks a solution that keeps the points in the same neighborhood close together while allowing the points not in the same neighborhood be very far.

In addition to the fact that TKE outperforms related methods like KPCA, KLE, etc., another advantageous feature of TKE is that it only relies on the pairwise similarities in the input data. The reason is that it only requires the kernel Gram matrix of the input data in its objective fuction during optimization. Using TKE, any kind of data (structured or not) can be visualized in a lower dimensional space as long as an appropriate kernel is defined on them. Thus, only the kernel Gram matrix on the input data $\mathbf{K_y}$ and an initialization for $\mathbf{X}$ will be adequate for TKE to find the optimal embeddings.

4   From Distance to Kernel

In the last section, we noted that TKE requires a kernel defined on the input data. [12] proposed a kernel function based on the edit distance which is applicable to both string and graph representations of patterns including fingerprints. However, what is widely adopted in matching biometric data is distance metric instead of kernel, like the enhanced shape context distance (ESC) we mentioned earlier. A similar problem exists in other fields such as bioinformatics. [14] presented several similarity measures for protein sequences other than kernels.

When we analyze the objective function of TKE in (2), several observations become obvious. Firstly, the kernel on the input data can actually be replaced by other similarity measures on condition that they can provide pairwise similarities represented in a matrix form. The special properties of kernel such as positive definiteness are not mandatory in TKE. Secondly, on the output side of TKE, an optimal kernel (k_x) can be learned from the input data as well as the embeddings since not only do we have the form of the kernel (for instance, the RBF kernel) in (3), but we also have all the optimal hyperparameters that enable us to reconstruct the kernel Gram matrix on the embeddings. Therefore, TKE can be seen as a learning machine for kernels. If we use other similarity measures on the input side such as those mentioned above, we are actually forcing TKE to learn a kernel Gram matrix from the similarity matrix. Thirdly, followed from the second observation, the kernel learned by TKE preserves most of the information in the input kernel matrix or similarity matrix. In other words, $\mathbf{K_x}$ is truly the mimic of $\mathbf{K_y}$ or the kernelized similarity matrix. Given $\mathbf{K_x}$ being a kernel Gram matrix that faithfully preserves the information contained in $\mathbf{K_y}$ or a similarity matrix that reflects the relations among the input data, we will be more confident in the subsequent learning algorithms since the information loss is visible and controllable. By varying the k in filtering, we can decide how much of $\mathbf{K_y}$ will be retained in learning $\mathbf{K_x}$. Furthermore, if we use $\mathbf{K_x}$ directly in learning procedures like SVM, etc., we no longer need to choose appropriate kernels for the embeddings. k_x is the right one that resembles the original kernel k_y as close as possible. This fills the gap we discussed in section 2.

Finally, we can convert a distance metric to kernel or roughly speaking a quasi-kernel which is not strictly a kernel mathematically but applicable in TKE because of the first observation if we simply regard the quasi-kernel as a similarity measure. Even the concept of distance metric here needs not to be strict. Although we could relax the necessity of a kernel on the input data, here we require at a minimum positivity and symmetry for the entries in the similarity matrix to ensure that the learning result will be correct. This observation also applies to Kernel Laplacian Eigenmap (KLE) since to make a Laplacian valid, the weights on the adjacency graph should be positive and symmetric according to the spectral graph theory [4]. This characteristic provides a solution to another hard problem, that is the initialization of embeddings in the latent space. Now we can use KLE with the quasi-kernel to give TKE an approximate initial state.

Now we return to our discussion on the input fingerprints. As described in [9], the ESC distance additionally incorporates the domain specific information like minutiae type and angle details, which were not considered in the original paper. Here, we will not explain its formulation in details other than showing that we can turn it from being a semi-metric into a metric distance as follow.

$$d(\mathbf{y}_i, \mathbf{y}_j) = \frac{1}{2}(d_{esc}(\mathbf{y}_i, \mathbf{y}_j) + d_{esc}(\mathbf{y}_j, \mathbf{y}_i)). \tag{4}$$

$d_{esc}(\mathbf{y}_i, \mathbf{y}_j)$ denotes the ESC distance between $\mathbf{y}_i$ and $\mathbf{y}_j$. Interested readers are referred to the original paper and references therein. We just use $d(\mathbf{y}_i, \mathbf{y}_j)$ to denote the metric distance between $\mathbf{y}_i$ and $\mathbf{y}_j$. It satisfies the following properties:

- $d(\mathbf{y}_i, \mathbf{y}_j) \geq 0$, equality holds if and only if $i = j$;
- Symmetry: $d(\mathbf{y}_i, \mathbf{y}_j) = d(\mathbf{y}_j, \mathbf{y}_i)$;
- Triangle inequality: $d(\mathbf{y}_i, \mathbf{y}_j) + d(\mathbf{y}_j, \mathbf{y}_l) \geq d(\mathbf{y}_i, \mathbf{y}_l)$.

Then, the following equation can convert $d(\mathbf{y}_i, \mathbf{y}_j)$ to a (quasi-)kernel [16]:

$$k(\mathbf{y}_i, \mathbf{y}_j) = e^{-d(\mathbf{y}_i, \mathbf{y}_j)} \tag{5}$$

The k defined above apparently satisfies the positivity and symmetry properties. Though it may not strictly be a kernel, according to our analysis above, it can be used in TKE. What remains is to use the (quasi-)kernel defined in (5) as the kernel on the input data to compute the kernel Gram matrix and then substitute it in TKE. Through this, the input fingerprints can be clustered and visualized in a 2-dimensional space that can be interpreted with ease. Note that TKE can use either KLE, KPCA or other methods applicable to non-vectorial data to initialize the embeddings. Because whether the k in (5) is a valid kernel or not is still in question, we decide to use KLE as the initialization for the reason we explained earlier.

5   Experimental Results

Experiments were conducted on the same fingerprint database used in [9]. The database we used contains 21 different fingers, each having 8 impressions totalling 168 fingerprint images. We randomly selected 10 subjects and processed the corresponding fingerprints images. Although the size of our database might be small, it is adequate for illustrating our idea. It is a common problem with biometric information processing in which most biometric data sets have large pool of subjects but limited samples for each subject. Figure 2 shows one finger impression for each of the 10 randomly selected subjects.

Fig. 2. Fingerprints of 10 subjects

The raw images of the fingerprints were put through six pre-processing steps: normalization, ridge orientations superimposing, filtering, binarizing, thinning and minutiae filtering. In the end, a set of minutiae including their coordinates, types and related information like tangent angles for each finger is extracted from each fingerprint image. Because the set of minutiae is highly structured, it creates problem for most DR methods. ESC is employed to evaluate the distance between each pair of fingers with the optimal parameters as stated in [9]. By using equation (4) and equation (5), the metric distance is converted to a (quasi-)kernel and as a result a kernel Gram matrix is constructed which serves as $\mathbf{K_y}$ in TKE.

The fingerprints from the same finger are anticipated to be close to each other in the 2-dimensional space with overlappings indicating similar prints actually from different fingers. The RBF kernel is used as k_x for the embedded data in TKE for its simple derivative formulation and ability to capture the nonlinear structure in the data. The regularization parameters are set to be 0.005 and 0.001 for kernel regularization λ_k and variable regularization λ_x respectively, and k in k nearest neighbor filtering of $\mathbf{K_y}$ is set to 13. As mentioned above, KLE provides the initialization for $\mathbf{X}$.

Figure 3 shows the visualization result of the fingerprints that we are seeking. Different symbols stand for different fingers. From the result of TKE, we can see that the nonlinear structure among the clusters are very clear. The result of KLE is not as good as that of TKE as can be seen by the overlappings in the middle that causes the embeddings to be indistinguishable. Observing the visualization result of TKE, we notice that fingerprints of some fingers have relatively high nonlinear structure like the plus, hollow pentagon and upright triangle in Figure 3 (a). This explains why simple linear models cannot correctly classify them.

Figure 4 demonstrates the kernel Gram matrices on the input data and the embeddings. If we simply observe the patterns of the images of these Gram matrices, we can easily detect that the rightmost one, that is $\mathbf{K_x}$ learned by TKE from $\mathbf{K_y}$ faithfully preserves the block property showing enhanced contrast. This reflects what we have discussed earlier regarding the ability of TKE to reproduce the kernel Gram matrix of the input data in the embeddngs as well as preserving the locality and non-locality at the same time.

As mentioned, the kernel $\mathbf{K_x}$ and the embeddings can serve as inputs to subsequent learning algorithms. To illustrate this point, $\mathbf{K_x}$ is put into SVM

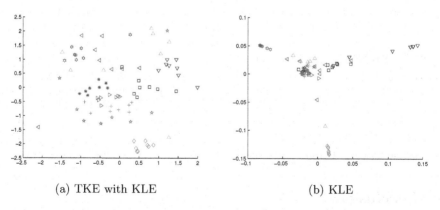

(a) TKE with KLE   (b) KLE

Fig. 3. Visualization of fingerprints using TKE and KLE

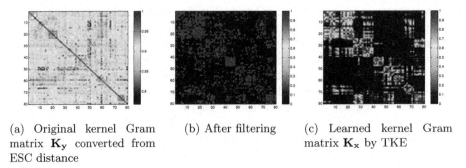

(a) Original kernel Gram matrix $\mathbf{K_y}$ converted from ESC distance

(b) After filtering

(c) Learned kernel Gram matrix $\mathbf{K_x}$ by TKE

Fig. 4. Kernel Gram matrices

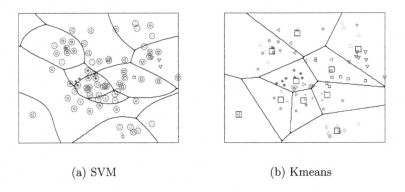

(a) SVM

(b) Kmeans

Fig. 5. SVM and Kmeans on the embedded data

directly. It is a multi-class problem, and we adopted a one-against-all strategy and set $C = 8$ in SVM. Since the embeddings are in a 2-dimensional space, we can visualize the result of SVM on a plane. Figure 5 (a) shows the optimal separating hyperplanes (black solid curves) betweem fingers while the red circles indicate the support vectors. Due to the high nonlinearity, there are 69 support vectors altogether and the training error is 23.75% which is not optimal. However, we want to emphasize that here our goal is not so much in classification but in illustrating the whole DR process visually. The error rate, we believe, can be further improved. Figure 5 (b) shows the result of K-means (that is, the original K-means using the Euclidean distance) on the embeddings. The black square in each area separated by the black boundaries is the median of the class. It is obvious that this simple linear method results in numerous mistakes. It is impossible to find straight lines that can separate those fingerprints correctly.

6   Conclusions

In this paper, we apply TKE to clustering and visualizing fingerprints in a 2-dimensional space. Twin Kernel Embedding (TKE) is dimensionality reduction method proposed recently that can process structured or non-vectorial data directly without vectorization. In this work, it is applied in learning an optimal kernel in the lower dimensional embedding or latent space from a distance metric defined on the input fingerprints instead of a kernel. The output are the embeddings of the fingerprints and a kernel Gram matrix in the latent space that can be used in subsequent learning procedures like Support Vector Machine (SVM) for classification or recognition. Experimental results supported positively the usefulness of the proposed method.

Several issues still need to be investigated further. Though we can use the images of the kernel Gram matrices as shown in Figure 4 to distinguish their differences, a method for quantifying the amount of information loss is necessary to determine how much information is actually preserved by the DR methods. Another problem is the initialization for TKE. If a similarity matrix other than a kernel Gram matrix is the only available information about the input, it is still unclear how to give an initial value for the embeddings since in general TKE only has local minimum solutions, thereby rendering random initialization infeasible.

Acknowledgements

This work is supported by the National Natural Science Foundation of China (NSFC 60373090), the ARC DP Development Grant from Charles Sturt University and the University Research Grant from the University of New England.

References

1. Belkin, M., Niyogi, P.: Laplacian eigenmaps for dimensionality reduction and data representation. Neural Computation 15(6), 1373–1396 (2003)
2. Buchala, S., Davey, N., Frank, R.J., Gale, T.M.: Dimensionality reduction of face images for gender classification. In: Proceedings of 2nd International IEEE Conference on Intelligent Systems, vol. 1, pp. 88–93 (June 2004)
3. Cherkassky, V., Mulier, F.: Learning from Data: Concepts, Theory, and Methods. In: Adaptive and Learning Systems for Signal Processing, Communications, and Control, John Wiley & Sons, Chichester (1998)
4. Chung, F.R.K.: Spectral Graph Theory. Number 92 in Regional Conference Series in Mathmatics. AMS (1997)
5. Guo, Y., Gao, J., Kwan, P.W.: Kernel Laplacian eigenmaps for visualization of non-vectorial data. In: Sattar, A., Kang, B.-H. (eds.) AI 2006. LNCS (LNAI), vol. 4304, pp. 1179–1183. Springer, Heidelberg (2006)
6. Guo, Y., Gao, J., Kwan, P.W.: Visualization of non-vectorial data using twin kernel embedding. AIDM 0, 11–17 (2006)
7. Jolliffe, M.: Principal Component Analysis. Springer, New York (1986)
8. Kwan, P.W., Gao, J.: A multi-step strategy for approximate similarity search in image databases. In: Proceedings of The Seventeenth Australasian Database Conference (ADC), Hobart, TAS, Australia, pp. 139–147 (January 2006)

9. Kwan, P.W., Gao, J., Guo, Y.: Fingerprint matching using enhanced shape context. In: Proceedings of the Image and Vision Computing New Zealand, pp. 115–120 (2006)

10. Mekuz, N., Bauckhage, C., Tsotsos, J.K.: Face recognition with weighted locally linear embedding. In: Proceedings of the 2nd Canadian Conference on Computer and Robot Vision, pp. 290–296 (May 2005)

11. Monro, D.M., Rakshit, S., Zhang, D.: Dct-based iris recognition. IEEE Transactions on Pattern Analysis and Machine Intelligence 29(4), 586–595 (2007)

12. Neuhaus, M., Bunke, H.: Edit distance-based kernel functions for structural pattern classification. Pattern Recognition 39(10), 1852–1863 (2006)

13. Pan, Z., Adams, R., Bolouri, H.: Dimensionality reduction of face images using discrete cosine transforms for recognition. In: IEEE Conference on Computer Vision and Pattern Recognition (2000)

14. Popescu, M., Keller, J.M., Mitchell, J.A.: Fuzzy measures on the gene ontology for gene product similarity. IEEE Transactions on Computational Biology and Bioinformatics 3(3), 263–274 (2006)

15. Roweis, S.T., Saul, L.K.: Nonlinear dimensionality reduction by locally linear embedding. Science 290(22), 2323–2326 (2000)

16. Schölkopf, B., Smola, A.J.: Learning with Kernels: Support Vector Machines, Regularization, Optimization, and Beyond. The MIT Press, Cambridge, MA (2002)

17. Sirovich, L., Kirby, M.: Low-dimension procedure for the characterization of human faces. Journal of the Optical Society of America 4(3), 519–524 (1987)

18. Tenenbaum, J.B., de Silva, V., Langford, J.C.: A global geometric framework for nonlinear dimensionality reduction. Science 290(22), 2319–2323 (2000)

19. Wang, P., Ji, Q., Wayman, J.L.: Modeling and predicting face recognition system performance based on analysis of similarity scores. IEEE Transactions on Pattern Analysis and Machine Intelligence 29(4), 665–670 (2007)

20. Yang, J., Zhang, D., Yang, J.Y., Niu, B.: Globally maximizing, locally minimizing: Unsupervised discriminant projection with applications to face and palm biometrics. IEEE Transactions on Pattern Analysis and Machine Intelligence 29(4), 650–664 (2007)

21. Gao, J., Harris, C.J., Gunn, S.R.: On a class of support vector kernels based on frames in function Hilbert spaces. Neural Computation 13(9), 1975–1994 (2001)

Efficiently Monitoring Nearest Neighbors
to a Moving Object

Cheqing Jin and Weibin Guo

Dept. of Computer Science, East China University of Science and Technololy, China
130 Meilong RD, Shanghai, 200237, China
{cqjin,gweibin}@ecust.edu.cn

Abstract. Continuous monitoring k nearest neighbors in highly dynamic scenarios appears to be a hot topic in database research community. Most previous work focus on devising approaches with a goal to consume litter computation resource and memory resource. Only a few literatures aim at reducing communication overhead, however, still with an assumption that the query object is *static*. This paper constitutes an attempt on continuous monitoring k nearest neighbors to a *dynamic* query object with a goal to reduce communication overhead. In our RFA approach, a Range Filter is installed in each moving object to filter parts of data (e.g. location). Furthermore, RFA approach is capable of answering three kinds of queries, including precise kNN query, non-value-based approximate kNN query, and value-based approximate kNN query. Extensive experimental results show that our new approach achieves significant saving in communication overhead.

1 Introduction

Finding k nearest neighbors (kNN) to a query object is one of the most critical operations in the field of spatial databases. The primary focus of spatial database research till recently has been on static spatial data which are updated infrequently, such as buildings, roads, .etc[7,13,14]. Nowadays, there has been an increasing interest in processing objects in motion, which change location frequently, such as vehicles, mobile networks, .etc. In a typical scenario, each moving object continues to report its location to a special site frequently, where the answer is calculated and output in real-time. Most previous work focus on devising various solutions to generate qualified results with few memory resource and computation resource (e.g., [8,9,11]).

The network communication resource is a critical resource in many real-world scenarios, especially distributed environments. For example, if the amount of objects becomes larger and larger, the network resource appears to be the bottle-neck for the processing system. Consequently, it is necessary to devise communication-efficient solutions for such scenarios. Unfortunately, to our best knowledge, only a few literatures[1,2,10] take this factor into account and give out

R. Alhajj et al. (Eds.): ADMA 2007, LNAI 4632, pp. 239–251, 2007.
© Springer-Verlag Berlin Heidelberg 2007

solutions. Babcock et al. propose a method to monitor k objects with largest numeric value over distributed environments, which can be viewed as a solution for finding kNN objects in 1-dimensional space with a special query object (∞)[1]. Recently, Cheng et al. present a solution to find k nearest neighbors with non-value tolerance over distributed environments[2]. Mouratidis et al. also propose one threshold-based approach to monitor k-nearest neighbors[10].

One common weakness of the above methods is that they mainly focus on handling *static* query object, i.e, the value of the query object being unaltered with time going on (e.g., the query object is ∞ in [1] and a constant in [2]). An example query can be described like: *what are k nearest taxies to a school?* However, there still exist some situations requiring *dynamic* query object, i.e, an object in motion. For example, in the query like: *what are k nearest clients to a free taxi?*, the *free taxi* is a moving object. Previous methods (e.g. [1,2]) cannot be easily adapted to solve such problem, it is necessary to seek new solutions. A simple way to handle *dynamic* query object is: some objects (more than k) are forced to report current location whenever the query object moves[10].

This paper focuses on devising novel approach on continuous monitoring k nearest neighbors to a *dynamic* query object with a goal to reduce the communication overhead. We assume an environment containing n moving objects sending their locations to a central site frequently. The central site continues to process the query without any knowledge about the velocities and the trajectories of objects.

The main contribution is that we have proposed a novel approach, the Range-Filter-based Approach (RFA), to cope with the problem. We sketch the approach as follows. Initially, each object is affiliated with a range filter whose purpose is to transmit new location to the central site if the location exceeds a specified range. The filter is initialized by the central site and maintained by the cooperation of the central site and the object itself. Simultaneously, the central site also reserves a copy of all filters. During the running time, when an object moves, it detects new location v, compares v with the range, and sends v to the central site once v is out of the range. The central site also updates filter settings of some objects if necessary. At any time, the central site can output the answer based on the a copy of filter settings in the central site.

The most significant characteristic of a range filter, the core structure in the RFA approach, is the self-adaptivity. In previous filter-based approaches, (e.g., [1,2,12]), when remote filters are outdated, the central site calculates the new setting, and sends them to remote objects. It may result in great network transmission overhead if such events frequently happen. But in RFA approach, when encountering such situation, the central site and the remote object are capable of calculating a same filter setting simultaneously. If the new setting satisfies the querying condition, no additional transmission occurs.

Some scenarios prefer approximate answer because it can be calculated easily even though the rate of stream is rapid and the volume of data is huge. Besides providing precise answer, RFA approach is capable of processing two kinds of approximate queries, including value-based approximate kNN query and non-value-based approximate kNN query[2]. The value-based approximate kNN query introduces a numeric value to guarantee the answer, whereas the non-value-based approximate kNN query expresses the error tolerance in terms of a *rank*.

The rest of the paper is organized as follows. Section 2 describes the environment in brief and defines the query formally. Section 3 depicts Range Filter structure and RFA approach in detail. Section 4 evaluates the performance of the new approach by a series of experiments. Section 5 reviews related work of this paper. Finally, Section 6 concludes the paper with a summary and directions for future work.

2 Preliminaries

2.1 Environment

We consider an environment containing n moving objects and 1 central site. When an object (say, O_i) moves, it sends its identity and new location to the central site through wireless network. Let S denote a set containing all objects, $S = \{O_1, O_2, \cdots, O_n\}$; let $V_{i,t}$ denote the location of the object O_i at time t. Each object O_i generates a stream of trace: $\{V_{i,0}, V_{i,1}, \cdots, \}$. Although an object can send(/receive) data to(/from) the central site, no direct communication routine exists between any pair of objects. Based on the data received from objects, the central site is capable of answering a kNN query in real-time, as demonstrated in Figure 1.

The location of each object is in a d-dimensional metric space. The distance between two arbitrary locations can be described by a distance function $dist$ with following properties, where V_1, V_2 and V_3 are locations of three objects.

1. $dist(V_1, V_2) = dist(V_2, V_1)$
2. $dist(V_1, V_2) > 0(V_1 \neq V_2)$ and $dist(V_1, V_2) = 0(V_1 = V_2)$
3. $dist(V_1, V_2) \leq dist(V_1, V_3) + dist(V_2, V_3)$

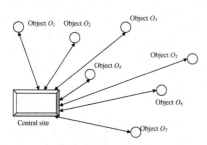

Fig. 1. Environment

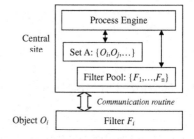

Fig. 2. Architecture

Table 1. An example of source data and query results

Time	O_1	O_2	O_3	O_4	O_5	$kNN(O_1,2)$	$nvakNN(O_1,2,1)$	$vakNN(O_1,2,1.5)$
0	1	3	7	10	12	$\{O_2,O_3\}$	$\{O_2,O_3,O_4\}$	$\{O_2,O_3\}$
1	4	5	10	14	9	$\{O_2,O_5\}$	$\{O_2,O_3,O_5\}$	$\{O_2,O_3,O_5\}$
2	3	4	15	14	9	$\{O_2,O_5\}$	$\{O_2,O_4,O_5\}$	$\{O_2,O_5\}$
3	1	3	11	12	8	$\{O_2,O_5\}$	$\{O_2,O_3,O_5\}$	$\{O_2,O_5\}$
4	2	8	15	16	7	$\{O_2,O_5\}$	$\{O_2,O_3,O_5\}$	$\{O_2,O_5\}$

2.2 Query Definition

This paper considers three kinds of kNN queries, including precise kNN query, non-value-based approximate kNN query and value-based approximate kNN query. The query object q can be either *static* (e.g, school, landmark), or *dynamic* (e.g, vehicle). Let q_t denote the location of q at time t.

Precise kNN query $(kNN(q,k))$**:** q is a query object, $q \in S$; k is the number of neighbors, $k \in N^+$. At any time point t, return a set A satisfying following conditions: (1) $q \notin A$; (2) $\forall O_i \in A, O_j \in S - A - \{q\}$, we have: $dist(V_{i,t}, q_t) \le dist(V_{j,t}, q_t)$; (3) $|A| = k$.

Non-value-based approximate kNN query $(nvakNN(q,k,r))$**:** q is a query object; k is the number of neighbors, $k \in N^+$; r is an error parameter, $r \in N^+$. At any time point t, return a set A satisfying following conditions: (1) $A \subseteq kNN(q, k+r)$; (2) $|A| = k$.

Value-based approximate kNN query $(vakNN(q,k,e))$**:** q is a query object; k is the number of neighbors, $k \in N^+$; e is an error parameter, $e \in R^+$. At any time point t, return a set A satisfying following conditions: (1) $q \notin A$; (2) $\forall O_i \in A, dist(V_{i,t}, q_t) \le e + \max_{O_j \in kNN(q,k)}(dist(V_{j,t}, q_t))$; (3) $|A| = k$.

Example 1. *Table 1 demonstrates a small example on processing above queries. The locations (in 1-dimensional space) of 5 objects in first 5 time points are illustrated in the left 6 columns. The seventh column shows the answer for* $kNN(O_1, 2)$. *The right 2 columns show the maximum result set for* $nvakNN(O_1, 2, 1)$ *and* $vakNN(O_1, 2, 1.5)$ *respectively, which implies that any subset containing two objects is a legal answer. For instance, at time point 4, any subset of* $\{O_2, O_3, O_5\}$ *(i.e.,* $\{O_2, O_3\}$, $\{O_2, O_5\}$, $\{O_3, O_5\}$*) is legal for the query* $nvakNN(O_1, 2, 1)$.

3 Range Filter-Based Approach (RFA)

This section describes our novel Range-Filter-based Approach (RFA) in detail. Figure 2 illustrates the architecture. A *range filter* (say, F_i) is installed in one object (say, O_i) to reduce communication overhead by filtering parts of new locations within a range. The central site consists of three components, such as a *range filter*, an answer set A and *process engine*. The *filter pool* reserves a copy

Algorithm 1. $newRange(F, v)$ /* $F = (c, l, u, b)$ */

1: $\Delta = dist(c, v) - b$;
2: **if** $(\Delta > 0)$ **then**
3:     $l = b + \frac{\Delta}{2}$;
4:     $u = b + \frac{3\Delta}{2}$;
5: **else**
6:     $l = \max(0, b + \frac{3\Delta}{2})$;
7:     $u = b + \frac{\Delta}{2}$;

---

of all filters of moving objects. At any time point, the *process engine* is capable of calculating an *answer set A* based on the data in *filter pool*. Another task of the *process engine* is to reset filters in remote objects if current settings cannot satisfy the query requirement.

Section 3.1 describes the structure of *Range Filter*. Section 3.2 describes a way to calculate answers from filters. Finally, Section 3.3 introduces the overall algorithm.

3.1   Range Filter

A Range Filter F is defined as (c, l, u, b). Field c is a location value in d-dimensional space, representing the central point of the filter's range. Fields l and u are the *lower diameter* and *upper diameter* respectively, representing the minimum and maximum distance from the location c, $0 \leq l \leq u$. The range of a filter F is the collection of locations whose distances to c are within $[l, u]$. In other words, a new location v is claimed in the range of filter F only if $dist(v, c) \in [l, u]$. In RFA approach, an object sends current location to the central site only when (1) its location exceeds range, (2) receives a request from the central site.

One important characteristic of a range filter is that it can generate a new range when current location v exceeds the range, as shown in Algorithm 1. With the help of field b (*backup diameter*). Algorithm `newRange` (Algorithm 1) alters the lower diameter l and upper diameter u, but retains the central location c unchanged. Clearly, after processing, we still have: $l \leq dist(c, v) \leq u$. Note that the value of b is only determined by the central site (see Algorithm 3).

Figure 3 demonstrates how to reset the range by invoking Algorithm `newRange`. When the new value v (the shadowed circle) jumps out of the range (covered by the dotted curve in the left part of Figure 3), Algorithm `newRange` calculates new lower and upper diameter (l', u'), as shown in the right part of Figure 3. Figure 3(a) shows situations when $dist(c, v) < b$, while Figure 3(b) shows situations when $dist(c, v) > b$.

3.2   Finding Nearest Neighbors from *Filter Pool*

One task of the *process engine* is to seek the nearest neighbors based on a copy of objects' filters reserved in *filter pool*. The first step is to determine the minimum

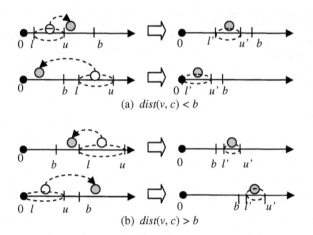

(a) $dist(v, c) < b$

(b) $dist(v, c) > b$

Fig. 3. Resetting the filter by invoking Algorithm `newRange`

and maximum distance between $V_{i,t}$ and q_t (q_t is the current value of query object q). Remember that $dist(V_{i,t}, c_i) \in [l_i, u_i]$. Consequently, $dist(V_{i,t}, q_t)$ is influenced by two factors, including the filter setting and the value of q_t. According to the definition of subroutine $dist$, the minimum and the maximum possible values of $dist(V_{i,t}, q_t)$, denoted as L_i and U_i, are calculated as follows.

$$L_i = \begin{cases} l_i - dist(q_t, c_i) & : & dist(q_t, c_i) < l_i \\ 0 & : & (dist(q_t, c_i) < u_i) \wedge (dist(q_t, c_i) > l_i) \\ dist(q_t, c_i) - u_i & : & dist(q_t, c_i) > u_i \end{cases} \quad (1)$$

$$U_i = dist(q_t, c_i) + u_i \quad (2)$$

Let R_i denote the maximum possible number of objects nearer to q_t than O_i. The value of R_i is calculated when assuming O_i is at the furthest possible point to q_t, and other objects are at the nearest possible points to q_t, as shown in Equ. (3).

$$R_i = |X|, \quad where \quad X = \{O_j | O_j \in (S - \{O_i, q\}), L_j \leq U_i\} \quad (3)$$

Lemmas (1)-(3) show how to answer three kinds of queries from the *filter pool*.

Lemma 1. *Let* $A = \{O_i | R_i < k\}$. *If* $|A| = k$, *then* A *is the answer for Precise kNN query* $kNN(q, k)$.

The correctness of Lemma 1 comes from the definition of R_i (Equ. (3)).

Lemma 2. *Let* $A = \{O_i | R_i < k + r\}$. *If* $|A| \geq k$, *then any subset of* A *containing* k *objects is an answer for non-value-based approximate kNN query* $nvakNN(q, k, r)$.

The correctness of Lemma 2 also comes from the definition of R_i (Equ. (3)).

Algorithm 2. RFA_client(i)

1: **loop**
2:     **if** *Sees a new location $V_{i,t}$* **then**
3:         **if** *Object O_i itself is the query object q* **then**
4:             Sends $V_{i,t}$ to central site
5:         **else if** $dist(V_{i,t}, c_i) \notin [l_i, u_i]$ **then**
6:             newRange($F_i, V_{i,t}$);
7:             Sends $V_{i,t}$ to central site;
8:     **else if** *Receives (c'_i, l'_i, u'_i, b'_i) from central site* **then**
9:         $(c_i, l_i, u_i, b_i) = (c'_i, l'_i, u'_i, b'_i)$;
10:    **else if** *Receives a request* SEND *from central site* **then**
11:        Sends $V_{i,t}$ to central site;

Lemma 3. *Let $A = \{O_i | R_i < k\} \cup \{O_i | U_i \leq \hat{L} + e, R_i \geq k\}$, where $\hat{L}$ is the kth smallest value of L_i. If $|A| \geq k$, any subset of A containing k objects is an answer for a value-based approximate kNN query $vakNN(q, k, e)$.*

We sketch the proof here. The set A consists of two parts. The first part is $\{O_i | R_i < k\}$. According to the definition of R_i, all objects belong to k nearest neighbors. The second part is $\{O_i | U_i \leq \hat{L} + e, R_i \geq k\}$. Because $\hat{L}$ is the minimum possible distance between the kth nearest neighbor and the query object, and U_i is the maximum possible distance between the object O_i and q_t, any object in the second part also meets the requirement according to the definition of value-based approximate kNN query.

3.3   Algorithm Description

This section introduces RFA approach in detail. RFA approach consists of two parts: (1) Algorithm RFA_client (Algorithm 2), running in moving objects; and (2) Algorithm RFA_server (Algorithm 3), running in the central site.

   Running in each moving object, the goal of Algorithm RFA_client is to handle new location value and to communicate with the central site. When object O_i 'sees' a new location $V_{i,t}$, and the object O_i is just the query object q, O_i sends $V_{i,t}$ to the central site immediately. Otherwise, O_i begins to check whether $V_{i,t}$ stays in the range or not. Only when O_i exceeds the range, subroutine newRange is invoked to update filter F_i and send $V_{i,t}$ to the central site (lines 2-7). Each remote object may receive two kinds of messages from the central site. The first kind is new filter setting (c'_i, l'_i, u'_i, b'_i), and the second kind is a SEND request. When receiving a first kind message, object O_i updates the local filter F_i accordingly. When receiving a second kind message, O_i sends $V_{i,t}$ to the central site immediately(lines 8-11).

   Algorithm RFA_server contains two phases, *initialization* phase and *maintaining* phase. In *initialization* phase (lines 1-7), the central site receives all locations of objects, based on which it creates set A containing k objects nearest to q_0. And then, it calculates filter settings to update all objects. The field c_i of all filters is set to q_0. The field b_i is set to the average value of the maximum distance

---

Algorithm 3. RFA_server()

1: Receives all locations $\{V_{1,0}, V_{2,0}, \cdots V_{n,0}\}$ from moving objects;
2: Creates set A, containing k objects with minimum $dist(V_{i,0}, q_0)$;
3: $V' = \min_{O_i \in S-A-\{q\}}(dist(V_{i,0}, q_0)); \quad V'' = \max_{O_i \in A}(dist(V_{i,0}, q_0))$;
4: $B = \frac{V'+V''}{2}$;
5: **foreach** filter F_i
6: $\quad c_i = q_0; \quad b_i = B; \quad$ newRange($F_i, V_{i,0}$);
7: $\quad$ Sends F_i to object O_i;
8: **loop**
9: $\quad$ **if** Receives $V_{i,t}$ from object O_i **then**
10: $\quad\quad$ newRange($F_i, V_{i,t}$);
11: $\quad\quad$ **if** Can't find answer according to Lemma (1)-(3) **then**
12: $\quad\quad\quad$ adjust();
13: $\quad\quad$ Output result;

Algorithm 4. adjust()

1: Calculates a set A satisfying $A = \{O_i | R_i < k\}$;
2: **while** $(|A| < k)$
3: $\quad$ Finds O_i with min L_i in $S - A - \{q\}$;
4: $\quad$ **if** O_i has not send $V_{i,t}$ to the central site **then**
5: $\quad\quad$ receives $V_{i,t}$ from O_i by sending SEND signal to O_i;
6: $\quad\quad$ $L_i = dist(V_{i,t}, q_t); \quad U_i = dist(V_{i,t}, q_t)$;
7: $\quad$ **else**
8: $\quad\quad$ $A = A + \{O_i\}$;
9: $B = \frac{L'+U'}{2}$, where $L' = \max(U_i | O_i \in A), U' = \min(L_i | O_i \in S - A - \{q\})$;
10: **forall** objects with $L_i = U_i$
11: $\quad c_i = q_t; \quad b_i = B; \quad$ newRange($F_i, V_{i,t}$);
12: $\quad$ Sends (c_i, l_i, u_i, b_i) to object O_i;

---

in A and the minimum distance in $S - A - \{q\}$, so that Lemmas (1)-(3) can be satisfied after invoking Algorithm newRange. In *maintaining* phase (lines 8-13), the central site begins to process query when a new location $V_{i,t}$ arrives. First, it invokes Algorithm newRange to update the filter setting in *filter pool* (lines 9-10). Second, it continues to check whether Lemmas (1)-(3) are satisfied. Once these lemmas cannot be satisfied, it would invoke Algorithm adjust(Algorithm 4) to update filter settings, and output new data.

The goal of Algorithm adjust (Algorithm 4) is to find a set of k objects nearest to the query object q and update filters accordingly. In fact, the initialization phase of Algorithm 3 has implied a simple method to cope with it. However, it requires all objects sending their locations to the central site, which results in heavy network transmission burden. Algorithm 4 illustrates a more efficient way. First, it calculates a set $A = \{O_i | R_i < k\}$. Clearly, if $|A| = k$, set A contains all k nearest neighbors. Otherwise, we should continue to add a neighbor object into A to make $|A| = k$ by iterations. For every an iteration, we check an object with smallest L_i (because this object is a candidate) (lines 1-8). Second, it continues to reset filter settings for parts of objects just sending new locations to the central site. Similar to the initialization phase in Algorithm 3, the local variable B is calculated as the average value of the maximum possible value in A and minimum possible value in $S - A - \{q\}$. Finally, it updates filter settings for objects with $L_i = U_i$ by invoking Algorithm newRange, and sends new settings to corresponding objects (lines 10-12).

Analysis: RFA approach is capable of answering a kNN query at any time point. Initially, Algorithm RFA_server creates filters for all objects. During the maintaining phase, if no new location is transmitted from remote objects, we can always answer the query because all current locations are within the range. Otherwise, if the central site receives a new data from any object, it invokes newRange to create new setting, and checks the validation by Lemmas (1)-(3). Algorithm adjust is then invoked to generate new settings satisfying Lemma 1 on condition that the above exam fails.

4 Experiments

This section begins to evaluate the performance of RFA approach through a series of experiments. Section 4.1 compares the performance between RFA approach and RTP approach[2]. Section 4.2 continues to analyze the network communication overhead in RFA approach. Finally Section 4.3 reports the performance of RFA approach upon different error tolerances.

All experiments are based on a dataset containing the location traces (in 2-dimensional space) of hundreds of vehicles. Each vehicle is initialized with (1) a location (x_0, y_0) random selected from [-1000, 1000], (2) a velocity s, $s \in [5, 15]$ and (3) a moving direction α, $\alpha \in [-\pi, \pi]$. At any time point t, the new location (x_t, y_t) is calculated as:$(x_t, y_t) = (x_{t-1} + s \cdot \cos(\alpha), y_{t-1} + s \cdot \sin(\alpha))$. The velocity and the direction are changed randomly for every a minute. Totally, 100 vehicles will generate 100*(60*60*24)=8,640,000 locations.

4.1 Handling a Static Query Object

The RTP approach is a filter-based solution to answer kNN queries over distributed environments[2]. However, this method only focuses on handling static query object, such as school, landmark, and so on.

Figure 4 compares the performance between the RTP approach and RFA approach. The query point is fixed at $(0, 0)$. Figures 4(a) and (b) report the number of messages transfered via network when running $kNN((0,0), k)$ and $nvakNN((0,0), k, 10)$ respectively. The x-axis represents the number of neighbors k, and the y-axis represents the number of messages raised. In all situations, RFA approach outperforms RTP approach significantly. The main reason is that all filters share same width in RTP approach, so that nearly all objects are forced to send their new locations to the central site once $|A| > k$. But in RFA

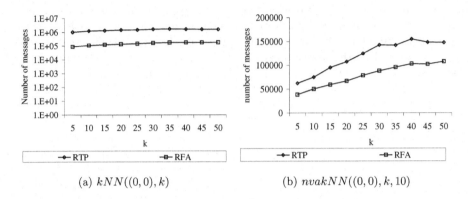

(a) $kNN((0,0),k)$ (b) $nvakNN((0,0),k,10)$

Fig. 4. Comparison between RTP approach and RFA approach

approach, each filter owns a different range, so that when $|A| < k$ or $|A| > k$ occurs, only a small fraction of objects send new locations to the central site.

4.2   Analysis on Communication Cost

RFA approach involves three kinds of network transmission costs, including (1) T_1: moving objects transmit new location to the central site when it moves out of the range; (2) T_2: the central site sends a SEND message to a moving object for new location; (3) T_3: the central site updates filter settings of moving objects. Figure 5 demonstrates the number of messages transmitted via network when running three different queries, such as $kNN(q,k)$, $nvakNN(q,k,10)$ and $vakNN(q,k,100)$. The number of neighbors changes from 5 to 50. When k increases, all kinds of costs increase. In all situations, T_2 and T_3 are smaller than T_1. We can also observe that the amount of messages transfered is still only a small fraction of total messages $(8,640,000)$.

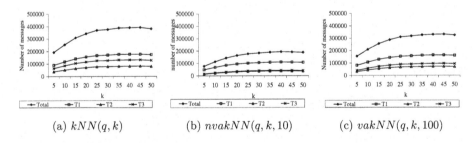

(a) $kNN(q,k)$ (b) $nvakNN(q,k,10)$ (c) $vakNN(q,k,100)$

Fig. 5. All kinds of communication costs

4.3   The Impact of Tolerance

RFA approach can support two kinds of approximate kNN queries, including non-value-based approximate kNN query and value-based approximate kNN query. Here, we implement experiments to evaluate the communication cost under different tolerances. Figure 6(a)and (b) examine queries $nvakNN(q, k, r)$ and $vakNN(q, k, e)$ respectively. The x-axis represents the error tolerance (e.g. r in Figure 6(a) and e in Figure 6(b)); the y-axis represents the number of neighbors; the z-axis represents the total number of messages transfered. In all situations, the communication cost is reduced when given larger tolerance.

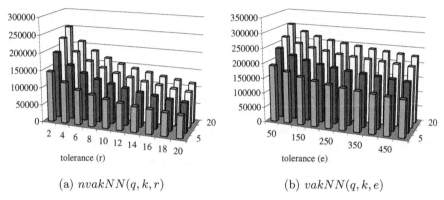

(a) $nvakNN(q, k, r)$ (b) $vakNN(q, k, e)$

Fig. 6. The performance under different error tolerance

5   Related Work

Finding k nearest neighbors to a query object has been widely studied for a long period. Traditional approaches mainly focus on creating and maintaining various indexes over static objects for optimization with a premise that all data have resided in the disks, and can be accessed multiple times[7,13,14]. Nowadays, there has been an increasing interest in continuous monitoring objects in motion[8,9,10,11,15]. Iwerks et al. invented CW approach to monitor PKO objects (point kinematic object)[8]. Koudas et al. proposed a system for approximate kNN queries over streams of multi-dimensional points[9]. Yu et al. gave out grid-base algorithms to index objects and queries to reduce processing cost[15]. Mouratidis et al. pioneered the work on continuous monitoring objects in road networks[11]. The common goal of such work is to provide qualified results with small memory resource and computation resource.

One of the critical goals for applications over distributed environments is to minimize the usage of network communication cost. Example algorithms consist of cardinality of set-expressions monitoring[6], quantile monitoring[4], general-purpose approximate query monitoring[3], distinct count estimate and distinct sample estimate[5], .etc. However, only a few literatures on monitoring k nearest neighbors consider network communication factors[1,2,10]. Babcock et al. give a solution for top-k monitoring, which can be treated as a special case of kNN monitoring. Cheng et al. propose solution for kNN monitoring with non-value tolerance[2]. These work only consider a static query object, while the work in [10] gives a simple method to handle dynamic query object, which is the target of this paper.

6   Conclusions

This paper constitutes an attempt in finding k nearest neighbors to a *dynamic* query object with a goal to reduce the communication overhead, which is critical for distributed monitoring applications. The main contribution is a novel approach, the Range-Filter-based Approach(RFA). Each moving object is installed with a range filter, so that only a small part of data is transmitted via network. At any time point, the central site can find k nearest neighbors to the query object by merely checking the information in its filter pool. Our approach supports three kinds of queries: precise kNN query, non-value-based approximate kNN query, and value-based approximate kNN query. Experimental results show that our approach only consumes small amount of network transmission. One challenging research direction is the kNN monitoring in road network, which is more common in real-life.

References

1. Babcock, B., Olston, C.: Distributed top-k monitoring. In: Proc. of SIGMOD (2003)
2. Cheng, R., Kao, B., Prabhakar, S., Kwan, A., Tu, Y.: Adaptive stream filters for entity-based queries with non-value tolerance. In: Proc. of VLDB (2005)
3. Cormode, G., Garofalakis, M.: Sketching streams through the net: Distributed approximate query tracking. In: Proc. of VLDB (2005)
4. Cormode, G., Garofalakis, M., Muthukrishnan, S., Rastogi, R.: Holistic aggregates in a networked world: Distributed tracking of approximate quantiles. In: Proc. of ACM SIGMOD (2005)
5. Cormode, G., Muthukrishnan, S., Zhang, W.: What's different: Distributed, continuous monitoring of duplicate-resilient aggregates on data streams. In: Proc. of ICDE (2006)
6. Das, A., Ganguly, S., Garofalakis, M., Rastogi, R.: Distributed set-expression cardinality estimation. In: Proc. of VLDB (2004)
7. Hjaltason, G., Samet, H.: Ranking in spatial databases. In: Egenhofer, M.J., Herring, J.R. (eds.) SSD 1995. LNCS, vol. 951, Springer, Heidelberg (1995)
8. Iwerks, G., Samet, H., Smith, K.: Continuous k-nearest neighbor queries for continuously moving points with updates. In: Proc. of VLDB (2003)
9. Koudas, N., Ooi, B., Tan, K., Zhang, R.: Approximate nn queries on streams with guaranteed error/performance bounds. In: Proc. of VLDB (2004)
10. Mouratidis, K., Papadias, D., Bakiras, S., Tao, Y.: A threshold-based algorithm for continuous monitoring of k nearest neighbors. IEEE Transactions on Knowledge and Data Engineering, 17(11) (November 2005)
11. Mouratidis, K., Yiu, M.L., Papadias, D., Mamoulis, N.: Continuous nearest neighbor monitoring in road networks. In: Proc. of VLDB (2006)

12. Olston, C., Jiang, J., Widom, J.: Adaptive filters for continuous queries over distributed data streams. In: Proc. of SIGMOD (2003)
13. Weber, R., Schek, H.-J., Blott, S.: A quantitative analysis and performance study for similarity-search methods in high-dimensional spaces. In: Proc. of VLDB (1998)
14. Yu, C., Ooi, B., Tan, K.-L., Jagadish, H.V.: Indexing the distance: An efficient method to knn processing. In: Proc. of VLDB (2001)
15. Yu, X., Pu, K.Q., Koudas, N.: Monitoring k-nearest neighbor queries over moving objects. In: Proc. of ICDE (2005)

A Novel Text Classification Approach Based on Enhanced Association Rule

Jiangtao Qiu[1], Changjie Tang[1], Tao Zeng[2], Shaojie Qiao[1], Jie Zuo[1], Peng Chen[1], and Jun Zhu[3]

[1] School of Computer Science, Sichuan University, Chengdu, China
[2] Computer and Information Engineering College, Tianjin Normal University, Tianjin, China
[3] National Center for Birth Defects Monitoring, Chengdu, China
qjt163@163.com,tangchangjie@cs.scu.edu.cn

Abstract. The current research on association rule based text classification neglected several key problems. First, weights of elements in profile vectors may have much impact on generating classification rules. Second, traditional association rule lacks semantics. Increasing semantic of association rule may help to improve the classification accuracy. Focusing on the above problems, we propose a new classification approach. This approach include: (1) Mining frequent item-sets on item-weighted transactions; (2) Generating enhanced association rule that has richer semantics than traditional association rule. Experiments show that new approach outperforms CMAR, S-EM and NB algorithms on classification accuracy.

Keywords: Association Rule, Text Classification, Enhanced Association Rule.

1 Introduction

Text classification is an important field on data mining and machine learning. Many studies on text classification have been conducted in the past. Existing techniques include SVM, K-NN, ANN, Naïve Bayes and etc. Association rule based classification is a new classification approach. In [2,9,10,11], association classification has been proved to have higher classification accuracy than other approaches.

Generally, designing an association rules based text classifier includes the following steps: (a) Extract profile vectors of texts. (b) Mine frequent item-sets from profile vectors and generate association rules. (c) Prune rules, and then build text classifier. To build high accurate classifier, however, two important problems must be thought out. First, weights of elements in profile vectors may have much impact on mining classification rules. Second, traditional association rule lacks semantics. Increasing the semantics of association rules may help to improve the classification accuracy.

Focusing on the above problems, we propose a new text classification approach based on enhanced association rule. Main contributions are summarized as follow. (i) Regarding profile vector as item-weighted transaction, an algorithm is proposed to

R. Alhajj et al. (Eds.): ADMA 2007, LNAI 4632, pp. 252–263, 2007.
© Springer-Verlag Berlin Heidelberg 2007

mine frequent item-sets on item-weighted transaction. The algorithm highlights those items having large weight. (ii) A method is proposed to generate enhanced association rule that have richer semantic than traditional association rule. Experiments show that enhanced association rule based text classifier may improve the classification accuracy apparently.

The remaining of this paper is organized as follows. Section 2 gives a brief introduction to related works. Section 3 revisits the description of the problems. Section 4 gives an extensive introduction to our works on building enhanced association based text classifier. Section 5 proposes an algorithm for enhanced association rule based text classification. Section 6 gives a thorough study on classification performance in comparison with other algorithms. Section 7 summarizes our study.

2   Related Works

CBA (Classification Based on Association rule)[2] algorithm first generates candidate rules satisfied supports and confidence threshold, and then prunes these rules by choosing rules with the highest confidence. Experiments show that CBA has better performance than C4.5. CMAR[9] employed a CR-Tree to mine association rules. Then CMAR uses a weighted χ^2 analysis and multiple strong association rules to perform classification. CPAR[10] directly generates classification rules from training set without generating candidate rules. The above algorithms do not take aim at text classification. Thus they do not involve preprocessing of profile vectors, but focus on pruning rules. For the text classification, training set is text collection. Therefore preprocessing of profile vectors must be thought out carefully.

The researchers have proposed some frequent item-sets mining methods on item-weighted transactions. In [6], author defines a domain for each item's weight, and then use two steps to generate weighted association rules. First, ignores the weight and uses traditional algorithm to find frequent item-sets. Second, searches frequent item-sets that satisfy support, confidence and density threshold by space partition. In [7], author gives a fixed weight to each item, and constructs the weighted FP-Tree. In [11], author mine frequent item-sets with weighted item from texts collection. However, the above methods are not suitable for mining frequent item-sets from text profile vectors.

Predicate Association Rule, which is defined and studied in [8], is a kind of enhanced association rule. Compared with traditional association rule, predicate association rule adds more logic connectives in the proposition formula. Thus predicate association rule has richer semantics than traditional association rules. With its advantages, we use predicate association rule to build text classifier so that better classification performance can be reached.

3   Problems Description

3.1   Predicate Association Rule

Let $I=\{I_1, I_2,..., I_n\}$ be a set of variables. X and Y be two subsets of I ($X \subset I$, $Y \subset I$, and $X \neq \varnothing$, $Y \neq \varnothing$, $X \cap Y = \varnothing$). A Traditional Association Rule (TAR hereafter) is an implication expression $P \rightarrow Q$. P is a proposition formula made of variable $I_i \in X$ and conjunction $\wedge$; Q is a proposition formula made of variables in Y and conjunction $\wedge$. For example, $I_2 \wedge I_3 \wedge I_4 \rightarrow I_7 \wedge I_9$ is a TAR.

Predicate Association Rule (PAR hereafter), as a kind of enhanced association rule, was defined in [8]. PAR adds conjunction $\wedge$, disjunction $\vee$ and negation $\neg$ in the proposition formula. Thus PAR has richer semantics and stronger expressing strength than TAR.

Let $I=\{I_1, I_2,...,I_n\}$ be a set of variables, $K=\{\wedge,\vee,\neg\}$ be a set of logic connectives and f $(I_1, I_2,..., I_n)$ be a proposition formula that consists of I and K. Then PAR is an implication expression $P \rightarrow Q$ ($P=f(X)$, $X \subset I$, $Q=f(Y)$, $Y \subset I$, $X \neq \emptyset$, $Y \neq \emptyset$, $X \cap Y=\emptyset$). For example, $(I_2 \wedge I_3) \vee (\neg I_4) \rightarrow I_7 \wedge I_9$ is a PAR.

3.2 Association Rule Based Text Classification

Let $S=\{s_1,..., s_j,..., s_n\}$ be a collection of texts. Each text has a pattern $(A_1,..., A_k)$ that is a collection of attribute-value. In the association rule (called as classification rule while association rule are used for classification) $P \rightarrow c$, P is a proposition formula that consists of attribute-values and logic connective and c is a class label. Association rules based text classifier is a function that maps a pattern $(A_1,...,A_k)$ to a class label.

Given a text object s and its pattern $(A_1,...,A_k)$, if there exists one classification rule r: $P \rightarrow c$ in classifier $TC(c)$ and P is subset of pattern of s, $P \subseteq \{A_1,..., A_k\}$, we call rule r **recognize** sample s, and then s will be mapped into c. If there exists no rule that recognize s, s will be mapped into $\neg c$.

3.3 Mining Frequent Item-Set on Item-Weighted Transactions

In our study, a profile vector will be regarded as an item-weighted transaction.

Definition 1 (item-weighted transaction): Let $I=\{I_1,..., I_i\}$ be a set of items, w_i be weight of I_i. A 2-tuple $<I_i, w_i>$ is called weighted item. An item-weighted transaction, IWT for short, is a set of weighted items $T=\{<I_1, w_1>,...,<I_i, w_i>\}$.

Intuitively, IWT emphasize effect of items by their weight. *frequent item-sets* mined from IWTs should highlight items that have large weights.

Obviously, traditional definitions of *support* and *frequent item-set* are not suitable to generate frequent item-sets on item-weighted transactions. Therefore, we propose new definitions.

Definition 2 (support of item-set): Let $A=\{<I_1, w_1>,...,<I_i, w_i>\}$ be an item-weighted item-set. *support* of A is $MIN(w_{1,...,}w_i)$.

Definition 3 (frequent item-set): Let $L=\{<I_1, w_1>,...,<I_n, w_n>\}$ be the set of all items in a IWT database D. Weight of item I_i, w_i, in L is the sum of weights of the I_i in D. *MaxWeight* is $MAX(w_{1,...,} w_n)$, *th* is support threshold, and then minimum support $Sup=MaxWeight \times th$. An item-weighted item-set A is called *frequent item-set* when *support* of A is not less than Sup.

The **Example 1** in section 4.1 may help to understand the new definitions. Obviously, traditional definition of *support* and *frequent item-set* [8] is a special case of the new definition where weight of all items in D is 1 and there is an item at least occurring in all transactions in D.

4   Designing PAR Based Text Classifier

In this study, PAR based text classifier is designed by the following steps. (1) Retrieval profile vector of texts and build VSM of training text set. (2) Use mean normalization to preprocess profile vectors. (3) Mine frequent item-sets on the set of IWT, and then generate association rules. (4) Generate PARs. (5) Prune negative rules in PARs. In Section 4.1 and 4.2, we will introduce two important parts in our works, mining frequent item-sets on IWTs and generating PARs.

4.1   Mining Frequent Item-Sets on Item-Weighted Transactions

In this study, we use VSM[3] to represent a collection of texts. Each text is regarded as a vector of terms. Terms are extracted from text in n-gram.

Frequency of each n-gram occurring in the text will be counted when extracting n-grams from a text. Then a text may be converted to a profile vector. Element of profile vector is a 2-tuple *<n-gram,weight>* where *weight* are computed using TFIDF[3] method. In order to reduce dimensions of profile vector, terms whose weights are smaller than a threshold will be deleted from vector.

Algorithm 1. Generating Frequent item-sets from IWTS

Input: *VSM*, Threshold *min_sup*

Output: frequent item-sets L

Method:

1 L_1=find_frequent_1-itemsets(*VSM*);

2 For (k=2; $L_{k-1} \neq \emptyset$; k++){

3   C_k=Apriori_gen (L_{k-1}, *min_sup*); // generating candidate k-itemset C_k from C_{k-1}

4   For each vector *vec*$\in$ *VSM*{ // scanning VSM

5   For each candidate $c \in C_k$ {

6   If subset(c, *vec*) // c is subset of *vec*

7   Add(c, *vec*); // weight of each item in c is added by weight of same
                               //item in *vec*

8   Else del (c); //deleting c

9   }

10   }

11   L_k={$c \in C_k \mid c.\ weight \geq min\_sup$}

12   ClearWeight ($c \in L_k$)}// set weight of each item of c to zero

13 return $L = \bigcup_k L_k$;

According to samples in the training set belonging or not belonging to class c when we build classifier $TC(c)$, training set will be divided into the positive training set and the negative training set. Then each profile vector in VSM is regarded as an IWT and association rules are generated by mining frequent item-sets on positive training set. However, neglecting weight of elements in IWT when mining frequent item-set will play down effect of the most representative item on classification. For example, there are two profile vectors $V_1=\{<t_1,15>,<t_2,10>,<t_3,2>\}$, $V_2=\{<t_2,3>,<t_3,1>\}$. Frequent item-sets whose support are not less than 2 are $\{t_2\},\{t_3\}$ and $\{t_2,t_3\}$ when neglecting weights of term. Although weight of t_1 is 15 that is greater than weight of other terms, $\{t_1\}$ is not frequent item-set. This problem often is overlooked, but it must be thought about carefully when designed association rules based text classifier.

In this study, frequent item-sets are generated from IWTs, and then classification rules are derived. In order to mine frequent item-sets from IWTs, we propose Algorithm 1, the improved apriori algorithm.

Example 1. There are three profile vectors $\{a:15, b:4, c:1\}$, $\{a:1, c:1\}$, $\{b:3, c:2\}$. Fig.1 shows the process using Algorithm 1 to mine frequent item-sets on profile vectors. Let support threshold be th=15%. From item collection L1, the greatest weight in collection is 16. By definition 3, minimum support is $16\times15\%=2.4$. Then frequent item-sets $\{a\}:16, \{b\}:7, \{c\}:4, \{a, b\}:4, \{b, c\}:3$ may be derived.

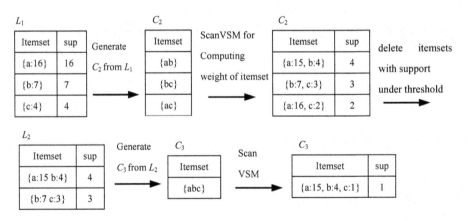

Fig. 1. Mining frequent Itemsets on feature vectors

4.2 Generate PARs

In Section 4.1, we have introduced how to generate classification rules from VSM. These rules, however, are traditional association rules and they lack semantics. Thus we need to further generate PARs based on these classification rules.

Let $TC(c)$ be a association rule based classifier, S be training set and $S_c \subseteq S$ be training subset in which samples have class label c. $\neg S_c$, called *self* set in this paper, is supplementary set of S_c. $\neg S_c$ represent samples in the training set that do not have class label c. If a rule r in TC(c) recognizes sample $s(s \in \neg S_c)$, we call r recognizes *self*.

According to *precision* formula in classification

$$\text{Precision} = \frac{\text{number of samples correctly labeled as positive } n}{\text{number of samples labeled as positive } m} \qquad (1)$$

If classifier recognizes more *self*s, m will increase while n keep unchanged. Thus precision of classifier will decrease.

If we simply delete those rules recognizing *self*, it will not be helpful for improving classification precision. For example, a rule r wrong recognize i samples and correctly recognize j samples, $j>i$. If we delete rule r from classifier, *precision* is $(n-j)/(m-i)$ that is less than n/m. It can be a good idea for improving classification *precision* that reduces recognizing negative samples while keep recognizing positive samples unchanged.

Suppose a rule $r: P \rightarrow c$ recognizes a sample A in S_c and a sample B in $\neg S_c$. It can be concluded that P is subset of both pattern of A and pattern of B, $P \subseteq A$ and $P \subseteq B$. Because B is in $\neg S_c$, A new rule $Q \rightarrow \neg c$, $Q \subseteq B\text{-}A$, can be derived. From expression $P \rightarrow c \wedge Q \rightarrow \neg c$, we can further infer the expression $P \wedge \neg Q \rightarrow c$ in which P, $P \rightarrow c$, Q and $Q \rightarrow \neg c$ are called positive logical formula, positive rule, negative logical formula and negative rule, respectively. For sample A in S_c, $P \subseteq A$, classification rule $P \wedge \neg Q \rightarrow c$ may label c to A correctly. For sample B in $\neg S_c$, $P \subseteq B$ and $Q \subseteq B\text{-}A$, although P is subset of pattern of B, $P \subseteq B$, Q is also subset of pattern of B, $Q \subseteq B$. Thus rule $P \wedge \neg Q \rightarrow c$ will label $\neg c$ to B.

In [12], it was proved that, for texts belonging to same class, there always exists a set of n-grams occurring in these texts in high frequency. Two texts belonging to same class will have roughly same n-gram frequency distribution. Therefore, if text D and B belong to same class, they will have roughly same n-gram frequency distribution. If rule $r: P \rightarrow c$ wrong recognize B ($B \in \neg S_c$), $Q \subseteq B\text{-}P$, and then PAR $r: P \wedge \neg Q \rightarrow c$ may be generated, there will be great probability on Q being subset of pattern of text D, $Q \subseteq D$, because D and B belong to same class. Thus rule r will label $\neg c$ to text D in high probability. By the above analysis, we can conclude that PAR may reduce wrong recognized samples in $\neg S_c$.

Let E be a text belonging to class c. According to the above n-gram frequency distribution, negative logical formula of rule r, Q, is generated from text B. B and E do not share same label. Therefore, $Q \subseteq E$ is in a small probability, i.e. rule r assign label $\neg c$ to E in a small probability. By this way, PAR may assure that number of wrong labeled positive sample reduce dramatically while number of correctly labeled sample reduce slightly. Thereby classification precision is improved.

Algorithm 2. generating predicate association rule

Input: set of classification rules *Rule*, set of selfs *self*, support threshold *Supp*
Output: Predicate Association Rules *H*
Methods:
1 For each rule *r* in *Rule*{
2   *W*=null;
3   For each sample *s* in *Self*
4   If (*Recog(r,s)*)
5   *Insert(s, W)*; //inserting sample recognized by *r* into *W*
6   *NegRule=CreateNegRules*(W, *Supp*); //generating negative rule of *r* from *W*
7   DelRepeateItems(*NegRule*);
8   Pick (*NegRule*) ;   //picking up less-general negative rules
9   *NewRule*=GenerateNewRule(*r*, *NegRule*);
10   Insert(*H,NewRule*);   //inserting PARs to *H*
11 }

In Algorithm 2, function GreateNegRules will mine negative rules satisfying support threshold by using method in Section 4.1. In function DelRepeateItems, if there is the intersection of items between left part of a negative rule, P_1, and a positive rule, P_2, P= $P_1 \cap P_2$, then item $I \in$ P will be deleted from P_1.

Example 2. Let *r*: $A_1 \wedge A_2 \rightarrow c$ be a classification rule, the collection of profile vectors in negative training set recognized by r denoted as W; $\{A_1, A_1A_2A_5A_6, A_1A_3, A_3A_4\}$ are part of frequent item-sets mined from W. After deleting items occurring in left part of rule *r* from the two frequent item-sets, $\{A_5A_6, A_3, A_3A_4\}$ can be derived. Picking up less-general rule $\{A_5A_6, A_3A_4\}$, we derive proposition formula $A_5 \wedge A_6$ and $A_3 \wedge A_4$. New proposition formula $\neg (A_5 \wedge A_6 \vee A_3 \wedge A_4)$ can be derived when connecting proposition formulas using disjunction $\vee$ and appending a negation $\neg$. Finally, a new PAR $A_1 \wedge A_2 \wedge \neg (A_5 \wedge A_6 \vee A_3 \wedge A_4) \rightarrow c$ is generated after adding the new proposition formula to rule *r*. The PAR means that class label *c* may be assigned to a text when profile vector of the text include elements of $\{A_1A_2\}$ and do not include $\{A_5A_6\}$ and $\{A_3A_4\}$. In other words, if a profile vector of text includes $\{A_5A_6\}$ or $\{A_3A_4\}$, class label $\neg c$ will be assigned to the text.

5 PARs Based Text Classification

From algorithm 2, we can learn that a PAR in classifier TC(*c*) includes two parts: positive logical formula and negative logical formula. Negative logical formula can be divided into several parts by disjunction. Positive logical formula and No.j part of negative logical formula of PAR r_i is denoted as PosRule(r_i) and NegRule($r_i^{\ j}$), respectively. Both PosRule (r_i) and NegRule($r_i^{\ j}$) may be reduced to the integer. Reduction of PosRule (r_i) includes the following steps. (1) Generate a set of items from all positive logical formulas in classifier, denoted as PV. (2) A unique integer number *id*, starting from 1, is assigned to each item in PV, and then items are ranked by their *id* in ascend order. (3) Let an integer be zero, for each item $A_i \in$ PosRule(r_i)

and its *id*, No.*id* bit of the integer is set to 1. Then PosRule(r_i) of a PAR may be represented by an integer, denoted as INT(PosRule(r_i)). (4) Reduced PosRule(r_i) of all PARs in TC(c) are stored in an array Pos(c). The array is the collection of positive logical formula of PARs. An element of array is a 2-tuple <INT(PosRule(r_i^j)), RID> where RID is number of a rule in TC(c).

After generating a set of items from all negative logical formulas in classifier, denoted as NV, NegRule(r_i^j) of PAR in TC(c) also may be reduced into an integer number, denoted as INT(NegRule(r_i^j)), by the above method. All reduced NegRule(r_i^j) of PARs in TC(c) are stored in an array Neg(c). The array is the collection of negative logical formula of PARs in TC(c). Structure of Neg(c) is same with Pos(c).

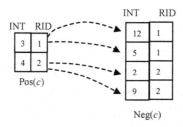

Fig. 2. Storing Reduced PARs

Example 3. There are two PARs in TC(c) r_1: $A_1 \wedge A_2 \wedge \neg(A_5 \wedge A_6 \vee A_3 \wedge A_5) \rightarrow c$ and r_2: $A_3 \wedge \neg(A_4 \vee A_3 \wedge A_6) \rightarrow c$. The set of items in positive logical formula is PV={ A_1, A_2, A_3 }; the set of items in negative logical formula is NV={ A_3, A_4, A_5, A_6 }. Items A_1, A_2, A_3 are assigned integer number 1, 2, 3 in PV, respectively. Items A_3, A_4, A_5, A_6 are assigned integer number 1,2,3,4 in NV, respectively. After PosRule(r_1)= $A_1 \wedge A_2$ is reduced, its binary integer is 011 and INT(PosRule(r_1))=3. NegRule(r_1^1)= $A_5 \wedge A_6$ also may be reduced into a binary integer 1100 and INT(NegRule(r_1^1))=12. By the same way, INT(NegRule(r_1^2))=5, INT(PosRule(r_2))=4, INT(NegRule(r_2^1))=2 and INT(NegRule(r_2^2))=9 also can be derived. Pos(c) and Neg(c) are shown in Fig.2.

An unlabeled text may get a class label by using Algorithm 3.

Algorithm 3(PARC). Predicate Association Rule Based Text Classifying

Input: unlabeled text s, PAR based Classifier TC(c), pruning threshold t
Output: label (c or $\neg c$)
Methods:
1 *vec*=GetVector(s); // extracting profile vector of s
2 *vec*=Prune(t, *vec*);//deleting elements in *vec* whose frequency under t
3 *pv*=GetPV;// getting the set of items in positive logical formula of TC(c)
4 *posInt*=Reduce(*pv*, *vec*);// reducing PV $\cap$ *vec* to a integer
5 *nv*=GetNV;// getting the set of items in negative logical formula of TC(c)
6 *negInt*=Reduce(*nv*, *vec*);// reducing NV $\cap$ *vec* to a integer
7 If (Match(*posInt*, Pos(c))) // get *RID* of positive rule matching with *posInt*
8 get *posInt.RID*;
9 else return $\neg c$;

---

10 If (Match(*negInt*, *Neg*(*c*)) and *posInt.RID*=*negInt.RID*)
11   return ¬ *c*;
12 else return *c*

In this algorithm, function GetVector extract profile vector of the unlabeled text in n-gram. However, number of n-grams may be large. By [12], n-grams with the high frequency in a text may represent feature of the text. Thus it is rational to delete n-grams whose frequencies do not satisfy the threshold because they cannot represent feature of the text. Deleting n-grams with low frequency may help to reduce their disturbance for classification. Function Match(*a*, *b*) return matching result of *a* and *b*. Two binary integers *a*, *b* are operated in logical OR. If the result is equal to *a*, we call *b* matching *a*.

6   Experiments and Analysis

The experiments aim at (1) studying how Algorithm 1 improve classification performance, (2) comparing PARC with other classification algorithms, such as CMAR[9], S-EM[1] and Naïve Bayes. Our experiments use 3-fold cross-validation. *precision*(Formula 1), *recall*(Formula 2) and *accuracy*(Formula 3) measures are used in our experiments.

$$\text{Recall}=\frac{\text{number of samples labeled correctly as positive}}{\text{number of all positive samples}} \qquad (2)$$

$$\text{Accuracy}=\frac{\text{number of samples labeled correctly}}{\text{number of all samples}} \qquad (3)$$

6.1   Dataset and Environment

Our experiments use Chinese Web Classification Training Set CCT2006[1] as dataset. This dataset include 1200 web pages distributing in eight classes. There are 150 web pages in each class. All web pages in each class are divided into three parts. Each part includes 50 pages. In 3-fold cross-validation, each time of experiment will use two parts of pages in one class as training set and another part as test set. Because there are eight classes, our experiment builds classifiers for each class respectively. For example, building a classifier TC(*c*) for class *c*, we use training set of class *c* as positive training set for classifier TC(*c*) and use test set of class *c* as positive test set, use all prior two parts pages of others classes as negative training set and other parts as negative test set.

Experiments were performed on an Intel C3 1.0G PC with 512M main memory and running Windows 2000 sever. PARC and CMAR are implemented in JAVA, and S-EM and Naïve Bayes(NB) are downloaded from website[2]

---

[1] http://www.cwirf.org

[2] http://www.cs.uic.edu/~liub/S-EM/S-EM-download.html

6.2 Experiment

6.2.1 Test 1

Test 1 compare classification performance on two situations where profile vectors are regarded as item-weighted transaction and common transaction. We build classifiers $TC(i)\{i=1,...,8\}$ by generating TARs on training set of eight classes respectively, and then classifiers are tested on their own positive and negative test set. *posNum* and *negNum*, in each row in Table 1, are average number of samples recognized as positive sample on positive and negative test set in 3-fold cross-validation respectively. In each fold validation, number of positive test samples is 50 and number of negative test samples is 350. *minimum support* is set at 10 in test.

Table 1. The Comparison of mining frequent item-sets on IWT and common transaction

Class	With weight		Without weight	
	posNum	*negNum*	*posNum*	*negNum*
1	32.67	54	37	17
2	21.33	53	26	17
3	31.67	52	34	20.33
4	21.67	54	26.33	32.33
5	44.67	50	44	7.33
6	30.33	54	44	44
7	19.67	51	29.33	45.67
8	42.33	11	43	12.33
Avg	30.54	43.38	35.46	24.5

From Table 1, we can observe that classifier that is built by mining item-sets with weight may increase recognizing positive samples (35.46-30.54) /50≈10% and reduce recognizing negative samples (43.38-24.5) /350≈5.4% than classifier neglecting item weight. It can be concluded that better classification accuracy can be reached when mining frequent item-sets with weight than without weight.

6.2.2 Test 2

Test 2 compare algorithm PARC with CMAR, S-EM and NB(Naïve Bayes) on classification performance.

Table 2 lists average *precision (Prec)*, *recall (Reca)* and *accuracy (Accu)* in 3-fold cross-invalidation on eight classes.

From table 2, we can observe that PARC has same *precision* with CARM, but 15% higher *recall* and 7% higher *accuracy* than CARM.

Table 2. The comparison of PARC, CMAR, S-EM and NB on classification performance

class	PARC			CMAR			S-EM			NB		
	Prec	Reca	Accu	Prec	Reca	Accu	Prec	Reca	Accu	Prec	Reca	Accu
1	0.80	0.65	0.93	0.87	0.55	0.93	0.6	0.87	0.79	0.65	0.87	0.92
2	0.87	0.5	0.93	0.93	0.41	0.92	0.26	0.81	0.7	0.35	0.7	0.81
3	0.75	0.44	0.91	0.82	0.29	0.90	0.26	0.65	0.77	0.33	0.72	0.82
4	0.68	0.51	0.91	0.67	0.32	0.89	0.49	0.73	0.84	0.59	0.73	0.89
5	0.95	0.87	0.98	0.93	0.84	0.97	0.86	0.91	0.97	0.86	0.91	0.97
6	0.82	0.78	0.95	0.57	0.7	0.89	0.62	0.89	0.92	0.62	0.89	0.92
7	0.75	0.63	0.93	0.75	0.25	0.90	0.5	0.76	0.85	0.54	0.75	0.88
8	0.86	0.83	0.96	0.94	0.61	0.95	0.63	0.91	0.92	0.72	0.92	0.94
Avg	0.81	0.65	0.94	0.81	0.5	0.92	0.53	0.82	0.85	0.58	0.81	0.89

Although S-EM and NB have 17% and 16% higher *recall* than PARC, their *precision* is 28% and 23% lower and their *accuracy* are 9% and 5% lower than PARC.

From test 2, we can conclude that PARC have better performance on text classification than CAR, CMAR, S-EM and NB in an overall evaluation.

7 Conclusions

In this paper, a new text classification approach is proposed. We highlight the effect of elements with high frequency on classification by mining frequent item-sets on item-weighted transactions. The new approach generates predicate association rules that have richer semantics than traditional association rule. Experiments show that the new approach outperforms CMAR, S-EM and NB on classification accuracy.

References

1. Liu, B., Lee, W.S., Yu, P.S., Li, X.: Partially Supervised Classification of Text Document[C]. In: Proceeding of ICML- 2002 (2002)
2. Liu, B., Hsu, W., Ma, Y.: Integrating classification and association rule mining[C]. Proceeding of KDD'98, New York (August 1998)
3. Salton, G., Wong, A., Yang, C.S.: A Vector Space Model for Automatic Indexing [J]. Communication of the ACM (1) (1995)
4. Sebastiani, F.: Machine Learning in Automated Text Categorization [J]. ACM computing Surveys, 34(1) 11–12, 32–33 (2002)
5. Han, J., Pei, J., Yin, Y.: Mining frequent patterns without candidate generation [C]. In: Proceeding of SIGMOD'00, Dallas, TX (2000)

6. Wang, W., Yang, J., Yu, P.: Efficient mining of weighted Association Rules (WAR) [J]. IBM Research Report RC 21692(97734) (March 2000)
7. Zhang, Z., et al.: Enabling Personalization Recommendation With WeightedFP for Text information Retrieval Based on User-Focus[C]. In: Proceeding of ITCC'04 (2004)
8. Jie, Z., Changjie, T., Tianqing, Z.: Mining Predicate Association Rule by Gene Expression Programming. In: Meng, X., Su, J., Wang, Y. (eds.) WAIM 2002. LNCS, vol. 2419, Springer, Heidelberg (2002)
9. Li, W., Han, J., Pei, J.: CMAR: accurate and efficient classification based on multiple class-association rules[C]. In: Proceeding of ICDM (2001)
10. Yin, X., Han, J.: CPAR: Classification based on Predictive Association Rules[C]. In: Proceeding of International Conference on Data Mining (SDM'03) (2003)
11. Xiaoyun, C., Wei, C.: Text Categorization Based on Classification Rules Tree by Frequent Patterns [J]. Journal of Software, 17(5) (May 2006)
12. Cavnar, W.B., Trenkle, J.M.: N-Gram-Based Text Categorization. In: Proceeding of Third Annual Symposium on Document Analysis (1994)
13. Zhou, S.G., et al.: A Chinese document categorization system without dictionary support and segmentation processing [J]. Journal of Computer Research and Development, 38(7) (2001)

Applications of the Moving Average of nth-Order Difference Algorithm for Time Series Prediction

Yang Lan[1] and Daniel Neagu[2]

[1] Department of Computing, University of Bradford, Bradford, BD7 1DP, UK
Y.LAN@bradford.ac.uk
[2] Department of Computing, University of Bradford, Bradford, BD7 1DP, UK
D.NEAGU@bradford.ac.uk

Abstract. Currently, as a typical problem in data mining, Times Series Analysis and Prediction are facing continuously more applications on a wide variety of domains. Huge data collections are generated or updated from science, military, financial and environmental applications. Prediction of the future trends based on previous and existing values is of a high importance and various machine learning algorithms have been proposed. In this paper we discuss results of a new approach based on the moving average of the nth-order difference of limited range margin series terms. Based on our original approach, a new algorithm has been developed: performances on measurement records of sunspots for more than 200 years are reported and discussed. Finally, Artificial Neural Networks (ANN) are added for improving the precision of prediction by addressing the error of prediction in the initial approach.

Keywords: Time Series Analysis, Pseudo-periodical Time Series Prediction, n^{th}-order Difference.

1 Introduction

From the early days of human civilization, time series analysis has become an important and indispensable part of human activity. The analysis of historical data is applied to agriculture (sun and moon cycle), industrial (quantity and quality of products), economic (resources consumption), finance (stock markets), meteorology, natural disasters (volcanic activity, earthquakes, river floods) and solar activity forecasting [Calvo et al 1995]. Today, with the increasing computing power and memory storage largely available, the complex time series data analysis and processing methods are accessible for all domains of human activity.

Time series prediction defines a problem requiring data mining analysis methods and proposes algorithms for which past data (mainly finite observation sequences of data points related to uniform time intervals) are used to generate models to forecast future data points of the series. The time series analysis was based originally on tools including mathematical modeling [Hathaway et al 1993], time-frequency analysis (Fourier, wavelet transformations) but started using in the last years machine learning methods: artificial neural networks

R. Alhajj et al. (Eds.): ADMA 2007, LNAI 4632, pp. 264–275, 2007.

(time-delay networks [Saad et al 1996], recurrent networks [Lee Giles et al 2001], Self-Organizing Maps [Simon et al 2005]). From a procedural perspective, using machine learning or computational approaches requires a first mathematical analysis to describe and breakdown the initial time series problem in simpler sub-problems for further computational models. A well-known approach in time series understanding and prediction is Auto-Regressive Moving Average (ARMA) [Box and Jenkins 1976] which comes with the advantage of addressing both auto-regressive terms as well as moving average terms.

A historical main constraint in using mathematical series models for prediction was the fact that the performance of the model is related to the length of data series, but currently, it is not an issue any more from the precessing of computing power and memory storage. However, most machine learning methods face the difficulty of requiring a priori knowledge about the problem at hand. On the other hand, some traditional methods in time series analysis were only able to relate the mathematics model results, but not available to satisfy the demand of specific applications. We intend to address these drawbacks for the restricted problem of pseudo-periodical series with limited boundaries by a two-step approach: we propose hereby a new algorithm to approximate the time series terms using the moving average of nth-order difference of already known values, then intend to address the specific order level of difference and its value of index to reduce the error, and train them to Artificial Neural Network (ANN) for the problem of error of approximation by a hybrid model. Therefore future work is proposed to identify as accurately as possible a general approximation by use of a supervised-learning model to forecast a further approximation error if found necessary.

We describe herewith an algorithm for efficient prediction of pseudoperiodical time series by identifying the number of necessary past values necessary to predict forthcoming value(s) with a good accuracy. We use for this aim some interesting properties related to the nth-order difference for bounded time series for which we're applying a "Short Selling" approach. A further step based on supervised-learning neural network for the prediction of the error of the average sum of nth-order difference to increase forecast performances of the time series is also considered. Therefore we prove that using an a priori identified number of previous values of time series within a hybrid model can provide good prediction performances for time series, even for noisy (natural) series. The results obtained using sunspot data series show the performance of our approach in mining bounded pseudo-periodical patterns in time series.

The paper is structured as follows: in the following section we introduce definitions on bounded pseudo-periodical time series used in our approach. Section 3 describes our proposed method and discusses the algorithm based on identification of the minimum historical values required for a good prediction. Section 4 describes the application data from Sunspot Number (also known as Wolf Number) time series. Results from this case study and further discussions on our algorithm come in section 5. Conclusions and further work about implementation issues of the proposed method for a hybrid approach are presented in the last section.

2   Time Series Prediction

We will introduce in this section some definitions necessary for the proof of our method and the algorithm proposed in following sections of the paper for time series prediction.

Notations 1

a_m :   *initial data series, $m > 0$ is its index (serial number);*

D_m^n :   *n^{th}-order difference, "n" is the nth-order, "m" is the series number*

E_m^n :   *moving average of nth-order difference, where "n" is the order of*
the difference terms, "m" is the series number.

Definition 1. *Time Series represents an ordered sequence of data values related to the same variable, measured typically at successive times, spaced apart at uniform time intervals:*

$$A = \{a_1, a_2, a_3, \ldots\} \text{ or } A = \{a_t\}, \ t = 1, 2, 3, \ldots \tag{1}$$

where series A is indexed by natural numbers.

Definition 2. *Time Series Prediction represents the use of a model to predict future events (data points) based on known past events, before they are measured. In other words, given first t measured data points of time series A, the aim of prediction is to develop a model able to advance the value*

$$a_{t+1} = f(a_1, a_2, \ldots, a_t) \tag{2}$$

Definition 3. *Pseudo-Periodical Time Series are series of which values repeat over a pseudo-fixed time interval:*

$$a_t \cong a_{t+d} \cong a_{t+p}, \ \ t > 0, \ d > 0, \ p > 0, \ p/d = k > 1, \ k \in N \tag{3}$$

For real applications time series, there are values showing a pattern of pseudo-periodical time series, where values show a repetition over a finite and almost constant time interval; a consequence for periodical series is that for a finite value d and initial values, the series values are bounded:

$$a_t \in [\min(a_1, a_2, \ldots, a_d), \max(a_1, a_2, \ldots, a_d)], \ t \geq 1 \tag{4}$$

The aim in time series data analysis and prediction is to find a model to provide with a good accuracy a future value of the series. Some research directions indicate as important the immediate future value (see eq.(2)) whereas other problems may indicate a further interval of interest, that it is, the prediction of the following m values of time series:

$$a_m = f(a_1, a_2, \ldots, a_n), \ m \in [n+1, n+k], \ k > 0 \tag{5}$$

3   The Approach of Moving Average of nth-Order Difference for Bounded Time Series Prediction

Our approach is based on the notion of the difference operator of stationary time series, and take a nth-order difference array, then analysis the moving average to predict data next period.

Definition 4. The Difference Operator for a given functional f with real values a calculates:

$$\Delta f(a) = f(a+1) - f(a) \tag{6}$$

$\Delta f(a)$ is called first-order difference (or simply difference) of $f(a)$. Is easy to induce that the nth-order difference is defined by:

$$\Delta^n f(a) = \sum_{i=0}^{n} (-1)^{n-i} C_n^i f(a+i) \tag{7}$$

$$\text{where } C_n^i = \binom{i}{n} = \frac{n!}{i!(n-i)!} \quad 0 \le i \le n$$

is the so-called the binomial coefficient.

Considering the last equation on n, we can find out that the same rule applies for $n+1$: Therefore, based on the induction principle (Peano) equation (number) is valid for any natural value of n. If $f(m)$, with m a natural number, generates a discrete series a_m, then the previous result can be written:

$$D_m^n = D_{m+1}^{n-1} - D_m^{n-1} \quad \text{where } D_m^n \text{ means } \Delta^n f(m) \tag{8}$$

The nth-order difference is used in the binomial transform of a function, the Newton forward difference equation and the Newton series. These are useful prediction relationships with the main drawback of difficult numerical evaluation because of rapid grow of the binomial coefficients for large n.

In order to avoid a complex calculus and also to provide a relationship for time series prediction, our idea starts from the fact that applying the difference operator we generate another series from the initial one, which has the property of pseudo-periodicity.

Let D_m^n represents the nth-order difference of a_m, then:

$$D_m^n = \sum_{i=0}^{n} (-1)^{n-i} C_n^i a_{m+i} = D_{m+1}^{n-1} - D_m^{n-1}, \quad m \ge 1 \tag{9}$$

The nth-order difference time series shows a pseudo-periodical bounded shape with amplitude modulated in time. The moving average of nth-order difference D_m^n time series for initial data $\{a_m\}$ based on the nth-order difference of a_m is:

$$E_m^n = \frac{1}{m}(D_1^n + D_2^n + D_3^n + \ldots + D_m^n) = \frac{1}{m}\sum_{i=1}^{m} D_i^n \tag{10}$$

At the same time, the average sum E_m^n can be expressed in terms of nth-order difference as:

$$E_m^n = \frac{1}{m}(D_{m+1}^{n-1} - D_1^{n-1}) \tag{11}$$

where, based on eq.(10) and eq.(11):

$$
\begin{aligned}
E_m^n &= \frac{1}{m} \sum_{i=1}^{m} D_i^n \\
&= \frac{1}{m}(D_m^n + D_{m-1}^n + \cdots + D_1^n) \\
&= \frac{1}{m}((D_{m+1}^{n-1} - D_m^{n-1}) + (D_m^{n-1} - D_{m-1}^{n-1}) + \cdots \\
&\quad \cdots + (D_3^{n-1} - D_2^{n-1}) + (D_2^{n-1} - D_1^{n-1})) \\
&= \frac{1}{m}(D_{m+1}^{n-1} - D_1^{n-1}) \tag{12}
\end{aligned}
$$

Since the initial series are bounded, the series generated by the difference operator is also bounded and their average converges to zero. We calculate below the limit of time series E_m^n: (for an easy calculation we can consider the following)

$$\lim_{m \to \infty} E_m^n = \lim_{m \to \infty} \frac{1}{m} \sum_{i=1}^{m} D_i^n \tag{13}$$

And therefore, based on eq.(13):

$$
\begin{aligned}
\lim_{m \to \infty} E_m^n &= \lim_{m \to \infty} \frac{1}{m}(D_1^n + D_2^n + \cdots + D_m^n) \\
&= \lim_{m \to \infty} \frac{1}{m}(D_{m+1}^{n-1} - D_1^{n-1}) \\
&= \lim_{m \to \infty} \frac{D_{m+1}^{n-1}}{m} - \lim_{m \to \infty} \frac{D_1^{n-1}}{m} \tag{14}
\end{aligned}
$$

For a large m, since D_1^{n-1} is a limited value, the second term in eq.(14) becomes negligible. And if D_{m+1}^{n-1} is also a limited value given the initial constraints on bounded time series, the first term in eq.(14) has a null limit also:

$$\lim_{m \to \infty} E_m^n = \left(\lim_{m \to \infty} \frac{D_{m+1}^{n-1}}{m} - \lim_{m \to \infty} \frac{D_1^{n-1}}{m} \right) \to 0 \tag{15}$$

Indeed, one can easily see this result is verified for our practical example in Fig.3. Based on the result in eq.(15) as depicted in Fig.3, given an initial time series $\{a_m\}$, with n^{th}-order difference $\{D_m^n\}$, our aim is to determine the value for $\{a_{m+1}\}$ (prediction) based on previous data measurements (and some negligible error). The moving average for n^{th}-order difference (term $n - 1$) values is easy to calculate:

$$E_{m-1}^n = \frac{1}{m-1}(D_1^n + D_2^n + \ldots + D_{m-1}^n) = \frac{1}{m-1} \sum_{i=1}^{m-1} D_i^n \tag{16}$$

Since $E_m^n \to 0$ for large values of m, then:

$$E_m^n = E_{m-1}^n + \varepsilon \tag{17}$$

where $\varepsilon > 0$ is a negligible error for large m. Replacing in eq.(17) E_m^n from eq.(11) and E_{m-1}^n from eq.(16):

$$\frac{1}{m} \sum_{i=1}^{m} D_i^n = \frac{1}{m-1} \sum_{i=1}^{m-1} D_i^n + \varepsilon \tag{18}$$

And therefore, from eq.(18)

$$\frac{1}{m} \left(\sum_{i=1}^{m-1} D_i^n + D_m^n \right) = \frac{1}{m-1} \left(\sum_{i=1}^{m-1} D_i^n \right) + \varepsilon$$

$$D_m^n = m \left(\frac{1}{m-1} \sum_{i=1}^{m-1} D_i^n + \varepsilon \right) - \sum_{i=1}^{m-1} D_i^n$$

$$D_m^n = \frac{1}{m-1} \sum_{i=1}^{m-1} D_i^n + m\varepsilon \tag{19}$$

As a result, when $n = 1$ (then $D_m^1 = a_{m+1} - a_m$), we obtain:

$$a_{m+1} - a_m = \frac{1}{m-1} \sum_{i=1}^{m-1} D_i^1 + m\varepsilon \tag{20}$$

The prediction precision for a_m depends on the nth-order difference D_m^n; in other words, precision of prediction increases in accuracy with the value of m and the order of the difference n. For simplicity, consider the first-order difference of the original bounded pseudo-periodical time series from eq.(20):

$$a_{m+1} = a_m + \frac{1}{m-1} \sum_{i=1}^{m-1} (a_{i+1} - a_i) + m\varepsilon$$

$$= a_m + \frac{1}{m-1} (a_m - a_1) + m\varepsilon$$

$$= \frac{1}{m-1} (ma_m - a_1) + m\varepsilon \tag{21}$$

Eq.(21), obtained by considering the average series of first-order difference, suggests a practical way to approximate the prediction a_{m+1} based on previous values a_m and a_1. For large numbers, although the error value is negligible (see eq.(15) and (17) and Fig.3) the accuracy of prediction may still be affected.

Now, we have two formulas here:

$$D_m^n = \sum_{i=0}^{n} (-1)^{n-i} C_n^i a_{m+i} = D_{m+1}^{n-1} - D_m^{n-1} \tag{22}$$

$$\sum_{j=1}^{m} D_j^n = D_{m+1}^{n-1} - D_1^{n-1} \tag{23}$$

Based on the above results, we have:

$$E_m^n = \begin{cases} E_{m-1}^n + \varepsilon & \text{if } \varepsilon \neq 0 \\ E_{m-1}^n & \text{if } \varepsilon \cong 0 \end{cases} \tag{24}$$

Then, take eq.(22) and eq.(23) into eq.(24):

$$E_m^n = E_m^n + \varepsilon$$

$$\Longleftrightarrow \frac{1}{m} \sum_{k=1}^{m} D_k^n = \frac{1}{m-1} \sum_{k=1}^{m-1} D_k^n + \varepsilon$$

$$\Longleftrightarrow \frac{1}{m} (\sum_{k=1}^{m-1} D_k^n + D_m^n) = \frac{1}{m-1} \sum_{k=1}^{m-1} D_k^n + \varepsilon$$

$$\Longleftrightarrow (m-1) \sum_{k=1}^{m-1} D_k^n + (m-1)D_m^n = m \sum_{k=1}^{m-1} D_k^n + m(m-1)\varepsilon$$

$$\Longleftrightarrow (m-1)D_m^n = \sum_{k=1}^{m-1} D_k^n + m(m-1)\varepsilon$$

$$\Longleftrightarrow (m-1)(D_{m+1}^{n-1} - D_m^{n-1}) = (D_m^{n-1} - D_1^{n-1}) + m(m-1)\varepsilon$$

$$\Longleftrightarrow (m-1)D_{m+1}^{n-1} = mD_m^{n-1} - D_1^{n-1} + m(m-1)\varepsilon$$

$$\Longleftrightarrow \frac{m}{m-1}(D_m^{n-1} + (m-1)\varepsilon) - \frac{1}{m-1}D_1^{n-1} \tag{25}$$

And let $n = n - 1$ in eq.(25) so that the general expression becomes:

$$D_{m+1}^n = \frac{m}{m-1}(D_m^n + (m-1)\varepsilon) + \frac{-1}{m-1}D_1^n \tag{26}$$

Thus, the coefficients can be seen as special "weights" related to two terms of the same order difference level, and they depend on the "start" and the "end" values. The case when $\varepsilon \neq 0$, will be processed using an Artificial Neural Network.

We aim now to find a suitable value for m as the prediction of forthcoming value to be well approximated. Since the second "weight" value is negative, and its condition number is so high, the eq.(26) is not a "normal" weighted function but a so known "Ill-conditioned Function" in "Short Selling" framework. As a result, with $m \to \infty$, its variance increases and the variation of function solution(s) is bigger, therefore the predicted precision is not good enough [Golub et al 1996].

Thus, k has been given by: $k \in [1, m]$, where m is the length of the initial data series; and let:

$$F(k) = \frac{k}{k-1}(D_k^n + (k-1)\varepsilon) + \frac{-1}{k-1}D_1^n \tag{27}$$

then calculate the D_{m+1}^n:

$$D_{k+1}^n = \frac{k}{k-1}\left(D_k^n + (k-1)\varepsilon\right) + \frac{-1}{k-1}D_1^n \tag{28}$$

next, we suppose Θ represents their Manhattan Distance:

$$\Theta(k) = |F(k) - D_{k+1}^n| \tag{29}$$

where $\Theta(k_{min}) = \min$, $F(k_{min})$ is the closest value to the real difference D_{m+1}^n. Our aim is to determine the value m for which a_{m+1} is approximated based on the previous data measurements $a_1, a_2, \ldots, a_m$. Thus starting from of "k_{min}", let $n = k_{min} + 1$ so that we determine the value of a_{m+1}.

The algorithm below implements the method described above for a general time series $\{a_m\}$ (Table 1).

Table 1. A Moving Average of nth-order Difference Algorithm (MAonDA)

Input: An Initial Time Series Data;
Method: Moving Average of nth-order Difference Algorithm;
Output: A Predicted Time Series Data;

---
01. A[ ] ← INPUT(An Initial Time Series Data);
02. $L \leftarrow length(A[\ ])$;   // Let L records the size of data sequence;
03. **while** $i < L - n$ {   // Calculate the nth-order difference D[ ] of A[ ];
04. $D[i] \leftarrow \sum_{i=0}^{n}(-1)^{n-i}C_n^i A[\ ])$;
05. }
06. **for** $m \leftarrow 1$ to $L - 2$ {   // Calculate the average sum E[ ] of D[ ];
07. $sum \leftarrow 0$;
08.   **for** $n \leftarrow 1$ to m {
09. $sum \leftarrow sum + D[\ ]$;
10. $E[\ ] \leftarrow sum/m$;
12. } }
13. $\varepsilon \leftarrow$ ANN(E[ ]); // Get the ε value by ANN (Fig.8);
14. (for) $k \leftarrow 1$ to K {   // K is a large number by given;
15. $F[k] \leftarrow \frac{k}{k-1}\left(D[k] + (k-1)\varepsilon\right) + \frac{-1}{k-1}D[1]$;
16. $\Theta[k] \leftarrow |F[k] - D[k]|$;
17. }
18. $m \leftarrow find(\Theta == \min(\Theta))$; $n \leftarrow m + 1$;
19. $D[m+1] \leftarrow \frac{m}{m-1}\left(D[m] + (m-1)\varepsilon\right) + \frac{-1}{m-1}D[1]$
20. OUTPUT($D[m+1]$)

---

4   Description of Sunspots Data Set

In this section, notions of sunspots and sunspots time series are introduced. On Sun's surface some regions could be characterized by a decrease in temperature simultaneously with an increase of magnetic activity, which inhibits convection, forming areas of low surface temperature called sunspots. Parameters of sunspots in terms of temperature and magnetic activity can vary for relatively close neighbourhoods, determining sunspots to organize in groups. There has been highlighted a correlation between the number of sunspots and the intensity of solar radiation. Given their lower temperature, sunspots appear darker in satellite images: then an increase in sunspots number may determine a decrease of solar radiation. This is a reason sunspot measurement and prediction is significant research objectives of solar activities. Solar radiation affects in various ways human activities (including telecommunications' quality for example).

Time Series Sunspot Data Set is an ordered data set of sunspot numbers based on observations, which can be treated as a tracking record of solar activities. Of course from the point of view of Definition 3, sunspot data is a pseudo-periodical time series, since there is not a fixed value d, but a series of values with an average of about 23 years.

The original time series of sunspot number used in this paper as a generic data set to describe the application of our algorithm has been generated by the National Geophysical Data Center (NGDC). NGDC provides stewardship, products and services for geophysical data describing the solid earth, marine, and solar-terrestrial environment, as well as earth observations from space [NGDC 2006]. Its data bases currently contain more than 300 digital and analog databases, which include Land, Marine, Satellite, Snow, Ice, Solar-Terrestrial subjects. NGDC's sunspots databases contain multiform original data from astronomical observatories; the sunspot data sets list various sunspots' attributes in time order, even others solar activities related to sunspots.

5   Prediction Results for the Sunspots Data Case Study

Our case study uses monthly average of sunspot number time series from January 1881 A.D to December 1980 A.D; therefore we use a time series data with 1200 data points to generate the times series $A = \{a_m\}, m \geq 1$ of their monthly average values. Table 2 lists the configuration of data. (Monthly average value reported to Julian date in format YYYY.MM) and Fig.5 depicts these values.

Table 2. Monthly Average of Sunspot Number Data Set Organization

Time	1881.01	1881.02	1881.03	$\cdots$	1980.10	1980.11	1980.12
Average	0.5	1.5	2.5	$\cdots$	12.4	11.0	12.1

Fig.1 shows an initial data of sunspot number; Fig.2 presents the nth-order difference (when n = 1); Fig.3 shows the moving average of nth-order difference;

As discussed in section 3, the results for the prediction of monthly average of sunspot number are based on values of the seventh-order difference for the original time series:

$$D_n = a_{n+7} - 7a_{n+6} + 21a_{n+5} - 35a_{n+4} + 35a_{n+3} - 21a_{n+2} + 7a_{n+1} - a_n \quad (30)$$

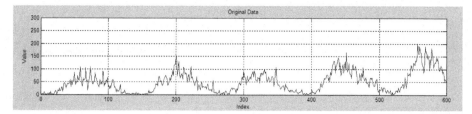

Fig. 1. The Original Value of Monthly Average of Sunspot Numbers

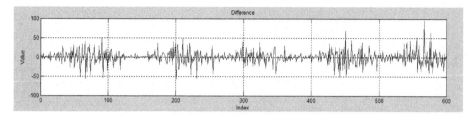

Fig. 2. The Seventh-order Difference of Monthly Average of Sunspot Number

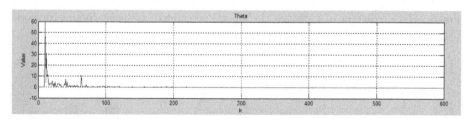

Fig. 3. The Moving Average of Monthly Average of Sunspot Number

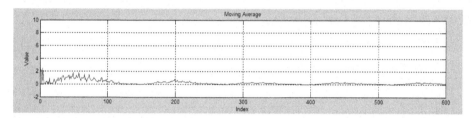

Fig. 4. An example of Manhattan Distance $\Theta(k) = |F(k) - D_{k+1}^n|$ (where $n = 7$)

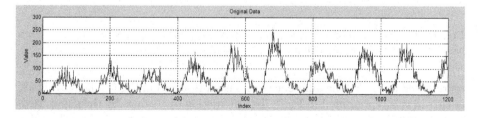

Fig. 5. The Original Value of Monthly Average of Sunspot Numbers

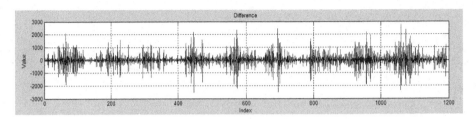

Fig. 6. The Seventh-order Difference of Monthly Average of Sunspot Number

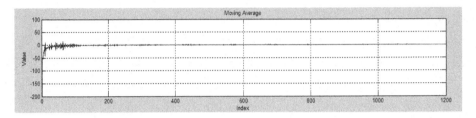

Fig. 7. The Moving Average of Monthly Average of Sunspot Number

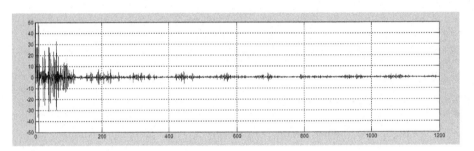

Fig. 8. The Prediction of ε Values by Supervised-trained Artificial Neural Network (ANN): whole set, including train and test value

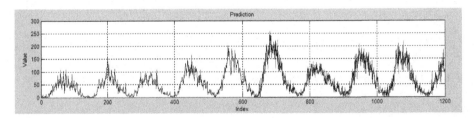

Fig. 9. The Prediction of Values for Monthly Average of Sunspot Number Time Series based on the proposed algorithm

represented in Fig.5. The difference series is represented in Fig.6. The moving average series is represented in Fig.7. The prediction of ε values by the supervised-trained Artificial Neural Network (ANN) is represented in Fig.8. And the value of prediction series is represented in Fig.9.

6   Conclusions and Future Work

The moving average of nth-order difference algorithm proposes a simple approach to determine the range of values necessary for a good prediction of the following time series terms in cases of bounded pseudo-periodical time series. We consider worth for further investigation the speed and also the complexity of our solution in comparison with traditional algorithms, such as Auto-Regress Moving Average (ARMA) and also performance issues in comparison with machine learning solutions.

Therefore, the moving average of nth-order difference algorithm generates an accurate prediction for the "shape" of pseudo-periodical series, but the precision of prediction (amplitude) suffers because of dependence on how many orders (i.e. value of n) difference have been considered, which increases the complexity calculus though and introduces a tuning parameter of the order of difference. The main disadvantage is the dependency of the prediction error to the index "m" which is consequently dependent on the difference's order "n". A small order of difference reduces the complexity but also the prediction precision, whereas a big order of difference increases the computing effort.

Further research proposes to reduce the error in eq.(25), eq.(27) and eq.(29) using machine learning techniques (see differences comparing Fig.1 and Fig.6). Some results to approximate the error using a neural network model are presented in Fig.4: the connectionist model is trained with the error values for the first period values.

References

1. Hathaway, D.H., Wilson, R.M., Reichmann, E.J.: The Shape of the Sunspot Cycle. Solar Physics 151, 177–190 (1994)
2. Calvo, R.A., Ceccatto, H.A., Piacentini, R.D.: Neural Network Prediction of Solar Activity. The Astrophysical Journal 444(2), 916–921 (1995)
3. Box, G., Jenkins, F.M.: Time Series Analysis: Forecasting and Control, 2nd edn. Holden-Day, Oakland, CA (1976)
4. Van Golub, L.: Matrix Computations, 3rd edn. Johns Hopkins University Press (1996)
5. Simon, G., Lendasse, A., Cottrell, M., Fort, J.C., Verleysen, M.: Time series forecasting: Obtaining long term trends with self-organizing maps. Pattern Recognition Letters 26, 1795–1808 (2005)
6. Saad, E.W., Prokhorov, D.V., Wunsch II, D.C.: Comparative study of stock trend prediction using time delay, recurrent and probabilistic neural networks, Neural Networks. IEEE Transactions on 9(6), 1456–1470 (1998)
7. Lee Giles, C., Steve, L., Tsoi, A.C.: Noisy Time Series Prediction using a Recurrent Neural Network and Grammatical Inference. Machine Learning 44(1/2), 161–183 (2001)
8. National Geophysical Data Center (NGDC) (2006), http://www.ngdc.noaa.gov/
9. Wikipedia (2006), http://en.wikipedia.org/wiki

Inference of Gene Regulatory Network by Bayesian Network Using Metropolis-Hastings Algorithm

Khwunta Kirimasthong[1], Aompilai Manorat[1], Jeerayut Chaijaruwanich[1], Sukon Prasitwattanaseree[2], and Chinae Thammarongtham[3]

[1] Department of Computer Science, Faculty of Science,
Biomedical Engineering Center, Chiang Mai University,
Chiang Mai, 50200, Thailand
jeerayut@science.cmu.ac.th
[2] Department of Statistics, Faculty of Science,
Chiang Mai University, Chiang Mai, 50200, Thailand
[3] National Center for Genetic Engineering and Biotechnology
113 Thailand Science Park, Phahonyothin Road, Klong 1,
Klong Luang, Pathumthani, 12120, Thailand

Abstract. Bayesian networks are widely used to infer genes regulatory network from their transcriptional expression data. Bayesian network of the best score is usually chosen as genes regulatory model. However, without the hint from biological ground truth, and given a small number of transcriptional expression observations, the resulting Bayesian networks might not correspond to the real one. To deal with these two constrains, this paper proposes a stochastic approach to fit an existing hypothetical gene regulatory network, derived from biological evidence, with few available amount of transcriptional expression levels of the genes. The hypothetical gene regulatory network is set as an initial model of Bayesian network and fitted with transcriptional expression data by using Metropolis-Hastings algorithm. In this work, the transcriptional regulation of gene CYC1 by co-regulators HAP2 HAP3 HAP4 of yeast (*Saccharomyces Cerevisiae*) is considered as example. Due to the simulation results, ten probable gene regulatory networks which are similar to the given hypothetical model are obtained. This shows that Metropolis-Hastings algorithm can be used as a simulation model for gene regulatory network.

Keywords: Bayesian network, Gene regulatory network, Metropolis-Hastings algorithm, Transcriptional expression analysis.

1 Introduction

Each gene of the living cells has its specific functions in the cellular processes. Gene is regulated to transcribe RNA, which is further translated to form certain proteins for some particular functions [1]. The transcriptional expression level, which could be measured at genomic scale by DNA microarray technology, indicates how much the gene reacts, or responds to the environment. The transcription of a gene could be occurred only under some suitable conditions. There must be a set of proteins such as transcriptional factors, activators, and co-activators to initiate, elongate, and terminate

R. Alhajj et al. (Eds.): ADMA 2007, LNAI 4632, pp. 276–286, 2007.

the transcription process. Gene regulatory network is therefore one of the most basic study issues to understand the complex behaviors of the living cells.

In general, without biological knowledge background, the cluster analysis of DNA microarray data is widely applied to group the genes which have similar expression patterns. These genes are inferred as co-regulated by some common transcriptional factors, activators, and co-activators [2]. Nevertheless, it is hard to biologically justify the results from cluster analysis. In the other ways, many statistical inference methods have been proposed for inferring gene regulatory networks such as graphical Gaussian model [3], probabilistic Boolean networks [4], and Bayesian networks[5, 6]. Again, these methods could not be verified whether their predicted model truly corresponds to the biological phenomena. These machine learning approaches seem not reliable when no biological ground truth is given. In some cases, the regulatory networks of certain genes might be proposed by the evidences from biological experiments. This paper aims to extend the justification of the proposed hypothetical gene regulatory network by fitting with the available transcriptional expression data using Bayesian network and Metropolis – Hastings algorithm which is a part of Markov Chain Monte Carlo techniques.

The application of Bayesian network has been recently proposed by [5] for inferring gene regulatory network from the transcriptional expression data. In [5], the hypothetical regulatory network must be given or all possible networks will be enumerated to find the optimal one with the highest score. In practical way, the microarray experiments can be preformed only in a small number comparing with the number of genes being studied. This small number of transcriptional expression observations consequently provides a low reliability of the resulting Bayesian networks [6].

In this work, Markov Chain Monte Carlo (MCMC) technique, or precisely Metropolis – Hastings algorithm, is combined with Bayesian network to simulate the set of most probable gene regulatory networks. The given hypothetical gene regulatory network is set as an initial model of Bayesian network. The regulatory relationship between two genes in the new network is randomly generated from the binomial distribution. The success probability of binomial distribution itself follows conjugated Dirichlet distribution which is parameterized by the number of such regulation occurred in all previous accepted networks. Our network generator is different from [6]. The acceptance probability of new network is defined by Metropolis – Hastings algorithm, basing on the proposal and posterior distributions of the current and the new network models. The proposal distribution used in this paper is similar to the one used in [6]. The posterior distribution is inferred from the conditional probability of transcriptional expression data in discrete form. When the simulation converges, last hundred of accepted networks are selected to represent the probable gene regulatory networks. In this paper, the transcriptional regulation of gene CYC1 by co-regulators HAP2 HAP3 HAP4 of yeast (*Saccharomyces Cerevisiae*) is considered as example.

2 Material

The transcriptional regulation of gene CYC1 (Cytochrome c, isoform 1) by co-regulators HAP2, HAP3, HAP4 of yeast *Saccharomyces Cerevisiae* is considered in this paper. CYC1 involves in electron carrier of the mitochondrial inter-membrane space that transfers electrons from ubiquinone-cytochrome c oxidoreductase to cytochrome c oxidase during cellular respiration in *Saccharomyces Cerevisiae* [7]. HAP2/3/4 complex is reported in [8] as the co-regulators of CYC1. HAP2 and HAP3 subunits are primarily responsible for site-specific DNA binding by the complex, whereas HAP4 subunit provides the primary transcriptional activation domain.

The transcriptional expression log-ratios of these four genes have been generated by [2] for 7 time points, see Table 1. The log-ratios of the transcriptional expressions are then discretisized into three levels; up-, stationary- and down-regulated. The log-ratios of expressions which are larger than the upper bound threshold is considered as up-regulated, denoted by '↑' symbol; smaller than the lower bound threshold is considered as down-regulated, denoted by '↓' symbol; and between the upper to lower bound thresholds are considered as stationary, denoted by '-' symbol. Transformed transcriptional expression levels are shown in Table 2. As there is no biological information available on how sensitive Hap2/3/4 complex regulates CYC1, the upper and lower bound thresholds are arbitrarily set to 0.2 and - 0.2 respectively.

Table 1. Transcriptional expression log-ratios of genes CYC1, HAP2, HAP3 and HAP4

CYC1	HAP2	HAP3	HAP4
-0.14	-0.15	-0.29	0.24
-0.22	0.03	0.24	0.58
-0.01	-0.12	0.12	1.04
-0.09	-0.32	-0.14	0.66
0.31	-0.23	-0.36	0.62
1.18	-0.04	-0.15	2.54
1.75	0.51	-0.09	3.13

Table 2. Transcriptional expression levels of genes CYC1, HAP2, HAP3 and HAP4

CYC1	HAP2	HAP3	HAP4
–	–	↓	↑
↓	–	↑	↑
–	–	–	↑
–	↓	–	↑
↑	↓	↓	↑
↑	–	–	↑
↑	↑	–	↑

3 Methods and Experiment

This paper uses Metropolis – Hastings algorithm and Bayesian network model to infer gene regulatory networks from an existing hypothetical gene regulatory network and the transcriptional expression data of genes in question. From the principle of Bayesian network modeling, the network structure must be a directed acyclic graph (DAG). Nodes of the network represent the genes and edges represent the regulatory relationship between two genes. In this work, the network matrix M of size $n \times n$, where n is the number of genes being studied is used to represent the regulatory relationship between genes. $M[i,j]$ is equal to one if gene i regulates gene j, and zero otherwise. For example, the hypothetical regulation of gene CYC1 by co-regulators HAP/2/3/4 complex can be shown in Fig. 1(a) and can be represented by the network matrix in Fig. 1(b). Rows of the network matrix represent the parent nodes or regulators and the columns represent the children nodes or regulated genes.

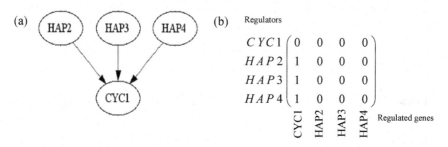

Fig. 1. Bayesian regulatory network of genes CYC1, HAP2, HAP3, and HAP4, (a) the graphical model (b) the network matrix of model

3.1 Metropolis – Hastings Algorithm

Metropolis – Hastings algorithm (MH) [9], which is a part of MCMC method, is used for sampling the gene regulatory networks. First, the regulatory network is initialized by the hypothetical one. Then, the MH algorithm repeatedly and randomly simulate new network M_{new}.

The binomial distribution [10] as shown in the equation (1) is used in this work to randomly generate the edges of the new network M_{new}.

$$M_{new}[i,j] \sim binomial(edge\_prob[i,j]) \tag{1}$$

$M_{new}[i,j]$ indicates the relationship between the regulator i and the regulated gene j in the new network. The parameter $edge\_prob[i,j]$ of the equation (1) is the binomial success probability for generating the edge (i, j) in M_{new}. As the success probability itself follows Dirichlet distribution [11], the values of $edge\_prob$ are then randomly generated from Dirichlet distribution as shown in the equation (2):

$$edge\_prob[i,j] \sim dirichlet(a) \tag{2}$$

The super-parameter a of Dirichlet distribution in the equation (2) is the shape parameter of Dirichlet which consists of ($alpha_1$, $alpha_2$) where $alpha_1$ is the parameter for generating edge and $alpha_2$ is the parameter for generating no edge. The values of these super-parameters are proposed by this paper to be the frequencies of having edge and having no edge in all previous accepted networks.

After generating M_{new}, the acceptance probability of M_{new} is computed by the equation (3).

$$\alpha(M_{old}, M_{new}) = min\left\{1, \frac{P(M_{new}|D)}{P(M_{old}|D)} * \frac{Q(M_{old}|M_{new})}{Q(M_{new}|M_{old})}\right\} \tag{3}$$

where D is the transcriptional expression levels of the genes. The stationary distribution $P(.|D)$ is computed by Bayesian network that will be described in section 3.2. The proposal distribution $Q(.|.)$ is similar to the one used in [6] and computed from the neighbor networks of M_{old} and M_{new} at each iteration. The neighbor networks which must be DAG, are the networks which have one edge modified. The edge modifying modes are deleting, reversing and adding. The examples of finding neighbor networks are shown in Fig. 2.

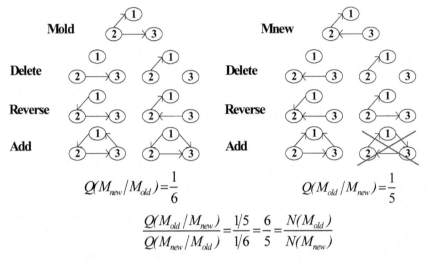

$$Q(M_{new}/M_{old}) = \frac{1}{6}$$

$$Q(M_{old}/M_{new}) = \frac{1}{5}$$

$$\frac{Q(M_{old}/M_{new})}{Q(M_{new}/M_{old})} = \frac{1/5}{1/6} = \frac{6}{5} = \frac{N(M_{old})}{N(M_{new})}$$

Fig. 2. Examples of neighbor networks of M_{old} and M_{new} are modified by deleting, reversing and adding an edge [6]. The neighbor networks must be DAG therefore the last neighbor network of M_{new} must be ignored.

For the example in Fig. 2, $Q(M_{old}|M_{new})$ is equal to 1/5 and $Q(M_{new}|M_{old})$ is equal to 1/6. Therefore, the proposal distribution $Q(.|.)$ in the equation (3) is computed by the Hastings ratio of $N(M_{old})$ - the number of neighbor networks of M_{old} - and $N(M_{new})$ - the number of neighbor networks of M_{new} - that shown in the equation (4):

$$\frac{Q(M_{old}|M_{new})}{Q(M_{new}|M_{old})} = \frac{N(M_{old})}{N(M_{new})} \tag{4}$$

Finally, the new simulated network M_{new} is accepted if a uniform random variable U is smaller than the acceptance probability from the equation (3):

```
U ~ uniform (0, 1)
If (U <= α(M_old, M_new)) {
    M_old[t+1] = M_new[t]
    t = t + 1 }
```

3.2 Bayesian Network

From Baye's theorem [12], the stationary distribution $P(.|D)$ of the acceptance probability in the equation (3) is the posterior distribution of the gene regulatory networks, given the observable gene expression levels D. This stationary distribution could be derived by the equation (5) using Baye's theorem:

$$\frac{P(M_{new}|D)}{P(M_{old}|D)} = \frac{P(D|M_{new})P(M_{new})}{P(D|M_{old})P(M_{old})} \tag{5}$$

where $P(M)$ is the prior probability of network M, $P(D|M)$ is the likelihood computed by using Bayesian belief network approach; which is the conditional probability of the genes in the network [13] . Fig. 3 gives an example of Bayesian belief network of four genes.

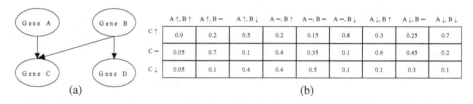

	A↑,B↑	A↑,B−	A↑,B↓	A−,B↑	A−,B−	A−,B↓	A↓,B↑	A↓,B−	A↓,B↓
C↑	0.9	0.2	0.5	0.2	0.15	0.8	0.3	0.25	0.7
C−	0.05	0.7	0.1	0.4	0.35	0.1	0.6	0.45	0.2
C↓	0.05	0.1	0.4	0.4	0.5	0.1	0.1	0.3	0.1

(a)   (b)

Fig. 3. Belief network, (a) graphical model shows the relation of genes in question, and (b) conditional probability table or CPT table of gene C where the expression of gene C are conditional on the expression of parent nodes: gene A and B

3.3 Learning of Bayesian Network by Metropolis-Hastings Algorithm for Gene Regulatory Network Inference

The learning of Bayesian network by Metropolis-Hastings algorithm for gene regulatory network inference can be summarized as follow:

Algorithm. MH-based Bayesian Network for Gene Regulatory Network Inference
Input: D: Transcriptional expression data
M_{Hypo}: Hypothetical regulatory network
Output: *Network*: A set of all accepted networks by Metropolis – Hastings

1. Transform D into discrete transcriptional expression levels
2. Initialize M_{old} by the given hypothetical regulatory network M_{Hypo}
3. Repeat step 4 to step 9 for long enough time
4. Sampling a new acyclic network M_{new} by
    If (it is the first sampling) Then
        For each pair of genes (i, j)
$$M_{new}[i,j] \sim binomial(0.5)$$
    Else
        For each pair of genes (i, j)
$$edge\_prob[i,j] \sim dirichlet(a)$$

$$M_{new}[i,j] \sim binomial(edge\_prob[i,j])$$

5. Find $N(M_{old})$ the number of acyclic neighbor networks of M_{old}
6. Find $N(M_{new})$ the number of acyclic neighbor networks of M_{new}
7. Compute acceptance probability: $\alpha(M_{old}, M_{new})$

$$\text{where } \alpha(M_{old}, M_{new}) = min\left\{1, \frac{P(M_{old}|D)}{P(M_{new}|D)} * \frac{N(M_{old})}{N(M_{new})}\right\}$$

8. Random a number U from $uniform(0,1)$
9. If($U <= \alpha(M_{old}, M_{new})$)) Then

> Accept M_{new} and keep it into *Network* set
> Set M_{new} to be M_{old} for the next sampling
> Update Dirichlet parameter $a = (alpha_1, alpha_2)$:
> $alpha_1[i,j] = alpha_1[i,j]+1$ if $M_{new}[i,j]=1$
> $alpha_2[i,j] = alpha_2[i,j]+1$ if $M_{new}[i,j]=0$

4   Results

Our algorithm is implemented by R language. After running the algorithm with given hypothetical transcriptional regulation and transcriptional expression data of genes HAP2, HAP3, HAP4 and CYC1 by 2,000 iterations repeatedly, 1,853 generated networks are accepted. The probability of all 16 edges, i.e. *edge_prob* that used to simulate all accepted networks, are sequentially converged and shown in Fig. 4. At the beginning, the *edge_prob* are set to be 0.5, then following Dirichlet distribution, they converge to their optimum regarding to Bayesian fit.

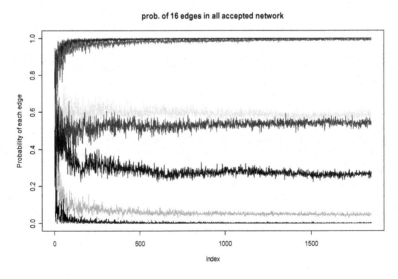

Fig. 4. Convergence of *edge_prob* probabilities of all edges

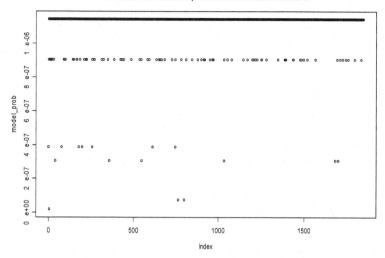

Fig. 5. Probabilities of all 1,853 accepted networks

When considering only the accepted networks, the structures of these networks are found to be similar to the hypothetical regulatory network. There is at least one edge from HAP2, HAP3, or HAP4 to CYC1. Furthermore, as shown in Fig. 5, the posterior probability of 1,853 accepted networks are mostly equal to 1.143835e-06 which is the highest posterior probability of the accepted networks found by the experiment. Although some accepted networks have a low posterior probability, it is found that the number of networks which have low posterior probabilities decreases in the last samplings. In other words, the accepted networks would likely converge to the ones with the highest posterior probability.

The last 100 accepted networks, long enough after burn-in, are selected to be the representative of the most probable regulatory networks. It is found that there are ten common patterns of network structures, and shown in Fig. 6. Each pattern in Fig. 6 shows the regulatory relationship, the frequency of network occurrences (frq.), the posterior probability of network (prob.), and the occurrence probability of each edge. The occurrence probability of each edge approaches to the value of *edge_prob* as in Fig. 4.

In Fig. 6, the regulation of CYC1 by HAP2 and HAP3 appear in all ten patterns. This corresponds to the biology evidence that the subunits HAP2 and HAP3 of the complex, primarily responsible for site-specific DNA binding of CYC1. On the other hand, HAP4 has no certain regulation with the others. This might support the fact that HAP4 itself provides the primary transcriptional activation domain which has role only at the beginning of the transcription and its expression levels are up-regulated all the time.

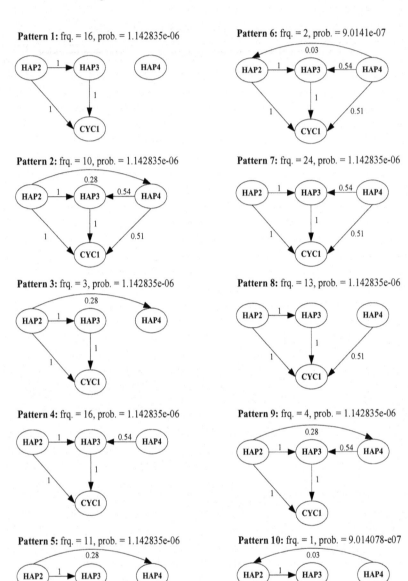

Fig. 6. Ten regulatory patterns of the last 100 accepted networks

5   Conclusion

This paper has proposed the usage of Metropolis – Hastings algorithm to simulate transcriptional gene regulatory Bayesian networks from an existing hypothetical network by fitting with the transcriptional expression data of genes in question. Our experimental simulation on the regulation of gene CYC1 by co-regulators HAP2, HAP3, and HAP4 of yeast (*Saccharomyces Cerevisiae*) shows that the biological evidence on regulatory network fits very well with the available transcriptional expression data. This shows that the gene regulatory networks could be reliably inferred by Bayesian network in combining with Metropolis - Hastings algorithm.

Acknowledgments. This study is supported by Thailand Graduate Institute of Science and Technology (TGIST), National Center for Genetic Engineering and Biotechnology (BIOTEC), National Science and Technology Development Agency (NSTDA), Thailand, and BioMedical Engineering Center (BMEC), Chiang Mai University, Thailand.

References

1. Alberts, B., Johnson, A., Lewis, J., Raff, M., Roberts, K., Walter, P.: Molecular Biology of the Cell, 4th edn. Garland Science, New York (2002)
2. Derisi, J.L., Iyer, V.R., Brown, P.O.: Exploring the Metabolic and Genetic Control of Gene Expression on a Genomic Scale. Science, Sciencemag, vol. 278, pp. 680–686 (1997)
3. Wu, X., Ye, Y., Subramanian, K.R.: Interactive Analysis of Gene Interactions Using Graphical Gaussian Model. In: BIOKDD03: 3rd ACM SIGKDD Workshop on Data Mining in Bioinformatics, pp. 63–69 (2003)
4. Shmulevith, I., Dougherty, E.R., Zhang, W.: From Boolean to Probabilistic Boolean Networks as Models of Genetic Regulatory Networks. Proceedings of the IEEE 90(11), 1778–1792 (2002)
5. Hartemink, A.J., Gifford, D.K., Jaakkola, T.S., Young, R.A.: Using Graphical Models and Genomic Expression Data to Statistically Validate Models of Genetic Regulatory Networks. Pac. Symp. Biocomputing, Hawaii, pp. 422–433 (January 2001)
6. Husmeier, D.: Sensitivity and Specificity of Inferring Genetic Regulatory Interactions from Microarray Experiments with Dynamic Bayesian Networks. In: Bioinformatics, vol. 19(17), pp. 2271–2282. Oxford, England (2003)
7. Hortner, H., Ammerer, G., Hartter, E., Hamilton, B., Rytka, J., Bilinski, T., Ruis, H.: Regulation of Synthesis of Catalases and iso-1-cytochrome c in Saccharomyces Cerevisiae by Glucose, Oxygen and Heme. Eur. J. Biochem. 128(1), 179–184 (1982)
8. Olesen, J.T., Guarente, L.: The HAP2 Subunit of Yeast CCAAT Transcriptional Activator Contains Adjacent Domains for Subunit Association and DNA Recognition: Model for the HAP 2/3/4 Complex. Genes Dev. 4, 1714–1729 (1990)
9. Gilk, W.R., Richardson, S., Spiegelhalter, D.: Markov Chain Monte Carlo in Practice. Chapman & Hall, London (1996)
10. Milton, J.S., Tsokos, J.O.: Statistical Methods in the Biological and Health Sciences, pp. 123–131. McGraw-Hill, Japan (1983)

11. Balding, D.J., Bishop, M., Cannings, C.: Handbook of Statistical Genetics, p. 730. Wiley, England (2001)
12. Gamerman, D.: Markov Chain Monte Carlo Stochastic Simulation for Bayesian Inference. Chapman & Hall, London (1997)
13. Han, J., Kamber, M.: Data Mining: Concepts and Techniques, pp. 348–354. Academic Press, San Francisco (2001)

A Consensus Recommender for Web Users

Murat Göksedef and Şule Gündüz Öğüdücü

Department of Computer Engineering
Istanbul Technical University
Maslak, Istanbul TR34469
Turkey
{goksedef, sgunduz}@itu.edu.tr

Abstract. In this paper, we propose a new hybrid recommendation model for web users which is based on multiple recommender systems working in parallel. With the rapid growth of the World Wide Web (www), it becomes a critical issue to find useful information from the Internet. Web recommender systems help people make decisions in this complex information space where the volume of information available to them is huge. Recently, a number of approaches have been developed to extract the user behavior from her navigational path and predict her next request as she visits Web pages. Some of these approaches are based on non-sequential models such as association rules and clustering, and some are based on sequential patterns. In this paper, we present a hybrid recommender model which combines the results of multiple recommender systems in an effective way. We have conducted a detailed evaluation on four different web usage data. Our results show that combining recommendation algorithms effectively leads a better recommendation accuracy. The experimental evaluation shows that our method can achieve a better prediction accuracy compared to standard recommendation systems while still guaranteeing competitive time requirements.

1 Introduction and Related Work

Most of the web users complain about finding useful information on web sites. Web recommender systems predict the information needs of users and provide them with recommendations to facilitate their navigation. Web recommender systems on the Internet accept a user model and a set of items to be recommended as input and generate a subset of these items as output. Given a user's (who may, for example, be a customer in an e-commerce site) current actions, the goal is to determine which web pages (items) will be accessed (bought) in the near future. Web usage mining is one of the main approaches for building a user model. Web usage mining refers to the application of data mining techniques to discover usage patterns from secondary data, such as web server access logs, proxy server logs, browser logs, user profiles, registration data, user sessions[1], cookies, user queries, and bookmark data, in order to understand and better serve the needs of Web-based applications. Recently, a number of approaches have been developed dealing with specific aspects of web usage mining like automatically

---

[1] The term *user session* is defined as the click stream of page views for a single visit of a user to a web site [17]. In this paper we will use this term interchangeably with "server session".

R. Alhajj et al. (Eds.): ADMA 2007, LNAI 4632, pp. 287–299, 2007.

discovering user profiles, recommender systems, web prefetching, design of adaptive web sites, etc. In all of these applications the goal is the development of an effective prediction algorithm. The most successful approach towards this goal has been the exploitation of the user's access history to derive prediction.

Various web usage mining techniques have been used to develop efficient and effective recommendation systems. For example, there have been attempts to use association rules [15], sequential patterns [5], and Markov models [7,10,16] in recommender systems. These techniques work well for web sites which do not have a complex structure, but experiments on complex, highly interconnected sites show that the storage space and runtime requirements of these techniques increase due to the large number of patterns for sequential pattern and association rules, and the large number of states for Markov models. It may be possible to prune the rule space, enabling faster on-line prediction. Except higher order Markov models, all of these techniques do not capture the entire behavior of a user in a session. Because the number of parameters for higher order Markov models are high, it is not feasible to learn higher order Markov models where the number of web pages in a site (i.e. the number of states for the Markov model) is big.

In summary, recommender systems have been extensively explored in web mining. However, the quality of recommendations and the user satisfaction with such systems are still not optimal. All recommender systems based on web usage mining techniques have strengths and weaknesses. Therefore, the need for hybrid approaches that combine the benefits of multiple algorithms has been introduced [6]. However, most of the hybrid recommender systems switch between recommendation algorithms which work independently, or combining different algorithms in one algorithm.

We concentrate in this study on designing a novel hybrid web recommender model to improve the recommendation quality while maintaining the time required for generating recommendation reasonable. In recent years, there has been an increasing interest in applying web content mining techniques to build web recommender systems. However, the web content mining techniques are unable to handle constantly changing web sites, such as news sites, and dynamically created web pages. Thus, using web content mining techniques in a recommender model leads to update the model frequently. For this reason, in this work we try to increase the prediction accuracy with web usage mining techniques. We have investigated methods for building a consensus (or hybrid) web recommender system which attempts to combine the information obtained from different recommender models. We call these models as the modules of the consensus recommender. Thus, our hybrid model is capable to present the recommendation results from more than one model together. Every time a Web user requests a web page, a recommendation set is generated for that user. We propose three different methods for combining the results coming from the modules of the hybrid recommender model. We adjusted the weights of the modules in such a way, that the more useful recommendation are presented to the active web user.

We conducted detailed comparitive evaluation of different recommender models which can be used as modules of the hybrid recommender system. Four recommender models which use different data mining approaches based on different characteristics of user sessions are implemented [7,11,12,13]. The reason for selecting these recommender systems for candidate modules of our hybrid recommender model is that they can be classified into three groups according to the data structure they use for representing user sessions:

1. Those that represent user sessions using only the time that a user spends on each page during her visit [12];
2. Those that represent user sessions by using only the visiting order of the web pages [7,11];
3. Those that represent user sessions by association rules which capture the relationships among pages based on their patterns of co-occurrence across user sessions [13].

Our experimental results show that using a recommender model as a module of hybrid recommender system, which has a lower accuracy comparing to the other modules of the hybrid model, decreases the final recommedation accuracy. For this reason, we have build the hybrid recommender model combining two different recommender systems, which have a better performance than the others.

The rest of the paper is organized as follows. In Section 2, we introduce the distinct models which are integrated to build a new recommendation method. In Section 3, we present the design of the consensus recommender. Section 4 provides detailed experimental results. Finally, in Section 5 we conclude and discuss future work.

2 Hybrid Recommender System Modules

In this section, we present briefly the distinct recommender models that we combine in our recommender system. Our overall approach can be summarized as follows. As with most recommender systems, our hybrid recommender system is composed of two parts: an off-line part and an on-line part. The off-line part preprocesses the web log data and extracts usage patterns. Since the focus of our paper is on building a correct Web recommender system, rather than on cleaning and preprocessing web log data, it is sufficient to assume that the data is a set of sequences, one sequence for each user session. A user session consists of a set of ordered distinct web pages on the web site that the user requests in her single visit. The off-line component models usage patterns from the web server access log data and builds a predictive model based on the extracted usage patterns. The on-line component considers the active user session and makes recommendations based on the discovered patterns. To make predictions in the on-line part, we aim to generate recommendations for the active user from different recommender models to which we refer as the modules of the hybrid recommender system. As stated in Section 1 we tried and tested four different recommender models which can serve as the modules of the hybrid recommender system. However, our detailed experiments show that using a recommender model as a module of hybrid recommender system, which has a lower accuracy comparing to the other modules of the hybrid model, decreases the final recommedation accuracy. For this reason, the hybrid recommender model consists of two modules: A web Page Prediction Model Based on Click-Stream Tree Representation of User Behavior (CST-Model) [11] and Model-Based Clustering and Visualization of Navigation Patterns on a web site (Markov-Model) [7]. The recommendations from these modules are then presented together to the active user. Each of these modules is described in the following subsections.

2.1 Click-Stream Tree Module

In the click-stream tree module (CST-Model), the recommendations are generated using the recommender model proposed in [11]. In this model, sequence alignment techniques are used to calculate pairwise similarities between user sessions. Using these pairwise similarity values, a graph is constructed whose vertices are user sessions. An edge connecting two vertices in the graph has a weight equal to the similarity between these two user sessions. Using an efficient graph-based clustering algorithm the user sessions are clustered, and each cluster is then represented by a click-stream tree (CST) whose nodes are pages of user sessions of that cluster. When a request is received from an active user, a recommendation set consisting of four different pages that the user has not yet visited, is produced using the best matching user session. The user session that has the highest similarity to the active user session is defined as the best session. For the first two requests of an active user session all clusters are explored to find the one that best matches the active user session. For the remaining requests, the best matching user session is found by exploring the top-N clusters that have the highest N similarity values computed using the first two requests of the active user session. The rest of the recommendations for the same active user session are made by using the top-N clusters.

2.2 Markov Module

This module is composed of the model proposed in [7]. In this model (Markov-Model), the user sessions are partitioned into clusters according to the order of web pages in each session A model based clustering approach is employed to cluster user session. In particular, user sessions are clustered by learning a mixture of first order Markov models using a standard learning technique, the Expectation-Maximization (EM) algorithm. Each cluster has a different Markov model which consists of a (sparse) matrix of state transition probabilities, and the initial state probability vector. The proportion of user sessions assigned to each cluster as well as the parameters of each Markov model is learned using EM algorithm. In this module, the user sessions are modeled as follows: (1) a user session is assigned to a particular cluster with some probability, (2) the order of web pages being requested in that session is generated from a Markov model with parameters specific to that cluster. Since the study in [7] focuses on visualization of navigational patterns rather than on predicting the next request of web users, the proposed model has not a recommendation part. For this reason, we developed a recommendation engine for Markov-Model. The details of the recommendation engine will be given in the following section.

3 Web Page Prediction Model

In this section, we present our consensus recommender model and its algorithm. Figure 1 depicts the overall process of our recommendation model. The details of the whole process is given below.

3.1 Data Preparation and Cleaning

In the off-line part of the recommender model, the raw Web access log data is cleaned and prepared for mining the usage patterns. Fundamental methods of data cleaning

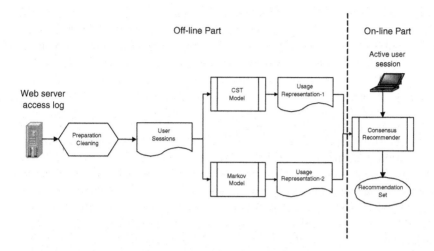

Fig. 1. Consensus Recommender Architecture

and preparation have been well studied in [8,18]. After data preparation (see [11] for details), the log entries are converted into a set of user sessions, where each user session of a length m is in the form of: $s_i = (p_i^1, p_i^2, ..., p_i^m)$, where s_i is a unique session number, $(p_i^1, p_i^2, ..., p_i^m) \subset P$ is the ordered web pages requested in session s_i and $P = \{p_1, ..., p_n\}$ is a set of pages on the web site.

3.2 Representation of Usage Patterns

After identifying user sessions, the CST-Model and Markov-Model are used to extract usage patterns. We advocate the use of these two models in a collaborative way for the following reasons:

1. These models are compatible to each other since both of the models consider the order of requested web pages in a user session to extract usage patterns. This enables the consensus model to generate the final recommendation set easier and faster in the on-line part.
2. CST-Model generates recommendations for a user session by looking all of the past requests made by the user in that session. Assume that two user session inserted to the CST are as below:

   $p_1\ p_2\ p_3$

   $p_2\ p_4$

   If a new user's first request is p_2, then the CST-Model generates p_4 as a recommendation. However, if the model generates a recommendation after visiting two pages p_1 and p_2, then the recommendation set consists of page p_3. Thus, this model complements the first order Markov model by treating as a mixed order Markov model.
3. When used as a stand-alone model both of the models have a high prediction accuracy.

In addition to these reasons, CST-Model has an advantage of representing user sessions as a single CST without using clusters of user sessions, thus increasing the prediction accuracy. In that case, the entire tree is searched to find the best session matching to the active user session. Since this causes a runtime overhead, it can be only used if the tree has a compact structure and the time spent for recommendation is acceptable.

In our model, we use two different representations of usage patterns, namely the representation in CST-Model and the representation in Markov-Model. The representation in CST-Model (Usage Representation-1 in Figure 1) is as follows: After clustering user sessions each cluster is represented by a tree which is called as click-stream tree (CST). Each user session in a cluster is a branch of the corresponding CST. Each CST has a *root* node, which is labeled as "null". Each node except the *root* node consists of three fields: *data*, *count* and *next_node*. Data field consists of page information. *Count* field registers the number of sessions represented by the portion of the path arriving at that node. *Next_node* links to the next node in the CST that has the same *data* field or null if there is any node with the same *data* field. Each CST has a *data_table*, which consists of two fields: *data* field and *first_node* that links to the first node in the CST that has the *data* field.

The second representation of usage patterns is the one produced by the Markov-Model (Usage Representation-2 in Figure 1). The parameters of the Markov-Model consists of: (1) the probabilities of assigning user sessions to various clusters ($p(c_k)$ where c_k is the kth cluster), (2) the parameters of each cluster. The parameters of each cluster are composed of a set of states called as state space, initial state probabilities, and transition probabilities t_{ij} between two adjacent states x_i and x_j. Each transition entry t_{ij} corresponds of moving to state x_j where the process is in state x_i. In our case, the state space of the Markov model is the set of pages making up the web site. A transition probability t_{ij} between state x_i and state x_j corresponds to the probability of visiting page p_j after visiting page p_i ($p(p_j|p_i)$). The Markov-Model of our recommender system uses a first order Markov model and predicts the probability of the next action of a user by only considering the user's previous action. Let $s_i = (p_i^1, p_i^2, ..., p_i^m)$ be user session of length m. The Markov-Model assumes that the user session s_i as being generated by a mixture of Markov models as follows:

$$p(s_i) = \sum_{k=1}^{K} p(s_i|c_k)p(c_k)$$

$$p(s_i|c_k) = p(p_i^1|c_k) \prod_{j=2}^{m} p(p_i^j|p_i^{j-1}, c_k) \qquad (1)$$

$$p(c_k|s_i) = \frac{p(s_i|c_k)p(c_k)}{\sum_j p(s_i|c_j)p(c_j)}$$

where K is the number of clusters, and $p(p_i^1|c_k)$ is the initial state probability of the kth cluster for page p_i^1.

3.3   Algorithm for Generating Recommendations

The previously described process consists of work done off-line. The recommendation engine is the real time component of the consensus recommender that predicts the next request of an active user given her previous actions in that session. Given a user's current request both CST-Model and Markov-Model predict the next request of the user according to their usage patterns extracted as mentioned in the previous subsection. However, there is a trade-off between the prediction accuracy of the next request and the time spent for recommendation. The speed of the recommendation engine is of great importance in on-line recommendation systems. Thus, both of the models are working parallel and are generating two recommendation sets consisting of four pages. The final recommendation set, which consists of four pages also, is produced by combining the results of both of the models in an effective way. For combining the results of the models we use an adaptive weighted sum method. The algorithm is as follows:

Input: an active user session s_a, usage pattern representation build by CST-Model, usage pattern representation build by Markov-Model.

Output: a recommendation set consisting of four pages.

1. Generate a recommendation set consisting of four pages from CST-Model as described in [11] (RS1). Generate a recommendation set consisting of four pages from Markov-Model (RS2) as follows: Assign the active user session one of the clusters (c_a) that has the highest probability calculated using Equation 1. Generate the recommendation set for the current user using the transition matrix of c_a. Sort all the transition entries t_{ij} of c_a in descending order, where the state x_i is equal to the last visited page in the current user session. Pick the top four pages to get a recommendation set from Markov-Model. In this step, both of the models are working parallel.

2. Combine RS1 and RS2 to get a final recommendation set (RS) as follows:

$$RS = \{p_i | p_i \in RS1 \ and \ i = 1, .., w_1\}$$
$$\cup \{p_j | p_j \in RS2 \ and \ j = 1, .., w_2\}$$

where $w_1, w_2 \in \{1, 2, 3\} \wedge w_1 + w_2 = 4$ which is the number of the recommended pages.

3. Update the weights before generating recommendations for a new user session according to the selected method.

The weights for combining the recommendation sets are updated by evaluating the performance of the models separately. If the active user requests a page which is generated by one model but not the other then the prediction accuracy of that model is increased. If the active user requests a page that is produced from both of the models then the prediction accuracies of the models are increased together. We use three different methods for updating the weights of the final recommendation set.

Method 1. If one of the models outperforms in the last active user session then the recommendation set of the next session consists of three pages from that model, and one page from the other model.

Method 2. After recommending two pages for the active user session the performances of the models are evaluated. The final recommendation set for the rest of the active user session is generated by picking three pages from the model that has higher prediction accuracy and one page from the other model.

Method 3. The performances of the models are evaluated until the current user session. The final recommendation set for the active user session is generated by picking three pages from the model that has higher prediction accuracy and one page from the other model.

4   Performance Evaluation

4.1   Experimental Setup

We report our experiments conducted on four different public data sets. The first data set is from the NASA Kennedy Space Center (NASA) server [2](with 92 pages and 15359 user sessions). The second log is from ClarkNet (C.Net) web server which is a full Internet access provider for the Metro Baltimore-Washington DC area [1](with 67 pages and 6846 user sessions). The third server log is from the web server at the University of Saskatchewan (UOS) [3](with 171 pages and 7452 user sessions). The last server log is from the web server of one of the portals of Turkey with 4,5 million of members (TISC)(with 212 pages and 6914 user sessions). Approximately 30% of the cleaned sessions extracted from each data set are randomly selected as the test set, and the remaining part as the training set. For each web page in a session in the test set, our recommendation model generates a recommendation list consisting of four pages. The two-processor computer used in the experiments was an Intel Xeon 3 GHz PC running Linux 2.6. The programs are coded in Java without code optimization.

Our evaluation metric is called as Hit-Ratio and is defined as follows: A hit is declared if any one of the four recommended pages is the next request of the user. The Hit-Ratio is the number of hits divided by the total number of recommendations made by the system.

We conducted the experiments with a single CST for C.Net, TISC and UOS data sets without clustering user sessions. Although using a single CST extends the time period required to generate the recommendation set, it is acceptable for those data sets which have a small number of sessions. The results obtained by using a single CST gives us the upper bound of the prediction accuracy of the CST-Model. In that case we do not have any side effects of the clustering algorithm nor the assumptions we made for assigning the active user session to a cluster since the entire tree is searched. For the NASA data set, the number of clusters is chosen to be five, as proposed in [11].

To determine the parameters of the Markov-Model, we conducted the experiments with different number of clusters and choose the one with highest Hit-Ratio. We perform ten runs (each with different initial parameter settings for the EM algorithm) for each different number of clusters and report the results with the highest Hit-Ratio.

4.2   Effects of Using Different Recommendation Models

Our consensus recommender can use also other recommender systems except CST-model and Markov-model as its module. We tested our recommender system with Apriori-Model [13] and Kmeans-Model [14].

The work in [13] presents a recommender system using association rule mining from user sessions. In this work, the feature weights $w(p_j, s_i)$ $j = 1, 2, 3, \ldots, n$ of the session vector s_i will be binary values: 0 if the page p_j is not visited during the session, 1 otherwise. Association rules capture the relationships among items based on their patterns of co-occurrence across transactions. In the case of web sessions, association rules capture relationships among visited pages. For the current paper, Apriori algorithm [4] is used to find groups of pages occurring frequently together in many user sessions. The recommendation engine uses the resulting frequent items (pages) to make a recommendation according to the user's actions. A fixed-size sliding window over the current active session is used to capture the current user's behavior. For example, if the current session (with a window size of 3) is $< A, B, C >$, and the user references the pageview D, then the new active session becomes $< B, C, D >$. The recommendation engine matches the current user session window with frequent pages to find candidate pageviews for giving recommendations. Given an active session window w, all frequent itemsets of size $|w| + 1$ which contain the active user session window are considered. The recommendation score of each candidate pageview is calculated using the confidence value of the corresponding association rule whose consequent is the singleton containing the pageview to be recommended. If the rule satisfies a user specified confidence threshold, then the candidate pageview is added to the recommendation set.

The model proposed in [14] clusters user sessions according to the visiting time of pages. After the cleaning step, a user sessions s_i is represented by an n-dimensional vector of visited pages over the space of page references:

$$s_i =< w(p_1, s_i), \ldots, w(p_j, s_i), \ldots, w(p_n, s_i) >$$

where n is the total number of unique pages and where p_j is a page reference represented by a unique ID. The feature weight $w(p_j, s_i)$ of the session s_i corresponds to the normalized time of jth page. If this page is not viewed during the visit, its weight will be 0. The session data is clustered by a simple k-means algorithm based on vector distances. The usage pattern for each cluster is represented by the center of that cluster. The center of a cluster c_t can be computed easily by calculating the mean vectors of the sessions assigned to the cluster:

$$ClusterCenter_t =< w(p_1), w(p_2), \ldots, w(p_n) >$$

where $w(p_j)$ is given by

$$w(p_j) = \frac{1}{|c_t|} \cdot \sum_{s \in c_t} w(p_j, s)$$

In the recommendation step, the similarity between each cluster and the active user session is calculated using the cosine similarity metric. The cluster with the highest similarity value is selected as the best matching cluster. To recommend pages, the recommendation algorithm uses the center vector of the best matching cluster. A recommendation score is calculated by multiplying each weight in the cluster center vector by the similarity value of that cluster. The pages that have a recommendation score greater than a user defined threshold are added to the recommendation set.

The test results of the consensus recommender with Kmeans-Model[14] and Apriori-Model[13] are given in Table 1 and Table 2 respectively. Although Apriori-Model and Kmeans-Model generate recommendations based on different aspects (time spent on page, association rules), they couldn't increase the recommendation accuracy of the consensus recommender. This may be due to the fact that recommender models considering the order of visiting pages have a better performance comparing the other models that represent user sessions in a different way(like time spent on page)[9]. This is why using two recommenders that recommend pages according to the page alignment, is more suitable to be a module in our system. Table 2 shows that using CST-model and Markov-model as a module of our consensus recommender, increase the prediction accuracy comparing to the individual modules of the consensus recommender.

Table 1. Hit-Ratio in %

Data Set	Kmeans-model	Markov-model	CST-model	Consensus Recommender(Kmeans-Markov)	Consensus Recommender(Kmeans-CST)
NASA	54	59	62	54	53
C.Net	46	60.5	60.3	57	52

Table 2. Hit-Ratio in %

Data Set	Apriori-model	Markov-model	CST-model	Consensus Recommender Apriori-Markov	Consensus Recommender Apriori-CST	Consensus Recommender Markov-CST
NASA	56	59	62	51	57	**66.4**
C.Net	55	60.5	60.3	51	53	**61.9**
UOS	56	67.3	67.5	60	63	**69.9**
TISC	18	51.6	46.8	47	38	**52.4**

4.3   Effects of the Modules

In our next experiment, we illustrate the effects of the hybrid recommender system modules to the recommendation accuracy of the consensus recommender. To do so, we present the number of successful recommendations produced by the modules (Table 3). The weights of the modules are updated by using method 2 since this method outperforms the other two proposed in this study. The result of the experiments for evaluating the proposed methods will be given in the next subsection. For each module, we show the percentage of successful recommendations produced only from that module. The last column in the table corresponds to the percentage of successful recommendations produced by both of the modules. This experiment shows that in most cases the

Table 3. Hit-Ratio in % of the hybrid system modules

Data Set	CST Model	Markov Model	CST+Markov
NASA	19.9	8.9	37.7
C.Net	15.7	8.5	37.7
UOS	11.6	4.9	53.5
TISC	8.4	6.4	37.6

recommendation sets generated by the distinct models consist of same pages. Using recommender models as the modules of the hybrid system that have high prediction accuracy, but have less pages in common in their recommendation sets may increase the final prediction accuracy of the consensus recommender.

4.4 Effects of the Methods for Updating the Weights

Table 4 indicates Hit-Ratio of the consensus recommender when using different methods for updating the weights of the final recommendation set. Although using method 1 or method 2 does not make a big difference in terms of Hit-Ratio, method 2 works better among the others. This is likely due to the fact that the weights are updated only considering the active user session. This part of the consensus recommender is the only part that causes a run-time overhead. However, this combination algorithm meets the real-time constraint of an on-line recommendation algorithm since it extends the time period for recommendation set generation only 3%. In our experiments, the CST-Model requires 20 ms. to generate a recommendation set which is longer than the Markov-Model requires. Using this combination algorithm increases the time required for generating a recommendation set only 6 μs.

Table 4. Hit-Ratio in %. The weights are updated using different methods.

Data Set	Method 1	Method 2	Method 3	Method 2 + Method 3
NASA	66.0	66.4	63.7	64.2
C.Net	61.8	61.9	61.1	61.2
UOS	69.9	69.9	69.6	69.4
TISC	52.2	52.4	52.3	52.1

4.5 Analysis of the Recommendation Set

The hybrid recommender system supplies the user with four pages it thinks the user will visit as the next page in her session. The system also ranks the pages according to how well they match to the current user session (this is the system's prediction). Table 5 shows the testing results, that evaluates the rank of popularity of the accessed pages in the recommendation set. As can be seen in the Table, in most of the cases the first page of the recommendation set is selected as the next page of the current user session.

Table 5. The rank of popularity of accessed pages in the recommendation set

Data Set	First Page	Second Page	Third Page	Forth Page
NASA	48.9	20.4	17.6	12.9
C.Net	44	25.2	16.5	14.1
UOS	64.3	16	12.1	7.4
TISC	62.2	17	11.3	9.3

However, the probability of visiting the fourth page of the recommendation set is less than 15% in all our experiments. Thus, there is no need to generate a recommendation set that consists of more than four pages.

The overall experimental results show that combining multiple recommender models in an effective way increase the prediction accuracy.

5   Conclusion and Future Work

In this paper, we proposed a new method to generate recommendations for web users. Our consensus recommender integrates the results from two distinct recommender models, which we call as the modules of our recommender system, and generates a single recommendation set for a new user. We tried three different methods to combine the results produced by the modules of the consensus recommender. Experiments show that this consensus model achieves a better prediction accuracy compared to the individual modules CST-Model and Markov-Model. The other recommender systems with low prediction accuracy are not suitable for consensus recommender. The modules of the consensus recommender have to complement each other. According to the experimental results, even though CST-Model and Markov-Model sometimes generate similar pages, they complement each other sufficiently to increase the prediction accuracy.

This paper presents the preliminary results of a work in progress. The current promising results promote further study. We are now extending the hybrid recommender system in several ways. We are working now on the combination methods of the recommender systems for improving the prediction accuracy. We will use more than two recommender systems as part of the hybrid recommendation system in order to represent different user behavior.

Acknowledgments. The authors were supported by the Scientific and Technological Research Council of Turkey (TUBITAK) EEEAG project 105E162.

References

1. Clarknet www server log:
   `http://ita.ee.lbl.gov/html/contrib/ClarkNet-HTTP.html`
2. Nasa kennedy space center log:
   `http://ita.ee.lbl.gov/html/contrib/NASA-HTTP.html`
3. The university of saskatchewan log:
   `http://ita.ee.lbl.gov/html/contrib/Sask-HTTP.html`
4. Agrawal, R., Srikant, R.: Fast algorithms for mining association rules. In: VLDB 1994. Proc. 20th Int. Conf. Very Large Data Bases, pp. 487–499. Morgan Kaufmann, San Francisco (1994)
5. Agrawal, R., Srikant, R.: Mining sequential patterns. In: ICDE. Proceedings of the International Conference on Data Engineering, Taipei, Taiwan, March 1995 (1995)
6. Burke, R.D.: Hybrid recommender systems: Survey and experiments. User Model. User-Adapt. Interact. 12(4), 331–370 (2002)

7. Cadez, I., Heckerman, D., Meek, C., Smyth, P., White, S.: Model-based clustering and visualization of navigation patterns on a web site. Data Min. Knowl. Discov. 7(4), 399–424 (2003)
8. Cooley, R., Mobasher, B., Srivastava, J.: Data preparation for mining world wide web browsing patterns. Journal of Knowledge and Information Systems 1(1), 5–32 (1999)
9. Demir, G.N., Goksedef, M., Uyar, A.S.: Effects of session representation models on the performance of web recommender systems. In: Proceedings of the Workshop on Data Mining and Business Intelligence (to appear, 2007)
10. Deshpande, M., Karypis, G.: Selective markov models for predicting web-page accesses. In: SDM'2001. Proceedings of the First SIAM International Conference on Data Mining (2001)
11. Gündüz, S., Özsu, M.T.: A web page prediction model based on click-stream tree representation of user behavior. In: KDD. Proceedings of Ninth ACM International Conference on Knowledge Discovery and Data Mining, Washington, DC, August 2003, pp. 535–540 (2003)
12. Mobasher, B., Dai, H., Luo, T., Nakagawa, M.: Discovery of aggregate usage profiles for web personalization. In: WebKDD'2000. Proceedings of the Web Mining for E-Commerce Workshop (2000)
13. Mobasher, B., Dai, H., Luo, T., Nakagawa, M.: Effective personalization based on association rule discovery from web usage data. In: Web Information and Data Management, pp. 9–15 (2001)
14. Mobasher, B., Dai, H., Luo, T., Nakagawa, M., Witshire, J.: Discovery of aggregate usage profiles for web personalization. In: Proceedings of the WebKDD Workshop (2000)
15. Nanopoulos, A., Katsaros, D., Manolopoulos, Y.: Effective prediction of web-user accesses: a data mining approach. In: Proceedings of WEBKDD workshop, San Francisco, CA (2001)
16. Sarukkai, R.R.: Link prediction and path analysis using markov chains. In: Proceedings of the Ninth International World Wide Web Conference, Amsterdam (2000)
17. Srivastava, J., Cooley, R., Deshpande, M., Tan, P.N.: Web usage mining: Discovery and applications of usage patterns from web data. SIGKDD Explorations 1(2), 12–23 (2000)
18. Zaïane, O.R.: Web usage mining for a better web-based learning environment. In: Proc. Conference on Advanced Technology for Education, Bannf, Alberta, June 27–28, 2001, pp. 27–28 (2001)

Constructing Classification Rules Based on SVR and Its Derivative Characteristics

Dexian Zhang[1], Zhixiao Yang[1], Yanfeng Fan[2], and Ziqiang Wang[1]

[1] College of Information Science and Engineering, Henan University of Technology,
Zhengzhou 450052, China
zdx@haut.edu.cn
[2] Computer College, Northwestern Polytecnical University, Xi'an 710072, China

Abstract. Support vector regression (SVR) is a new technique for pattern classification , function approximation and so on. In this paper we propose an new constructing approach of classification rules based on support vector regression and its derivative characteristics for the classification task of data mining. a new measure for determining the importance level of the attributes based on the trained SVR is proposed. Based on this new measure, a new approach for clas-sification rule construction using trained SVR is proposed. The performance of the new approach is demonstrated by several computing cases. The experimen-tal results prove that the approach proposed can improve the validity of the extracted classification rules remarkably compared with other constructing rule approaches, especially for the complicated classification problems.

1 Introduction

The goal of data mining is to extract knowledge from data. Data mining is an inter-disciplinary field, whose core is at the intersection of machine learning, statistics and databases. There are several data mining tasks, including classification, regression, clustering, dependence modeling, etc. Each of these tasks can be regarded as a kind of problem to be solved by a data mining approach. In classification task, the goal is to assign each case (object, record, or instance) to one class, out of a set of predefined classes, based on the values of some attributes (called predictor attributes) for the case. In this paper we propose an new constructing approach of classification rules based on support vector regression and its derivative characteristics for the classification task of data mining. The classification rule extraction has become an important aspect of data mining, since human experts and corporate managers are able to make better use of the classification rules for making decision and easily discover unknown relationships and patterns from a large data set than other expression forms of knowledge.

The existing approaches for constructing the classification rules can be roughly classified into two categories, data driven approaches and model driven approaches. The main characteristic of the data driven approaches is to extract the symbolic rules completely based on the treatment with the sample data.

R. Alhajj et al. (Eds.): ADMA 2007, LNAI 4632, pp. 300–309, 2007.

The most representative approach is the ID3 algorithm and corresponding C4.5 system proposed by J.R.Quinalan. This approach has the clear and simple theory and good ability of rules extraction, which is appropriate to deal with the problems with large amount of samples. But it still has many problems such as too much dependence on the number and distribution of samples, excessively sensitive to the noise, difficult to deal with continuous attributes effectively and etc. The main characteristic of the model driven approaches is to establish a model at first through the sample set, and then extract rules based on the relation between inputs and outputs represented by the model. Theoretically, these rule extraction approaches can overcome the shortcomings of data driven approaches mentioned above. Therefore, the model driven approaches will be the promising ones for rules extraction. The representative approaches are rules extraction approaches based on neural networks [1-8]. Though these methods have certain effectiveness for rules extraction, there exist still some problems, such as low efficiency and validity, and difficulty in dealing with continuous attributes etc.

There are two key problems required to be solved in the classification rule extraction, i.e. the attribute selection and the discretization to continuous attributes. Attribute selection is to select the best subset of attributes out of original set. The attributes that are important to maintain the concepts in the original data are selected from the entire attributes set. How to determine the importance level of attributes is the key to attribute selection. Mutual information based attribute selection [9-10] is a common method of attribute selection, in which the information content of each attribute is evaluated with regard to class labels and other attributes. By calculating mutual information, the importance levels of attributes are ranked based on their ability to maximize the evaluation formula. Another attribute selection method uses entropy measure to evaluate the relative importance of attributes [11]. The entropy measure is based on the similarities of different instances without considering the class labels. In paper [12], the separability-correlation measure is proposed for determining the importance of the original attributes. The measure includes two parts, the intra-class distance to inter-class distance ratio and an attributes-class correlation measure. Through attributes-class correlation measure, the correlation between the changes in attributes and their corresponding changes in class labels are taken into account when ranking the importance of attributes. The attribute selection methods mentioned above can be classified into the sample driven method. Their performance depends on the numbers and distributions of samples heavily. It is also difficult to use them to deal with continuous attributes. Therefore, it is still required to find more effective heuristic information for the attribute selection and the discretization to continuous attribute in the classification rule extraction.

In this paper, we use trained SVR to obtain the position and shape characteristics of the classification hypersurface. Based on the analysis of the relations among the position and shape characteristics of classification hypersurface, the partial derivative distribution of the outputs of trained SVR to its corresponding inputs and the importance level of attributes to classifications, this paper mainly studies on the measure method of the classification power of attributes on the basis of differential information of the trained SVR and develops new approach for the rule extraction.

The rest of our paper is organized as follows. Section 2 discusses the representation of the classification rules . Section 3 describes the classifier construction based on SVR. Section 4 presents the measure method for attribute importance ranking. The rules extraction method is presented in section 5. Experimental results and analysis are reported in Section 6. Finally, we give the conclusion in Section 7.

2   Representation of the Classification Rules

Classification rules should be not only accurate but also comprehensible for the user. Comprehensibility is important whenever classification rules will be used for supporting a decision made by a human user. After all, if classification rules is not comprehensible for the user, the user will not be able to interpret and validate it. In this case, probably the user will not trust enough the classification rules to use it for decision making. This can lead to wrong decisions.

In this paper, the classification rule is often expressed in the form of IF-THEN rules which is commonly used, as follows: IF <conditions > THEN < class>.The rule antecedent (IF part) contains a set of conditions, connected by a logical conjunction operator (AND). In this paper we will refer to each rule condition as a term, so that the rule antecedent is a logical conjunction of terms in the form: IF term 1 AND term 2 AND ... Each term has two kind of forms. One kind of the form is a triple <attribute, operator, value>. The operator can be $<$,$\geq$ or $=$. Another kind of the form is a triple <attribute, $\in$, value range>. The rule consequent (THEN part) specifies the class label predicted for cases whose attributes satisfy all the terms specified in the rule antecedent.

This kind of classification rule representation has the advantage of being intuitively comprehensible for the user, as long as the number of discovered rules and the number of terms in rule antecedents are not large.

3   Constructing the Classifier Based on SVR

SVR is a new technique for pattern classification , function approximation and so on. The SVR classifiers have the advantage that we can use them for classification problem with more than 2 class labels. In this paper, we use an SVR classifier to determine the importance level of attributes and construct classification rules.

Given a set of training sample points,$\{(x_i, z_i), i = 1, .., l\}$,such that $x_i \in R^n$ is an input and $z_i \in R^1$is a target output, The primal form of Support vector Regression is

$$min_{w,b,\xi,\xi^*,\varepsilon} \frac{1}{2} w^T w + C(\nu\varepsilon + \frac{1}{l} \sum_{i=1}^{l} (\xi_i + \xi_i^*)) \tag{1}$$

subject to

$$(w^T \phi(x_i) + b) - z_i \geq \varepsilon + \xi_i \tag{2}$$

$$z_i - (w^T \phi(x_i) + b) \geq \varepsilon + \xi_i^* \tag{3}$$

$$\xi_i, \xi_i^* \geq 0, i = 1, .., l, \varepsilon > 0 \tag{4}$$

Here sample vector x are mapped into a higher dimensional space by the function ϕ. $C > 0$ is the penalty parameter of the error item. In this paper we usually let $C = 10 \sim 10^5$. The ν and ε are two parameters. The parameter $\nu, \nu \in (0, 1]$, control the number of support vectors. In this paper we usually set $\nu = 0.5, \varepsilon = 0.1$. Furthermore, $K(x_i, x_j) \equiv \phi(x_i)^T \phi(x_j)$. is the kernel function. In this paper, we use the following radial basis function (RBF) as kernel function.

$$K(x_i, x_j) = exp(-\gamma||(x_i - x_j)^2||), \gamma > 0 \tag{5}$$

Here, γ is kernel parameter. In this paper we usually let $\gamma = (0.1 \sim 1)/n$, n is the number of attributes in training sample set. The formula (1) dual is

$$min_{\alpha,\alpha^*} \frac{1}{2}(\alpha - \alpha^*)^T Q(\alpha - \alpha^*) - z^T(\alpha - \alpha^*) \tag{6}$$

subject to

$$e^T(\alpha - \alpha^*) = 0, e^T(\alpha + \alpha^*) \leq C\nu \tag{7}$$

$$0 \leq \alpha_i, \alpha^* \leq C, i = 1, \ldots, l \tag{8}$$

Where e is the vector of all ones, Q is a l by l positive semidefinite matrix. The output function of SVR classifier is

$$Z(x) = \sum_{i=1}^{l}(-\alpha + \alpha^*)K(x_i, x) + b \tag{9}$$

Assuming the attribute vector of classification problems is $x=[X_1, X_2, \ldots, X_n]$, where n is the number of attributes in sample set, and the corresponding classification label is z and $z \in R^1$, then the sample of classification problems can be represented as $< X, z >$. In order to constructing SVR classifiers as the form shown by formula (9), quan-tification and normalization of the attribute values and classification labbels will be carried out as follows.

The quantification is performed for the values of discrete attributes and classification labels. In this paper, values of discrete attributes and classification labels are quantified as integer numbers in some order, for example, 0,1,2,3,....

The normalization is performed to adjust the of SVR input ranges. For a given attribute value space Ω, utilizing the following linear transformation to map the attribute value X of sample to the SVM input x, making every elements in the x with the same range of $[\Delta, -\Delta]$.

$$x = bX + b_0 \tag{10}$$

where $b = (b_{ij})$ is a transformation coefficient matrix. Here

$$b_{ij} = \begin{cases} \frac{2\Delta}{MaxX_i - MinX_i} : j = i \\ 0 : \qquad\qquad otherwise \end{cases} \qquad (11)$$

$b_0 = (b_{0i})$ is a transformation vector.Here,

$$b_{0i} = \Delta - a_{ii} MaxX_i \qquad (12)$$

The parameter Δ affects the generalization of trained SVM, in this paper we usually set $\Delta = 0.5 \sim 2$.

During the construction of classification rules, only the attribute space covered by the sample set should be taken into account. Obviously according to formula (9), when the kernel function of SVR is the radial basis function shown by formula (5), any order derivatives of network output $Z(x)$ to each SVR input x_k exist.

4   Measure for Attribute Importance Ranking

Without losing the universality, next we will discuss the classification problems with two attributes and two class labels.

For a 2-dimension classification problem, assuming the shape of classification hypersurface in the given area Ω is as shown in Fig.1, in which the perpendicular axis is attribute $Z(x)$is class label, the area A and B are the distribution area of different classes. In the cases (a) and (b),the importance level of x_1 for classification is obviously higher, so area should be divided via attribute x_1.In the case (c), attribute x_1 and attribute x_2 have the equal classification powers. Therefore, for a given attribute value space Ω , the importance level of each attribute depends on the mean perpendicular degree between each attribute axis and classification hypersurface in space Ω or its adjacent space. The higher is the mean perpendicular degree, the higher is the importance level.

For a given sample set, the attribute value space Ω is defined as follows.

$$\Omega = \{x | Minx_k \le x_k \le Maxx_k, k = 1, \dots, n\} \qquad (13)$$

Where $Minx_k$ and $Maxx_k$are the minimal and maximal value of k-th attribute in the given sample set, respectively.

For a given trained SVM and the attribute value $\Gamma, \Gamma \subset \Omega$, the perpendicular level between classification hypersurface and attribute axis x_k is defined as follows.

$$P_{x_k}(x) = \frac{|\frac{\partial Z(x)}{\partial x_k}|}{\sqrt{\sum_k [(\frac{\partial Z(x)}{\partial x_k})^2 + 1]}} \qquad (14)$$

According to formula (14), the value of the perpendicular level $P_{x_k}(x)$, mainly depends on the value $\frac{\partial Z(x)}{\partial x_k}$. Therefore for the convenience of computing, we can use the following formula to replace formula (14).

$$P_k(x) = |\frac{\partial Z(x)}{\partial x_k}| \qquad (15)$$

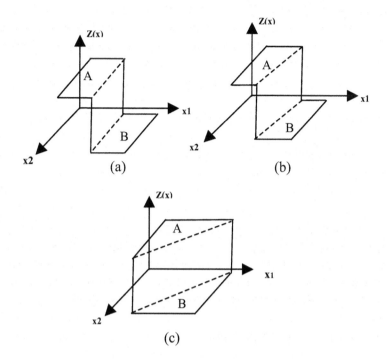

Fig. 1. Typical shapes of classification hypersurface

From the formula (9) and (5), we can get

$$\frac{\partial Z(x)}{\partial x_k} = \sum_{i=1}^{l} 2(-\alpha + \alpha^*)\gamma(x_{ik} - x_k)K(x_i, x) \tag{16}$$

Here the x_{ik} and the x_k are the k-th attribute values of the j-th Support vector and the sample point x, respectively.

For a given attribute value space Γ, $\Gamma \subset \Omega$,the measurement of classification power of attribute x_k is defined as follows.

$$JP(x_k) = \begin{cases} \frac{\int_{V_\Gamma} P_k(x)dx}{\int_{V_\Gamma} dx} : \int_{V_\Gamma} dx \neq 0 \\ 0 : \qquad otherwise \end{cases} \tag{17}$$

The importance level measure $JP(x_k)$ of the attribute x_k represents the influence degree of attribute x_k to classification. So in the process of rules extraction, the value $JP(x_k)$ is the important instruction information for selecting attributes and dividing attribute value space.

The typical classification problem of weather type for playing golf is employed to demonstrate the performance of the new measure method. The computing results are shown as table 1. The attributes and their values are as follows: Outlook has the value of sunny, overcast and rain, quantified as 0, 1, 2. Temperature has the value of 65 $\sim$ 96. Humidity has the value of 65 $\sim$ 96. Windy has the value of true, false, quantified as 0, 1. The size of the training sample set is 14.

Table 1. Computing Results of Measurement Value $JP(x_k)$

Attributes	Whole Area	Outlook=sunny	Outlook=rain
Outlook	0.0547	–	–
Temperature	0.0292	0.0551	0.0334
Humidity	0.0353	0.0861	0.0552
Windy	0.0362	0.0457	0.0929

From table 1, in the whole space the measure value of importance level of attribute outlook is the biggest, therefore attribute outlook should be selected as the root node of the decision tree, and the attribute value space should be divided by its values. While in the subspace of outlook= rain, measure value of attribute windy is the biggest. Thus according to this information, the optimal decision tree and corresponding classification rules can be generated.

5 Rules Extraction Method

The algorithm for classification rule construction from trained SVR proposed in this paper is described as follows.

Step 1: Initializing.
a) Divide the given sample set into two parts, the training sample set and the test set. According to the training sample set, generate the attribute value space Ω by formula (13).
b) Set the interval number of attributes and the predefined value of error rate.
Step 2: Rule generating.
a) Generate a queue R for finished rules and a queue U for unfinished rules.
b) Select attribute x_k with the biggest value of $JP(x_k)$ computed by formula (17) as the extending attribute out of the present attributes. Divide the attribute x_k into intervals according to the chosen interval number. Then for each interval, pick attribute x_j with the biggest $JP(x_k)$ as the extending attribute for each interval. Merge the pairs of adjacent intervals with the same extending attribute and same class label with the largest proportion in all of the class labels. A rule is generated for each merged interval. If the class error of the generated rule is less than the predefined value, put it into the queue R, otherwise put it into the queue U.
c) If U is empty, the extraction process terminates, otherwise go to d).
d) Pick an unfinished rule from the queue U by a certain order, and perform division and mergence. A rule is generated for each merged interval. If the class error of the generated rule is less than the predefined value, then put it into the queue R, otherwise put it into the queue U. Go to c).

Step 3: Rule Processing.

Check the rule number of each class label. Let the rules of the class label with the largest number of rules be default rules.

6   Experiment and Analysis

The spiral problem [13] and congressional voting records(voting for short), hepatitis, iris plant(iris for short), statlog australian credit approval(credit-a for short) in UCI data sets [14] are employed as computing cases, shown in table 2. The attribute value distribution of spiral problem is shown as Fig.2, in which solid points are of Class C0, empty points are of Class C1.

Table 2. Computing Cases

	Spiral	Voting	Hepatitis	Iris	Credit-A
Total Samples	168	232	80	150	690
Training Samples	84	78	53	50	173
Testing Samples	84	154	27	100	517
Classification Numbers	2	2	2	3	2
Total Attributes	2	16	19	4	15
Discrete Attributes	0	16	13	0	9
Continuous Attributes	2	0	6	4	6

Table 3. Experimental Results Comparison between New Approach(NA) and C4.5R

	#Rules(NA: C4.5R)	Err.Train(NA: C4.5R)	Err.Test(NA: C4.5R)
Spiral	8: 3	0: 38.1%	1.1: 40.5%
Voting	3: 4	2.5%: 2.6%	2.5%: 3.2%
Hepatitis	3: 5	7.5%: 3.8%	11.1%: 29.6%
Iris	4: 4	0%: 0%	4%: 10%
Credit-A	5: 3	12.1%: 13.9%	14.5%: 14.9%

Since no other approaches extracting rules from SVR are available, we include a popular rule learning approach i.e.C4.5R for comparison. The experimental results are tabulated in Table 3. For the spiral problem and the Iris plant problem, the rules set extracted by the new approach are shown in Table 4 and Table 5, respectively. Table 3 shows that the rules extraction results of the new approach are obviously better than that of C4.5R, especially for spiral problem. For the case of spiral problem, C4.5R is difficult to extract effective rules, but the new approach has so impressive results that are beyond our anticipation. This means that the new approach proposed can improve the validity of the extracted rules for complicated classification problems remarkably. Moreover, the generalization ability of those rules extracted by the new approach is also better than that of rules extracted by the C4.5R.

Table 4. Rules Set of Spiral Problem Gemerated by the Algorithm Proposed

R1 $x_0 \geq 2.22 \longrightarrow C1$
R2 $x_0[-2.22, -1.33) \wedge x_1 < 1.68 \longrightarrow C1$
R3 $x_0[-1.33, 0) \wedge x_1 < -1.75 \longrightarrow C1$
R4 $x_0[-1.33, 0) \wedge x_1[0.76, 1.2) \longrightarrow C1$
R5 $x_0[0, 1.33) \wedge x_1 < -2.2 \longrightarrow C1$
R6 $x_0[0, 1.33) \wedge x_1[-0.76, 1.75] \longrightarrow C1$
R7 $x_0[1.33, 2.22) \wedge x_1 < -1.68 \longrightarrow C1$
R8 $Default \longrightarrow C0$

Table 5. Rules Set for the Iris Plant Problem Generated by the Algorithm Proposed

R1 $petalwidth < 0.72 \longrightarrow Iris - setosa$
R2 $petalwidth \geq 1.66 \longrightarrow Iris - virginica$
R3 $petalwidth[1.34, 1.66] \wedge petallength \geq 4.91 \longrightarrow Iris - virginica$
R4 $Default \longrightarrow Iris - versicolor$

7 Conclusions

In this paper, based on the analysis of the relations among the characteristics of position and shape of classification hypersurface, the partial derivative distribution of the trained SVR output to the corresponding inputs, a new measure for determining the importance level of the attributes based on the differential information of trained SVR is proposed, which is suitable for both continuous attributes and discrete attributes, and can overcome the shortcomings of the measure method based on information entropy. On the basis of this new measure, a new approach for rules extraction from trained SVR is presented, which is also suitable for classification problems with continuous attributes. The performance of the new approach is demonstrated by several typical examples, the computing results prove that the new approach can improve the validity of the extracted rules remarkably compared with other rule extracting approaches, especially for complicated classification problems.

References

1. Fu, L M.: Rule generation from neural network. IEEE Trans on Sys, Man and Cybernetics 8, 1114–1124 (1994)
2. Towell, G., Shavlik, J.A.: The extraction of refined rules from knowledge-based neural networks. Machine Learning 1, 71–101 (1993)
3. Lu, H.J., Setiono, R., Liu, H.: NeuroRule: a connectionist approach to data mining. In: Proceedings of 21th International Conference on Very Large Data Bases, Zurich, Switzerland, pp. 81–106 (1995)
4. Zhou, Z.H., Jiang, Y., Chen, S.F.: Extracting symbolic rules from trained neural network ensembles. AI Communications 6, 3–15 (2003)

5. Sestito, S., Dillon, T.: Knowledge acquisition of conjunctive rules using multilayered neural networks. International Journal of Intelligent Systems 7, 779–805 (1993)
6. Craven, M.W., Shavlik, J.W.: Using sampling and queries to extract rules from trained neural networks. In: Proceedings of the 11th International Conference on Machine Learning, New Brunswick, NJ, pp. 37–45 (1994)
7. Maire, F.: Rule-extraction by backpropagation of polyhedra. Neural Networks 12, 717–725 (1999)
8. Setiono, R., Leow, W.K.: On mapping decision trees and neural networks. Knowledge Based Systems 12, 95–99 (1999)
9. Battiti, R.A.: Using mutual information for selecting featuring in supervised net neural learning. IEEE Trans on Neural Networks 5, 537–550 (1994)
10. Bollacker, K.D., Ghosh, J.C.: Mutual information feature extractors for neural classifiers. In: Proceedings of 1996 IEEE international Conference on Neural Networks, Washington, pp. 1528–1533 (1996)
11. Dash, M., Liu, H., Yao, J.C.: Dimensionality reduction of unsupervised data. In: Proceedings of the 9th International Conference on Tools with Artificial Intelligence, Newport Beach, pp. 532–539 (1997)
12. Fu, X.J., Wang, L.P.: Data dimensionality reduction with application to simplifying RBF network structure and improving classification performance. IEEE Transactions on Systems, Man and Cybernetics, Part B - Cybernetics. 33, 399–409 (2003)
13. Kamarthi, S.V., Pittner, S.: Accelerating neural network training using weight extrapolation. Neural Networks 12, 1285–1299 (1999)
14. Blake, C., Keogh, E., Merz, C.J.: UCI repository of machine learning databases, Department of Information and Computer Science, University of California, Irvine, CA(1998), [http://www.ics.uci.edu/~meearn/MLRepository.htm]

Hiding Sensitive Associative Classification Rule by Data Reduction

Juggapong Natwichai, Maria E. Orlowska, and Xingzhi Sun

School of Information Technology and Electrical Engineering
The University of Queensland, Brisbane, Australia
{jpn, maria, sun}@itee.uq.edu.au

Abstract. When data sharing becomes necessary, there is a dilemma in preserving privacy. On one hand sensitive patterns such as classification rules should be hidden from being discovered. On the other hand, hiding the sensitive patterns may affect the data quality. In this paper, we present our studies on the sensitive classification rule hiding problem by data reduction approach, i.e., removing the whole selected records. In our work, we focus on a particular type of classification rule, called canonical associative classification rule. And, the impact on data quality is evaluated in terms of the number of affected non-sensitive rules. We present the observations on the data quality based on a geometric model. According to the observations, we can show the impact precisely without any re-computing. This helps to improve the hiding algorithms from both effectiveness and efficiency perspective. Additionally, we present the algorithmic steps to demonstrate the removal of the records so that the impact on the data quality is potentially minimal. Finally, we conclude our work and outline future work directions for this problem.

1 Introduction

Privacy Preservation Data Mining (PPDM) is a new emerging research area of interest. In general, it primarily focuses on developing algorithms to modify the original data, such that the private information remains private after the data being mined. In PPDM, there are two types of privacy which are regarded, the privacy of individual or personal information, and the privacy of some sensitive patterns [1]. For the first type, individual privacy, the problem is how to mine the data without any access to the sensitive data directly. Many attempts have been proposed to address the problem by adding artificial noise [2,3] or applying cryptography-based technique [4,5] to the original data before it is used for data mining. The challenge is how can the mining result (from noise-added-data) be accurate as the result from the original data.

The other type of privacy considered in PPDM is the privacy of sensitive patterns. Generally, it occurs when the data owner wants to share the data set with the collaborators or release the data set publicly. However, some discoverable patterns in the data are considered by the data owner as the "sensitive" patterns, for example, the patterns which could give too much useful information

R. Alhajj et al. (Eds.): ADMA 2007, LNAI 4632, pp. 310–322, 2007.

to the data recipient [6]. Hence, in order to hide the sensitive patterns, the data owner needs to modify the data before the sharing takes place.

To address the pattern hiding problem, many approaches have been proposed. Most of them use data perturbation technique, i.e., modifying some values in the data such that the patterns can not be discovered when the same pattern discovery algorithm is used (with the same parameter setting). For example, in the context of association rules, given specified support and confidence values, the data modification should guarantee that the sensitive association rules will not be discovered with the same support and confidence values [6,7,8].

One specific pattern for the above problem is classification rule. Generally, given the training data which are records and their class labels, the classification is formulating a model to classify new record to the correct class label. The set of derived classification rules is one of such models, with which, we can assign the class label to the record correctly. Particularly, we focus on associative classification rules [9,10] in this paper. A scenario for classification rule hiding problem can be: Given a set of records with class labels, the associative classification rules can be derived from the data set. The data owner wishes to hide some sensitive rules before sharing the data with a collaborator. Naturally, the most trivial way to hide the rules is simply not to disclose data itself, but show what we wish to share (e.g., non-sensitive rules). This is not an option in our consideration – we wish to share the data but only the data that is "minimally" corrupted in order to hide the selected rules.

In this paper, we address the data modification problem which impacts on the data quality with great details. Basically, instead of arbitrary perturbation on the particular data values, we simply apply data reduction, i.e., to remove the whole selected record(s) to hide the classification rules. In this way, the released data is still part of the original data, however, with smaller size. Our main task is to hide the given sensitive rules by data reduction and meanwhile, minimize the impact on the data quality. Here, a sensitive classification rule is considered hidden successfully when its support falls below the minimal support threshold. The impact on data quality is evaluated in terms of the number of affected non-sensitive classification rules. Importantly, we introduce a special type of classification rule, called *canonical form rule*. Based on this concept, given a set of rules of the data set, when a record is removed during the modification process, we can derive the new set of classification rules without re-applying the classification algorithm on the modified data set. We call this property "no re-computing". At the end, we demonstrate an algorithmic way to choose the potentially best selection of records towards the hiding rule goal.

For the sake of clarity, let us state here the advantage of "no re-computing" in the process of rule hiding. We discuss the benefit from both efficiency and effectiveness. First, usually, when an algorithm is applied to hide the rules, it modifies the data set by following a particular heuristic, e.g., selecting the records which contain most frequent patterns to reduce the number of modifications [7].

The algorithm will proceed by following such heuristic until the sensitive rules are successfully hidden. The impact, e.g. the number of affected non-sensitive rules, of the hiding process can be determined at the end by redoing the pattern discovery process. If the data owner does not satisfy with the impact of the modification, usually, the try-and-error of the whole process including the computational costly patterns discovery must be done repetitively. Hence, being able to determine the impact without the patterns discovery process can improve the efficiency. For the effectiveness issue, ideally, taking the intermediate impact into consideration can help to chose right candidate to remove in every step, which improves the data quality at the end.

The organization of the paper is as follows. The basic notation is presented in Section 2. In Section 3, we define the canonical form classification rule and present the observations for the problem of hiding classification rules. Based on these observations, we propose algorithms to demonstrate how to hide the sensitive rules in Section 4. The conclusion and future work are presented in Section 5.

2   Basic Notation

In this section we introduce the basic notation for our problem.

Definition 1 (Data Set). *Let a data set D be a set of records, $D = \{d^1, d^2, \ldots, d^n\}$, and $I = \{1, \ldots, n\}$ be a set of identifiers for elements of D.*

The data set D is defined on a schema $\boldsymbol{A} = \{A_1, A_2, \ldots, A_k\}$, and $J = \{1, \ldots, k\}$ be a set of identifiers for elements of $\boldsymbol{A}$.

For each $j \in J$, domain of A_j, denoted as $dom(A_j) \subseteq N$, where N is a set of natural number.

For each $i \in I$, $d^i(\boldsymbol{A}) = (d^i(A_1), d^i(A_2), \ldots, d^i(A_k))$, denoted as $(d_1^i, d_2^i, \ldots, d_k^i)$.

Let C be a set of class labels, denoted as $C = \{c_1, c_2, \ldots, c_o\}$, and $M = \{1, \ldots, o\}$ be a set of identifiers for elements of C. For all $m \in M$, $c_m \in N$, where N is a set of natural numbers.

The label is just an identifier of a class. A class which is labelled as c_m defines a subset of records which is described by data assigned to the class. The classification problem is to establish a mapping from D to C.

Definition 2 (Classification). *A literal p is a pair, consisting of an attribute A_j and a value v in $dom(A_j)$. A record d^i will satisfy the literal $p(A_j, v)$ iff $d_j^i = v$.*

Given a data set D, and a set of class labels C, let R be a set of classification rules, such that $R = \{r_1, r_2, \ldots, r_q\}$, and $L = \{1, \ldots, q\}$ be a set of identifiers for elements of R.

For any $l \in L$, a rule r_l is in the form of $\bigwedge p \to c_m$, where p is the literal, and c_m is a class label. The left hand side (LHS) of the rule r_l is the conjunction of the literals, denoted as $r_l.LHS$. The right hand side (RHS) is a class label of the rule r_l, denoted as $r_l.RHS$.

A record d^i satisfies the rule r_l iff it satisfies all literals in $r_l.LHS$, and has a class label c_m same as $r_l.RHS$.

A record d^i which satisfies the classification rule r_l is called a **supporting record** of r_l. The **support** of the rule r_l, denoted as $Sup(r_l)$, is the number of supporting records of r_l. The **confidence** of rule r_l, denoted as $Conf(r_l)$, is the ratio between $Sup(r_l)$ and the total number of records which satisfy all literals in LHS of r_l.

To illustrate the basic concepts, we present here the example data, which will be used through out this paper. Suppose that we are dealing with the data set with three attributes A_1, A_2, and A_3, $J = \{1, 2, 3\}$, and their domains are $\{0, 1\}$. The set of class labels is given as $C = \{1, 2, 3\}$. The simplicity of the example with restricted domain $\{0, 1\}$ for each attribute will not limit our general considerations, but will assist with presentation in this paper. Table 1 shows the data set where duplication is not allowed, and we also have the complete data set, $I = \{1, 2, \ldots, 8\}$.

Table 1. The example data set

I	A_1	A_2	A_3	C
1	0	0	0	1
2	0	0	1	1
3	0	1	0	1
4	0	1	1	1
5	1	0	0	2
6	1	0	1	1
7	1	1	0	1
8	1	1	1	3

From the example data set, three classes define three disjoint subsets of data: label-1-class defines subset where $I = \{1, 2, 3, 4, 6, 7\}$, and label-2-class and label-3-class define subsets of one record where $I = \{5\}$ and $I = \{8\}$ respectively. From the data in Table 1, we can derive a set of classification rules as shown in Table 2. The first rule is one-literal rule which classifies the data into class label 1, its support is 4. While the second and third rules with two literals also classify the data record into class label 1, their supports are both 2. The fourth and fifth rules has support 1, classify the data record into class label 2 and 3 respectively. All the rules have 100% confidence value.

Considering a rule $r : (A_2, 1) \rightarrow 1$ in the example data set, we have $Sup(r) = 3$ and confidence $Conf(r) = 75\%$. In general, this rule could be a classification rule. However, in this paper, for precise classification, we are only interested in the rule with 100% confidence. In other words, any derived classification rule must have 100% confidence.

Table 2. The example classification rules

Rule No. (L)	Content
1.	$(A_1, 0) \rightarrow 1$
2.	$(A_2, 1) \wedge (A_3, 0) \rightarrow 1$
3.	$(A_2, 0) \wedge (A_3, 1) \rightarrow 1$
4.	$(A_1, 1) \wedge (A_2, 0) \wedge (A_3, 0) \rightarrow 2$
5.	$(A_1, 1) \wedge (A_2, 1) \wedge (A_3, 1) \rightarrow 3$

3   Modelling of Hiding Classification Rules

For the brevity of expressions, from now on when we refer to the impact of data modification, we always refers to the impact on the data quality in term of the classification rules on a data set, in particular the data set after removal of some records.

Also, for each rule r_l where $l \in L$, we define its **supporting set** D_l as $\{d^i \in D | d^i$ satisfies $r_l\}$. According to Definition 2, we have $|D_l| = Sup(r_l)$. Because D is non-duplication data set, there is no class-conflict between records (i.e., same attribute values, but different class labels). Also, in our problem, every classification rule has 100% confidence. Therefore, for each rule r_l, D_l can also be defined as the set of records in D that satisfy all literals in the LHS of r_l.

Example 1. Considering rule 1 in Table 2, we have its supporting set $D_1 = \{d^i | d^i(A_1) = 0\} = \{d^1, d^2, d^3, d^4\}$.

In this paper, we illustrate the impact based on a geometric model. Note that the geometric model is only applied to facilitate our discussion. Essentially our proposal on hiding the classification rules is not necessary to be presented based the geometric model. However, applying the geometric model can help to explain some key concepts / observations more clearly.

In the running example, we have three attributes A_1, A_2 and A_3. Hence, we can represent the data by a 3-dimensional geometric model.

In the 3-dimensional geometric model, a data record is represented as a point. The rules which correspond to the labelled subsets of D can be represented as one of the following categories: points, lines or faces. Generally, a face in the model can represent a one-literal-rule, on the other hand, a point can represent a k-literal-rule where k is the cardinality of $\mathbf{A}$.

For our example data set, we have 8 records and 3 attributes. Hence we have 8 points at the corner of the 3-dimension cube as shown in the Figure 1a). For the set of rules in Table 2, we can represent them in geometric model in Figure 1b), where rule r_1 is represented as a face with supporting set $\{d^1, d^2, d^3, d^4\}$, rule r_2 and r_3 as lines with $\{d^3, d^7\}$ and $\{d^2, d^6\}$ as the supporting set respectively. Finally, rule r_4 and r_5 are represented as points with $\{d^5\}$ and $\{d^8\}$ as supporting set respectively. Based on the model, it can be seen that the data set is partitioned into non-overlapping sets based on class labels. While, for the rules with the same class label, their supporting sets may be overlapped.

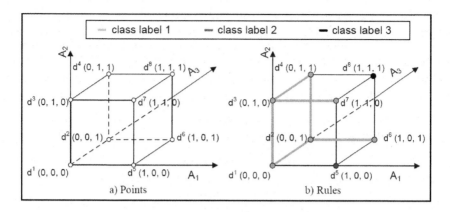

Fig. 1. The geometric model

Observation 1. *Classification deliver a partition on data set.*

Example 2. In our running example, we can deliver a partition of three sets $\{d^1, d^2, d^3, d^4, d^6, d^7\}$, $\{d^5\}$ and $\{d^8\}$.

Observation 2. *Each class is associated with a set of rules.*

Example 3. From the example, class 1 is associated with set of rules r_1, r_2, and r_3. The class 2 and 3 are associated with one-literal set of rules r_4 and r_5 respectively.

In general, we have different representations of rules for one classification. For example, considering the data records with class label 1 in Table 1, we can either represent them as a set of one face-rule and two line-rules as in Table 2 (r_1, r_2 and r_3), or, alternatively, represent them as a set of six point-rules which are $\{(A_1, 0) \wedge (A_2, 0) \wedge (A_3, 0) \rightarrow 1, (A_1, 0) \wedge (A_2, 0) \wedge (A_3, 1) \rightarrow 1, (A_1, 0) \wedge (A_2, 1) \wedge (A_3, 0) \rightarrow 1, (A_1, 0) \wedge (A_2, 1) \wedge (A_3, 1) \rightarrow 1, (A_1, 1) \wedge (A_2, 1) \wedge (A_3, 0) \rightarrow 1$ and $(A_1, 1) \wedge (A_2, 0) \wedge (A_3, 1) \rightarrow 1\}$.

Table 3. The example data set with only 3 records

I	A_1	A_2	A_3	C
1	0	0	0	1
2	0	0	1	1
3	0	1	0	1

On the other case, when dealing with non-complete data set, for example, the 3-record data set in Table 3, the set of rules can also be represented in many different ways as shown in Table 4 (note that every listed rule has 100% confidence).

First, we can represent it as a set of single most general rule shown in Table 4a), or shown as a face-rule in the geometric model in Figure 2a). However, this representation is not **precise** because a record $(0, 1, 1)$ with unknown class label is also covered by r_1. The second option is to choose the set of line-rules as shown in Table 4b) and Figure 2b). The third representation is using the detail set of point-rules, which is shown in Table 4c) and Figure 2c).

Table 4. The example classification rules from 3-record data set

Rule No. (L)	Content
1.	$(A_1, 0) \rightarrow 1$

a) The set of one-literal rule

Rule No. (L)	Content
1.	$(A_1, 0) \wedge (A_3, 0) \rightarrow 1$
2.	$(A_1, 0) \wedge (A_2, 0) \rightarrow 1$

b) The set of two-literals rules

Rule No. (L)	Content
1.	$(A_1, 0) \wedge (A_2, 0) \wedge (A_3, 0) \rightarrow 1$
2.	$(A_1, 0) \wedge (A_2, 0) \wedge (A_3, 1) \rightarrow 1$
3.	$(A_1, 0) \wedge (A_2, 1) \wedge (A_3, 0) \rightarrow 1$

c) The set of three-literals rules

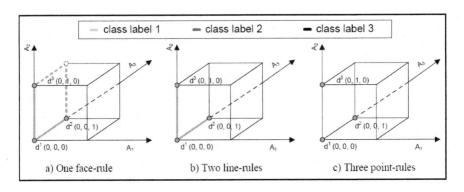

Fig. 2. The example classification rules from 3-record data set in the geometric model

Obviously, we need the regulation on the form of classification rules. Here, we introduce a concept of canonical form in Definition 3. Intuitively, the canonical form requires a rule to be "precise" and "as general as possible".

Definition 3 (Canonical Form). *Given a data set D and a classification rule r_l, let $D_{rhs} \subseteq D$ be the set of points (records) whose class label is $r_l.RHS$. In addition, let $\overline{D_l}$ be the set of points in the geometric model that satisfy all the literals in $r_l.LHS$ (note that $\overline{D_l}$ is not necessary a subset of D). The rule r_l is in canonical form if 1) $\overline{D_l} \subseteq D_{rhs}$, 2) and there does not exist a more general rule $r_{l'}$ such that $r_{l'}$ is derived by removing a literal from r_l and $\overline{D_l} \subset \overline{D_{l'}} \subseteq D_{rhs}$.*

Example 4. If we re-consider the data set in Table 3 and its derived sets of rules in Table 4, it can be seen that only the set of rules in Table 4b) are in canonical form. We can see from the geometric model in Figure 2b) that this set of rules has all points it covers. While the rule r_1 from Table 4a) fails to satisfy the first condition because the point $(0, 1, 1)$ is in $\overline{D_1}$, but does not have the class label 1. On the other hand, the set of rules in Table 4c) fails to satisfy the second condition because there exists more general rules satisfying the first condition.

Based on the concept of canonical form, we have the following two observations.

Observation 3. *For any given data set D, we can derive a unique set of canonical form classification rules in D.*

Observation 4. *First, for a non-canonical form rule which only violates the condition 1 in Definition 3, we call it **imprecise** rule. Based on simple facts from set theory and topology, any imprecise rule can be transformed into a set of canonical form rules.*

Example 5. From the data set in Table 3, we can transform the imprecise rule "$(A_1, 0) \rightarrow 1$" into the set of canonical form rules $\{(A_1, 0) \wedge (A_2, 0) \rightarrow 1, (A_1, 0) \wedge (A_3, 0) \rightarrow 1\}$. In the geometric model, this can be seen as the transformation from Figure 2a) to Figure 2b).

Now let us consider the data reduction process for hiding classification rules. When a data set is given, the set of unique canonical form classification rules can be derived. Suppose that a record is removed, aside from the support values of the sensitive rules are decreased, some rules could be affected by the removal and become non-canonical form rules. However, based on the Observation 4, we are able to transform each affected rule (which is the imprecise rule) into a set of canonical form rules. It means that the impact of any removal can be determined precisely. Also, the transformation process can be done without any re-computing. Based on this fact, rule hiding algorithms can greatly benefit from the impact determination. In the next section, the algorithmic steps will be presented to demonstrate how to hide the sensitive classification rules.

4   Hiding Sensitive Classification Rules

In this section, we propose the algorithms to solve the problem below.

Problem 1. Given a threshold *minsup* and a data set D with a set of class labels C, let R be the set of canonical form classification rules from D and for any rule $r \in R$, $Sup(r) > minsup$. In addition, let $R_s \subset R$ be a set of sensitive classification rules. The problem is to transform D into D' by removal of some records from D such that 1) any rule $r_s \in R_s$ is invalid in D' in terms of the threshold *minsup* and 2) the impact of removal on other rules is minimized.

Based on the discussions in Section 3, we hide sensitive classification rules by judiciously removing the records (points) in the data set. Importantly, according to Definition 3, the impact of a removal on the canonical form classification rules can be easily tracked. So, after each removal, we could transform every affected rule into the set of new canonical form rules without any re-computing on the data set. As a result, when the hiding process completes, besides the result data set D', we can also output the canonical form classification rules in D'.

Before giving the algorithm, let us clarify some basic issues for the classification rule hiding. First, considering a single sensitive rule r_s, r_s is hidden if $Sup(r_s)$ drops below the threshold $minsup$. Let D_s be the set of points that support the rule r_s. Clearly, we need to remove $Sup(r_s) - minsup + 1$ points from D_s. Now, we discuss the impact of removing a single points $d^i \in D_s$ on the set of classification rules. Obviously, the support of r_s decreases by 1. Considering others rules $r \in R$ and $r \neq r_s$, according to the assumption of our problem, only the rule r with $r.RHS = r_s.RHS$ is possible to be affected. For any such rule r_a, if the point d^i supports r_a, i.e., $d^i \in D_a$, r_a will be affected and it needs to be transformed into a set of canonical form classification rules, as mentioned in Observation 4. For the sensitive rule r_s itself, we could also re-write it into a set of canonical form classification rules. However, it is unnecessary because for any canonical form rule generated by r_s, eventually its support must be less than $minsup$.

Now, we propose two demonstrative algorithms for the problem of hiding classification rules. Algorithm 1 is the main algorithm, giving the process of hiding a set of sensitive classification rules with the concern of minimizing the side-effect on other rules. Algorithm 2 performs an important task of hiding process, which is to transform a set of affected classification rules into canonical form after a point is removed.

Figure 3 shows the pseudo code of Algorithm 1. For any sensitive rule r_s, we first find the set D_s of points that support r_s and the set $R_{possible}$ of possible affected rules. The key step is to find a set of $(Sup(r_s) - minsup + 1)$ points with minimal impact on other rules. To do this, we define the *Impact* of removing a point d^i as the number of affected rules, i.e., $Impact(d^i) = |\{r_a \in R_{possible} | d^i \in D_a\}|$. Note that a rule r_a is affected by removing d^i if d^i satisfies all the literals in $r_a.LHS$ (it is unnecessary to compute D_a). We select the first $(Sup(r_s) - minsup + 1)$ points with minimal impact to be removed. For each point d^i in the selected set, if the set $R_{affected}$ of affected rules of d^i is not empty, after removing d^i, we need to update the set R of current valid rules by transforming $R_{affected}$ into the canonical form R_{can}. Finally, for each sensitive rule, we update the data set once after selecting all points to be removed.

In Figure 4, Algorithm 2 gives the procedure of transforming affected rules into canonical form. According to Definition 3, after a point $(d_1^i, d_2^i, \cdots, d_k^i)$ is removed, the affected rule r_a needs to be decomposed into less general rules, as shown in Example 5. It is not difficult to see that after decomposing r_a, the left

---

Algorithm 1
Input:
D: a data set
$minsup$: a support threshold
R: the set of canonical form classification rules in D (satisfying $minsup$)
$R_{sensitive}$: the set of sensitive classification rules, $R_{sensitive} \subset R$
Output:
D': the output data set, from which $R_{sensitive}$ can not be derived
R': the set of canonical form classification rules in D'
Method:
for each $r_s \in R_{sensitive}$ **do**
  $R = R - \{r_s\}$;
  Find D_s;
  $R_{possible} = \{r \in R | r.RHS = r_s.RHS\}$;
  Sort D_s in ascending order by their $Impact(d^i), d^i \in D_s$
  where $Impact(d^i) = \left| \{r_a \in R_{possible} | d^i \in D_a\} \right|$;
  /* select points with minimum impact on other rules */
  Select first $(Sup(r_s) - minsup + 1)$ points from D_s,
  remove them from D, put them into D_Remove
  for each $d^i \in D\_Remove$ **do**
    $R_{affected} = \{r_a \in R_{possible} | d^i \in D_a$ where $d^i \in D\_Remove\}$;
    if $(R_{affected} \neq \phi)$ /* Update rules */
    $R_{rest} = R - R_{affected}$;
    Transform $R_{affected}$ into canonical form R_{can};/* Algorithm 2 */
    $R = R_{rest} \cup R_{can}$;
  Update data set D;
Output $D' = D, R' = R$;

Fig. 3. Demonstrative data reduction algorithm

hand side of every derived canonical form rule should be $r_a.LHS$ plus *one* literal on the attribute which is not included in $r_a.LHS$. Therefore, in the algorithm, we first find the set S_A of the attributes that do not appear in $r_a.LHS$. For each such attribute A_j, we discuss the corresponding canonical form rule(s) as below. First, given a canonical form rule r_a with support $Sup(r_a)$ and an attribute A_j, let $r_{new} = r_a.LHS \wedge p \rightarrow r_a.RHS$ be a new generated canonical form rule, where p is a literal on A_j. From the geometric model, we can easily see that the support of r_{new} must be $\frac{Sup(r_a)}{|dom(A_j)|}$, where $|dom(A_j)|$ is the cardinality of attribute A_j. This means that for any new canonical form rule formed by adding a literal on attribute A_j to $r_a.LHS$, its support is a fixed value $\frac{Sup(r_a)}{|dom(A_j)|}$, denoted as Sup. If Sup is less than the threshold $minsup$, we do not need to consider the attribute A_j. Otherwise, for each value v in $dom(A_j)$ except for d_j^i, we form a literal $p = (A_j, v)$ and generate the canonical form rule $r_a.LHS \wedge p \rightarrow r_a.RHS$. After the canonical form rules are generated for all affected rules, we need to remove redundant rules from the result set R_{can}. A rule r_r in R_{can} is redundant if there exists another rule $r_c \in R_{can}$ which is more general than r_r, meaning that $r_r.LHS$ contains $r_c.LHS$.

Algorithm 2
Input:
$d^i = (d_1^i, d_2^i, \cdots, d_k^i)$: a point to be removed
$R_{affected}$: a set of affected canonical form rules
$minsup$: a support threshold
Output:
R_{can}: the canonical form of $R_{affected}$ after removing d^i
Method:
$R_{can} = \phi$;
for each $r_a \in R_{affected}$ **do**
$\quad S_A = \{A_j | \ (A_j, d_j^i) \text{ not in } r_a.LHS \}, j = 1 \cdots k$;
$\quad$ **for each** $A_j \in S_A$
$\quad\quad Sup = \frac{Sup(r_a)}{|dom(A_j)|}$;
$\quad$ **if** $(Sup \geq minsup)$
$\quad\quad$ **for each** $v \in dom(A_j)$ and $v \neq d_j^i$
$\quad\quad\quad$ /* Generate new rule r_{new} */
$\quad\quad\quad r_{new}.LHS = r_a.LHS \wedge (A_j, v)$;
$\quad\quad\quad r_{new}.RHS = r_a.RHS$;
$\quad\quad\quad Sup(r_{new}) = Sup$;
$\quad\quad\quad R_{can} = R_{can} \cup \{r_{new}\}$;
/* Remove redundant rules */
for each $r_c \in R_{can}$ **do**
$\quad R_r = \{r_r \in R_{can} | r_r.LHS \text{ contains } r_c.LHS\}$;
$\quad R_{can} = R_{can} - R_r$;
Output R_{can};

Fig. 4. Transform affected rules into canonical form

Table 5. The example classification rules with $minsup = 2$

Rule No. (L)	Content	Support value
1.	$(A_1, 0) \rightarrow 1$	4
2.	$(A_2, 1) \wedge (A_3, 0) \rightarrow 1$	2
3.	$(A_2, 0) \wedge (A_3, 1) \rightarrow 1$	2

Example 6. Consider the data set in Table 1, suppose that the $minsup$ is set at 2, the set of rules shown in Table 5 will be shown to the data owner. Suppose that the data owner decides to hide rule r_1, $R_{sensitive} = \{r_1\}$, D_s is $\{d^1, d^2, d^3, d^4\}$. According to the $minsup$, we need to remove three points from the data set. The first two out of the three points are d^1 and d^4. Both of the removals have zero impact. For the last point, either d^2 or d^3 can be chosen because both options have the same impact ($impact(d^2) = impact(d^3) = 1$). Suppose that d^3 is chosen. Subsequently, rule r_2 becomes the affected rule, and the transformation process in the Algorithm 2 is needed. Because the support of the new canonical form rule is less than $minsup$, we do not need to generate such the rule. Note here that,

if we do not consider *minsup* value, the new rule, $(A_1 = 1) \land (A_2 = 1) \land (A_3 = 0) \rightarrow 1$, can be derived from the data set with support value 1. The output data set, after the reduction process, $D' = \{d^2, d^5, d^6, d^7, d^8\}$. The output set of canonical form rule R' is $(A_2, 0) \land (A_3, 1) \rightarrow 1$, its support value is 2.

5 Conclusion

In this paper, we address the problem of hiding sensitive classification rules by data reduction. We consider the data reduction as a reasonable way to deal with this problem. A key concept, called canonical form classification rule, is introduced into the hiding problem. Also, we demonstrate how to select the data record to be removed with potential minimal impact on other rules. Importantly, the impact of removal can be presented precisely without any re-computing on the data set. Finally, the algorithmic steps to hide the rules are presented.

In the future work, we will focus on the problem which duplication in the data set is allowed. The richer domain attribute such as [0,1] will be considered. Also, the other kind of impacts which can be applicable to our proposed way e.g. number of non-sensitive lost rules (due to the supports fall below the minimal support) or number of new rules which do not exist in the original data set [6], will be considered.

References

1. Verykios, V.S., Bertino, E., Fovino, I.N., Provenza, L.P., Saygin, Y., Theodoridis, Y.: State-of-the-art in privacy preserving data mining. SIGMOD Rec. 33(1), 50–57 (2004)
2. Evfimievski, A., Srikant, R., Agarwal, R., Gehrke, J.: Privacy preserving mining of association rules. Inf. Syst. 29(4), 343–364 (2004)
3. Zhang, N., Wang, S., Zhao, W.: A new scheme on privacy-preserving data classification. In: KDD '05. Proceeding of the eleventh ACM SIGKDD international conference on Knowledge discovery in data mining, pp. 374–383. ACM Press, New York (2005)
4. Vaidya, J., Clifton, C.: Privacy preserving association rule mining in vertically partitioned data. In: Proceedings of the eighth ACM SIGKDD international conference on Knowledge discovery and data mining, pp. 639–644. ACM Press, New York (2002)
5. Kantarcioglu, M., Clifton, C.: Privacy-preserving distributed mining of association rules on horizontally partitioned data. IEEE Transactions on Data and Knowledge Engineering 16(9), 1026–1037 (2004)
6. Verykios, V.S., Elmagarmid, A.K., Bertino, E., Saygin, Y., Dasseni, E.: Association rule hiding. IEEE Trans. Knowl. Data Eng. 16(4), 434–447 (2004)
7. Oliveira, S.R.M., Zaïane, O.R.: Protecting sensitive knowledge by data sanitization. In: Proceedings of the 3rd IEEE ICDM International Conference on Data Mining, pp. 613–616. IEEE Computer Society Press, Los Alamitos (2003)
8. Sun, X., Yu, P.S.: A border-based approach for hiding sensitive frequent itemsets. In: Proceedings of the 5th IEEE ICDM International Conference on Data Mining, pp. 426–433. IEEE Computer Society Press, Los Alamitos (2005)

9. Li, W., Han, J., Pei, J.: Cmar: Accurate and efficient classification based on multiple class-association rules. In: Proceedings of the 2001 IEEE ICDM International Conference on Data Mining, Washington, DC, pp. 369–376. IEEE Computer Society Press, Los Alamitos (2001)
10. Liu, B., Hsu, W., Ma, Y.: Integrating classification and association rule mining. In: Proceedings of the fourth ACM SIGKDD international conference on Knowledge discovery and data mining, pp. 80–86. AAAI Press, New York (1998)

AOG-ags Algorithms and Applications*

Lizhen Wang[1,2], Junli Lu[1], Joan Lu[2], and Jim Yip[2]

[1] Department of Computer Science and Engineering, School of Information, Yunnan University, Kunming, 650091, P.R. China
Lzhwang@ynu.edu.cn
[2] Department of Informatics, School of Computing and Engineering, University of Huddersfield, Huddersfield, UK, HD1 3DH

Abstract. The attribute-oriented generalization (AOG for short) method is one of the most important data mining methods. In this paper, a reasonable approach of AOG (AOG-ags, attribute-oriented generalization based on attributes' generalization sequence), which expands the traditional AOG method efficiently, is proposed. By introducing equivalence partition trees, an optimization algorithm of the AOG-ags is devised. Defining interestingness of attributes' generalization sequences, the selection problem of attributes' generalization sequences is solved. Extensive experimental results show that the AOG-ags are useful and efficient. Particularly, by using the AOG-ags algorithm in a plant distributing dataset, some distributing rules for the species of plants in an area are found interesting.

Keywords: Attribute-oriented generalization (AOG); Concept hierarchy trees; Attributes' generalization sequences (AGS); Equivalence partition trees; Interestingness of AGS.

1 Introduction

The general idea of Attribute-oriented generalization (AOG) is to abstract each attribute of each record in a relation from a relatively low conceptual level to higher conceptual levels by using domain knowledge(Concept Hierarchy Trees), in order to discover rules among attributes from multilevel or higher lever. Two AOG algorithms are well-known: AOI [1] and LCHR [2]. Both are not incremental and also don't allow fast re-generalization. An AOG method possessing fast re-generalization was proposed in literature [3]. However, it is not perfect in efficiency and occupies too much memory space.

C. L. Carter and H. J. Hamilton proposed two new AOG algorithms (the GDBR and the FIGR) in literature [4]. The GDBR is an online algorithm, while the FIGR has characteristics of incremental and fast re-generalization. One important thing is that the runtime of the GDBR and the FIGR is less than the AOI and the LCHR. But there is a supposition in the FIGR. That is that the size of attributes and the number of the possible values in an attribute are relatively small. In addition, the four algorithms control generalization levels by using attribute thresholds. That is not so practical in

---

* Supported by the National Natural Science Foundation of China under Grant No.60463004.

R. Alhajj et al. (Eds.): ADMA 2007, LNAI 4632, pp. 323–334, 2007.

some applications, because it is impossible to try every possible combination of thresholds for every attribute, and the size of attributes and the number of the possible values of an attribute are not small in some practical applications.

Considering the generalization threshold, besides the attribute thresholds, Chen [5] and Han [6] discussed the degree of tuple (record) generalization (i.e. tuple thresholds), and the problem of combining the attribute thresholds and the tuple thresholds. Under the tuple thresholds, it is convenient to control the process of AOG. But the low-efficiency of AOG algorithms becomes the main problem in its applications, for we have to consider all combinations of attributes which satisfy the tuple threshold in an attribute-oriented generalization process, and explain the results of generalization (or rank them for user).

In this paper, by introducing the concept of the attributes' generalization sequence, the attribute thresholds and the tuple thresholds are unified, and a reasonable approach of AOG—AOG-ags (Attribute-Oriented Generalization based on Attributes' Generalization Sequences), which expands the traditional AOG efficiently, is proposed. Some technologies, for example, partitions, equivalence partition trees, prune optimization strategies and interestingness, are used to improve the efficiency of the algorithm. It is shown that the AOG-ags algorithm has special advantages.

The rest of the paper is organized as following. Section 2 formally defines the concept of the degree of tuple generalization, and introduces the method of AOG based on attributes' generalization sequence (AOG-ags). In Section 3, by introducing the equivalence partition trees, an optimization algorithm of AOG-ags is devised. Interestingness of attributes' generalization sequences is discussed in Sect. 4. Section 5 discusses correctness, completeness and complexity of our algorithms. Performance and application results of algorithms are evaluated in Sect. 6. The last Section is conclusions.

2 AOG-ags

In traditional AOG algorithms, the generalization process is controlled by setting a threshold for each attribute. But in some applications, user does not want to consider each attribute for generalization threshold, so the degree of tuple generalization is introduced.

Definition 1. Given a relation $r(r_1,...,r_n)$ and the generalization relation $r'(r_1',\cdots,r_n')$, then the rate of reserved tuples is defined as $\overline{Z}=n'/n$, so the degree of tuple generalization is defined as $Z=1-\overline{Z}=1-(n'/n)$.

Z is a measure for the degree of tuple generalization. The higher is the value of Z, the greater the degree of generalization. The value Z meets $0 \le Z \le (n-1)/n$.

Z cannot confirm certain generalization result. That is to say, given a tuple threshold Z, we will get some generalization relations that satisfy this threshold Z. But analyzing the process of AOG, we find that generalization for each attribute is independent, that is, an attribute is generalized earlier or latter will not affect the generalization result. Further to say, generalization result is the same no matter that it is obtained by generalizing gradually or directly up to the k-th level, so attributes' generalization sequences (AGS for short)is introduced in this research. One AGS confirms certain generalization relation.

The Depth of certain node V in the tree is defined as the path length from root to V, the Height of V is the length of the longest path in the tree whose root is V. The Height of the tree is the Height of its root. The Level of V is its Depth more 1.

Definition 2. Given a relation pattern $R(A_1, \cdots, A_m)$, attributes' concept hierarchy trees $h_1, \cdots, h_m$, the Heights of trees $l_1, \cdots, l_m$, sequence $A_1^{g_1} \cdots A_i^{g_i} \cdots A_m^{g_m}$ is called an AGS of AOG, where ($1 \le g_i \le l_i + 1$).

Property 1. The number of all AGS in a relation pattern is $\prod_{i=1}^{m}(l_i + 1)$。

    Proof. $\because$ One sequence $g_1 \cdots g_i \cdots g_m$ can only confirm one AGS $A_1^{g_1} \cdots A_i^{g_i} \cdots A_m^{g_m}$.

    Meanwhile, $\because 1 \le g_i \le l_i + 1$ $\therefore |g_i| = l_i + 1$ $(1 \le i \le m)$

    $\therefore$ The number of attributes' generalization sequences is:

$$(l_1 + 1) \times \cdots \times (l_i + 1) \times \cdots \times (l_m + 1) = \prod_{i=1}^{m}(l_i + 1) \qquad \square$$

Definition 3. Given the tuple threshold Z. If the generalization relation $r'(r_1', \cdots r_n')$ which are confirmed by the AGS $A_1^{g_1} \cdots A_i^{g_i} \cdots A_m^{g_m} (1 \le g_i \le l_i + 1)$ satisfies $1 - (n'/n) \ge Z$, and if increasing any $g_i (1 \le i \le m)$, it will not satisfy $1 - (n'/n) \ge Z$, then $A_1^{g_1} \cdots A_m^{g_m}$ is called an AGS which satisfies the Z, and $r'(r_1', \cdots r_n')$ is called a generalization result under $A_1^{g_1} \cdots A_m^{g_m}$.

From Definition 3, we can conclude that the AOG method of using attribute thresholds is a special case of using the tuple thresholds. That is to say, the attribute thresholds and the tuple thresholds are unified under the concept of AGS. The AOG based on AGS (for short, call it AOG-ags) is an efficient extension to the traditional approach.

An ordinary AOG-ags algorithm is devised as follows.

```
Algorithm 1: The ordinary AOG-ags algorithm
Input: The un-generalized dataset (relation) r, which has m
attributes{ A₁,A₂….Aₘ },Attributes' concept hierarchy
```
trees $\{h_1, \cdots, h_m\}$ and the height of trees $\{l_1, \cdots, l_m\}$,The tuple
```
threshold Z
Output: Generalization rules which meet the Z
Algorithm description▯
1) Gen_seq(relation,1,m,L₁,S,Gs);
2) Selecting a sequence from the set Gs of AGS and returning
a generalization relation;
3) Producing generalization rules from the generalization re-
lation.
```

```
Procedure Gen_seq(r,i,m,Li,S,Gs);
(1)    For k=Li+1 downto 1 Do
(2)        Begin If k< Li+1 then
(3)               Gen_r ← generalize (r,i,k)
(4)           Else Gen_r ← r    Endif;
(5)           If i<m then
(6)               Gen_seq (Gen_r,i+1,m,Li+1,S∪ Aki,Gs)
(7)           Else
(8)             If |Gen-r|≤ n(1-Z) then
(9)                 Gs ← Gs∪ {S∪Aki}
                  endif
                Endif
        End
```

When the number of attributes (m) is larger, in order to obtain all attributes' generalization sequences which meet the Z in algorithm, we must search $\prod_{i=1}^{m}(l_i+1)$ times, and it will waste much time. So, how to efficiently compute all AGS which meet the Z is the chief problem in this algorithm. Further more, how to quickly calculate generalization relations that are related to AGS is another solving problem. To solve these problems efficiently, an optimization AOG-ags algorithm is presented by introducing equivalence partition trees according to the property of AGS.

3 An Optimization AOG-ags Algorithm

3.1 AOG-ags and Partition

Let $r(r_1,\cdots,r_n)$ is a relation in relation pattern $R(A_1,\cdots,A_m)$, $X \subseteq R$, $\forall r_i, r_j \in r$, r_i and r_j are equal with respect to X if and only if $r_i[A_k]=r_j[A_k]$ for $\forall A_k \in X$, which is denoted as $r_i =_x r_j$. X partitions the rows of r into equivalence classes. We can denote the equivalence class of $r_i \in r$ with respect to $X \subseteq R$ by $[r_i]_{=x}$. The quotient set $\pi_x = \{[r_i]_{=x} \mid r_i \in r\}$ of equivalence classes is an equivalence partition of r under X.

Given two partitions $\pi_x = \{\tau_1,\cdots,\tau_k\}$, $\pi_y = \{\tau_1',\cdots,\tau_{k'}'\}$, $\pi_x \otimes \pi_y = \{\tau_i \cap \tau_j' \mid \tau_i \in \pi_x, \tau_j' \in \pi_y\}$ is called intersection partition of π_x and π_y, $\pi_x \otimes \pi_y$ is a partition of r. We know that $\pi_{X \cup Y} = \pi_x \otimes \pi_y$ holds.

In fact, the equivalence class $\pi_R = \pi_{(A_1,\cdots A_m)}$ is the relation r. According to the property of $\pi_{X \cup Y} = \pi_x \otimes \pi_y$, there is a one-one correspondence from the records of r to equivalence classes of $\pi_{(A_1)} \otimes \cdots \otimes \pi_{(A_m)}$. If $\pi_{A_i,j} (1 \le i \le m, 1 \le j \le l_i+1)$ denotes equivalence partition which attribute A_i generalizes up to the j-th level along with the concept hierarchy tree, then the generalization relation and AGS $A_1^{g_1} \cdots A_m^{g_m}$ is corresponding one by one. This leads to a new partition-based approach of AOG-ags:

(1) Compute all $\pi_{A_i, g_i}(1 \le i \le m, 1 \le g_i \le l_i + 1)$.

(2) Obtain all AGS which meet the Z.

(3) Select a sequence $A_1^{g_1} \cdots A_m^{g_m}$, and then calculate generalization relation $r' = \pi_{A_1 g_1} \otimes \cdots \otimes \pi_{A_m g_m}$.

(4) Produce generalization rules from the generalization relation.

3.2 Pruning Strategies

Definition 4. The Grid that is constituted by $\prod_{i=1}^{n}(l_i + 1)$ possible AGS and satisfies the following properties is called the search space.

(1) There are AGS that satisfy $g_1 + \cdots + g_m = k + 1$ in the k-th level.

(2) Each sequence is connected to any sequence $A_1^{g_1} \cdots A_i^{g_i - 1} \cdots A_m^{g_m}$ of the (k-1)-th level.

The search space will increase with the increasing of m and l_i. By introducing the concept "refinement", some pruning is executed for reducing search space.

Definition 5. Given a relation and its two partitions $\pi_x = \{\tau_1, \cdots, \tau_k\}$, $\pi_y = \{\tau_1', \cdots, \tau_{k'}'\}$, if $\forall \tau_i \in \pi_x$, $\exists \tau_j' \in \pi_Y$, $\tau_i \subseteq \tau_j'$ holds, then π_x is called as a refinement of π_Y .

Obviously, $\pi_x \otimes \pi_y$ is the refinement of π_x and π_Y , and $\pi_{A,j}$ refines $\pi_{A,k}$, $1 \le i \le m, 1 \le k < j \le l_i + 1$. If π_x refines π_Y, then $|\pi_x| \ge |\pi_Y|$ holds.

Definition 6. Given two sequences $A = A_1^{g_1} \cdots A_m^{g_m}$ and $A' = A_1^{g_1'} \cdots A_m^{g_m'}$, if $\forall A_i$, $g_i \ge g_i'(1 \le i \le m)$ holds, then A is called a sub-sequence of A', denote as $A = sq(A')$, and A' is the parent-sequence of A , denote as $A' = fq(A)$.

If $A = sq(A')$, then π_A refines $\pi_{A'}$. Therefore, what the pruning strategies we can get are the followings.

(1) If there exists a π_{A_i, g_i} , and $|\pi_{A_i, g_i}| > n(1 - Z)$ holds, then any sequence which includes $A_i^{g_i}$ or its sub-sequence $A_i^{g_k}$ ($g_k \ge g_i$) cannot meet the Z.

(2) If there is a sequence $A_i^{g_i} \cdots A_j^{g_j}$, and $|\pi_{A_i, g_i} \cup \cdots \cup \pi_{A_j, g_j}| > n(1 - Z)$ holds, then any sequence which includes $A_i^{g_i} \cdots A_j^{g_j}$ or its sub-sequence may not meet the Z.

(3) If a sequence $A = A_1^{g_1} \cdots A_m^{g_m}$ meets the Z, then all parent-sequences of A will not meet the Z, so it can be pruned.

3.3 Calculating π_{A_i, g_i}

By introducing the concept of equivalence partition trees, π_{A_i, g_i} can be calculate efficiently. At first, we assign each node (i.e., concept) in a concept hierarchy tree to a concept's code. It is shown as the numbers of bracket in Fig. 1. The unary concept's codes represent the first level in the concept hierarchy tree (i.e., the root of the tree),

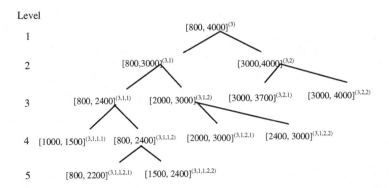

Level

Fig. 1. A concept hierarchy tree of an attribute "elevation" and its concept's codes

("3" is the code of the root, which only represents the difference from other attributes). The binary codes represent the second level, etc.

Definition 7. The equivalence partition tree of the attribute A, each branch is a concrete value of A, which is concept's code with respect to the value in the concept hierarchy tree, each node in the tree is a value in concept's code.

The equivalence partition tree of the attribute "elevation", which values are from the Table 1 in Section 6.2, is showed as Fig. 2.

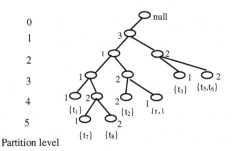

Fig. 2. The equivalence partition tree of the attribute "elevation" in Table 1

An equivalence partition tree can be constructed by the following steps: (1) create the root of the tree, labeled with "null". (2) for each value of the attribute, a branch is created. (3) note down the corresponding Row-Id under the corresponding leaf node. The set of the Row-Id noted down is called the identity of leaf-node.

The partition level of an equivalence partition tree is defined as follows. The root node "null" is the level 0, the following is level 1, ..., till the leaf nodes. The identity of a node in an equivalence partition tree is the union of the identity of all leaf-nodes of the sub-trees with this node as their root. The set of equivalence partition class with respect to attribute A at level l is the set of the identity of nodes of the l-th partition level in equivalence partition tree with respect to the attribute A.

Considering the equivalence partition tree in Figure 3, the partition result in the third level is $\{\{t_1,t_7,t_8\}, \{t_2,t_4\}, \{t_3\}, \{t_5,t_6\}\}$, which is the same as the equivalence partition result after "elevation" is generalized to the third level.

3.4   An Optimization AOG-ags Algorithm

Introducing partition and refinement, we can efficiently speed the AOG process up. Introducing equivalence partition trees, we can quickly obtain equivalence partition results of attributes in any level. So we have an optimization algorithm of AOG-ags as below:

```
Algorithm 2: An optimization algorithm of AOG-ags
Input: The un-generalized dataset r, which has m attrib-
```
utes $\{A_1, A_2...A_m\}$; Attribute's concept hierarchy tree $\{h_1,\cdots,h_m\}$ and their heights $\{l_1,\cdots,l_m\}$; The tuple threshold Z
```
Output: generalization rules which meet the Z
Algorithm description:
1) creation_partition_tree (r);
2) computing lower bound L(A) which attribute A is
generalized;
```
3) Gen$(\pi_{A_i,l(A_i)}, 1, m, L(A_i), S, Gs)$; //the initial values are S="null", Gs=Φ. Obtain all AGS (Gs) which meet the Z
4) Selecting a generalization sequence $A_1^{g_1}\cdots A_m^{g_m}$ from Gs, computing generalization relation $r' = \pi_{A_1g_1} \otimes \cdots \otimes \pi_{A_mg_m}$;
```
5) Producing generalization rules from the generaliza-
tion relation.

      Procedure Gen(r,i,m,L(A),S,Gs);
      (1)     For k= L(A) downto 1 Do

      (2)         Begin  If i=1 and k < L(A)    then
```
      (3)   Gen_r $\leftarrow \pi_{A_i,k}$
```
      (4)                 Else  If i=1 and k= L(A) then
      (5)                         Gen_r ← r
```
      (6)   Else Gen_r $\leftarrow r \otimes \pi_{A_i,k}$ Endif
```
                    Endif;
      (7)             if | Gen_r| > n(1-Z) then
      (8)                exit for
                    endif
      (9)             If i<m then
```
      (10)   Gen (Gen_r,i+1,m,L(A_{i+1}),S$\cup$ A$_i^k$,Gs)
```
      (11)           Else  If |Gen-r|≤ n(1-Z) then
```
      (12)   Gs ← Gs$\cup$ {S$\cup$ A$_i^k$};
```
      (13)                   Exit for
                        Endif
                    Endif
              End
```

In the optimization algorithm of AOG-ags, we efficiently control recursive times by computing each attribute's lower limit L(A$_i$), and consider whether AGS can be pruned or not in a recursive process according to Pruning Strategies. In fact, many recursive steps will be jumped over using pruning strategies in algorithm 2.

4 Interestingness of AGS

Motivation example: For the plant "Magnolia sieboldii" in a plant distributing dataset, suppose the following rules have been obtained.

(1) Plant "Magnolia sieboldii" ⇒ 50% grows in the conifer forest and scrub whose elevation is from 2600 to 4100 meter of Lijiang, and 50% grows in the forest, scrub and meadow whose elevation is from 2400 to 3900 meter of Weixi.

(2) Plant "Magnolia sieboldii" ⇒ 90% grows in the conifer forest and scrub whose elevation is from 2600 to 4100 meter of Lijiang, and 10% grows in the forest, scrub and meadow whose elevation is from 2400 to 3900 meter of Weixi.

The rule (2) is more meaningful than the rule (1), because the growth characteristics of plant "Magnolia sieboldii" are more obvious in the rule (2).

Definition 8. In generalization relation, the t-weight of the ith generalization record t_i is defined as formula (1).

$$t_i = \frac{count\,(i)}{\sum_{j=1}^{n'} count\,(j)} \tag{1}$$

In formula (1), count (i) is the number of repeated records of the ith generalization record in generalization relation, n' is the number of records in generalization relation.

Definition 9. Given $r'(r_1', \dots r_{n'}')$ is a generalization relation under $A_1^{g_1} \cdots A_m^{g_m}$, $(1 \le g_i \le l_i + 1)$, then interestingness $I_{g_1 \cdots g_m}$ of $A_1^{g_1} \cdots A_m^{g_m}$ is defined as formula (2).

$$I_{g_1 \cdots g_m} = \sqrt{\sum_{i=1}^{n'} (t_i - \frac{1}{n'})^2} \tag{2}$$

When the number of repeated records for each generalization record in a generalization relation $r'(r_1', \dots r_{n'}')$ gets average value, $I_{g_1 \cdots g_m}$ achieves the minimum 0. The farer t-weight of generalization records in a generalization relation from average value, the larger the contribution to interestingness. The larger the value of $I_{g_1 \cdots g_m}$, the more interesting the rule expressed by the attributes' generalization sequence $A_1^{g_1} \cdots A_m^{g_m}$.

Therefore, after obtaining sequences which meet the Z, computing their interestingness, and ranking the generalization sequences with the decline of interestingness, we can produce generalization relation and rules.

5 Analysis

In this section, we analyze our algorithms for completeness, correctness and computational complexity. Correctness means that the generalization rules meet the user specified threshold. Completeness implies that no AGS that satisfies the given threshold Z is missed.

– The algorithm 1 is correct

Proof. The algorithm 1 uses very simple way to get generalization rules. It is obvious that this algorithm is correct if we can prove the recursive procedure Gen_seq is correct. That means the Gen_seq will return the AGS that satisfy the Z. It is guaranteed by step (8) in Gen_seq, because this step will check whether every sequence satisfies the Z or not.

– The algorithm 2 is correct

Proof. The pruning strategy (1) guarantee the step 2) in algorithm 2 will return the low boundary of every attributes. If the return of step 3) is correct, then the Section 3.2 ensures the generalization relation computed by the step 4) is correct.

For step 3) (i.e., the recursive procedure Gen), the property of $\pi_{X \cup Y} = \pi_X \otimes \pi_Y$ and Section 3.2 ensure the correctness of the step (3)-(6) in Gen. The step (3)-(6) ensures the correctness of the step (11). And the step (11) guarantees every AGS satisfies the threshold Z.

– The algorithms are complete

Proof. We prove if a sequence satisfies the Z, it is found by our algorithms. In the recursive procedure Gen_seq of algorithm 1, the step (1) iterates all generalization levels of an attribute and the step (6) recursively perform Gen_seq. So the combine of step (1) and step (6) ensures the Gen_seq will check all possible candidate sequences.

For algorithm 2, the pruning strategy (1) guarantee the step 2) of algorithm will return the low boundary of every attributes. In the recursive procedure Gen, The step (7) and (8) is because the pruning strategy (2) and the step (11) and (13) are just because the pruning strategy (3). The combine of step (1) and step (10) guarantee the Gen will check all possible candidate sequences.

– Computational Complexity Analysis

Suppose the number of records in the relation T is N, the number of attributes in T is m, and the height of attribute i-th concept hierarchy tree is l_i. In the worse case, the computational complexity of the algorithm 1 will be $O(N * \prod_{i=1}^{m}(l_i + 1))$. And the algorithm 2 will be $O(\prod_{i=1}^{m}(l_i + 1))$. Theoretically speaking, the computational complexity of the two algorithms seems not very distinctive. But we will jump over many recursive steps using pruning strategies in algorithm 2, so the complexity will be much lower than algorithm 1. We will show the real execution results in the experiments.

6 Performance Evaluation and Applications

The performance of our algorithms is evaluated by synthetic datasets and a real data-set (a plant distributing dataset in "The Three Parallel Rivers in Yunnan Area" zone). The experiments are performed on a Celeron computer with a 2.40 GHz CPU and 256 Mbytes memory running the Windows XP operating system.

6.1 Evaluation Using Synthetic Datasets

The experiments using synthetic data sets are aimed at answering the following ques-tions. (1) How does the size of dataset affect the two algorithms? (2) How do the Algorithm 1 and the Algorithm 2 behave with the Z is changed?

We ran a series of experiments with increasing number of spatial data points. The results are showed in Fig. 3. (a). We can see that the algorithm 2 is almost linear and much faster than the algorithm 1.

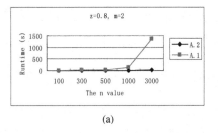

(a)

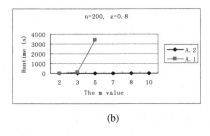

(b)

Fig. 3. Performance of algorithms using synthetic datasets

Fixed on the number of records, the number of attributes is an important parameter. The detailed comparative results are showed as Fig. 3. (b). We can see that the per-formance of Algorithm 1 is very bad when m=5. The results indicate that the pruning strategies and equivalence partition trees used in Algorithm 2 are very effective.

Now we look at the characters of fast re-generalization of the two algorithms. The results are shown in Fig. 4(a) (b). We used 6 different settings of the thresholds Z. We can see the Algorithm 2 possesses the character of fast re-generalization.

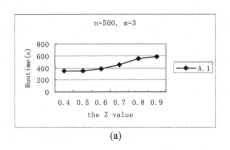

(a)

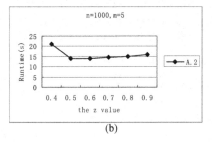

(b)

Fig. 4. Characters of fast re-generalization for the two algorithms

6.2 Application in a Real Dataset

A plant distributing dataset, which involves 29 plant species which are very valuable and rare in "The Three Parallel Rivers in Yunnan Area" zone and 319 instances (tuples), is used in our experiments. Table 1 is some tuples of this dataset.

Table 1. Some tuples of a plant distributing dataset

Tuple-ID	Plant-name	Veg-name	Elevation /m	Location
t_1	Orchid	meadow	[1000, 1500]	Lijiang
t_2	Fig	scrub	[2400, 3000]	Weixi
t_3	Magnolia	scrub	[3000, 3700]	Lijiang
t_4	Calligonum	taiga	[2000, 3000]	Jianchuan
t_5	Magnolia	meadow	[3000, 4000]	Lanping
t_6	Agave	taiga	[3000, 4000]	Lanping
t_7	Yucca	forest	[1500, 2400]	Weixi
t_8	Waterlily	meadow	[800, 2200]	Jianchuan

The experiments using this dataset are aimed at checking the usefulness of the AOG-ags algorithms. Can they discover valuable patterns? Are the rules discovered by our algorithms interesting towards geographers and botanists?

34 AGS are obtained when the threshold Z is set 0.8, and 57 plant distributing rules are discovered when one of the 34 AGS is chosen according to their interestingness. When the threshold Z is set 0.85, the number of AGS is 28 and the number of rules is 19. When the Z is set 0.9, the number of AGS and rules is 22 and 16 respectively.

Some rules discovered by our algorithms are really attractive to geographers and botanists. The followings are some examples:

- "Tricholoma matsutake" $\Rightarrow$ 40% grows in the forest and meadow whose elevation is from 3300 to 4100 meter of Lijiang.
- "Angiospermae" $\Rightarrow$ 80% grows in the forest, scrub and meadow whose elevation is from 2400 to 3900 meter of Lijiang and Weixi.
- Lijiang $\Rightarrow$ There are a plenty of plants species in severe danger such as "Tricholoma matsutake", "Angiospermae", "Gymnospermae", and so on.

7 Conclusions

Related approaches for mining the associations of attributes can be divided into the clustering-based approach, the association rule-based method and the approach of AOG. Clustering-based approach treats every attribute as a layer and considers clusters of point-data in each layer as candidates for mining associations [7, 8]. The complexity and the low-efficiency are the crucial problems of this method. The association rule-based approach is divided into the transaction-based method and distance-based method again. The transaction-based method computes the mining transaction (two-dimension table) by a reference-object centric model, so one can use the method which is similar to *Apriori* for mining the rules [9,10]. The problem of this method is that a suitable reference-object is required to be specified. The distance-based method was proposed by Morimoto [11], and Shekhar together with Huang [12-14] did further research. Because of doing a plenty of join operations, executing efficiency is the key problem of this method. The approach of AOG is presented firstly by Cai [1]. It is a simple and understandable method. But it is inconvenient because setting each attribute threshold is required.

The AOG-ags proposed in this paper can obtain automatically rules under setting a threshold Z. Particularly, by using the AOG-ags algorithm in a plant distributing dataset, some distributing rules for the species of plants in "Three Parallel Rivers of Yunnan Protected Areas" zone are found interesting. The advantage of AOG method is that expert knowledge (concept hierarchy trees) is used in the process of data mining.

References

1. Cai, Y., Cercone, N., Han, J.: Attribute-Oriented Induction in Relational Databases. In: Piatetsky-Shapiro, G., Frawley, W.J. (eds.) Knowledge Discovery in Databases, Menlo Park, Calif., pp. 213–228. AAAI/MIT Press, Cambridge (1991)
2. Han, J.: Towards Efficient induction mechanisms in database systems. Theoretical Computering Science 133, 361–385 (1994)
3. Wang, L.: A method of the abstract generalization on the bases of the semantic proximity. Chinese J. Computers 23(10), 1114–1121 (2000)
4. Carter, C.L., Hamilton, H.J.: Efficient attributed-oriented generalization for knowledge discovery from large databases. IEEE Trans. on Knowledge and Data Eng. 10(2), 193–208 (1998)
5. Chen, H., Wang, L.: Quantifiable Attribute-Oriented Generalization. Journal of Computer Research & Development 38(2), 150–156 (2001)
6. Han, J., Kamber, M.: Data mining: concepts and techniques. Morgan Kaufmann publishers, San Francisco (2001)
7. Estivill-Castro, V., Lee, I.: Data Mining Techniques for Autonomous Exploration of Large Volumes of Geo-Referenced Crime Data. In: Proc. Sixth Int'l Conf. Geocomputation (2001)
8. Estivill-Castro, V., Murray, A.: Discovering Associations in Spatial Data—An Efficient Medoid Based Approach. In: Wu, X., Kotagiri, R., Korb, K.B. (eds.) PAKDD 1998. LNCS, vol. 1394, Springer, Heidelberg (1998)
9. Koperski, K., Han, J.: Discovery of Spatial Association Rules in Geographic Information Databases. In: Egenhofer, M.J., Herring, J.R. (eds.) SSD 1995. LNCS, vol. 951, pp. 47–66. Springer, Heidelberg (1995)
10. Wang, L., Xie, K., Chen, T., Ma, X.: Efficient discovery of multilevel spatial association rule using partition. Information and Software Technology (IST) 47(13), 829–840 (2005)
11. Morimoto, Y.: Mining Frequent Neighboring Class Sets in Spatial Databases. In: Proc. ACM SIGKDD Int'l Conf. Knowledge Discovery and Data Mining, pp. 353–358 (2001)
12. Xiong, H., Shekhar, S., Huang, Y., Kumar, V., Ma, X., Yoo, J.S.: A Framework for Discovering Co-location Patterns in Data Sets with Extended Spatial Objects. In: SDM. Proc. 2004 SIAM International Conference on Data Mining, pp. 1–12 (2004)
13. Huang, Y., Shekhar, S., Xiong, H.: Discovering Colocation Patterns from Spatial Data Sets: A General Approach. IEEE Transactions on Knowledge and Data Engineering 2004, 1472–1485 (2004)
14. Yoo, J.S., Shekhar, S.: A partial Join Approach for Mining Co-location Patterns. In: Proc. of the 12th annual ACM international workshop on Geographic information systems, pp. 241–249 (2004)

A Framework for Titled Document Categorization with Modified Multinomial Naivebayes Classifier

Hang Guo and Lizhu Zhou

Computer Science & Technology Department
100084, Tsinghua University,Beijing,China
guohang@mails.tsinghua.edu.cn,
dcszlz@mail.tsinghua.edu.cn

Abstract. Titled Documents (TD) are short text documents that are segmented into two parts: Heading Part and Excerpt Part. With the development of the Internet, TDs are widely used as papers, news, messages, etc. In this paper we discuss the problem of automatic TDs categorization. Unlike traditional text documents, TDs have short headings which have less useless words comparing to their excerpts. Though headings are usually short, their words are more important than other words. Based on this observation we propose a titled document classification framework using the widely used MNB classifier. This framework puts higher weight on the heading words at the cost of some excerpt words. By this means heading words play more important roles in classification than the traditional method. According to our experiments on four datasets that cover three types of documents, the performance of the classifier is improved by our approach.

1 Introduction

TDs (Titled Documents) are short text documents composed by Heading Part and Excerpt Part. The former is the title or heading of the document. TDs are used in many applications such as papers, news, messages and so on. Their numbers are increasing very fast these years and they are getting more and more important. TDs are widely used in the Web. People are reading on-line news at home, searching for papers at college and posting messages to the Newsgroups. Now TDs have been important Internet resources for business communities. They are also valuable for data mining researchers.

The Multinomial Naivebayes [4](MNB) classifier is widely used in text categorization because it is fast and easy to implement. [6] shows that its performance is competitive with the state-of-the-art models like SVM [9] with simple modifications. MNB model follows the assumption that every word in the documents is independent with each other. [7] discusses why it performs well on such a severe assumption.

Traditional text classification models can be used in titled document classification. These classifiers are built on the bag-of-words model. They do not

R. Alhajj et al. (Eds.): ADMA 2007, LNAI 4632, pp. 335–344, 2007.
© Springer-Verlag Berlin Heidelberg 2007

distinguish heading and excerpt words. These classifiers are developed for plain text documents, not for TDs. Usually the headings are more important. They are shorter but more important than other words. As shown in Section 2, the headings have much less useless words than the excerpts. In our previous work [11], we found that putting different weights on the words of different sections could improve classification performance of the classifiers. Based on this observation we consider the possibility to replace some excerpt words with heading words so that heading words can be "emphasized" in classification. Following this idea, we propose a titled document classification framework. In the training phase, we remove less words from the heading vocabulary than the excerpt vocabulary. In the classifying phase, the weights of heading words are increased. By this means more heading words are used in classification at the cost of some excerpt words. According to the experiments on four datasets, the performance of the classifiers is improved.

The reminder of the paper is organized as follows: in Section 2 we show that why heading words should be emphasized. Section 3 introduces our feature reduction approach. The classification framework is shown in Section 4. In Section 5 we present our experiments on four datasets. Section 6 glances at the related work. Finally the paper is concluded in Section 7.

2   Motivation

Intuitively heading words are more important than other words. The heading (title) words of a document are usually the keywords. Heading words should play more important roles than others. To test this idea, we select four real life text collections. They are: OHSUMED[1], Reuters-21578[2], 20-Newsgroups[3] and Cite-Seer papers[4]. Every document in these collections has a short title(headline) and a longer abstract(excerpt). The four datasets cover three types of documents: papers, news and messages. We use Information Gain[5] – a popular dimensionality reduction method, to remove the less important words in documents. According to [5], the words left after reduction are considered to be more informative for classifiers. Before reduction there are about 3,000 words in titles and excerpts. Then we reduce the vocabulary size to 2000, 1000, 500, 200 and 100 words. Each time we compare the words left in excerpts and headings. The results are shown in Table 1.

The result shows that the number of heading words are much less than that of the excerpt words before feature reduction. However, the heading words are much more important than excerpt words in terms of the rate of informative words. Therefore heading words should play more important roles than others in training and classification.

---

[1] http://trec.nist.gov/data/t9_filtering/README

[2] http://www.ics.uci.edu/ kdd/databases/reuters21578/reuters21578.html

[3] http://people.csail.mit.edu/ jrennie/20Newsgroups/

[4] http://citeseer.ist.psu.edu/directory.html

Table 1. Comparison between Title Words and Other Words

Word Number		3000	2000	1000	500	300	100
OHSUMED	Excerpt	83.46	66.63	36.95	26.11	17.98	6.46
	Heading	11.43	8.37	5.87	4.58	3.74	1.99
	Excerpt/Heading	8.49	7.96	6.29	5.69	4.81	3.25
Reuters-21578	Excerpt	51.90	43.57	36.79	30.85	26.20	17.94
	Heading	6.11	5.17	4.68	4.26	3.81	2.76
	Excerpt/Heading	7.24	8.42	7.87	7.24	6.89	6.5
20-Newsgroup	Excerpt	87.79	54.11	27.87	18.52	13.26	4.91
	Heading	4.70	3.48	2.52	2.07	1.80	1.29
	Excerpt/Heading	18.68	15.55	11.05	8.95	7.36	3.81
CiteSeer	Excerpt	54.03	29.22	14.37	8.35	6.22	3.53
	Heading	7.24	4.49	2.79	2.35	2.19	1.70
	Excerpt/Heading	7.47	6.51	5.15	3.43	2.84	2.08

3 Feature Reduction

Before building the classifier, the size of the vocabulary must be reduced because the training documents usually have over 10,000 words. It is a great challenge for training a feasible classifier. Traditional feature reduction methods do not distinguish heading words and excerpt words in documents. We have shown in Table 1 that heading words are usually more informative than excerpt words. Therefore we should be more careful when a heading word is discarded.

Though heading words are less likely to be removed in feature reduction, sometimes they are incorrectly removed because they are mixed with other words. For instance, if the word "database" appear in the title of a document, this document is probably about database technology. Whereas "database" may also appear in the excerpts of articles on other categories since databases are widely used in different applications like machine learning, information retrieval, etc.. As a result, word "database" is likely to be removed according to Information Gain algorithm.

The solution to the problem is to separate heading words and excerpt words in feature reduction. Therefore we need to build two vocabularies: heading vocabulary and excerpt vocabulary. In feature reduction, we employ Information Gain algorithm on the two vocabularies independently. At last they are united as one vocabulary. The "compress rates" of the two vocabularies are different in feature reduction. Since heading words are more important, we remove less words from the heading vocabulary.

According to our feature reduction method, many less informative heading words are kept. In the classification phrase, the average length of the headings of the testing documents are increased. We make the heading words more "affective" in classification so that they can take the place of the removed excerpt words.

4   Classifying

As mentioned in Section 3, the idea of the titled document classifying method is to increase the "affects" of heading words in classification. In our previous work [11], we have found that putting higher weight on the more important words would increase the performance of the classifiers. According to the most popular weighting function– TFIDF, doubling the weight of the title words is doubling the occurrences of these words. Following this idea, we assume that

Assumption. When classifying a titled document $i = (h, e)$, it is equal to classify a plain text document $(\theta * h, e)$.
where h denotes the heading words, e denotes the excerpt words. $\theta(\theta > 1)$ is a prior.

In this way we can transform a titled document into a plain text document. Then we can use the well-developed plain text classification technology to classify titled document. Substantial efforts have been made in the literature to develop good plain text classification models, which are general enough to be applied to diverse document representations. In this study, we take one of the most popular classification model–Multinomial NaiveBayes (MNB) [4], as a case in point.

4.1   Traditional MNB Model

Multinomial Naivebayes is a widely used document classification model, whose performance is acceptable for many corpora [4]. It follows the NaiveBayes Independence Assumption that *"the probability of each word event in a document is independent of the word's context and position in the document"*.

Suppose document d has words $\{w_1, w_2, \ldots, w_{|W|}\}$, the assumption is :

$$p(d|c) = \prod_{i=1}^{|W|} p(w_i|c) \qquad (1)$$

here W denotes the vocabulary. According to the Bayes rule, we have

$$p(c|d) = p(c) * p(d|c)/p(d) = p(c) * (\prod_{i=1}^{|W|} p(w_i|c))/p(d)$$

$$\approx p(c) * (\prod_{i=1}^{|W|} \frac{n(w_i, c)}{n(c)})/p(d)$$

where $n(w_i, c)$ is the number of occurrences of word w_i ($w_i \in W$) in the training examples labeled with category c, and $n(c)$ is the total number of occurrences of all the words in W in the training examples associated with category c.

The goal of the classification problem is to find class j that maximizes $p(c_j|d)$. The complexity of the training procedure is $O(|D_{train}| * |W|)$ and the complexity of the classification is $O(|W| * |\mathcal{C}|)$, where $|D_{train}|$ is the total number of training examples and $|\mathcal{C}|$ is the number of categories(classes).

4.2   Our Algorithm

Following the NaiveBayes Independence Assumption in Formula 1 , we can extend the scope of MNB model for titled document classification. When classifying document $(\theta * h, e)$, here is:

$$
\begin{aligned}
p(c_i|\theta * h, e) &= \frac{p(\theta * h, e|c_i) * p(c_i)}{p(\theta * h, e)} \\
&= \frac{p(\theta * h|c_i) * p(e|c_i) * p(c_i)}{p(\theta * h, e)} \\
&= \frac{p(h|c_i)^\theta * p(e|c_i) * p(c_i)}{p(\theta * h, e)} \\
&= \frac{(p(c_i|h) * p(h))^\theta * p(c_i|e) * p(e)}{p(c_i)^\theta * p(\theta * h, e)} \\
&\propto \frac{p(c_i|h)^\theta * p(c_i|e)}{p(c_i)^\theta}
\end{aligned}
\tag{2}
$$

Using Formula 2, we can easily calculate the classification result of titled document $(\theta * h, e)$ by the classification results of its heading and excerpt. When classifying document $(\theta * h, e)$, we first classify its heading and excerpt respectively. Then we have $p(c|e)$ and $p(c|h)$. At last we combine the results according to Formula 2.

Because $\theta > 1$, $p(c|h)$ is more affective than $p(c|e)$ for the classification result. However, in most cases it is hard to calculate $p(c|h)$ because headings are too short. If the approximation of $p(c|h)$ is far from the real possibility, the classification results of $p(c|\theta * h, e)$ will be incorrect.

The idea is to weight $p(c|h)$ by its classification error rate. We propose a heuristic weighting function to evaluate the classification result with different θ. The weight function of $p(c|\theta * h, e)$ is:

$$
w_\theta = \begin{cases}
10^{-10*(e_h/e_e)-(\theta-1)} & : \quad e_h < 2 * e_e, \theta > 1 \\
0 & : \quad e_h \geq 2 * e_e, \theta > 1 \\
1 & : \quad \theta = 1
\end{cases}
\tag{3}
$$

here e_h is the error rate of $p(c|h)$ and e_e is the error rate of $p(c|e)$. In this paper we classify the training set to get an optimistic approximation of e_h and e_e. If e_h is too high, the weight of $p(c|\theta * h, e)$ will be dropped. According to Formula 2, the errors of $p(c|h)$ greatly affect the predication of $p(c|\theta * h, e)$ with the increase of θ. Then when θ increases, the weight of $p(c|\theta * h, e)$ will decrease.

We set an upper bound of θ, namely θ_{max} (set to 3), and combine all the classification results using different θ. The classifying algorithm is shown in Figure 1. The classification cost is $O(|C| * \theta_{max})$.

Input: a titled document (t, b), θ_{max}
Output: classification result vector v

(a) Preprocessing
1   Classify the training set to get e_e and e_h.
2   Calculate weight $w_1, \ldots, w_{\theta_{max}}$ by formula 3;

(b) Calculate $p(c_i|\theta * h, e)$;
2   **foreach** $1 \leq i \leq |\mathcal{C}|$
3   Calculate $P(c_i)$, $P(c_i|h)$ and $P(c_i|e)$;
4   **foreach** $1 \leq \theta \leq \theta_{max}$
5   Calculate $p(c_i|\theta * h, e)$ by formula 2;

(c) Combine Results
6                $v_i = v_i + p(c_i|\theta * h, e) * w_\theta$;
7   **endfor**
8   **endfor**

Fig. 1. Pseudo Codes for classifying algorithm

5   Experiment

5.1   Datasets

To test our framework on different type of titled documents, we collect four real life datasets from a variety of applications: scientific papers search engines, the on-line news agency, Newsgroups, etc. CiteSeer and OHSUMED are widely-used paper collections. The abstracts of papers are used as Excerpt Parts. Reuters-21578 and 20-Newsgroup are often used as benchmarks in text categorization problems. We use the body section of news or messages as their Excerpt Part. Due to the size of these collections, a small portion of data are selected and used in our experiments.

CiteSeer. CiteSeer[5] provides on-line scientific materials on computer science. We use its classified papers[6] as our training and testing set. There are 17 top categories in its hierarchy. To avoid possible overlap among those categories, we only use 8 categories in our experiments. They are Database, Agents, Compression, Hardware, Networking, Programming, Security, Software Engineering and Theory.

---

[5] http://citeseer.ist.psu.edu/
[6] http://citeseer.ist.psu.edu/directory.html

OHSUMED. OHSUMED[7] document collection is a set of 348,566 references from MEDLINE, the on-line medical information database, consisting of titles and/or abstracts from 270 medical journals over a five-year period (1987-1991). These papers are categorized into 4904 topics. Two of the largest 10 categories are randomly selected for our experiments.

Reuters-21578. Reuters-21578 [8] is a collection of documents that appeared on Reuters newswire in 1987. The documents were assembled and indexed with categories by personnel from Reuters Ltd and and Carnegie Group, Inc.. We use ModApte Split in our experiments.

20-Newsgroup. 20-Newsgroup[9] is a collection of 20,000 messages, collected from 20 different netnews newsgroups. One thousand messages from each of the twenty newsgroups were chosen at random and partitioned by newsgroup name. We chose four newsgroups from the collection because there are over-laps between different categories. The selected newsgroups are "alt.atheism", "talk.politics.guns", "comp.os.ms-windows.misc" and "rec.sport.hockey".

5.2 Training and Testing

Environment and Library. We perform the experiments on a 1.5GHz work-station with 512M memory. We select Naivebayes[4] classifier in our experiment. It is implemented by Weka[10] and Judge[11]. They are both open source classifi-cation toolkits in Java.

Training. We use the popular Information Gain algorithm [5] to reduce the feature space. In the excerpt vocabulary we select the top 50% words out of 4,000 words. And in the heading vocabulary we select the top 30% words out of 2,000 words. Then the two reduced vocabularies are united as the training vocabulary. Based on this vocabulary we build a MNB classifier.

Classifying. We compare the results of our framework with the results of tra-ditional plain text classification method. The results are listed in Table 2. We use the overall error rate as the criteria in our experiments. The error rates are affected by θ_{max}.

5.3 Discussion

Table 2 shows that the titled document classification framework declines the clas-sification error rate. Generally speaking, we get more improvement in CiteSeer and OHSUMED. That is probably because the titles of papers are longer and

---

[7] http://trec.nist.gov/data/t9_filtering/README

[8] http://www.ics.uci.edu/ kdd/databases/reuters21578/reuters21578.html

[9] http://www.cs.cmu.edu/afs/cs.cmu.edu/project/theo-20/www/data/news20.html

[10] http://www.cs.waikato.ac.nz/ml/weka/

[11] http://www3.dfki.uni-kl.de/judge/

Table 2. Classification Error Rates

θ_{max}	CiteSeer	OHSUMED	Reuters	20-Newsgroup
2	24.7%	5.2%	3.3%	5.5%
3	24.9%	5.4%	3.4%	5.8%
4	25.1%	5.6%	3.5%	6.1%
tradition	27.0%	6.0%	3.4%	5.8%

more formal. Researchers usually choose titles carefully. Therefore title words are usually the keywords of the whole documents. They deserve to be emphasized. However the headlines of news and messages are relatively shorter and more informal. Many news headlines use abbreviates and numbers such as "MGM/UA COMMUNICATIONS 2ND QTR FEB 28 LOSS". These headlines are meaningless for our classifier. Things are even worse in 20-Newsgroups. Newsgroup users always use titles like "I agree", "You got it" and so on. These titles are actually meaningless words that have little use in classification. Our framework keeps these title words at the cost of some excerpt words. If many title words are useless in classification, our method fails.

The performance of the classifier declines with the increase of θ_{max}, especially in 20-Newsgroups and Reuters. If θ_{max} is too large, the error rate increases. That is because $p^\theta(c|h)$ are used in Formula 2. It is clear that these possibilities given by MNB is not the real distributions. When θ goes larger, we have more errors. To get the optimized θ_{max} on a given corpus, it is better to run the algorithm on a small portion of documents first. For those collections that documents have long and meaningful headings (titles), we can set larger θ_{max}.

6   Related Work

Conventional automated documents categorization models have been developed for years. [5] has introduced many frequently used models, Naivebayes[4], SVM[9][10], Decision Tree, etc. SVM performs best in many datasets. However, even with the linear kernel, its training and classifying costs are higher than simple models like NB. Moreover, other simple models still can be improved in terms of accuracy. [7] shows that with simple modification, the performance of Naive Bayes Multinomial model approaches SVM.

The use of titles in document classification is first explored in [12]. Recently many experiments have shown that titles are very useful when classifying various documents. [8] uses titles and other features in Web page classification with support vector machine. [15] improves the performance of classifiers by combining the main body text, anchor text and titles when classifying Web pages. [13] proposes an intelligent document title classification method based on information theory. [14] proves that weighting titles in life science publications improves the performance of classifiers.

7   Conclusion and Future Work

Titled documents such as papers, news and messages are widely used these days. The headings words of these documents are usually more important than other words. Traditional document classification methods usually ignore the differences of heading words and other words. In this paper we propose a titled document classification framework. According to this framework, we remove less heading words in feature reduction at the cost of some excerpt words. In classification all the heading words are put more weight. By this means heading words play more important roles in classification than the traditional method. According to the experiments on four real life datasets, the error rates are dropped by our method.

We are working on new methods to determine θ_{max}. In the next step we will develop a new framework based on the state-of-art classifiers like SVM.

Acknowledgement

This work is supported by National Nature Science Foundation of China under Grant No.60520130299.

References

1. Hull, D.P., Schutze, J., Method, H.: Combination for document filtering. In: Proc. the 19th ACM SIGIR Conference on Research and Development in Information Retrieval, Switzerland, pp. 279–287 (1996)
2. Tumer, K., Ghosh, J.: Linear and order statistics combination for pattern classification. In: Sharkey, A. (Ed.) Combining Artificial Neural Networks, pp. 127–162. Springer-Verlag (1999)
3. Merz, C.J., Pazzani, M.J.: Combining neural network regression estimates with regularized linear weights. In: Advances in Neural Information Processing Systems, vol. 9, pp. 564–570. MIT Press, Cambridge (1997)
4. Mccallum, A., Nigam, K.: A Comparison of Event Models for Naive Bayes Text Classification. In: Proc. AAAI workshop on Learning for Text Categorization, Wisconsin, pp. 41–48 (1998)
5. Fabrizio, S.: Machine Learning in Automated Text Categorization. ACM Computing Surveys 34, 1–47 (2002)
6. Rennie, J.D.M., et al.: Tackling the Poor Assumptions of Naive Bayes Text Classifiers. In: Proc. International Conference on Machine Learning, Washington, DC (2003)
7. Domingos, P., Pazzani, M.: Beyond independence: conditions for the optimality of the simple Bayesian classifier. In: Proc. International Conference on Machine Learning, Italy (1996)
8. Sun, A., Lim, E., Ng, W.: Web Classification Using Support Vector Machine. In: Proc. Workshop on Web Information and Knowledge Management, Virginia (2002)
9. Joachims, T., Sebastiani, F.: Guest editors's categorization. J. Intell. Inform. Syst. 18(2/3), 103–105 (2002)

10. Joachims, T.: Text Categorization with Support Vector Machines: Learning with Many Relevant Features. In: Nédellec, C., Rouveirol, C. (eds.) Machine Learning: ECML-98. LNCS, vol. 1398, pp. 137–142. Springer, Heidelberg (1998)

11. Guo, H., Zhou, L.: Segmented Document Classification: Problem and Solution. In: Bressan, S., Küng, J., Wagner, R. (eds.) DEXA 2006. LNCS, vol. 4080, Springer, Heidelberg (2006)

12. Hamill, K., Zamora, A.: The use of titles for automatic document classification. In J. of the American Society for Information Science (1980)

13. Song, D., Bruza, P., Huang, Z., Lau, R.: Classifying Document Titles Based on Information Inference. In: Zhong, N., Raś, Z.W., Tsumoto, S., Suzuki, E. (eds.) ISMIS 2003. LNCS (LNAI), vol. 2871, Springer, Heidelberg (2003)

14. Hakenberg, J., Rutsch, J., Leser, U.: Tuning Text Classification for Hereditary Diseases with Section Weighting. In: Proc International Symposium on Semantic Mining in Biomedicine (2005)

15. Kaist, I., Kim, G.: Query type classification for web document retrieval. In: Proc. of ACM SIGIR, ACM Press, New York (2003)

Prediction of Protein Subcellular Locations by Combining K-Local Hyperplane Distance Nearest Neighbor*

Hong Liu, Haodi Feng, and Daming Zhu

School of Computer Science and Technology, Shandong University, Jinan 250061,
Shan-dong Province, People's Republic of China
{Hong-liu, Fenghaodi, Dmzhu}@sdu.edu.cn

Abstract. A huge number of protein sequences have been generated and collected. However, the functions of most of them are still unknown. Protein subcellular localization is important to elucidate protein function. It would be worthwhile to develop a method to predict the subcellular location for a given protein when only the amino acid sequence of the protein is known. Although many efforts have been done to accomplish such a task, there is the need for further research to improve the accuracy of prediction. In this paper, with K-local Hyperplane Distance Nearest Neighbor algorithm (HKNN) as base classifier, an ensemble classifier is proposed to predict the subcellular locations of proteins in eukaryotic cells. Each basic HKNN classifiers are constructed from a separated feature set, and finally combined with majority voting scheme. Results obtained through 5-fold cross-validation test on the same protein dataset showed an improvement in pre-diction accuracy over existing algorithms.

Keywords: Protein, Subcellular Location, Ensemble Classifier.

1 Introduction

As a result of the Human Genome Project and related efforts, protein sequence data accumulate at an accelerating rate. This raises the challenge of understanding the functions of proteins from high throughput sequencing projects. Protein subcellular localization is a key functional characteristic of proteins [1] and correct prediction of protein subcellular localization will greatly help in understanding its functions. However, experimental determination of subcellular location is time-consuming and costly. Therefore, a reliable and efficient computational method is highly required to construct prediction systems to predict the subcellular location for a given protein when only the amino acid sequence of the protein is known.

Several machine learning techniques have been applied to construct such prediction systems, for example, Support Vector Machines (SVM) [2-4], Neural Network [5,6], Naïve Bayesian [7] and Fuzzy KNN [8], using different sets of

---

* Supported by National Science Foundation of China under grant No. 60603007 and Science and Technology Development Foundation of Shandong Province, China under grant No. 2006GG2201005.

R. Alhajj et al. (Eds.): ADMA 2007, LNAI 4632, pp. 345–351, 2007.
© Springer-Verlag Berlin Heidelberg 2007

features extracted from amino acid composition [2,3,6,9], amino acid pair composition[9], gapped amino acid composition [2], pseudo amino acid composition [10], evolutionary and structural information [11] and motif information [12]. While SVM, Neural Network, Naïve Bayesian need a separate training procedure, Fuzzy KNN not. Given the various sets of features, one approach to make use of them is combining (or ensembling) different classifiers constructed from different set of features. By this means, reduction in variance caused by the peculiarity of a single feature set and consequently more reliable and stable prediction system could be obtained.

In this paper, we present a new method based on an ensemble of K-local Hyper-plane Distance Nearest Neighbor algorithm (HKNN) [13] for the prediction of protein subcellular locations. Experimental results obtained through 5-fold cross-validation tests on the same protein dataset showed an improvement in prediction accuracy over existing algorithms.

The rest of this paper is organized as follows. Section 2 presents the methods proposed. Experimental results and comparison with existing methods are presented in Section 3. Conclusion is given in Section 4

2 Methods

2.1 Feature Presentation

Protein sequences are composed of amino acids, which are denoted by letters from the alphabet {A, C, D, E, F, G, H, I, K, L, M, N, P, Q, R, S, T, W, V, Y}. In order to be able to perform computation on these sequences, non-numerical amino acids should be represented by numerical values.

From amino acid composition, we extracted the first set of features. In detail, for each protein sequence, the occurrence frequency of each amino acid (letter) was calculated and normalized (i.e. divided by the length of the sequence minus one). Each of the number obtained corresponds to an element of a 20-dimension vector (see Fig. 1). That is, a protein sequence was mapped as a point in a feature space with dimension 20.

Prediction based on only amino acid composition features would lose sequence order information. Thus, to capture this kind of information, amino acid pair composition and gapped amino acid composition [2] were considered. While the former corresponds to two adjacent amino acids, the latter corresponds to two amino acids separated by one or more intervening residue positions. In fact, amino acid pair composition can be seen as a special case of gapped amino acid composition with zero gaps. For four different gap values (i.e. 0, 1, 2, 3), we extracted four different set of features separately. In detail, since there are 20 different amino acids, we considered $20 \times 20 = 400$ amino acid pairs for each gap value. For each protein sequence, the occurrence frequency of each gapped pair was calculated and normalized (i.e. divided by the length of the sequence minus 3). Thus, a protein sequence was converted to other four different 400-dimension vectors one for each gap value (see Fig. 1) or, in other words, was mapped as a point in other four different feature spaces with dimension 400.

Since amino acids have different biochemical and physical properties that influence their relative replace-ability in evolution, we re-substituted the 20-letter amino acids by 9-letter amino acids according to their physicochemical properties, as illustrated in Table 1[14]. Based on this new encoding scheme, each protein sequence was converted to other five different vectors using the similar process as in 20-letter encoding scheme case (see Fig. 1).

So far, for each protein sequence, we got ten feature vectors. In other words, each protein sequence was mapped as a point in ten different feature spaces.

Table 1. The 9-letter encoding scheme for the 20 amino acids based on their physical-chemical proper-ties

Group	Residues	Description
1	C	Highly conserved
2	M	Hydrophobic
3	N, Q	Amides, polar
4	D, E	Acids, positive, polar
5	S, T	Alcohols
6	P, A, G	Aliphatic, small
7	I, V, L	Aliphatic
8	F, Y, W	Aromatic
9	H, K, R	Bases, charged

2.2 K-Local Hyperplane Distance Nearest Neighbor Algorithm (HKNN)

HKNN is a modified k-nearest neighbor algorithm (KNN). By building a (non-linear) decision surface, separating different classes of the data, directly in the original feature space, it is intended to improve the classification performance of the conventional KNN to a level of SVM.

Suppose the number of different classes in the training set is c, HKNN computes distances of a test point $\mathbf{x}$ to c local hyperplanes, where each hyperplane is composed of k nearest neighbors of $\mathbf{x}$, belonging to the same class, in the training set. Then the test point $\mathbf{x}$ is assigned to the class whose hyperplane is closest to $\mathbf{x}$ (see [13] for details).

HKNN has two parameters, k and λ, a penalty term introduced to find the hyperplane.

2.3 Voting Scheme

In each of the ten feature space described above, a HKNN classifier was constructed. To combine the prediction results of all the ten classifiers, majority voting scheme [15] was used, in which the final prediction class was the most voting one. Ties were randomly resolved.

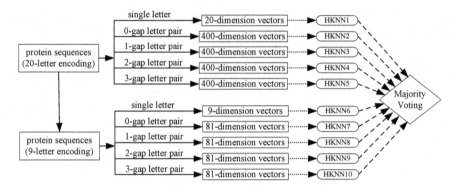

Fig. 1. Framework of the method used in this paper

3   Experimental Results and Discussion

3.1   Protein Dataset

For comparison, the protein dataset (downloadable at "http://web.kuicr.kyoto-u.ac.jp/~park/Seqdata/") studied in previous investigations [2] were used in this study. In this dataset, all protein sequences were collected from the SWISS-PROT database release 39.0. Totally 7579 protein sequences of eukaryotic cells for 12 subcellular locations were contained in this dataset. The number of protein sequences in each location is shown in Table 2.

3.2   Performance Measurement

The prediction performance was evaluated by 5-fold cross-validation test. In detail, proteins in the dataset were separated into five balanced sets. Each of these sets contained almost the same number of proteins. In each round of cross-validation, four sets were used to construct the ensemble of HKNNs while one set was set aside for evaluating the method. This procedure was repeated five times, once for each set. In order to evaluate the prediction performance of our method, two measures were used. The first measure, total accuracy (TA), is defined as

$$TA = \frac{\sum_{i=1}^{c} T_i}{N} \qquad (1)$$

The second measure, local accuracy (LA), is defined as

$$LA = \frac{\sum_{i=1}^{c} P_i}{c} \qquad (2)$$

In Equation (1) and (2), c is the number of subcellular locations, N is the total number of proteins in the dataset, T_i is the number of proteins correctly predicted in location i, and $P_i=T_i/n_i$, where n_i is the number of proteins in location i. These two measures were also used in [2].

Table 2. The 9-letter encoding scheme for the 20 amino acids based on their physical-chemical proper-ties

Subcellular Location	Number of protein sequences
Chloroplast	671
Cytoplasmic	1241
Cytoskeleton	40
Endoplasmic reticulum	114
Extracellular	861
Golgi apparatus	47
Lysosomal	93
Mitochondrial	727
Nuclear	1932
Peroxisomal	125
Plasma membrane	1674
Vacuolar	54
Total	7579

Table 3. Prediction performance for the 12 subcellular locations

Subcellular Location	Accuracy (%)
Chloroplast	78.1
Cytoplasmic	73.2
Cytoskeleton	67.5
Endoplasmic reticulum	71.9
Extracellular	76.1
Golgi apparatus	42.6
Lysosomal	71.0
Mitochondrial	55.7
Nuclear	92.2
Peroxisomal	44.0
Plasma membrane	94.5
Vacuolar	46.3
LA	67.8
TA	80.9

3.3 Results

In this paper, ten HKNNs with parameters $k=4$ and $\lambda=0.8$ were ensembled and test on the dataset. The total accuracy (TA) and location accuracy (LA) were calculated by 5-fold cross-validation. The prediction accuracy for each subcellular location is shown in Table 3.

3.4 Comparison with Previous Methods

In order to examine the performance of our method, we made comparison with previous methods, especially the methods by Park and Kanehisa[2] who had used the same dataset and 5-fold cross-validation test. The comparison results are shown in Table 4. Our method improved the LA significantly, from 57.9% to 67.8%, together with a small increase in the TA, from 78.2% to 80.9%. Our method is more balanced than their method, that is, our method performs better than theirs for all small groups, such as Cytoskeleton (58.5% vs. 67.5%), Endoplasmic reticulum (46.5% vs. 71.9%), Golgi apparatus (14.6% vs. 42.6%), Lysosomal (61.8% vs. 71.0%), Peroxisomal (25.2% vs. 44.0%) and Vacuolar (25.0% vs. 46.3%).

Table 4. Comparison our method with a previous method

Subcellular Location	Park and Kanehisa[2]	Our Method
Chloroplast	72.3	**78.1**
Cytoplasmic	72.2	**73.2**
Cytoskeleton	58.5	**67.5**
Endoplasmic reticulum	46.5	**71.9**
Extracellular	78.0	76.1
Golgi apparatus	14.6	**42.6**
Lysosomal	61.8	**71.0**
Mitochondrial	57.4	55.7
Nuclear	89.6	**92.2**
Peroxisomal	25.2	**44.0**
Plasma membrane	92.2	**94.5**
Vacuolar	25.0	**46.3**
LA	57.9	67.8
TA	78.2	80.9

4 Conclusion and Future Work

In this paper, a new method based on an ensemble of K-local Hyperplane Distance Nearest Neighbor algorithm is proposed to predict protein subcellular locations of eukaryotic cells. The experimental results on the same dataset showed an improvement in prediction accuracy over existing algorithms. Other advantages of our method include: (1) it has relatively fewer adjustable parameters compared with SVM and Neural Network and (2) like KNN, it does not need a training process.

In the future, we will develop the ensembling classifier as a web service for public usage.

References

1. Chou, K.C.: Review: prediction of protein structural classes and subcellular locations. Current Protein and Peptide Science 1, 171–208 (2000)
2. Park, K.J., Kanehisa, M.: Prediction of Protein Subcellular Locations by Support Vector Machines using Compositions of Amino Acids and Amino Acid Pairs. Bioinformatics 19(13), 1656–1663 (2003)
3. Hua, S.J., Sun, Z.R.: Support vector machine approach for protein subcellular localization prediction. Bioinformatics 17(8), 721–728 (2001)

4. Matsuda, S., et al.: A novel representation of protein sequences for prediction of subcellular location using support vector machines. Protein Science 14, 2804–2813 (2005)
5. Cai, Y.D., et al.: Artificial neural network model for predicting protein subcellular location. Computers and Chemistry 26, 179–182 (2002)
6. Emanuelsson, O., et al.: Predicting subcellular localization of proteins based on their N-terminal amino acid sequence. Journal of Molecular Biology 300(4), 1005–1016 (2000)
7. Lu, Z., et al.: Predicting subcellular localization of proteins using machine-learned classifiers. Bioinformatics 20(4), 547–556 (2004)
8. Huang, Y., Li, Y.: Prediction of protein subcellular locations using fuzzy k-NN method. Bioinformatics 20(1), 21–28 (2004)
9. Nakashima, H., Nishikawa, K.: Discrimination of intracellular and extracellular proteins using amino acid composition and residue-pair frequencies. Journal of Molecular Biology 238(1), 54–61 (1994)
10. Chou, K.C.: Prediction of protein cellular attributes using pseudo-amino-acid-composition. Proteins 43(3), 246–255 (2001)
11. Nair, R., Rost, B.: Better prediction of sub-cellular localization by combining evolutionary and structural information. Proteins 53, 917–930 (2003)
12. Cai, Y.D., et al.: Support vector machines for predicting membrane protein types by using functional domain composition. Biophysical Journal 84(5), 3257–3263 (2003)
13. Vincent, P., Bengio, Y.: K-local hyperplane and convex distance nearest neighbor algorithms. In: Dietterich, T.G., Becker, S., Ghahramani, Z. (eds.) NIPS. Advances in Neural Information Processing Systems, vol. 14, pp. 985–992. MIT Press, Cambridge (2002)
14. Yang, M.Q., Yang, J.Y.: Identification of Intrinsically Unstructured Regions in Proteins Using Primary Structure. In: Arabnia, H.R., Valafar, H. (eds.) BIOCOMP'06. Proceedings of the 2006 International Conference on Bioinformatics & Computational Biology, pp. 303–309. CSREA Press (2006)
15. Freund, Y.: Boosting a weak learning algorithm by majority. Information and computation 121(2), 256–285 (1995)

A Similarity Retrieval Method in Brain Image Sequence Database

Haiwei Pan[1], Qilong Han[1], Xiaoqin Xie[1], Zhang Wei[2], and Jianzhong Li[2]

[1] Dept. of Computer Science, Harbin Engineering University, Harbin, P.R. China
{panhaiwei, hanqilong, xiexiaoqin}@hrbeu.edu.cn
[2] Dept. of Computer Science, Harbin Institute of Technology, Harbin, P.R. China
{lijzh, wzhang74}@hit.edu.cn

Abstract. Image mining is more than just an extension of data mining to image domain but an interdisciplinary endeavor. Very few people have systematically investigated this field. Similarity Retrieval in medical image sequence database is an important part in domain-specific application because there are several technical aspects which make this problem challenging. In this paper, we introduce a notion of image sequence similarity patterns (ISSP) for medical image database. These patterns are significant in medical images because the similarity for two medical images is not important, but rather, it is the similarity of objects each of which has an image sequence that is meaningful. We design the new algorithms with the guidance of the domain knowledge to discover the possible Space-Occupying Lesion (PSO) in brain images and ISSP for similarity retrieval. Our experiments demonstrate that the results of similarity retrieval are meaningful and interesting to medical doctors.

1 Introduction

Advances in image acquisition and storage technology have led to tremendous growth in very large and detailed image databases [1]. A vast amount of image data is generated in our daily life and each field, such as medical image (CT images, ECT images and MR images etc), satellite images and all kinds of digital photographs. These images involve a great number of useful and implicit information that is difficult for users to discover.

Image mining can automatically discover these implicit information and patterns from the high volume of images and is rapidly gaining attention in the field of data mining. Image mining is more than just an extension of data mining to image domain. It is an interdisciplinary endeavor that draws upon computer vision, image processing, image retrieval, machine learning, artificial intelligence, database and data mining, etc. While some of individual fields in themselves may be quite matured, image mining, to date, is just a growing research focus and is still at an experimental stage. Research in image mining can be broadly classified to two main directions: (1) domain-specific applications; (2) general applications [2]. The focus in the first direction is to extract the most relevant image features into a form suitable for data mining [4,8,9] and the latter is to generate image patterns that may be helpful in understanding

R. Alhajj et al. (Eds.): ADMA 2007, LNAI 4632, pp. 352–364, 2007.
© Springer-Verlag Berlin Heidelberg 2007

of the interaction between high level human perceptions of image and low level image features [1,10,14,15]. Data mining in medical images belongs to the first direction.

Brain tissue is human's advanced nerve center, so its function is particularly important. The disease affecting the brain has received much attention in the domain of medicine. That is why data mining in medical images for assisting medical staff is so significant. Computerized Tomography (CT) is one of the most important techniques that are used to diagnose by medical doctors. Brain CT scan of each patient (as an object below) is an image sequence in which each one is an image of a layer every a few millimeters from calvaria. There exists a certain spatial relationship between images in the sequence.

This paper presents a new method to retrieve similar objects each of which includes an image sequence. The novelty includes three directions. The first is to make use of medical domain knowledge efficiently to guide data mining. The second is that we utilize two different clustering algorithms on pixels to discover the possible brain diseases that are called Space-Occupying Lesion (SO) by doctors, as shown in figure 1(c). The third is that we introduce a notion of image sequence similarity patterns ISSP for similarity retrieval.

The rest of the paper is organized as follows: section 2 is the statement of problem. Pixel clustering for detecting PSO is introduced in section 3. Section 4 presents similarity retrieval method based on ISSP. Section 5 is experimental result. Conclusions and future research are presented in section 6.

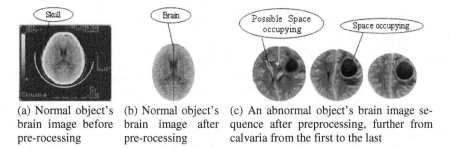

(a) Normal object's brain image before pre-rocessing   (b) Normal object's brain image after pre-rocessing   (c) An abnormal object's brain image sequence after preprocessing, further from calvaria from the first to the last

Fig. 1. An example of normal and abnormal brain image

2   Statement of Problem

At present, the main work of data mining on medical images [3,4,5,6,7,12,13] has two characteristics: (1) research content is the images in the medical image database, not the objects with medical images. For example, it is possible to classify images in the same object into the different class because they always have different morphological SO. This determines the type of knowledge that will be mined; (2) research method is to extract features from images to form feature attributes and use data mining on these attributes, not to consider the fundamental element – pixel's significance. In fact, medical doctors make a diagnosis mainly according to medical knowledge and the tone of pixels. Figure 2 shows the framework of these precious works, where IM_i is an image and F_{ij} is a feature in IM_i. Also, these work paid little attention to the guidance effect of domain knowledge to data mining.

Our research content is objects each of which contains a series of images and the images from different objects maybe have different intensity, see figure 3. The images from one object have the same structure but come from the different layers of a brain.

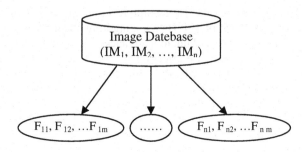

Fig. 2. Framework of Precious Works. IM_n is an image and F_{nm} is one of the features extracted from IM_n.

They have similar pixel density distribution and it is possible for two spatial-adjacent images to be similar in one object, see figure 1(c). Each image contains some PSOs that also have a spatial relationship in an image. Therefore, the objects with the similar image sequence patterns (ISP) should have similar clinical manifestations. It is very helpful to assist medical doctors to make a diagnosis.

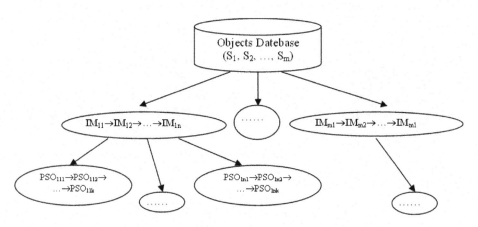

Fig. 3. Framework of Our Work

3 Pixel Clustering for Detecting Possible SO (PSO)

In preprocessing, our work uses domain knowledge effectively to remove the noisy data. A brain CT image mainly consists of three parts: noisy data, skull and cerebrum. The noisy data includes the black background and some additive information on it, such as CT identification, date and patient's name etc. This information is not only helpless but revealing patient's privacy. We are only interested in cerebrum. So we use domain knowledge (DK1) and cropping technique of image processing to gain it.

DK1 --- human's brain skull has the highest density and surrounds the cerebrum. That is, the skull is a cricoid area with the whitest pixels that separate cerebrum from the noisy data, see figure 1(a). It becomes easy to remove the noisy data by using cropping technique and keep the interesting region with the guidance of DK1.

After image preprocessing, all the objects are formed as follows, see table 1. Each object has a unique identification (ID) and its image part (IM) is a preprocessed image sequence where every code is composed of id and the image sequence number. The arrow represents the spatial relationship.

Table 1. Each object is formed as a record in the table after its images are preprocessed

ID	IM
001	$001.01 \rightarrow 001.02 \rightarrow 001.03 \rightarrow \ldots \rightarrow 001.07$
002	$002.01 \rightarrow 002.02 \rightarrow 002.03 \rightarrow \ldots \rightarrow 002.13$
...	......
n	$n.01 \rightarrow n.02 \rightarrow n.03 \rightarrow \ldots \rightarrow n.10$

The following domain knowledge (DK2) is used to direct clustering algorithm.

(1) The normal persons have nearly the same brain structure that is evident to be lateral symmetry. That is, the distribution of density in the left hemisphere of the brain is almost identical with the right, see figure 1 (b). If there is SO in either side, its density will change and destroy the symmetry;

(2) If one object has SO, it is more possible that this SO will be shown in some continuous images, see figure 1 (c).

3.1 Basic Definitions

Let $S=\{S_i \mid i=1\ldots m\}$ be object set; Let $S_i=\{IM_{i1}, IM_{i2}, \ldots, IM_{in}\}$ be object or ordered image set, where (1) IM_{i1} and IM_{in} are the nearest and farthest image from calvaria; (2) For any $IM_{i1}, IM_{i2}, \ldots, IM_{ip}$, IM_{ip} must be the farthest image from calvaria;

For any preprocessed image IM_p, it is halved by brain midline (shown as figure 1(b)) and is composed of two parts: $IM_p(L)$ and $IM_p(R)$ present the left and right hemisphere image respectively. For any IM_p and IM_j, they are adjacent if (1) $IM_p \in S_i$ and $IM_j \in S_i$; (2) p=j+1 or p=j-1;

Definition 1. Pixel set of IM_p is defined as $P=\{p_i \mid p_i$ is the pixel with coordinate (x_i, y_i) in the image $IM_p\}$, $P(L)$ and $P(R)$ are pixel set of $IM_p(L)$ and $IM_p(R)$ respectively.

According to the symmetry of the brain structure in DK2, we assume that the number of pixels in $IM_p(L)$ and $IM_p(R)$ are equal. That is, $|P(L)|=|P(R)| = |P|/2$. For any $p_{li} \in P(L)$, $p_{rj} \in P(R)$, they are symmetric pixel if the line between p_{li} and p_{rj} is halved vertically by brain midline. They are denoted as p_{li} and p_{ri} below.

Definition 2. We partition all pixels in pixel set P into m blocks. Pixels in the same block have the same grey level and pixels in the different blocks have the different grey level. Let $G(P)=\{g_1, g_2, \ldots, g_m\}$ be P's grey-scale (GS) set if $G(P)$ is an ascending sort set of $g_1', g_2', \ldots, g_m'$ and g_i' is grey level of pixels in the ith block, where

g_i (i=1,…,m) is the ith GS, g_1' and g_m' are P's minimum and maximum GS respectively. The GS of pixel p_i is denoted as $g(p_i)$. we call $g_{mean}(P)$ the mean GS if

$$g_{mean}(P) = \sum_{i=1}^{|P|} g(p_i)/|P|$$

Definition 3. For any P and distance function DisA=$|g_k - g_{mean}(P)|$, mid-value GS is a middle value in the GS set that minimizes DisA.

Theorem 1. Mid-value GS set includes not more than two values.

Proof. The value of DisA has two possibilities:

(1) If $g_k - g_{mean}(P) < 0$, then DisA= $g_{mean}(P) - g_k$;
(2) If $g_k - g_{mean}(P) \geq 0$, then DisA= $g_k - g_{mean}(P)$;

In the first case, if there exist more than two GS that make the value of DisA be a certain minimum ϒ: g_1', g_2', … , g_k'(k>2), then they must satisfy the equation g_1'=g_2'= … =g_k'. This doesn't agree with the definition of GS set. Therefore, it is only one GS that minimizes DisA in the first case, denoted as g_{mida}.
Case 2 is proved similarly and thus it is also only one GS that minimizes DisA, denoted as g_{midb}.
If $g_{mean}(P) - g_{mida} = g_{midb} - g_{mean}(P)$, then mid-value set includes two elements. Otherwise, it only includes one of g_{mida} and g_{midb} that minimizes DisA.
For any P, g_s is called Benchmark GS and another one is denoted as g_s', if
(1) Mid-value GS set includes one element g_{mid}, and g_s is the minimum value between g_{mean} and g_{mid}; (2) Mid GS set includes two elements, g_s is the minimum value between these two values.

Definition 4. For pixel set P, let $g^{(l)}$={g_i | $g_1 \leq g_i \leq g_1 + |g_1 - g_s|/2$} be low bound GS; $g^{(h)}$={g_i | $g_m - |g_m - g_s'|/2 \leq g_i \leq g_m$} be high bound GS; $g^{(b)}= g^{(l)} \cup g^{(h)}$ be bound GS.
For pixel set P, let
$P^{(l)}$={p_i | $g(p_i) \in g^{(l)}$} be low bound pixel set;
$P^{(h)}$={p_i | $g(p_i) \in g^{(h)}$} be high bound pixel set;
$P^{(b)}= P^{(l)} \cup P^{(h)}$ be bound pixel set and pixels in $P^{(b)}$ are bound pixels.

Definition 5. $\Delta g(P)$ is IM_p's difference set if for any symmetrical pixel p_{li} and p_{ri} in P, $\Delta g(P)$={ $\Delta g_i | \Delta g_i = g(p_{li}) - g(p_{ri})$, i=1,2,…,|p|/2 }.

Definition 6. For any image sequence <IM_{ij}, …, IM_{ik}>, if

(1) only the first and last image IM_{ij} and IM_{ik} have one adjacent IM;
(2) other IM_{ip} (if existed) has two adjacent IM;
we called that it has the property of continuity. Image sequence <IM_{ij}, …, IM_{ik}> has the property of discontinuity if it can't satisfy the property of continuity.

Definition 7. For any pixel p_i and a certain integer $\mathcal{E}$, the assemble of pixels is called ε-adjacent area (ε-AA) if distance between p_i and pixels in the assemble is not more than $\mathcal{E}$. P_i is core pixel (c-pixel) if its ε-AA involves no less than MP pixels that satisfy some conditions. P_i is immediate density reachable from p_j if p_i is in the ε-AA of

p_j and p_j is c-pixel. P_1 and p_k are *density reachable* if for some pixels $p_1, p_2, \ldots, p_k$, any pixel p_{i+1} is immediate density reachable from p_i. P_i and p_j are *density connective* if there exists pixel p_k that is density reachable from not only p_i but also p_j.

3.2 Pixel Clustering with the Guidance of DK2

It is very crucial step for medical doctors to determine whether there is a Space-Occupying Lesion or not in the brain images. In this paper, we use clustering method on the pixels of images to detect the PSO. Firstly, for any IM_p, we compute to get the IM_p's difference set $\Delta g(P)$, then sort the absolute value of each element in $\Delta g(P)$ to yield the set $\Delta g'(P) = \{ | \Delta g_i | \ | \Delta g_i \in \Delta g(P)$ and for any $|\Delta g_i|$, it must be maximal in the former i elements$\}$. Each element of $\Delta g'(P)$ is regarded as an atomic cluster and hierarchical clustering from bottom to top will not stop in the light of the following similarity function of difference between pixels' GS until the number of clusters is equal to a specified value k.

Similarity$(C_i, C_j) = \min|T_i - T_j|$, where T_i and T_j are mean value of all $|\Delta g_i|$ in cluster C_i and C_j respectively. The algorithm is as follows:

Clustering algorithm I:
Input: the set $\Delta g'(P)$ and the number of clusters k
Output: k clusters that satisfy the similarity function similarity(C_i, C_j)
1.   Each element of $\Delta g'(P)$ is regarded as an atomic cluster and compute $|T_i - T_j|$ of the adjacent clusters;
2.   Clustering in terms of similarity(C_i, C_j);
3.   While (the number of clusters is not equal to k) {
4.   Compute $|T_i - T_j|$ of the adjacent clusters;
5.   Clustering in terms of similarity(C_i, C_j);}

According to "If there is SO in either side, its density will change and destroy the symmetry", we can deduce that if there exists SO in the image, then pixels' GS of SO should change and the value of the elements corresponding to these pixels in $\Delta g'(P)$ will be much greater than zero. Otherwise, the value of these corresponding elements in $\Delta g'(P)$ will approximate zero. Therefore, we set the number of clusters to 2 as the termination condition. The first step of Clustering algorithm I scans $|P|/2$ elements in the set $\Delta g'(P)$ one time and time complexity is $O(|P|/2)$. The second step is to select the minimum and time complexity is $O(|P|/2)$, too. In the third step, the loop times is related to the speed of clustering. The worst case, only two clusters is clustered to a bigger cluster in each loop. The number of time is $|P|/2-3$ and time complexity of the 4th and 5th step is $O(|P|/2-i)$, where i is the ith loop. Accordingly, the 3rd and 5th step for the worst case need $(n-3)(n+2)/2$ operations and time complexity is $O(n^2)$.

According to the symmetry of the brain structure, we single out the cluster with the greater $|\Delta g_i|$ from clustering algorithm I (denoted as high difference cluster) to be the main study objects of the next step. It means that data size to be processed may be reduced. For Each Δg_i, there are two corresponding symmetric pixels p_{li} and p_{ri}. All symmetric pixels $g(p_{li})$ and $g(p_{ri})$ in the high difference cluster are judged to be whether bound GS or not, then bound pixel set is generated.

Next, the based-on density clustering method is utilized to re-cluster these bound pixel set and determine the location and size of SO in each brain image. The algorithm is as follows:

Clustering algorithm II:
Input: all bound pixel set, ε and MP
Output: k clusters
1. Assume that the pixel count of any bound pixel set is bn and examine ε-AA of these bn pixels;
2. If (ε-AA of p_i involves more than MP bound pixels)
3. Then mark p_i as the c-pixel;
4. While (all c-pixels) {
5.   Clustering all *density reachable* pixels; }

For ε and MP in clustering algorithm II, we specify their value through learning on the normal object's brain images. The learning process is as follows: (1) run the first clustering on the normal object's IM_p and achieve the bound pixel set; (2) count (not clustering) on the bound pixel sets which, in fact, are noisy data but not SO, and compute the maximum of the radius and the count of all bound pixel set which are the greatest lower bound of ε and MP. Time complexity of this algorithm is O(nlogn). The k clusters generated from this algorithm are k PSOs.

4   Similarity Retrieval Method

In this section, we will: (1) discover the image sequence pattern (ISP) of one object; (2) discover the image sequence similarity patterns (ISSP) of two objects.

4.1   Discovering ISP Algorithm

The whole PSO in an image can be found using the above algorithm. For these PSOs, we denote each PSO as the following: $<H(L), (x_i, y_i), (x_{a1}, x_{a2}), (y_{b1}, y_{b2})>$, where H (L) represents the high (low) bound GS, (x_i, y_i) is the coordinate of the center of this PSO, (x_{a1}, x_{a2}) and (y_{b1}, y_{b2}) are the max and min x and y coordinate of the PSO. They are computed by the following formula:

$$x_i = \frac{1}{k}\sum_{j=1}^{k} x_j \qquad (1) \qquad\qquad y_i = \frac{1}{k}\sum_{j=1}^{k} y_j \qquad (2)$$

$$x_{a1} = \max_{j=1}^{k}(x_j) \qquad (3) \qquad\qquad x_{a2} = \min_{j=1}^{k}(x_{ij}) \qquad (4)$$

$$y_{b1} = \max_{j=1}^{k}(y_j) \qquad (5) \qquad\qquad y_{b2} = \min_{j=1}^{k}(y_j) \qquad (6)$$

We take the center of the brain as the origin of coordinates. If $x_i \leq 0$, then this PSO is in IM(L). Otherwise, it is in IM(R).

Definition 8. PSO_k is prior to PSO_j if for $PSO_k = <H(L), (x_k, y_k), (x_{ka1}, x_{ka2}), (y_{kb1}, y_{kb2})>$ and $PSO_j = <H(L), (x_j, y_j), (x_{ja1}, x_{ja2}), (y_{jb1}, y_{jb2})>$, they satisfy one of the three prior conditions:

(a) $x_k < x_j$; (b) $x_k = x_j$ and $y_k > y_j$; (c) $x_{ka1} \leq x_{ja1}$ and $x_{ka2} \geq x_{ja2}$; (d) $y_{kb1} \leq y_{jb1}$ and $y_{kb2} \geq y_{jb2}$;
We denote it as $PSO_k >> PSO_j$.

According to this priority, we describe PSO pattern (PSOP) of each image as the following form:

$$PSOP(IM_i) = <L_{i1}, L_{i2}, ..., L_{im}; R_{i1}, R_{i2}, ..., R_{in}>$$

Where L_{im} and R_{in} represent a PSO in $IM_i(L)$ and $IM_i(R)$ respectively.

Definition 9. PSOP of two images is complete similar if $|PSOP(IM_i)| = |PSOP(IM_j)|$, and the corresponding L_{ik} and L_{jk} (or R_{ik} and R_{jk}) have the same bound GS, locate the same part of the image (IM(L) or IM(R)) and satisfy the same prior condition. That is, if L_{ik} and $L_{i(k+1)}$ satisfy the prior condition (a) or (b), then L_{jk} and $L_{j(k+1)}$ must satisfy (a) or (b).

Definition 10. PSOP of two images is incomplete similar if cut one or more discontinuous PSO from one or both of these two images, then $|PSOP(IM_i)| = |PSOP(IM_j)|$, and the corresponding L_{ik} and L_{jk} (or R_{ik} and R_{jk}) have the same bound GS, locate the same part of the image (IM(L) or IM(R)) and satisfy the same prior condition.

For each image of one object, it has a PSOP and maybe is different from the PSOP of the adjacent image. We compare the PSOP of all adjacent images and get a pattern sequence in which two adjacent patterns are not same. This pattern sequence is called image sequence pattern (ISP).

The algorithm of discovering ISP is as follows.

DISP Algorithm:
Input: m images of one object
Output: Image sequence patterns (ISP)
1. Initialization: j=1, n_j=1, k=1;
2. For i = 1 to m {
3. Compare the pattern $PSOP(IM_i)$ and $PSOP(IM_{i+1})$ of i_{th} and $(i+1)_{th}$ image;
4. If complete similar
5. Then record the pattern as $< PSOP(IM_k), n_j = n_j +1>$;
6. Else if j = i
7. Then k=i and record the pattern as $< PSOP(IM_k), n_j = 1>$; j=j+1;
8. Else k=i+1 and record the pattern as $< PSOP(IM_k), n_j = 1>$; j=j+1;}

We use an example to illustrate ISP from DISP algorithm, see figure 4. Convenient for description, we denote ISP as the following form:

$$ISP(S_i) = \{<M_i(IM_{i1}), n_1>, <M_i(IM_{i2}), n_2>, ..., <M_i(IM_{ik}), n_k>\}, \text{ or }$$
$$\{<M_i, n_1>, <M_i, n_2>, ..., <M_i, n_k>\}$$

Where for j=1...k, M_j and M_{j+1} are the distinct PSOP. IM_{ij} is the first image with the PSOP M_j and n_j is the number of the adjacent images with the same PSOP M_j.

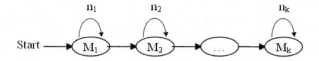

Fig. 4. Patterns from Discovering ISP Algorithm

4.2 Discovering ISSP Algorithm

For ISP(S_i) and ISP(S_j), ISSP of two objects refers to the longest similar and continuous sub-patterns that belongs to the ISP of each object. For example, ISP(S_i)= {< M_1, 1>, < M_2, 2>, < M_3, 2>, < M_4, 1>, < M_5, 1>} and ISP(S_j)= {< M_1', 1>, < M_2', 3>, < M_3', 3>, < M_4', 2>, < M_5', 1>, < M_6', 1>}, if M_1 is similar to M_1', M_3 similar to M_3', M_4 similar to M_4' and M_5 similar to M_6', then ISSP of these two objects is < M_1', M_3', M_4', M_6' >. Here, M_3' and M_4' are called continuous sup-pattern, and M_1' and M_3' are discontinuous sup-pattern.

Since there are spatial relationship between all images in S_i, that is, for any IM_{i1}, IM_{i2}, ..., IM_{in}, $IM_{i(j+1)}$ must be farther from calvaria than IM_{ij}, it is not necessary to retrieve the whole PSOPs of one object to find the similar pattern of a given M_j. For example, M_1 is PSOP of the farthest image from calvaria of one object and M_p is PSOP of the nearest image of another different object, it is not meaningful to compare M_1 and M_p to find whether they are similar or not because M_1 and M_p show the different parts of the brain. According to this, two rules are introduced to reduce the retrieval space and enhance the efficiency. For two objects S_i and S_j, assumed that ISP(S_i) = m and ISP(S_j) = n,

(1) if m=n, that is, the number of the PSOPs in these two objects is equal, then we only need to retrieve M_{i-1}', M_i' and M_{i+1}' in ISP(S_i) to discover the similar patterns of M_i in ISP(S_i).

(2) If m<n, then we only need to retrieve M_i', M_{i+1}', ..., M_{i+n-m}' in ISP(S_i) to discover the similar patterns of M_i in ISP(S_i).

The algorithm of discovering ISSP is as follows.

DISSP Algorithm:
Input: ISP of two objects
Output: Image sequence similarity patterns (ISSP)

1. Initialization: C=NULL, NC=NULL;
2. Assumed that ISP(S_i)=m and ISP(S_j)=n, if m≤n, start the following steps:
3. For i = 1 to m {
4. if m=n, then goto step (5); if m<n, then goto step(7);
5. Compare M_i in ISP(S_i) with M_{i-1}', M_i' and M_{i+1}' in ISP(S_j). If discover a complete similar PSOP, C=C∪ (M_i→M_j). If there is no PSOP to compare in ISP(S_j), return step (3) to start the next iteration. If there is no complete similar PSOP to be discovered, goto step (6);
6. Compare M_i in ISP(S_i) with M_{i-1}', M_i' and M_{i+1}' in ISP(S_j). If discover a incomplete similar PSOP, NC=NC∪ (M_i→M_j). If there is no PSOP to compare in ISP(S_j), return step (3) to start the next iteration;
7. Compare M_i in ISP(S_i) with M_i', M_{i+1}', ..., M_{i+n-m}' in ISP(S_j). If discover a complete similar PSOP, C=C∪ (M_i→M_j). If there is no PSOP to compare in ISP(S_j), return step (3) to start the next iteration. If there is no complete similar PSOP to be discovered, goto step (8);
8. Compare M_i in ISP(S_i) with M_i', M_{i+1}', ..., M_{i+n-m}' in ISP(S_j). If discover a incomplete similar PSOP, NC=NC∪ (M_i→M_j). If there is no PSOP to compare in ISP(S_j), return step (3) to start the next iteration; }
9. Order the whole patterns in C∪NC by the spatial relationship of M_i and use the exhausting method to discover the ISSP;

Since the above rules reduce the number of the PSOPs to be retrieved efficiently, the time cost of this algorithm mainly lies on the procedure of comparing the similarity of two PSOPs that depends on the count of the PSOs. Therefore, the time complexity of this algorithm is O(km), where k is the number of the PSOs.

Definition 11. Two objects are similar if there exists ISSP in two objects.

We give an example to illustrate this algorithm. For two objects S_i and S_j, ISP(S_i)= $\{< M_1, 1>, < M_2, 2>, < M_3, 2>, < M_4, 1>, < M_5, 1>, < M_6, 1>\}$ and ISP(S_j)= $\{< M_1', 1>, < M_2', 3>, < M_3', 1>\}$, we use DISSP algorithm to get the results: C=($M_2' \rightarrow M_2,$ M_4) and NC=($M_1' \rightarrow M_1,$ M_3) $\cup$ ($M_3' \rightarrow M_6$), where C and NC includes the complete and incomplete similar patterns respectively. Notice that |ISP(S_j)|<|ISP(S_i)|, we order C$\cup$NC by the spatial relationship of M_j in ISP(S_j). The result is ($M_1' \rightarrow M_1,$ M_3) $\cup$ ($M_2' \rightarrow M_2,$ M_4) $\cup$ ($M_3' \rightarrow M_6$). By means of exhausting method, the final ISSP is $<M_1, M_2, M_6>$ and $<M_3, M_4, M_6>$. Therefore, S_i is similar to S_j.

5 Experiments

The dataset utilized in our experiments was real data from hospital. The main reason that we study on real brain CT images instead of any simulative data is to avoid reducing the accuracy of the similarity retrieval and the reliability of the discovered knowledge. To have access to real medical images is a very difficult undertaking due to legally privacy issues and management of hospital. But with some specialists' help and support, we got 103 pieces of precious data, which included 11 normal objects' CT scans and 92 abnormal objects data including CT scans and clinical data. Up to now, we have not found the method used on this kind of object dataset where each object has a brain image sequence. Therefore, we only give our experimental results.

We randomly sample 10 percent of normal objects and 10 percent of abnormal objects as the targets from the dataset and use our algorithm to retrieve the similar objects of each target from the whole dataset. The formulas for precision and recall are given below to evaluate our algorithm:

$$P(p)= \frac{TP}{TP + FP} \qquad\qquad R(p)= \frac{TP}{TP + FN}$$

$$P(n)= \frac{TN}{TN + FN} \qquad\qquad R(n)= \frac{TN}{TN + FP}$$

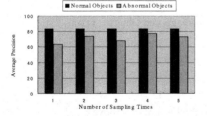
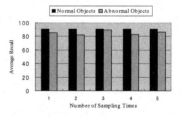

Fig. 5. Average Precision when the Norman and Abnormal Objects as Targets

Fig. 6. Average Recall when the Norman and Abnormal Objects as Targets

Where TP stands for true positives, FP for false positives, FN for false negatives, TN for true negatives, P(p) and R(p) for the precision and recall of similarity retrieval of abnormal objects, P(n) and R(n) for the precision and recall of similarity retrieval of normal objects. Figure 5 and figure 6 shows the experimental results of randomly sampling 5 times. We can observe that the precision and recall are very high for normal targets. Only one abnormal object is retrieved as the similar object of normal targets for each time because the tumor in the brain of this abnormal object is laterally symmetrical. While our clustering algorithms calculate the difference set, they ignore the tumor. One normal object is not retrieved for each time because his (her) brain images are too bright and some noisy dark pixels cause the false results.

The average precision of the abnormal targets is more than 60% but not very high. All retrieved objects are certainly the abnormal objects with tumor in their brain and the image sequence of them is similar to that of the target. The reason why not very high is that their concrete tumor kind is not the same as the target and the medical doctors made a different detailed diagnosis. But this demonstrates that our similarity retrieval algorithm based on ISSP can gain the similar image sequence of the target. The average recall, however, is very high. This illustrates that the results of our similarity retrieval method based on ISSP can include most of the similar objects with the targets. For example, there are 7 objects including the target itself that are actually similar to the target in the dataset and our method can find 10 similar objects in which 6 objects belong to the actually similar objects.

In addition, there are two abnormal objects in the dataset who were misdiagnosed in the first visit to the doctor. Their final diagnosis had to resort to the biopsy techniques that are costly and need to remove the tissue from the patient. Our algorithm succeeds in retrieving the similar objects whose clinical diagnosis are the same as these two abnormal objects. Figure 7 shows one of these two misdiagnosed objects and one of the similar objects with it. This result can provide significant information to the medical doctors to make a right diagnosis.

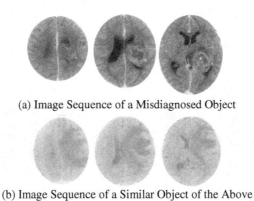

(a) Image Sequence of a Misdiagnosed Object

(b) Image Sequence of a Similar Object of the Above

Fig. 7. Two objects with the similar image sequence pattern have a same clinical diagnosis

6 Conclusions

In this paper, we firstly use two clustering algorithms to generate the possible Space-Occupying Lesion, and then discover ISP for each object. Next, we introduce a notion of image sequence similarity patterns (ISSP) for medical image database. ISSP refer to the longest similar and continuous sub-patterns hidden in two objects each of which contains an image sequence. These patterns are significant in medical images because the similarity for most medical images is not important, but rather, it is the similarity of objects each of which has an image sequence that is meaningful. We design a new algorithm with the guidance of the domain knowledge to generate ISSP for similarity retrieval. Our experiments demonstrate that the guidance of domain knowledge is meaningful and the results are interesting to medical doctors.

References

[1] Zaiane, O.R., et al.: Mining MultiMedia Data. Meeting of Minds, CASCON (1998)
[2] Hsu, W., Lee, M.L., Zhang, J.: Image Mining: Trends and Developments. Journal of Intelligent Information Systems 19(1), 7–23 (2002)
[3] Megalooikonomou, V., Davatzikos, C., Herskovits, E.H.: Mining Lesion-Deficit Associations in a Brain Image Database. In: KDD-99, San Diego, CA (1999)
[4] Hsu, W., Lee, M.L., Goh, K.G.: Image Mining in IRIS: Integrated Retinal Information System. In: Proceedings of the ACM SIGMOD, Dellas, Texas, May 2000, p. 593 (2000)
[5] Liu, Y., Dellaert, F., Rothfus, W.E., Moore, A., Schneider, J., Kanade, T.: Classification-Driven Pathological Neuroimage Retrieval Using Statistical Asymmetry Measures. In: Niessen, W.J., Viergever, M.A. (eds.) MICCAI 2001. LNCS, vol. 2208, Springer, Heidelberg (2001)
[6] Antonie, M.-L., Zaiane, O.R., Coman, A.: Application of Data Mining Techniques for Medical Image Classification. In: Perner, P. (ed.) MLDM 2001. LNCS (LNAI), vol. 2123, Springer, Heidelberg (2001)
[7] Zaiane, O.R., Antonie, M.-L., Coman, A.: Mammography Classification by an Association Rule-based Classifier. In: Zaïane, O.R., Simoff, S.J., Djeraba, C. (eds.) MDM/KDD 2002 and KDMCD 2002. LNCS (LNAI), vol. 2797, Springer, Heidelberg (2003)
[8] Fayyad, U.M., Djorgovski, S.G., Weir, N.: Automating the Analysis and Cataloging of Sky Surveys. In: Advances in Knowledge Discovery and Data Mining, pp. 471–493 (1996)
[9] Kitamoto, A.: Data Mining for Typhoon Image Collection. In: Perner, P. (ed.) MLDM 2001. LNCS (LNAI), vol. 2123, Springer, Heidelberg (2001)
[10] Ordonez, C., Omiecinski, E.: Discovering Association Rules Based on Image Content. In: IEEE Advances in Digital Libraries Conference, IEEE Computer Society Press, Los Alamitos (1999)
[11] Burl, M.C., et al.: Mining For Image Content. In systems, Cybernetics, and Informatics / Information Systems: Analysis and Synthesis (1999)
[12] Soltanian-Zadeh, H., Nezafat, R., Windham, J.P.: Is There Texture Information in Standard Brain MRI? In: Proceedings of SPIE Medical Imaging 1999: Image Processing conference, San Diego, CA (1999)

[13] Barra, V., Boire, J-Y.: Tissue segmentation on MR Images of the Brain by Positivistic Clustering on a 3D Wavelet Representation. J. of Magnetic resonance Imaging 11, 267–278 (2000)

[14] Yanai, K.: Image Classification by a Probabilistic Model Learned from Imperfect Training Data on the Web. In: Seventh International Workshop on Multimedia Data Mining, pp. 75–81 (2006)

[15] Walter, J.A., WeBling, D., Essig, K., Ritter, H.: Interactive Hyperbolic Image Browsing-Towards an Integrated Multimedia Navigator. In: MDM2006. Seventh International Workshop on Multimedia Data Mining, pp. 101–108 (2006)

A Criterion for Learning the Data-Dependent Kernel for Classification

Jun-Bao Li[1], Shu-Chuan Chu[2], and Jeng-Shyang Pan[3]

[1] Department of Automatic Test and Control, Harbin Institute of Technology, Harbin, China
junbaolihit@hotmail.com
[2] Department of Information Management, Cheng Shiu University, Kaohsiung, Taiwan
jspan@cc.kuas.edu.tw
[3] Department of Electronic Engineering, National Kaohsiung University of Applied Sciences,
Kaohsiung, Taiwan

Abstract. A novel criterion, namely Maximum Margin Criterion (MMC), is proposed for learning the data-dependent kernel for classification. Different kernels create the different geometrical structures of the data in the feature space, and lead to different class discrimination. Selection of kernel influences greatly the performance of kernel learning. Optimizing kernel is an effective method to improve the classification performance. In this paper, we propose a novel kernel optimization method based on maximum margin criterion, which can solve the problem of Xiong's work [1] that the optimal solution can be solved by iteration update algorithm owing to the singular problem of matrix. Our method can obtain a unique optimal solution by solving an eigenvalue problem, and the performance is enhanced while time consuming is decreased. Experimental results show that the proposed algorithm gives a better performance and a lower time consuming compared with Xiong's work.

1 Introduction

Pattern classification is one hot research topic in the pattern recognition area. In the recent years, many method were developed in the past years. In the recent years, kernel-based learning machines, e.g., Kernel Principal Component Analysis (KPCA) [1], Kernel Discriminant Analysis (KDA) [2] and Support Vector Machine (SVM) [3], have attracted much attention in the areas of pattern recognition and machine learning [4]. KPCA was originally developed by Scholkopf et al. in 1998 [5], while KFD was firstly proposed by Mika et al. in 1999 [6]. In the past research work, researchers developed a series of KDA algorithms (Juwei Lu[7], Baudat and Anouar [8], Liang and Shi [9],[10],[11], Yang [1],[13], J. Lu [12], Zheng [14], Huang [15], Wang [16] and Chen [17]). KDA has been applied in many real-world applications owing to its excellent performance on feature extraction. However, KDA endures a crucial problem, i.e., kernel selection. Kernel selection is one of the key problems in many other kernel-based methods because the geometrical structure of the data in the kernel mapping space has significant impact on the performance of these methods, while the geometrical structure of the data in the kernel mapping space is totally determined by

R. Alhajj et al. (Eds.): ADMA 2007, LNAI 4632, pp. 365–376, 2007.

the kernel function. In the kernel-machine, three kinds of kernels, i.e., polynomial kernels, Gaussian kernels, and sigmoid kernels are widely used. But all the above kernels have only their own kernel structure, and different kernels create different geometrical structures of the data in the feature space, and lead to different class discrimination. Selection of kernel influences greatly the performance of kernel learning. So the separability of the data in the feature space could be even worse if an inappropriate kernel is used. Researchers have paid increasing attention to improve KDA through optimizing the kernel parameters, and many algorithms are developed (Huang [15], Wang [16] and Chen [17]). However, the kernel optimization method of choosing the parameters of kernel function just from a set of discrete values of the parameters is not able to change the geometrical structures of the data in the kernel mapping space. A data-dependent kernel is suitable to all data sets as a general kernel function. We can optimize the kernel structure through seeking optimal combination coefficients for a data-dependent kernel. Xiong [18] optimized kernel in the empirical feature space by seeking the optimal combination coefficients of data-dependent kernel based on Fisher criterion and a good performance was reported, but it endures a singular matrix problem during seeking the optimal solution. In this paper, we propose a novel kernel optimization method based on maximum margin criterion, which can solve the matrix singular problem.

The rest of paper is organized as follows. In section 2, we introduce the data-dependent kernel, and the proposed algorithm is presented in Section 3. Experimental results and conclusion are given in Section 4 and Section 5 respectively.

2 Data-Dependent Kernel

Data-dependent kernel with a general geometrical structure is applied to create a new kernel in this paper. Given a basic kernel $k(x, y)$, its data-dependent kernel $\tilde{k}(x, y)$ can be defined as follows [21].

$$\tilde{k}(x, y) = f(x)f(y)k(x, y) \tag{1}$$

where $f(x)$ is a positive real valued function x, Amari and Wu [11] expanded the spatial resolution in the margin of a SVM by using $f(x) = \sum_{i \in SV} a_i e^{-\delta \|x - \tilde{x}_i\|^2}$, where $\tilde{x}_i$ is the ith support vector, SV is a set of support vector, a_i is a positive number representing the contribution of $\tilde{x}_i$, and δ is a free parameter. In this paper, we generalize Amari and Wu's method as follows.

$$f(x) = b_0 + \sum_{n=1}^{N_{xv}} b_n e(x, \tilde{x}_n) \tag{2}$$

where $e(x, \tilde{x}_n) = e^{-\delta \|x - \tilde{x}_n\|^2}$, and δ is a free parameter, and $\tilde{x}_n$, $1 \leq n \leq N_{xv}$, are called the "expansion vectors (XVs)" in this paper, N_{xv} is the number of XVs, and

$b_i \in R$ is the "expansion coefficient" associated with $\tilde{x}_n$. The $\tilde{x}_n$, $1 \le n \le N_{XV}$, have different notations in the different kernel learning algorithms. For example, they are called support vectors in the SVM, and relevance vectors in the relevance vector machine. In this paper we uniformly call them "expansion vectors".

From the above discussion, given the δ and $\tilde{x}_n$, $1 \le n \le N_{XV}$, we can obtain a different kernel with different geometrical structure with a different set of expansion coefficients. In this paper, we select the centers of classes as XVs as follows.

$$e(x, \tilde{x}_n) = e(x, \overline{x}_n) = e^{-\delta \| x - \overline{x}_n \|^2} , (n = 1, 2, ..., L) \tag{3}$$

where , $\overline{x}_n$ is the mean of each class and $N_{XV} = L$. After we select the expansion vectors and the free parameter, our goal is to find the expansion coefficients varied with the input data to optimize the kernel. According to the equation (2), given one free parameter δ and the expansion vectors $\{ \tilde{x}_i \}_{i=1,2,...,N_{XVs}}$, we create a matrix as follows.

$$E = \begin{bmatrix} 1 & e(x_1, \tilde{x}_1) & \cdots & e(x_1, \tilde{x}_{N_{XVs}}) \\ \vdots & \vdots & \ddots & \vdots \\ 1 & e(x_M, \tilde{x}_1) & \cdots & e(x_M, \tilde{x}_{N_{XVs}}) \end{bmatrix} \tag{4}$$

Let $\beta = [b_0, b_1, b_2, ..., b_{N_{XVs}}]^T$ $(i = 0, 1, 2, ..., N_{XVs})$ and $\Lambda = diag(f(x_1), f(x_2), ..., f(x_M))$, and according to the equation (2), we obtain

$$\Lambda 1_n = E\beta \tag{5}$$

where 1_n is a n-dimensional vector whose entries equal to unity.

Proposition 1. Let K and $\tilde{K}$ denote the basic kernel matrix and data-dependent kernel matrix respectively, then $\tilde{K} = \Lambda K \Lambda$.

Proof. Since $K = [k(x_i, x_j)]_{n \times n}$ and $\tilde{K} = [\tilde{k}(x_i, x_j)]_{n \times n}$, according to equation (1), we can obtain

$$\tilde{K} = [\tilde{k}(x_i, x_j)]_{n \times n} = [f(x_i)f(x_j)k(x_i, x_j)]_{n \times n} \tag{6}$$

Hence $\tilde{K} = \Lambda K \Lambda$. ☐

Now our goal is to create a constrained optimization function to seek an optimal expansion coefficient vector β. In this paper, we apply the maximum margin criterion to solve the expansion coefficients.

3 Learning the Data-Dependent Kernel for Classification

Xiong [18] proposed a kernel optimization method based on Fisher criterion in the empirical feature space and reported a good performance. Let's review Xiong's kernel optimization method firstly, and then introduce the proposed maximum margin criterion based kernel optimization method.

3.1 Xiong's Method

In [18], Xiong applied Fisher criterion to optimize kernel in the empirical feature space. Fisher criterion is to measure the class separability in the high-dimensional feature space to optimize the expansion coefficients. The Fisher criterion can be defined as follows.

$$J_{Fisher} = \frac{tr\left(S_B^\Phi\right)}{tr(S_W^\Phi)} \tag{7}$$

where S_B^Φ is the between-class scatter matrix, S_W^Φ is the within-class scatter matrix, and tr denotes the trace of a matrix. J_{Fisher} is the well-known Fisher scalar for measuring the class linear separabiltiy.

Proposition 2. Assume Kc_{ij} $(i, j = 1, 2, ..., L)$ is a kernel matrix calculated with the ith and jth class samples and kernel matrix K_{total} with its elements k_{ij} , Let

$$B = diag\left(\tfrac{1}{n_1} Kc_{11}, \tfrac{1}{n_2} Kc_{22}, ..., \tfrac{1}{n_L} Kc_{LL}\right) - \tfrac{1}{n} K_{total} \quad , \qquad W = diag\left(k_{11}, k_{22}, ..., k_{nn}\right)$$

$$-diag\left(\tfrac{1}{n_1} Kc_{11}, \tfrac{1}{n_2} Kc_{22}, ..., \tfrac{1}{n_L} Kc_{LL}\right), \text{ then } tr\left(S_B^\Phi\right) = 1_n^T B 1_n \text{ and } tr\left(S_W^\Phi\right) = 1_n^T W 1_n$$

According to Proposition 2, The Fisher criterion in high-dimensional kernel space can be described as follows.

$$J_{Fisher} = \frac{1_n^T B 1_n}{1_n^T W 1_n} \tag{8}$$

When a data-dependent kernel is selected as the general kernel simultaneously according to Proposition 1, we can acquire

$$\tilde{B} = \Lambda B \Lambda \tag{9}$$

$$\tilde{W} = \Lambda W \Lambda \tag{10}$$

Then we can obtain Fisher criterion in a high-dimensional kernel space, which is described as follows.

$$J_{Fisher} = \frac{1_n^T \Lambda B \Lambda 1_n}{1_n^T \Lambda W \Lambda 1_n} \tag{11}$$

According to Equation (9) and (10), the Equation (11) can be rewritten as follows.

$$J_{Fisher} = \frac{\beta^T E^T BE\beta}{\beta^T E^T WE\beta} \tag{12}$$

$E^T BE$ and $E^T WE$ are the constant matrices, so J_{Fisher} is a function with its variable β. We can create an optimization function constrained by the unit vector β, i.e., $\beta^T \beta = 1$, to maximize J_{Fisher}, which can be described as follows.

$$\max \quad \frac{\beta^T E^T BE\beta}{\beta^T E^T WE\beta}$$

$$subject \quad to \quad \beta^T \beta - 1 = 0 \tag{13}$$

Based on the Fisher criterion the optimal solution must be obtained by iteration updating algorithm [20] owing to the singular problem of matrix, and detailed algorithm can be found in [20]. But based on this method, we can not obtain the unique optimal solution and it needs high time consuming. In order to solve it, we apply maximum margin criterion to find the optimal expansion coefficients.

3.2 Our Method

We maximize the class discrimination in the high-dimensional feature space by maximizing the average margin between different classes which is widely used as maximum margin criterion for feature extraction [20]. The average margin between two classes c_i and c_j in feature space can be defined as follows.

$$Dis = \frac{1}{2n} \sum_{i=1}^{L} \sum_{j=1}^{L} n_i n_j d\left(c_i, c_j\right) \tag{14}$$

where $d\left(c_i, c_j\right) = d\left(m_i^\Phi, m_j^\Phi\right) - S\left(c_i\right) - S\left(c_j\right)$, $i, j = 1, 2, ..., L$, denotes the margin between any two classes, and $S\left(c_i\right)$, $i = 1, 2, ..., L$, is the measure of the scatter of the class c_i, $i = 1, 2, ..., L$, and $d\left(m_i^\Phi, m_j^\Phi\right)$, $i, j = 1, 2, ..., L$, is the distance between the means of two classes. Let S_i^Φ, $i = 1, 2, ..., L$, denote the within-class scatter matrix of class i, which is defined as follows.

$$tr(S_i^\Phi) = \frac{1}{n_i} \sum_{p=1}^{n_i} \left(\Phi\left(x_i^p\right) - m_i^\Phi\right)^T \left(\Phi\left(x_i^p\right) - m_i^\Phi\right) \tag{15}$$

and $tr\left(S_i^\Phi\right)$ measures the scatter of the class i, that is, $S\left(c_i\right) = tr\left(S_i^\Phi\right)$, $i = 1, 2, ..., L$.

Proposition 3. Let $tr(S_B^\Phi)$ and $tr(S_W^\Phi)$ denote the trace of between classes scatter matrix and within classes scatter matrix respectively, then $Dis = tr(S_B^\Phi) - tr(S_W^\Phi)$. □

The proof of Proposition 3 is shown in Appendix.

Let $M = 2 * diag\left(\frac{1}{n_1} Kc_{11}, \frac{1}{n_2} Kc_{22}, ..., \frac{1}{n_L} Kc_{LL}\right) - diag\left(k_{11}, k_{22}, ..., k_{nn}\right) - \frac{1}{n} K_{total}$,

then we can obtain $tr\left(S_B^{\Phi}\right) - tr\left(S_W^{\Phi}\right) = 1_n^T M 1_n$ (Proof refer to Appendix). Accordingly we can obtain

$$Dis = 1_n^T M 1_n \tag{16}$$

Simultaneously according to Proposition 1, we can acquire

$$\tilde{M} = \Lambda M \Lambda \tag{17}$$

where $\tilde{M} = 2 * diag\left(\frac{1}{n_1} \tilde{K}_{11}, \frac{1}{n_2} \tilde{K}_{22}, ..., \frac{1}{n_L} \tilde{K}_{LL}\right) - diag\left(\tilde{k}_{11}, \tilde{k}_{22}, ..., \tilde{k}_{nn}\right) - \frac{1}{n} \tilde{K}_{total}$, and

$\tilde{K}_{ij}$ $(i, j = 1, 2, ..., L)$ is calculated by the ith and jth class of samples with the data-dependent kernel and $\tilde{K}_{total}$ represents the kernel matrix with its elements $\tilde{k}_{pq}$ ($p, q = 1, 2, ..., n$) which is calculated by pth and qth samples with adaptive data-dependent kernel. Thus when a data-dependent kernel is selected as the general kernel, $\widetilde{Dis}$ is obtained as follows.

$$\widetilde{Dis} = 1_n^T \Lambda M \Lambda 1_n^T \tag{18}$$

The above equation can be written as follows.

$$\widetilde{Dis} = \beta^T E^T M E \beta \tag{19}$$

Given a basic kernel $k(x, y)$ and relative data-dependent kernel coefficients, $E^T M E$ is a constant matrix, so $\widetilde{Dis}$ is a function with its variable β. So it is reasonable to seek the optimal expansion coefficient vector β by maximizing $\widetilde{Dis}$. Now we create an optimization function constrained by the unit vector β, i.e., $\beta^T \beta = 1$ as follows.

$$\max \quad \beta^T E^T M E \beta$$

$$subject \quad to \quad \beta^T \beta - 1 = 0 \tag{20}$$

The solution of the above constrained optimization problem can often be found by using the so-called Lagrangian method. We define the Lagrangian as

$$L(\beta, \lambda) = \beta^T E^T M E \beta - \lambda\left(\beta^T \beta - 1\right) \tag{21}$$

with the parameter λ. The Lagrangian L must be maximized with respect to λ and β, and the derivatives of L with respect to β are must vanish, that is,

$$\frac{\partial L(\beta, \lambda)}{\partial \beta} = \left(E^T M E - \lambda I\right)\beta \tag{22}$$

And

$$\frac{\partial L(\beta, \lambda)}{\partial \beta} = 0 \tag{23}$$

Hence

$$E^T ME\beta = \lambda\beta \tag{24}$$

So the problem of solving the constrained optimization function is transformed to the problem of solving eigenvalue equation shown in (24). We can obtain the optimal expansion coefficient vector β^*, that is, the eigenvector of $E^T ME$ corresponding to the largest eigenvalue. It is easy to see that the data-dependent kernel with β^* is adaptive to the input data, which leads to the best class discrimination in feature space for given input data.

4 Experimental Results

After we present our detail theoretical arguments, we implement some experiments on one real dataset (ORL) to test the feasibility of the proposed algorithm. Firstly we select the procedure parameters, δ for similarity measure, kernel parameters, with cross-validation method, and secondly we evaluate the performance of proposed algorithm on computation efficiency and recognition accuracy.

4.1 Dataset Description and Experiment Setting

ORL face database [19], developed at the Olivetti Research Laboratory, Cambridge, U.K., is composed of 400 grayscale images with 10 images for each of 40 individuals. The variations of the images are across pose, time and facial expression. To reduce computation complexity, we resize the original ORL face images sized 112×92 pixels with a 256 gray scale to 48×48 pixels, and some examples are shown in Figure 1.

Fig. 1. Example cropped face images from the ORL face database in our experiments (cropped to the size of 48×48 to extract the facial region)

After describing the two data set used in our experiments. It is worthwhile to make some remarks on the experiment setting as follows: 1) For ORL face database, we randomly select 5 images from each subject, 200 images in total for training, and the rest 200 images are used to test the performance. 5 images of each person randomly selected from YALE database are used to construct the training set, and the rest 5 images of each person are used to test the performance of the algorithms. 2) We run experiments for 10 times, and the average rate is used to evaluate the classification performance. 3) The experiments are implemented on a Pentium 3.0 GHz computer with 512MB RAM and programmed in the MATLAB platform (Version 6.5). 4) The procedural parameters, i.e., kernel parameters and the free parameter δ of a data-dependent kernel are chosen with cross-validation method. 5) The number of projection vectors in each dimensionality reduction method is set to $C-1$ in all experiments.

Polynomial kernel $k(x, y) = (x \cdot y)^d$ $(d \in N)$ and Gaussian kernel $k(x, y) = \exp(-\dfrac{\|x - y\|^2}{2\sigma^2})$ $(\sigma > 0)$ are selected as the basic kernels to compute the kernel matrix. And $d = 2$ and $d = 3$ are selected as polynomial kernel parameter, and $\sigma^2 = 4 \times 10^9$ and $\sigma^2 = 4 \times 10^{10}$ for Gaussian kernel parameter, and the free parameters $\delta = 1 \times 10^{-6}$ and $\delta = 1 \times 10^{-8}$ for the data-dependent kernel parameter.

4.2 Performance Evaluation

Recognition accuracy and time consuming are considered to evaluate the performance of the algorithm. We evaluate the performance of Xiong's method and the algorithm to solve expansion coefficients. Time consuming and recognition accuracy are considered to evaluate the performance. We find the optimal parameters, kernel parameters and free parameters, to evaluate the performance of two methods. As shown in Table 1 and 2, our method can give higher recognition accuracy and lower time consuming compared with Xiong's method. Certainly time consuming of Xiong's method is changeable with different iteration number. But it is apparent that our method costs less time than Xiong's method.

Table 1. Recognition accuracy performance on ORL face databases

Kernel function	Polynomial kernel (d=2)	Polynomial kernel (d=3)	Gaussian kernel ($\sigma^2 = 4 \times 10^9$)	Gaussian kernel ($\sigma^2 = 4 \times 10^{10}$)
Xiong's method	0.8870	0.8770	0.9185	0.8950
Our method	0.9075	0.8890	0.9450	0.9205

Table 2. Time consuming performance on ORL face databases

Kernel function	Polynomial kernel (d=2)	Polynomial kernel (d=3)	Gaussian kernel ($\sigma^2 = 4 \times 10^9$)	Gaussian kernel ($\sigma^2 = 4 \times 10^{10}$)
Xiong's method (seconds)	0.3962	0.3695	0.2625	0.2695
Our method (seconds)	0.0221	0.0204	0.0781	0.0218

5 Conclusion

A novel kernel optimization method based on maximum margin criterion is proposed in this paper. The method gives a high classification accuracy and low time consuming for classification problem. The kernel optimization is expected to be applied to other kernel based learning methods.

References

1. Yang, J., Frangi, A.F., Yang, J.-y., Zhang, D., Jin, Z.: KPCA Plus LDA: A Complete Kernel Fisher Discriminant Framework for Feature Extraction and Recognition. IEEE Trans. Pattern Analysis and Machine Intelligence 27(2), 230–244 (2005)
2. Liu, Q., Lu, H., Ma, S.: Improving kernel Fisher discriminant analysis for face recognition. IEEE Trans. Pattern Analysis and Machine Intelligence 14(1), 42–49 (2004)
3. Vapnik, V.: The Nature of Statistical Learning Theory. Springer, New York (1995)
4. Müller, K.R., Mika, S., Rätsch, G., Tsuda, K., Schölkopf, B.: An introduction to kernel-based learning algorithms. IEEE Trans. Neural Networks 12, 181–201 (2001)
5. Scholkopf, B., Smola, A., Mu'ller, K.R.: Nonlinear Component Analysis as a Kernel Eigenvalue Problem. Neural Computation 10(5), 1299–1319 (1998)
6. Mika, S., Ratsch, G., Weston, J., Scholkopf, B., Mu'ller, K.-R.: Fisher Discriminant Analysis with Kernels. In: Proc. IEEE Int'l Workshop Neural Networks for Signal Processing IX, August 1999, pp. 41–48. IEEE Computer Society Press, Los Alamitos (1999)
7. Lu, J., Plataniotis, K.N., Venetsanopoulos, A.N.: Face recognition using kernel direct discriminant analysis algorithms. IEEE Transactions on Neural Networks 14(1), 117–226 (2003)
8. Baudat, G., Anouar, F.: Generalized Discriminant Analysis Using a Kernel Approach. Neural Computation 12(10), 2385–2404 (2000)
9. Liang, Z., Shi, P.: Uncorrelated discriminant vectors using a kernel method. Pattern Recognition 38, 307–310 (2005)
10. Liang, Z., Shi, P.: Efficient algorithm for kernel discriminant anlaysis. Pattern Recognition 37(2), 381–384 (2004)
11. Liang, Z., Shi, P.: An efficient and effective method to solve kernel Fisher discriminant analysis. Neurocomputing 61, 485–493 (2004)
12. Lu, J., Plataniotis, K.N., Venetsanopoulos, A.N.: Face Recognition Using Kernel Direct Discriminant Analysis Algorithms. IEEE Trans. Neural Networks 14(1), 117–126 (2003)

13. Yang, M.H.: Kernel Eigenfaces vs. Kernel Fisherfaces: Face Recognition Using Kernel Methods. In: Proc. Fifth IEEE Int'l Conf. Automatic Face and Gesture Recognition, May 2002, pp. 215–220 (2002)
14. Zheng, W., Zou, C., Zhao, L.: Weighted maximum margin discriminant analysis with kernels. Neurocomputing 67, 357–362 (2005)
15. Huang, J., Yuen, P.C, Chen, W.-S., Lai, J H: Kernel Subspace LDA with Optimized Kernel Parameters on Face Recognition. In: Proceedings of the Sixth IEEE International Conference on Automatic Face and Gesture Recognition, IEEE Computer Society Press, Los Alamitos (2004)
16. Wang, L., Chan, K.L., Xue, P.: A Criterion for Optimizing Kernel Parameters in KBDA for Image Retrieval. IEEE Trans. Systems, Man and Cybernetics-Part B: Cybernetics 35(3), 556–562 (2005)
17. Chen, W.-S., Yuen, P.C., Huang, J., Dai, D.-Q.: Kernel Machine-Based One-Parameter Regularized Fisher Discriminant Method for Face Recognition. IEEE Trans. Systems, Man and Cybernetics-Part B: Cybernetics 35(4), 658–669 (2005)
18. Huilin Xiong, Swamy, M.N.S., Omair Ahmad, M.: Optimizing the Kernel in the Empirical Feature Space. IEEE Trans. Neural Networks 16(2), 460–474 (2005)
19. Samaria, F., Harter, A.: Parameterisation of a Stochastic Model for Human Face Identification. In: Proceedings of 2nd IEEE Workshop on Applications of Computer Vision, Sarasota, FL (December 1994)
20. Li, H., Jiang, T., Zhang, K.: Efficient and Robust Feature Extraction by Maximum Margin Criterion. IEEE Trans. Neural Networks 17(1), 157–165 (2006)
21. Amari, S., Wu, S.: Improving support vector machine classifiers by modifying kernel functions. Neural Netw. 12(6), 783–789 (1999)

Appendix

1 Proof of Proposition 3

$$Dis = \frac{1}{2n} \sum_{i=1}^{L} \sum_{j=1}^{L} n_i n_j \left[d\left(m_i^\Phi, m_j^\Phi\right) - tr\left(S_i^\Phi\right) - tr\left(S_j^\Phi\right) \right]$$

$$= \frac{1}{2n} \sum_{i=1}^{L} \sum_{j=1}^{L} n_i n_j d\left(m_i^\Phi, m_j^\Phi\right) - \frac{1}{2n} \sum_{i=1}^{L} \sum_{j=1}^{L} n_i n_j \left[tr\left(S_i^\Phi\right) + tr\left(S_j^\Phi\right) \right] \tag{25}$$

Firstly we use Euclidean distance to calculate $d\left(m_i^\Phi, m_j^\Phi\right)$ as follows.

$$\frac{1}{2n} \sum_{i=1}^{L} \sum_{j=1}^{L} n_i n_j d\left(m_i^\Phi, m_j^\Phi\right) = \frac{1}{2n} \sum_{i=1}^{L} \sum_{j=1}^{L} n_i n_j \left(m_i^\Phi - m_j^\Phi\right)^T \left(m_i^\Phi - m_j^\Phi\right)$$

$$= \frac{1}{2n} \sum_{i=1}^{L} \sum_{j=1}^{L} n_i n_j \left[\left(m_i^\Phi - m^\Phi\right)^T \left(m_i^\Phi - m^\Phi\right) + \left(m^\Phi - m_j^\Phi\right)^T \left(m^\Phi - m_j^\Phi\right) \right]$$

$$+ \frac{1}{2n} \sum_{i=1}^{L} \sum_{j=1}^{L} n_i n_j \left[\left(m_i^\Phi - m^\Phi\right)^T \left(m^\Phi - m_j^\Phi\right) + \left(m^\Phi - m_j^\Phi\right)^T \left(m_i^\Phi - m^\Phi\right) \right] \tag{26}$$

It is easy to see $m^\Phi = \frac{1}{n} \sum_{i=1}^{L} n_i m_i^\Phi$, so

$$\frac{1}{2n} \sum_{i=1}^{L} \sum_{j=1}^{L} n_i n_j \left[\left(m_i^\Phi - m^\Phi \right)^T \left(m^\Phi - m_j^\Phi \right) \right]$$

$$= \frac{1}{2n} \left[\sum_{i=1}^{L} n_i \left(m_i^\Phi - m^\Phi \right)^T \right] \left[\sum_{j=1}^{L} n_j \left(m^\Phi - m_j^\Phi \right) \right] \qquad (27)$$

Then

$$\frac{1}{2n} \sum_{i=1}^{L} \sum_{j=1}^{L} n_i n_j \left[\left(m_i^\Phi - m^\Phi \right)^T \left(m_i^\Phi - m^\Phi \right) \right]$$

$$= \frac{1}{2} \sum_{i=1}^{L} n_i \left[\left(m_i^\Phi - m^\Phi \right)^T \left(m_i^\Phi - m^\Phi \right) \right] = \frac{1}{2} tr(S_B^\Phi) \qquad (28)$$

So

$$\frac{1}{n} \sum_{i=1}^{L} \sum_{j=1}^{L} n_i n_j d \left(m_i^\Phi, m_j^\Phi \right) = tr(S_B^\Phi) \cdot \qquad (29)$$

$$\frac{1}{2n} \sum_{i=1}^{L} \sum_{j=1}^{L} n_i n_j tr \left(S_i^\Phi \right) = \frac{1}{2n} \sum_{j=1}^{L} n_j \left(\sum_{i=1}^{L} n_i tr \left(S_i^\Phi \right) \right)$$

$$= \frac{1}{2} \sum_{i=1}^{L} \sum_{p=1}^{n_i} \left(\left(\Phi \left(x_i^p \right) - m_i^\Phi \right)^T \left(\Phi \left(x_i^p \right) - m_i^\Phi \right) \right) = \frac{1}{2} tr(S_W^\Phi) \qquad (30)$$

So

$$\frac{1}{2n} \sum_{i=1}^{L} \sum_{j=1}^{L} n_i n_j \left[tr \left(S_i^\Phi \right) + tr \left(S_j^\Phi \right) \right] = tr(S_W^\Phi) \cdot \qquad (31)$$

We can acquire $Dis = tr(S_B^\Phi) - tr(S_W^\Phi)$ ☐

2 Proof of $tr \left(S_B^\Phi \right) - tr \left(S_W^\Phi \right) = 1_n^T M 1_n$

$$tr(S_B^\Phi) = \sum_{i=1}^{L} n_i \left(m_i^\Phi - m^\Phi \right)^T \left(m_i^\Phi - m^\Phi \right) = \sum_{i=1}^{L} n_i m_i^{\Phi T} m_i^\Phi - n m^{\Phi T} m^\Phi \qquad (32)$$

$$tr(S_W^\Phi) = \sum_{i=1}^{L} \sum_{j=1}^{n_i} \left(\Phi \left(x_i^j \right) - m_i^\Phi \right)^T \left(\Phi \left(x_i^j \right) - m_i^\Phi \right)$$

$$= \sum_{l=1}^{n} \Phi \left(x_l \right)^T \Phi \left(x_l \right) - \sum_{i=1}^{L} n_i m_i^{\Phi T} m_i^\Phi \qquad (33)$$

where $\sum_{i=1}^{L} n_i m_i^{\Phi T} m_i^\Phi = \sum_{i=1}^{L} n_i \left(\frac{1}{n_i} \sum_{j=1}^{n_i} \Phi \left(x_i^j \right) \right)^T \left(\frac{1}{n_i} \sum_{k=1}^{n_i} \Phi \left(x_i^k \right) \right)$

$$
= 1_n^T
\begin{bmatrix}
\frac{1}{n_1} K_{11} & & & \\
& \frac{1}{n_2} K_{22} & & \\
& & \ddots & \\
& & & \frac{1}{n_L} K_{LL}
\end{bmatrix}
1_n
\tag{34}
$$

and

$$
nm^T m = n \left(\frac{1}{n} \sum_{i=1}^{L} \sum_{j=1}^{n_i} \Phi\left(x_i^j\right) \right)^T \left(\frac{1}{n} \sum_{p=1}^{L} \sum_{q=1}^{n_i} \Phi\left(x_p^q\right) \right)
$$

$$
= \sum_{i=1}^{L} \sum_{p=1}^{L} \sum_{j=1}^{n_i} \sum_{q=1}^{n_i} \left(\frac{1}{n} \Phi\left(x_i^j\right)^T \Phi\left(x_p^q\right) \right)
$$

$$
= 1_n^T
\begin{bmatrix}
\frac{1}{n} K_{11} & \frac{1}{n} K_{12} & \cdots & \frac{1}{n} K_{1L} \\
\frac{1}{n} K_{21} & \frac{1}{n} K_{22} & \cdots & \frac{1}{n} K_{2L} \\
\vdots & \vdots & \ddots & \vdots \\
\frac{1}{n} K_{L1} & \frac{1}{n} K_{L2} & \cdots & \frac{1}{n} K_{LL}
\end{bmatrix}
1_n
\tag{35}
$$

and

$$
\sum_{l=1}^{n} \Phi\left(x_l\right)^T \Phi\left(x_l\right) = 1_n^T
\begin{bmatrix}
k_{11} & & & \\
& k_{22} & & \\
& & \ddots & \\
& & & k_{nn}
\end{bmatrix}
1_n
\tag{36}
$$

Let $M = 2 * diag\left(\frac{1}{n_1} K_{11}, \frac{1}{n_2} K_{22}, ..., \frac{1}{n_L} K_{LL}\right) - diag\left(k_{11}, k_{22}, ..., k_{nn}\right) - \frac{1}{n} K_{total}$ then

$tr\left(S_B^\Phi\right) - tr\left(S_W^\Phi\right) = 1_n^T M 1_n$.   □

Topic Extraction with AGAPE

Julien Velcin and Jean-Gabriel Ganascia

Université de Paris 6 – LIP6
104 avenue du Président Kennedy, 75016 Paris
{Julien.Velcin, Jean-Gabriel.Ganascia}@lip6.fr

Abstract. This paper uses an optimization approach to address the problem of conceptual clustering. The aim of AGAPE, which is based on the tabu-search meta-heuristic using split, merge and a special "k-means" move, is to extract concepts by optimizing a global quality function. It is deterministic and uses no *a priori* knowledge about the number of clusters. Experiments carried out in topic extraction show very promising results on both artificial and real datasets.

Keywords: conceptual clustering, global optimization, tabu search, topic extraction.

1 Introduction

Conceptual clustering is an unsupervised learning problem the aim of which is to extract concepts from a dataset [1,2]. This paper focuses on concept extraction using optimization techniques. The concepts are descriptions that label the dataset clusters. They are automatically discovered from the internal structure of the data and not known *a priori*.

AGAPE is a general approach that solves the clustering problem by optimizing a global function [3]. This quality function q must be given and computes the correspondence between the given dataset E and the concept set C. Hence, the objective is to find the solution C^* that optimizes q. The function q is rarely convex and the usual methods are trapped in local optima. AGAPE therefore uses the meta-heuristic of tabu search, which improves the basic local search [4]. The main contribution of AGAPE is to discover better solutions in a deterministic way by escaping from local optima. In addition, an original operator inspired by the classical k-means algorithm is proposed. Note that the number of concepts to be found during the search process is not fixed in advance.

This approach has been implemented for the problem of topic extraction from textual datasets [5]. These datasets are binary (presence or absence of words), high dimensional (more than 10,000 variables) and sparse (each description contains only a few words), and require a specific computation of the tabu-search meta-heuristic. Experiments have been done using this kind of dataset to demonstrate AGAPE's validity.

The paper is organized as follows: section 2 presents the general framework of AGAPE and its implementation for the topic extraction problem; section 3 details

R. Alhajj et al. (Eds.): ADMA 2007, LNAI 4632, pp. 377–388, 2007.

the experiments based on both artificial and real datasets. Artificial datasets are used to compare the results to the score of the "ideal" solution. The experiments show the effectiveness of our approach compared to other clustering algorithms. The real dataset is based on French news and very convincing topics have been extracted. The conclusion proposes a number of lines of future research.

2   AGAPE Framework

2.1   Logical Framework

The aim of the AGAPE framework is to extract concept sets from the dataset E. Each example e of this dataset is associated to a description $\delta(e)$. Let $\mathcal{D}$ be the set of all the possible descriptions. Let us define the concept set C as a $\mathcal{D}$-element set including the empty-description $d_\emptyset$. The latter is the most possible general description which is used to cover the examples uncovered by the other concepts of C.

In the AGAPE approach, the function that evaluates the quality of the concept set is given. Let q be this function, which uses as input the concept set C and the dataset E. This function provides a score $q(C, E) \in [0, 1]$ which has to be optimized. The greater the value of $q(C, E)$, the better the concept set corresponds to the dataset. For our experiments, we have defined the following quality function q:[1]

$$q(C, E) = \frac{1}{|E|} \sum_{e \in E} sim(\delta(e), R_C(e)) \tag{1}$$

where sim is a similarity measure, $\delta(e)$ is the description of the example e in $\mathcal{D}$ and $R_C(e)$ is the function that relates the example e to the closest concept c in C (relative to the similarity measure sim). The similarity measure sim can be of any kind and will be defined in our experiments (see section 3.1). The ideal objective of the clustering algorithm is to find the best concept set C^* such that:

$$C^*(E) = \arg\max_{C \in \mathcal{H}} q(C, E) \tag{2}$$

where $\mathcal{H}$ is the hypothesis space, i.e. the set of all the possible concept sets. This task is known to be NP-complete and it is very difficult to find a perfect solution. A good approximation $\tilde{C}(E)$ of $C^*(E)$ is often enough and can be discovered through optimization techniques, which is why we have chosen a local search-based algorithm to solve this clustering problem. The algorithm uses a meta-heuristic called "tabu search" to escape from local optima.

2.2   Tabu-Based Algorithm

This global optimization approach for conceptual clustering is based on tabu search [6,7], which extends the classical local search to go beyond the local

---

[1] Note that q is similar to the classical "squared-error criterion"; this is just an example used for the experiments presented in this paper.

optima. The objective is to optimize the function q in order to find a solution close to the optimal one. Remember that each concept set $C \in \mathcal{H}$ is a potential solution to the clustering problem and that the best set C^* optimizes q. The hypothesis space is explored by computing at each step a *neighborhood* $\mathcal{V}$ of the current solution. The neighborhood contains new solutions computed from the current solution using *moves*, i.e. operators that change one solution into another. The solution of $\mathcal{V}$ optimizing q becomes the new current solution and the process is iterated.

To create these new potential solutions, three kinds of moves are considered: (1) splitting one existing concept into p ($p > 1$) new concepts, (2) merging two existing concepts to form one new concept, (3) performing a *k-means* step. Split and merge moves can be seen as similar than those used in the COBWEB framework [8]. The original "k-means move" is inspired from the classical k-means algorithm [9]. In fact, it is really important to "recenter" the search in order to better fit the dataset and this requires two steps: an *allocation* step followed by an *updating* step. The allocation step associates each example e to the closest concept in C relative to *sim* and builds clusters around the different set items. The updating step computes a new description for each concept in line with the examples covered.

Tabu search enhances the performance of a basic local search method by using memory structures. Short-term memory implements a *tabu list* T that records the lastly chosen moves and therefore prevents the search process from backtracking. If the move that lead to the new current solution has an opposite move (e.g. splitting a concept after merging it), the latter is marked *tabu*. This move remains in T for a fixed number of iterations (called the *tabu tenure*). The moves in T cannot be chosen temporarily to go through the hypothesis space. Hence, the search process avoids backtracking to previously-visited solutions. The search is stopped when a predefined iteration number $MAXiter$ is reached.

The k-means move needs special attention because, contrary to split and merge moves, there is no opposite move. Furthermore, our preliminary experiments show that once it has been used it cannot be marked tabu. We have therefore chosen to mark it tabu as soon as the current solution can no longer be improved. In a way, the k-means move is its "own" opposite move.

2.3   AGAPE for Topic Extraction

The general AGAPE framework has been implemented for the problem of topic extraction from textual datasets. Here, a *topic* is just a set of words or phrases (similarly to the "bag-of-words" representation) that label a cluster of texts covering the same theme. Let $\mathcal{L}$ be the word dictionary and a description d be a set of $\mathcal{L}$ elements. Therefore, the description set $\mathcal{D}$ corresponds to all the possible $\mathcal{L}$ elements. A constraint is added, that one word or phrase cannot belong to more than one topic at the same time. This is similar to the co-clustering approach [10] and is particularly useful since it is possible to restrain the hypothesis space and to obtain more intelligible solutions. Here is a minimalistic example of topic set: { { *ball, football player* }, { *presidential election, France* }, {} }. The whole clustering algorithm is detailed as shown below:

1. An initial solution S_0 is given, which is also the current solution S_i ($i = 0$). We choose the topic set $S_0 = \{d_T, d_\emptyset\}$, where d_T is the description that contains all the words or phrases of $\mathcal{L}$. The tabu list T is initially empty.
2. The neighborhood $\mathcal{V}$ is computed from the current solution S_i with the assistance of authorized moves relative to T. This moves are split, merge and k-means moves.
3. The best topic set B is chosen from $\mathcal{V}$, i.e. the solution that optimizes the quality function q.
4. The aspiration criterion is applied: if the k-means move leads to a solution A better than all previously discovered solutions, then B is replaced by A.
5. The new current solution S_{i+1} is replaced by B.
6. If the chosen move has an opposite move (e.g. splitting a topic after merging it), the latter is marked *tabu*. This move remains in a tabu list T for a fixed number of iterations (called the *tabu tenure*). The moves in T cannot be chosen temporarily to go through the hypothesis space. Hence, the search process avoids backtracking to previously-visited solutions.
7. The tabu list T is updated: the oldest tabu moves are removed according to the tabu tenure t_T.
8. If a local optimum is found then an intensification process is executed temporarily *without taking into account* the tabu list. It means that the neighborhood $\mathcal{V}$ is built with *all* the possible moves (even if they are marked tabu) and that the best quality solution is chosen until a local optimum is reached. Otherwise, better solutions can be missed because of the tabu list.
9. The best up-to-now discovered solution $\tilde{B}$ is recorded.
10. The search is stopped when a predefined iteration number $MAXiter$ is reached and $\tilde{B}$ is returned.

Note that the moves used to compute the neighborhood, especially split and merge moves, must be adapted to high-dimensional data. These moves are not the focus of this paper and are therefore not detailed here.

The overall complexity can be estimated to $\mathcal{O}(|\mathcal{L}| \times |E| \times max_\delta \times max_C^2 \times MAXiter)$, where max_δ is the maximum length of example description and max_C is the maximum number of topics. Note that this algorithm is linear relative to the number of examples $|E|$ and almost linear relative to the number of words $|\mathcal{L}|$. In fact, the factor $|\mathcal{L}| \times max_\delta$ is greater, but far lower than the squared one $|\mathcal{L}|^2$. The term max_C^2 is very overestimated because a low proportion of topic combinations is considered in practice (see fig. 6). The common drawback of local search is the factor $MAXiter$ which inevitably increases runtime, but this extra time is the price to pay for better solutions, as we will see below.

3   Experiments and Results

This section presents the experiments carried out using AGAPE on topic extraction. The first subsection describes the whole methodology, including the evaluation measure and the datasets chosen. The second subsection concerns the experiments carried out on artificial datasets where an ideal topic set can be used in the evaluation. The third subsection presents the results obtained from a real news dataset.

3.1 Evaluation Methodology

Quality Function. As said in the introduction, the clustering problem is solved using a given quality function q. In fact, the algorithm is general and can use any kind of function as input, its purpose primarily being to find a solution that optimizes this function as well as possible. The function we propose here was already defined in equation 1 and computes the average homogeneity between the topics of C and the associated clusters of E. The similarity measure s_α we use is inspired by works on adaptive distances [11] and is defined as follows:

$$s_\alpha(d_1, d_2) = \frac{a}{a + b + \alpha.c} \quad (3)$$

where a is the number of common words (or phrases) in the descriptions d_1 and d_2, b is the number of words appearing only in d_1 and c is the number of words appearing only in d_2. This measure is not symmetrical and gives as output a non negative real number. The value 0 means that no word is shared by the two descriptions, whereas the value 1 means two equal descriptions. d_1 stands for the example description $\delta(e), e \in E$, and d_2 stands for the topic $t \in C$. Note that $\alpha = 1$ corresponds to the classical jaccard measure used for binary data. After investigation, it was decided to set α to 10%, which corresponds to a good trade-off for our experiments.

Internal Criteria. The evaluation is mainly based on internal criteria [12] and on the number of topics extracted. The internal criteria are based on a compactness-separation trade-off, where compactness computes the within-class homogeneity while separation computes the dissimilarity between the different classes. The compactness measure is that of [13] and relies on the correspondence[2] between the topic set and the example descriptions covered. A low value means a better average homogeneity within the clusters. A new separation measure adapted to topic extraction is proposed below:

$$\sigma(C, E) = \frac{1}{|E|} \sum_{\substack{e \in E/ \\ s_\alpha(\delta(e), R_C(e)) > 0}} \frac{s_\alpha(\delta(e), D_C(e))}{s_\alpha(\delta(e), R_C(e))} \quad (4)$$

where $D_C(e)$ relates the example e to the second best topic in C. This new measure is more adapted to topic set search from binary datasets than the usual separation measure [13]. A low value means a better separation in E: each example is associated to one topic with no ambiguity. A high value means that the examples could more easily be associated to other topics.

---

[2] Diday et al. (1979) uses the term of "adequation".

Attention must also be paid to the number of topics extracted, which may be very high (several hundred) depending on the dataset considered. This is the reason why we divide the topic set into three types of topic depending the number of examples they cover: *main topics* (over 1% of the examples covered), *weak topics* (between 2 examples and 1% of the dataset) and *outliers* (only 1 example covered). The 1% parameter may appear arbitrary and will be discussed in section 3.3.

General Methodology. The methodology requires two steps. First, experiments are carried out on artificial datasets, which were generated from an original topic set INI in $\mathcal{V}$. Each topic in INI is the seed of examples with a similar theme. These examples are mixed and form the new dataset E, which can be analyzed using clustering algorithms. Then, it is possible to compare the results obtained using these algorithms to the "optimal" solution INI. The second step compares the topic sets extracted from a real dataset, containing French news available online between the 1st and the 8th September 2006. The process is described in terms of q-score evolution in order to highlight the advantages of tabu search.

3.2 Experiments on Artificial Datasets

Artificial Datasets. 40 datasets were generated artificially, corresponding to 5 datasets for 8 different noise levels from 0% to 70%. The generation process is as follows:

- k topics are generated over $\mathcal{L}$ with an homogeneous distribution. These k descriptions and $d_\emptyset$ stand for the initial topic set INI.
- Each topic $t \in INI$ is the seed of δ example descriptions. μ% of the words or phrases are drawn from these seeds to create a dataset E of $k \times \delta$ examples. This corresponds to the sparseness of real textual data.
- E is degraded by a percentage ν of noise (between 0% and 70%), done by swapping the location of two well-placed words or phrases.

In our experiments, the number k of initial topics was set to 5, the duplication dup to 100 (E therefore contains 500 examples) and μ to 99%. The number of elements in the dictionary $\mathcal{L}$ was set to 30,000, which is not that much higher than the dictionary size of our real dataset (17,459). The results could thus be averaged over 5 similar datasets for each noise level.

Comparison with two Algorithms. AGAPE is compared with the initial topic set INI and with two other clustering algorithms. The first of these algorithms is the classical *k-means*, adapted to binary co-clustering, and is very similar to the k-modes algorithm [14]. The best topic set was chosen from 30 runs with three values of k (respectively 5, 30 and 300). The second algorithm is a variation called *bisecting k-means*, known to obtain very good results in text (or document) clustering [15]. It was implemented in order to be adapted to binary co-clustering. 30 k-means were executed at each step to discover the best splitting of the biggest cluster. The process was stopped when the quality function showed no more improvements.

Figure 1 shows the evolution of the q-score and the topic number relative to a noise between 0% and 70%. The tabu tenure t_T was set to 20 and the maximum number of iterations $MAXiter$ to 100. Even if AGAPE does not always reach the maximum score, it seems to be the most stable algorithm. Furthermore, its results are the best for noise over 30% and the number of main topics discovered remains close to the 5 original topics. Note that the k-means algorithm can find fewer than k topics because the empty clusters are automatically removed.

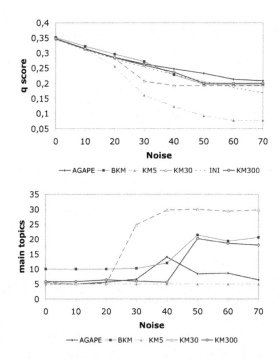

Fig. 1. Compared evolution of the q-scores and of the main topic number

Figure 2 shows the dual evolution of compactness and separation. Lower scores entail more compact and well-separated clusters. The compactness curves reflect the q-score exactly because the two formulas rely on the similarity between the topics and the example descriptions. The separation score obtained by AGAPE is very good, even if the initial topic set INI is better. It is interesting to note that as noise increases the topic sets discovered are more compact but less separated than the initial topic set. Hence, the q-score shows that the added noise makes it impossible to find the initial solution using the function q.

These results show the effectiveness of the AGAPE-based algorithm, whatever the rate of noise added to the datasets. Runtime is not very long (between 90 and 400 seconds on a pentium II with 2Gh-memory) and similar to the time taken by the bisecting k-means algorithm. Although the simple k-means algorithm is

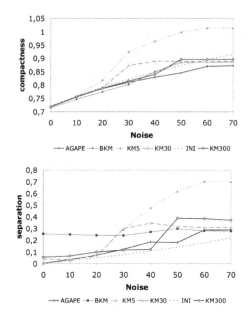

Fig. 2. Compared evolution of the compactness-separation scores

much faster (between 12 and 50 seconds), its results are much worse. We decided not to consider the 5-means algorithm in the following experiments.

3.3   Experiments on AFP News

AFP News. This dataset was automatically extracted from the French web news by the research team under F. Chateauraynaud. The news was available on a public site linked to the French press agency AFP and was collected between the 1st and the 8th September 2006. It contains 1566 news items described using a vocabulary of 17, 459 words or phrases. These words (or word sequences) were identified by a semi-automatic indexation tool and are the basis of a French dictionary supervised by researchers. In this work, only nouns and noun phrases were considered (e.g. "president", "presidential election", "information society"). For more details about the word extraction technique, see [16]. If you wish to receive the complete dataset, please contact us by email.

Results. Table 1 compares the topic sets obtained using three clustering algorithms on the French news dataset. The 5-means algorithm was not considered because the results on the artificial datasets were poor. The tabu tenure t_T was set to 20 and the maximum number of iterations $MAXiter$ to 500. Here, the results are even clearer than in the previous experiments: using the AGAPE-based algorithm it is possible to extract extremely compact and well-separated topics.

Table 1. Comparisons with the AFP dataset

	AGAPE	BKM	KM30	KM300
Q-SCORE	0.3122	0.2083	0.1832	0.2738
COMPACTNESS	0.7171	0.824	0.8496	0.7583
SEPARATION	0.1511	0.1827	0.2464	0.2519
MAIN TOPICS	14	27	29	23
WEAK TOPICS	186	3	1	157
OUTLIERS	301	0	0	57

The price to pay is a higher (but still tractable) runtime: AGAPE needs 6654 seconds, which is ten times more than with the BKM algorithm (694 seconds).

The number of topics extracted, 501, may seem rather high, though they can be subdivided into 14 main topics, 186 weak topics and 301 outliers. The boundary between main and weak topics is not a hard one and Figure 3 shows that a small variation of the cut parameter (1% here) has little impact on this partition. Besides, remember that the quality function to optimize is given. q could thus take into account the number of topics as a penalty, similarly to what is done in the BIC criterion [17].

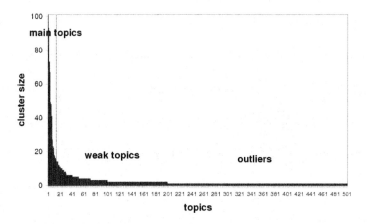

Fig. 3. Example distribution through the topics

Fig. 4 describes the most frequent words and phrases of the 7 main topics of the collection. Their meaning was clear and corresponded to newspaper headlines. Figure 5 gives term clarifications for non-French readers.

Figure 6 presents the join evolution of the q-score and of the neighborhood size $|\mathcal{V}|$. The neighborhood size is computed both with (grey bars) and without (dark bars) the tabu list. Note that the iterations without dark bars corresponds to intensification steps (step 8 in our algorithm). The temporary degradation of the current solution (see **(1)**) leads to better solutions (see **(2)**).

t_1 (101 news): students, education, de Robien, academic, children, teachers, minister, parents, academy, high school ...
t_2 (93 news): AFP, Royal, UMP, Sarkozy, PS, president, campaigner, French, Paris, party, debate, presidential election ...
t_3 (73 news): project, merger, French deputy, bill, GDF, privatization, government, cluster, Suez ...
t_4 (67 news): nuclear, UN, sanction, USA, Iran, Russia, security council, China, Countries, uranium, case, enrichment ...
t_5 (49 news): rise, market, drop, Euros, analyst, index
t_6 (48 news): tennis, Flushing Meadows, US Open, chelem, year, tournament ...
t_7 (41 news): health minister, health insurance, health, statement, doctors, Bertrand, patients ...

Fig. 4. The first 7 main topics of the AFP dataset

-*Sarkozy, Royal:* candidates for the French presidential election in 2007; they were the heads of the *UMP* and *PS* party (respectively).
-*De Robien, Bertrand:* members of the French government (minister of education and Health minister respectively).
-*GDF, Suez:* two big French companies.

Fig. 5. Clarification of some French terms

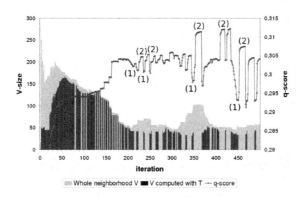

Fig. 6. Join evolution of the neighborhood size and of the q-score

4   Conclusion and Future Work

This paper presents a general framework for concept extraction. The learning problem is centered on concepts and addressed using optimization techniques. The AGAPE-based algorithm can be adapted to any kind of function q and similarity measure *sim*. In addition, the number or nature of the concepts can be constrained. This approach was successfully applied to topic extraction from textual data.

Textual data are ubiquitous, especially on the internet (blogs, forums, RSS feeds, emails). It seems natural to develop algorithms capable of summarizing such data, which is why we choose to implement AGAPE to extract topics from high-dimensional data. Results on both artificial datasets and the real AFP news dataset seem to be better than those obtained using classical text clustering algorithms, and gives better solutions in the whole hypothesis space, at the cost of slightly longer runtime.

The AGAPE framework offers future research possibilities. Different quality functions [18] and similarity measures can be compared thanks to the generality of our approach. Furthermore, the different optima discovered during the search process can be recorded and used to return not just one, but several good solutions. These solutions can either be redundant or offers specific interesting features, which can be computed using a measure such as mutual information [19]. They can then generate ensembles [20] which can be used in robust clustering [21].

References

1. Michalski, R.S., Stepp, R.E., Diday, E.: A recent advance in data analysis: clustering objects into classes characterized by conjunctive concepts. In: Pattern Recognition, vol. (1), pp. 33–55 (1981)
2. Mishra, N., Ron, D., Swaminathan, R.: A New Conceptual Clustering Framework. In: Machine Learning, vol. 56 (1-3), pp. 115–151. Kluwer Academic Publishers, Dordrecht (2004)
3. Sherali, H.D., Desai, J.: A Global Optimization RLT-based Approach for Solving the Hard Clustering Problem. In: Journal of Global Optimization, vol. 32(2), pp. 281–306. Kluwer Academic Publishers, Dordrecht (2005)
4. Glover, F., Laguna, M.S.: Tabu Search. Kluwer Academic Publishers, Dordrecht (1997)
5. Newman, D.J., Block, S.: Probabilistic Topic Decomposition of an Eighteenth-Century American Newspaper. Journal of the American Society for Information Science and Technology 57(6), 753–767 (2006)
6. Ng, M.K., Wong, J.C.: Clustering categorical data sets using tabu search techniques. In: Pattern Recognition, vol. 35(12), pp. 2783–2790 (2002)
7. Velcin, J., Ganascia, J.-G.: Stereotype Extraction with Default Clustering. In: Proceedings of the 19th International Joint Conference on Artificial Intelligence, Edinburgh, Scotland (2005)
8. Fisher, D.H.: Knowledge Acquisition Via Incremental Conceptual Clustering. In: Machine Learning, vol. (2), pp. 139–172 (1987)
9. MacQueen, J.: Some methods for classification and analysis of multivariate observations. In: Proceedings of the Fifth Berkeley Symposium on Mathematical Statistics and Probability, vol. 1, pp. 281–297. University of California Press, Berkeley, Califonia (1967)
10. Dhillon, I.S., Mallela, S., Modha, D.S.: Information-Theoretic Co-Clustering. In: KDD-2003. Proceedings of The Ninth ACM SIGKDD International Conference on Knowledge Discovery and Data Mining, pp. 89–98. ACM Press, New York (2003)
11. Aggarwal, C.: Re-designing distance functions and distance-based applications for high dimensional data. In: ACM SIGMOD Record, vol. 30(1), pp. 13–18. ACM Press, New York (2001)

12. Halkidi, M., Batistakis, Y., Vazirgiannis, M.: Cluster Validity Methods: Part I - Part II. In: Special Interest Groups on Management Of Data (2002)
13. He, J., Tan, A.-H., Tan, C.-L., Sung, S.-Y.: On Qualitative Evaluation of Clustering Systems. In: Information Retrieval and Clustering, Kluwer Academic Publishers, Dordrecht (2002)
14. Huang, Z.: A Fast Clustering Algorithm to Cluster Very Large Categorical Data Sets in Data Mining. In: DMKD, vol. 8 (1997)
15. Steinbach, M., Karypis, G., Kumar, V.: A comparison of document clustering techniques. In: Proceedings of the KDD Workshop on Text Mining (2000)
16. Chateauraynaud, F.: Prospéro: une technologie littéraire pour les sciences humaines. CNRS Editions (2003)
17. Kass, R.E., Raftery, A.E.: Bayes factors. Journal of American Statistical Association 90, 773–795 (1995)
18. Zhao, Y., Karypis, G.: Empirical and Theoretical Comparisons of Selected Criterion Functions for Document Clustering. In: Machine Learning, vol. 55, pp. 311–331. Kluwer Academic Publishers, Dordrecht (2004)
19. Gondek, D., Hofmann, T.: Non-redundant clustering with conditional ensembles. In: Proceedings of the eleventh ACM SIGKDD international conference on knowledge discovery and data mining, Chicago, Illinois, pp. 70–77 (2005)
20. Dimitriadou, E., Weingessel, A., Hornik, K.: A cluster ensembles framework. In: Design and application of hybrid intelligent systems, pp. 528–534. IOS Press, Amsterdam (2003)
21. Fred, A., Jain, A.: Robust data clustering. In: Proceedings of IEEE Computer Society Conference on Computer Vision and Pattern Recognition, pp. 128–133. IEEE Computer Society Press, Los Alamitos (2003)

Clustering Massive Text Data Streams by Semantic Smoothing Model

Yubao Liu[1], Jiarong Cai[1], Jian Yin[1], and Ada Wai-Chee Fu[2]

[1] Department of Computer Science of Sun Yat-Sen University, Guangzhou, 510275, China
liuyubao@mail.sysu.edu.cn, kelvin2004_cai@163.com,
issjyin@mail.sysu.edu.cn
[2] Department of Computer Science and Engineering, the Chinese University of Hong Kong,
Hong Kong
adafu@cse.cuhk.edu.hk

Abstract. Clustering text data streams is an important issue in data mining community and has a number of applications such as news group filtering, text crawling, document organization and topic detection and tracing etc. However, most methods are similarity-based approaches and use the TF*IDF scheme to represent the semantics of text data and often lead to poor clustering quality. In this paper, we firstly give an improved semantic smoothing model for text data stream environment. Then we use the improved semantic model to improve the clustering quality and present an online clustering algorithm for clustering massive text data streams. In our algorithm, a new cluster statistics structure, cluster profile, is presented in which the semantics of text data streams are captured. We also present the experimental results illustrating the effectiveness of our technique.

Keywords: Semantic Smoothing, Text Data Streams, Clustering.

1 Introduction

Clustering text data streams is an important issue in data mining community and has a number of applications such as news group filtering, text crawling, document organization and TDT (topic detection and tracing) etc. In such applications, text data comes as a continuous stream and this presents many challenges to traditional static text clustering [1].

The clustering problem has recently been studied in the context of numeric data streams [2, 3]. But, the text data streams clustering research is only on the underway stage. In [4], an online algorithm framework based on traditional numeric data streams clustering approach is presented for categorical and text data streams. In [4], the concept of cluster droplet is used to store the real-time condensed cluster statistics information. When a document comes, it would be assigned to the suitable cluster and then the corresponding cluster droplet is updated. This framework also distinguishes the historical documents with the presents by employing a fading function to describe the temporal locality attribute. The single pass clustering method for online event

R. Alhajj et al. (Eds.): ADMA 2007, LNAI 4632, pp. 389–400, 2007.
© Springer-Verlag Berlin Heidelberg 2007

detection of text data streams is presented in [5] that is also an extension of traditional numeric data streams clustering. Different from [4], in [5], the time window is used to describe the changes of text data streams. Different from [4, 5], the idea of [6] uses the static spherical k-means text clustering method for online text data streams clustering and it only keeps cluster centers instead of cluster droplets. When a document comes, it would not be clustered at once. Instead, it is accumulated as a portion of a segment. When a segment (a fixed number of documents) forms, the online spherical k-means algorithm would be employed and then update the cluster centers. In [9], a feature-pivot clustering technique is presented for the detection of a set of bursty events from a text stream.

Recently, [7] argues that a text document is often full of class-independent "general" words (such as stop words that may shared by different classes of two documents) and short of class-specific "core" words (such as the related topic words occur in the document), which often leads to poor document clustering quality. In [7], model-based clustering approaches based on semantic smoothing that is widely used in information retrieval (IR) [8] is presented for efficient text data clustering. Actually, most existing clustering algorithms for text data streams are similarity-based approaches and often employ the heuristic TF*IDF scheme to discount the effect of "general" words. As shown in [7], semantic smoothing model is often better than TF*IDF scheme in improving the clustering quality. Inspired by semantic smoothing model, in this paper, we extend semantic smoothing model for text data streams environment and use the semantic smoothing model to improve the clustering quality. We also present an online clustering algorithm (short for *OCTS*) for clustering massive text data streams. In our algorithm, a new cluster statistics structure, cluster profile (short for *CP*), is presented in which the semantics of text data streams are captured. Different from the cluster droplet [4], cluster profile is build by the semantic smoothing model but the TF. We also present the experimental results illustrating the effectiveness of the technique.

2   The Basic Concepts for the Clustering of Text Data Streams

In text data streams environment, text document data comes as a continuous stream. In order to account for the evolution of the data stream, we assign a time-sensitive weight to each document data point. It is assumed that each data point has a time-dependent weight defined by the function $f(t)$. The function $f(t)$ is also referred to as the *fading function*. The fading function $f(t)$ is a non-monotonic decreasing function which decays uniformly with time t. In order to formalize this concept, we will define the *half-life* of a point in the data stream.

Definition 1 (Half life). The half life t_0 of a point is defined as the time at which $f(t_0)$ = $(1/2) f(0)$.

The aim of defining a half life is to define the rate of decay of the weight associated with each data point in the stream. The decay-rate is defined as the inverse of the half life of the data stream. Similar to [4], we denote the decay rate by $\zeta = 1/t_0$ and the fading function is defined as follow.

Definition 2 (Fading Function). Consider the time t, the fading function value is defined as $f(t) = 2^{-\zeta t}$, here $\zeta = 1/t_0$ and t_0 is the half life of the stream.

3 The Semantic Smoothing Model

Many previous approaches use word extraction method and single word vector as the document features. However, they suffer from the context-insensitivity problem. The terms in these models may have ambiguous meanings. In contrast, the semantic smoothing model uses the multiword phrases as topic signatures (document features). For example, the multiword phrase "fixed star" (denotes planet) has clearer meaning than the single word "star" (denotes either a celestial body or a pop star).

After phrase extraction, the training process determines the probability of translating the given multiword phrase to terms in the vocabulary. For example, if the word "planet" frequently appears in the documents whose topic contains "fixed star", then "fixed star" and "planet" must have some specific relationship. The translation model finds out such relationship and assigns a degree to describe it. In the following process (e.g. clustering), if we encounter a document contains the topic signature "fixed star" (but not "planet"), we can also assign a rational probability count to the word "planet" for the document.

For each phrase t_k, it would have a set of documents (D_k) containing that phrase. Since not all words in D_k center on the topic signature t_k, we assume D_k is generated by a mixture language model (i.e. all terms in the document set are either translated by the given topic signature model $p(w|t_k)$ or generated by the background collection model $p(w|C)$). The formulas (1) (2) (3) are used to iteratively compute the translation probabilities [7, 8].

$$p(w \mid D_k) = (1 - \beta) p(w \mid t_k) + \beta p(w \mid C) \tag{1}$$

$$\hat{p}^{(n)}(w) = \frac{(1-\beta)^{(n)} p(w|t_k)}{(1-\beta) p^{(n)}(w|t_k) + \beta p(w|C)} \tag{2}$$

$$p^{(n+1)}(w \mid t_k) = \frac{c(w, D_k) \hat{p}^{(n)}(w)}{\sum_i c(w_i, D_k) p^{(n)}(w_i)} \tag{3}$$

Here, β is a coefficient accounting for the background noise, t_k denotes the translation model for topic signature t_k and $c(w, D_k)$ is the frequency count of term w in document set D_k (means the appearance times of w in D_k), and C denotes the background collection, which is the set of all word occurrences in the corpus. In practice, the EM algorithm is used to estimate the translation model in the formulas (2) (3).

The cluster model with semantic smoothing (or referred to as semantic smoothing model) is estimated using a composite model $p_{bt}(w|c_j)$, which means the likelihood of each vocabulary word w generated by a given document cluster c_j after smoothing. It has two components: a simple language model $p_b(w|c_j)$ and a topic signature (multiword phrase) translation model $p_t(w|c_j)$. The influence of two components is controlled by the translation coefficient (λ) in the mixture model.

$$p_{bt}(w \mid c_j) = (1 - \lambda) p_b(w \mid c_j) + \lambda p_t(w \mid c_j) \tag{4}$$

$$p_b(w \mid c_j) = (1 - \alpha) p_{ml}(w \mid c_j) + \alpha p(w \mid C) \tag{5}$$

$$p_t(w \mid c_j) = \sum_k p(w \mid t_k) p_{ml}(t_k \mid c_j) \tag{6}$$

In simple language model (5), α is a coefficient controlling the influence of the background collection model $p(w|C)$ and $p_{ml}(w|c_j)$ is a maximum likelihood estimator cluster model. They can be computed using formulas (7) and (8). In translation model (6), t_k denotes the topic signatures (multiword phrases) extracted from documents in cluster c_j. The probability of translating t_k to individual term (word) is estimated using formulas (1) (2) (3). The maximum likelihood estimator of t_k in cluster c_j can be estimated with (9), where $c(w, c_j)$ denotes the frequency count of word w in cluster c_j, and $c(w, C)$ is the frequency count of word w in the background corpus. $c(t_k, c_j)$ is the frequency count of topic signature t_k (multiword phrase) in cluster c_j. The function of the translation model is to assign reasonable probability to core words in the cluster.

$$p_{ml}(w \mid c_j) = \frac{c(w, c_j)}{\sum_{w_i \in c_j} c(w_i, c_j)} \tag{7}$$

$$p(w \mid C) = \frac{c(w, C)}{\sum_{w_i \in C} c(w_i, C)} \tag{8}$$

$$p_{ml}(t_k \mid c_j) = \frac{c(t_k, c_j)}{\sum_{t_i \in c_j} c(t_i, c_j)} \tag{9}$$

Due to the likelihood of word w generated by a given cluster $p(w|c_j)$ can be obtained by the cluster model with semantic smoothing, the remaining problem for the clustering of text document is how to estimate the likelihood of a document d generated by a cluster. The log likelihood of document d generated by the j-th multinomial cluster model is described in formula (10), where $c(w, d)$ denotes the frequency count of word w in document d and V denotes the vocabulary.

$$\log p(d \mid c_j) = \sum_{w \in V} c(w, d) \log p(w \mid c_j) \tag{10}$$

Compared to semantic smoothing model, traditional similarity-based approaches just uses the technique of frequency count of words (which is similar to the simple language model $p_b(w|c_j)$) and does not takes into account the translation model $p_t(w|c_j)$. As shown in [7], semantic smoothing model is efficient to improve the clustering quality for traditional static text document clustering. However, it can not be directly used in the dynamical text data streams environment. The key reason is that, in text data stream, the text data comes as a continuous stream and it is hard to get the background collection model of all document data point of text data streams in advance. That is, it is hard to determine $p(w|C)$ in $p_b(w|c_j)$.

In this paper, we present an improved semantic smoothing model in which the background model is not included and set coefficient α as zero. Then we define the semantic smoothing model as follows.

$$p_{bt}(w \mid c_j) = (1 - \lambda) p_{ml}(w \mid c_j) + \lambda p_t(w \mid c_j) \qquad (11)$$

From the formula (11), we can see that the improved semantic smoothing model also consists of two components, one is $p_{ml}(w|c_j)$ (which consists of frequency count of words) and the other is $p_t(w|c_j) = \sum_k p(w \mid t_k) p_{ml}(t_k \mid c_j)$.

4 The Proposed Online Clustering Algorithm

4.1 Cluster Statistics Structure

In order to achieve greater accuracy in the clustering process, we also maintain a high level of granularity in the underlying data structures. We refer to such cluster statistic structure as cluster profile in which the semantics are captured by the improved semantic smoothing model. Similar to the formula (11), in cluster profile, we maintain two kinds of weighted sums of the components $p_{ml}(w|c_j)$ and $p_t(w|c_j)$.

Definition 3 (Weighted Sum of Frequency Count). The weighted sum of frequency count for word w_i in cluster c is defined as $w\_c(w_i, c) = \sum_{d \in c} f(t - T_d) c(w_i, d)$.

Here, $c(w_i, d)$ denotes the frequency count of w_i in document d, T_d is the arrival time of document d and $f(t - T_d)$ is the weight for word w_i in document d. Similarly, the weighted sum of frequency count for topic signature t_k in the cluster c is defined as $w\_c(t_k, c) = \sum_{d \in c} f(t - T_d) c(t_k, d)$, where $c(t_k, d)$ denotes the frequency count of t_k in document d.

Definition 4. (Weighted Sum of Translation). The weighted sum of topic signature translation probability in cluster c for word w_i is defined as $w\_t(w_i, c) = \sum_k p(w_i \mid t_k) w\_c(t_k, c) = \sum_k p(w_i \mid t_k)(\sum_{d \in c} f(t - T_d) c(t_k, d))$. Here, $c(t_k, d)$ denotes the frequency count of topic signature t_k in document d, $p(w_i|t_k)$ denotes the probability of translating topic signature t_k to word w_i, T_d is the arrival time of document d and $f(t - T_d)$ is the weight for topic signature t_k in document d.

Definition 5. (Cluster profile, CP). A cluster profile $D(t, c)$ for a document cluster c at time t is defined to as a tuple ($\overline{DF2}, \overline{DF1}, s, l$). Consider wb denotes the number of distinct words in the dictionary V, then each tuple components is defined as follows.

- The vector $\overline{DF2}$ contains wb entries and the i entry $\overline{DF2}_i$ is defined as $w\_c\,(w_i, c)$.
- The vector $\overline{DF1}$ contains wb entries and the i entry $\overline{DF1}_i$ is defined as $w\_t\,(w_i, c)$.
- The entry s is defined as $\sum_k w\_c(t_k, c)$, which denotes the summation of $w\_c\,(t_k, c)$

    for all the topic signature t_k in the cluster c.
- The entry l denotes the last time when cluster c is updated.

Then we can estimate the cluster model with semantic smoothing using formula (12) in which $\dfrac{\overline{DF2}_i}{\sum_i \overline{DF2}_i}$ and $\dfrac{\overline{DF1}_i}{s}$ denotes the weighted form of the components

$p_{ml}(w|c_j)$ and $p_t(w|c_j)$ respectively.

$$p'_{bt}(w_i \mid c_j) = (1-\lambda)\frac{\overline{DF2}_i}{\sum_i \overline{DF2}_i} + \lambda\frac{\overline{DF1}_i}{s} \tag{12}$$

Interestingly, we also find CP structure also has some similar properties for clustering process as the cluster droplet [4].

Property 1 (Additivity). Additivity describes the variation of cluster profile after two clusters c_1 and c_2 are merged as $c_1 \cup c_2$. Consider two cluster profiles as $D(t,c_1)= (\overline{DF2}(c_1), \overline{DF1}(c_1),\ s_{c_1},\ l_{c_1})$ and $D(t,c_2)= (\overline{DF2}(c_2), \overline{DF1}(c_2),\ s_{c_2},\ l_{c_2})$. Then $D(t,c_1 \cup c_2)= (\overline{DF2}(c_{12}),\ \overline{DF1}(c_{12}),\ s_{c_{12}},\ l_{c_{12}})$ can be defined by tuple $(\overline{DF2}(c_1)+\overline{DF2}(c_2), \overline{DF1}(c_1)+\overline{DF1}(c_2), s_{c_1}+s_{c_2}, max(l_{c_1}, l_{c_2}))$.

Proof

(1) For the i entry of $\overline{DF2}(c_{12})$, $\overline{DF2}(c_{12})_i = w\_c\ (w_i, c_1 \cup c_2) =$
$\sum_{d \in c1 \cup c2} f(t-T_d)c(w_i,d) = \sum_{d \in c1} f(t-T_d)c(w_i,d) + \sum_{d \in c2} f(t-T_d)c(w_i,d) = w\_c\ (w_i, c_1) + w\_c$
$(w_i, c_2) = \overline{DF2}(c_1)_i + \overline{DF2}(c_2)_i$.

(2) For the i entry of $\overline{DF1}(c_{12})$, $\overline{DF1}(c_{12})_i = w\_t\ (w_i, c_1 \cup c_2) =$
$\sum_k p(w|t_k)w\_c(t_k, c_1 \cup c_2) \quad = \quad \sum_k p(w_i \mid t_k)(\sum_{d \in c1 \cup c2} f(t-T_d)c(t_k,d)) \quad =$
$\sum_k p(w_i \mid t_k)(\sum_{d \in c1} f(t-T_d)c(t_k,d) + \sum_{d \in c2} f(t-T_d)c(t_k,d)) = \sum_k p(w|t_k)(w\_c(t_k,c_1)+w\_c(t_k,c_2)) =$
$\sum_k p(w|t_k)w\_c(t_k,c_1) + \sum_k p(w|t_k)w\_c(t_k,c_2) = w\_t\ (w_i,\ c_1) + w\_t\ (w_i,\ c_2) =$
$\overline{DF1}(c_1)_i + \overline{DF1}(c_2)_i$.

(3) For s_{c12}, $s_{c12} = \sum_k w\_c(t_k, c_1 \cup c_2) = \sum_k \sum_{d \in c1 \cup c2} f(t-T_d)c(t_k,d) =$
$\sum_k (\sum_{d \in c1} f(t-T_d)c(t_k,d) + \sum_{d \in c2} f(t-T_d)c(t_k,d)) = \sum_k w\_c(t_k,c_1) + \sum_k w\_c(t_k,c_2) =$
$s_{c1} + s_{c2}$.

(4) For l_{c12}, the proof is trivial since the last updated time of the merged cluster is the later of the original two ones.

Property 2 (Updatability). Updatability describes the variation of cluster profile after a new document is added into the clusters. Consider a new document d is merged into a cluster c, the current cluster profile $D_b(t,c_b)= (\overline{DF2}(c_b), \overline{DF1}(c_b)$, s_{cb}, l_{cb}). Then the updated cluster profile is denoted as $D_a(t,c_a)=$ ($\overline{DF2}(c_a)$, $\overline{DF1}(c_a)$, s_{ca}, l_{ca}). Here $\overline{DF2}(c_a)_i = \overline{DF2}(c_b)_i + c(w_i,d)$, $\overline{DF1}(c_a)_i = \overline{DF1}(c_b)_i + \sum_k p(w_i \mid t_k)c(t_k,d)$, $s_{ca}=s_{cb}+ \sum_k c(t_k,d)$ and $l_{ca}= t$.

Actually, property2 can also be viewed as the special case of property 1 in which one of the two clusters to be merged only consists of one document.

Property 3 (Fading Property). Fading Property describes the variation of cluster profile with time. Consider the cluster profile at the time t_1 is $D(t_1, c_{t1})= (\overline{DF2}(c_{t1})$, $\overline{DF1}(c_{t1})$, s_{ct1}, l_{ct1}) and no document is added to the cluster c_{t1} during $[t_1, t_2]$. Then the cluster profile at the time t_2 is defined as $D(t_2, c_{t2})= (\overline{DF2}(c_{t2}), \overline{DF1}(c_{t2})$, s_{ct2}, l_{ct2}), where $c_{t2}=c_{t1}$, $\overline{DF2}(c_{t2}) = \overline{DF2}(c_{t1}) *2^{-\zeta(t2-t1)}$, $\overline{DF1}(c_{t2}) = \overline{DF1}(c_{t1}) *2^{-\zeta(t2-t1)}$, $s_{ct2} = s_{ct1}*2^{-\zeta(t2-t1)}$ and $l_{ct2}= l_{ct1}$.

Proof

(1) For the i entry of $\overline{DF2}(c_{t2})$, $\overline{DF2}(c_{t2})_i = w\_c$ $(w_i, c_{t2}) =$
$$\sum_{d\in ct2} f(t2-T_d)c(w_i,d) = \sum_{d\in ct2} f(t2-t1+t1-T_d)c(w_i,d) = \sum_{d\in ct1} 2^{-\zeta(t2-t1+t1-T_d)} *c(w_i,d) =$$
$$\sum_{d\in ct1} 2^{-\zeta(t2-t1)} *2^{-\zeta(t1-T_d)} *c(w_i,d) = 2^{-\zeta(t2-t1)} * \sum_{d\in ct1} 2^{-\zeta(t1-T_d)} *c(w_i,d) = \overline{DF2}(c_{t1})_i *2^{-\zeta(t2-t1)}.$$

(2) For the i entry of $\overline{DF1}(c_{t2})$, $\overline{DF1}(c_{t2})_i = w\_t$ $(w_i, c_{t2}) =$
$$\sum_k p(w\mid t_k)w\_c(t_k,c_{t2}) = \sum_k p(w_i \mid t_k)(\sum_{d\in ct2} f(t2-T_d)c(t_k,d))$$
$$\sum_k p(w_i \mid t_k)(\sum_{d\in ct2} f(t2-t1+t1-T_d)c(t_k,d)) = \sum_k p(w_i \mid t_k)\sum_{d\in ct1} 2^{-\zeta(t2-t1)} *2^{-\zeta(t1-T_d)} *c(w_i,d) =$$
$$\overline{DF1}(c_{t1})_i *2^{-\zeta(t2-t1)}.$$

(3) For s_{ct2}, $s_{ct2}= \sum_k w\_c(t_k,c_{t2}) = \sum_k \sum_{d\in ct2} f(t2-T_d)c(t_k,d) =$
$$\sum_k \sum_{d\in ct1} f(t2-t1+t1-T_d)c(t_k,d) = \sum_k \sum_{d\in ct1} 2^{-\zeta(t2-t1)} * 2^{-\zeta(t1-T_d)} * c(t_k,d) = s_{ct1}*2^{-\zeta(t2-t1)}.$$

(4) For l_{ct2}, the proof is trivial since no document is added to cluster c_{t1} during $[t_1, t_2]$.

4.2 The Online Clustering Algorithm

Our online clustering algorithm *OCTS* includes two phases: (1) offline initialization process, (2) online clustering process. The detailed description of *OCTS* algorithm framework is given in Fig.1.

The offline initialization process corresponds to line 1-2. In detail, *OCTS* first reads in the retrospective documents stored in disk as the training text data set. From the training document set, *OCTS* generates the topic signature translation model. In our algorithm implementation, the topic signature translation model is estimated by matrix M(wb*tb), where *wb* denotes the number of vocabulary words and *tb* denotes the number of topic signatures, and the data element M_{ik} of matrix represents $p(w_i|t_k)$. Then *OCTS* reads in the first *k* documents from the text data stream and build the initial cluster profiles *CPs* (each for one cluster) using definition 5.

The online clustering process corresponds to line 3-20. In this process, as a new text document arrives, firstly all the *CPs* would be updated using property 3. Next, the probability $p'_{bt}(w_i | c_j)$ is computed by formula (12). Then the similarity between document *d* and each cluster is estimated using formula (10). Then if the similarity between document *d* and its most similar cluster is less than specified threshold *MinFactor*, the most inactive cluster is deleted, and a new cluster is created. We associate the new cluster with the ID same to the deleted cluster's ID. Then the document *d* is assigned to the new cluster using property 2. Otherwise, document *d* would be assigned to the most similar cluster using property 2.

Algorithm: *OCTS*
Inputs: A stream of text data, a training data set *D*, *k* is the number of clusters, *MinFactor* is the threshold.
Output: A set of *k* cluster profiles (D(*t*, c_1),...., D(*t*, c_k))
Method:
1. Extract words and multiword phrases, and build the translation model for the training data set *D*. /* Notice that the extraction process of words and phrase are the same as in [7].*/
2. Read in the first *k* documents from the text data stream and generate *k* cluster profiles (D(*t*,c_1),....,D(*t*,c_k));
    /*The while loop is the online clustering process*/
3. While (the stream is not empty) do
        /*Step4-5 is to update all *CP* using the fading property */
4.   *t*= GetCurrentTimestamp();
5.   Update all cluster profiles using property 3;
        /*Step 6-8 is the online building of the cluster model */
6.   For each cluster *j*
7.   For each word w_i
8.   The cluster model with semantic smoothing $p'_{bt}(w_i | c_j)$ is estimated using formula (12);
        /*Step 9-12 is to find the most similar cluster to document *d* */

Fig. 1. The description of OCTS algorithm framework

9.   Read in the next document d of text data stream and extract words and multiword phrases from d;

10.   For each cluster j

11.   The similarity p $(d \mid c_j)$ between document d and c_j is estimated with cluster model $p'_{bt}(w_i \mid c_j)$ using formula (10);

12. $AssignID$ = argmax$_j$ p $(d \mid c_j)$ //Get the ID of the most similar cluster to d

                /* Step 13-17 is to delete the most inactive cluster and create a new cluster*/

13.   If (p $(d \mid c_{AssignID}$) < $MinFactor$)   //$MinFactor$ is the threshold

14. NID = argmin$_j$ $D(t,c_j).l$;   /*Get the most inactive cluster's ID, if there are more than one inactive clusters then we randomly choose one)*/

15.   Delete the most inactive cluster c_{NID};

16.   Create a new empty cluster and associate it with ID=NID and build the cluster profile of the new cluster using definition 5;

17.   Assign document d to the new cluster profile using property 2;

18.   Else

                /*Step 19 is to assign document d to its most similar cluster */

19.   Assign document d to cluster profile with ID = $AssignID$ using property 2;

20. End while

Fig. 1. (*continued*)

5   The Experimental Results

In the experimental studies, we compare our text streams clustering method with the framework proposed in [4] (denoted as *Ostc*) and the single pass clustering method in [5] (denoted as *SPstc*) in terms of clustering accuracy. All clustering algorithms are implemented by java 1.4. Our experiment is performed on AMD 1.60G, 240M memory, 40G hard disk and Window XP OS. The normalized mutual information (*NMI*) evaluation function is used to evaluate the clustering quality, and 20ng-newsgroups (*20NG*) (20000 documents) corpus [7] is used as the test data set. To simulate the real stream environment, we randomly give every document a time stamp to stand for the arrival time and create three different text data streams with different document sequence (denoted as Stream1, Stream2 and Stream3). By default, in our experiments, we randomly choose 500 documents as the training set and set the translation coefficient λ=0.6, cluster number k=20, stream speed is 100doc/s, half life=1s and $MinFactor$=0.05.

The first set of experiments is about the clustering quality comparison with different streams and the results are given in Fig.2-Fig.4. In this set of experiments, there are two kinds of results with different test granularities. In the left figures, the NMI values are compared by every 1 second interval (fine granularity). In the right figures, the NMI value is estimated by every 50 seconds interval (coarser granularity).

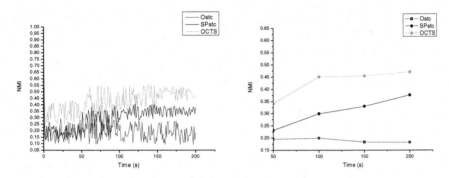

Fig. 2. Clustering Quality Comparison of Stream 1

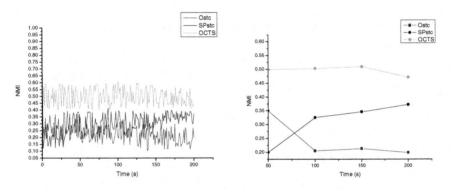

Fig. 3. Clustering Quality Comparison of Stream 2

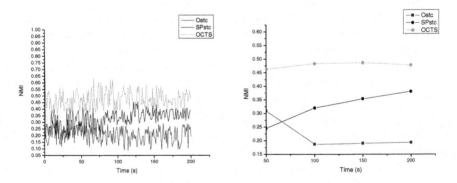

Fig. 4. Clustering Quality Comparison of Stream 3

From the above results, it is easy to know our proposed method *OCTS* obviously outperform *Ostc* and *SPstc* in which both TF [4] and incremental IDF [5] schemas are used. The results also show the effectiveness of semantic smoothing model in the aspect of improving the clustering quality.

The second set of experiments about the clustering quality of *OCTS* with different translation coefficient (λ) and different text data streams is also studied. The results are given in Fig.5-Fig.8. From the results, it is easy to know that the NMI values of *OCTS* with different λ are also better than the NMI values of *Ostc* and *SPstc* (referred to Fig.2, Fig.3 and Fig.4). From these results, we can also know that NMI values of *OCTS* will increase with the increase of the translation coefficient till the peak point (between 0.2 and 0.6) and then go downward. That is, our method is most effective as λ is between 0.2 and 0.6.

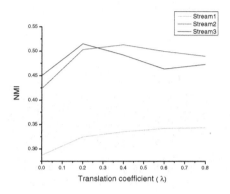

Fig. 5. The variance of the clustering quality with λ (test time: 50 seconds)

Fig. 6. The variance of the clustering quality with λ (test time: 100 seconds)

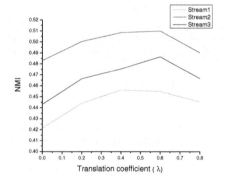

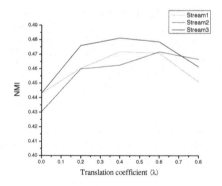

Fig. 7. The variance of the clustering quality with λ (test time: 150 seconds)

Fig. 8. The variance of the clustering quality with λ (test time: 200 seconds)

6   Conclusions

In this paper, we give an improved semantic smoothing model which is suitable for text data stream environment. Then we use the improved semantic model to improve the clustering quality and present an online clustering algorithm for clustering massive text data streams. In our algorithm, a new cluster statistics structure, cluster profile, is presented in which the semantics of text data streams are captured by the semantic smoothing model. We also present the experimental results illustrating the effectiveness of our technique.

Acknowledgments. This work was supported by the National Natural Science Foundation of China (60573097), Natural Science Foundation of Guangdong Province (05200302, 06104916), Research Foundation of Science and Technology Plan Project in Guangdong Province (2005B10101032), Specialized Research Fund for the Doctoral Program of Higher Education (20050558017), and Program for New Century Excellent Talents in University of China (NCET-06-0727).

References

1. Aggarwal, C.C.: A Framework for Diagnosing Changes in Evolving Data Streams. In: Proc. ACM SIGMOD 2003, pp. 575–586 (2003)
2. Agrawal, C.C., Han, J., Wang, J., Yu, P.S.: A framework for clustering evolving data streams. In: Aberer, K., Koubarakis, M., Kalogeraki, V. (eds.) Databases, Information Systems, and Peer-to-Peer Computing. LNCS, vol. 2944, pp. 81–92. Springer, Heidelberg (2004)
3. Agrawal, C.C., Han, J., Wang, J., Yu, P.S.: A framework for projected clustering of high dimensional data streams. In: Jonker, W., Petković, M. (eds.) SDM 2004. LNCS, vol. 3178, pp. 852–863. Springer, Heidelberg (2004)
4. Aggarwal, C.C., Yu, P.S.: A Framework for Clustering Massive Text and Categorical Data Streams. In: Proc. SIAM conference on Data Mining, pp. 477–481 (2006)
5. Yang, Y., Pierce, T., Carbonell, J.: A study of retrospective and on-line event detection. In: Proc. ACM SIGIR 1998, pp. 28–36. ACM Press, New York (1998)
6. Zhong, S.: Efficient Streaming Text Clustering. Neural Networks 18(5-6), 790–798 (2005)
7. Xiaodan, Z., Zhou, X., Hu, X.: Semantic Smoothing for Model-based Document Clustering. In: Perner, P. (ed.) ICDM 2006. LNCS (LNAI), vol. 4065, pp. 1193–1198. Springer, Heidelberg (2006)
8. Zhou, X., Hu, X., Zhang, X., Lin, X., Song, I.-Y.: Context-Sensitive Semantic Smoothing for the Language Modeling Approach to Genomic IR. In: Proc. ACM SIGIR 2006, pp. 170–177 (2006)
9. Fung, G.P.C., Yu, J.X., Yu, P.S., Lu, H.: Parameter Free Bursty Events Detection in Text Streams. In: Draheim, D., Weber, G. (eds.) TEAA 2005. LNCS, vol. 3888, pp. 181–192. Springer, Heidelberg (2006)

GraSeq: A Novel Approximate Mining Approach of Sequential Patterns over Data Stream

Haifeng Li and Hong Chen

School of Information, Renmin University, Beijing, 100872, P.R. China
`mydlhf@126.com,`
`chong@ruc.edu.cn`

Abstract. Sequential patterns mining is an important data mining approach with broad applications. Traditional mining algorithms on database were not adapted to data stream. Recently, some approximate sequential pattern mining algorithms over data stream were presented which solved some problems except the one of wasting too many system resources in processing long sequences. According to observation and proof, a novel approximate sequential pattern mining algorithm is proposed named *GraSeq*. *GraSeq* uses directed weighted graph structure and stores the synopsis of sequences with only one scan of data stream; furthermore, a subsequences matching method is mentioned to reduce the cost of long sequences' processing and a conception *validnode* is introduced to improve the accuracy of mining results. Our experimental results demonstrate that this algorithm is effective and efficient.

Keywords: sequential pattern, data stream, directed weighted graph.

1   Introduction

A sequential pattern is a subsequence that appears frequently in the sequence database. The sequential patterns mining has shown its importance in many applications include business analysis, web mining, security, and bio-sequences analysis. For instance, a website wants to make users find their favorite contents conveniently, so they will get users' visit orders from web log. These visit orders should be seen as sequences and could be mined to sequential patterns so that the website's structure is improved according to these sequential patterns.

Before data stream appears, almost all of sequence pattern mining algorithms use the accurate matching method which waste a lot of system resources, and moreover, the results have to be got by multiple scans. There are two main kinds of accurate matching algorithms so far.

The first kinds are the appriori-like algorithms such as *GSP* [1] and *SPADE* [2], which need to create the candidate set and use multiple scans over database.

The second kinds are the projection-based algorithms such as *PrefixSpan*[3], *FreeSpan*[4] and *SPAM*[5], which avoid the process of creating candidate set and extend the local frequent itemsets to long sequential patterns according to the projection of sequential patterns.

R. Alhajj et al. (Eds.): ADMA 2007, LNAI 4632, pp. 401–411, 2007.

Data Stream is fast, unlimited, dynamic and continuous, so data can't be wholly stored in memory and can't be scanned for multiple times. Furthermore, the mining results always bring noises from data stream, so traditional mining methods are absolutely not adapted to data stream. The adaptation and approximation will be mainly considered in this environment. Two kinds of approximate sequential patterns mining methods have been developed: One is block data processing method, another is tuple processing method.

ApproxMap [6] uses the block data processing method, it clusters data stream into a series of blocks according to the similarity among sequences, and then compresses the similar sequences with multiple alignment method to reduce memory usage. The minimum support threshold is set to ignore the items which are not frequent so that noise is filtered. Finally the compression results are stored within a tree to make query convenient. *ApproxMap* can't get the real-time results because it can't compute until a group of data has arrived.

Hoong Hyuk Chang proposed an algorithm names *eISeq* [7] that uses the tuple data processing method. *eISeq* regards data stream as continuous transaction tuples and processes each tuple at once when it comes so that the real-time results are achieved. Five steps are in this algorithm: parameter update, count update, sequence insertion, sequential patterns selection and data pruning. *eISeq* computes the sequential patterns efficiently, but on the other hand, it wastes many time in scanning the tree to decide whether a new sequence can be inserted into the monitor tree, and also *eISeq* can't process long sequences because the longer a sequence is, the more system resources to create all its subsequences is used. For example, if $\langle a_1, \cdots, a_{20} \rangle$ is a sequence, there are $(2^{20} - 1)$ subsequences in total must be created. It is obviously difficult to compute and store all these subsequences.

In this paper, a novel sequential patterns mining algorithm names *GraSeq* is presented which uses the directed weighted graph structure so that memory usage is reduced a lot. *GraSeq* increasingly stores the whole information of the coming data from data stream. In this algorithm, a new subsequences matching method is proposed to create graph, and some relational data rules are introduced to filter most of the redundancy data. *GraSeq* has four steps to finish mining task:

1. Subsequences generating, Without creating all subsequences, only 1-subsequences set and 2-subsequences set of each sequence are created.
2. Sequence insertion and update. In this step, 1-subsequences and 2-subsequences of each sequence are inserted into graph as vertices and edges if they do not exist in graph, whereas the weight of vertices and edges are updated.
3. Sequential pattern mining. Users can traverse the directed weighted graph to acquire approximate sequential patterns with setting the minimum support threshold.
4. Data pruning. When system resources are not enough to support the running of *GraSeq*, some data will be erased according to given rules.

The main contributes of this paper are shown as follows: Firstly, a new directed weighted graph structure is presented to stored the synopsis of the sequences of data stream; Secondly, in allusion to the characteristic of directed weighted graph, a new approximate sequential pattern mining method is proposed where a sequence is regarded as sequential pattern when all the 1-subsequences and 2-subsequences of which are frequent; Finally, to reduce the cost of system resources, a non-reclusive depth first search algorithm is introduced.

The rest of paper is organized as follows. Section 2 introduces the preliminaries of this algorithm and section 3 describes the data structure and implementation of *GraSeq* in detail. In section 4, a series of experiments are finished to show the performance of the proposed method. Finally, section 6 concludes this paper.

2   Preliminaries

Sequential pattern mining is the constraint of frequent patterns mining in data item's order which is presented in [8] firstly in 1995. There is a detailed description of sequential patterns definition:

An itemset is a non-empty set of items. Let $I = \{i_1, \cdots, i_l\}$ be a set of items. An itemset $X = \{i_{j_1}, \cdots, i_{j_k}\}$ is a subset of *I*. Conventionally, itemset $X = \{i_{j_1}, \cdots, i_{j_k}\}$ is also written as $\{x_{j_1}, \cdots, x_{j_k}\}$. A sequence $S = \langle X_1, \cdots, X_n \rangle$ is an ordered list of itemsets. A sequence database SDB is a multi-set of sequences.

A sequence $S_1 = \langle X_1, \cdots, X_n \rangle$ is a subsequence of sequence $S_2 = \langle Y_1, \cdots, Y_m \rangle$, and S_2 is a super-sequence of S_1, if $n \leq m$ and there exist integers $1 \leq i_1 < \cdots < i_n \leq m$ such that $X_j \subseteq Y_{i_j} (1 \leq j \leq n)$.

Given a sequence database SDB, the support of a sequence *P*, denoted as $sup(P)$, is the number of sequences in SDB that are super-sequences of *P*. Conventionally, a sequence *P* is called a sequential pattern if $sup(P) \geq S_{min}$, where S_{min} is a user-specified minimum support threshold.

In this paper, directed weighted graph is the data structure whose vertices denote itemsets and edges denotes order between itemsets. Some definitions are given as follows:

Definition 1. *For a sequence $S = \langle s_1, \cdots, s_n \rangle (n \geq 1)$, a 1-subsequence of* S *is $oss_i = \langle s_i \rangle (1 \leq i \leq n)$, which denotes a vertex in graph. A 1-subsequence set of* S *is $OA = \{oss_i | i = 1 \cdots n\}$.*

Definition 2. *For a sequence $S = \langle s_1, \cdots, s_n \rangle (n \geq 2)$, a 2-subsequence of* S *is $tss_i = \langle s_i, s_j \rangle (1 \leq i < j \leq n)$, which denotes an edge in graph. A 2-subsequence set of* S *is $TA = \{tss_i | i = 1 \cdots n - 1\}$.*

Definition 3. *For a 2-subsequence $tss_i = \langle s_i, s_j \rangle (i = 1 \cdots n - 1, j = i + 1 \cdots n)$, s_i is parent of s_j and s_j is child of s_i.*

In data stream, a sequence support achieves its maximum value when all its subsequences happen in as many transactions as possible, so the support of a sequence must be not higher than the ones of all its subsequences.

Proposition 1. *For a sequence $S = \langle s_1, \cdots, s_n \rangle (n \geq 2)$, C(S) is support threshold of* S *and $C_{max}(S)$ is maximum support threshold of* S, *the approximate estimate formula of $C_{max}(S)$ is shown as follows:*

$$C_{max}(S) = min(\{C(a)|a \subseteq S \wedge |a| = n-1\}) \,. \tag{1}$$

Theorem 1. *For a sequence $S = \langle s_1, \cdots, s_n \rangle (n \geq 1)$, the approximate estimate formula of $C_{max}(S)$ is shown as follows:*

$$C_{max}(S) = \left\{ \begin{array}{ll} C(S) & |S| = 1 \\ min(\{C(a)|a \subseteq S \wedge |a| = 1\}) & |S| = 2 \\ min(\{C(a)|a \subseteq S \wedge |a| = 2\}) & |S| > 2 \end{array} \right. \tag{2}$$

Proof. For a sequence $S = \langle s_1, \cdots, s_n \rangle (n \geq 1)$, the results are obvious as $n \leq 2$. When $n > 2$, if $C_{max}(S) = min(\{C(a)|a \subseteq S \wedge |a| = n-1\})$, then $C_{max}(S_n) = min(C_{max}(S_{n-1}))$, and also $C_{max}(S_{n-1}) = min(C_{max}(S_{n-2})), \cdots, C_{max}(S_3) = min(C_{max}(S_2))$, so $C_{max}(S_n) = min(min(\cdots C_{max}(S_2) \cdots)) = min(C_{max}(S_2))$. Then, when $n > 2$, $C_{max}(S) = min(\{C(a)|a \subseteq S \wedge |a| = 2\})$, proof done.

3 *GraSeq* Method

From Theorem 1 we can find that if the support of all the 1-subsequences and 2-subsequences of one sequence S are acquired, the approximate support of this sequence S is acquired too. In other words, if the support of all the 1-subsequences and 2-subsequences of a sequence is higher than the minimum support threshold, the support of this sequence is higher than the minimum support threshold, and it means this sequence is frequent. So the main task in this paper is to store all the information of 1-subsequences and 2-subsequences of all sequences, and this guarantees the mining is almost valid. Users can traverse the graph to find all frequent sequences as sequential patterns on condition that their 1-subsequences and 2-subsequences are frequent.

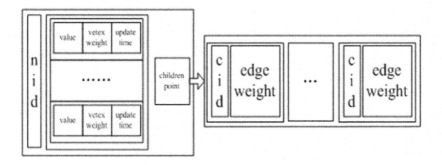

Fig. 1. Data structure of graph

3.1 Data Structure and Meaning

GraSeq uses directed weighted graph structure to share data information. Each vertex in graph denotes one itemset of sequence, and each edge denotes the order of two different itemsets. Each vertex is a 3-tuple $\langle nid, itm, ch \rangle$, in which nid is the identification of vertex which is fixed to make it possible for quick visit; *itm* denotes the main information of vertex which is a collection of 3-tuple $\langle va, wt, ut \rangle$, where *va* is the real value of the vertex, *wt* is the weight of the vertex and *ut* is the latest update time stamp of the vertex; *ch* is a pointer to the children of vertex named *childrenlist* . The element in *childrenlist* is a 2-tuple $\langle cid, lwt \rangle$, *cid* is the identification of each child vertex and *lwt* denotes the edge weight between current vertex and its child vertex. Figure 1 shows the data structure of graph.

Example 1. After a series of sequences $\{\langle a, c \rangle$, $\langle a, d \rangle$, $\langle a, b, c \rangle$, $\langle a, b \rangle$, $\langle b, c \rangle$, $\langle b, d \rangle$, $\langle a, b, d \rangle\}$ have arrived, the data storage in memory is shown in figure 2.

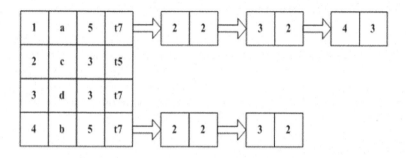

Fig. 2. Example of data storage in memory

3.2 Algorithm Description

GraSeq includes 4 steps to deal with each sequence: subsequences generating, sequence insertion and update, sequential patterns mining and data pruning. First two steps are graph establishment phase, step three is data mining phase, and they are parallel in running.

3.2.1 Subsequences Generating

In this section, the 1-subsequences set and 2-subsequences set are created from a sequence. For example, a sequence $\langle a, c, d, e \rangle$ has 1-subsequences set $\{\langle a \rangle, \langle c \rangle, \langle d \rangle, \langle e \rangle\}$ and 2-subsequences set $\{\langle a, c \rangle, \langle a, d \rangle, \langle a, e \rangle, \langle c, d \rangle, \langle c, e \rangle, \langle d, e \rangle\}$. From definition 1 and definition 2, it is clear to see that a n-sequence has only n 1-subsequences and $\sum_{i=1}^{n} i$ 2-subsequences. The count is much smaller than that of all subsequences $2^n - 1$ when sequence is longer.

3.2.2 Sequence Insertion and Update

The old sequence information has weakly affection with the coming of the new sequence. To differentiate the information of recently generated data elements from the obsolete information of old data elements which may be no longer useful or possibly invalid at present to make result reflect recent rule of data stream, a decay rate *d* [9] is used as follows:

$$d = b^{-(1/h)} \qquad (b \geq 1, h \geq 1) .$$

(3)

In this formula, b is decay-base and h is decay-base-life.

If decay-base is set to 1, frequent sequences are found as a mining result set as in the other mining algorithms of recent sequences, on the other hand, if a decay-base is set to be greater than 1, recently sequences are effective in mining. To avoid fluctuation in the set of recently frequent sequences, a decay-base-life h of a decay rate should be set to be greater than or equal to its lower bound h^{LB} , found as follows [9]:

$$h^{LB} = \lceil -\{\gamma/\log_b(1 - S_{min})\}\rceil . \tag{4}$$

In this formula, γ is safety factor and S_{min} is minimum support threshold.

Sequence insertion and update are to combine all 1-subsequences and 2-subsequences of one sequence with graph in the form of adding weight of vertices and edges. Two steps are in this section: The first step is to add weight of vertices, if a vertex corresponding one of 1-subsequences is not in graph, create and insert this vertex into graph, the real value is 1-subsequence's value and the initial weight $wt=1$; otherwise update the weight $wt=wt+1/d$. The second step is to add weight of edge as the same method as in step one. The optimized algorithm of sequence insertion and update is shown as follows.

```
Function createGraph(seq){
  create1subsequences(seq);
  create2subsequences(seq);
  for each item in 1-subsequences{
    findVertex(item);
    if(findVertex)
      getId(updateVertex(item));
    else
      getId(addNewVertex(item));
  }
  for each <preItem, nextItem> in 2-subsequences{
    findEdge(preItem,nextItem);
    if(findEdge)
      updateEdge(preItem,nextItem);
    else
      addNewEdge(preItem,nextItem);
  }
}
```

3.2.3   Sequential Pattern Mining

This step is independent from previous two steps. When the graph is constructing, users can provide the minimum support threshold S_{min} to get the sequential patterns anytime. Sequential patterns are obtained by depth first search over the graph. The mining process is to recursively traverse the graph with every vertex as beginner, and finally the real-time sequential patterns are acquired in which all the vertices are *validnode*.

Considering the efficiency of algorithm, a non-recursive traverse method is used. A stack named *nodestack* is introduced to store each *validnode*.

If there are no other rules in traverse, the results should be a superset of accurate sequential patterns. To eliminate most of the redundant sequential patterns, a concept *validnode* is introduced. For a vertex on the top of stack, if once there is no *validnode* in its children, output all vertices from the bottom to the top of stack as a sequential pattern.

Definition 4. *A vertex is* validnode *when this vertex satisfies the follow rules after it is pushed into stack.*

1. Every vertex in stack is greater than minimum support threshold S_{min}.

2. There are no repeated vertices in stack.

3. If the vertices from the bottom to the top of stack are regarded as a sequence, then every 2-subsequence of this sequence is frequent, i.e., the edge weight between each two vertices in stack is greater than the minimum support threshold S_{min}.

In traverse the current vertex may be searched through its parent vertex or ancestor vertex, the relationship between them will be recorded so that when all the parent and ancestor of the current vertex is popped from stack, the current vertex is surely visited. so a 32 bits binary integer *sign* of *validnode* in *nodestack* is imported to avoid data lose.

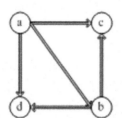

Fig. 3. Example of graph

Example 2. The sequences in example 1 is used here. The minimum support S_{min} is set to 2, then the graph where the weights of each vertex and edge greater than or equal to 2 is shown in Figure 3. They are denoted in data structure as $(a \rightarrow c \rightarrow d \rightarrow b, b \rightarrow c \rightarrow d, c, d)$, and Table 1 shows the process of sign operation with a as beginner.

As shown in Table 1, c is firstly signed by a because a is parent of c, and a is the *first* vertex in stack, so $Sign(c) = 1$, here c has no children, so output $\langle\{\langle a\rangle, \langle c\rangle\}$ in row 3 as sequential pattern and pop c in row 4. Next b is pushed in row 5 because it is the second child of a and also $Sign(b)=1$, Then c is pushed and signed as the first child of b in row 6, now b is the second vertex in stack, so $Sign(c)=11$. c is not unsigned as $Sign(c)=1$ until b which is one of c's parent is popped from stack in row 9. The same as a's being popped.

The sequential patterns mining algorithm is shown as follows.From this algorithm we can find each sequence built from the bottom to the top of stack is a sequential pattern whose 1-subsequences and 2-subsequences are all frequent.

Table 1. The process of sign operation with vertex a as beginner

	Stack	Sign(a)	Sign(b)	Sign(c)	Sign(d)	Output
1		0	0	0	0	
2	a	0	0	0	0	
3	a,c	0	0	1	0	$\langle a \rangle, \langle c \rangle$
4	a	0	0	1	0	
5	a,b	0	1	1	0	
6	a,b,c	0	1	11	0	$\langle a \rangle, \langle b \rangle, \langle c \rangle$
7	a,b,d	0	1	11	10	$\langle a \rangle, \langle b \rangle, \langle d \rangle$
8	a,b	0	1	11	10	
9	a	0	1	1	0	
10	a,d	0	1	1	1	$\langle a \rangle, \langle d \rangle$
11	a	0	1	1	1	
12		0	0	0	0	

```
Function querySequence(Graph g){
 for(each vertex in g){
  if(vertex is valid){
   push(vertex);
   while(stack is not empty){
    nextvertex = findnextValidvertex(vertex);
    if(find){
     push(nextvertex);
     sign(nextvertex);
    }else{
     out(stack);
     unsign(all nextvertex of vertex);
     pop(topvertexofstack);
    }
   }
  }
 }
}
```

3.2.4 Data Pruning

Generally, the smaller the weight of vertex is, the greater possibility a vertex will be erased, but in fact, some vertices with smaller weight may become frequent in future, so data can not be erased according to weight.

And on the other hand, the effect of old data will decay follows the new data's coming. If a vertex is not updated for a long time, we supposed that it would not be updated in future. Data pruning is to find the vertices that were not updated for a long time and erases them when system resources can't satisfy the running of algorithm.

4 Experimental Results

A set of experiments are designed to test the performance of this algorithm. The experiment uses synthetic supermarket business data created by the data generator (*http://www.almaden.ibm.com/cs/quest*) of IBM. *PrefixSpan* is a multiple scan algorithm and creates accurate sequential patterns, so a precision and recall compare is based on the result of *PrefixSpan*. *eISeq* is a one scan algorithm and creates approximate sequential patterns, and it has the best performance in processing time and memory usage to the best of our knowledge, so it is chose as the comparative algorithm. The machine in experiment has PIV3.2G CPU and 512M memory, C++ is the implementary language under Windows environment.

Four data sets are created named *T8I10KD100K*, *T8I80KD100K*, *T15I10KD100K* and *T15I80KD100K*,where *T* is the average sequence length, *I* is the count of distinct sequence itemsets and *D* is the number of sequences.

Figure 4 shows the memory usage of two algorithms *GraSeq* and *eISeq* in sequences insertion and update phase over data set *T8I10KD100K* with $S_{min} = 100$, and figure 5 shows the memory usage over *T8I80KD100K* with $S_{min} = 10$. For *eISeq*, another parameter significant support threshold [7] $S_{sig} = 1$. It can be seen that on condition that significant support threshold is small, the memory usage of *eISeq* obviously increases a lot. To compare two figures, it is clear to see that *GraSeq* uses less memory when the count of distinct itemsets is smaller, that is because sequence information is stored in vertex of graph, and the same itemsets are stored in the same vertex so no redundant information are stored.

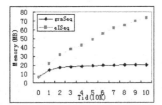

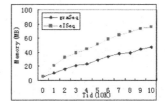

Fig. 4. Memory usage of *GraSeq* and *eISeq* over data set *T8I10KD100K*

Fig. 5. Memory usage of *GraSeq* and *eISeq* over data set *T8I80KD100K*

Figure 6 shows the data processing time of *GraSeq* and *eISeq* in sequences insertion and update phase over data set *T15I10KD100K* with $S_{min} = 100$, and figure 7 shows data processing time of two algorithms over *T15I80KD100K* with $S_{min} = 10$. We can see that *eISeq* needs more time because the analysis of sequence spends much time when average length of sequence increases.

To compare the correctness of results of *GraSeq* and *eISeq*, two concepts precision and recall are introduced. For two result sets R_1 and R_2, define precision and recall as follows:

$$Precision(R_1|R_2) = \frac{|R_1 \cap R_2|}{|R_2|}. \tag{5}$$

$$Recall(R_1|R_2) = \frac{|R_1 \cap R_2|}{|R_1|}. \tag{6}$$

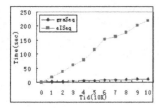
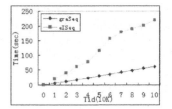

Fig. 6. Data processing time of *GraSeq* and *eISeq* over data set *T15I10KD100K*

Fig. 7. Data processing time of *GraSeq* and *eISeq* over data set *T15I80KD100K*

Figure 8 shows both $Precision(R_{GraSeq}|R_{PrefixSpan})$ and $Precision(R_{eISeq}|R_{PrefixSpan})$, and figure 9 shows both $Recall(R_{GraSeq}|R_{PrefixSpan})$ and $Recall(R_{eISeq}|R_{PrefixSpan})$ with different S_{min}, both figures use the same data set *T15I10KD100K*. In *eISeq*, the significant support threshold is also set as $S_{sig} = 1$. We can find both precisions and recalls of *GraSeq* and *eISeq* are almost same and become uniform follows the increasing minimum support threshold.

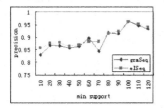
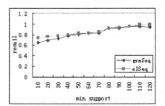

Fig. 8. Precision compare of *GraSeq* and *eISeq* over data set *T15I10KD100K*

Fig. 9. Recall compare of *GraSeq* and *eISeq* over data set *T15I10KD100K*

All of the above experiments show that when the average length of sequence is greater or the count of distinct itemsets are smaller, *GraSeq* can save more system resource with less lose of precision and recall than *eISeq*.

5   Conclusions

This paper investigates the problem of sequential patterns mining over data stream and proposes a novel algorithm named *GraSeq* based on directed weighted graph structure. *GraSeq* is a one scan algorithm, and it stores the synopsis of sequences. With subsequence matching method, final approximate sequential patterns in different minimum support threshold are obtained dynamically with deep traverse over graph. The experiments show that *GraSeq* has a better performance in processing time and in memory usage than *eISeq* with the accordant accuracy of mining results. In this algorithm, data sharing brings less resources usage but a little more imprecise results, more adaptive rules may further improve the data correctness. Moreover, graph compression is also an interesting topic for future research.

References

1. Srikant, R., Agrawal, R.: Mining sequential patterns: Generalizations and performance improvements. In: Apers, P.M.G., Bouzeghoub, M., Gardarin, G. (eds.)
   EDBT 1996. LNCS, vol. 1057, pp. 3–17. Springer, Heidelberg (1996)
2. Zaki, M.: SPADE: An Efficient Algorithm for Mining Frequent Sequences. Machine
   Learning 40, 31–60 (2001)
3. Pei, J., Han, J., Mortazavi-Asl, B., Pinto, H., Chen, Q., Dayal, U., Hsu, M.-C.: PrefixSpan: Mining Sequential Patterns Efficiently by Prefix-Projected Pattern Growth.
   In: ICDE'01. Proceeding of the International Conference on Data Engineering, pp.
   215–224 (2001)
4. Han, J., Pei, J., Mortazavi-Asl, B., Chen, Q., Dayal, U., Hsu, M.-C.: FreeSpan:
   Frequent Pattern-Projected Sequential Pattern Mining. In: KDD '00. Proceeding of
   ACM SIGKDD International Conference Knowledge Discovery in Databases, August 2000, pp. 355–359 (2000)
5. Ayres, J., Flannick, J., Gehrke, J., Yiu, T.: Sequential pattern mining using a bitmap
   representation. In: KDD'02. Proceeding of ACM SIGKDD International Conference
   on Knowledge Discovery and Data Mining, July 2002, pp. 429–235 (2002)
6. Kum, H.C., Pei, J., Wang, W., Duncan, D.: Approx-MAP: Approximate Mining of
   Consensus Sequential Patterns. Technical Report TR02-031, UNC-CH (2002)
7. Chang, J.H., Lee, W.S.: Efficient Mining method for Retrieving Sequential Patterns
   over Online Data Streams. Journal of Information Science, 31–36 (2005)
8. Agrawal, R., Srikant, R.: Mining Sequential Patterns. In: ICDE'95. Proceedings
   of the 11th International Conference on Data Engineering, March 1995, pp. 3–14
   (1995)
9. Chang, J.H., Lee, W.S.: Finding recent frequent itemsets adaptively over online
   data streams. In: Getoor, L., et al. (eds.) Proceedings of the 9th ACM SIGKDD
   International Conference on Knowledge Discovery and Data Mining, August 2003,
   pp. 487–492 (2003)

A Novel Greedy Bayesian Network Structure Learning Algorithm for Limited Data

Feng Liu[1], Fengzhan Tian[2], and Qiliang Zhu[1]

[1] Department of Computer Science, Beijing University of Posts and
Telecommunications,
Xitu Cheng Lu 10, 100876 Beijing, China
lliufeng@hotmail.com
[2] Department of Computer Science, Beijing Jiaotong University,
Shangyuan Cun 3, 100044 Beijing, China

Abstract. Existing algorithms for learning Bayesian network (BN) require a lot of computation on high dimensional itemsets, which affects accuracy especially on limited datasets and takes up a large amount of time. To alleviate the above problem, we propose a novel BN learning algorithm MRMRG, Max Relevance and Min Redundancy Greedy algorithm. MRMRG algorithm is a variant of K2 algorithm for learning BNs from limited datasets. MRMRG algorithm applies Max Relevance and Min Redundancy feature selection technique and proposes Local Bayesian Increment (LBI) function according to the Bayesian Information Criterion (BIC) formula and the likelihood property of overfitting. Experimental results show that MRMRG algorithm has much better efficiency and accuracy than most of existing BN learning algorithms when learning BNs from limited datasets.

1 Introduction

There are many problems in fields as diverse as medical diagnosis, weather forecast, fault diagnosis, where there is a need for models that allow us to reason under uncertainty and take decisions, even when our knowledge is limited. To model this type of problems, AI community has proposed Bayesian network which allows us to reason under uncertainty. [1] During the last two decades, many BN learning algorithms have been proposed. But, the recent explosion of high dimensional and limited datasets in the biomedical realm and other domains has induced a serious challenge to these BN learning algorithms. The existing algorithms must face higher dimensional and smaller datasets.

In general, BN learning algorithms take one of the two approaches: the constraint-based method and the search & score method. The constraint-based approach [2],[3],[15] estimates from the data whether certain condition independences hold between variables. Typically, this estimation is performed using statistical or information theoretical measure. The search & score approach [4],[5],[6],[9],[12],[13] attempts to find a graph that maximizes the selected score. Score function is usually defined as a measure of fitness between the graph and

R. Alhajj et al. (Eds.): ADMA 2007, LNAI 4632, pp. 412–421, 2007.
© Springer-Verlag Berlin Heidelberg 2007

the data. These algorithms use a score function in combination with a search method in order to measure the goodness of each explored structure from the space of feasible solutions. During the exploration process, the score function is applied in order to evaluate the fitness of each candidate structure to the data.

Although encouraging results have been reported, the two approaches both suffer some difficulties in accuracy on limited datasets. A high order statistical or information theoretical measure may become unreliable on limited datasets. At the same time, the result of selected score function may also be unreliable on limited datasets.

To further enhance learning efficiency and accuracy, this paper proposes Max-Relevance and Min-Redundancy Greedy BN learning algorithm. MRMRG algorithm applies Max-Relevance and Min-Redundancy feature selection technology to obtain better efficiency and accuracy on limited datasets, and proposes Local Bayesian Increment function according to BIC approximation formula and the likelihood property of overfitting for limited datasets.

This paper is organized as follows. Section 2 provides a brief review of some basic concepts and theorems. Section 3 describes K2 algorithm. In Section 4, we propose Local Bayesian Increment function. Section 5 represents the details of MRMRG algorithm. At the same time, we also analyze the time complexity of MRMRG. Section 6 shows an experimental comparison among K2 and MRMRG. Finally, we conclude and present future work.

2 Concepts and Theorems

2.1 Bayesian Network

A Bayesian network is defined as a pair $B = \{G, \Theta\}$, where G is a directed acyclic graph $G = \{V(G), A(G)\}$, with a set of nodes $V(G) = \{V_1, \ldots, V_n\}$ representing a set of random variables and a set of arcs $A(G) \subseteq V(G) \times V(G)$ representing causal independence/dependence relationships that exist among the variables. Θ represents the set of parameters that quantifies the network. It contains a parameter $\theta_{v_i | \pi_i} = P(v_i \mid \pi_i)$ for each possible value v_i of V_i, and π_i of Π_i. Here Π_i denotes the set of parents of V_i in G and π_i is a particular instantiation of Π_i.

2.2 Max-Dependence and MRMR

Definition 1. *In feature selection, **Max-Dependence scheme** [7] is to find a feature set S with m features, which jointly have the largest dependency on the target class C; $S = \arg \max\limits_{\{X_i, i=1,\ldots,m\}} I(\{X_i, i = 1, \ldots, m\}; C)$.*

Definition 2. *In feature selection, **Max-Relevance criterion** [7] is to select a feature set S with m features satisfying $S = \arg \max\limits_{S} \left(\frac{1}{|S|} \sum\limits_{X_i \in S} I(X_i; C) \right)$, which approximates $I(\{X_i, i = 1, \ldots, m\}; C)$ with the mean value of all mutual information values between individual features $X_i, i = 1, \ldots, m$ and class C.*

Definition 3. *In feature selection, **Min-Redundancy criterion** [7] is to select a feature set S with m features such that they are mutually minimally similar (mutually maximally dissimilar):* $S = \arg\min_S \left(\frac{1}{|S|^2} \sum_{X_i, X_j \in S} I(X_i; X_j) \right).$

Definition 4. *In feature selection, **Max-Relevance and Min-Redundancy criterion** [7] is to find a feature set S with m features obtained by optimizing the Max-Relevance criterion and the Min-Redundancy criterion simultaneously. Assume that the two conditions equally important, and consider the following criteria:* $S = \arg\max_S \left(\sum_{X_i \in S} I(X_i; C) - \frac{1}{|S|} \sum_{X_i, X_j \in S} I(X_i; X_j) \right).$

We select the feature set $S_m = \{X_1, X_2, \ldots, X_m\}$, the classification variable C. Using the standard multivariate mutual information $MI(X_1, \ldots, X_m) = \int \int p(x_1, \ldots, x_m) \log \frac{p(x_1, \ldots, x_m)}{p(x_1) \ldots p(x_m)} dx_1 \ldots dx_n$, we can get the following formula:

$$I(S_m; C) = \int \int p(S_m, c) \log \frac{p(S_m, c)}{p(S_m)p(c)} dS_m dc = MI(S_m, C) - MI(S_m). \quad (1)$$

Equation (1) is similar to the MRMR feature selection criterion: The second term requires that the features S_m are maximally independent of each other(that is, minimum redundant), while the first term requires every feature to be maximally dependent on C. In practice, the authors have shown that if one feature is selected at one time, then MRMR criterion is almost optimal implementation scheme of Max-Dependence scheme on limited datasets. [8]

3   K2 Algorithm

Given a complete dataset D, K2 searches for the Bayesian network G^* with maximal $P(G, D)$.

Let D be a dataset of m cases, where each case contains a value for each variable in V. D is sufficiently large. Let V be a set of n discrete variables, where x_i in V has r_i possible values $(v_{i1}, v_{i2}, \ldots, v_{ir_i})$. Let G denote a Bayesian network structure containing just the variables in V. Each variable x_i in G has the parents set π_i. Let $\phi_i[j]$ denote the jth unique instantiation of π_i relative to D. Suppose there are q_i such unique instantiation of π_i. Define N_{ijk} to be the number of cases in D in which variable x_i is instantiated as v_{ik} and π_i is instantiated as $\phi_i[j]$. Let $N_{ij} = \sum_{k=1}^{r_i} N_{ijk}$.

Given a Bayesian network model, cases occur independently. Bayesian network prior distribution is uniform. It follows that $g(i, \pi_i) = \prod_{j=1}^{q_i} \frac{(r_i-1)!}{(N_{ij}+r_i-1)!} \prod_{k=1}^{r_i} N_{ijk}!$.

It starts by assuming that a node has no parents, and then in every step it adds incrementally the node which can most increase the probability of the resulting BN, to the parents set. K2 stops adding nodes to parents set when the addition

Input: A set V of n variables, an ordering on the variables, a dataset D containing m cases, an upper bound u_{max}

Output: for each variable X_i ($i=1, \ldots, n$), a printout of the parents set π_i.

Procedure K2()

  For $i = 1$ to n do

    $\pi_i = $ NULL;

    $P_{old} = g(i, \pi_i)$;

    OK=TRUE;

    while OK and ($\| \pi_i \| < u_{max}$) do

      $Y = \underset{X_j \in Pre_i - \pi_i}{\arg\max} \left[g(i, \pi_i \cup X_j) \right]$;

      $P_{new} = g(i, \pi_i \cup X_j)$;

      If $P_{new} > P_{old}$ then

        $P_{old} = P_{new}$;

        $\pi_i = \pi_i \cup \{Y\}$;

      Else

        OK=FALSE;

      Endif

    Endwhile

    Output(X_i, π_i);

  EndFor

Endproc

Fig. 1. K2 algorithm

cannot increase the probability of the BN given the data. The pseudo-code of MRMRG algorithm sees Fig.1.

Pre_i denotes the set of variables that precede X_i. π_i denotes the current parents set of the variable X_i. u_{max} denotes an upper bound on the number of parents a node may have.

4   Local Bayesian Increment Function

Let X and Y be two discrete variables, $\mathbf{Z}$ be a set of discrete variables, and z be an instantiation for $\mathbf{Z}$. $X, Y \notin \mathbf{Z}$.

Definition 5. *According to Moore's recommendation [10] about the chi-squared test, the dataset D satisfying the following condition is **sufficiently large** for $\{X \cup Y\}$: All cells of $\{X \cup Y\}$ in the contingency table have expected value greater than 1, and at least 80% of the cells in the contingency table about $\{X \cup Y\}$ have expected value greater than 5.*

Definition 6. *According to Moore's recommendation [10] about the chi-squared test, the sub-dataset $D_{\mathbf{Z}=z}$ satisfying the following condition is **locally sufficiently large** for $\{X \cup Y\}$ given $\mathbf{Z} = z$: All cells of $\{X \cup Y\}$ in the contingency*

table conditioned on $\mathbf{Z} = z$ have expected value greater than 1, and at least 80%
of the cells in the contingency table about $\{X \cup Y\}$ on $\mathbf{Z} = z$ have expected value
greater than 5.

Learning on limited datasets, we loose the "locally sufficiently large" condition: If the number of cases in $D_{\mathbf{Z}=z}$ is much larger than the number of values for $\{X \cup Y\}$, for example $\|D_{\mathbf{Z}=z}\| \geq 4 \times (\|X\| \times \|Y\|)$; then we assume that the sub-dataset $D_{\mathbf{Z}=z}$ is "locally sufficiently large" for $\{X \cup Y\}$ given $\mathbf{Z} = z$.

Let D be a dataset of m cases. Let V be a set of n discrete variables, where X_i in V has r_i possible values $(v_{i1}, v_{i2}, \ldots, v_{ir_i})$. B_P and B_S denote BN structures containing just the variables in V. B_S exactly has one edge $Y \rightarrow X_i$ more than B_P. X_i has the parents set π_i in B_P and the parents set $\pi_i \cup Y$ in B_S. N_{ijk} is the number of cases in D, which variable X_i is instantiated as v_{ijk} and π_i is instantiated as $\phi_i[j]$. Let $N_{ijk} = \sum_y N_{i,\{j \cup y\},k}$, $N_{ij} = \sum_{k=1}^{r_i} N_{ijk}$. $\hat{\Theta}_i, \hat{\Theta}$ denote the maximum likelihoods of Θ_i, Θ. π_i^l denotes the instantiation of π_i in the lth case.

Cases occur independently. The prior distribution of possible Bayesian networks is uniform. Given a Bayesian network model, there exist two properties: Parameter Independence and Parameter Modularity. [5]

We apply the BIC formula also used by Steck in [11] : $BIC(B_S) = \log L\left(\hat{\Theta}\right) - \frac{1}{2}\log(m)dim\left(\hat{\Theta}\right) \approx \log(P(D \mid B_S))$ to control the complexity of BN model. BIC adds the penalty of structure complexity to LBI function to avoid overfitting.

Definition 7 (Local Bayesian Increment Function)

$$Lbi(Y, i, \pi_i) = \log\left(P(B_S, D)/P(B_P, D)\right) \approx BIC(B_S) - BIC(B_P)$$

$$= \log\left(L\left(\hat{\Theta}^{B_S}\right)/L\left(\hat{\Theta}^{B_P}\right)\right) - \frac{1}{2}\log(m)\left[dim\left(\hat{\Theta}^{B_S}\right) - dim\left(\hat{\Theta}^{B_P}\right)\right]$$

$$\log\left(L\left(\hat{\Theta}^{B_S}\right)/L\left(\hat{\Theta}^{B_P}\right)\right) = \log\left(P\left(D \mid \hat{\Theta}^{B_S}\right)\right) - \log\left(P\left(D \mid \hat{\Theta}^{B_P}\right)\right)$$

$$= \sum_{l=1}^{m} \log\left(P\left(x_i^l \mid \hat{\Theta}_i^{B_S}, \pi_i^l \cup y\right)/P\left(x_i^l \mid \hat{\Theta}_i^{B_P}, \pi_i^l\right)\right)$$

According to the likelihood property of overfitting(the marginal likelihood of overfitting for the training dataset is usually no less than non-overfitting), we assume that the log-likelihood does not change on the sub-dataset $D_{\pi_i = \phi_i[*]}$ which are not "locally sufficiently large" for $\{X \cup Y\}$ (that is, to assume that there is overfitting between X and the parents set $\pi_i \cup Y$ on the $D_{\pi_i = \phi_i[*]}$),

$$\sum_{d_l \in D_{\pi_i = \phi_i[*]}} \log P\left(x^l \mid \hat{\Theta}^{B_S}, \pi_l \cup y\right) = \sum_{d_l \in D_{\pi_i = \phi_i[*]}} \log P\left(x^l \mid \hat{\Theta}^{B_P}, \pi_l\right). \quad (2)$$

According to (2), we infer the following results:

$$\log\left(L\left(\hat{\Theta}^{B_S}\right)\right) - \log\left(L\left(\hat{\Theta}^{B_P}\right)\right)$$

$$= \sum_j N_{ij} \times I_j(X,Y), \text{for } j, D_{\pi_i=\phi_i[j]} \text{ is "locally sufficiently large"}$$

$$\dim\left(\hat{\Theta}^{B_S}\right) - \dim\left(\hat{\Theta}^{B_P}\right) = (r_y - 1)(r_i - 1)q_i$$

$$Lbi(Y, i, \pi_i) = \sum_j N_{ij} \times I_j(X,Y) - \frac{1}{2}(r_y - 1)(r_i - 1)q_i \log(m),$$

$$\text{for } j, D_{\pi_i=\phi_i[j]} \text{ is "locally sufficiently large"}.$$

Note: $I_j(X,Y)$ is the mutual information between X and Y on $D_{\pi_i=\phi_i[j]}$.

5   MRMRG Algorithm

MRMRG algorithm initializes the current parents set π_i of the variable X_i to NULL, and then adds the variables one by one, which acquire the maximal value for Local Bayesian Increment (LBI) function, into the parents set π_i from $Pre_i - \pi_i$, until the result of LBI function is no more than 0. Repeating the above steps for every variable, we can obtain an approximately optimal Bayesian network. The pseudo-code of MRMRG algorithm sees Fig.2.

Pre_i denotes the set of variables that precede X_i. π_i denotes the current parents set of the variable X_i. $(k < 5)$.

Given an ordering on the variables, MRMRG algorithm improves greedy BN learning algorithms (such as K2 algorithm [4]) in the following two ways in order to learn more accurately and efficiently on limited datasets.

Firstly, on limited datasets, the results of traditional scoring functions (such as K2 score [4],MDL score [6],BDe score [5], etc) $score(C, \pi_i \cup X_j)$ have less and less reliability and robustness with the dimension increase of $\pi_i \cup X_j$, so that the formula $Y = \arg \max\limits_{X_j \in Pre_i - \pi_i} score(C, \pi_i \cup X_j)$ cannot obtain the variable Y with the maximal score, even cannot acquire a variable with approximately maximal score sometimes. Since MRMR technology only uses 2-dimensional computation, it has much higher reliability and robustness than traditional scoring functions on limited datasets. Furthermore, according to the discussion in section 2.2, we know that if one feature is selected at one time (that is Greedy search), MRMR technology is nearly optimal implementation scheme of Max-Dependence scheme, which is equivalent to the maximal score method, on limited datasets. We consider that for some variable $X_j \in Pre_i - \pi_i$, if the value of $\{I(X_j; C) - \frac{1}{|\pi_i|+1} \sum\limits_{X \in \pi_i} I(X_j; X)\}$ is the largest, then it is the most probable that the value of the formula $score(C, \pi_i \cup X_j)$ is the largest. Thus, MRMRG algorithm applies Max-Relevance and Min-Redundancy (MRMR) feature selection technology and replaces $score(C, \pi_i \cup X_j)$ with the formula $\{I(X_j; C) - \frac{1}{|\pi_i|+1} \sum\limits_{X \in \pi_i} I(X_j; X)\}$ to obtain the variable Y which gets the maximal score. Firstly, MRMRG algorithm selects the top k variables from the sorted variables set $Pre_i - \pi_i$ according to the value of the formula $\{I(X_j; C) - \frac{1}{|\pi_i|+1} \sum\limits_{X \in \pi_i} I(X_j; X)\}$ by descendant order.

Input: A set V of n variables, an ordering on the variables, a dataset D containing m cases.

Output: for each variable X_i ($i=1, \ldots, n$), a printout of the parents set π_i.

Procedure MRMRG()

For $i = 1$ to n do

    Initialize π_i to NULL and OK to TRUE;

    while OK

        For every variable $X \in Pre_i - \pi_i$, compute the formula $\left\{ I(X;C) - \dfrac{1}{|\pi_i|+1} \sum_{X_j \in \pi_i} I(X;X_j) \right\}$ (1);

        Sort the variables in $Pre_i - \pi_i$ by descendant order according to the value of (1);

        Obtain the top k variables $\{Y_1, Y_2, \ldots, Y_k\}$ from the sorted variables set $Pre_i - \pi_i$;

        $Y_{max} = \underset{Y_j \in \{Y_1, Y_2, \ldots Y_k\}}{\arg \max} \left[Lbi(Y_j, i, \pi_i) \right]$;

        If $Lbi(Y_{max}, i, \pi_i) > 0$ then

            $\pi_i = \pi_i \cup \{Y_{max}\}$;

        Else

            OK=FALSE;

        Endif

    Endwhile

    Output (X_i, π_i);

Endfor

Endproc

Fig. 2. MRMRG Bayesian network learning algorithm

Then, it take the variable Y with the largest value of LBI function among the k variables as the variable with the maximal score.

Secondly, MRMRG algorithm proposes LBI function to replace traditional score increment functions (such as K2 [4],MDL [6],BDe [5]) to control the complexity of Bayesian network and to avoid overfitting. When the dataset D is "sufficiently large" for $\{X \cup Y \cup \pi_i\}$, LBI function is equivalent to K2 increment function. When the dataset D is not "sufficiently large" for $\{X \cup Y \cup \pi_i\}$, but there exist sub-datasets $D_{\pi_i = \phi_i[*]}$ are "locally sufficiently large" for $\{X \cup Y\}$ given $\pi_i = \phi_i[*]$, MRMRG algorithm can also apply LBI function to improve accuracy and avoid overfitting (see section 4). The technique also makes it unnecessary to set the maximal parents number u_{max} of a node.

5.1 The Time Complexity of MRMRG

$r = max(r_i), i = 1, \ldots, n$, where r_i is the number of values for the variable X_i. The complexity of the formula $I(X;C) - \frac{1}{|\pi_i|+1} \sum_{X_j \in \pi_i} I(X;X_j)$ is O(n). The complexity of computing $Lbi(Y, i, \pi_i)$ is O(mnr). The while statement loops at most O(n) times, each time it is entered. The for statement loops n times. So, in the worst case, the complexity of **MRMRG()** is O(kmn^3r). On the other hand, the worst-case time complexity of K2 algorithm is O(mn^4r). Therefore, if

$n >> k$, then MRMRG has much better efficiency (more than one magnitude) than K2 algorithm.

6   Experimental Results

We implemented MRMRG algorithm, K2 algorithm, TPDA algorithm [3] and presented the comparison of the experimental results for 3 implementations.

Tests were run on a PC with Pentium4 1.5GHz and 1GB RAM. The operating system was Windows 2000. These programs were developed under Matlab 7.0. 5 Bayesian networks were used. Table 1 shows the characteristics of these networks. The characteristics include the number of nodes, the number of arcs, the maximal number of node parents/children(Max In/Out-Degree), and the minimal/maximal number of node values(Domain Range).

From these networks, we performed these experiments with 200, 500, 1000, 5000 training cases each. For each network and sample size, we sampled 20 original datasets and recorded the average results by each algorithm. Let $u_{max} = 7$ in Fig.1 and $k = 3$ in Fig.2.

Table 1. Bayesian networks

BN	Nodes Num	Arcs Num	Max In/Out-Degree	Domain Range
Insur	27	52	3/7	2-5
Alarm	37	46	4/5	2-4
Barley	48	84	4/5	2-67
Hailf	56	66	4/16	2-11
Munin	189	282	3/15	1-21

6.1   Comparison of Runtime Among Algorithms

A summary of the time results of the execution of all the 3 algorithms is in Table 2. We normalized the times reported by dividing by the corresponding running time of MRMRG on the same datasets and reported the averages over sample sizes. [14] Thus, a normalized running time of greater than 1 implies a slower algorithm than MRMRG on the same learning task. A normalized running time of lower than 1 implies a faster algorithm than MRMRG.

From the results, we can see that MRMRG has better efficiency than other 3 algorithms K2 and TPDA. In particular, for smaller sample sizes (200, 500, 1000), MRMRG runs several times faster than K2 and TPDA. For larger sample sizes (5000), MRMRG performs nearly one magnitude faster than K2.

6.2   Comparison of Accuracy Among Algorithms

We compared the accuracy of Bayesian networks learned by these 3 algorithms according to the BDeu score. The BDeu score corresponds to the posteriori probability of the learned structure.[5] The BDeu scores in our experiments

Table 2. Normalized Runtime

Size	MRMRG	K2	TPDA
200	1.0	2.39	8.82
500	1.0	5.57	7.61
1000	1.0	9.18	4.57
5000	1.0	12.03	1.22

Table 3. Average BDeu(Insur)

Size	MRMRG	K2	TPDA
200	-21.915	-22.993	-31.507
500	-19.032	-19.583	-28.786
1000	-19.386	-19.533	-25.111
5000	-18.177	-18.152	-20.857

Table 4. Average BDeu(Alarm)

Size	MRMRG	K2	TPDA
200	-15.305	-16.069	-24.456
500	-13.858	-13.950	-18.097
1000	-13.319	-13.583	-15.429
5000	-13.021	-12.979	-14.496

Table 5. Average BDeu(Barley)

Size	MRMRG	K2	TPDA
200	-81.794	-83.972	-106.782
500	-76.327	-79.194	-103.783
1000	-76.846	-77.375	-111.069
5000	-75.710	-76.281	-116.892

Table 6. Average BDeu(Hailf)

Size	MRMRG	K2	TPDA
200	-72.138	-73.361	-106.135
500	-71.217	-72.662	-101.382
1000	-70.955	-71.734	-97.374
5000	-70.277	-71.105	-84.300

Table 7. Average BDeu(Munin)

Size	MRMRG	K2	TPDA
200	-65.971	-68.393	-135.103
500	-62.483	-63.250	-125.625
1000	-61.967	-62.837	-140.476
5000	-59.392	-60.943	-145.635

were calculated on a seperate test set sampled from the true Bayesian network containing 50000 samples. Table 3-7 reports the results.

From the results, we can see that MRMRG can learn more accurately than TPDA on limited datasets. In particular, MRMRG has better accuracy than K2 on limited datasets. The accuracy of MRMRG is almost the same as K2 on larger datasets relative to the true Bayesian network, such as Insur(5000), Alarm(5000).

7   Conclusion

Efficiency and accuracy are two main indices in evaluating algorithms for learning Bayesian network. MRMRG algorithm greatly reduces the number of high dimensional computations and improves scalability of learning. The experimental results indicate that MRMRG has better performance on efficiency and accuracy than most of existing algorithms on limited datasets.

We are interesting in incorporate ordering-based search method [9] into our MRMRG algorithm to implement MRMRG without the information of the order between nodes. In addition, much more experimentation is needed on different network structures.

References

1. Heckerman, D.: Bayesian Networks for Data Mining, 1st edn. Microsoft Press, Redmond (1997)
2. Spirtes, P., Glymour, C., Scheines, R.: Causation, Prediction and Search, 2nd edn. MIT Press, Massachusetts (2000)
3. Cheng, J., Greiner, R., Kelly, J., Bell, D., Liu, W.: Learning Belief Networks form Data: An Information Theory Based Approach. Artificial Intelligence 137(1-2), 43–90 (2002)
4. Cooper, G., Herskovits, E.: A Bayesian Method for Constructing Bayesian Belief Networks from Databases. In: Ambrosio, B., Smets, P. (eds.) UAI '91. Proceedings of the Seventh Annual Conference on Uncertainty in Artificial Intelligence, pp. 86–94. Morgan Kaufmann, San Francisco (1991)
5. Heckerman, D., Geiger, D., Chickering, D.M.: Learning Bayesian Networks: the Combination of Knowledge and Statistical Data. Machine Learning 20(3), 197–243 (1995)
6. Wai, L., Fahiem, B.: Learning Bayesian Belief Networks An approach based on the MDL Principle. Computational Intelligence 10(4), 269–293 (1994)
7. Hanchuan, P., Chris, D., Fuhui, L.: Minimum redundancy maximum relevance feature selection. IEEE Intelligent Systems 20(6), 70–71 (2005)
8. HanChuan, P., Fuhui, L., Chris, D.: Feature Selection Based on Mutual Information: Criteria of Max-Dependency, Max-Relevance, and Min-Redundancy. IEEE Transactions on PAMI 27(8), 1226–1238 (2005)
9. Teyssier, M., Koller, D.: Ordering-Based Search: A Simple and Effective Algorithm for Learning Bayesian Networks. In: Chickering, M., Bacchus, F., Jaakkola, T. (eds.) UAI '05. Proceedings of the 21st Conference on Uncertainty in Artificial Intelligence, pp. 584–590. Morgan Kaufmann, San Francisco CA (2005)
10. Moore, D.S.: Goodness-of-Fit Techniques, 1st edn. Marcel Dekker, New York (1986)
11. Steck, H.: On the Use of Skeletons when Learning in Bayesian Networks. In: Boutilier, C., Goldszmidt, M. (eds.) UAI '00. Proceedings of the 16th Conference on Uncertainty in Artificial Intelligence, pp. 558–565. Morgan Kaufmann, San Francisco (2000)
12. Moore, A.W., Wong, W.K.: Optimal Reinsertion: A New Search Operator for Accelerated and More Accurate Bayesian Network Structure Learning. In: Fawcett, T., Mishra, N. (eds.) ICML 2003 – Machine Learning. Proceedings of the Twentieth International Conference, pp. 552–559. AAAI Press, Washington DC (2003)
13. Friedman, N., Nachman, I., Peter, D.: Learning Bayesian Network Structure from Massive Datasets: The Sparse Candidate algorithm. In: Laskey, K.B. (ed.) UAI '99. Proceedings of the Fifteenth Conference on Uncertainty in Artificial Intelligence, pp. 196–205. Morgan Kaufmann, San Francisco (1999)
14. Ioannis, T., Laura, E.B.: The Max-Min Hill-Climbing Bayesian Network Structure Learning Algorithm. Machine Learning 65(1), 31–78 (2006)
15. Margaritis, D., Thrun, S.: Bayesian Network Induction via Local Neighborhoods. In: Solla, S.A., Leen, T.K., Muller, K. (eds.) NIPS '99. Advances in Neural Information Processing Systems 12, pp. 505–511. MIT Press, Cambridge (1999)

Optimum Neural Network Construction Via Linear Programming Minimum Sphere Set Covering

Xun-Kai Wei[1,2], Ying-Hong Li[1], and Yu-Fei Li[1]

[1] School of Engineering, Air Force Engineering University,
Shaanxi Province, Xian 710038, China
[2] Institute of Aeronautical Equipment, Air Force Equipment Academy,
100076 Beijing, China
xunkai.wei@gmail.com, yinghong_li@126.com,
horizontal_lyf@hotmail.com

Abstract. A novel optimum feedforward neural networks construction method was proposed. We first define a 0-1 covering matrix and propose an integer programming method for minimum sphere set covering problem. We then further define the extended covering matrix with smooth function, relax the objective and constraints to formulate a more general linear programming method for the minimum sphere set covering problem. We call this method Linear Programming Minimum Sphere set Covering (LPMSC). After that, we propose to apply the LPMSC method to neural network construction. With specific smooth functions, we can obtain optimum neural networks via LPMSC needless of prior knowledge, which is more objective and automatic like Support Vector Machines (SVM). Finally, we investigate the performances of the proposed method through UCI benchmark datasets compared with SVM. The parameters of SVM and our method are determined using 5-fold cross validation. Results shows that our method need less neurons than SVM while retain comparable even superior performances in the datasets studied.

1 Introduction

The neural network structure has extremely influence on its performance. Generally different structure has different performance. The neural network structure selection because of lacking solid theory foundation is still left open. There are many heuristic selection methods which were reported to be effective such as data structure preserving criterion [1], hybrid selection in evolutionary computation framework [2], orthogonal least square (OLS) [3], pruning and growing [4], regularization [5] etc. In fact, for most of these methods, we all need to specify some criterion for the structure selection problem, which is quite dependent on subjective experiences. So it may be wise to seek an alternative method to avoid the ambiguous criterion problem.

Recently, the minimum enclosing sphere approximation method [6] draws extensive attention in various research domains. The problem we interested is to enclose the class specific regions with unions of spheres. This problem is quite interested. Since each sphere acts just like the neuron in neural network especially RBFNN, so why not use sphere instead of hyperplane based neuron. This initial idea enlightens us a new way i.e. we may first construct a set of data dependent spheres, and then use some

R. Alhajj et al. (Eds.): ADMA 2007, LNAI 4632, pp. 422–429, 2007.
© Springer-Verlag Berlin Heidelberg 2007

methods to select a compact subset with minimum sphere number to obtain a compact classifier. The problem is also known as minimum set covering [7]. What's more, finding the smallest number of class-specific spheres can be justified in the VC theory and Structural Risk Minimization (SRM) framework. The VC dimension of the class of union of m spheres in R^d is bounded above by $2m(d+1)\ln[m(d+1)]$ [8]. Using this bound, we can easily conclude that a classifier with a smaller number of spheres owns better generalization performance. The same conclusion was proved to be true in general data-dependent structural risk minimization framework [9]. Due to the solid theory foundation, it is now very naturally to propose the minimum sphere set covering method for automatically constructing a new class of feedforward neural networks.

The paper is organized as follows. Section 2 will present an integer programming and an extended linear programming method for minimum sphere set covering problem. Section 3 will elaborate on how to build a compact feedforward neural network using different specific smooth functions. Experiments are studied in Section 4, and concluding remarks are given in Section 5.

2 Minimum Sphere Set Covering

Given training samples $D = \{(x_1, y_1), (x_2, y_2), \cdots, (x_n, y_n)\}$, the sphere set covering for binary classification problem is to find a set of class-specific spheres $\mathbb{Z} = \{S_1, S_2, \cdots, S_m\}$ such that $x_i \in \bigcup_{y(S_j)=y_i} S_j, x_i \notin \bigcup_{y(S_j)\neq y_i} S_j, \forall i = 1, 2, \cdots n$, where each sphere S_i is described by center $C(S_i)$, radius $r(S_i)$ and class label $y(S_i)$. We define following data dependent sphere S_i for each sample x_i as:

$$\begin{cases} r_i = \min_{j:y_j \neq y_i} d(x_i, x_j) - \varepsilon, \varepsilon > 0 \\ C(S_i) = x_i, r(S_i) = r_i, y(S_i) = y_i \end{cases} \tag{1}$$

The minimum sphere set covering is then to select a compact sphere subset $\mathbb{Z}$ from $\mathbb{S} = \{S_i, i = 1, 2, \cdots, n\}$.

2.1 Minimum Sphere Set Covering Via Integer Programming

For the given sphere set $\mathbb{S} = \{S_i, i = 1, \cdots, n\}$, we can define 0-1 covering matrix:

$$K_{ij} = \begin{cases} 1, if\ x_i \in S_j \Leftrightarrow d(x_i, x_j) \leq r_j \\ 0, otherwise \end{cases}, \tag{2}$$

$$or\ K_{ij} = K(x_i, x_j) = Heaviside(r_j - d(x_i, x_j)),$$

where, $d(x_i, x_j)$ is the Euclid distance between x_i and x_j, r_j is the radius of the sphere S_j centered on x_j. Thus, the entry K_{ij} is 1, only if the sphere centered on

x_j covers x_i. The minimum sphere set covering problem can be formulated as following integer programming form:

$$\min \sum_{i=1}^{n} z_i$$

$$s.t. \begin{cases} \sum_{j=1}^{n} K_{ij} z_j \geq 1, i = 1, 2, \cdots, n \\ z_i \in \{0,1\}, i = 1, 2, \cdots, n \end{cases} \tag{3}$$

where z_i indicates whether or not the sphere set S_i occurs in the sphere subset $\mathbb{Z}$. The sum $\sum_{i=1}^{n} z_i$ is the number of spheres in the sphere subset $\mathbb{Z}$. $\sum_{j=1}^{n} K_{ij} z_j$ is the number of spheres that cover the point x_i. The objective function minimizes the number of spheres in subset $\mathbb{Z}$ while the constraints ensure that a sphere covers at least one point.

For some problem, it might not be possible for each point to be covered by a sphere. To allow for some error, we can introduce slack variable $\xi_i \in \{0,1\}$ like Support Vector Machine (SVM) and reformulate the integer programming as:

$$\min \sum_{i=1}^{n} z_i + C \sum_{i=1}^{n} \xi_i$$

$$s.t. \begin{cases} \sum_{j=1}^{n} K_{ij} z_j \geq 1 - \xi_i, i = 1, 2, \cdots, n \\ z_i \in \{0,1\}, \xi_i \in \{0,1\}, i = 1, 2, \cdots, n \end{cases} \tag{4}$$

where $C > 0$ is a constant that controls the tradeoff between the training error and the number of spheres.

Solving the integer programming problem, we can obtain the resulting classification function:

$$f(x) = \operatorname{sgn} \left(\sum_{i=1}^{n} K(x, x_i) y_i z_i \right) \tag{5}$$

However, both these formulations are integer programming problems which are known to be NP-hard and so we would like to relax the constraints in order to form a Linear Programming (LP) formulation for minimum sphere covering problem just as [10] gives a LP extension for the set covering machine.

2.2 Extended Minimum Sphere Set Covering Via Linear Programming

As defined in equation (2), the covering matrix K is binary. Due to the element of set $\mathbb{S}$ is sphere, thus the decision boundary is piece-wise spherical. So as to further obtain a smooth decision boundary, we can re-define the covering matrix using smooth functions as:

$$K_{ij} = K\left(x_i, x_j\right) = f_{basis}(x_i, x_j, r_j) \tag{6}$$

where $f_{basis}(\bullet)$ is a base smooth function. What's more, we can further relax the constraints to be just $z_i \geq 0$ and $\xi_i \geq 0$ instead of rigid integers. Considering the class label information, we can get following linear programming formulation:

$$\min \sum_{i=1}^{n} z_i + C \sum_{i=1}^{n} \xi_i$$

$$s.t. \begin{cases} \sum_{j=1}^{n} y_i y_j K_{ij} z_j \geq 1 - \xi_i, i = 1, 2, \cdots, n \\ z_i \geq 0, \xi_i \geq 0, i = 1, 2, \cdots, n \end{cases} \tag{7}$$

where $C > 0$ is a constant that controls the tradeoff between the training error and the number of spheres.

Finally, we can apply any linear programming solver in order to solve the problem. Again we can get decision function for any unseen sample x:

$$f(x) = \text{sgn}\left(\sum_{i=1}^{n} K\left(x, x_i\right) y_i z_i\right) \tag{8}$$

3 Proposed Neural Network Construction Method

This section will elaborate procedures for neural networks construction using former linear programming sphere set covering method. Suppose the output layer activation function is linear and the hidden layer activation function is a nonlinear smooth function. Generally, the neural network for binary classification can be expressed via following form:

$$f(x) = \sum_{i=1}^{m} y_i w_i f_{basis}(x_i, x) \tag{9}$$

As is mentioned in section 2.2, the only work we need to do is to specify a certain smooth function type. There are many smooth functions available, such as RBF exponential function and sigmoid function. For RBF exponential function, we can define a smooth covering matrix:

$$K_{ij} = K(x_i, x_j) = \exp(-\frac{\|x_i - x_j\|^2}{sr_j^2}) \tag{10}$$

where $s > 0$ is a parameter controls the decay speed of exponential function. Another often used smooth function is sigmoid function. We can also define a smooth covering matrix:

$$K_{ij} = K(x_i, x_j) = \frac{1}{1 + \exp(-\frac{\|x_i - x_j\|^2 - r_j^2}{\sigma})} \tag{11}$$

where $\sigma > 0$ is a parameter controls the slope of the sigmoid function. The smaller $\sigma > 0$ is, the more closely the sigmoid function approximates the Heaviside function.

Now we can conclude following feedforward neural network construction algorithm (LPMSC-FNN):

Table 1. Proposed LPMSC method for feedforward neural network construction

LPMSC-FNN
Step 1, Generate data dependent sphere set $\mathbb{S}$ for training dataset D,
Step 2, Specify smooth function f_{basis} and define the covering matrix K,
Step 3, Solve the minimum sphere set covering problem via LP, and obtain compact sphere subset $\mathbb{Z}$,
Step 4, Output optimum neural network LPMSC-FNN.

The minimum number objective is to keep compact capacity of the classifier, which assures good generalization performance according to VC theory. For RBF exponential function, we get optimum RBF neural networks with compact linear combination of RBF exponential function. We denote the model as LPMSC-RBF.

$$f(x) = \mathrm{sgn}\left(\sum_{i=1}^{n} y_i z_i \exp(-\frac{\|x - x_i\|^2}{sr_i^2})\right) \tag{12}$$

The main virtue of this approach is that it does not use a prior knowledge to determine the neural network structure. We only need to specify the basis function term and thus define a smooth covering matrix. But our function is obviously different from standard RBF and sigmoid function, it is bounded and localized. This means our neural network is more data dependent i.e. the points fall into the data dependent spheres will have more influence on the classifier than the points far away from the spheres. Given a smooth function type, we can optimally determine the structure. The neuron of the proposed neural network construction is actually the selected spheres in the subset $\mathbb{Z}$, which covers as many samples of the same class as possible.

4 Experiments

We investigate the proposed neural network construction method via UCI benchmark datasets [11]. To show its superiority, a state-of-the-art machine learning algorithm SVM is adopted. In order to show that the proposed neural network construction is more compact, we investigate the number of effective computation unit i.e. Support Vectors (SV) for SVM and neurons for NN. Further in order to show generalization performance, we use RBF smooth function for LPMSC-RBFNN and RBF kernel

$$K(x_i, x_j) = \exp(-\gamma \|x_i - x_j\|^2) \text{ for SVM.}$$

Each algorithm was tested on the UCI data sets of Table 2. Each data set was randomly split in two parts [12]. The parameters C and γ for the SVM and parameters s and C for the LPMSC-RBF were determined by the 5-fold cross validation (CV) method performed on the training set.

The results are reported in Table 2. The support vector number and RBF basis number are presented in the final classifier and the classification errors obtained on the testing set is also reported. We observe that LPMSC-RBF needs smaller computation units than SVM. Generally, as for SVM, the Lagrange multipliers (or called dual variables) have many zero entries, which is called sparseness. As for LPMSC-RBF, the sparseness is more commendable than SVM, i.e. it needs less efficient computation units. Yet, we also observe that our proposed method is always much sparser than SVM but with roughly the same generalization error.

Table 2. SVM and LPMSC-RBFNN results

Dataset			SVM		LPMSC-RBFNN	
Name	train	test	SV	error	No.	error
breast cancer	341	342	89	0.039	32.5	0.023
diabetes	384	384	207	0.232	21.5	0.249
ionosphere	175	176	64	0.051	22.4	0.053
mushrooms	4062	4062	157	0.000	89.5	0.000
sonar	104	104	83	0.113	33.3	0.107

5 Conclusions

In this paper we proposed to solve an integer programming based minimum num sphere set covering problem to construct compact feedforward neural networks. We propose to use radial basis functions or sigmoid functions to further obtain an extended LP based sphere set covering solution. In contrast to traditional RBFNN and sigmoid MLP, in which neuron is specified a priori, the new method explicitly minimizes the number of basis units (which is the selected sphere in our method) in the resulting classifiers. Our proposed neural network construction method has rigid theory foundation, which is based on the structural risk minimization principle. Applications to the UCI Datasets show our proposed LPMSC method is commendably sparse and compact while owns comparable generalization performance compared with SVM.

However, as for our proposed LPMSC, due to the use of Euclidean distance metric in original sample space, there exists dimension curse problem, i.e. the integer (Equation 4) and linear (Equation 7) programming methods. But fortunately, the kernel trick works here easily. We just first map the original data to a high dimensional feature space using some nonlinear mapping $\Phi(\bullet)$ and then reformulate the distance in pairwise dot form

$$d_{\Phi(\cdot)}(x_i, x_j) = \left\| \Phi(x_i) - \Phi(x_j) \right\| = \sqrt{\Phi^T(x_i) \bullet \Phi(x_i) - 2\Phi^T(x_i) \bullet \Phi(x_j) + \Phi^T(x_j) \bullet \Phi(x_j)} \quad (13)$$

By using kernel trick $\Phi^T(x_i) \bullet \Phi(x_j) = k(x_i, x_j)$, we can obtain a totally kernelized distance form

$$d_{k(\cdot)}(x_i, x_j) = \sqrt{k(x_i, x_i) - 2k(x_i, x_j) + k(x_j, x_j)} \quad (14)$$

Using above distance metric, we could expect a kernelized LPSMC method to solve the problem of dimension curse. The kernelized LPSMC method will be investigated in future developments and related results will be reported elsewhere.

Acknowledgements. The authors would like to give special thanks to the anonymous reviewers for their invaluable suggestions. Special thanks also should go to Dr. Guang-bin Huang for his elaborate polish. This paper is jointly supported by NSFC and CAAC under grant #60672179, National High Technology Program Youth Foundation under grant #2005AA100200 and also supported by the Doctorate Foundation of the Engineering College, Air Force Engineering University of China under grant #BC0501.

References

1. Mao, K.Z., Huang, G.-B.: Neuron Selection for RBF Neural Network Classifier Based on Data Structure Preserving Criterion. IEEE Trans. Neural Networks 16(6), 1531–1540 (2005)
2. Vonk, E., Jain, L.C., Johnson, R.P.: Automatic Generation of Neural Network Architecture Using Evolutionary Computation. World Scientific Publishing, River Edge NJ (1997)
3. Rutkoski, L.: Adaptive Probabilistic Neural Networks for Pattern Classification in Time-varying Environment. IEEE Trans. Neural Networks 15(4), 811–827 (2004)
4. Huang, G.-B., Saratchandran, P., Sundararajan, N.: A Generalized Growing and Pruning RBF (GGAP-RBF) Neural Network for Function Approximation. IEEE Trans. Neural Networks 16(1), 57–67 (2005)
5. Girosi, F., Jones, M., Poggio, T.: Regularization Theory and Neural Networks Architectures. Neural Computation 7(2), 219–269 (1995)

6. Tsang, I.W., Kwok, J.T., Cheung, P.-M.: Core Vector Machines: Fast SVM Training on Very Large Data Sets. Journal of Machine Learning Research 6, 363–392 (2005)
7. Marchand, M., Shawe-Taylor, J.: The Set Covering Machine. Journal of Machine Learning Research 3, 723–746 (2002)
8. Floyd, S., Warmuth, M.: Sample Compression, Learnability, and the Vapnik-Chervonenkis Dimension. Machine Learning 21, 269–304 (1995)
9. Shawe-Taylor, J., Bartlett, P.L., Williamson, R.C., Anthony, M.: Structural Risk Minimization over Data-dependent Hierarchies. IEEE Trans. Information Theory 44, 1926–1940 (1998)
10. Hussain, Z., Szedmak, S., Shawe-Taylor, J.: The Linear Programming Set Covering Machine. In: PASCAL04 (2004), http://eprints.pascal-network.org/archive/00001210/01/lp_scm.pdf
11. Newman, D.J., Hettich, S., Blake, C.L., Merz, C.J.: Repository of Machine Learning Databases (1998), http://www.ics.uci.edu/m learn/MLRepository.html
12. Wei, X.-K, Löfberg, J., Feng, Y., Li, Y.-H, Li, Y.-F.: Enclosing Machine Learning for Class Description. In: Liu, D., et al. (eds.) ISNN 2007. LNCS, vol. 4491, pp. 428–437. Springer, Heidelberg (2007)

How Investigative Data Mining Can Help Intelligence Agencies to Discover Dependence of Nodes in Terrorist Networks

Nasrullah Memon, David L. Hicks, and Henrik Legind Larsen

Department of Computer Science and Engineering
Esbjerg Institute of Technology, Aalborg University
Niels Bohrs Vej 8, DK-6700, Esbjerg, Denmark

Abstract. A new model of dependence centrality is proposed. The centrality measure is based on shortest paths between the pair of nodes. We apply this measure with the demonstration of a small network example. The comparisons are made with betweenness centrality. We discuss how intelligence investigation agencies could benefit from the proposed measure. In addition to that we argue about the investigative data mining techniques we are using, and a comparison is provided with traditional data mining techniques.

Keywords: Destabilizing terrorist networks, dependence centrality, investigative data mining, social network analysis, subject-based link analysis

1 Introduction

Terrorists seldom operate in a vacuum but interact with one another to carry out terrorist activities. To perform terrorist activities requires collaboration among terrorists. Relationships between individual terrorists are essential for the smooth operation of a terrorist organization, which can be viewed as a network consisting of nodes (for example terrorists, terrorist camps, supporting countries, *etc.*) and links (for example, communicates with, or trained at, *etc.*). In terrorist networks, there may exist some group or cell, within which members have close relationships. One group may also interact with other groups. For example, some key nodes (key players) may act as leaders to control activities of a group, while others may serve as gatekeepers to ensure smooth flow of information or illicit goods.

In social network literature, researchers have examined a broad range of types of ties [1]. These include communication ties (such as who talks to whom or who gives information or advice to whom), formal ties (such as who reports to whom), affective ties (such as who likes whom, or who trust whom), material or work flow ties (such as who gives bomb making material or other resources to whom), proximity ties (who is spatially or electronically close to whom). Networks are typically multiplex, that is, actors share more than one type of tie. For example, two terrorists might have a formal tie (one is a foot-soldier or a newly recruited person in the terrorist cell and reports to the other, who is the cell leader) and an affective tie (they are friends); and

R. Alhajj et al. (Eds.): ADMA 2007, LNAI 4632, pp. 430–441, 2007.

may also have a proximity tie (they are residing in the same apartment and their flats are two doors away on the same floor).

Network researchers have distinguished between strong ties (such as family and friends) and weak ties such as acquaintances [2, 3]. This distinction will involve a multitude of facets, including affect, mutual obligations, reciprocity, and intensity. Strong ties are particularly valuable when an individual seeks socio-emotional support and often entail a high level of trust. Weak ties are more valuable when individuals are seeking diverse or unique information from someone outside their regular frequent contacts.

Ties may be non directional (for example, Atta attends a meeting with Nawaf Alhazmi) or vary in direction (for instance, Bin Laden gives advice to Atta vs. Atta gets advice from Bin Laden). They may vary in content (Atta talks with Khalid about the trust of his friends in using them as human bombs and his recent meeting with Bin Laden), frequency (daily, weekly, monthly, *etc.*), and medium (face-to-face conversation, written memos, email, fax, instant messages, *etc.*). Finally ties may vary in sign, ranging from positive (Iraqis like Zarqawi) to negative (Jordanians dislike Zarqawi).

A network may be divided into groups or subgroups (also known as cells), which is known as a structural feature of a network in the literature of social network analysis (SNA). Structural network patterns in terms of subgroups and individual roles are important in understanding the organization and operation of terrorist networks. Such knowledge can help law enforcement and intelligence agencies to disrupt terrorist networks and develop effective control strategies to combat terrorism. For example, capture of central members in a network may effectively upset the operational network and put a terrorist organization out of action [4, 5, 6]. Subgroups and interaction patterns between groups are helpful in finding a network's overall structure, which often reveals points of vulnerability [7, 8]. Networks come in basically three types [9]:

(1) The chain network, as in a smuggling chain, where people, goods, or information move along a line of separated contacts and where end-to-end communication must travel through the intermediate nodes. (2) The star, hub, or wheel network, as in a terrorist syndicate or a cartel structure, where a set of actors is tied to a central node or actor and all must go through that node to communicate and coordinate with each other. (3) The all-channel network, as in a collaborative network of small militant groups, in which every group or node is connected to every other node.

Each type of network may be suited to different conditions and purposes, and there may be any number of hybrids. The all-channel network has historically been the most difficult to organize and sustain, partly because of the dense communications required. Yet the all-channel network is the type that is gaining strength from the information revolution. The design is flat. Ideally, there is no single, central leadership or command or headquarters—no precise heart or head that can be targeted. Decision-making and operations are decentralized, allowing for local initiative and autonomy [10].

The remainder of the paper is organized as follows: Section 2 introduces investigative data mining and also provides a comparison between traditional data mining and the data mining used for counterterrorism. This section also introduces the importance of social network analysis in counterterrorism research. Section 3 describes the fundamentals of graph theory on which this study is based. Section 4 briefly discusses the centrality measures available in the literature and relates them this research. Section 5 provides a detailed discussion of a newly proposed centrality measure, that is, dependence centrality. This section also discusses the strengths of this new centrality measure using two examples and argues how the investigative world could benefit from this measure. Section 5 concludes the paper with a discussion of some limitations of this measure.

2 Investigative Data Mining (IDM)

Defeating terrorist networks requires a nimble intelligence apparatus that operates actively and makes use of advanced information technology. Data mining for counterterrorism (also known as investigative data mining[1]) is a powerful tool for intelligence and law enforcement officials fighting against terrorism [11]. Investigative data mining is a combination of data mining and subject-based automated data analysis techniques. "Data mining" actually has a relatively narrow meaning: the approach that uses algorithms to discover predictive patterns in datasets. Subject-based automated data analysis applies models to data to predict behaviour, assess risk, determine associations, or do other types of analysis [11]. The models used for automated data analysis can be used on patterns discovered by data mining techniques.

Although these techniques are powerful, it is a mistake to view investigative data mining techniques as a complete solution to security problems. The strength of IDM is to assist analysts and investigators. IDM can automate some functions that analysts would otherwise have to perform manually. It can help to prioritize attention and focus an inquiry, and can even do some early analysis and sorting of masses of data. Nevertheless, in the complex world of counterterrorism, it is not likely to be useful as the only source for a conclusion or decision.

In the counterterrorism domain, much of the data could be classified. If we are to truly get the benefits of the techniques we need to test with actual data. But not all researchers have the clearances to work on classified data. The challenge is to find unclassified data that is, representative of the classified data. It is not straightforward to do this, as one has to make sure that all classified information, even through implications, is removed. Another alternative is to find as good data as possible in an unclassified setting for researchers to work on. However, the researchers have to work not only with counterterrorism experts but also with data mining specialists who have the clearances to work in classified environments. That is, the research carried out in an unclassified setting has to be transferred to a classified setting later to test the

---

[1] The term is firstly used by Jesus Mena in his book Investigative Data Mining and Criminal Detection, Butterworth (2003) [18].

applicability of data mining algorithms. Only then do we get the true benefit of investigative data mining.

Investigative data mining is known as a data-hungry project for academia. It can only be used if researchers have good data. For example, in monitoring central banking data, in the detection of an interesting pattern, e.g., a person who is earning less and spending more continuously for the last 12 months, the investigative officer may send a message to an investigative agency to keep the person under observation. Further, if a person, (who is under observation of an investigative agency) has visited Newark airport more than 5 times in a week (a result received from video surveillance cameras), then this activity is termed a suspicious activity and investigative agencies may act. It was not possible for us to get this type of sensitive data. Therefore, in this research, we harvested unclassified data of the terrorist attacks that occurred, or were planned in the past, from a number of authenticated web resources [12].

IDM offers the ability to map a covert cell, and to measure the specific structural and interactional criteria of such a cell. This framework aims to connect the dots between individuals and to map and measure complex, covert, human groups and organizations [13]. The method focuses on uncovering the patterning of people's interaction, and correctly interpreting these networks assists in predicting behaviour and decision-making within the network [13]. The technique is also known as subject based link analysis. This technique uses aggregated public records or other large collections of data to find the links between a subject— a suspect, an address, or a piece of relevant information—and other people, places, or things. This can provide additional clues for analysts and investigators to follow [11].

IDM also endows the analyst with the ability to measure the level of covertness and efficiency of the cell as a whole, and the level of activity, ability to access others, and the level of control over a network each individual possesses. The measurement of these criteria allows specific counter-terrorism applications to be drawn, and assists in the assessment of the most effective methods of disrupting and neutralising a terrorist cell [13]. In short, IDM provides a useful way of structuring knowledge and framing further research. Ideally it can also enhance an analyst's predictive capability [13].

On the other hand, traditional data mining commonly refers to using techniques rooted in statistics, rule-based logic, or artificial intelligence to comb through large amounts of data to discover previously unknown but statistically significant patterns. However, in the application of IDM in the counterterrorism domain, the problem is much harder, because unlike traditional data mining applications, we must, find an extremely wide variety of activities and hidden relationships among individuals. Table 1 gives a series of reasons for why traditional data mining isn't the same as investigative data mining.

In this research we have chosen to use a very small portion of data mining for counterterrorism; and we have borrowed techniques from social network analysis and graph theory for connecting the dots. Our goal is to propose mathematical methods, techniques, and algorithms to assist law enforcement and intelligence agencies in destabilizing terrorist networks.

Table 1. Traditional data mining vs. investigative data mining

TRADITIONAL DATA MINING	INVESTIGATIVE DATA MINING
Discover comprehensive models of databases to develop statistically valid patterns	Detect connected instances of rare patterns
No starting points	Known starting points or matches with patterns estimated by analysts
Apply models over entire data	Reduce search space; results are starting points for human analysts
Independent instances (records)	Relational instances (networked data)
No correlation between instances	Significance autocorrelation
Minimal consolidation needed	Consolidation is key
Dense attributes	Sparse attributes
Sampling is used	Sampling destroys connections
Homogeneous data	Heterogeneous data
Uniform privacy policy	Non-uniform privacy policy

Traditionally, most of the literature in SNA has focussed on networks of individuals. Although SNA is not conventionally considered as a data mining technique, it is especially suitable for mining a large volume of association data to discover hidden structural patterns in terrorist networks. SNA primarily focuses on applying analytic techniques to the relationships between individuals and groups, and investigating how those relationships can be used to infer additional information about the individuals and groups [14]. There are a number of mathematical and algorithmic approaches that can be used in SNA to infer such information, including connectedness and centrality [15].

Law enforcement personnel have used social networks to analyze terrorist networks [16, 17] and criminal networks [6]. The capture of Saddam Hussein was facilitated by social network analysis: military officials constructed a network containing Hussein's tribal and family links, allowing them to focus on individuals who had close ties to Hussein [19].

After the 9/11 attacks, SNA has increasingly been used to study terrorist networks. Although these covert networks share some features with conventional networks, they are harder to identify, because they mask their transactions. The most significant complicating factor is that terrorist networks are loosely organized networks having no central command structure; they work in small groups, who communicate, coordinate and conduct their campaign in a network like manner. There is a pressing need to automatically collect data of terrorist networks, analyze such networks to find hidden relations and groups, prune datasets to locate regions of interest, detect key players, characterize the structure, trace the point of vulnerability, and find the efficiency of the network [20]. Hence, it is desirable to have tools to detect the hidden hierarchical structure of horizontal terrorist networks [21] in order to assist law enforcement agencies in the capture of the key players so that most of the network can be disrupted after their capture.

3 Graph Theory[2]

Throughout this article the topology of a network is represented by a *graph* $G = (V, E)$ which is an abstract object, formed by a finite set V of vertices ($m = |V|$) and a finite set E of edges ($n = |E|$). An edge $e = (u, v) \in E$ connects two vertices u and v. The vertices u and v are said to be *incident* with the edge e and *adjacent* to each other. The set of all vertices which are adjacent to u is called the neighbourhood $N(u)$ of u: ($N(u) = \{v: (u, v) \in E\}$). A graph is called *loop-free* if no edge connects a vertex to itself. An *adjacency matrix* A of a graph $G = (V, E)$ is a ($n \times n$) matrix, where $a_{uv} = 1$ if and only if $(u, v) \in E$ and $a_{uv} = 0$ otherwise. Graphs can be *undirected* or *directed*. The adjacency matrix of an undirected graph is symmetric. An undirected edge joining vertices $u, v \in V$ is denoted by $\{u, v\}$.

In directed graphs, each directed edge (arc) has an *origin* (*tail*) and a *destination* (*head*). An edge with origin $u \in V$ is represented by an ordered pair (u, v). As a shorthand notation, an edge $\{u, v\}$ can also be denoted by uv. It should be noted that, in a directed graph, uv is short for (u, v), while in an undirected graph, uv and vu are the same and both stand for $\{u, v\}$. Graphs that can have directed as well undirected edges are called *mixed graphs*, but such graphs are encountered rarely. Let $(e_1,...,e_k)$ be a sequence of edges in a graph $G = (V, E)$. This sequence is called a *walk* if there are vertices $v_0,...,v_k$ such that $e_i = (v_{i-1}, v_i)$ for $i = 1, ..., k$. If the edges e_i are pairwise distinct and the vertices v_i are pairwise distinct the walk is called a *path*. The *length* of a walk or path is given by its number of edges, $k = |\{e_i,...,e_k\}|$. The *shortest path* between two vertices u, v is a path with minimum length, all shortest paths between u and v are called *geodesics*.

The *distance* (dist (u, v)) between two vertices u, v is the length of the shortest paths between them. The vertices u, v are called *connected* if there exists a walk from vertex u to vertex v. If any pair of different vertices of graph $G = (V, E)$ is connected, the graph is called *connected*.

In the remainder of this article, we consider only non-trivial undirected loop-free connected graphs.

4 Centrality Measures

Centrality is one of the network properties that frequently has been used to study actors or events in terrorist social networks. The general notion of centrality encompasses a number of different aspects of the "importance" or "visibility" of actors within a network. A review of key centrality concepts can be found in the papers by Freeman *et al.* [23]. Their work has significantly contributed to the conceptual clarification and theoretical application of centrality. He provides three general measures of centrality termed "degree", "closeness", and "betweenness". His development was partially motivated by the structural properties of the center of a star graph. The most basic idea of degree centrality in a graph is the adjacency count of its constituent nodes.

Formally a centrality is a function C which assigns every vertex $v \in V$ of a given graph G a value $C(v) \in \mathbb{R}$. As we are interested in the ranking of the vertices of the

---

[2] Most of the concepts discussed in this section are taken from [22].

given graph G we choose the convention that a vertex u is more important than another vertex v, iff $C(u) > C(v)$. Now we explain different centrality measures.

"The *degree* of a node v is simply the count of the number of encounters with other nodes that are adjacent to it, and with which it is, therefore, in direct contact" [23]; it is known as a measure of activity. The degree centrality $C_d(v)$ of a vertex v is simply defined as the degree $d(v)$ of v if the considered graph is undirected. The degree centrality is, e.g., applicable whenever the graph represents something like a voting result. These networks represent a static situation and we are interested in the vertex that has the most direct votes or that can reach most other vertices directly. The degree centrality is a local measure, because the centrality value of a vertex is only determined by the number of its neighbors. The most commonly employed definition of degree centrality is:

$$C_d(u) = \sum r(u,v) \tag{1}$$

Where $r(u, v)$ is a binary variable indicating whether a link exists between nodes u and v.

The second measure relates to the *closeness* or the *distance* between the nodes. According to Freeman (1979), this closeness measure can be conceptualized as independence (the extent to which an actor can avoid the control of others) or efficiency (extent to which an actor can reach all other actors in the shortest number of steps). Thus, it measures independent access to others. A central actor can reach other actors through a minimum number of intermediary positions and is therefore dependent on fewer intermediary positions than a peripheral actor.

Suppose a terrorist organization wants to establish a new camp, for example, a human bomb training camp, such that the total distance to all persons interested to kill themselves, for a cause, in the region is minimal. This would make traveling to the camp as convenient as possible for most people who are living in that region and are willing to be used for human bombs in the near future.

We denote the sum of the distances from a vertex $u \in V$ to any other vertex in a graph $G = (V,E)$ as the total distance $\sum_{u \in V} d(u,v)$, where $d(u,v)$ is shortest distance between the nodes u and v The problem of finding an appropriate location can be solved by computing the set of vertices with minimum total distance.

In SNA literature, a centrality measure based on this concept is called closeness. The focus lies here, for example, on measuring the closeness of a person to all other people in the network. People with a small total distance are considered as most important as those with high total distance. The most commonly employed definition of closeness is the reciprocal of the total distance:

$$C_C(u) = \frac{1}{\sum_{v \in V} d(u,v)} \tag{2}$$

$C_C(u)$ grows with decreasing total distance of u, therefore it is also known as a structural index. Unlike degree centrality this measure is a global metric.

The third measure is *betweenness* which is defined as the frequency at which a node occurs on geodesics that connect pairs of nodes. Thus, any node that falls on the shortest path between other nodes can potentially control the transmission of information or effect exchange by being an intermediary; it is the potential for control that defines the centrality of these nodes [23]. Thus, if two persons A and C are connected only through person B, B would fall between A and C and would have control of any resources that flow between A and C. This measure easily discovers gatekeepers.

Let $\delta_{uw}(v)$ denote the fraction of shortest paths between u and w that contain vertex v:

$$\delta_{uw}(v) = \frac{\sigma_{uw}(v)}{\sigma_{uw}} \tag{3}$$

where σ_{uw} denotes the number of all shortest-paths between u and w. The ratio $\delta_{uw}(v)$ can be interpreted as the probability that vertex v is involved in any communication between u and w. Note, that this measure implicitly assumes that all communication is conducted along shortest paths. Then the betweenness centrality C_B (v) of a vertex v is given by:

$$C_B(v) = \sum_{u \neq v \in V} \sum_{w \neq v \in V} \delta_{uw}(v) \tag{4}$$

Any pair of vertices u and w without any shortest path from u to w will just add zero to the betweenness centrality of every other vertex in the network. This measure is also a global metric like closeness centrality.

These measures can help in finding key players in terrorist social networks. But if intelligence agencies have the data about terrorist cells, how do they detect who is depending on whom in the network. The next Section proposes and discusses a new centrality measure that can be useful in discovering who is depending on whom in terrorist networks.

5 Dependence Centrality

Dependence centrality represents how much a node is dependent on other nodes in a network. Consider a network representing a symmetrical relation, "communicates with" for a set of nodes. When a pair of nodes (say, u and v) is linked by an edge so that they can communicate directly without intermediaries, they are said to be adjacent. A set of edges linking two or more modes (u, v, w) such that node u would like to communicate with w, using node v. The dependence centrality can discover how many times node u uses node v to reach node w and how many shortest paths node u uses to reach node w. There can, of course, be more than one geodesic, linking any pair of nodes.

Let $\zeta_{(u,v)}(w)$ = dependence factor of the node u on node v to reach any other node (*i.e.*, node w) in the graph of communication as shown in (5):

$$\zeta_{(u,v)}(w) = \frac{occurence(u,v)}{d(u,v) \times path(u,w)} \qquad (5)$$

where $occurence(u,v)$ = the number of times (shortest paths), the node u uses node v in the communication with one another, $path(u,w)$ = the number shortest paths between node u and node w, and $d(u,v)$ = the geodesic distance between node u and node v.

As mentioned earlier, the dependence centrality of a node represents how much a node is dependent on other nodes. Usually the nodes which are adjacent to a node are always important for that node, as all activities of that node depend on the nodes which are adjacent to it (or directly connected to that node).

Now we define dependence centrality as the degree to which a node, u, must depend upon another, v, to relay its messages along geodesics to and from all other reachable nodes in the network. Thus, for a network containing n nodes, the dependence centrality of u on v can be found by using:

$$C_{dep(u,v)} = 1 + \sum_{w=1}^{n} \zeta_{(u,v)}(w), \qquad u \# v \# w \qquad (6)$$

We have used 1 in the above formula, because we expect every graph/ network we use is connected.

We can calculate the dependence centrality of each vertex on every other vertex in the network and arrange the results in a matrix $D = [C_{dep(u,v)}]$. The value of the dependence matrix can be normalized by dividing each value with $(n—1)$ where n represents the total number of nodes in the network.

Each entry in D is an index of the degree to which the node designated by the row of the matrix must depend on the vertex designated by the column to relay messages to and from others. Thus D captures the importance of each node as a gatekeeper with respect to each other node—facilating or perhaps inhibiting its communication.

The dependence matrix benefits an analyst by providing an even clearer picture than betweenness or closeness centrality, not only identifying how much a particular node is dependent on others, but also how much others depend on that particular node. Consider, for example, a fictitious small terrorist network as shown in Figure 1. The nodes a, b, and c show a relatively low score based on overall centrality. The (column) sum of each node (a, b, c) is approximately 1, in comparison to the sums of other nodes in the network. The following is the summary of the dependency matrix shown in Table 2.

The *lowest sum of values in a row* points out that the nodes that are most difficult to be deactivated, as its communications are least damaged with the capture of other nodes. These nodes are least dependent on others. Its communications are uniformly distributed. Whereas the *highest sum of values* in a row points out that the nodes that can be easily deactivated. These nodes are mostly dependent on others.

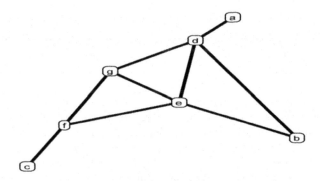

Fig. 1. A fictitious Terrorist Network

Table 2. Dependency matrix of the shown in Figure 1

Node	g	e	f	d	b	c	a	Sum
g		0.25	0.33	0.42	0.17	0.17	0.17	1.51
e	0.17		0.33	0.33	0.17	0.17	0.17	1.34
f	0.33	0.5		0.25	0.17	0.17	0.17	1.59
d	0.33	0.33	0.25		0.17	0.17	0.17	1.42
b	0.17	0.58	0.25	0.42		0.17	0.17	1.76
c	0.25	0.33	1.00	0.22	0.17		0.17	2.19
a	0.25	0.25	0.22	1.00	0.17	0.17		2.06
Sum	1.5	2.24	2.38	1.64	1.02	1.02	1.02	

The lowest sum of values in a column tells us that minimum communication takes place through these nodes. The capture of these nodes will be of least damage for a network. Whereas the *largest sum of values in a column* points out the nodes whose capture would be most disruptive the network.

The dependence matrix benefits the analyst by giving an even clearer picture than betweenness centrality. The matrix not only identifies how much a particular node is dependent on others, but also it discovers how much others depend on that particular node. Thus an analyst may be able to tell the intelligence agencies about the over all picture of the nodes in a network: keeping in view the strength and weaknesses of the node being considered for capture.

Looking at the values of betweenness, we note that the betweenness centrality for the nodes a and b, i.e., $C_B(a)$ and $C_B(b)$, is zero, as these nodes participate least in the paths of communication. One may only deduce that other nodes in the network are not much dependent on these nodes for communication.

If we view the dependence matrix which is tabulated in Table 2, row of the node a tells us how much node a depends on others. It is crystal clear that node a mostly depends on the node d (by removing node d, node a will automatically be isolated). If we analyze the row of node b, the row is more uniformly distributed as compared to the row of node a. The node b mostly depends on node e (which has the highest value of 0.58).

Table 3. Betweenness Centrality of the network shown in Figure 1

Node	g	e	f	d	b	c	a
Betweenness	0.13	0.30	0.33	0.37	0	0	0

The node b also has some other alternatives, for example, it also depends on node d (it has value 0.42). As these two nodes are not frequently in communication as resulting from betweenness, there is significant difference as indicated by the dependence matrix. It should be noted that if intelligence agencies desire to know which node should be removed or eradicated (either node a or node b), the analyst by looking at the dependence centrality matrix can assist in advising to capture node b in comparison to node a, because node b's capture will cause more harm than node a. These points have significant effects, which are not indicated by betweenness centrality.

6 Conclusions

A new measure of centrality, dependence centrality, is proposed, and its application in analysis of terrorist networks illustrated by use of an example. This measure reflects the information contained in the shortest paths of a network. We have demonstrated how this measure could be useful for the law enforcement and intelligence agencies in understanding the structure of the network.

We have attempted to illustrate the calculation of centralities to these prototypical situations. However we regard our efforts in this direction as only a beginning. We only considered shortest distances in this paper. It is quite possible that information will take a more circuitous route either by random communication or may be intentionally channeled through many intermediaries in order to "hide" or "shield" information in a way not captured by geodesic paths. These considerations raise questions as to how to include all possible paths in a centrality measure.

References

1. Monge, P.R., Contractor, N.: Theories of Communication Networks. Oxford University Press, New York (2003)
2. Granovetter, M.: The Strength of Weak Ties. American Journal of Sociology 81, 1287–1303 (1973)
3. Granovetter, M.: The Strength of Weak Ties: A Network Theory Revisited. In: Collins, R. (ed.) Sociological Theory, pp. 105–130 (1982)
4. Baker, W.E., Faulkner, R.R.: The Social Organization of Conspiracy: Illegal Networks in the Heavy Electrical Equipment Industry. American Sociological Review 58(6), 837–860 (1993)
5. McAndrew, D.: The Structural Analysis of Criminal Networks. In: Canter, D., Alison, L. (eds.) The Social Psychology of Crime: Groups, Teams, and Networks, Aldershot, Dartmouth. Offender Profiling Series, III, pp. 53–94 (1999)
6. Sparrow, M.: The Application of Network Analysis to Criminal Intelligence: An Assessment of the Prospects. Social Networks 13, 251–274 (1991)

7. Evan, W.M.: An Organization-set Model of Inter-organizational Relations. In: Tuite, M., Chisholm, R., Radnor, M. (eds.) Inter-organizational Decision-making, Aldine, Chicago, pp. 181–200 (1972)
8. Ronfeldt, D., Arquilla, J.: What Next for Networks and Netwars? In: Arquilla, J., Ronfeldt, D. (eds.) Networks and Netwars: The Future of Terror, Crime, and Militancy, Rand Press (2001)
9. Arquilla, J., Ronfeldt, D.: Swarming and a Future of Conflict. RAND National Defense Institute (2001)
10. Hoffman, B.: Terrorism evolves Toward Netwar. RAND Review 22(2) (1999)
11. DeRosa, M.: Data Mining and Data Analysis for Counterterrorism, CSIS Report (2004)
12. Nasrullah, M.: A First Look on iMiner's Knowledge Base and Detecting Hidden Hierarchy of Riyadh Bombing Terrorist Network. In: Proceedings of International Conference on Data Mining Applications, Hong Kong, March 21-23, 2007 (2007)
13. Memon, N., Larsen, H.L.: Structural Analysis and Mathematical Methods for Destabilizing Terrorist Networks. In: Li, X., Zaïane, O.R., Li, Z. (eds.) ADMA 2006. LNCS (LNAI), vol. 4093, pp. 1037–1048. Springer, Heidelberg (2006)
14. Degenne, A., Forse, M.: Introducing Social Networks. Sage Publications, London (1999)
15. Wasserman, S., Faust, K.: Social Network Analysis: Methods and Applications. Cambridge University Press, Cambridge (1994)
16. Krebs, V.E.: Uncloaking Terrorist Networks (2002), Accessed on 23/3/2005 http://www.firstmonday.org/issues/issue7_4/krebs
17. Stewart, T.: Six Degrees of Mohamed Atta (2001), Accessed on 24/1/2006 http://money.cnn.com/magazines/business2
18. Mena, J.: Investigative Data Mining for Security and Criminal Detection. Butterworth Heinemann (2003)
19. Hougham, V.: Sociological Skills Used in the Capture of Saddam Hussein (2005), Accessed on 22/2/2005 http://www.asanet.org/footnotes/julyaugust05/fn3.html
20. Memon, N., Kristoffersen, K.C., Hicks, D.L., Larsen, H.L.: Detecting Critical Regions in Covert Networks: A Case Study of 9/11 Terrorists Network. In: Proceedings of International Conference on Availability, Reliability, and Security 2007, Vienna, Austria, March 10-13, 2007 (2007)
21. Memon, N., Larsen, H.L.: Detecting Hidden Hierarchy from Terrorist Networks. Mathematical Models for Counterterrorism. Springer, Heidelberg (in Press) (2007)
22. West, B.D.: Introduction to Graph Theory, 2nd edn. Prentice Hall, Englewood Cliffs (2001)
23. Freeman, L.C., Freeman, S.C., Michaelson, A.G.: On Human Social Intelligence. Journal of Social and Biological Structures 11, 415–425 (1988)

Prediction of Enzyme Class by Using *Reactive Motifs* Generated from Binding and Catalytic Sites

Peera Liewlom, Thanawin Rakthanmanon, and Kitsana Waiyamai

Data Analysis and Knowledge Discovery Laboratory (DAKDL), Computer Engineering Department, Engineering Faculty, Kasetsart University, Bangkok, Thailand
{oprll, fengtwr, kitsana.w}@ku.ac.th

Abstract. The purpose of this research is to search for motifs directly at binding and catalytic sites called *reactive motifs*, and then to predict enzyme functions from the discovered reactive motifs. The main challenge is that the data of binding, or catalytic sites is only available in the range 3.34% of all enzymes, and many of each data provides only one sequence record. The other challenge is the complexity of motif combinations to predict enzyme functions.

In this paper, we introduce a unique process which combines statistics with bio-chemistry background to determine *reactive motifs*. It is consisting of *block scan filter*, *mutation control*, and *reactive site-group define* procedures. The purpose of *block scan filter* is to alter each 1-sequence record of binding or catalytic site, using similarity score, to produce quality blocks. These blocks are input to *mutation control*, where in each position of the sequences, amino acids are analyzed an extended to determine complete substitution group. Output of the *mutation control* step is a set of motifs for each 1-sequence record input. These motifs are then grouped using the *reactive site-group define* procedure to produce reactive motifs. Those reactive motifs together with known enzyme sequence dataset are used as the input to C4.5 learning algorithm, to obtain an enzyme prediction model. The accuracy of this model is checked against testing dataset. At 235 enzyme function class, the reactive motifs yield the best prediction result with C4.5 at 72.58%, better than PROSITE motifs.

Keywords: mutation control, sequence motif, reactive motif, enzyme function prediction, binding site, catalytic site, amino acid substitution group.

1 Introduction

An enzyme function or enzyme reaction mechanism is the combination of two main sub-functions: binding, and catalyzing. The parts in an enzyme sequence are called binding sites, and catalytic sites. A site is a short amino acid sequence. To perform one type of binding or catalyzing may be achieved by each of several short amino acid sequences. These sequences can be represented in one pattern (motif).

One of the most well-known collections of motif sequences is PROSITE [1]. PROSITE contains only 152 motifs of binding and catalytic sites, covering 396 out of 3,845 classes of enzyme functions. Therefore the insufficient of data is one of main

R. Alhajj et al. (Eds.): ADMA 2007, LNAI 4632, pp. 442–453, 2007.

challenges. In addition, one of the motifs can be a part of 46 enzyme functions, while 139 enzyme functions can have more than one of the motifs. These create complexity. Therefore, many methods [2,3,4,5] avoid the direct usage of motifs generated from binding and catalytic sites to predict enzyme functions. Those methods use other resources and need data in the form of blocks [6] or multiple sequence alignment [7], which contain very few sequences of binding and catalytic sites.

In this paper, we choose to develop the method to predict enzyme functions based on the direct usage of these binding and catalytic sites. Principal motivation is that information of enzyme reaction mechanism is very important for applied science, especially bioinformatics. We introduce a unique process to determine *reactive motifs* using *block scan filter, mutation control*, and *reactive site-group define*. The main step, *mutation control*, is a method based on motif patterns of PROSITE, which involve amino acid substitution, insertion-deletion, and conserved region, to generate amino-acid substitution group. For example, with the PROSITE motif [RK]-x(2,3)-[DE]-x(2,3)-Y, mutation in position 1 is a substitution [8,9] of amino acids R or K, while maintaining same function. The mutation in position 2 is a insertion-deletion (ins-dels/gap) of amino acids of 2 or 3 residues, and the last position is a conserved region Y necessary in most mutation sequences. In our work, only conserved region and substitution are used in the mutation control operation.

The amino acid substitution has been described in 2 paradigms; expert-based and statistic-based motifs. In the case of expert-based motifs such as PROSITE, the substitution is manually resulted from expert knowledge and bio-chemistry background. The main principle is that the different enzymes with the same reaction mechanism on binding sites and catalytic sites perform the same enzyme function [8]. Due to the need of expertise, motifs discovered by experts are in slow progress. In the case of statistic-based motifs such as EMOTIF [3], motifs are discovered using statistical methods. Therefore the fast predictions of enzyme functions can be achieved. Almost of the statistic-based motifs are not discovered directly from the binding sites or catalytic sites; but from the surrounding sites. Statistic-based motifs yield high enzyme function prediction accuracy. However, in certain applications, it is necessary to understand how motifs of these sites are combined to perform enzyme function. This is the reason why the statistic-based motifs cannot replace the expert-based motifs completely.

In this paper, we propose a method for searching motifs directly at binding and catalytic sites called *reactive motifs*. The proposed method combines statistics with bio-chemistry background in the similar way of expert working process. We develop a procedure called *block scan filter* to alter the 1-sequence record of binding or catalytic site to generate a block, which will be input to the *mutation control* step. As a result, 1-sequence record can produce one motif. The motifs generated from a set of input sequences are then grouped using the *reactive site-group define* procedure to produce reactive motifs. Those reactive motifs together with known enzyme sequence dataset are used as the input to C4.5 learning algorithm, to obtain an enzyme prediction model.

The following will be the details in reactive motifs discovery (phase I), and reactive motif-based prediction of enzyme class (phase II). The overall process is described in Fig. 1, and details are described in section 2 and 3. Experimental results, conclusion and discussion are given respectively in section 4 and section 5.

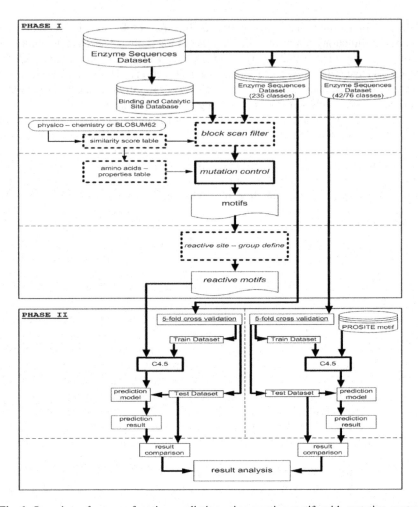

Fig. 1. Overview of enzyme function prediction using reactive motifs with mutation control

2   Reactive Motifs Discovery with Mutation Control (Phase I)

This phase consists of 4 steps; data preparation step, block scan filter step, mutation control step, and reactive site-group define step. The result is reactive motif representing each enzyme mechanism. These details are explained order in the next.

2.1   Data Preparation

We use the protein sequences data from the SWISSPROT part [10] in the UNIPROT database release 9.2, and the enzyme function class from ENZYME NOMENCLATURE [11] in the ENZYME NOMENCLATURE database release 37.0 of Swiss Institute of Bioinformatics (SIB). The enzyme protein sequences are grouped to be used as a database, called *Enzyme Sequence Dataset*. Some of the enzyme proteins in the Enzyme Sequence Dataset provide the information of the amino acid position, which is a part of a binding or catalytic site. Setting the position as the center, a binding or catalytic sequence with the length of 15 amino acids, forming a binding or catalytic site, is retrieved from the enzyme protein sequence (See Fig. 2). These binding and catalytic sites are grouped and used as another database called *Binding and Catalytic Site Database*.

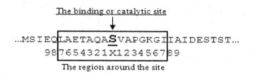

Fig. 2. Data preparation of the sequences around the binding and catalytic site

In this Binding and Catalytic Site Database, the sites are divided to subgroups. In case the sites are binding sites, the sites are in the same subgroup when they have the same reaction descriptions, which are the same substrate, the same binding method (i.e. via amide nitrogen), and the same type of amino acid(s). In case the sites are catalytic sites, the sites are in the same subgroup when they have the same mechanism (i.e. the same proton acceptor), and the same type of amino acid(s). There are in total 291 subgroups in this Binding and Catalytic Site Database. The sites in each subgroup will be used to scan to all related enzyme protein sequences in the *block scan filter* step in order to get quality blocks.

In a function class, if only one type of binding or catalytic site is found, the function class is also neglected. The reason is the enzyme classes having only one motif cannot represent the complexity of the sub-function combination. Only the function classes having enzyme members between 10 and 1000 are used. The rest of the classes are neglected. Therefore, the Enzyme Sequence Dataset covers 19,258 enzymes in 235 function classes. And the Binding and Catalytic Site Database covers 3,084 records of binding or catalytic sites with 291 enzyme reaction descriptions.

2.2 Block Scan Filter

Objective of the block scan filter step is to alter one record of binding or catalytic site data to form a block. This step is divided into two subtasks: the *similarity block scanning* and the *constraint filter*. The first subtask is to use only 1 record of binding or catalytic site to induce the related binding and catalytic sites in order to create a block. One record of binding site or catalytic site is used to scan over the related protein sequences, all protein sequences in enzyme functions that have the same site descriptions. Several similarity scores, such as BLOSUM62 [12], are given, while the record scans over. The part of the protein with the length of 15 amino acids, giving highest score will be stored in a block. The scanning is repeated to other related protein sequences. Therefore, the result is a block containing sets of highest score binding or catalytic sites.

From this block, some of the sites inside the block are filtered out using the second subtask, *constraint filter*. To achieve that, the sites in the block are sorted from highest scores to the lowest. Based on the works of Smith et.al. [13], a block has high quality when each site in the block having at least 3 positions presenting the same type of amino acids. This is the criteria to filter the block. Fig. 3. shows example of block members selection of the constraint filter subtask.

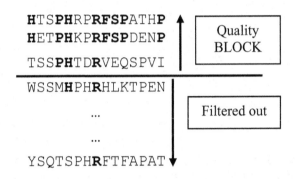

Fig. 3. Selection of members in the blocks using *constraint filter*

2.3   Mutation Control

An enzyme mechanism can be represented by several amino acid sequences of binding or catalytic sites. Therefore specific positions in sequences necessary for controlling the properties of the enzyme mechanism shall have common or similar properties. In some positions, they are of the same types of amino acids, which are called *conserved regions*. In some positions, there are many types of amino acids, however having similar properties. All amino acids in the same position are grouped with respect to the mutation in biological evolution and the resulting group is called *substitution group*. The characteristics of the substitution group can be of two types submitted from patterns of PROSITE motif:

(1)   The substitution group shall have some common properties representing by [], for example [ACSG].

(2)   The amino acids having prohibited properties shall not be included in the group. This prohibition is represented by { }, for example {P}, meaning any amino acids but P.

We call *mutation control* when the substitution group at each position of binding or catalytic sites is controlled by the above characteristics. Thus, mutation control regarding to biological evolution is important for enzyme mechanism to function. The objective of mutation control is to determine complete substitution group at each position in sequences. The mutation control is formalized using concept lattice theory is explained in [16]. In the following, we give an example to illustrate mutation control process.

Table 1. The physico-chemistry table representing the background knowledge of amino acids properties

	Small	Tiny	Proline	Polar	Charge	Positive	Negative	Hydrophobic	Aromatic	Aliphatic
A	X	X						X		
C	X	X		X				X		
D	X			X	X		X			
E				X	X		X			
F								X	X	
G	X	X						X		
H				X	X	X		X	X	
I								X		X
K				X	X	X		X		
L								X		X
M								X		
N	X			X						
P	X		X							
Q				X						
R				X	X	X				
S	X	X		X						
T	X			X						
V	X							X		X
W								X	X	
Y				X				X	X	

For the first pattern [ACSG], their common properties using background knowledge from physico-chemistry (see Table 1) is {small, tiny} which are necessary for enzyme mechanism to function. For the second pattern {P}, the prohibited property is {proline}, for which other amino acids do not have. The prohibited property blocks the enzyme mechanism to function. We call *boundary properties*, the complement of the prohibited property that do not block the enzyme mechanism.

For example (see Fig. 3), at position 1 of BLOCKS, the original substitution group is {H,T}. Their common properties are hydrophobic and polar. It follows that the amino acid group having the common properties is {H, T, W, Y, K, C}. All properties representing {H,T} are the boundary properties, or the properties - polar charge positive hydrophobic aromatic and aliphatic. Any amino acid having the other properties may be the prohibited properties which blocks the enzyme mechanism. Thus, the amino acids having the boundary properties are {H, T, F, K, M, N, Q, R, W, Y}. The complete substitution group controlled by the common properties and the boundary properties can be obtained by intersecting the amino acids generated from the common properties and the boundary properties. From this example, we obtain the complete substitution group as {H, T, W, Y, K}. This process is repeated with all other positions of the quality block. The result is a reactive motif from one binding or catalytic site.

However, the output motifs from using *block scan filter* and *mutation control* should be generated using the same background knowledge. In case of using physico-chemistry table in the *mutation control* step, similarity score table transformed from the physico-chemistry table should be used in the *block scan filter* step. Similarly, when using BLOSUM62 table as similarity score table in *block scan filter step*, we should use amino acids properties table transformed from BLOSUM62 table at the *mutation control* step.

The similarity score table transformed from Physico-Chemistry is given in Table 2. The score is given in relation to the number of the same properties, for example, if two amino acids have 3 same properties, the similarity score is 3. For example, amino acids A and C have properties {small, tiny, hydrophobic} and {small, tiny, polar, hydrophobic}, the similarity score is the shared properties weight by 1 = |{small, tiny, hydrophobic} ∩ {small, tiny, hydrophobic}| = 3. However in case of pairing the same amino acid type, the score is weighed more than one, in our case, it is weighed by 4.

The reason is to give higher score for conserved region. For example, the similarity score of amino acid A and itself is 4 x I{small, tiny, hydrophobic} ∩ {small, tiny, hydrophobic}I = 12.

Table 2. The similarity score table transformed from physico-chemistry table of amino acids properties

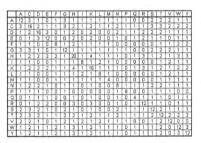

On the other hand, the amino acids properties table transformed from BLOSUM62 table consist of 3 steps. Firstly, BLOSUM62 is transformed to binary table using threshold at zero. The score more than or equal to zero is replaced by 1, and the score less than zero is set to zero. Then, a property is set by the greatest group of amino acids sharing binary 1. Last step, all the new properties are put a table in relation to the amino acids. All steps are described in Fig. 4.

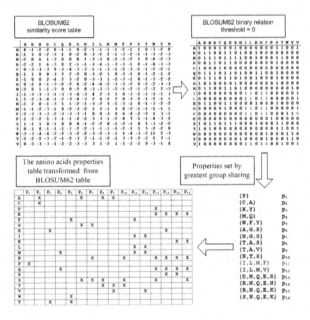

Fig. 4. The amino acids properties table transformed from BLOSUM62 table

These two background knowledge types give different potentials. The background knowledge based on BLOSUM62, in general, is better statistically, while the background knowledge based on physico-chemistry yields motifs closer to PROSITE.

From Fig. 3, we can discover two different reactive motifs from 1 binding or catalytic site with respect to the different background knowledge. Using physico-chemistry table, we obtain [HTWYK] [CDENQST] [CNST] P H [KNQRT] [DNP] R [FILMV] [DENQS] . [ACDGNST] . . . as output motif. Using BLOSUM62, we obtain . . [ST] P H . . R . [ENS] . . . . . as output motif.

2.4   Reactive Site – Group Define

From the previous step, motifs are produced from different records of the same binding or catalytic function, by definition, are redundant, and should be grouped together and represent as one motif, namely *reactive motif.* It means that the 291 subgroups in the Binding and Catalytic Site database will yield 291 reactive motifs.

Although motifs are retrieved from the same original binding or catalytic sites in the same subgroup of the Binding and Catalytic Site Database; they can have different binding structures to the same substrate. In other words, there are many ways to "fit and function". Therefore these motifs, in some cases, can be rearranged to several reactive motifs. The separate method called *conserved region group define* is based on the conserved region, where the motifs with the same amino acids at the same positions of conserved regions are grouped together. As the results, 1,328 reactive motifs are achieved by BLOSUM62 tool, and 1,390 by physico-chemistry table tool. This grouping process is called the *reactive site – group define.*

3   Reactive Motif-Based Prediction of Enzyme Class (Phase II)

In this phase, the problem is to construct an enzyme prediction model using reactive motifs together with known Enzyme Sequence Dataset (train data set). The efficiency of our prediction model is compared with the one of PROSITE, the original pattern subscribed by mutation control to create reactive motif automatically.

Concerning data preparation for phase II, the motifs and enzyme classes in PROSITE are selected for the comparison purpose. In PROSITE, there are 152 motifs of binding and catalytic sites. To be comparable, the same conditions used with reactive motif are applied. However using the condition, which the function classes having members between 10 and 100, covers very small number of motifs (36 motifs in 42 functions, 2,579 sequences) and yields very low accuracy. Therefore we use the function classes having members between 5 and 1000 instead, which covering 65 motifs in 76 Enzyme Classes (2,815 sequences).

To construct the prediction model, given a set of the motifs (reactive motifs or PROSITE motifs), we aim to induce classifiers that associate the motifs to enzyme functions. As suggested in [14], any protein chain can be mapped into a representation based attributes. Such a representation supports efficient function of data-driven algorithms, which represent instances as classified part of fixed set of attributes. In our case, an enzyme sequence is represented as a set of reactive motifs (or PROSITE motif for comparison purpose).

Suppose that from phase I, N reactive motifs have been obtained. Each sequence is encoded as an N-bit binary pattern where the ith bit is 1 if the corresponding reactive motif is present in the sequence; otherwise the corresponding bit is 0. Each N-bit, sequence is associated with an EC number (Enzyme Commission Number). A training set is simply a collection of N-bit binary patterns each of which has associated with it, an EC number. This training set can be used to train a classifier which can then be used to assign novel sequences to one of the several EC-numbers represented in the training set. The reactive motif-based representation procedure is given in Fidg. 5.

R1		R2					
			PROTEIN SEQUENCE, ID:1				

Protein	Reactive motif						Class (EC)
ID	1	2	3	4	....	N	
1	1	1	0	0	....	0	3.2.1.1
2	1	0	1	0	....	0	3.2.1.1
3	0	0	0	1	....	1	4.1.2.13
....	....	....	....	....	....	....	....
M	0	1	0	0	....	1	2.6.1.1

Fig. 5. Reactive Motif-Based Representation of Enzymes

In this paper, we use Weka [15], the machine learning suit, to compare the efficiency of different enzyme function prediction models. C4.5 decision tree (J4.8 Weka's implementation) has been used as a prediction learner in order to assess efficiency of the reactive motifs used for predicting enzyme functions.

4   Experimental Results

In this part, we present the results of the efficiency and the quality of the reactive motifs to predict enzyme functions. The results are divided to 2 sections 1) the prediction accuracy comparison between reactive motifs resulted from different background knowledge (BLOSUM62 or physico-chemistry table), 2) The quality of each reactive motif.

4.1   Prediction Accuracy Comparison Between Reactive Motifs Resulted from Different Background Knowledge

This section compares the accuracy of prediction between different enzyme function prediction models, which resulted from the reactive motifs, which are generated from different background knowledge. The reactive motif generated from BLOSUM62 is called *BLOSUM – reactive motif*. The reactive motif generated from Taylor's physico-chemistry table is called *physicochemistry – reactive motif*. The reactive motifs with out substitution group element are used as reference of reactive motif, that retrieve from conserved region of reactive motif generated from BLOSUM62, called *conserved amino acid – reactive motif*. In addition, the prediction accuracy of enzyme function prediction model from PROSITE motifs is presented. The dataset we used covers 235 enzyme function classes, 19,258 protein sequences. The enzyme function prediction models are created by learning algorithm C4.5 with 5 fold – cross validation. The results are presented in the table 3 and 4.

In case the *conserve region -group define* step is not applied, the prediction model with using BLOSUM – reactive motifs gives the best result: 68.69% accuracy. The prediction model with using physicochemistry – reactive motifs with application of conserve region-group define gives the best result: 72.58% accuracy, however, the accuracies of all models are very close.

Table 3. The maximum scale comparison among the enzyme function prediction systems (19,258 sequences, 235 functions)

Reactive site – group define	Reactive motif					
	Conserved amino acid		BLOSUM		Physicochemistry	
	# motif	C4.5 (%)	# motif	C4.5 (%)	# motif	C4.5 (%)
From Binding and Catalytic Site Database	291	60.84	291	**68.69**	291	64.38
Conserved region - group define	1324	70.57	1328	71.66	1390	**72.58**

Table 4. The maximum scale of the enzyme function prediction system with PROSITE motifs

Selected function class with condition of #Members	# Functions	# Motifs	# Sequences	C4.5 (%)
Between 10 and 1000	42	36	2579	37.15
Between 5 and 1000	76	65	2815	**67.25**

The accuracy of the prediction model retrieved from PROSITE motifs gives the best result of 67.25%.

4.2 Quality of Discovered Reactive Motifs

In case the learning algorithm is not used, the quality efficiency of motifs/reactive motifs to represent the sub-functions of binding or catalytic sites are measured and compared. The quality is represented by 2 values: *coverage value*, and *motifs found per enzyme sequence*. The coverage value is the percentage of the motifs that relevant to enzyme sequences in all related enzyme classes. From the sequences, which motifs/reactive motifs cover, each sequence is checked on how many motifs are matched, and the average value from all sequences is calculated, called motifs found per enzyme sequence.

From table 5, the higher the coverage value is the better. However, the motifs found per sequence, theoretically, should close to 2, because one enzyme at least has one type of binding site and one type of catalytic site. The reactive motifs using physico-chemistry background knowledge gives the result closest to PROSITE, both coverage value and motifs found per sequence.

Table 5. Show the quality values of the reactive motifs and the PROSITE motif

Motif type	# Class	# seq	# seq not match any motif	Coverage value (%)	Motif found per enzyme sequence
PROSITE	42	2579	1752	32.07	1.5562
PROSITE	76	2815	590	79.04	1.6431
Conserved amino acid - Reactive motif	235	19258	59	99.69	27.8416
BLOSUM - Reactive motif	235	19258	665	96.55	6.6293
Physicochemistry - Reactive motif	235	19258	2772	85.61	3.4724

5   Conclusion and Discussion

The process introduced here yields good results (~70% accuracy of enzyme function prediction), and can solve the main problems such as the insufficient of data: binding sites and catalytic sites (~5.8% in our dataset). The reactive motifs using physico-chemistry background knowledge gives the best results, although the coverage value is not satisfied, the reactive-motifs found per enzyme sequence is very good. It means the motifs are very specific. The improvement of accuracy caused from conserved region group define shows that the details in the mechanism descriptions are not complete. The quality of the descriptions of binding and catalytic sites should be improved.

The proposed reactive motif discovery process can be applied using other types of background knowledge. Using other background knowledge such as HMM profile to classify protein domain or family in another interesting future work.

References

1. Bairoch, A.: PROSITE: a dictionary of sites and patterns in proteins. Nucleic Acids Res. 19, 2241–2245 (1991)
2. Sander, C., Schneider, R.: Database of homology-derived protein structures and the structural meaning of sequence alignment. Proteins Struct. Funct. Genet. 9, 56–68 (1991)
3. Huang, J.Y., Brutlag, D.L.: The EMOTIF database. Nucleic Acids Res. 29, 202–204 (2001)
4. Eidhammer, I., Jonassen, I., Taylor, W.R.: Protein structure comparison and structure patterns. Journal of Computational Biology 7(5), 685–716 (2000)
5. Bennett, S.P., Lu, L., Brutlag, D.L.: 3MATRIX and 3MOTIF: a protein structure visualization system for conserved sequence. Nucleic Acids Res. 31, 3328–3332 (2003)
6. Henikoff, S., Henikoff, J.G.: Automated assembly of protein blocks for database searching. Nucleic Acids Res. 19, 6565–6572 (1991)
7. Barton, G.J.: Protein multiple sequence alignment and flexible pattern matching. Methods Enzymol (183), 403–428 (1990)
8. Taylor, W.R.: The classification of amino acid conservation. J. Theor. Biol. 119(2), 205–218 (1986)

9. Wu, T.D., Brutlag, D.L.: Discovering Empirically Conserved Amino Acid Substitution Groups in Databases of Protein Families. In: Proc. Int. Conf. Intell. Syst. Mol. Biol., vol. (4), pp. 230–240 (1996)
10. Bairoch, A., Apweiler, R.: The SWISS-PROT protein sequence database and its supplement TrEMBL in 2000. Nucleic Acids Res. 28, 45–48 (2000)
11. Nomenclature Committee of the International Union of Biochemistry and Molecular Biology (NC-IUBMB): Enzyme Nomenclature. Recommendations 1992. Academic Press (1992)
12. Henikoff, S., Henikoff, J.G.: Amino acid substitution matrices from protein blocks. Proc. Natl. Acad. Sci. USA (89), 10915–10919 (1992)
13. Smith, H.O., Annau, T.M., Chandrasegaran, S.: Finding sequence motifs in groups of functionally related proteins. Proc. Natl. Acad. Sci. 87(2), 826–830 (1990)
14. Diplaris, S., Tsoumakas, G., Mitkas, P.A., Vlahavas, I.: Protein Classification with Multiple Algorithms. In: Bozanis, P., Houstis, E.N. (eds.) PCI 2005. LNCS, vol. 3746, Springer, Heidelberg (2005)
15. Frank, E., Hall, M., Trigg, L., Holmes, G., Witten, I.H.: Data mining in bioinformatics using Weka. Bioinformatics 20(15), 2479–2481 (2004)
16. Liewlom, P., Rakthanmanon, M.P., Waiyamai, K.: Concept Lattice-based Mutation Control for Reactive Motif Discovery. DAKDL technical report, Faculty of Engineering, Kasetsart University, Thailand

Bayesian Network Structure Ensemble Learning

Feng Liu[1], Fengzhan Tian[2], and Qiliang Zhu[1]

[1] Department of Computer Science, Beijing University of Posts and
Telecommunications,
Xitu Cheng Lu 10, 100876 Beijing, China
lliufeng@hotmail.com
[2] Department of Computer Science, Beijing Jiaotong University,
Shangyuan Cun 3, 100044 Beijing, China

Abstract. Bayesian networks (BNs) have been widely used for learning
model structures of a domain in the area of data mining and knowledge
discovery. This paper incorporates ensemble learning into BN structure
learning algorithms and presents a novel ensemble BN structure learning
approach. Based on the Markov condition and the faithfulness condition
of BN structure learning, our ensemble approach proposes a novel sample decomposition technique and a components integration technique.
The experimental results reveal that our ensemble BN structure learning approach can achieve an improved result compared with individual
BN structure learning approach in terms of accuracy.

1 Introduction

Bayesian network is an efficient tool to represent a joint probability distribution and causal independence relationships among a set of variables. Therefore,
there has been great interest in automatically inducing Bayesian networks from
datasets. [6] During the last two decades, two kinds of BN learning approaches
have emerged. The first is the search & score method [2],[9], which uses heuristic
search methods to find the Bayesian network that maximizes some given score
function. Score function is usually defined as a measure of fitness between the
graph and the data. The second approach, which is called the constraint-based
approach, estimates from the data whether certain conditional independences
hold among the variables. Typically, this estimation is performed using statistical or information theoretical measures [1],[11].

Although encouraging results have been reported, both of the approaches done
so far suffer some computational difficulties in accuracy and cannot overcome
the local maxima problem. A statistical or information theoretical measure may
become unreliable on small sample datasets. At the same time, the computation of selected score function may also be unreliable on small sample datasets.
Moreover, the CI-testing space and structure-searching space are so vast that
heuristic methods have to be used. So, the two approaches are usually limited
to find a local maxima.

To further enhance the accuracy and to try to overcome the local maxima
problem in BN leaning, this paper proposes an ensemble BN structure learning

R. Alhajj et al. (Eds.): ADMA 2007, LNAI 4632, pp. 454–465, 2007.

approach that aims to achieve a better result in BN induction. In Section 2, we briefly introduces Bayesian network. In Section 3, we propose the overall process of our learning method. We present the details of our learning approach in sections 4, 5 and 6. In section 7, experimental results are compared and analyzed. Finally, we conclude our work in section 8.

2   Bayesian Network

A Bayesian network is defined as a pair $B = \{G, \Theta\}$, where G is a directed acyclic graph $G = \{V(G), A(G)\}$, with a set of nodes $V(G) = \{V_1, \ldots, V_n\}$ representing a set of random variables and a set of arcs $A(G) \subseteq V(G) \times V(G)$ representing causal independence/dependence relationships that exist among the variables. Θ represents the set of parameters that quantifies the network. It contains a parameter $\theta_{v_i | \pi_i} = P(v_i \mid \pi_i)$ for each possible value v_i of V_i, and π_i of Π_i. Here Π_i denotes the set of parents of V_i in G and π_i is a particular instantiation of Π_i.

For example, Fig.1 shows the Bayesian network called *World* and the parameter table $\Theta_{H | \{E,F\}}$ of the node H. In the *World* network, the nodes A, B and E are root nodes which have not inarcs in the *World*.

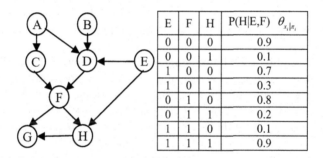

E	F	H	P(H|E,F) $\theta_{x_i|\pi_i}$
0	0	0	0.9
0	0	1	0.1
1	0	0	0.7
1	0	1	0.3
0	1	0	0.8
0	1	1	0.2
1	1	0	0.1
1	1	1	0.9

Fig. 1. An example of Bayesian network (*World*)

3   Ensemble BN Learning Overview

The overall process of our ensemble BN structure learning approach is shown in Fig.2.

Our approach belongs to the category of ensemble methods, "sub-sampling the training dataset".[3] Given the original training dataset D, our algorithm applies sample decomposition technique to generate several training sub-datasets D_i. From each generated training sub-dataset D_i, the component learner learns a component (Bayesian network) BN_i. Then, using components integration technique, these learned components (BNs) are combined into a result Bayesian network.

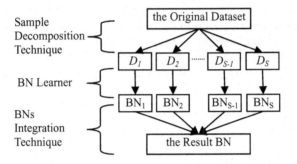

Fig. 2. Process of ensemble BN structure learning

4   Root Nodes Based Sample Decomposition Method

BN structure learning is to estimate conditional independences and dependences among the variables in the training dataset. The two most important sufficient conditions for BN structure learning are the Markov condition and the faithfulness condition. The joint probability distribution P over the training dataset and the true BN G satisfy the Markov condition if and only if under the distribution P, a node is independent of its non-descendant nodes given its parent nodes in the true BN G. The faithfulness condition means that all and only the conditional independence relations true in P are entailed by the Markov condition applied to G. [10]

Bootstrap sampling used by Bagging methods is a powerful tool for model selection. When learning the structure of a graphical model from limited datasets, such as the gene-expression datasets, Bagging methods [3] which use the bootstrap [4] sampling have been applied to explore the model structure [14],[13],[15], [16]. However, Bagging methods have several disadvantages over BN structure learning. On the one hand, the distributions generated from the sub-datasets obtained using Bagging methods and the true BN G may be unsatisfied with the Markov condition and the faithfulness condition. On the other hand, although Bagging methods may asymptotically converge to the true BN by re-sampling a large number of times, they require some convergence conditions. For example, the bagging method using non-parametric bootstrap sampling requires uniform convergence in the distribution of the bootstrap statistic as well as a continuity condition in the parameters.[14], [5]

To solve the above problems, we propose a novel sample decomposition method, which sub-samples the training dataset according to the values of root nodes in the true BN, for ensemble BN structure learning. The sample decomposition method is called Root Nodes based Sample Decomposition(RNSD).

4.1   Root Nodes Based Sample Decomposition

The detail of RNSD method is shown in Fig.3. The idea behind this decomposition method is based on the following 3 facts:

1. Learning BN from the sub-dataset sampled from the original training dataset by limitting the range of values for some root node to be a part of all the possible values for the root node, we expect to get all the Markov independences implied by the true BN.
2. Any arc, which is not connected with some root node in the true BN, can be learned on at least one sub-dataset sampled from the original training dataset by specifying values for the root node. So, learning on sampled sub-datasets that all the possible values of the specified root node are considered, we expect to learn all the "dependences" (arcs) of the true BN.
3. The joint marginal probability distributions of some nodes vary with the different sub-datasets sampled from the original training dataset by limitting the different ranges of the values for some root node.

Therefore, RNSD method can guarantee that all the Markov independences implied by the true BN can be learned from the sampled sub-datasets. Moreover, RNSD method can also hold the diversity in marginal probability distributions during components (BNs) learning.

---

1) Search root nodes H_i ($i=1,...,L$) on the original dataset D (Note: L is the number of found root nodes);

2) Compute marginal probability tables of every found root node $\{P(H_i)|1 \leq i \leq L\}$;

3) Construct probability tables T_s for all pairs and triples of the found root nodes $\{P(H_i,H_k), P(H_i,H_j,H_k) \mid 1 \leq i,j,k \leq L\}$, where $P(H_i,H_k) = P(H_i) * P(H_k)$, $P(H_i,H_j,H_k) = P(H_i) * P(H_j) * P(H_k)$;

4) For each probability table T_s, sort the probability values by ascendant order;

5) For each probability table T_s, obtain the group $Gr(H_i,...,H_k)$ of the sub-datasets $D_{\{H_i \neq h_i \wedge ... \wedge H_k \neq h_k\}}$, ..., by sub-sampling the dataset D given $\{H_i \neq h_i \wedge ... \wedge H_k \neq h_k\}$, where $(h_i,...,h_k) \in T_s$;

6) For each group $Gr(H_i,...,H_k)$ of the sub-datasets, prune the sub-datasets which sampling rate is smaller than σ;

7) Prune the groups of sub-datasets which have few sub-datasets (such as the groups which only have no more than 2 sub-datasets)

Fig. 3. Root Nodes based Sample Decomposition method

Take the *World* network in Fig.1 as an example. Let $\sigma = 0.8$. Assume that the algorithm found the root nodes A, B and E in step 1. The sorted probability tables after step 4 are shown in Fig.4. In step 5, the groups of the sampled sub-datasets were generated. Finally, 4 groups of sub-datasets obtained after pruning in steps 6 and 7 are shown in Fig.5.

4.2 Correctness Proof

Definition 1. *Given a joint probability distribution P, X and Y are **conditional independent** given Z, denoted as $Ind(X,Y |Z)$, if and only if the following statement holds: $P(x \mid y,z) = P(x \mid z)$, $\forall x,y,z$ such that $P(yz) > 0$, where x,y and z denote an instantiation of the subsets of variables X,Y and Z, respectively.[11]*

Definition 2. *Given a joint probability distribution P, X and Y are **conditional dependent** given Z, denoted as $Dep(X,Y |Z)$, if and only if the following statement holds: $P(x \mid y,z) \neq P(x \mid z)$, $\exists x,y,z$ such that $P(yz) > 0$,*

Gr(A, B)	Sampling rate
$D_{\{A\neq0\wedge B\neq1\}}$	0.98
$D_{\{A\neq1\wedge B\neq1\}}$	0.92
$D_{\{A\neq0\wedge B\neq0\}}$	0.82

Gr(A, E)	Sampling rate
$D_{\{A\neq0\wedge E\neq0\}}$	0.96
$D_{\{A\neq0\wedge E\neq1\}}$	0.96
$D_{\{A\neq0\wedge E\neq2\}}$	0.88
$D_{\{A\neq1\wedge E\neq0\}}$	0.84
$D_{\{A\neq1\wedge E\neq1\}}$	0.84

Gr(B, E)	Sampling rate
$D_{\{B\neq1\wedge E\neq0\}}$	0.98
$D_{\{B\neq1\wedge E\neq1\}}$	0.98
$D_{\{B\neq1\wedge E\neq2\}}$	0.94
$D_{\{B\neq0\wedge E\neq0\}}$	0.82
$D_{\{B\neq0\wedge E\neq1\}}$	0.82

Gr(A, B, E)	Sampling rate
$D_{\{A\neq0\wedge B\neq1\wedge E\neq0\}}$	0.996
$D_{\{A\neq0\wedge B\neq1\wedge E\neq1\}}$	0.996
$D_{\{A\neq0\wedge B\neq1\wedge E\neq2\}}$	0.988
$D_{\{A\neq1\wedge B\neq1\wedge E\neq0\}}$	0.984
$D_{\{A\neq1\wedge B\neq1\wedge E\neq1\}}$	0.984
$D_{\{A\neq0\wedge B\neq0\wedge E\neq0\}}$	0.964
$D_{\{A\neq0\wedge B\neq0\wedge E\neq1\}}$	0.964
$D_{\{A\neq1\wedge B\neq1\wedge E\neq2\}}$	0.952
$D_{\{A\neq0\wedge B\neq0\wedge E\neq2\}}$	0.892
$D_{\{A\neq1\wedge B\neq0\wedge E\neq0\}}$	0.856
$D_{\{A\neq1\wedge B\neq0\wedge E\neq1\}}$	0.856

A	P(A)
0	0.2
1	0.8

B	P(B)
1	0.1
0	0.9

E	P(E)
0	0.2
1	0.2
2	0.6

A	B	P(A,B)
0	1	0.02
1	1	0.08
0	0	0.18
1	0	0.72

A	B	E	P(A,B,E)
0	1	0	0.004
0	1	1	0.004
0	1	2	0.012
1	1	0	0.016
1	1	1	0.016
0	0	0	0.036
0	0	1	0.036
1	1	2	0.048
0	0	2	0.108
1	0	0	0.144
1	0	1	0.144
1	0	2	0.432

B	E	P(B,E)
1	0	0.02
1	1	0.02
1	2	0.06
0	0	0.18
0	1	0.18
0	2	0.54

A	E	P(A,E)
0	0	0.04
0	1	0.04
0	2	0.12
1	0	0.16
1	1	0.16
1	2	0.48

Fig. 4. Sorted probability tables

Fig. 5. Groups of sampled sub-datasets ($\sigma = 0.8$)

where x, y and z denote an instantiation of the subsets of variables X, Y and Z, respectively.[11]

Assume that the original training dataset D is data faithful to the true BN G.[11] We take the *World* network in Fig.1 as an example to prove the correctness.

Proposition 1. *Learning a BN from the sub-dataset sampled from the original training dataset by limitting the range of the values for some root node to be a part of all the possible values for the root node, we expect to obtain all the Markov independences implied by the true BN from the learned BN.*

Proof. Assume that we obtain the sub-dataset $D_{E\neq0}$ from the original training dataset D by limitting the range of the values for the root node E to be $E \neq 0 \Leftrightarrow \{(E = 1) \cup (E = 2)\}$.

Assume that P denotes the distribution faithful to the true BN and P' denotes the distribution over the sub-dataset $D_{E\neq0}$.

We take 2 cases to consider whether there exists the Markov independences implied by the true BN in the distribution P' over the sub-dataset $D_{E\neq0}$.

1. For the nodes of which the parents set contains the root node E, for example the node D, there exists the Markov independence $Ind(D, C \mid A, B, E)$ in the true BN.

   According to the definition of conditional independence, there exists:

$$P(d \mid c, a, b, E = 1) = P(d \mid a, b, E = 1)$$
$$P(d \mid c, a, b, E = 2) = P(d \mid a, b, E = 2)$$

We can infer $P'(d \mid c, a, b, E = 2) = P'(d \mid a, b, E = 2)$.

According to the definition of conditional independence, we can infer that there exists $Ind(D, C \mid A, B, E)$ in the distribution P' over the sub-dataset $D_{E \neq 0}$.

2. For the nodes of which the parents set does not contain the root node E, for example the node G, there exists the Markov independences $Ind(G, C \mid F, H)$ and $Ind(G, E \mid F, H)$ in the true BN.

According to the definition of conditional independence, there exists:

$$P(g \mid c, f, h, E = 1) = N_{gcfh1}/N_{cfh1} = P(g \mid f, h)$$
$$P(g \mid c, f, h, E = 2) = N_{gcfh2}/N_{cfh2} = P(g \mid f, h)$$
$$P(g \mid f, h, E = 1) = N_{gfh1}/N_{fh1} = P(g \mid f, h)$$
$$P(g \mid f, h, E = 2) = N_{gfh2}/N_{fh2} = P(g \mid f, h)$$

We can infer that $P'(g \mid c, f, h) = P'(g \mid f, h)$.

According to the definition of conditional independence, we can infer that there exists $Ind(G, C \mid F, H)$ in the distribution P' over the sub-dataset $D_{E \neq 0}$.

According to case 1, 2, we infer that Proposition 1 is correct.

Lemma 1. *Learning a BN from the sub-dataset sampled using RNSD method, we can obtain all the Markov independences implied by the true BN from the learned BN.*

Proof. Using the same way as Proposition 1, we can infer it.

Proposition 2. *Learning BNs from the sub-datasets which are sampled by by limiting the range of values for some root node to be a part of all the possible values for the root node, if all the possible values of the root node can be included in the sub-datasets, then we can obtain all the edges of the true BN from the learned BNs.*

Proof. There is an edge between the node C and the node F in Fig.1. We can get $Dep(C, F \mid A)$ and $Dep(C, F \mid A, E)$. According to the definition of conditional dependence, we can infer the following formula:

$$Dep(C, F \mid A, E) \Leftrightarrow \exists e \in E, Dep(C, F \mid A, E = e) \tag{1}$$

According to $Dep(C, F \mid A)_{D_{E \neq 0}} \Leftrightarrow Dep(C, F \mid A, E \neq 0)_D$ and the formula (1), we can infer the proposition.

According the above inferences, we can conclude that the BNs learned from the sub-datasets, which are sampled using our RNSD method, include all the Markov independences and edges implied by the true BN.

4.3   Root Nodes Search Method

The search method sees Fig.6.

---

1) Conduct order-0 CI tests for each pair nodes, then build the undirected graph UG_0;

2) Conduct order-1 CI tests for any three nodes X, Y and Z that in the undirected graph UG_0, X and Y, and Y and Z, are directly connected; and X and Z are not directly connected, if $Dep(X, Z \mid Y)$, then direct the edges $X \rightarrow Y$ and $Z \rightarrow Y$;

3) Find the maximal cliques $\{G_{a1}, ..., G_{ak}\}$ consisting of the nodes which have no inarcs, that every clique is undirected complete graph and has at least one outarc;

4) Order the maximal cliques by the number of nodes in ascendant order and prune the cliques G_{ai} that $\|G_{ai}\| > \delta$;

5) For every maximal clique, use the exhaustive search & Bayesian score function method to learn the root node in the maximal clique;

6) Detect and delete pseudo root nodes.

Fig. 6. Root nodes search method

The idea behind the search root nodes method is based on the following assumption, which is correct in most situations both for synthetic datasets and for real-life datasets:

Assumption 1. *If there exists a directed path $X \longmapsto Y$ between node X and node Y in a Bayesian network, then $Dep(X, Y \mid NULL)$.*

Under the above assumption, we can obviously infer that every clique obtained after the 4th step of the method has one and only one root node.

In most cases, Assumption 1 is satisfied. However, some exceptions may occur when there are many nodes (normally, the number of nodes including the two nodes on the path $\gg 5$) on the directed path between two nodes, that is, the two nodes may be independent conditional on NULL. Moreover, even if Assumption 1 is satisfied in any situation, some results after step 5 in Fig.6 may be pseudo root nodes on limited datasets.

We take one step to solve the pseudo root nodes problem. The step is to detect pseudo root nodes and delete them (see step 6 in Fig.6). The detection for pseudo root nodes is based on 2 kinds of independences. One kind of independences is the Markov independences given the obtained root node of other nodes in the maximal clique. The other kind of independences is the independences among root nodes. Firstly, if the first kind of independences given some found root node is not satisfied, then the found root node is pseudo root node and prune the pseudo root node. Secondly, if the second kind of independences is not satisfied by the two found root nodes, then at least one found root node is pseudo root node, and we prune the two root nodes.

For example, during the *World* network learning, the running result for every step of our root nodes searching method sees Fig.7.

Our root nodes search method does not have to find all the root nodes in the true BN, it is enough to find several root nodes in the true BN for our ensemble BN learning in terms of accuracy. We can also use other methods to search root nodes, such as the RAI algorithm [12].

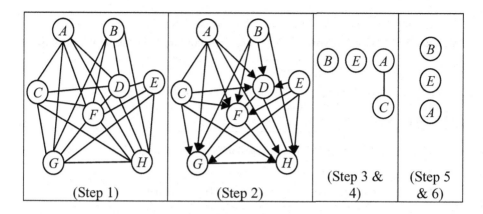

Fig. 7. Search root nodes in the *World* network

5   Bayesian Network Learner

Bayesian network learner is an individual BN learning algorithm and a building block of ensemble BN structure learning algorithm. Normally, it needs to be computationally efficient. Therefore, we selected OR algorithm [9], TPDA algorithm [1], and other algorithms using fast heuristic search (such as Greedy search) methods [8] as our BN learners. In our implementations for these algorithms, we applied partial nodes order information which was acquired by our root nodes search method.

6   Bayesian Networks Integration Method

Our integration method includes 2 parts: the integration of the Bayesian networks in the same group; the integration of the intergroup undirected networks.

6.1   Intragroup Bayesian Networks Integration

The method for intragroup Bayesian network integration is shown in Fig.8.

For any edge $\widetilde{e}$, ($\widetilde{e}$ is undirected edge of the arc e) in the BNs, consider the quantity:

$$P(\widetilde{e}) = \frac{1}{L} \sum_{i=1}^{L} 1\{\widetilde{e} \in BN_i\}$$

If $P(\widetilde{e}) > P(\widetilde{e}')$, then it is more probable that $\widetilde{e}$ exists in the true BN than $\widetilde{e}'$ does. Furthermore, if $P(\widetilde{e}) > 1/2$, we classify edge $\widetilde{e}$ as "true". Therefore, we can obtain the most probable BNs represented in the form of undirected network UG_i which every edge $\widetilde{e}$ has a probability table $g(e)$ of $\{\rightarrow, \leftarrow, \leftrightarrow\}$.

In the group Gr_i, assume we have learned L Bayesian networks $BN_1,\ldots,BN_L$:

1) For the Bayesian Networks $BN_1,\ldots,BN_L$, according to the principle of simple voting, compute the probability of every edge $\tilde{e}$ without considering orientation in the Bayesian Networks

2) Prune the edges whose probabilities are less than 1/2, and create a undirected network UG_i, which every edge $\tilde{e}$ has a probability table $g(e)$ about $\{\rightarrow,\leftarrow,\leftrightarrow\}$

Fig. 8. Intra-group Bayesian Networks Integration

6.2   Intergroup Undirected Networks Integration

Assume there are m groups of sub-datasets sampled using RNSD method. After intragroup Bayesian networks integration for every group $Gr_i(i = 1,\ldots,m)$, we obtained m undirected networks UG_i. For any possible arc e, consider the quantity:

$$P'(e) = \frac{1}{m} \sum_{i=1}^{m} p(e) 1\{\tilde{e} \in UG_i, i = 1,\ldots,m\}$$

Note: Weight $p(e)$ is the probability value of $\{\rightarrow,\leftarrow\}$, where $p(\rightarrow) = g(\rightarrow)+g(\leftrightarrow)$ and $p(\leftarrow) = g(\leftarrow) + g(\leftrightarrow)$.

Finally, we take search method and score function to generate the result Bayesian network of our ensemble BN structure learning method. The process sees Fig.9.

Assume we have learned m undirected networks UG_i $(i=1,\ldots,m)$:

1) For the undirected networks UG_i $(i=1,\ldots,m)$, according to the principle of weighted voting, compute the probability value of every possible arc which exists in the undirected networks

2) Order arcs by the probability values in ascendant order, and apply exhaustive or heuristic search method and Bayesian score function to learn the maximal score Bayesian network by 'adding arc', 'deleting arc' and 'reversing arc' operators

Fig. 9. Inter-group Bayesian Networks Integration

7   Experimental Results

We implemented OR algorithm, OR-BWV algorithm, OR-HNSD algorithm, TPDA algorithm, TPDA-BWV algorithm, and TPDA-HNSD algorithm. OR-BWV and TPDA-BWV algorithms use Bagging sampling method and weighted voting integration method. Tests were run on a PC with Pentium4 1.5GHz and 1GB RAM. The operating system was Windows 2000. 4 Bayesian networks were used. From these networks, we performed experiments with 500, 1000, 5000

Table 1. Bayesian Networks

BN	Nodes Num	Arcs Num	Roots Num	Max In/Out-Degree	Domain Range
Alarm	37	46	12	4/5	2-4
Barley	48	84	10	4/5	2-67
Insur	27	52	2	3/7	2-5

Table 2. Average BDe(Alarm-OR)

SIZE	OR	OR-BSW	OR-RNSD
500	-15.1000	-14.9235	-13.8901
1000	-14.8191	-14.1917	-13.7610
5000	-13.9041	-13.8941	-13.1002

Table 3. Average BDe(Alarm-TPDA)

SIZE	TPDA	TPDA-BSW	TPDA-RNSD
500	-18.0973	-17.8045	-17.3650
1000	-15.4286	-15.0012	-14.2455
5000	-14.4960	-14.4731	-13.2682

Table 4. Average BDe(Barley-OR)

SIZE	OR	OR-BSW	OR-RNSD
500	-82.3448	-81.4104	-80.1943
1000	-78.9655	-78.2544	-76.2538
5000	-76.7081	-75.2273	-73.3371

Table 5. Average BDe(Barley-TPDA)

SIZE	TPDA	TPDA-BSW	TPDA-RNSD
500	-103.7931	-117.5443	-110.5973
1000	-111.0690	-112.2151	-103.5482
5000	-116.8919	-106.7328	-99.8341

Table 6. Average BDe(Insur-OR)

SIZE	OR	OR-BSW	OR-RNSD
500	-24.0167	-23.3812	-24.0010
1000	-22.6077	-21.5445	-22.5077
5000	-19.4286	-18.9523	-19.2286

Table 7. Average BDe(Insur-TPDA)

SIZE	TPDA	TPDA-BSW	TPDA-RNSD
500	-28.7857	-28.3471	-28.5172
1000	-25.1111	-24.8157	-25.0111
5000	-20.8571	-20.7916	-20.5571

training cases each. For each network and sample size, we sampled 10 original datasets and record the average results by each algorithm. Moreover, we applied Bagging sampling with 200 times in OR-BWV and TPDA-BWV algorithms. Let $\sigma = 0.8$ in Fig.3 and $\delta = 5$ in Fig.6.

We compared the accuracy of Bayesian networks learned by these algorithms according to the average BDeu score. The BDeu score corresponds to the posteriori probability of the structure learned.[7] The BDeu scores in our experiments were calculated on a seperate testing dataset sampled from the true BN containing 50000 samples. Tables 2-7 report the results.

There are several noticeable trends in these results. Firstly, as expected, as the number of instances grow, the quality of learned Bayesian network improves, except to TPDA for Barley network (500, 1000, 5000). It is due to that constraint-based method is unstable for limited datasets (500, 1000, 5000) relative to Barley network. At the same time, we can see that TPDA-BWV algorithm and TPDA-RNSD algorithm improve the stability of TPDA, that is, the quality of learned Bayesian networks by TPDA-BWV and TPDA-RNSD improves with the increase of sample size. Secondly, our RNSD based ensemble algorithms OR-RNSD and TPDA-RNSD are almost better than or at least equal to the individual Bayesian network learning algorithms in terms of accuracy on limited datasets.

Thirdly, in most cases, our ensemble algorithms have better performance than BWV ensemble algorithms. Finally, for Bayesian networks with few root node (such as Insur network), our ensemble algorithms have little improvement on learning accuracy. So, they are ineffective for these Bayesian networks.

8   Conclusion

We proposed a novel sampling technique and a components integration technique to incorporate ensemble learning into BN structure learning. Our results are encouraging in that they indicate that the our method achieved a more accurate result BN than individual BN learning algorithms.

References

1. Cheng, J., Greiner, R., Kelly, J., Bell, D., Liu, W.: Learning Belief Networks form Data: An Information Theory Based Approach. Artificial Intelligence 137(1-2), 43–90 (2002)
2. Cooper, G., Herskovits, E.: A Bayesian Method for Constructing Bayesian Belief Networks from Databases. In: Ambrosio, B., Smets, P. (eds.) UAI '91. Proceedings of the Seventh Annual Conference on Uncertainty in Artificial Intelligence, pp. 86–94. Morgan Kaufmann, San Francisco (1991)
3. Dietterich, T.G.: Machine Learning Research: Four Current Directions. AI Magazine 18(4), 745–770 (1997)
4. Davison, A.C., Hinkley, D.V.: Bootstrap Methods and Their Application, 1st edn. Cambridge Press, New York (1997)
5. Efron, B., Tibshirani, R.J.: An Introduction to the Bootstrap, 1st edn. Chapman Hall, New York (1993)
6. Heckerman, D.: Bayesian Networks for Data Mining, 1st edn. Microsoft Press, Redmond (1997)
7. Heckerman, D., Geiger, D., Chickering, D.M.: Learning Bayesian Networks: the Combination of Knowledge and Statistical Data. Machine Learning 20(3), 197–243 (1995)
8. Chickering, D.M., Heckerman, D., Geiger, D.: Learning Bayesian Network: Search Methods and Experimental Results. In: Fisher, D., Lenz, H. (eds.): Learning from Data: Artificial Intelligence and Statistics 5, Lecture Notes in Statistics 112, 143–153. Springer, New York (1996)
9. Moore, A.W., Wong, W.K.: Optimal Reinsertion: A New Search Operator for Accelerated and More Accurate Bayesian Network Structure Learning. In: Fawcett, T., Mishra, N. (eds.) ICML 2003 – Machine Learning. Proceedings of the Twentieth International Conference, pp. 552–559. AAAI Press, Washington DC (2003)
10. Pearl, J.: Causality: Models, Reasoning, and Inference, 1st edn. Cambridge Press, London (2000)
11. Spirtes, P., Glymour, C., Scheines, R.: Causation, Prediction and Search, 2nd edn. MIT Press, Massachusetts (2000)
12. Yehezkel, R., Lerner, B.: Recursive Autonomy Identification for Bayesian Network Structure Learning. In: Cowell, R.G., Ghahramani, Z. (eds.) AISTATS05. Proceedings of the Tenth International Workshop on Artificial Intelligence and Statistics, pp. 429–436. Society for Artificial Intelligence and Statistics, London (2005)

13. Friedman, N., Goldszmidt, M., Wyner, A.: On the Application of the Bootstrap for Computing Confidence Measures on Features of Induced Bayesian Networks. In: Heckerma, D., Whittaker, J. (eds.) Learning from Data: Artificial Intelligence and Statistics VII. Proceedings of the Seventh International Workshop on Artificial Intelligence and Statistics, pp. 197–202. Morgan Kaufmann, San Francisco (1999)
14. Friedman, N., Goldszmidt, M., Wyner, A.: Data Analysis with Bayesian Networks: A Bootstrap Approach. In: Laskey, K.B. (ed.) UAI '99. Proceedings of the Fifteenth Conference on Uncertainty in Artificial Intelligence, pp. 196–205. Morgan Kaufmann, San Francisco (1999)
15. Friedman, F., Linial, M., Nachman, I., Pe'er, D.: Using Bayesian Networks to Analyze Expression Data. Journal of Computational Biology 7, 601–620 (2000)
16. Pe'er, D., Regev, A., Elidan, G., Friedman, N.: Inferring Subnetworks from Perturbed Expression Profiles. Bioinformatics 17(Suppl. 1), 1–9 (2001)

Fusion of Palmprint and Iris for Personal Authentication

Xiangqian Wu[1], David Zhang[2], Kuanquan Wang[1], and Ning Qi[1]

[1] School of Computer Science and Technology,
Harbin Institute of Technology (HIT), Harbin 150001, China
{xqwu, wangkq}@hit.edu.cn
http://biometrics.hit.edu.cn
[2] Biometric Research Centre, Department of Computing,
Hong Kong Polytechnic University, Kowloon, Hong Kong
csdzhang@comp.polyu.edu.hk

Abstract. Traditional personal authentication methods have many instinctive defects. Biometrics is an effective technology to overcome these defects. The unimodal biometric systems, which use a single trait for authentication, can result in some problems like noisy sensor data, non-universality and/or lack of distinctiveness of the biometric trait, unacceptable error rates, and spoof attacks. These problems can be addressed by using multi-biometric features in the system. This paper investigates the fusion of palmprint and iris for personal authentication. The features of the palmprint and the iris are first extracted and matched respectively. Then these matching distances are normalized. Finally, the normalized distances are fused to authenticate the identity. The experimental results show that combining palmprint and iris can dramatically improve the accuracy of the system.

1 Introduction

Security is becoming increasingly important in the information based society. Personal authentication is one of the most important ways to enhance the security. However, the traditional personal authentication methods, including token-based ones (such as keys and cards, etc.) and knowledge-based ones (such as PINs and passwords, etc.), suffer from some instinctive defects: the token can be stolen or lost and the knowledge can be cracked or forgotten. Biometrics, the technology automatically using the physiological (such as fingerprint, palmprint, iris, etc.) or behavioral characteristic (such as signature, gait, etc.) for personal recognition, is one of the effective techniques to overcome these problems [1,2,3]. Up to now, most of the biometric systems, which is called unimodal biometric systems, use single trait to recognize the identity. The unimodal biometric systems can result in some problems like noisy sensor data, non-universality and/or lack of distinctiveness of the biometric trait, unacceptable error rates, and spoof attacks [4]. To address these problems, the multimodal biometric systems, which combine multiple biometric features, are developped [3,5,6].

R. Alhajj et al. (Eds.): ADMA 2007, LNAI 4632, pp. 466–475, 2007.

The iris recognition is one of the most accuracy methods among the current available biometric techniques and many effective algorithms have been developed. Daugman [7, 8] encoded the iris into a 256-bytes IrisCode by using two-dimensional Gabor filters, and took the Hamming distance to match the codes. Wildes [9] matched the iris using Laplacian pyramid multi-resolution algorithms and a Fisher classifier. Boles et al. [10] extract iris features using a one-dimensional wavelet transform, but this method has been tested only on a small database. Ma et al. [11] construct a bank of spatial filters whose kernels are suitable for use in iris recognition. They have also developed a preliminary Gaussian-Hermite moments-based method which uses local intensity variations of the iris [12]. Later, they proposed an improved method based on characterizing key local variations [13]. The accuracy of these algorithms are heavily affected by the quality of iris images, so it is very difficult to obtain a high accuracy if the iris images are not good enough.

The palmprint is a relatively new biometric feature and has several advantages compared with other currently available features [3,14]: palmprint capture devices are much cheaper than iris devices; palmprints contain additional distinctive features such as principal lines and wrinkles, which can be extracted from low-resolution images; a highly accurate biometrics system can be built by combining all features of palms, such as palm geometry, ridge and valley features, and principal lines and wrinkles, etc. Many approaches have been developed for palmprint recognition in the last several years. Han [15] used Sobel and morphological operations to extract line-like features from palmprints. Similarly, for verification, Kumar [16] used other directional masks to extract line-like features. Wu [17] used Fisher's linear discriminant to extract the algebraic feature (called Fisherpalms). Zhang [18,19] used 2-D Gabor filters to extract the texture features (called PalmCode) from low-resolution palmprint images and employed these features to implement a highly accurate online palmprint recognition system. Wu [20] extract the palm lines and authenticate persons according to the line structure. However, the accuracies of these approaches are not high enough to meet the requirement of some very high security applications.

Combining iris and palmprint can resolve the problems of both iris and palmprint algorithms. This paper investigates the fusion of the palmprint and the iris.

The rest of this paper is organized as follows. Section 2 gives the feature extraction and matching. Section 3 presents several fusion strategies. Section 4 contains some experimental results and analysis. And Section 5 provides some conclusions and future work.

2 Feature Extraction and Matching

2.1 Palmprint Feature Extraction and Matching

When palmprints are captured, the position and direction of a palm may vary so that even palmprints from the same palm may have a little rotation and translation. Furthermore, palms differ in size. Hence palmprint images should be orientated and normalized before feature extraction and matching. In this paper, we use the preprocessing technique described in [18] to align and normalize the palmprints.

Let I denote a palmprint image and G_σ denote a 2D Gaussian filter with the variance σ. The palmprint is first filtered by G_σ as below:

$$I_f = I * G_\theta \tag{1}$$

where $*$ is the convolution operator.

Then the difference of I_f in the horizontal direction is computed as following:

$$D = I_f * b \tag{2}$$

$$b = [-1, 1] \tag{3}$$

where $*$ is the convolution operator.

Finally, the palmprint is encoded according to the sign of each pixel of D:

$$C(i, j) = \begin{cases} 1, & \text{if } D(i,j) > 0; \\ 0, & \text{otherwise.} \end{cases} \tag{4}$$

C is called DiffCode of the palmprint I. The size of the preprocessed palmprint is 128×128. Extra experiments shows that the image with 32×32 is enough for the DiffCode extraction and matching. Therefore, before compute the DiffCode, we resize the image from 128×128 to 32×32. Hence the size of the DiffCode is 32×32. Fig. 1 shows some examples of DiffCode. From this figure, the DiffCode preserves the structure information of the lines on a palm.

Because all DiffCodes have the same length, we can use Hamming distance to define their similarity. Let C_1, C_2 be two DiffCodes, their Hamming distance $(H(C_1, C_2))$ is defined as the number of the places where the corresponding values of C_1 and C_2 are different. That is,

$$H(C_1, C_2) = \sum_{i=1}^{32} \sum_{j=1}^{32} C_1(i, j) \otimes C_2(i, j) \tag{5}$$

where $\otimes$ is the logical **XOR** operation.

The matching distance of two DiffCodes C_1 and C_2 is defined as the normalized Hamming distance:

$$D(C_1, C_2) = \frac{H(C_1, C_2)}{32 \times 32} \tag{6}$$

Actually, $D(C_1, C_2)$ is the percentage of the places where C_1 and C_2 have different values. Obviously, $D(C_1, C_2)$ is between 0 and 1 and the smaller the matching distance, the greater the similarity between C_1 and C_2. The matching distance of a perfect match is 0. Because of imperfect preprocessing, there may still be a little translation between the palmprints captured from the same

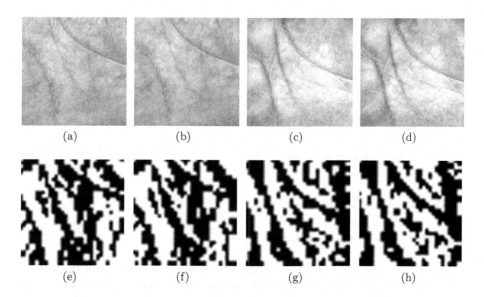

Fig. 1. Some examples of DiffCodes. (a) and (b) are two palmprint samples from a palm; (c) and (d) are two palmprint samples from another palm; (e)-(h) are the DiffCodes of (a)-(d), respectively.

palm at different times. To overcome this problem, we vertically and horizontally translate C_1 a few points to get the translated C_1^T , and then, at each translated position, compute the matching distance between C_1^T and C_2. Finally, the final matching distance is taken to be the minimum matching distance of all the translated positions.

2.2   Iris Feature Extraction and Comparison

As the iris images used in this paper contain some reflection (see Fig. 2), we first use a threshold to remove these reflection points from the image. Then the two circular boundaries of iris are searched by using the integrodifferential operators. Finally, the disk-like iris area is unwrapped to a rectangular region by using doubly dimensionless projection.

The IrisCode [7, 8, 1] is produced by 2-D complex Gabor filters. Let I denote a preprocessed iris image. The IrisCode C contains two parts, i.e. the real part C_r and the imaginary part C_i, which are computed as following:

$$I_G = I * G. \tag{7}$$

$$C_r(i,j) = \begin{cases} 1, \text{ if } \mathbf{Re}[I_G(i,j)] \geq 0; \\ 0, \text{ otherwise.} \end{cases} \tag{8}$$

$$C_i(i,j) = \begin{cases} 1, \text{ if } \mathbf{Im}[I_G(i,j)] \geq 0; \\ 0, \text{ otherwise.} \end{cases} \tag{9}$$

where G is a complex 2-D Gabor filter and $*$ is the convolution operation.

IrisCode comparison is based on Hamming distance. The Boolean operator (XOR) is used to compare each pair of IrisCodes C_1 and C_2 bit by bit. The Boolean operator equals 1 if and only if the two bits are different; otherwise, it equals to 0. The normalized Hamming distance equation is given below,

$$HD = \frac{1}{2048} \sum_{j=1}^{2048} [C_1(j) \otimes C_2(j)] \tag{10}$$

where $\otimes$ is the Boolean operator **XOR**.

3   Distance Fusion

3.1   Distance Normalization

Though the matching distances obtained from the palmprints and the iris are in the interval $[0, 1]$, they are not covered the whole interval. Their maximum and the minimum values are not same. In order to fuse these two distance, they should be normalized to cover this interval. Let $d_1, d_2, \ldots, d_n$ denote the matching distances obtained from palmprint or iris. These distance can be normalized as following:

$$d_i = \frac{d_i - min}{max - min} \tag{11}$$

where min and max are the minimum and the maximum values of these distances. That is, the minimum and maximum values of these distances are normalized to 0 and 1.

3.2   Fusion Strategies

Denote x_1 and x_2 as the distances obtained from palmprint matching and iris matching between two persons, respectively. To obtain the final matching distance x, we fuse these two distance by following simple strategies.

S_1: *Maximum Strategy:*

$$x = \max(x_1, x_2) \tag{12}$$

S_2: *Minimum Strategy:*

$$x = \min(x_1, x_2) \tag{13}$$

S_3: *Product Strategy:*

$$x = \sqrt{x_1 x_2} \tag{14}$$

S_4: *Sum Strategy:*

$$x = \frac{x_1 + x_2}{2} \tag{15}$$

4   Experimental Results And Analysis

4.1   Database

We have collected $21,240$ iris images and $21,240$ palmprint images from of 2124 people by CCD-based devices. Each eye/palm provides 5 images. The size of the iris images is 768×568 and the size of palmprint is 384×284. Using the iris or palmprint, most people can be effectively recognized. However, there are still some people cannot be recognized. We choose 120 persons whose right iris images are not good enough for the iris recognition algorithms to get a high accuracy. The right iris images and right palmprints of these 120 persons are used to build a database to investigate the proposed approach. From this database, 5 pairs of palmprint/iris can be obtained for each person. Therefore, there are total $120 \times 5 = 600$ pairs of palmprint and iris. Fig. 2 shows some samples in the database.

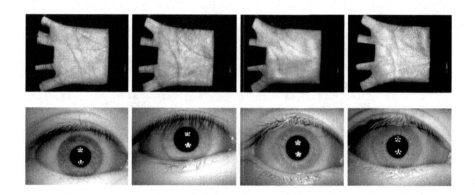

Fig. 2. Some Samples in the Database

4.2   Matching Tests

To test the performance of the different fusion strategies, each palmprint/iris pair in the database is matched against the other pairs using different fusion strategies. The matching distance distributions of each strategy are showed in Fig. 3. This figure shows that the distribution curves of each strategy have two peaks, which respectively correspond the genuine and impostor matching distances. The two peaks of all strategies are widely separated. However, the genuine matching curve and the impostor matching curve of different strategies overlapped with different degree. The less the overlapping area is, the better the matching performance is. This overlapping area is defined as minimum total error (MTR), and the MTR of each strategy is listed in Table 1. From this table, the Maximum and Minimum strategies improve the matching performance little, while the Sum and Product strategies can improve the performance dramatically, their MTRs decreased to 0.012%.

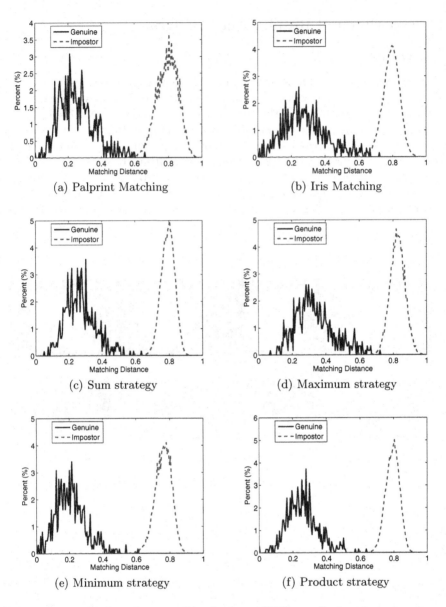

Fig. 3. Matching Distance Distributions of Different Fusion strategies

Table 1. MTR of Different Fusion Strategies

Strategy	Palmprint Matching	Iris Matching	Sum	Maximum	Minimum	Product
MTR (%)	0.376	1.161	0.012	0.191	0.165	0.012

4.3   Accuracy Tests

To evaluate the accuracies of the different fusion strategies, each pair of palm-print/iris in the database is again matched with the other pairs. For each strategy, the false accept rate (FAR) and false reject rate (FRR) at different distance thresholds is plotted in Fig. 4 and the equal error rate (EER) of them are listed in Table 2. This figure and table also demonstrate that the accuracies of the Maximum and the Minimum strategies are close to that of the palmprint matching, while the Sum and the Product strategies can improve the accuracy dramatically, their EERs can be reduce to 0.006%.

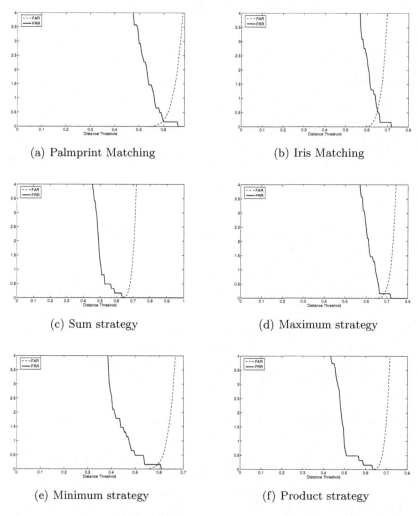

Fig. 4. FAR and FRR Curves of Different Strategies

Table 2. EER of Different Strategies

Strategy	Palmprint Matching	Iris Matching	Sum	Maximum	Minimum	Product
EER (%)	0.188	0.589	0.006	0.148	0.175	0.006

5   Conclusions and Future Work

This paper combines palmprint and iris for personal authentication. Several simple fusion strategies are investigated. And the results show that the Sum and the Product strategies can improve the accuracy dramatically the MTRs and EERs of which are respectively reduced to 0.012% and 0.006%.

In future, we will investigate some complex fusion strategies for the combination of palmprint and iris.

Acknowledgements

This work is supported by the National Natural Science Foundation of China (No. 60441005), the Key-Project of the 11th-Five-Year Plan of Educational Science of Hei Longjiang Province, China (No. HZG160) and the Development Program for Outstanding Young Teachers in Harbin Institute of Technology.

References

1. Zhang, D.: Automated Biometrics–Technologies and Systems. Kluwer Academic Publishers, Dordrecht (2000)
2. Jain, A., Bolle, R., Pankanti, S.: Biometrics: Personal Identification in Networked Society. Kluwer Academic Publishers, Dordrecht (1999)
3. Jain, A., Ross, A., Prabhakar, S.: An introduction to biometric recognition. IEEE Transactions on Circuits and Systems for Video Technology 14, 4–20 (2004)
4. Jain, A.K., Ross, A.: Multibiometric systems. Communications of the ACM, Special Issue on Multimodal Interfaces 47, 34–40 (2004)
5. Wang, Y., Tan, T., Jain, A.K.: Combining face and iris biometrics for identity verification. In: Kittler, J., Nixon, M.S. (eds.) AVBPA 2003. LNCS, vol. 2688, pp. 805–813. Springer, Heidelberg (2003)
6. Ross, A., Jain, A.K.: Information fusion in biometrics. Pattern Recognition Letters 24, 2115–2125 (2003)
7. Daugman, J.: High confidence visual recognition of persons by a test of statistical independence. IEEE Transactions on Pattern Analysis and Machine Intelligence 15, 1148–1161 (1993)
8. Daugman, J.: How iris recognition works. IEEE Transactions on Circuits and Systems for Video Technology 14, 21–30 (2004)
9. Wildes, R.: Iris recognition: an emerging biometric technology. Proceedings of the IEEE 85, 1348–1363 (1997)
10. Boles, W., Boashash, B.: A human identification technique using images of the iris and wavelet transform. IEEE Transactions on Signal Processing 46, 1185–1188 (1998)

11. Ma, L., Tan, T., Wang, Y., Zhang, D.: Personal identification based on iris texture analysis. IEEE Transactions on Pattern Analysis and Machine Intelligence 25, 1519–1533 (2003)
12. Ma, L., Tan, T., Wang, Y., Zhang, D.: Local intensity variation analysis for iris recognition. Pattern Recognition 37, 1287–1298 (2004)
13. Ma, L., Tan, T., Wang, Y., Zhang, D.: Efficient iris recognition by characterizing key local variations. IEEE Transactions on Image Processing 13, 739–749 (2004)
14. Wu, X., Zhang, D., Wang, K.: Palmprint Recognition. Scientific Publishers, China (2006)
15. Han, C., Chen, H., Lin, C., Fan, K.: Personal authentication using palm-print features. Pattern Recognition 36, 371–381 (2003)
16. Kumar, A., Wong, D., Shen, H., Jain, A.: Personal verification using palmprint and hand geometry biometric. In: Kittler, J., Nixon, M.S. (eds.) AVBPA 2003. LNCS, vol. 2688, pp. 668–678. Springer, Heidelberg (2003)
17. Wu, X., Wang, K., Zhang, D.: Fisherpalms based palmprint recognition. Pattern Recognition Letters 24, 2829–2838 (2003)
18. Zhang, D., Kong, W., You, J., Wong, M.: Online palmprint identification. IEEE Transactions on Pattern Analysis and Machine Intelligence 25, 1041–1050 (2003)
19. Kong, W., Zhang, D.: Feature-level fusion for effective palmprint authentication. In: Zhang, D., Jain, A.K. (eds.) ICBA 2004. LNCS, vol. 3072, pp. 761–767. Springer, Heidelberg (2004)
20. Wu, X., Wang, K., Zhang, D.: Palm-line extraction and matching for personal authentication. IEEE Transactions on Systems, Man, and Cybernetics—Part A: Systems and Humans 36, 978–987 (2006)

Enhanced Graph Based Genealogical Record Linkage

Cary Sweet[1], Tansel Özyer[2], and Reda Alhajj[1,3]

[1] Department of Computer Science, University of Calgary, Calgary, Alberta
[2] TOBB Ekonomi ve Teknoloji Üniversitesi, Ankara, Turkey
[3] Department of Computer Science, Global University, Beirut, Lebanon
alhajj@cpsc.ucalgary.ca

Abstract. Record linkage is an essential social problem that has received considerable attention for over two centuries. With the popularity of computers, automated systems started to be realized for better record linkage systems. In this paper, we look at improving the latest techniques in record linkage, from the perspective of genealogy, to obtain results that require decreased human intervention. The increased benefit will be measured against the mainstream string and entity comparison techniques.

1 Introduction

With the Internet being utilized in more homes around the world, information on the web is becoming more accessible. As a result, the new generation of genealogy software, which allows the user to load multiple genealogies, search and merge duplicate individuals, is becoming popular in the current market; most of it is likely to be on-line and be used interactively. However, the complex queries for searching are still restricted to the boundaries defined by the software developer and all have GUI interfaces that run the queries; much of the searching in genealogy has suffered from the limitation of a GUI interface. So, one of the objectives of this research work is to allow users to type in a command that will resemble the English language structure, and return the result from genealogical database. The major reason for developing this system is to give users the added flexibility of potential relationships (cousins, ancestries) in the database.

Family history is generally represented by a genetic family tree (also called a "pedigree"), which shows the past and present members of the family joined together by a series of links that help in ascertaining their relationship to each other, and the location, documentation and recording of a family history. An example of a genetic family tree is shown in Figure 1.

The main idea in using genetic family tree is that it is easy to view and understand by users. By using chart diagram, users can easily track and transfer the relationships including sound, video, and text files - along with a wide variety of graphic formats in the trees. When the multiple records identified reside within the same family tree, a smaller tree with no duplicates will be produced. When

R. Alhajj et al. (Eds.): ADMA 2007, LNAI 4632, pp. 476–487, 2007.

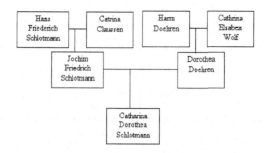

Fig. 1. Example genetic family tree

the identified records reside in two or more disjoint trees, a larger tree will result. This process is also referred to as duplicate reduction or merge/purge.

Graph-matching [19] is an essential technique in genealogical record linkage that takes advantage of the hierarchical nature of genealogical data. In the abstract form, trees represent genealogical data whose nodes are the real world entities and edges are relationships between the entities. A graph-matching algorithm then uses the identified records as focal points and extends to all nodes that overlap between the trees. The record linkage evaluation will be based on the extended coverage of the trees.

A set of potentially matching records goes through a ranking stage. At the record level, attributes of records are compared for corresponding accuracy. The aggregate ranking of a set of identified records is determined by both these attribute matches and the subsequent node matches. Once all the identified records have been sorted by ranking, they are grouped into three categories: Linked, Unlinked and Unknown. Any record that falls into the "Unknown" category will require manual research to determine which category the record finally belongs. The decision of which ranking will act as a cutoff is also made manually.

This process seems to be able to produce effective results. Addressed in this research work are several areas that could be improved upon that would produce a more user-friendly set of final data. In particular, the following issues are handled.

- The algorithms used to evaluate the record attributes (names, dates, etc.)
- The calculation of both the entity and aggregate rankings
- Knowing the structure of the overlapping tree
- Using inferred dates in replacement of missing ones
- A smaller and better-organized "Unknown" category would reduce the amount of time necessary for manual evaluation of the records.

The conducted analysis demonstrate the power of the developed process.

The rest of this paper is organized as follows. Section 2 is a brief overview of the related work. Section 3 highlights the aspects of the developed data miner for handling record linkage information. Section 4 reports on the analysis of the ranking process. Section 5 is summary and conclusions.

2   Related Work

Record linkage is one of the most essential social problems studied in history, and several attempts tried to handle the problem from different perspectives, e.g., [1,3,9,12,16,20] . Heuristics based on matching Soundex codes on names and exact matching on date and place fields for some event, including birth, marriage, or death, were the drivers for the first initiatives in record linkage algorithms. Once a pair of duplicates was linked, additional, less-stringent rules were used to link related family members. The next milestone in record linkage could be identified as statistical based on a Fellegi-Sunter algorithm [7,13]. The weighted features used in the statistical algorithm were based on name and event information for various world regions.

In general, there are two main areas of interest dealing with genealogical databases. They are family reconstruction and the work of projects like those undertaken by the GRANDMA project [8] of the California Mennonite Historical Society. The Minnesota Population Center [15] is Family reconstruction organization that is trying to link the records from 5 different censuses ranging from 1860 through 1910. Considering these crosses over approximately 3 generations, record linkage would be occurring with the new families of the children as well as within the same family.

Taking multiple genealogical databases and merging them together help produce a larger, more complete database. The algorithms utilized in this task have been progressing in their level of complexity. The early attempts involved only comparing two records [11]. Post-processing rules were used to handle merging the relationships of the selected records [14]. Current work in this area by the Family History department involves graph-matching [19,14].

While graph-matching generates accurate results, Wilson's algorithm [19] focused on the resultant tree only as a benefit to the merging process. Wilson's algorithm also focused on finding only high-ranking matches while building the tree. This meant that the content of the tree could not be utilized in determining the ranking of the individuals within the tree. This has been the basic motivation to conduct the work described in this paper.

3   Developing the Data Miner

The DataMiner project involved the creation of a query language that would be used to assist in the determination for candidate objects to be merged within a hierarchical data model. Specifically, a genealogical database usually contains one or more entries of the same person. In this data model, only the individual, parents and spouse are used for comparison and ranking purposes. It covers several stages of the record linkage process.

3.1   Find All the Matching Individuals Within the Database

This involved taking a list of groups where the only grouping criterion was the first and last names had to match. This list was then manually verified as a possibility match with secondary and tertiary sources [17,4,5,8,18]. This has been tested on a database of 22696 persons. While this procedure identified 362 duplicates and 42 potential duplicates as not a match, it did not guarantee that *all* matches were found. Great pains were made, however, to ensure that most reasonable combinations of the 22696 person database were evaluated.

3.2   Query Language

DataMiner has the capability to describe when a field does not contain null. Previously, when an attribute was placed within the query, the resultant grouping allowed all the nulls to match one another. To help reduce the level of undesirable information, the attribute modifier not_null was added to the query language.

Date(not_null:birth:indiv)
Father_modified_given(not_null)

3.3   Analyze the Ranking Function

It was important to determine the maximum efficiency of the implemented algorithm. This will allow the comparisons with the following enhancements to be better understood. A hill-climbing algorithm was used to drive the search for the best values for each attribute that was used within the ranking function. When analyzing the values for the attributes three interests were being considered:

- The highest level of accuracy for the current algorithm
- List of attributes that had no impact on the outcome of the results
- List of attributes that had the highest impact on the outcome

Attributes of the ranking function. Three relationships were exploited and the attributes of each person involved were combined to reach a single ranking value. The relationships were: Current individual, Spouse, Mother and Father. For each of these the aggregate attributes of name, birth date, baptismal date and death date were selected. In total, there are 98 attributes (23 for each individual, plus the 6 for only the spouse, mother and father). New attributes were added in to determine how they would impact the results. These included: Entity Comparison results in zero ranking; The individual does not exist; Sex; Location; First and last name both match; First and last name both do not match; Year, month and day all match; and Fuzzy logic date matching.

It was previously assumed that Sex and Location would have zero impact on the results and were removed, for this reason, from the ranking function. Since a full analysis of the ranking function was identified essential, the basic attributes of Sex and Location were added to ensure completeness.

Each attribute is counted only once. For instance, if year, month and day all match, only the full date match attribute is applied to the ranking function. The individual component evaluations are ignored in this case.

Individual level attributes: These attributes are applied to only the spouse, mother and father. The individual is assumed to always be involved in the equation and thus exists with at least a partial match of the name. This assumption makes these attributes unnecessary for the ranking of the primary individual of interest.

Comparison results in a rank of zero. None of the following attributes match between the individuals of interest. This is useful when a person has married more than once. The first individual could represent the spouse of the first marriage, while the second individual could represent the individual of the second marriage.

The individual does not exist. If an individual were missing both parents, this situation would be more probable for a match than if both parents existed and had a very low individual level ranking.

Name and sex attributes

First name and last name. These are the basic constructs for comparing an individual. These rankings only apply if only one and both attributes are matched.

Comparison results in a rank of zero. Both the first and last names did not match. This attribute was added to see what the hill-climbing algorithm would do with it.

Both match. This attribute gives the capability of enhancing the ranking for a better than partial match.

Sex. Some names are unisexual, such as Kim. This attribute may play a role in obtaining better results. This attribute was added to test its significance within the ranking function. The hill-climbing algorithm was designed with this in mind.

Birth, baptismal and death events

Location. Most of the data does not have locations and some have mixed level of detail for the same location: Swift Current, Canada; Swift Current, Saskatchewan. This attribute will help test some of the string comparison functions.

Year, month and day. This set of attributes are processed the same as the name attributes. Their ranking is applied only if some, but not all, of the attributes match.

Year, month and day all match. This attribute gives the capability of enhancing the ranking for a better than partial match.

Fuzzy logic matching. While looking at the data it was determined that there were several focal points surrounding the perfect date match. There were peaks at 0, 7, 14, and 21 days as the absolute difference between two entities that were determined by research to be an exact match.

Calculation of the ranking function. The ranking function is created in several phases. First, each of the individuals - primary individual, spouse, mother and father - are evaluated separately by adding the attribute values together. Each individual is then identified as a percentage of the maximum rank for that individual. These percentages get one added to them and then are multiplied together. This gives us an overall ranking for the individual.

All the relationships for that individual are then evaluated. Moving recursively along the parents, children and spouse lines, the tree is built up with individual matches. These may be good or bad matches. The only criterion is that both individuals have the same movement (e.g. both have a father). The ranking for the tree is the average of all the individuals within the tree. The final ranking for the individual is obtained by finding out what percentage of the maximum individual ranking the current individual and multiplying that to the tree ranking (e.g. 45 % × 450).

Hill-climbing Algorithm. The hill-climbing algorithm looks at marginal changes to the original ranking and decides which values to keep for all the different attributes. Each attribute is evaluated over a range of values (0 thru 128). The rank that achieves the highest level of accuracy is kept and the next attribute is evaluated. Once all the attributes have been evaluated, the algorithm continues with the first one. This is repeated until the level of accuracy at the end of evaluating the 98th attribute remains the same.

Evaluating the points 0 thru 128 results in either a convex, concave or linear curve. Since all of these only have at most one maximum point, a more optimized algorithm can be employed by using a bisection line search algorithm. Using this algorithm results in an average of 5 evaluations instead of the original 129.

Determining the accuracy of the current algorithm. Before any modification could be made to the algorithms, a baseline level of accuracy had to be defined.

When the program is run, a list of groups is obtained. Each group of two individuals is then ranked using the ranking function and the set of ranks configured by the user, or the hill-climbing algorithm. This list of groups will then be sorted with the highest ranking being first in the list. This sorted list was then compared against the list of known matches. This comparison had to consider three things:

1. The first group of 2 should have a higher score than the last group of 2. This will imply that an incorrect mismatch occurring at the highest likely duplicate will have a more significant impact than the least likely duplicate.
2. A score of 100% should be theoretically attainable
3. The scoring mechanism should be consistent across differing data sets

The algorithm decided upon has the following attributes:

- The first group has a rank of n. Where n is the total number of matched records, which were manually identified earlier. In this case: 362
- The last group has a rank of 1
- The total for all the groups is $\frac{n \times (n+1)}{2}$
- For each group that exists within the matched record list, add that group's rank to the total

- Only evaluate the first n groups
- Divide the final total by the maximum total to obtain a percentage of accuracy

3.4   Date Comparisons

People who are entering dates into data entry system can always make mistakes. Another discrepancy in dates could arise when changing from the Julian to the Gregorian calendars. This discrepancy could be anywhere from 10 to 13 days depending on the country that the data originated from.

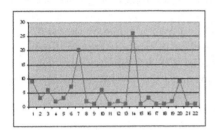
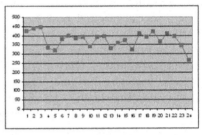

(a) Differences within matching data

(b) Differences within non-matching data of almost the same number of records

Fig. 2. Differences within matching and non-matching

Figure 2(a) shows the results of differences between the dates on the records that were manually identified as matches (see Section 3.1). Several peaks can be identified, the tallest peaks occurring at a difference of 7 and 14 and lower peaks at 1 and 20. Looking at the unmatched data (Figure 2(b)), we can see that no identifiable peaks exist. For this set of data, another unique characteristic of the data can be used as a focus for the ranking function. Under normal date comparisons, these differences could not be utilized. Fuzzy logic along multiple user-defined focus points helps solve these issues. There are three aspects for each focal point: 1) Offset from the date of interest (e.g. 14 days); 2) Range that the fuzzy logic will be applied (e.g. +/-3 days); 3) Type of curve that will be used: concave, convex, or linear.

Table 1. Example focal points identified from Figure 2(a)

Offset	Range	Curve
7	2	Concave
14	2	Concave
20	2	Concave

3.5   Entity Comparisons

Entity comparisons always involve weighting the individual evaluations and obtaining an aggregate ranking value. Part of this project is to develop an automated evaluation of the different attributes in the data and obtain ideal weightings for each of the attributes. In this way, we can evaluate the importance of a death date over a birth date. An important factor for this analysis will be the availability of the attribute within the data.

Entity comparisons will utilize the graph-matching algorithm [14]. Modifications to this algorithm will involve Identifying potential matches within the main database, Fully evaluating all the nodes within the overlapping sub tree, and Using the final tree ranking to assist in the individual ranking.

Table 2. Highest and lowest Standard Deviations of the attributes

Standard Deviation	Preferred Value	Attribute
8.27	0	Father No name given
7.26	0	Father Last Name
5.89	0	Mother No name given
2.68	1	Individual Birth Month
1.97	4	Mother First Name
1.76	0	Father sex
1.62	0 - 15	Father Birth Location
1.48	14	Mother Last Name
1.45	14	Father Both first and last name
1.43	0	Spouse No name given
1.36	0	Father Birth Month
1.20	1	Mother Sex
1.06	6	Individual Birth Year
0	0 – 99	All information regarding Baptism and Death for all individuals
0	0 – 99	The overall rank of the Mother when zero attributes match
0	0 – 99	Individual Both first name and last name match
0	0 – 99	Spouse Birth Location

3.6   Query Analysis

The blocking technique allows the user to define the minimum requirement of how records will be grouped together for further evaluation and ranking. Sometimes the user will create a blocking definition that creates groups that are larger than necessary (10 or more records within a single group). Evaluating the group could result in suggestions being made to the user as to how to make a more restrictive definition.

By looking at what attributes would match, a histogram can be created that would say which attributes have a higher success ratio and which ones do not.

4   Analysis of the Ranking Function

The highest percentage of accuracy achieved was 96.28%. The standard deviation was used to determine the effect that different values of rank had on an attribute. Attributes that had standard deviations greater than one were considered highly influential in the level of accuracy. These attributes fell into two categories: Those that affected the rank and those that did not. According to Table 2, half of the attributes had a preferred ranking value of zero. Any value higher than this adversely affected the level of accuracy for the ranking function. Attributes that had a standard deviation equal to zero were determined to have no impact on the level of accuracy. The effect of these attributes was felt more in the ranking function, where the more attributes matched the higher the probability of a successful match. It is interesting to note that the highest ranking exists when

Table 3. Attributes with the highest preferred value

Preferred Value	Attribute
99	Father not exist
0 - 99	Mother zero rank
98	Mother not exist
81 – 94	Spouse not exist
82	Mother First Name
0 – 99	Individual both first and last name match
97	Individual birth date all matches
94	Spouse both first and last name match
0 – 99	All information regarding Baptism and Death for all individuals

both the mother and father do not exist and the spouse does. This is one of the tree structures that might prove to have a higher possibility of a match.

4.1   Further Analysis

The final accuracy of the tree-ranking algorithm with fuzzy dates was 96.56%. The final ranking for the non-tree algorithm was 94.12%. The main improvement in the final ranking was due to adding in newly found matches by following the graph-matching algorithm. Table 4 shows that keeping all the attributes at the same ranking value, the resulting accuracy is quite diminished compared to the results obtained from before this project began. The ranking function analysis, however, was able to converge upon an ideal ranking function more easily with the new algorithm. This would seem to imply that there are more optimal ranking functions available by using the new algorithm.

Some attributes should never be used in ranking the data. Father's last name, which should always match the primary individual, has a very adverse effect on the level of accuracy when used. Some attributes do not add value to the matching process. While the hill-climbing algorithm says that the information

Table 4. Attributes with the highest matches

Initial Accuracy	Iter.s before Convergence	Matches Found	Algorithm	Fuzzy Dates
9.55	417	303	Tree Algorithm focus on tree average	Yes
9.57	421	300	Tree Algorithm focus on tree average	No
2.77	916	303	Tree Algorithm focus on tree size	Yes
91.55	955	258	Non-tree algorithm	Yes
91.65	1017	254	Non-tree algorithm	No

about the baptism and death events can be ignored by the ranking function, they are still important to help determine which groups are more likely to be a real match.

Subjectively, we would think that the sex of an individual should have no bearing on the outcome. For out primary individual the sex attribute had a standard deviation of 0.26 and a preferred value of three. Odd results like this only mean that the data entered always has a chance of being incorrect. The source of the inaccuracy could be the result of the person recording the information in the registry, the person transcribing the registry or even the person reading the transcription and entering the data into the computer. It is because of these possible sources for errors that any algorithm for genealogical record linkage should be flexible and unrestricted.

In our quest for producing results that are useful for the user, inaccuracies can produce unwanted results. These unpredictable results need to be taken into consideration when creating new algorithms.

The fuzzy dates did not appear to enhance the accuracy of the query process. It might work better with data that focuses on a specific time and/or location that a change in calendar occurred.

This proved to be quite effective when the ranking function was not biased on the size of the tree. At one point, the tree's rank was the sum of the entities in the tree. This produced less then desirable results. Once the ranking focused on the average of the tree's entities, the hill-climbing algorithm was able to converge on an optimal solution more quickly. This would imply that there were more optimal solutions for this new approach to the tree's ranking. On small datasets, the hill-climbing algorithm reached an optimal ranking with only one pass through the data. As the datasets became larger, the number of passes through the data also became larger.

The focus has been on how a single query could be used to produce the best possible results. However, the data being considered might be too much for the computer/user to absorb. Being able to reduce the amount of data returned and only look at higher quality data can be advantageous. The best way to do this would be adding an additional attribute to the query. Having a list of attributes and how they could affect the outcome is very handy.

Table 5. The larger the number of groups(# of grps), the more iterations required before convergence(Iter.s Before Conv.)

# of pos. grps found out of 362	# of grps evaluated	# of attribs used out of 98	Iter.s Before Conv.	Query
254	267	49	417	modified_given, modified_surname, Date(not_null:birth:indiv), MaxGroups(100000)
279	3446	73	916	modified_given, modified_surname, Date(birth:indiv), MaxGroups(100000)
354	256943	95	13212	modified_given, modified_surname, MinRank(5000), MaxGroups(100000)

Table 5 shows that the more initial data that the grouping query obtains, the longer it takes to obtain a satisfactory result for the ranking function analysis.

5   Summary and Conclusions

This paper discussed the importance of using graph matching in handling record linkage. As a result of this study, it was found that using a single query over a large number of data produces good results. However, we found out that we need to investigate the accuracy of combining the results of multiple queries running against smaller cross sections of the data. It would be interesting to find out if combining the results of one phone and one string comparison algorithm would produce a better result. Strings to be compared include: First name, last name and location. We also decided to investigate the effect of adding inferred dates when some, but not all the dates are missing within the tree of individual being ranked. Focusing on the tree structure can help arrange the unknown category of records by using probabilistic reasoning. Two individuals have a high match. The parents, siblings, spouses along the tree structure originating at these individuals also produce a high probability of a match. It would then make sense to keep these matches close together. Comparisons could be made based on spatial closeness.

References

1. Bengtsson, T., Lundh, C.: Name-standardization and automatic family reconstitution. Lund Papers in Economic History. Department of Economic History, Lund University 29, 1-24 (1993)
2. Bloothooft, G.: Multi-Source Family Reconstruction. History and Computing 7(2), 90–103 (1995)

3. Bouchard, G.: Current issues and new prospects for computerized record linkage in the province of Québec. Historical Methods 25, 67–73 (1992)
4. Dyck, J., William, H.: Reinlaender Gemeinde Buch (1994)
5. Ens, A., Jacob, E.P., Otto, H.: 1880 Village Census of the Mennonite West Reserve (1998)
6. Family History Department, CJCLS (2002) http://www.familysearch.org/Eng/Home/FAQ/faq_gedcom.asp
7. Fellegi, I.P., Sunter, A.B.: A Theory for Record Linkage. Journal of the American Statistical Association 64, 1183–1210 (1969)
8. GRANDMA, Genealogical Project Committee of the California Mennonite Historical Society, vol. 3 (2000)
9. Jaro, M.A.: Advances in Record-Linkage Methodology as Applied to Matching the 1985 Census of Tampa, Florida. Journal of the American Statistical Association 89, 414–420 (1989)
10. Katz, M., Tiller, J.: Record-linkage for everyman: A semi-automated process. Historical Methods Newsletter 5, 144–150 (1972)
11. NeSmith, N.P.: Record Linkage and Genealogical Files. Record Linkage Techniques, pp. 358–361. National Academy Press, Washington, DC (1997)
12. Nygaard, L.: Name standardization in record linking: An improved algorithmic strategy. History & Computing 4, 63–74 (1992)
13. Newcombe, H.B., Kennedy, J.M., Axford, S.J., James, A.P.: Automatic Linkage of Vital Records. Science 130, 954–959 (1959)
14. Quass, D., Paul, S.: Record Linkage for Genealogical Databases. In: ACM SIGKDD 2003 Workshop on Data Cleaning, Record Linkage, and Object Consolidation, August 2003 (2003)
15. Ruggles, S.: Linking Historical Censuses: A New Approach. In: IMAG Workshop (2003)
16. Schofield, R.: Automatic family reconstitution - the Cambridge experience. Historical Methods 25, 75–79 (1992)
17. Unger, H., Martha, M., Ens, A.: Sommerfelder Gemeinde Buch (2004)
18. Unpublished genealogical records Swift Current Gemeinde Buch. vol. 1, 2
19. Wilson, R.: Graph-Based Remerging of Genealogical Databases. In: Provo UT. Proceedings of the 2001 Family History Technology Workshop, pp. 4–6 (2001)
20. Winkler, W.E.: Advanced Methods of Record Linkage. American Statistical Association. In: Proceedings of the Section of Survey Research Methods, pp. 467–472 (1994)
21. Winchester, I.: The linkage of historical records by man and computer: Techniques and problems. Journal of Interdisciplinary History 1, 107–124 (1970)

A Fuzzy Comprehensive Clustering Method

Shuliang Wang[1,2] and Xinzhou Wang[2]

[1] International School of Software, Wuhan University, Wuhan 430072, China
[2] The State Key Laboratory for Information Engineering in Surveying, Mapping and Remote Sensing, Wuhan University, Wuhan 430072, China
slwang2005@whu.edu.cn

Abstract. Fuzzy comprehensive evaluation cannot reasonably differentiate the close membership values, e.g. 0.70 and 0.69. When the results have to be decided on the basis of maximum fuzzy membership value, some related information among similar objects may be neglected. At the same time, supervised fuzzy clustering analysis selects the threshold according to subjective experience. But different users may give different thresholds, and different thresholds may further get different clustering results. Integrating both fuzzy comprehensive evaluation and fuzzy clustering analysis in a unified way, this paper proposes a fuzzy comprehensive clustering method based on the maximum remainder algorithms and maximum characteristics algorithms. First, the principle of fuzzy comprehensive clustering is given. Based on the membership matrix of fuzzy comprehensive evaluation, fuzzy similar matrix is generated. Then a fuzzy equivalent matrix is produced from the fuzzy similar matrix. According to the fuzzy equivalent matrix, fuzzy clustering is implemented via the maximum remainder algorithms on the basis of fuzzy confidence level. And the grades of the resulting clusters are computed by using the maximum characteristics algorithms. Finally, a case study is given on land grading in Nanning city, the results of which show the proposed fuzzy comprehensive clustering method is able to overcome the disadvantages of either fuzzy comprehensive evaluation or fuzzy clustering analysis.

1 Introduction

When an object in the real world is approached, there is very often no sharp boundary. Furthermore, the boundary of its qualitative concept or classification is so vague that an element of the data set will not be uniquely assigned to one subset (Burrough, Frank, 1996). As a consequence of the fuzzy characteristics of natural classes and concepts, fuzzy sets characterize the fuzziness via a fuzzy membership function instead of the crisp characteristics function, and map the uncertainty to a numerical value on the interval [0, 1] instead of a set {0,1} (Zadeh, 1965). The fuzzy membership function is the relationship between the values of an element and its degree of membership in a set (Wang, Klir, 1992). Fuzzy sets allow elements to be partially in a set, and each element is given a fuzzy membership value ranged from 0 to 1, e.g. the membership of an element belonging to a concept. If one only allowed the extreme membership values of 0(not an element of the set) and 1(a member of the set),

R. Alhajj et al. (Eds.): ADMA 2007, LNAI 4632, pp. 488–499, 2007.
© Springer-Verlag Berlin Heidelberg 2007

this would actually be equivalent to crisp sets (Vaught, 1995). Simultaneously, fuzzy sets deal with the similarity of an element to a class. The possibilistic approach of uncertainty offered by fuzzy sets also forms a useful complement to the measures of probability theory for the probability is an indicator of the frequency or likelihood that an element is in a class (Arthurs, 1965; Zadeh, 1978). In fuzzy sets, fuzzy comprehensive evaluation and fuzzy clustering analysis are two essential techniques.

During the process of fuzzy comprehensive evaluation, an element may be assigned to a series of membership values in relation to the respective subsets of the discourse universe for the membership values indicate the possibility of an element to belong to a specific subset, and to another identified subset, and to yet another, and so on (Wang, 2002). For example, a land parcel can be depicted with grade I, II, III, IV memberships together. Traditionally, fuzzy comprehensive evaluation considers various factors of alternatives, and treats the grade with the maximum membership value as the grade of the land parcel (Wang et al., 2003). But it cannot reasonably differentiate two close membership values, e.g. 0.70 (grade I) and 0.69 (grade II), and it further neglects the influences of obstructers such as river and railway.

Ruspini (1969, 1970) first applied the theory of fuzzy sets to cluster analysis. Then, a number of algorithms were described to derive optimal fuzzy partitions of a given set of sample points,

At the same time, fuzzy clustering analysis is to partition a number of observed data into a few clusters on the basis of the characteristics of the spatial entities concerned. And it is to apply fuzzy membership values between zero and one instead of crisp assignments of the data to clusters (Frank, 1999). Similar data are assigned to the same cluster whereas dissimilar data should belong to different clusters (Grabmeier, Rudolph, 2002). However, supervised fuzzy clustering analysis selects the threshold according to subjective experience. The threshold is often decided by users, which are greatly affected by subjective wish (Zadeh, 1977). And the clustering result is not unique because of human errors to decide thresholds. Based on the same fuzzy equivalence matrix, different users may give different thresholds, the same user may also select different thresholds in different contexts, and different thresholds may further get different clustering results. And what is a suitable threshold? Moreover, how will it be if fuzzy comprehensive evaluation and fuzzy clustering are integrated together in the context of fuzzy sets?

In order to distinguish two close membership values with the help of obstructers, and avoid the subjective thresholds, this paper proposes a fuzzy comprehensive clustering that integrates both fuzzy comprehensive evaluation and fuzzy clustering analysis. In the following sections, section 2 will give the detailed principles. A case study on land grade in Nanning city with the proposed methods will be presented in section 3. Finally, section 4 will draw a conclusion on the whole paper.

2   Principles of Fuzzy Comprehensive Clustering

Fuzzy comprehensive clustering firstly acquires the fuzzy evaluation matrix on each influential factor. Then all fuzzy evaluation matrices multiply the corresponding weight matrices, the product matrix of which is the comprehensive matrix on all factors. Third, the comprehensive matrix is further used to create a fuzzy similar matrix, on the basis of which a fuzzy equivalent matrix is obtained. Fourth, fuzzy clustering is implemented via the proposed Maximum Remainder Algorithms. Finally, the resulting clusters are graded by using Maximum Characteristics Algorithms.

2.1 DHP to Select and Weight Factors

There are many factors to influence a spatial entity in the real world, e.g. land quality, and they are hierarchical in logics. Each factor shows different weight according to its contribution. However, it is impossible to consider all the factors when the quality of the spatial entity is evaluated. Therefore, we put forward a DHP (Delphi Hierarchical Processing) method to select the important influential factors and determine their weights.

The DHP method is an integration of the Delphi method and the analysis hierarchy process. In the contexts, the Delphi method is employed to select influential factors and decide their primary weights, and then the analysis hierarchy process is used to determine their weights. Fig.1 shows the whole flow chart.

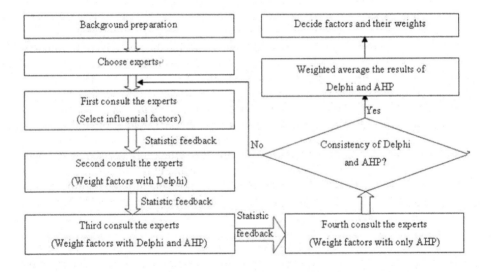

Fig. 1. A DHP method to select and weight influential factors

The Delphi method takes good advantage of group information feedback and control. It synthesizes the decentralized experiences of experts in different fields and gets the final consistent result via evaluating the average and variance of influential factor probability (Arthurs, 1965). The experts have wide representation, high authority and suitable number, and the consultative form strongly centralizes influential factors and is easy to fill in. Generally, the experts are consulted three or four times. The cycle of the process is consultation, evaluation, feedback and consultation. As the Delphi method pays more attention to qualitative analysis, the analysis hierarchy process is used to calculate the weights again. It firstly systemizes the selected influential factors and makes them hierarchical, then the experts are asked to fill in the decision array. Based on the decision array, the individual hierarchy is ordered and its consistency is also tested. When the consistency is satisfactory, the sum hierarchy is ordered ascending and its consistency is tested. If the decision array does not reach the required consistency, the experts will have to revise it (Li, Wang, Li, 2006).

2.2 Fuzzy Equivalent Matrix

Here take a model with two hierarchies for example. Suppose that the set of influential factors is $U = \{u1, u2, ..., um\}$ with the matrix of weights $A = (a1, a2, ..., am)T$, and the set of sub-factors $ui = \{ui1, ui2, ..., ui\ ki\}$ with the matrix of weights $Ai = (ai1, ai2, ..., aiki)T$ ($i = 1, 2, ..., m$) simultaneously (Fig.2). The set of grades is $V = \{v1, v2, v3, ..., vn\}$, including n grades.

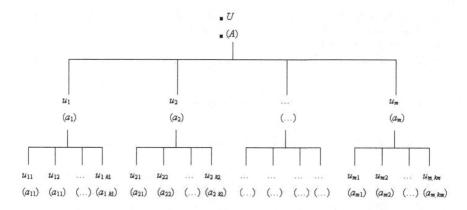

Fig. 2. Factors and their weights

In the context of the given grades, the fuzzy membership matrix on the sub-factors of factor Ui may be described as equation (1).

$$\underset{\sim}{X}_i = \begin{pmatrix} x_{11} & x_{12} & ... & x_{1n} \\ x_{21} & x_{22} & ... & x_{2n} \\ ... & ... & ... & ... \\ x_{k_i 1} & x_{k_i 2} & ... & x_{k_i n} \end{pmatrix} \tag{1}$$

Multiply the fuzzy evaluation matrix and its weight matrix, and the fuzzy evaluation matrix of ui is got.

$$\underset{\sim}{Y}_i = A_i \cdot \underset{\sim}{X}_i \tag{2}$$

The fuzzy evaluation matrix of U, i.e., all factors, on the spatial entity p is computed via equation (3).

$$Y^{(p)} = A \cdot Y = A \cdot (Y_1, Y_2, ..., Y_i, ..., Y_m)^T \tag{3}$$

When p (p≥1) spatial entities are evaluated at the same time, we can get a total matrix of fuzzy comprehensive evaluation. Let the number of spatial entity be l, then , i.e.,

$$Y_{l \times n} = \begin{pmatrix} y_{11} & y_{12} & \cdots & y_{1n} \\ y_{21} & y_{22} & \cdots & y_{2n} \\ \cdots & \cdots & \cdots & \cdots \\ y_{l1} & y_{l2} & \cdots & y_{ln} \end{pmatrix} \tag{4}$$

where, p = 1, 2, …, l.

Take the element yij (i = 1, 2, …, l; j = 1, 2, …, n) of matrix as original data, the fuzzy similar matrix can be created via equation (5), i.e., = (rij) l×l, which indicates the fuzzy similar relationships among the entities.

$$r_{ij} = \frac{\sum_{k=1}^{n}(y_{ik} \times y_{jk})}{\left(\sum_{k=1}^{n} y_{ik}^2\right)^{\frac{1}{2}} \times \left(\sum_{k=1}^{n} y_{jk}^2\right)^{\frac{1}{2}}} \tag{5}$$

The fuzzy similar matrix = (rij) l×l will have to be changed into the fuzzy equivalent matrix t( ) when clustering. The fuzzy matrix for clustering is a fuzzy equivalent matrix t( ), and it shows a fuzzy equivalent relationship among the entities instead of the fuzzy similar relationships. The fuzzy equivalent matrix has three characteristic conditions of self-reverse, symmetry, and transfer. However, the fuzzy similar matrix = (rij) l×l frequently satisfies two characteristic conditions of self-reverse and symmetry, but not transfer. The fuzzy similar matrix = (rij) l×l may be changed into the fuzzy equivalent matrix t( ) via self-squared method(Frank, 1999), for there must be a minimum natural number k(k = 1, 2, …, l, and k≤ l) that makes equivalent matrix if and only if (Zadeh, 1965). The algorithms are

$$R \to R \cdot R = R^{2^1} \to R^{2^1} \cdot R^{2^1} = R^{2^2} \to R^{2^2} \cdot R^{2^2} = R^{2^3} \to ... \to R^{2^k} \tag{6}$$

where, i = 1, 2, …, l; j = 1, 2, …, l; i≠ j; = max (min(ri1, rj1), min(ri2, rj2),…, min(rik, rjk),…, min(ril, rjl)). The algorithms of max(.), min(.) are to choose the maximum, minimum of two elements respectively. The calculus complexity is 2k -1 $<$ l ≤2k, that is, k$<$log2n + 1. For example, if there are l=30 entities, then the maximum times of calculus is only 5. The fuzzy equivalent matrix t( ) is a matrix of equation (7)

$$t(\underset{\sim}{R})_{l \times l} = \begin{pmatrix} t_{11} & t_{12} & \cdots & t_{1l} \\ t_{21} & t_{22} & \cdots & t_{2l} \\ \cdots & \cdots & \cdots & \cdots \\ t_{l1} & t_{l2} & \cdots & t_{ll} \end{pmatrix} \tag{7}$$

2.3 Maximum Remainder Algorithms-Based Clustering

Fuzzy confidential level α is a fuzzy probability that two or more than two entities belong to the same cluster. It is a prior determination from data. Under the umbrella of the fuzzy confidential level α, the Maximum Remainder Algorithms is described as the following.

1. Select fuzzy confidential level α.
2. Summarize the elements column by column in the fuzzy equivalent matrix $t(R)$, excluding the diagonal elements.

$$T_j = \sum_{i=1}^{l} t_{ij} \tag{8}$$
$, (i \neq j, i, j = 1, 2, \ldots, l)$

3. Compute the maximum and the ratio

$$T_{max}^{(1)} = \max (T_1, T_2, \ldots, T_l), \qquad K_j^{(1)} = \frac{T_j}{T_{max}^{(1)}} \tag{9}$$

4. Put the $K_j^{(1)} \geq \alpha$ of entity j into the first cluster.
5. Repeat the above steps [3][4] in the remained Tj until the end.

2.4 Maximum Characteristics Algorithms-Based Grading

The resulting cluster only gives is a group of entities with similar or the same characteristics, e.g. the grades of entities. However, it may not show obviously which grade the entities in the cluster belong to. In the context of fuzzy sets, we propose an algorithm of maximum characteristics to decide the grades.

In all the l spatial entities, suppose that there are h (h $\leq$ l) spatial entities that are grouped into ClusterZ.

$$\text{ClusterZ} = \{ \text{Entity\_1, Entity\_2, } \ldots, \text{Entity\_h} \} \tag{10}$$

Then, the fuzzy evaluation values yij in (euation (4)) of each cluster are added respectively column by column, i.e., grade by grade in the set V = {v1, v2, v3, …, vn }. And the grade of ClusterZ is the grade with the maximum sum of every column, i.e., all the entities in ClusterZ belonging to the same grade.

$$Grade\_ClusterZ = \max(Sum\_v_1, Sum\_v_2, \ldots, Sum\_v_n) = \max(\sum_{i=1}^{h} y_{i1}, \sum_{i=1}^{h} y_{i2}, \ldots, \sum_{i=1}^{h} y_{in}) \tag{11}$$

The resulting grade is further the grade of all the entities in ClusterZ.

3 A Case Study on Land Grade

It is essential to evaluate land grade in the urban development. In order to set up a good system of land grades, people have tried fuzzy comprehensive evaluation. It considers the fuzziness of land quality, but the related information among similar land parcels may be lost when the results are decided on the principle of maximum membership value. Then fuzzy clustering analysis has also been attempted. Though the related information is considered more in fuzzy clustering analysis than in fuzzy comprehensive evaluation, it is more subjective to select and decide the threshold of fuzzy clustering analysis on the basis of human experience. And the supervised clustering results are not unique. Moreover, it is not suitable to respectively use these two methods, and then compare whether their results are consistent to each other, which not only increases the workload, but also leads to a dilemma when the results of the two techniques are very different. In fact, the dilemma often happens (Huang, 1997).

In order to verify the feasibility and effectiveness of our proposed fuzzy comprehensive clustering, this section studies a case on land grade evaluation in Nanning city. In the case study, 20 pieces of land parcels with different characteristics are used (Table 1), and they stochastically distribute in Nanning city (Fig.2). The set of grades is decided as $V = \{I, II, III, IV\}$.

At the beginning of land evaluation, the influential factors are selected and weighted by using the DHP method (Fig.1) in the context of local reality and the experience of experts. And the selected and weighted factors are got under the proposed DHP method.

Thank to the advice of Nanning experts, trapezoidal fuzzy membership function (Lowen, 1996) is chosen to compute the membership value that each influential factor belongs to every grade. Then the matrix of fuzzy comprehensive evaluation can be got on the basis of principles of fuzzy comprehensive clustering (Table 1).

After the matrix of is got, the fuzzy comprehensive evaluation traditionally determines the land grades on the basis of maximum fuzzy membership value (Table 2).

Contrast to the traditional fuzzy comprehensive evaluation, the proposed fuzzy comprehensive clustering takes the elements y_{ij} ($i=1,2,\ldots,20$; $j=1,2,3,4$) of (Table 1)as the original data of clustering analysis. And the fuzzy similar matrix is created on the basis of . Then, in the context of the matrix , the fuzzy equivalent matrix $t()$ is computed via Equation (6). According to on fuzzy probability (Zadeh, 1984), the clustering threshold is selected as 0.99, which is the probability that the entities belong to the same cluster. The clustering process is shown in Table 3 by using the Maximum Remainder Algorithms based clustering. In the end, the grades of each land parcels are got via Maximum Characteristics Algorithms (Table 4), and they are also mapped in Fig.2.

Table 1. Land parcels and their evaluation matrix

Land Parcel		Fuzzy Evaluation Matrix $Y_{20\times4}$			
No.	Name	Grade I	Grade II	Grade III	Grade IV
1	Nanning railway station	0.671	0.001	0.328	0.005
2	Nanhua building	0.885	0.099	0.016	0.021
3	Train department	0.176	0.162	0.662	0.300
4	Yong river theatre	0.787	0.057	0.155	0.041
5	Medical school of women and children health care	0.141	0.654	0.204	0.834
6	Nanning shop building	0.508	0.002	0.491	0.001
7	Yongxin government division	0.009	0.594	0.404	0.006
8	Minsheng shop	0.723	0.135	0.142	0.213
9	Provincial procuratorate	0.454	0.423	0.121	0.210
10	No.1 Middle School of Nanning railway station	0.180	0.193	0.626	0.232
11	Xinyang road primary school	0.265	0.215	0.519	0.002
12	No.25 middle school	0.125	0.419	0.455	0.013
13	Municipal engineering corporation	0.335	0.558	0.092	0.034
14	Dragon palace hotel	0.058	0.117	0.704	0.902
15	Provincial school of broadcasting and TV	0.160	0.041	0.798	0.030
16	Provincial architecture institute	0.210	0.123	0.666	0.041
17	Training school of water supply corporation	0.023	0.335	0.641	0.102
18	No.2 Tingzi Middle School	0.065	0.441	0.523	0.210
19	Jiangnan government division	0.016	0.038	0.945	0.060
20	Baisha paper mill	0.031	0.622	0.424	0.604

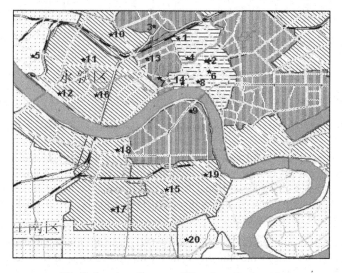

Fig. 3. Land grade map of Nanning city (parts)

Table 2. The results of traditional fuzzy comprehensive evaluation

Grade	I	II	III	IV
Land parcel	1,2,4,6,8,9	7,13,20	3,10,11,12,15,16,17,18,19	5,14

Table 3. The process of Maximum Remainder Algorithms -based clustering

Land parcel	$T^{(1)}$	$K^{(1)}$	Cluster 1	$T^{(2)}$	$K^{(2)}$	Cluster 2	$T^{(3)}$	$K^{(3)}$	Cluster 3	$T^{(4)}$	$K^{(4)}$	Cluster 4
1	16.253	0.954		16.253	0.982		16.253	0.998	1			
2	16.280	0.956		16.280			16.280	1.000	2			
3	17.005	0.998	3									
4	16.280	0.956		16.280	0.984		16.280	1.000	4			
5	16.516	0.969		16.516	0.998	5						
6	16.220	0.952		16.220	0.980		16.220	0.996	6			
7	16.104	0.945		16.104	0.973		16.104	0.989		16.104	1.000	7
8	16.267	0.955		16.267	0.983		16.267	0.999	8			
9	15.928	0.935		15.928	0.963		15.928	0.978		15.928	0.991	9
10	17.005	0.998	10									
11	17.005	0.998	11									
12	16.967	0.996	12									
13	15.858	0.931		15.858	0.958		15.858	0.974		15.858	0.990	13
14	16.182	0.950		16.182	0.978		16.182	0.994	14			
15	17.036	1.000	15									
16	17.036	1.000	16									
17	16.939	0.994	17									
18	16.939	0.994	18									
19	17.033	0.999	19									
20	16.545	0.971		16.545	1.000	20						
$T^{(1)}_{max}$	17.036		$T^{(2)}_{max}$	16.545		$T^{(3)}_{max}$	16.280		$T^{(4)}_{max}$	16.104		

Note : [1] $T^{(1)} = \sum_{-1}^{20} t_{ij}$, $i \neq j$, $i, j = 1, 2, \cdots, 20$;

[2] $T^{(2)} = T^{(1)}_{remainder}$, $T^{(3)} = T^{(2)}_{remainder}$, $T^{(4)} = T^{(3)}_{remainder}$;

[3] $K_j^p = \dfrac{T_j^{(p)}}{T^p_{MAX}}$, $j = 1, 2, \cdots, 20, p = 1, 2, 3, 4$.

Now, compare the results of our proposed comprehensive clustering (Table 4, Fig.2) with the results of single fuzzy comprehensive evaluation (Table 2). Seen from the contrast, the results of traditional fuzzy comprehensive evaluation do not match the reality of Nanning city on some pieces of land parcels, for example, land parcel 14, land parcel 9, land parcel 20.

1. Mistake the land parcel 14 of grade I for grade II. Land parcel 14 is Dragon Palace Hotel that locates the city center with good shops, infrastructures, and big population density.

Table 4. The results of Maximum Characteristics Algorithms -based grading

Grade	I	II	III	IV
Cluster	Cluster 3	Cluster 4	Cluster 1	Cluster 2
Land parcel	1, 2, 4, 6, 8, 14	7, 9, 13	3, 10, 11, 12, 15, 16, 17, 18, 19	5, 20
Property	Old urban districts, with good shops, infrastructures, Nanning commercial center, golden land	Near city center with perfect shops infrastructures, but most are obstructed by Yong river, railway from grade I land parcels	Locate at the neighbor of the land parcels with grade I, grade II Mainly resident, culture, and education land	Distribute in the periphery of land grade III. All infrastructures are bad except the ratio of green cover is big.

2. Mistake the land parcel 9 of grade II for grade I. Land parcel 9 is Provincial Procuratorate that is obstructed away from the city center by Yong River.
3. Mistake the land parcel 20 of grade IV for grade II. Land parcel 20 is Baisha Paper Mill that is in suburban suburb, with bad infrastructures.

The reasons of these errors are that fuzzy comprehension evaluation cannot differentiate the close membership value, especially when the difference between the first maximum membership value and the second maximum membership value is small. When the grade with maximum membership value is chosen as the grade of the entity, some important information on land quality, which is hidden in the associations of the influential factors, may be lost, e.g. the obstructer of river, lake, road, railway, the spread of land position advantage, etc. Because the subordinate grade will be cut off once the maximum membership value is chosen though the maximum membership may represent the essential grade that a land parcel belongs to. For example, seen from the matrix (Table 2), the land parcel 20 of Baisha Paper Mill with the membership value of grade II 0.622, grade IV 0.604, they are both close to each other. This dilemma cannot be overcome by the fuzzy comprehensive evaluation itself.

Let look at the results of our method. It is shown in Table 4 and Fig.4 that the grade of each land parcel is unique. The results include the information from the essential factors and subordinate factors. Further the shortcoming of fuzzy maximum membership value is complemented via the quantitative fuzzy clustering without subjective thresholds. The obstructer of river, lake, road, railway, the spread of land position advantage, etc. are considered because the results come from manipulating and analyzing data. The results obviously give the grades of land parcels 1~20 in the context of Nanning land properties. That is, encircling the center of city, the quality of land in Nanning city decreases with the increase of the distance away from city center like radiation. From the center of city to the fringe, land grade changes from high down to low gradually. The land with high grade centralizes the city center with good shops, infrastructures, etc., while the land with low grade distributes around the fringe. This reflects the mapping relationships between land quality and land location. Because of the obstructer of railway, Yong River, the grade of land parcel on one side of the obstructers, i.e., land parcel is close to the city center, is obviously different from the grade of land parcel on the other side of the obstructers, i.e., land parcel is away from the city center. For example, Dragon Palace Hotel, Provisional Procuratorate.

The transportation is one of the essential factors that impact land quality. On the two sides of main item road, the land grade shows the decreasing trend with the increasing distance away from the road. For example, Nanning Railway Station, Baisha Paper Mill. The resulting rules match the reality value and characteristics of Nanning land.

4   Conclusions and Discussion

This paper proposed a new fuzzy comprehensive clustering. It included two algorithms, one was Maximum Remainder Algorithms for clustering, the other was Maximum Characteristics Algorithms for grading. Qualitative analysis and quantitative calculation were combined in this method, which considered the objective reality and subjective whishes at the same time. Moreover, the proposed method integrated fuzzy comprehensive evaluation and fuzzy clustering analysis according to their advantages, and might further avoid their shortcomings to a certain extent. When it was applied to evaluate land grades in Nanning city, it considered multiple factors. And the results thought of the obstructers, i.e., railway, river. When the characteristics of two land parcels are similar or the same, the relationship between the obstructers and land grade can be used as a rule to guide land quality evaluation.

Nevertheless, there are still some problems to be further studied in fuzzy sets, e.g. How to establish, measure, and estimate the shape of fuzzy membership function? How to deal with other uncertainties besides fuzziness when there exist more than fuzziness at the same time? How to describe the distribution of spatial uncertainties with fuzzy sets in spatial analysis? How to propagate and inherit the uncertainties in the spatial reasoning? How to measure the error accumulation when uncertainties are propagated and inherited? If really there is a fixed value to represent how a specific element belong a subset of given discourse universe? They will be our continuous study in the future.

Acknowledgements

This study is supported by the funds from 973 Program (2006CB701305), National 100 best doctoral theses Fund (2005047), the State Key Laboratory of Software Engineering Fund (WSKLSE05-15), and the Ministry Key GIS Lab Fund (wd200603). And it is further in memory of Professor Xinzhou WANG.

References

1. Arthurs, A.M.: Probability theory. Dover Publications, London (1965)
2. Burrough, P.A., Frank, A.U.: Geographic Objects with Indeterminate Boundaries. Taylor and Francis, Basingstoke (1996)
3. Dunn, J.C.: A fuzzy relative of the ISODATA process and its use in detecting compact well separated clusters. Journal of Cybernetics 3, 32–57 (1974)
4. Frank, H., et al.: Fuzzy Cluster Analysis: Methods for Classification, Data Analysis, and Image Recognition. John Wiley, New York (1999)
5. Grabmeier, J., Rudolph, A.: Techniques of Cluster Algorithms in Data Mining. Data Mining and Knowledge Discovery 6(4), 303–360 (2002)
6. Huang, C.F.: Principle of information diffusion. Fuzzy Sets and Systems 91, 69–90 (1997)
7. Li, D.R., Wang, S.L., Li, D.Y.: Spatial Data Mining Theories and Applications. Science Press, Beijing (2006)
8. Ruspini, E.R.: A new approach to clustering. Information and Control 15, 22–32 (1969)
9. Ruspini, E.R.: Numerical methods for fuzzy clustering. Information Science 2, 319–350 (1970)
10. Vaught, R.L.: Set Theory: An Introduction, 2nd edn. Birkhäuser, Boston (1995)
11. Wang, Z.Y., Klir, G.J.: Fuzzy measure theory. Plenum Press, New York (1992)
12. Wang, S.L.: Data Field and Cloud Model -Based Spatial Data Mining and Knowledge Discovery. Ph.D. Thesis, Wuhan University, Wuhan (2002)
13. Wang, S.L., et al.: A method of spatial data mining dealing with randomness and fuzziness. In: Shi, W., Goodchild, M.F, Fisher, P.F, Kong, H. (eds.) Proceedings of the Second International Symposium on spatial Data Quality, March 19th – 20th, 2003, pp. 370–383 (2003)
14. Zadeh, L.A.: Fuzzy Sets. Information and Control 8(3), 338–353 (1965)
15. Zadeh, L.A.: Fuzzy sets and their application to pattern classification and clustering analysis. In: Van Ryzin, J. (ed.) Classification and Clustering, pp. 251–299. Academic Press, New York (1977)
16. Zadeh, L.A.: Fuzzy sets as basis for a theory of probability. Fuzzy sets and Systems 1(1), 3–28 (1978)
17. Zadeh, L.A.: Fuzzy probabilities. Information Processing and Management 20(3), 363–372 (1984)

CACS: A Novel Classification Algorithm Based on Concept Similarity

Jing Peng[1], Dong-qing Yang[1], Chang-jie Tang[2], Jing Zhang[3], and Jian-jun Hu[2]

[1] School of Electronics Engineering and Computer Science; Peking University; Beijing 100871, China
[2] School of computer Science and Engineering, Sichuan University, Chengdu 610065, China
[3] Chengdu Jiuheyuan Industry Company, Chengdu 610015, China
pj@pku.edu.cn

Abstract. This paper proposes a novel algorithm of classification based on the similarities among data attributes. This method assumes data attributes of dataset as basic vectors of m dimensions, and each tuple of dataset as a sum vector of all the attribute-vectors. Based on transcendental concept similarity information among attributes, this paper suggests a novel distance algorithm to compute the similarity distance of each pairs of attribute-vectors. In the method, the computing of correlation is turned to attribute-vectors and formulas of their projections on each other, and the correlation among any two tuples of dataset can be worked out by computing these vectors and formulas. Based on the correlation computing method, this paper proposes a novel classification algorithm. Extensive experiments prove the efficiency of the algorithm.

Keywords: Classification; Concept Similarity; Correlation Computing; k-NN.

1 Introduction

Data mining [1] is a process of extracting effective, connotative, and potentially valuable information from large database. It is a front line research in a lot of scientific fields especially in database research. In data mining tasks, classification [1,2] which widely used in a lot of domains is one of the most branch. It is used to discover same characters among same type of data objects, to construct classifier, and to classify unknown samples.

Nowadays, there are a lot of methods and technologies used in constructing classify models, such as decision tree, decision table, nearest neighbors [1], bayesian method, and support vector machine. In general, a classification algorithm needs three elements - known training dataset, test dataset and domain knowledge the problem belongs to. Domain knowledge does not directly exist in sample data, but in the relations among the data attributes potentially. Take text classification for example, words which are text attributes are often considered as the base of classification; however, there are actually potential similar relationships among words. The classification models aforementioned do not consider this kind of knowledge. So, these models do not have ideal results in many practical classification applications.

R. Alhajj et al. (Eds.): ADMA 2007, LNAI 4632, pp. 500–507, 2007.

This paper proposes a novel Classification Algorithm based on Concept Similarity (CACS). The main point of the algorithm is beginning from the issue of the k-nearest neighbor, then exploring on expressions of the domain potential knowledge. The k-nearest neighbor here means searching the top-k nearest or the approximate top-k nearest tuples to a given tuple in a certain database D, which has n tuples. By using a measure space, the method of k-nearest neighbor can easily turn to issues of similarity searching or classification based on models into issues of dimensional vector distance. Theory researches on these k-nearest neighbor methods refer to references [6], applications refer to references [3,4,5]. The kernel of top-k nearest neighbor is distance computing, Kleinberg proposed a method of projection on random linearity in d dimension space, D Achlioptas [7,8] applied the projection method into database environments. C.Dwork [4] proposed method of Rank Aggregation. In these methods, the tuples in database are considered as dimensional vectors, and the dimensional coordinate is the database attributes set.

CACS considers tuples in database as dimensional vectors, but it differs from the existed k-nearest neighbor classification model. It does not consider the database attributes set as coordinate. However, it considers the attributes of every tuple as a vector based on an unknown dimension space, and every tuple is considered as sum-vector of attribute vectors. According to the transcendent domain similarity information, standard attribute vector distance and projection method are defined. Through similarity formulas, the issue of distance computing is turned into attribute vectors and their projection on each other. Extensive experiments show that compared with classification without domain similarity information, the precision of CACS is greatly improved.

2 Attribute Vector Distance and Projection Algorithm

After the similarity information among attributes are known, we need to find out a method to express the relationship among data attributes, and this method should measure data attributes and their relationships in a uniform frame. Only by this way, can we really import domain knowledge. This paper assumes:

Every attribute in database is defined as a unit vector in a multi-dimension space.

Every tuple in the database is considered as a weight sum vector of these unit vectors, and the weight is the tuple value on each attribute.

According the similarity, define the known similarity information among attributes as unit distance between 2 vectors.

Through these assumptions, the attributes and computing formulas of dimensional vectors can be expediently used to express the correlativity among attributes or among tuples.

The detail is described as follow:

2.1 Definitions

Definition 1 (Attribute-vector). Assume $r_1, r_2, \ldots, r_m$ are attributes ranges, $R^m = (r_1 \times r_2 \ldots \times r_m)$ is a m-dimension vector set.

Assume $f_1, f_2, .. f_d$ are vectors in dimension R^m, namely, $f_i = (v_{i,1}, v_{i,2}, ... v_{i,m})$, $v_{i,k} \in r_k$. If $|f_i| = 1$, and c is non-negative real number, then f_i is considered as unit attribute-vector in dimension R^m, and $c \cdot f_i$ are considered as attribute-vectors, $i = 1, ..., d$.

Assume Range $A_i = Dom(f_i)$, then the maximum relationship based on the mode $F = (f_1, f_2, .. f_d)$ is denoted as $D^d = (A_1 \times A_2 ... \times A_d)$.

Definition 2. Under the same environment and symbol system as definition 1:

For $\forall v \in D^d$, assume the corresponding attribute-set's value is $(u_1, u_2, ... u_d)$, and u_i is non-negative real number, $i = 1..d$. Then $u = \sum_{i=1}^{d} u_i \cdot f_i$ is a tuple-vector corresponding to v, and $u \in R^m$.

(1) For $\forall r_i, r_j \in R^m$, $\xi(r_i, r_j)$ is named the vector projection of r_i on r_j. Then there is $\xi(r_i, r_j) = |r_i| \cdot \cos(r_i, r_j)$, and $\cos(r_i, r_j)$ denotes the cosine of angle between vector r_i and r_j.

(2) Assume u, v, w $\in$ F, c1, c2 are non-negative real numbers, <u, v, c1> denotes a path from u to v, and the length of the path is c1. Assume path0 = {<u, v, c1>, <v, w, c2>} is a known path-set, then path <u, w, c1+c2> is the derived path of path0.

(3) Assume P = {<u, v, c> | u, v $\in$ F, c is a non-negative real number, and denotes the path length between u,v } is a known path-set, then the set of paths in P and P's derived paths is called path closure, denoted as P+.

(4) Assume u,v are attribute-vectors in Rm, P is a known path-set. Then d(u, v) = Min{h| <u, v, h> in P+} is called attribute-vector distance from u to v, namely, the shortest path length sum from u to v.

2.2 Shortest Attribute Distance Computing

At first, the similar information among data attributes are converted into path set P, and then the similarity degree is expressed by the path length. According to the definition 1 and 2, an algorithm to compute the attribute vector distance between $\forall r_i, r_j \in F, (i, j = 1, ..., d)$ is proposed, and the main steps of this algorithm are follows.

First, sort the known attribute-vector path set P, namely, known similarity rules among attributes;

Second, create initial vector distance matrix $M(d \times d)$ among $\forall r_i, r_j \in F, (i, j = 1, ..., d)$. Browse every known path in path-set $p \in P$, then estimate if there is a shorter path between p.u or p.v among other attributes. If there is, then replace it, and add it into a modified attribute-set.

Third, for each path of the created attribute-set in previous step, seek whether there is a shorter distance existing after this path is added between every two attributes, if there is , then replace the old path between these two attributes; repeat this seeking-estimating-replacing action until all paths are finished.

From the method of computing distance between any pairs of attribute-vectors above, an attribute-vector projection computing formula in m-dimension space can be created. According to the definition 2, there is vector projection $\xi(r_i, r_j) = |r_i| \cdot \cos(r_i, r_j)$. However, $\cos(r_i, r_j)$ can not be got directly, so, it can be approximately expressed by computed distance among vectors:

$$\cos(r_i, r_j) = \begin{cases} 1 & \text{if } d(r_i, r_j) = 0 \\ (\varphi - d(r_i, r_j))/\varphi & \text{if } 0 < d(r_i, r_j) < \varphi \\ 0 & \text{if } d(r_i, r_j) \geq \varphi \end{cases} \quad (1)$$

3 Tuple-Vectors Correlation Computing

In above section, the distance between any pairs of attribute-vectors is computed, and a formula of projection between any two attribute-vectors is gained according to the distance between two attribute-vectors.

From the definition 2, tuple-vector $u = \sum_{i=1}^{d} u_i \cdot f_i$ and $u \in R^m$. From this formula, the tuple-vector's length can be got as:

$$|u| = \sqrt{\sum_{i=1}^{d}\sum_{j=1}^{d}(\cos(f_i, f_j)\, u_i \cdot |f_i| \cdot u_j \cdot |f_j|)} = \sqrt{\sum_{i=1}^{d}\sum_{j=1}^{d}(\cos(f_i, f_j) \cdot u_i \cdot u_j)} \quad (2)$$

From the quality of vectors, the projection of tuple-vector u on any attribute-vector $f_k \in F$ (k=1,..,d) equals the sum of all of its part-vectors' projections on f_i, namely:

$$\xi(u, f_k) = \sum_{j=1}^{d}\xi(u_j \cdot f_j, f_k) = \sum_{j=1}^{d}u_j \cdot \xi(f_j, f_k) = \sum_{j=1}^{d}u_j \cdot \cos(f_j, f_k) \quad (3)$$

From formula (2),(3), the cosine of the angle of tuple-vector u and any attribute-vector $f_k \in F$, (k=1,..,d) can be got as:

$$\cos(u, f_k) = \frac{\xi(u, f_k)}{|u|} = \frac{\sum_{j=1}^{d}u_j \cdot \cos(f_j, f_k)}{\sqrt{\sum_{i=1}^{d}\sum_{j=1}^{d}(\cos(f_i, f_j)\, u_i \cdot u_j)}} \quad (4)$$

From formula (4), the cosine value of the angle of tuple-vector u and each attribute-vector $f_k \in F$, (k=1,..,d) can be got. When every attribute-vector is orthogonal each other, and $i \neq j$, there must be $\cos(f_i, f_j) = 0$, when i=j, there must be $\cos(f_i, f_j) = 1$, and $f_i, f_j \in F$. From formula (2), the length of tuple-vector u on this condition is:

$$|u| = \sqrt{\sum_{i=1}^{d}\sum_{j=1}^{d}(\cos(f_i, f_j) \cdot u_i \cdot u_j)} = \sqrt{\sum_{i=1}^{d} u_i^2} \qquad (5)$$

Then, formula (3) transform to $\xi(u, f_k) = u_k$. Put this into formula (4), it becomes:

$$\cos(u, f_k) = \frac{\xi(u, f_k)}{|u|} = \frac{u_k}{\sqrt{\sum_{i=1}^{d} u_i^2}} \qquad (6)$$

Formula (5) is the vector-length formula in d-dimension space, and formula (6) is the cosine formula of the angle of dimension-vector and each dimension. From this, it can be figure out that formula (2),(3) are respectively expansions of the formulas of length and projection of m-dimension space vector, and formula (4) is an expansion of the formula of cosine of the angle of m-dimension space vector and each dimension.

In addition, when each attribute-vector f_k completely superposes each other, $\cos(f_i, f_j) = 1$, and $f_i, f_j \in F$, and formula (2) becomes:

$$|u| = \sqrt{\sum_{i=1}^{d}\sum_{j=1}^{d}(\cos(f_i, f_j) \cdot u_i \cdot u_j)} = \sqrt{\sum_{i=1}^{d}\sum_{j=1}^{d} u_i \cdot u_j} = \sqrt{(\sum_{i=1}^{d} u_i)^2} = \sum_{i=1}^{d} u_i$$

From formula (2),(3),(4), the correlation of $\forall r, s \in D^d$ is defined as:

$$\lambda(r, s) = \frac{\sqrt{2}}{2} \cdot \sqrt{\sum_{k=1}^{d}\left(\frac{r_i}{|r|} + \frac{s_i}{|s|}\right)\Big|\cos(r, f_k) - \cos(s, f_k)\Big|} \qquad (7)$$

and $f_k \in F$ (k=1,...,d)

In many practical applications, the most parts of attribute-vectors of tuple-vectors in high-dimension space are empty. For example, in text classification, the words in a certain text are a tiny part in all words set. We will discuss the correlation computing on this condition.

Assume $r \in D^d$, $r = \sum_{i=1}^{x} r_i \cdot f_i^r$, $f_i^r \in F$ denotes all non-null standard attribute-vectors in r; and assume $s \in D^d$, $s = \sum_{i=1}^{y} s_i \cdot f_i^s$, $f_i^s \in F$ denotes all non-null standard attribute-vectors in s, and get rid of all the zero items in formula (7), there must be:

$$\lambda(r, s) = \frac{\sqrt{2}}{2} \cdot \sqrt{\sum_{i=1}^{x}\frac{r_i}{|r|} \cdot \Big|\cos(r, f_i^r) - \cos(s, f_i^r)\Big| + \sum_{i=1}^{y}\frac{s_i}{|s|} \cdot \Big|\cos(s, f_i^s) - \cos(r, f_i^s)\Big|} \qquad (8)$$

Through above discussions, it is discovered that the correlation formula $\lambda(r, s)$ among tuple-vectors completely represents the similarity between vectors. It is based the inductive offset: the tuples in database can be represented by an m-dimension space vector, although the definition of this m-dimension space and size of m are not known. Through the observation of formula (7) and (8), it can conclude that all the computing can be represented by the angles of attribute-vectors and the angles can be got from the distance of vectors. So, every item in formula (7), (8) is computable, and there is not any unknown variant in these two formulas.

After compared formula (7) with formula (8), it can be found out that formula (8) can effectively solve the tuple computing issues under the conditions of attribute dimension is very high and actual tuple contain few attribute-vectors.

4 Classification Method Based on Concept Similarity

According to the above discussions, this paper proposes a classification algorithm based on concept similarity-CACS (Classification Method based on Concept Similarity). The object of this algorithm is to compute the tuple-vector which has smallest correlation to new data, and to classify the new data by the class this tuple-vector belongs to. The class set $C(r)$ that any $r \in R^m$ corresponding to is:

$$C_r = \{Classify(p_i) \mid p_i \in D^d, d(p_i, r) \le \theta, i = \min\{k \mid d(q_k, r) = \min\{d(q_j, r) \mid q_j \in D^d\}\}\}$$

and, Classify(p) denotes the class the tuple-vector belongs to, D^d denotes sample database.

The algorithm has 2 steps:

1) At preprocessing step, it gets the distance between any 2 attributes from transcendent attribute similarity database.

2) According to the correlation computing formula, compute the correlation of each data, which is to be classified, to every sample data, and get the tuple-vector r whose value is the smallest (P.S. smaller the value is, the more similar they are). If the correlation is less than threshold value, then classifies the data to the classification which r corresponds to, or takes it as an isolated point.

If the pretreatment processing is ignored, the classification time complexity of the CACS algorithm on each tuple is: $O(n*d)$, and n is the size of training set, d is the number of attribute-vectors.

5 Experiments

The hardware for the experiments environment is Athlon 64 3000+, 1G memory. Sample data is Chinese traditional medicine spleen-stomach type prescriptions database, and the information in database includes: spleen-stomach Chinese traditional medicine prescriptions, similar action table, and basic medicament table. The information in Chinese traditional medicine prescriptions includes not only basic information such as name of the prescription, symptom, pathogeny, and pathology, but also medicaments contained in the prescription. In Chinese traditional medicine prescription, every medicament includes information such as nature, flavors, Meridian Tropism, classification and action, and each medicament has different kinds of action.

The process of the experiment is: First, getting the medicaments which compose the prescription; second, inducting the action of the prescription according to the action of component medicaments; finally, estimate the classification of the prescription according to the action of the prescription. The result of the experiment is shown as the following:

Experiment 1. According to the action of a prescription, estimate its symptom classification. Three classification algorithms are implemented in this study, CACS, CACS without considering similar information and traditional Euclidean distance method respectively, 1060 records of classified prescription action were taken as samples in the experiment, and the action of these 3 methods are tested by 3 groups of data (there are 400 records of data in every group). The results of the experiment are shown in follows.

Table 1. Comparison of the 3 Methods' presicion

	CACS	CACS (without similar information)	Euclidean distance
Experiment 1	92.8%	68.0%	37.0%
Experiment 2	85.0%	65.0%	48.3%
Experiment 3	89.0%	70.3%	55.3%
Average	88.9%	67.8%	46.8%

Table 2. Comparison of 3 Methods' Time Consumption

	CACS	CACS (without similar information)	Euclidean distance
Experiment 1	24940	24611	23442
Experiment 2	24903	24917	23775
Experiment 3	27097	25515	23916

Trough experiment, we can find out that the 3 methods are little different in time consumption, but the precision of CACS algorithm with attribute similar information is much better than that of Euclidean distance. In real practice, efficiency of CACS algorithm can be improved more by some pretreatment, such as pre-compute the angle of tuple-vector and every attribute vector.

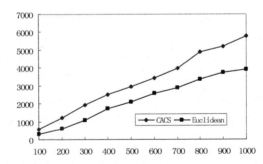

Fig. 1. Capability Comparison of CACS and classification algorithm based on Euclidean distance

Experiment 2. Estimate the difference of capability between CACS algorithm and Euclidean distance classification algorithm under different size of samples. In the experiment, the size of samples was set from 100 to 1000, and number of test data was set to 80. The detail of experiment result is shown in the following figure, and horizontal coordinate denotes numbers of samples, vertical coordinate denotes time consumption, unit is millisecond.

Figure 1 shows that same as Euclidean distance, the capability of CACS algorithm are approximately linear with increasing of sample size, and the time consumptions of these 2 algorithms are almost same.

6   Conclusions

Using the transcend similar information among attributes, this paper adopts shortest attribute distance algorithm to compute the distance between any 2 attribute-vectors, and deduces the formula of any tuple-vector correlation based on the distance between vectors, and through relative analysis, it proves this formula primely expresses the similar degree between tuple-vectors. Using this formula, this paper proposes a classification algorithm based on correlation: CACS. Through experiment to compare with classification method based on Euclidean distance, this algorithm is proved to have high precision in classification.

Acknowledgments. This research was supported by the National Natural Science Foundation of China under Grant No.60473051, 60473071 and 60503037, the China Postdoctoral Science Foundation under Grant No.20060400002, the Sichuan Youth Science and Technology Foundation of China under Grant No.07ZQ026-055, the National High-tech Research and Development of China under Grant No.2006AA01Z230 and the Natural Science Foundation of Beijing Natural Science Foundation under Grant No. 4062018.

References

1. Han, J., Kamber, M.: Data Mining: Concepts and Techniques, 2nd edn. Morgan Kaufmann Publishers, San Francisco (2006)
2. Zhou, C., Xiao, W., Nelson, P.C., Tirpak, T.M.: Evolving Accurate and Compact Classification Rules with Gene Expression Programming. IEEE Transactions on Evolutionary Computation 7(6), 519–531 (2003)
3. Aggarwal, C.: Hierarchical subspace sampling: A unified framework for high dimensional data reduction, selectivity estimation, and nearest neighbor search. In: Proceedings of the ACM SIGMOD Conference, pp. 452–463. ACM Press, New York (2002)
4. Dwork, C., Kumar, R., Naor, M., Sivakumar, D.: Rank aggregation methods for the web. In: Proceedings of the 10th International World Wide Web Conference, pp. 613–622 (2001)
5. Goldstein, J., Ramakrishnan, R.: Contrast plots and P-sphere trees: space vs. time in nearest neighbour searches. In: Proceedings of the 28th International Conference on Very Large Databases, pp. 429–440 (2002)
6. Kushilevitz, E., Ostrovsky, R., Rabani, Y.: Efficient search for approximate nearest neighbor in high dimensional spaces. SIAM Journal on Computing 30(2), 451–474 (2000)
7. Achlioptas, D.: Database-friendly random projections: Johnson-Lindenstrauss with binary coins. Journal of Computer and System Sciences 4(66), 671–687 (2003)
8. Achlioptas, D.: Database-friendly random projections. In: 20th Annual Symposium on Principles of Database Systems, Santa Barbara, CA, pp. 274–281 (2001)

Data Mining in Tourism Demand Analysis: A Retrospective Analysis

Rob Law[1], Henry Mok[2], and Carey Goh[2]

[1] School of Hotel & Tourism Management, The Hong Kong Polytechnic University,
Hung Hom, Kowloon, Hong Kong
[2] Department of Decision Sciences and Managerial Economics The Chinese University of
Hong Kong, Shatin, N.T., Hong Kong
careygoh@cuhk.edu.hk

Abstract. Despite numerous studies have applied various forecasting models to tourism demand analysis, data mining techniques have been largely overlooked by academic researchers in tourism forecasting prior to 1999. Based on our review of published articles in tourism journals that applied data mining techniques to tourism demand forecasting, we find that the application of data mining techniques are still at their infancy. This paper concludes with practical implications and future research areas.

Keywords: tourism, demand forecasting, data mining, business applications.

1 Introduction to Tourism Demand Analysis

Ever since the recognition of the important contribution of tourism to economic development, researchers, policy makers, planners, and industrial practitioners have been trying to forecast tourism demand. Likewise, the perishable nature of tourism products and services, the information-intensive nature of the tourism industry, and the long lead-time investment planning of equipment and infrastructures all render accurate forecasting of tourism demand necessary. Accurate forecast facilitates accurate planning. According to Frechtling [7] and McIntosh et al. [18], accurate planning can reduce the risk of decision making regarding the future. At the macro level, accurate forecasting results can help a destination predict the contributions and consequences of visitors to the local economy, culture, and environment. As well, the government bodies can project public revenues from tourism, and ensure that appropriate capacity and infrastructures can be maintained [26]. At the micro level, practitioners can use accurate forecasting results to set up operational requirements while investors can study project feasibility [26]. In all cases, accurate planning can minimize if not totally avoid the economic loss due to either excessive or inadequate supply.

An increased number of papers have studied tourism demand forecasting, however, these past studies have predominantly applied statistical or econometric techniques to measure and predict the future market performance [15] in terms of the number of

R. Alhajj et al. (Eds.): ADMA 2007, LNAI 4632, pp. 508–515, 2007.

tourist arrivals in a specific destination. Econometric forecasting techniques are very much highly exploited in empirical studies but only margin improvement could be attained with a substantial amount of efforts because the development of such techniques might have reached a plateau for the time being. In view of this situation, academic researchers have attempted to incorporate data mining techniques to tourism demand forecasting and have achieved some ground breaking outcomes [6].

Generally speaking, data mining refers to the process of discovering useful patterns, correlations, and rules, which are previously unknown, by filtering through a large amount of data stored in some repositories [3]. The central idea for data mining is to perform an automatic or semi-automatic identification to determine the hidden relationships and patterns which are beyond humans' manual capacity. In a business environment, data mining would be useful for managers to analyze and explore market opportunities and threats, and particularly those inherent in growing or declining markets. In the context of tourism, Law [16] stated that one of the intrinsic problems that managers have is the large amount of raw data carried in the industry, and these data are basically not comprehensible to the non-technical practitioners. Law [16] further commented that desk reviewing, dynamic tracing with keyword recognition, and *ad hoc* searching for various patterns simply cannot meet the present industrial need. Delen and Sirakaya [6] echoed such a comment, and stated that data mining methods have been rarely applied to tourism and hospitality research. Among various types of business areas, trend identification or forecasting appears to be a crucial application field for data mining [3, 6].

In view of the growing importance of data mining in business applications, we review and critically analyze the relevant articles that adopt data mining techniques to forecast tourism demand in tourism research journals. The objectives are to identify research gaps and to provide future research suggestions. Section 2 presents our methodology for searching published papers that adopted data mining techniques. We will critically review those studies in this section. Section 3 discusses the key observations and research gaps. Section 4 concludes the paper, and offers suggestions for future research directions. This paper will make a contribution for academic researchers, industrial practitioners, and official policy makers by drawing attention to the importance and necessity of integrating two prominent fields, namely data mining and tourism demand forecasting.

2 Methodology and Findings

According to McKercher, Law, and Lam [19], there are presently 70 research journals in tourism. To reflect the development of research efforts on data mining in tourism demand forecasting, several database searches were conducted on ScienceDirect and the Hospitality & Tourism Complete index on EBSCOhost Web using appropriate keywords from mid-2006 to early-2007. We restrict our search on publications in tourism journals. At the end of the database search process, we identified 174 papers on tourism demand forecasting published in a 28-year span (1980 to early 2007). Among these 174 papers, only 14 used data mining techniques, and the first was published in 1999. As reflected on Table 1, nine out of the 14 papers were published in the last 5 years (2003-early 2007). In comparison, Sami [22] found that there were 15 studies applying data mining techniques in medical fields in the nine-year span between 1997 and 2005. Notwithstanding the fact that the secondary data used in tourism demand analysis are relatively easy to obtained compared to the primary data collected by medical researchers, data mining application in tourism field seems to be worse.

Table 1. Application of Data Mining Techniques in Tourism Demand Analysis

Reference No.	Type of Analysis [#]	Authors	Year of Publication	data Frequency [**]	Data Mining Techniques [*]										
					NN (SFF)	NN (BP)	RS	NN (Elman's)	NN (MLP)	NN (RBF)	NN (BAYN)	SVR	GMDH	FTS	GFM
24	C	Uysal and Roubi	1999	Q:80-94	1										
9	C	Goh and Law	2003	A:85-00			1								
13	C&TS	Law and Au	1999	A:67-96	1										
15	C&TS	Law	2000	A:67-96	1	1									
17	C&TS	Law	2001	A:67-98	1										
14	C&TS	Law et al	2004	A:67-96			1								
1	TS	Burger *et al*	2001	M:92-98	1								1		
4	TS	Cho	2003	M:74-00				1							
25	TS	Wang	2004	A:89-00										1	1
20	TS	Pai and Hong	2005	A:56-87								1			
12	TS	Kon and Turner	2005	Q:85-01					1	1	1				
27	TS	Yu and Schwartz	2006	A:85-02										1	1
21	TS	Palmer et al	2006	Q:86-00					1						
2	TS	Chen and Wang	2007	Q:85-01		1						1			
		Frequency of model application:			**5**	**2**	**2**	**1**	**2**	**1**	**1**	**2**	**1**	**2**	**2**

Note:
* NN(SFP)=Supervised Feed Forward Neural Network; NN(BP)=Back propagation Neural Network; RS=Rough Sets; NN(Elman's)=Elman's ANN; NN(MLP)=Multilayer Perception Neural Network; NN(RBF)=Radius Basis Function Neural Network; NN(BAYN)=Bayesian Neural Network; SPR=Support Vector Regression; GMDH=Group Method of Data Handling; FTS=Fuzzy Time Series; GFM=Grey Forecasting Model
** A=Annual data; Q=Quarterly data; M=Monthly data
# C=Causal Analysis; TS=Time Series Analysis

In its broadest terms, the reviewed tourism demand forecast papers fall into two major categories, namely time series methods and causal methods. Both models were then used to project future values based on historical values. While the time series methods extrapolate future values purely based on the variable's past values, the causal approaches model the relationship between the dependent variable and the independent variables. The following subsections discuss these papers.

2.1 Causal Methods

Six papers were found in this category, and the first two data mining papers were published in 1999. In the first paper, Uysal and El Roubi [24] developed a preliminary neural network that used Canadian tourism expenditures in the United States as dependent variable; whereas per capita income of Canada, consumer price index, lagged expenditure by one period, and seasonal dummy variables were used as independent variables. Their findings showed that the neural network achieved highly accurate results with high adjusted correlations and low error terms. In addition, Law and Au [13] used a supervised feed-forward neural network model to forecast Japanese tourist arrivals to Hong Kong. Using six independent variables, including relative service price, average hotel rate in Hong Kong, foreign exchange rate, population in Japan, marketing expenses by the Hong Kong Tourist Association, and gross domestic expenditure in Japan, the neural network achieved the lowest mean average percentage error (MAPE), and highest acceptance percentage and normalized correlation coefficient compared to four other commonly used econometric models, including multiple regression, naïve I, moving average and single exponential smoothing. Further, Law [15] built a back-propagation neural network to analyze the non-linearly separable Taiwanese visitor arrivals to Hong Kong from 1967 to 1996 using the similar variables in Law and Au [13]. Their results showed that the calibrated neural network model outperformed all other approaches, attaining the highest forecasting accuracy.

Asia was badly affected by the regional financial crisis in 1997, which had resulted in many turbulent travel behaviors. Law [17] tested the forecasting accuracy of seven forecasting techniques for Japanese demand for travel to Hong Kong. These seven models comprised naïve 1, naïve II, moving average, single exponential smoothing, Holt's exponential smoothing, multiple regression, and feed-forward neural network. Using the same variables as in Law and Au [13] and Law [15] but with updated data, neural network, again, outperformed other techniques in three of the five accuracy measurements, including MAPE, trend change accuracy, and the closeness between actual and estimated values. Law [17] then concluded that no single forecasting method could outperform others in all situations when there was a sudden environmental change but neural network appeared to perform reasonably well in terms of predicted values.

There were also two causal regression papers that incorporated the rough sets theory to tourism demand forecasting. In the first paper, Goh and Law [9] built a rough sets based model using independent variables of real gross domestic product, relative consumer price index, population, trade volume, and foreign exchange rate to form six decision rules for forecasting tourist arrivals. Their findings achieved 87.2% accuracy rate in six samples. Furthermore, Law, Goh, and Pine [14] categorized and discretized the variables in Law and Au [13] with the most updated data, and the empirical findings showed that the rough sets based model correctly classified 86.5% of the testing data.

2.2 Time Series Methods

Eight time series papers were published in the field. In the first paper, Burger et al. [1] investigated eight forecasting methods. These methods included Naïve I, moving average, decomposition, single exponential smoothing, ARIMA, autoregression, genetic regression, and feed-forward neural network for the demand of American travelers to Durban, South Africa, and found that neural networks with 1-month-ahead forecast generated the lowest MAPE values. After that, Cho [4] established an Elman's ANN to forecast the number of visitors from different origins to Hong Kong. Research findings revealed that neural networks performed the best in five of the six origins. Applying the fuzzy time series and hybrid grey theory to tourist arrivals to Taiwan from Hong Kong, the United States, and Germany, Wang [25] showed that the approach did not need to have a large sample size, and that the approach did reach a low error rate. Pai and Hong [20] introduced a support vector machine model with genetic algorithm, applied the model to forecast tourist arrival data in Bardados, and claimed that the approach could accurately forecast the arrival data with MAPE values of less than 4%.

Kon and Turner [12] performed a large scale study for visitors from Australia, China, India, Japan, the United States, and the United Kingdom to Singapore in 1985 to 2001. Research findings shown the correctly structured neural network models could outperform basic structural models (BMS). Palmer et al. [21] developed neural network models for forecasting tourist expenditures in the Balearic Islands, Spain. Although the models did not perform well with raw data, they attained very low MAPE values for preprocessed data. Such findings strongly hint the necessity for data preprocessing.

Moreover, Yu and Schwartz [27] established a grey theory based fuzzy time series forecasting model for tourist arrivals to the United States from eight origins. Unlike other studies, experimental findings indicated that the sophisticated data mining technique did not produce more accurate forecasting results. The authors thus suggested that tourism researchers should proceed with care when dealing with innovative data mining techniques. Lastly, Chen and Wang [2] presented a genetic algorithm-support vector machine model to forecast tourist arrivals in China in the period 1985 to 2001, and the presented model achieved an MAPE value of 2.457%, which was considered as highly accurate.

3   Analysis and Discussion

3.1   Key Observations and Analysis

Some interesting patterns were revealed from the reviewed tourism demand studies that adopted data mining techniques. For instance, these articles were predominantly published in first-tier and second-tier research journals in tourism. According to a recent survey with worldwide scholars, two of these journals were rated at 3.6 and 3.8, and the other journals were all rated above 4 on a 5-point Likert scale [19]. Also, both time series and causal regression approaches were used in these studies. Although six studies were to investigate causal tourism demand relationship, their results were either not compared with econometric models or the comparisons were done with simple regression models. Other than Uysal and El Roubi's [24] study, causal studies only analyzed annual data under which seasonality, an important phenomenon in tourism, was basically neglected.

Another interesting observation for tourism demand forecasting studies using data mining is the dominance of artificial neural network approach. Looking at this more closely, the earlier studies [1, 13, 17, 24] all applied supervised feed-forward neural networks to discover patterns in independent and dependent variables. Later, researchers extended the applicability of neural networks by incorporating back-propagation learning process into non-linearly separable tourism demand data [2, 15]. Elman's concurrent network was also used to fit time series tourism data [4]. In a more comprehensive approach, Kon and Turner [12] fitted the tourism time series data with three different network architectures, including the MLP (Multi-Layer Perceptron), RBF (Radial Basis Function Network), and BAYN (Bayesian Network), and compared the models' performance with a number of time series econometric models.

The data mining techniques adopted in tourism demand forecasting studies were mainly used to analyze tourist arrivals to a specific destination from either one specific origin [1, 13, 14, 15, 17, 24], a number of selected origins [2, 20, 21], or from an aggregate total of certain countries [4, 9, 12].

3.2 Research Gaps

The previously analyzed publications exhibit several gaps. First, the forecasting accuracy of the adopted data mining techniques was often compared with the traditional, and often non-sophisticated, forecasting techniques. It remains largely unknown whether data mining techniques are able to outperform the advanced econometric techniques such as non-stationary regression approaches, system of equation techniques, models for structural stability like Time Varying Parameter (TVP) models. The comparison between forecasting quality of data mining techniques and advanced econometric techniques is crucial for tourism data mining researchers to be confident of their research findings. Another noticeable gap is the non-existence of examining non-economic variables in the published data mining articles. With the exception of Goh, Law, and Mok, (forthcoming) who have introduced a leisure time index and a climate index into the rough sets forecasting framework, none of the reviewed articles has investigated the impact of non-economic factors on tourism demand. Given the fact that strict statistical assumptions are not required in data mining techniques as they are in econometric models, non-economic variables which are found in many tourism studies [5, 8, 11, 23] that affect tourists' decision making could also be investigated using data mining techniques.

Apparently, prior studies on tourism demand forecasting using data mining techniques largely followed the approach that has long been used by economists. This is manifested by the same variables of demand measurements instead of innovative variables like tourism receipts in aggregate or in per capita term or number of overnight stays. Moreover, hybrid data mining systems such as fuzzy neural networks were still unpopular in tourism demand studies. The last observable gap related to the use of data for model calibration and model testing. Virtually all prior studies on data mining in tourism demand forecasting relied on regional instead of truly international data. It is understood that global data are difficult to collect but we also advocate the importance of knowledge generalization from research findings.

Compared to other business or scientific fields that have conducted numerous studies on forecasting using data mining techniques, the number of published articles in tourism falls short. Such a lagging behind phenomenon is probably due to the late commencing of data mining applications in tourism forecasting. In spite of the general claim that data mining techniques could generate more accurate forecasting results, there does not exists sufficient evidence that can prove the generalization of these claims.

4 Conclusions and Future Research

In conclusion, it is encouraging to note that interest and research efforts have been made on adopting data mining into tourism demand forecasting. The interest in this area is evident by the fact that ever since the initial attempt of utilizing artificial neural networks to tourism forecasting in 1999, articles on data mining and tourism demand forecasting have been regularly published in leading tourism journals. Still, comparing to the traditional econometric or statistical modeling techniques, data mining is at its infancy stage. What are needed in the immediate future, in addition to the involvement of more researchers, would be the endeavors to assemble and integrate different data mining approaches into a commonly agreeable process that can be applied to the tourism industry at large. In the meantime, efforts should be devoted to data mining development, which can uniquely capture the features of different tourism sectors. The tourism industry, as well, needs the introduction of more rigorously developed data mining techniques with sufficient tests on both primary and secondary data (instead of purely using secondary data). Also, it would be desirable to have more solid and concrete business solutions rather than general suggestions, which most of the published tourism demand forecasting articles offered. To measure the effectiveness of data mining in tourism demand analysis, the forecasting quality could be validated by human experts using qualitative approaches, such as jury of executive opinion and Delphi method.

Although this study is of use to provide an overview of recent research on data mining applications to tourism demand forecasting, its scope of coverage is limited by the selection channels. Future research efforts can certainly go beyond this limitation in order to have a more comprehensive review of related literature. The gaps that were discussed in the previous section also deserve further investigations by researchers in data mining and tourism forecasting.

Acknowledgement. This research was supported by Research Grant: CUHK4631/06H.

References

1. Burger, C.J.S.C., Dohnal, M., Kathrada, M., Law, R.: A practitioners guide to a time-series methods for tourism demand forecasting – a case study of Durban, South Africa. Tourism Management 22, 403–409 (2001)
2. Chen, K.Y., Wang, C.H.: Support vector regression with genetic algorithms in forecasting tourism demand. Tourism Management 28, 215–226 (2007)
3. Chen, L.D.: A review of the Data Mining literature. In: Proceedings of the Hong Kong International Computer Conference '97, pp. 23–31 (1997)
4. Cho, V.: A comparison of three different approaches to tourist arrival forecasting. Tourism Management 24, 323–330 (2003)

5. Decrop, A.: Tourists' decision-making and behavior processes. In: Consumer Behavior in Travel and Tourism, Haworth Hospitality Press, New York (1999)
6. Delen, D., Sirakaya, E.: Determining the efficacy of Data-mining methods in predicting gaming ballot outcomes. Journal of Hospitality & Tourism Research 30(3), 313–332 (2006)
7. Frechtling, D.C.: Forecasting Tourism Demand: Methods and Strategies. Butterworth-Heinemann, Oxford (2001)
8. Galarza, M.G., Saura, I.G., Garcia, H.C.: Destination image: towards a conceptual framework. Annals of Tourism Research 29(1), 56–78 (2002)
9. Goh, C., Law, R.: Incorporating the rough sets theory into travel demand analysis. Tourism Management 24, 511–517 (2003)
10. Goh, C., Law, R., Mok, H.M.K.: Analyzing and forecasting tourism demand: A Rough Sets approach. Journal of Travel Research (forthcoming)
11. Heung, V.C.S., Qu, H., Chu, R.: The relationship between vacation factors and socio-demographic and characteristics: the case of Japanese leisure travelers. Tourism Management 22(3), 259–269 (2001)
12. Kon, S.C., Turner, L.W.: Neural network forecasting of tourism demand. Tourism Economics 11(3), 301–328 (2005)
13. Law, R., Au, N.: A neural network model to forecast Japanese demand for travel to Hong Kong. Tourism Management 20, 89–97 (1999)
14. Law, R., Goh, C., Pine, R.: Modeling tourism demand: A decision rules based approach. Journal of Travel & Tourism Marketing 16(2/3), 61–69 (2004)
15. Law, R.: Back-propagation learning in improving the accuracy of neural network-based tourism demand forecasting. Tourism Management 21, 331–340 (2000)
16. Law, R.: Hospitality data mining myths. FIU Hospitality Review 16(1), 59–66 (1998)
17. Law, R.: The impact of the Asian Financial Crisis on Japanese demand for travel to Hong Kong: A study of various forecasting techniques. Journal of Travel & Tourism Marketing 10(2/3), 47–65 (2001)
18. McIntosh, R.W., Goeldner, C.R., Ritchie, J.R.B.: Tourism: Principles, Practices, Philosophies. John Wiley & Sons, New York (1995)
19. McKercher, B., Law, R., Lam, T.: Rating tourism and hospitality journals. Tourism Management 27(6), 1235–1252 (2006)
20. Pai, P.F., Hong, W.C.: An improved neural network model in forecasting arrivals. Annals of Tourism Research 32(4), 1138–1141 (2005)
21. Palmer, A., Montaño, J.J., Sesé, A.: Designing an artificial neural network for forecasting tourism time series. Tourism Management 27, 781–790 (2006)
22. Sami, A.: Obstacles and misunderstandings facing medical data mining. In: Li, X., Zaïane, O.R., Li, Z. (eds.) ADMA 2006. LNCS (LNAI), vol. 4093, pp. 856–863. Springer, Heidelberg (2006)
23. Um, S., Crompton, J.L.: Attitude determinants in tourism destination choice. Annals of Tourism Research 17(3), 432–448 (1990)
24. Uysal, M., El Roubi, S.E.: Artificial neural networks versus multiple regression in tourism demand analysis. Journal of Travel Research 38, 111–118 (1999)
25. Wang, C.H.: Predicting tourism demand using fuzzy time series and hybrid grey theory. Tourism Management 25, 367–374 (2004)
26. Witt, S.F., Witt, C.A.: Forecasting tourism demand: A review of empirical research. International Journal of Forecasting 11(3), 447–475 (1995)
27. Yu, G., Schwartz, Z.: Forecasting short time-series tourism demand with Artificial Intelligence models. Journal of Travel Research 45, 194–203 (2006)

Chinese Patent Mining Based on Sememe Statistics and Key-Phrase Extraction

Bo Jin[1], Hong-Fei Teng[2,*], Yan-Jun Shi[2], and Fu-Zheng Qu[2,3]

[1] Department of Computer Science, Dalian Univ. of Tech., P.R. China
[2] School of Mechanical Engineering, Dalian Univ. of Tech., P.R. China
[3] Key Laboratory for Precision and Non-traditional Machining Technology,
Dalian Univ. of Tech., P.R. China
tenghf@dlut.edu.cn

Abstract. Recently, key-phrase extraction from patent document has received considerable attention. However, the current statistical approaches of Chinese key-phrase extraction did not realize the semantic comprehension, thereby resulting in inaccurate and partial extraction. In this study, a Chinese patent mining approach based on sememe statistics and key-phrase extraction has been proposed to extract key-phrases from patent document. The key-phrase extraction algorithm is based on semantic knowledge structure of HowNet, and statistical approach is adopted to calculate the chosen value of the phrase in the patent document. With an experimental data set, the results showed that the proposed algorithm had improvements in recall from 62% to 73% and in precision from 72% to 81% compared with term frequency statistics algorithm.

Keywords: Patent, Data Mining, Key-Phrase, Sememe Statistics.

1 Introduction

Patent information is one of the most crucial sources of key technologies for industrial research and development. The World Intellectual Property Organization (WIPO) has predicted that judicious use of patent information could reasonably lead to prevention of duplication of research, which could save as much as 60% in time and 40% in funding. On the basis of an estimate by WIPO, patent publications cover approximately 90–95% of the results of scientific research worldwide, probably greater than the percentage that all scientific journals can cover. Because patent authorities periodically publish patent data for public inquiry, enormous data have been accumulated for many years. However, the tools and concepts used in patent research are not on par with that in scientific research. One of the reasons is the lack of key-phrases.

Key-phrase extraction is considered one of the major barriers to text retrieval, especially for Asian languages [1]. It is commonly known as indexing and finding the phrases that are representative of a document [2]. The needs to harness the

---

[*] Corresponding author.

R. Alhajj et al. (Eds.): ADMA 2007, LNAI 4632, pp. 516–523, 2007.

tremendous amount of information available from Internet and to achieve semantic retrieval have progressively prompted more interest and research to advance the state-of-the-art of key-phrase extraction [3]. Besides, key-phrase extraction is also the sticking point for automatic classification and text data mining application.

Traditionally, key-phrases have been extracted using various information-process related-methods, such as term frequency statistics [4], machine learning [5], and natural language processing [6]. However, only few examples of key-phrase extraction, particularly in full context, are available. Furthermore, most of the available examples may result in inaccurate and partial extraction.

In this article, we first introduce the semantic structure used in our key-phrase extraction algorithm and then provide the details of our algorithm. The main finding in this study is that we realized a semantic understanding algorithm using traditional statistics approach. Finally, we provide and discuss the obtained results and propose a conclusion and some perspectiveideas.

2 Semantic Structure

The knowledge of semantic structure in this study is obtained from HowNet. HowNet [7] is an online common-sense knowledge base that unveils inter-conceptual and inter-attribute relations of concepts as connotations in lexicons of the Chinese and their English equivalents. The philosophy behind the design of HowNet is its ontological view that all physical and nonphysical matters undergo a continual process of motion and change in a specific space and time. The motion and change are usually reflected by a change in state that in turn, is manifested by a change in value of some attributes. In HowNet, a set of sememes, the most basic set of semantic units that are non-decomposable, is extracted from approximately 6,000 Chinese characters. A total of more than 1,400 sememes are found and organized hierarchically. This is a closed set from which all concepts are defined. The set of sememes is stable and robust enough to describe all kinds of concepts by deriving the set of sememes in a bottom-up fashion. The fact that HowNet has more than 65,000 concepts provides a good proof for its robustness.

Table 1. Description of knowledge specimen in HowNet

ID	Phrase	Sememe
061554	gentleman|男人	human|人,family|家,male|男
005241	must必须	{modality|语气}
114646	plagiarism|剽窃	steal|偷,*copy|抄写
011940	outflank|抄袭	attack|攻打,military|军

Table 1 shows the description of knowledge specimen in HowNet. The method of description and the related rules are based on the Knowledge Dictionary Mark-up Language (KDML), which comprises the following components: (i) over 1400 features and event roles, (ii) pointers and punctuation, and (iii) word order. We set up knowledge description model for key-phrase extraction based on KDML.

3 Key-Phrase Extraction Algorithm

As the key-phrase extraction approach needs to access all the available patent specifications, a highly efficient extraction algorithm, both in terms of space and time, must be employed, or else it becomes too large or too slow for practical use. There has been a tremendous amount of researches in this area [4-6]. In this study, we propose a sememe statistics algorithm (SSA) for key-phrase extraction. Our algorithm adopted statistics approach to deal with the above-mentioned semantic structure. We consider key-phrase extraction as a classification task: each phrase in a document is either a key-phrase or not, and the problem is to correctly classify a phrase into one of these two categories.

Generally, the more frequently a word occurs, the more important is. Therefore, the first factor influencing the chosen function of word is word frequency. To this extent, we have used the relative frequency ratio (RFR) [8]. For a given word, RFR is the ratio of the frequency of the word in the document to the frequency of the word in a general corpus. The idea is that words that occur relatively frequently in a certain document compared to how frequently they occur in a general document are more likely to be good key-phrases.

The second criterion we have used to find important key-phrases relies on a measure of semantic similarity derived from the available lexical resources. There are some related research work of concept counting using the WordNet lexical database [9]. As mentioned above, HowNet is adopted as our lexical resource. For HowNet, since sememe is attached to each concept associated with a particular word, we can measure the semantic similarity with the collection of sememes.

```
Initialization:
    DocSet←{w₁,w₂,...,wₙ}
    ThresholdVal←0.3
    ChosenNum←5
Main:
While (DocSet≠∅) do
    w←pop(DocSet)
    fw[w]←fw(w)
    SemSet[w]←{ s₁,s₂,...,sₘ }
    while (SemSet[w]≠∅) do
        s←pop(SemSet[w])
        fs[w][s]←fs(s)
        fc[w][s]←fc(s)
    endwhile
    S[w]←fw[w]; fs[w]; fc[w]
    If S[w]>ThresholdVal then
        w→ChosenSet
    endif
endwhile
KeyPhraseSet←ChosenSet(ChosenNum)
```

Fig. 1. SSA: Sememe Statistics Algorithm

Let d be a patent document. We note that document d consists of the set of words w: $d=\{w_1,w_2,...,w_n\}$ and that word w consists of the set of sememes s: $w=\{s_1,s_2,...,s_m\}$. Thus, $d=\{(s_{11},s_{12},...,s_{1m}), (s_{21},s_{22},...,s_{2m}), ..., (s_{n1},s_{n2},...,s_{nm})\}=\{s_1,s_2,...,s_k\}$. We provide a chosen function for this kind of situation. The function $S(w)$ is defined as:

$$S(w) = \alpha \log(fw(w)+1) + \beta \sum_{i=1}^{m} c_i \qquad (1)$$

where α and β are the weighting factors. $fw(w)$ is the relative frequency ratio (RFR) of word w. c_i is the center value of sememe.

Figure 1 showed the instantiation of SSA for key-phrase extraction. As in the untimed case, the algorithm is based on a document set, DocSet, containing all the words of a patent document, and a sememe set, SemSet, containing all the sememes of a word.

3.1 Word-Sememe Conversion

Basically, we have a static corpus that can be preprocessed to convert word to sememe. The corpus includes a vocabulary (covered over 65,000 concepts) and a sememe structure (covered over 1400 sememes).

Firstly, the word segmentation has to be done and the part-of-speech should be tagged because there are no word boundaries in Chinese text. In HowNet, the concept of word or phrase and its description form one entry. Therefore, the coherent sememe of each word can be found with the knowledge dictionary. The word is represented by several sememes and the sememe has various descriptive ways based on the knowledge structure. A "type" parametric should then be defined to describe the sememe type after transferring word into sememe. The word and the sememe may frequently occur during document processing. The more frequently it occurs, the more important is the word or the sememe.

3.2 Relative Frequency Ratio

RFR of word is based upon the fact that the significant words will appear frequently in specific collection of document (treated as foreground corpus) but rarely or even not in other quite different corpus (treated as background corpus). The higher of RFR values of the words, the more informative of the words will be in foreground corpus than in background one.

However, selection of background corpus is an important problem. Degree of difference between foreground and background corpus is rather difficult to measure and it will affect the values of RFR of terms. Commonly, large and general corpus will be treated as background corpus for comparison. In this paper, for our foreground corpus (patent document), we select the PFR People's Daily corpus as compared background corpus. The corpus contains newspaper text from People's Daily.

The relative frequency ratio (RFR) of word w is defined as:

$$fw(w) = \frac{f(d(w))/t_d}{f(PFR(w))/t_{PFR}} \tag{2}$$

where $f(d(w))$ is the frequency at which word w appears in the document, t_d is the total number of words in the document. $f(PRF(w))$ is the frequency at which word w appears in the PFR People's Daily corpus, t_{PFR} is the total number of word in the PFR People's Daily corpus.

3.3 Central Value

The central value is used to review the importance of sememe of word in a document. There are two parameters that influence the central value from different aspects: (i) sememe frequency, and (ii) conclusion degree. The central value is defined as:

$$c(s) = \left[\gamma_1 \cdot \log(fs(s)+1)\right] \cdot \left[\gamma_2 \cdot fc(s) + \gamma_3\right] \tag{3}$$

where $fs(s)$ is the sememe frequency, and $fc(s)$ is the conclusion degree. γ_1, γ_2, and γ_3 are the weighting factors.

We believe that the more words a sememe covers in a document, the more important that the sememe would be. Therefore, the most important factor influencing central value is the sememe frequency. Suppose the set of words which associated with sememe s is $\{w_1, w_2, ..., w_n\}$. The sememe frequency of sememe s is defined as:

$$fs(s) = \sum_{i=1}^{n} f(w_i) \tag{4}$$

where $f(w_i)$ is the frequency at which word w_i appears in the document.

In HowNet, the sememes are hierarchically organized with tree structure. In this tree structure, the upper sememes may cover more concepts, but sometimes too general. Whereas the lower sememes may focus on the detailed concepts, but sometimes too limited. Therefore, we consider that the crucial sememe should have certain ability of conclusion and should locate in the middle of sememe tree structure. The conclusion degree is used to detect conclusion ability of sememe. Let the child nodes set of sememe s be $\{s'_1, s'_2, ..., s'_n\}$, the ancestor nodes set of sememe s be $\{s''_1, s''_2, ..., s''_m\}$. Then the whole nodes set associated with sememe s is $\{s'_1, s'_2, ..., s'_n, s, s''_1, s''_2, ..., s''_m\} = \{s_1, s_2, ..., s_k\}$ $(k=n+m+1)$. The conclusion degree $fc(s)$ reflects the relative importance of sememe s in the tree structure. The higher $fc(s)$ is, the more disperse and equal the content covered, and the more conclusive ability the sememe would have. The conclusion degree $fc(s)$ is defined as:

$$fc(s) = 1 - \frac{fs(s)}{\left[\sum_{i=1}^{n} fs(s'_i)\right] \cdot \left[\sum_{i=1}^{m} fs(s''_i)\right] + Max_{i=1}^{k}(fs(s_i))} \tag{5}$$

where $fs(s)$ is the sememe frequency of sememe s.

3.4 Threshold Value

Formula (1) to (5) can be used to calculate the chosen value of each word in document. The chosen threshold value and selection number should then be set up. If the chosen value of a word is larger than threshold value, the word can be added into the chosen key-phrases set. After calculating all the words in the document, we will chose the largest n (n is the chosen number) chosen value words as key-phrases, if the size of chosen key-phrases is larger than the chosen number. In this study, the threshold value is *0.3* and the chosen number is *5*.

4 Evaluation

We carried out an empirical evaluation of SSA using patent documents from the State Intellectual Property Office (SIPO) of P.R. China. Our goals were to assess SSA's overall effectiveness. As is known, there are no author-assigned key-phrases in patent specification. So the current evaluation of key-phrase extraction from patent document is mainly dependent on manual evaluation. Experts in the relative domain are invited to assign key-phrases for patent document. A total of six evaluators responded to our invitation. The evaluators belonged to professors, researchers, or graduate students who are familiar with patent.

We measured key-phrase quality by quality metrics such as recall and precision. Recall and precision have long been used to assess the quality of literature searches. In this study, recall and precision could be used as indicators of the quality of key-phrase extraction. Precision is the proportion of the returned key-phrases that are correct answer; recall is the proportion of the correct answers that are returned.

5 Experimental Results

We selected randomly 200 patent documents from State Intellectual Property Office (SIPO) of P.R. China as our test data. We extracted 921 phrases after performing the aforementioned procedure. Table 2 shows a simple document and the result of key-phrase extraction.

We compared two key-phrase extraction techniques for Chinese patent mining, conventional term frequency statistics method and proposed sememe statistics method. Experimental results are shown in Table 3. The measure to evaluate the key-phrase extraction is recall and precision. Our algorithm was able to recover 73% of the key-phrases with 81% precision. In comparison, the term frequency statistics algorithm obtained only 62% precision when recall is 72%. From Table 3, it can be said that the sememe statistics using HowNet knowledge in the test data is effective in key-phrase extraction.

Table 2. Result of the algorithm

Test Document	**Title:** 薄膜执行机构			
	Abstract: 本发明涉及一种控制阀的薄膜执行机构，其中齿条设置成可以使齿条/齿轮系统的接触点位于实现运动的压力空间的可移动壁的中心连接直线上，所以薄膜本身或薄膜基板上都没有扭转应力作用，实现往复移动的部件的运动的力的方向基本上和运动路径平行而且相互对称。另外，是齿条和齿轮之间的间隙可以通过带有轴承的支撑辊子被基本消除。根据本发明，提供一种可以轻快移动的执行机构，间隙或在活塞-缸系统内的壁摩擦力引起的延迟作用效果可以基本上被消除掉，这种执行机构可以是双作用类型，也可以是单作用类型。			
	Assigned Key-Phrases: 控制阀, 薄膜, 执行机构, 齿轮, 齿条			
Result Key-Phrases	运动	0.907	齿轮	0.382
	薄膜	0.654	齿条	0.382
	执行机构	0.641		

Table 3. Experimental result for different algorithms

Methods	Recall	Precision
Term Frequency Statistics Algorithm	62%	72%
Sememe Statistics Algorithm (SSA)	73%	81%

6   Conclusions and Perspectives

We propose a simple algorithm for key-phrase extraction, called sememe statistics algorithm (SSA). Our algorithm adopted the natural language understanding approach and performed better than the term frequency statistics algorithm in our experiments. We also showed how HowNet can be used by similarity calculation for key-phrase extraction. Experiments on a collection of patent documents showed that our algorithm significantly improved the quality of the key-phrase extraction.

Frequently used key-phrases in a particular domain such as patent would have the additional advantage. This property makes it easier to categorize patent documents using the key-phrase extraction and may benefit to patent search and patent analysis. Furthermore, the proposed method of key-phrase extraction in this paper can be used not only for patent specification, but also for other documents.

Acknowledgments. This work is supported by the National High-tech R&D Program (863 Program) of P.R. China (Grant No. 2006AA04Z109) and National Natural Science Foundation of P.R. China (Grant Nos. 60674078, 50575031).

References

1. Chien, L.F., Pu, H.T.: Important Issues on Chinese Information Retrieval. Computational Linguistics and Chinese Language Processing 1, 205–221 (1996)
2. Schatz, B., Chen, H.: Digital Libraries: Technological Advancements and Social Impacts. IEEE Computer 2, 45–50 (1999)
3. Chen, H., Houston, A.L., Sewell, R.R., Schatz, B.R.: Internet Browsing and Searching: User Evaluation of Category Map and Concept Space Techniques. Journal of the American Society for Information Science 7, 582–603 (1998)
4. Wang, H., Li, S., Yu, S.: Automatic Keyphrase Extraction from Chinese News Documents. In: Wang, L., Jin, Y. (eds.) FSKD 2005. LNCS (LNAI), vol. 3614, pp. 648–657. Springer, Heidelberg (2005)
5. Freitag, D.: Machine Learning for Information Extraction in Informal Domains. Journal Machine Learning 39, 169–202 (2000)
6. Ong, T.H., Chen, H.: Updateable PAT-Tree Approach to Chinese Key Phrase Extraction using Mutual Information: A Linguistic Foundation for Knowledge Management. In: Proceedings of the Second Asian Digital Library Conference, Taiwan, pp. 63–84 (1999)
7. Dong, Z.D.: Bigger Context and Better Understanding: Expectation on Future MT Technology. In: Proceedings of the International Conference on Machine Translation & Computer Language Information, Beijing, pp. 17–25 (1996)
8. Damerau, F.J.: Generating and Evaluating Domain-Oriented Multi-word Terms from Texts. Information Processing & Management 4, 433–447 (1993)
9. Ji, H., Luo, Z., Wan, M., Gao, X.: Research on Automatic Summarization Based on Concept Counting and Semantic Hierarchy Analysis for English Texts. Journal of Chinese Information Processing 2, 14–20 (2003)

Classification of Business Travelers Using SVMs Combined with Kernel Principal Component Analysis*

Xin Xu[1], Rob Law[2], and Tao Wu[1]

[1] Institute of Automation, National University of Defense Technology,
410073, Changsha, P.R. China
xuxin_mail@263.net
[2] School of Hotel & Tourism Management, The Hong Kong Polytechnic University,
Hung Hom, Kowloon, Hong Kong

Abstract. Data mining techniques for understanding the behavioral and demographic patterns of tourists have received increasing research interests due to the significant economic contributions of the fast growing tourism industry. However, the complexity, noise and nonlinearity in tourism data bring many challenges for existing data mining techniques such as rough sets and neural networks. This paper makes an attempt to develop a data mining approach to tourist expenditure classification based on support vector machines (SVMs) with kernel principal component analysis. Compared with previous methods, the proposed approach not only makes use of the generalization ability of SVMs, which is usually superior to neural networks and rough sets, but also applies a KPCA-based feature extraction method so that the classification accuracy of business travelers can be improved. Utilizing the primary data collected from an Omnibus survey carried out in Hong Kong in late 2005, experimental results showed that the classification accuracy of the SVM model with KPCA is better than other approaches including the previous rough set method and a GA-based selective neural network ensemble method.

1 Introduction

Nowadays, data mining and knowledge discovery techniques have become more and more important for business organizations to remain competitive in the global economy. The knowledge thus obtained, in whatever format, can help decision makers to establish empirical rules of past performance. More importantly, future business performance can be forecasted by utilizing the knowledge found in historical data, and appropriate planning activities can then be carried out. As one of the world's largest and fastest growing industries, the tourism industry particularly needs accurate planning at all levels. Gunn [4] stated the importance of demand-supply match in tourism planning. The planning process can extend from the micro level of internal

* Supported by the School Research Funding of School of Hotel & Tourism Management, The Hong Kong Polytechnic University and partly supported by the National Natural Foundation of China under Grant 60303012.

R. Alhajj et al. (Eds.): ADMA 2007, LNAI 4632, pp. 524–532, 2007.

planning in hotels such as staffing and yield management to macro planning for a destination including transportation infrastructure and theme park development. Accurate planning, which largely relies on the business pattern generated from past performance, can reduce the risks associated with over (or under) supply and thus the associated costs. This, in turn, leads to better and more acceptable returns on investment.

In recent years, various studies have shown the economic contributions of international travelers in general, and business travelers in particular, to a destination [7, 8]. For example, the tourism figures [5, 6] of Hong Kong, a Special Administrative Region of the People's Republic of China (HKSAR), highlight the importance of the business traveler market and its contribution to the economy. As a result, the high-yield business traveler segment has attracted the attention of policy makers, practitioners, and, to a much lesser extent, academic researchers in general and in particular researchers in data mining. Nevertheless, few attempts, if any, have been made to understand the behavioral patterns of business travelers. The statistical data from the Hong Kong Tourism Board (HKTB) only showed the number and percentages of business visitors from different major source markets but very few efforts have been made to mine the profile of these business travelers [6]. In the existing data mining and tourism literature, the study of data mining of business travelers has been almost entirely overlooked by academic researchers. In other words the demographic and trip profiles of this high yield group of visitors remain largely unknown. In a recent work [12], a model based on rough sets was developed to extract useful information and knowledge rules from business travelers to Hong Kong using primary tourist expenditure data collected in a survey. Although the work in [12] provided very promising results for data mining applications in business data analysis, the classification accuracy still needs to be improved. The main difficulties for improving the classification accuracies of tourist expenditure include several aspects. One aspect is the complex and nonlinear relationship between attributes of tourist behaviors such as mode of travel, region of residence, etc. The other aspect is due to unknown noise in the survey data which were collected under different questionnaire situations. Therefore, the classification and prediction analysis of tourist profiles provides new challenges for data mining and machine learning techniques.

To construct classification and regression models with high accuracy and good generalization ability, support vector machines (SVMs), which were originated from the research work on statistical learning theory and kernel learning machines [1, 11], have received lots of research interests in the past decade and there have been many successful applications of SVMs in pattern recognition and data mining problems, where SVM-based models usually obtain state-of-the-art results. However, the applications of SVMs to tourism data analysis have only received very little attention in the literature and previous works mainly focused BP (back-propagation) neural network models [3], and rough sets [12], etc. Recently, SVM-based regression models were studied for tourism demand analysis [2] and very promising results were obtained. However, as a class of popular kernel-based learning machines, SVM models' performance still relies on the selection of feature representations, which is still an open problem in the literature. In this paper, a multi-class SVM model with kernel principal component analysis (KPCA) is studied to solve the problem of tourist expenditure analysis. Compared with previous methods, the approach not only makes use of the generalization ability of SVMs, which is usually superior to BP neural networks and rough sets, but also utilizes the KPCA method [10] to realize nonlinear feature extraction for better performance. Based on the primary data collected from an Omnibus survey carried out in Hong Kong in late 2005, experimental results showed that the classification accuracy of the SVM model with KPCA is better than other approaches including the previous rough set method and a GA-based selective neural network ensemble method.

This paper is organized as follows. In Section 2, a brief introduction on the problem of tourist profile analysis is given. In Section 3, the multi-class SVM algorithm with KPCA for feature extraction is presented in detail. In Section 4, experimental results on a set of survey data carried out in Hong Kong are provided to illustrate the effectiveness of the proposed method. Some conclusions and remarks on future work are given in Section 5.

2   The Problem of Business Traveler Classification

As discussed in the previous section, tourist profile analysis plays an important role in the tourism industry since the business traveler market contributes a lot to the economy. In addition, the data analysis of tourist profiles presents some research challenges for data mining techniques. In this Section, the problem of tourist profile analysis is introduced in general and the task of tourist expenditure classification is discussed in particular. To facilitate discussion, a set of survey data on tourists visiting Hong Kong during a certain period is used as an example. The data were collected in an Omnibus Survey carried out by the School of Hotel & Tourism Management at the Hong Kong Polytechnic University during the period from October 3, 2005 to October 22, 2005. During this period, a total of 1,282 non-transit visitors from seven major tourist generating regions were interviewed face-to-face in the restricted departure lounge of the Hong Kong International Airport. The regions included: Mainland China, Taiwan, Singapore, Malaysia, the United States, Australia, and Western Europe. A quota sampling method was used by a team of nine interviewers. Upon the completion of an interview, a souvenir was presented to the respondent as a token of appreciation for participation.

Following the practice of the Omnibus Survey [9], the questionnaire was developed in English and then translated into Chinese. The final version of the questionnaire was pilot-tested in September 2005 to ensure that questions were clearly understood. The questionnaire consisted of a common set of questions for demographic and trip profile. These demographic and trip profile data were utilized in this research for classification of business travelers. Among the 1,282 respondents, 303 identified themselves as business travelers and provided usable demographic and trip profile data. All attributes (variables) were grouped as condition attributes whereas expenses per night excluding accommodation and airfare were used as the decision attribute. An equal percentile approach was adopted to split the values in the decision attribute into three categories of High (H), Medium (M), and Low (L). For detailed discussion on the profile data, please refer to [12].

Based on the above data, the tourist data analysis problem becomes a pattern classification task. The input attributes may include various behavior attributes of the tourists and the decision variable is the three types of expenditure amounts. The distribution of the data has unknown nonlinearities and there are also potential noises during the data collection process. Thus, the business traveler classification problem presents a new challenge for data mining techniques. In [12], a rough set method was developed to classify the business travelers according to their expenditures, but the overall classification accuracy still needs to be improved.

3   Traveler Classification Using Multi-class SVMs with KPCA

Based on the results in statistical learning theory, SVMs have been developed to construct classifiers or regression models according to the principle of structural risk minimization (SRM) [11], which is superior to the empirical risk minimization (ERM) in most traditional pattern recognition and data mining algorithms such as neural networks, etc. For classification problems, the basic rule of SVMs is to generate a linear hyper-plane to maximize the margin between two classes. However, since in most real applications, the features of two classes cannot be separated linearly, it is necessary to choose an adequate nonlinear feature mapping ϕ to make the data points become linearly separable or mostly linearly separable in a high-dimensional space. Thus, the key problem remained for SVMs is to select appropriate feature mappings for given samples. In order to avoid computing the mapped patterns $\phi(x)$ explicitly, which may usually lead to huge computational costs, the 'kernel trick' was commonly used for SVMs in nonlinear classification problems. Due to the kernel trick, the dot products between mapped patterns are directly available from the kernel function which generates $\phi(x)$. Although kernel functions in SVMs can implement nonlinear feature mappings for pattern classification problems, how to select appropriate kernel functions for given samples is still an open problem in the literature. Therefore, it is necessary to combine SVMs with other feature extraction methods so that the classification performance can be improved for difficult problems such as the traveler classification problem studied in this paper.

In this paper, a multi-class SVM classification method combined with KPCA is employed for the forecasting problem of business traveler expenditures so that the performance of forecasting can be optimized both in terms of accuracy and statistical significance.

3.1   Kernel PCA for Feature Extraction

As an efficient feature selection and dimension reduction method, principal component analysis (PCA) has been widely applied in many pattern recognition and data analysis areas. In PCA, the aim is to extract a set of basis vectors which is, in some sense, adapted to the data by searching for directions where the variance is maximized. Since the vectors found by PCA can be viewed as linear transformations from the original features, in some applications, it may be inadequate to use linear PCA when nonlinear transformations are needed. Therefore, KPCA, which is to implement a linear form of PCA in a kernel-induced nonlinear feature space, has received increasing research interests in recent years.

.

In KPCA, let $\phi(X)$ be the feature vector introduced by a kernel function $k(.\,,.)$, i.e.

$$k(x, y) =< \phi(x), \phi(y) > \tag{1}$$

To simplify the case, we will also suppose that

$$\sum_{i=1}^{n} \phi(x_i) = 0 \tag{2}$$

It is obvious that this condition is easy to be satisfied (Centralizing data can make this equation to be valid). Thus, in the feature space, the covariance matrix will be obtained as follows:

$$\overline{C} = \frac{1}{n} \sum_{i=1}^{n} \phi(x_i) \phi^T(x_i) \tag{3}$$

Let λ_i be the i-th eigen-value of the covariance matrix and V^i be the corresponding eigen-vector, we can have:

$$\lambda_i V^i = \overline{C} V^i \tag{4}$$

Note that V^i can be represented as the combination of $\{\phi(x_i)\}_{i=1,2,...,n}$, i.e.

$$V^i = \sum_{j=1}^{n} a_j^i \phi(x_j) \tag{5}$$

So we have:

$$\lambda_i < \phi(x_j), V^i >=< \phi(x_j), \overline{C} V^i > \tag{6}$$

Thus,

$$\lambda_i \sum_{k=1}^{n} a_k^i <\phi(x_j), \phi(x_k) >= \frac{1}{n} \sum_{k=1}^{n} a_k^i < \phi(x_k), \sum_{j=1}^{n} \phi(x_j) k(x_j, x_k) > \tag{7}$$

$$n\lambda_i K a^i = K^2 a^i \tag{8}$$

where K is the kernel matrix introduced by kernel function $k(.,.)$.

Then, the solution to (4) can be determined by the following equation:

$$n\lambda_i a^j = K a^j \tag{9}$$

If V^i is the corresponding eigen-vector of λ_i, for any vector $\phi(x)$ in the feature space, we can obtain its' projection to V^i as follows:

$$<V^i, \phi(x) >= \sum_{j=1}^{n} a_j^i k(x, x_j) \tag{10}$$

Therefore, the extracted features using KPCA can be obtained by computing the projected values based on the eigen-vectors of m largest eigen-values. However, how to select the appropriate dimensions of projected vectors is determined by different applications. In subsection 3.3, according to the task of business traveler classification, we will discuss the performance criteria for the dimension selection of feature vectors derived by KPCA.

3.2 The Multi-class SVM Algorithm for Classification of Business Travelers

As is well known, SVMs were originally proposed for binary classification problems. Nevertheless, most real world pattern recognition applications are multi-class classification cases. Thus, multi-class SVM algorithms have received much attention over the last decade and several decomposition-based approaches were proposed [13]. There are several strategies for the implementation of multi-class SVMs using binary algorithms, which include one-vs-all, one-vs-one, and error correcting output coding [13], etc. To solve the analysis problem of tourist expenditure, we will use the one-vs-all strategy for the multi-class SVMs, which has been studied in our previous work and applied successfully to classification problems of network information [14]. More importantly, the multi-class SVMs will be integrated with the KPCA method for feature extraction so that better generalization ability can be obtained.

For multi-class SVMs based on the one-vs-all strategy, the classification problem is transformed to m binary sub-problems, where m is the total number of classes. Every binary classifier separates the data of one class from the data of all the other classes. In the training of binary SVM classifiers, an optimal hyperplane is considered to separate two classes of samples. Based on the SRM principle in statistical learning theory and after using the kernel trick as in KPCA [11], the optimal separating hyperplane can be constructed by the following optimization problem

$$\max_{\alpha} \sum_{i=1}^{N} \alpha_i - \frac{1}{2} \sum_{i,j=1}^{N} \alpha_i \alpha_j y_i y_j k(\vec{x}_i \cdot \vec{x}_j) \tag{11}$$

subject to

$$0 \le \alpha_i \le C, \ i = 1,2,..., N \text{ and } \sum_{i=1}^{N} \alpha_i y_i = 0 \tag{12}$$

To solve the above quadratic programming problem, various decomposition-based fast algorithms have been proposed, such as the SMO algorithm [15], etc. For details on the algorithmic implementation of SVMs, please refer to [15].

After constructing the binary SVM classifiers, the decision function of each binary SVM is

$$f_k(\vec{x}) = \text{sgn}(\sum_{i=1}^{N} \alpha_{ki} y_{ki} k(\vec{x}_{ki}, \vec{x}) + b_k) \quad k = 1,2,...,m \tag{13}$$

where $f_k(\vec{x})$ is the decision function of classifier k and ($\vec{x}_{ki}$, y_{ki}) (k=1,2,...,m) are corresponding training samples. The output of the whole multi-class SVMs can be determined by the decision outputs of every binary classifier, where various voting strategies can be used.

3.3 Performance Criteria for Data Analysis of Business Travelers

As a typical application in social science, the data analysis of business travelers has more performance criteria than conventional classification and regression problems. In addition to the classification or regression precision, the statistical coincidence between the estimated values and the real values is another performance criterion, which can be computed by various statistical testing tools such as the Wilcoxon Signed Rank Test, t-Test, Sign Test, etc. Among these tools, the Wilcoxon Signed Rank Test has been considered to be one of the most efficient non-parametric testing methods for non-Gaussian data. By making use of the Wilcoxon Signed Rank test, two possibly related sets of sample data can be tested and we can make conclusions on whether there is any significant difference between the two data sets.

In our studies, we use both classification accuracies and the p-values in the Wilcoxon Signed Rank test for performance evaluations of different data mining and pattern recognition methods. Based on the above performance criteria, parameters of data mining methods can be selected and optimized.

4 Experimental Results

In this section, experimental results on the expenditure classification problem of business travelers are provided to compare the performance of different classification methods. The data set was constructed by randomly selecting 80% (N=243) of all the cases for model calibration and the remaining ones (N=60) for model testing. In addition to the model testing of the proposed multi-class SVMs combined with KPCA, we also tested other four popular data mining methods including rough sets, ensemble neural networks using genetic algorithms (GAs) [15], SVMs without feature extraction, and SVMs using PCA. The classification approach using rough sets was proposed for the data analysis problem of business travelers in [12], where the accuracy of the rough sets classification was determined in two quality terms of percentage of successfully classified cases (i.e., with a decision) and percentage of correctly classified cases (the estimated value matches the actual value). Although the rough set method can derive symbolic rules from the data and the Wilcoxon Signed Rank test showed that there is no statistical difference between estimated values and the real values, it can only classify 82% of the testing cases and 41% of these classified cases were correctly estimated. So the overall classification accuracy is 33%.

In the experiments, parameters of different data mining methods were all selected according to the performance criteria discussed above. For the multi-class SVMs with KPCA, since radius basis function (RBF) kernel functions were used, there are two main parameters to be selected, which are the width parameter of the RBF kernel function and the dimension of features projected by the KPCA method. After manual selection and optimization, the width parameters for RBF kernels are all set to 10 for all the methods. For KPCA and PCA, the dimension of extracted features was set to 6. The population size for GA selected neural network ensemble was 40.

Experimental results on performance comparisons of different methods are shown in Table 1 and Table 2. The parameters were selected as above and similar results were also obtained for other parameter settings. In Table 1, the testing accuracies of the five data mining methods are listed. It is illustrated that the proposed multi-class SVMs with KPCA has the highest test precision among all the methods. In Table 2, the results of Wilcoxon Signed Rank tests also show that the estimated values of SVMs combined with KPCA also have the highest probability (0.835) for coming from the same distribution of the actual data. While the SVMs with PCA for feature extraction have the lowest probability (pr=0.011<0.05) and it can be concluded that there are differences between the real values and estimated values. From the experimental results, it is shown that the multi-class SVMs combined with KPCA have the best performance than other methods both in terms of testing accuracy and statistical coincidence.

Table 1. Performance comparison among different classifiers

Classification methods	Test precision
Rough sets [12]	33%
GA selected NN ensemble	52.7%
SVMs with original features	51.4%
SVMs with PCA	56.6%
SVMs with KPCA	**58.4%**

Table 2. Wilcoxon Signed Ranks Test

	Estimated - Actual		
	SVM+KPCA	Rough sets	SVM+PCA
p-value 1	**-0.209^a**	-0.807^a	-2.554 ^{a}
pr (2-tailed)	**0.835**	0.420	0.011

a. Based on negative ranks.

5 Conclusions

In view of the challenges that the Hong Kong inbound business traveling sector is facing, there is an urgent need for policy makers and practitioners to better understand the demand for business visitor arrivals, and to carry out more accurate planning at strategic, tactical, and operational levels. Due to the complex relationships and sampling noises, the analysis of tourism data presents many research challenges for data mining methods. In this paper, we propose a new way to the data analysis of business travelers, which is based multi-class SVMs combined with KPCA for feature extraction. The performance criteria and parameter selection for data analysis were also discussed. The experimental results showed that the proposed method not only has better classification accuracy but also has advantages in statistical coincidence between estimated and real values. Future work may include developing automatic parameter selection and optimization methods for the multi-class SVMs with KPCA.

References

1. Schölkopf, B., Smola, A.: Learning With Kernels. MIT Press, Cambridge, MA (2002)
2. Pai, P.-F., Hong, W.-C.: An Improved Neural Network model in Forecasting Arrivals. Annals of Tourism Research 32(4), 1138–1141 (2005)
3. Chena, K.-Y., Wang, C.-H.: Support vector regression with genetic algorithms in forecasting tourism demand, Tourism Management (2006)
4. Gunn, C.A.: Tourism Planning. Routledge, London (2002)
5. Hong Kong Tourism Board: Statistics on Conventions & Exhibitions 2004. Accessed on February 3, 2006. (2005a) Available online at http://partnernet.hktourismboard.com/
6. Hong Kong Tourism Board: Visitor Profile Report 2004. Accessed on February 3, 2006. (2005b) Available online at http://partnernet.hktourismboard.com/
7. Braun, B.M., Rungeling, B.: The relative economic impact of convention and tourist on a regional economy: a case study. International Journal Hospitality Management 11(1), 65–71 (1992)
8. Lawson, F.R.: Trends in business tourism management. Tourism Management 3(4), 298–302 (1982)
9. Hui, E.L.L., McKercher.: Operational Issues in Marketing Research: An Example of the Omnibus Tourism Survey. Pacific Tourism Review 5(1/2), 5–13 (2001)
10. Scholkopf, B. et al.: Input space vs. feature space in kernel-based methods. IEEE Transctions on Neural networks (1999)
11. Vapnik, V.: The nature of statistical learning theory. Springer, New York (1995)
12. Law, R., Bauer, T., Weber, K., Tse, T.: Towards a Rough Classification of Business Travelers. In: Li, X., Zaïane, O.R., Li, Z. (eds.) ADMA 2006. LNCS (LNAI), vol. 4093, Springer, Heidelberg (2006)
13. Dietterich, T.G., Bakiri, G.: Solving multiclass learning problems via error-correcting output codes. Journal of Artificial Intelligence Research 2, 263–286 (1995)
14. Xu, X., Wang, X.N.: Adaptive network intrusion detection method based on PCA and support vector machines. In: Li, X., Wang, S., Dong, Z.Y. (eds.) ADMA 2005. LNCS (LNAI), vol. 3584, pp. 696–703. Springer, Heidelberg (2005)
15. Zhou, Z.-H., Wu, J., Tang, W.: Ensembling neural networks: many could be better than all. Artificial Intelligence 137(1-2), 239–263 (2002)

Research on the Traffic Matrix Based on Sampling Model

Fengjun Shang

College of Computer Science and Technology, Chongqing University of Posts and
Telecommunications, Chongqing 400065, China
shangfj@cqupt.edu.cn

Abstract. Traffic matrix information is very important to networks. In this paper, a traffic matrix model is proposed based on passive measurement that can
be used to high-speed IP network. The core of model has three parts as follows:
1) measuring traffic at the edge node of network. The passive measurement
method is introduced to measure the node traffic based on software measurement. Because the software is based on flow measurement, the flow matching,
that is, packet classification is a key problem. In packet classification, the dual
hash algorithm is proposed. The algorithm is introduced based on the non-
collision hash and XOR hash. 2) introducing non-intrusive measurement
method to acquire path information and then the sampling method is introduced. In this method, the path information is writen in the flag field. 3) deduc-
ing sampling probability so that the point of optimization is selected. Simulation results prove the effectiveness of this algorithm.

Keywords: non-collision hash; sampling; traffic matrix.

1 Introduction

Many decisions that IP network operators make depend on how the traffic flows in
their network. *Traffic Matrix* (TM) reflects the volume of traffic that flows between
all possible pairs of origins and destinations in a network. Many of the decisions that
IP network operators make depend on how the traffic flows in their network. A traffic
matrix describes the amount of data traffic transmitted between every pair of ingress
and egress points in a network. When used together with routing information, the
traffic matrix gives the network operator valuable information about the current network state and is instrumental in traffic engineering, network management and provi-
sioning [1].

Traffic matrix is important for many network design, engineering, and management functions. However they are often difficult to measure directly. Because networks are dynamic, analysis tools must be adaptive and computationally light weight.

The traffic model mainly depict the traffic distribution in the network, in this paper, we use traffic matrix to depict the traffic distribution. Traffic matrix includes two
parts: *sd*(source-destination) pairs traffic and *sd* pair. We introduce interrelated traffic
model as follows.

R. Alhajj et al. (Eds.): ADMA 2007, LNAI 4632, pp. 533–544, 2007.
© Springer-Verlag Berlin Heidelberg 2007

The origin-destination estimation problem for telephone traffic is a well-studied problem in the telecom world. For instance, already in 1937, Kruithof suggested a method for estimation of point-to-point traffic demands in a telephone network based on a prior traffic matrix and measurements of incoming and outgoing traffic. The literature [2] uses active method to measure LSP and update by delay and the dropping rate of packet. The literature [3] synthesizes user connect information, SLA, traffic forecasting, historical data, network topology etc. to estimate traffic matrix.

Estimating a traffic matrix from aggregated traffic data such as link byte counts is an inverse problem that resembles tomography in medical imaging, so it was named *network tomography* by Vardi [4]. Vardi assumes a Poisson model for the traffic demands and covariances of the link loads is used as additional constraints. The traffic demands are estimated by Maximum Likelihood estimation.

Vanderebei and Iannone (1994) assumed independent Poisson traffic counts for the entries of the *sd* matrix, developed three equivalent formulations of the *EM* (Expectation Maximization) algorithm and studied the fixed points of the *EM* operator. Cao et al. (2000) assumed independent Gaussian *sd* traffic flows, and used *EM* algorithm [5] to derive estimates for the parameters, that depended on *t*. The Poisson model is also used by Tebaldi and West [6], they use a Bayesian approach. Since posterior distributions are hard to calculate, the authors use a Markov Chain Monte Carlo simulation to simulate the posterior distribution. The Bayesian approach is refined by Vaton et al. [7], who propose an iterative method to improve the prior distribution of the traffic matrix elements. An evaluation of the methods in [6] together with a linear programming model is performed by Medina et al. [8]. The gravity model is introduced by Zhang et al. [9]. In its simplest form the gravity model assumes a proportionality relation between the traffic entering the network at node *i* and destined to node *j* and the total amount of traffic entering at node *i* and the total amount of traffic leaving the network at node *j*. In a paper by Nucci et al. [10] the routing is changed and shifting of link load is used to infer the traffic demands. Feldmann et al. [11] uses a somewhat different approach to calculate the traffic demands. Instead of estimating from link counts they collect flow measurements from routers using Cisco's NetFlow tool and derive point-to-multipoint traffic demands using routing information from inter- and intradomain routing protocols.

In the traffic mapping, the most measurement systems based on flow, [12][13] works on network boundary. The traffic matrix can be made up according to the measuring result of boundary and synthesizing user information, network topology, forecasting information of traffic demand.

Direct model need additional hardware in router. In [14], traffic matrix is gotten by sampling method, which adopt hash function and acquire measurement packet. No information about network topology and route is needed in this method. Shortcoming of model need acquire right hash sampling function, but this can't apply in the reality.

Traffic measurement, especially traffic matrix measurement is the most basic demand [15]. In this paper, we study acquiring traffic matrix by passive measurement. Summing up, when measuring traffic matrix in an AS system, we may measure synthetical traffic at the network edge. At the same times, we also know how to transmit traffic. In traditional measurement, it is difficult to get traffic matrix, but it is feasible to measure traffic matrix by our sampling method.

In the paper, we have four contributions as follows:

1. We propose a dual-hash algorithm for packet classification so that the model may measure Gbit node traffic on network edge node.
2. We introduce edge measurement model so that measurement structure is convenient to software realization.
3. In route probing, we introduce sampling method based on packet content so that simulation and realization are simple.
4. We design a non-intrusive model. In this model, the additive traffic isn't introduced so as to reduce the effect on the network and then we verify the availability of system

In this paper, we present a dynamic traffic matrix measurement model based on the sampling method.

This paper is organized as follows. We discuss the Problem Descriptions in Section 2. In Section 3, the Passive Measurement Model is introduced. In Section 4, the Sampling based on Packet Content is introduced. We present performance evaluation and conclusion in Section 5. The Conclusions are introduced in Section 6.

2 Problem Descriptions

Our model measures every path at the edge, and then uses measurement result to compute traffic matrix. Firstly, route matrix is measured only in the boundary and not involving network nuclear. Secondly, we measure edge node traffic. At last, we may compute sd node pairs traffic by route matrix measured. We use non-intrusive measurement method to acquire route information.

The measurement period T *is* logically selected, for example 5 minutes, the active path numbers is few among edge nodes. Measurement model may measure path and traffic by means of non-intrusive method, so element numbers of path aggregate and measurement result aggregate is limited and few. The mathematics formula is as follows, where P_l is a path and Φ_l is the path traffic.

$$y_{ij} = \begin{cases} y_{ij}, e_{ij} \not\subset P_l \\ y_{ij} + \Phi_l, e_{ij} \subseteq P_l \end{cases}$$

Where y_{ij} is path traffic.

In this paper, we proposed a novel traffic computation method by collecting fine-grain packet or flow-level measurement at the edge of the network. In measuring sd traffic, the packet classification algorithm is very important to classify packet for special sd pairs, so we introduce the dual hash algorithm as follows. We use non-intrusive measurement to acquire the route probability.

3 Passive Measurement Model

The model structure includes two parts: *EMU* (edge measure unit) and *NCU* (network convergence unit). Figure 1 is shown the measurement model structure.

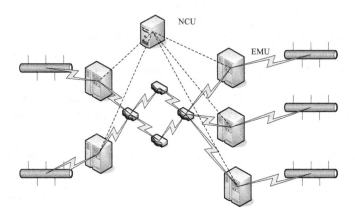

Fig. 1. Measurement Model Structure

EMU lies in edge router and includes traffic measurement and route probability matrix probing. EMU measures the route matrix and path and transmit result to *NCU* . *NCU* is a server or special processor, which collects and save *EMU* measurement result, and then computing traffic matrix. According to different system need, it may transmit result to all *EMU* or make result use network computing, for example, restriction route, load balance, network plan or short period TE traffic controlling.

3.1 Edge Node Traffic Measurement

We introduce passive measurement method to measure the node traffic. Because using the hardware measurement techniques outlined is expensive and complex, we use software measurement. The monitor software has four modules. They are network interface module, classification module, flow record collecting and storage module, analyzing and data mining module.

A careful study of the use of software measurement techniques shows that it is possible to significantly improve them. However, the process is far from trivial. Most of our work has taken place in the context of the Linux operating system where the source of the code is available. Without this it would be impossible to understand the operating system effects introduced into the measurement process let alone do anything about them. (Incidentally, Linux is fundamentally no better suited to measurement than any other operating system, but its transparency and modifiability make it by far the best choice if serious work is to be done).

Our idea is that our exploiting measurement software acting as a most useful part of operation system so that it may avoid to switch kernel to application software to improve measuring speed.

Creating a Minimal Kernel. This Linux kernel is still a fully functional operating system at this point, stripped only of unused code for a given machine. The next step is to further remove services and functionality unused by the monitor that would reduce performance.

Now we are left with an absolutely minimal kernel to manage low-level hardware tasks and provide a useful Application Programming Interface (API), which we use to actually program the monitor.

Architecture of the Monitor. Our measurement system includes foreground system and background system. The foreground system has four modules. They are network interface module, classification module, flow record collecting and storage module and analyze and mine data module. Classification module is used as packet classification.

Network interface module, classification module and flow record collecting and storage module are executed one server. At the same times, analysis module is executed at another server.

Network Interface Module. Network interface module is executed, it obtains the buffer in memory that buffer the burst traffic in link and pass the address of the buffer to the customed driver of NIC (network interface card) make NIC to deposit captured packets at this buffer. A customed driver labels timestamp for packet and transfers the packet information from NIC to memory accord to defined format beforehand.

Classification Module. This information is in our opinion the minimal but most useful subset of data available. As above, five domains are TCP or UDP and IP Addresses of Source and Destination and Ports of Source and Destination define a flow.

How to look up a specific record in so many flow records maintaining in memory is a trouble. How to match the packets to a specific flow is one of key issues in the specific flows-based measurement system, because the performance of the flow matching algorithm remarkably affects the results of the traffic measurement in high-speed network. This paper presents a method for this purpose, namely non-collision hash and XOR hash algorithm.

In this article, the authors survey the recent advances in the research of IP classification and introduce some of the typical algorithms. At last, a novel IP classification is proposed based the non-collision hash and XOR hash algorithm, which is based on non-collision hash Trie-tree Algorithm and XOR hash algorithm. The core of algorithm has four parts: The algorithm scheme has four parts: 1) structuring the non-collision hash function, which is constructed mainly based on destinationsource port and protocol type field so that the hash function usually can avoid space explosion problem; 2) introducing the XOR hash algorithm. At first, we concatenate the source/destination IP pairs, and then divide the packet header into 4 chunks and each chunk has 16bits. Then one of chunks is mapped into stochastic space, and is applied XOR operation on the rest 3 chunks and mapped random number. Because the stochastic space is even distribution after XOR operation, its collision is limitary;

3) introducing multibit trie-tree based key value of XOR hash in order to reduce time complexity; 4) lookuping every rule index in order to ensure the validity that we get the final rule index. It decreases the time complexity of algorithm after expanding normally because we introduce the hash algorithm. Space complexity consumed and space requirement of this algorithm are limitary. The test results show that the classification rate of dual-hash algorithm is up to 2 million packets per second and the maximum memory consumed is 6MB for 10,000 rules.

In storing traffic, we use SRAM chip to store traffic, and then unload to hard disk timely. We introduce software method to realize the node traffic measurement model based on edge survey method.

Flow Record Collecting and Storage Module. Flow record collecting and storage module has two important functions. Firstly it collects the overtime records and inactive records, overtime record is no packet arrival in destine interval and inactive record has set 1 of the FIN domain of the packet belongs to the flow record, to release the memory. Secondly it stores the overtime records and inactive records. If flow record collecting and storage module is connecting with analyze and mine data module by the network, it transfers the overtime and inactive records to analyze and mine data module real-time by IPFIX protocol. If not, it stores them to local-storage at hard disk, on condition that connection gets, then it read them from hard disk and transfer them to analyze and mine data module.

3.2 Route Measurement

After acquiring node traffic, we must know traffic how to pass network. In this paper, we introduce non-intrusive measurement method to acquire route information, because we use IP identification field to store route information.

As we reuse of the IP identification field, we must address issues of backwards-compatibility for IP fragment traffic. Ultimately, there is no perfect solution to this problem and we are forced to make compromises that disadvantage fragmented traffic. Fortunately, recent measurements suggest that less than 0.25% of packets are fragmented. Moreover, it has long been understood that network-layer fragmentation is detrimental to end-to-end performance so modern network stacks implement automatic MTU discovery to prevent fragmentation regardless of the underlying media. Consequently, we believe that our encoding will interoperate seamlessly with existing protocol implementations in the vast majority of cases.

In the basic approach, we use this approach where we use an explicit *flag* field to indicate which hash function the router has used for the marking. In particular, we divide the overloaded IP Identification field into a w-bit flag field, *fID*, a $(11-w)$ bits edge field, and a 5-bit distance field. Note that 5 bits can represent 32 hops which is sufficient for almost all Internet paths. Figure 2 shows an example of this approach for $w=3$. With a given fID, the encoding of a router R_i is simply $h(<fID, R_i>)$. Thus different fIDs indicate different independent hash functions. When a router R_i decides to mark a packet, it chooses a random number x of w bits and write it in the flag field and use $g(<x, R_i>)$ as its IP address encoding, as figure 2 shows [16].

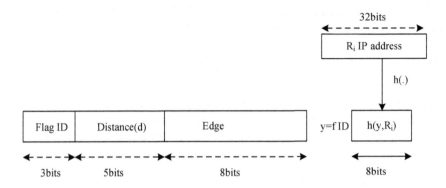

Fig. 2. Encoding in Non-intrusive Measurement

Marking. Note that we actually use one independent hash functions, h, in the encoding of the routers' IP addresses. h has 8-bit outputs. In network entrance, the router marks a packet with a probability q when forwarding the packet. Moreover, it writes $h(y, R_i)$ into the edge field and 1 into the distance field in packet P. Otherwise, the router always makes the distance field be 0 if it decides not to mark the packet. In inner network, if the distance field is nonzero which implies its previous router has marked the packet, it XORs $h(y, R_i)$ with the edge field value and overwrites the edge field with the result of the XOR. The router always increments the distance field if it decides to mark the packet. The XOR of two neighboring routers encode the edge between the two routers of the upstream router map. The edge field of the marking will contain the XOR result of two neighboring routers, except for samples from routers one hop away from the destination. Because $a \oplus b \oplus a = b$, we could start from markings from the routers one hop away from the destination, and then hop-by-hop, decode the path.

The Marking Scheme in the case of $w=3$ at router R_i
for sampled packet P
let x be a random integer from [0,4)
P.fID $\leftarrow$ x
P.distance $\leftarrow$ 1
P.edge $\leftarrow$ g($\langle$x, Ri$\rangle$)
if (P.distance $\neq$ 0) then
P.edge $\leftarrow$ P.edge $\oplus$ g($\langle$P.fID,Ri$\rangle$)
P.distance $\leftarrow$ P.distance + 1

Reconstruction the path
Let destination/source address and fID are same for packet P,
Decompose P.distance for router IP
output path

We explain our idea by using Figure 3. A packet from SubNet2 to SubNet4 has several paths. Firstly, we map entrance router IP into 8bits shown in Figure 2. Secondly, the IP identification field will store the value of XORing each 8bits of passed route IP. If there is collision, we may select the mapped some router IP low 8bits in order to reduce collision rate so that the XORed value is non-collision. We take SubNet2 to SubNet4 for an example and use formula to explain the process. We only consider same distance field, because different distance is direct distinguished.

From SubNet2 to SubNet4, the path aggregate includes:

{Router2 → Router1 → Router4, Router2 → Router5 → Router4, Router2 → Router3 → Router4}

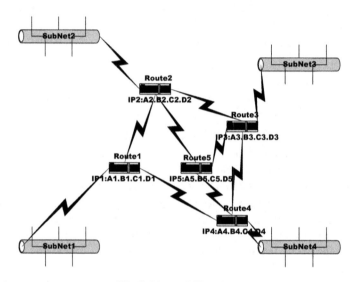

Fig. 3. Network Topology

Reconstruction.. Intuitively, in order to reconstruct the pass paths, we may know the path from the 8bits XORed result. When XORing operation, it has collision, so we must dispose collision. Our idea is that 8bits is mapped a stochastic space and then the mapped number XORed other router IP. The analyse is as follows.

Theorem 1. Random number $x \in [0, m]$ follows even distribution, that is, $x \sim \pi[0, m]$.

Theorem 2. After random number XORing random number, the result of XOR operation is random number.

Proof: Given $b \in [0, m]$ is random number, $x \sim \pi[0, m]$.

$$\therefore \forall x, P(x) = \frac{1}{m}$$

$$\because \forall x_1, x_2 \quad and \quad x_1 \neq x_2 \quad \therefore x_1 \oplus x_2 \neq 0$$

$$\because \forall y_1, y_2 \quad y_1 = x_1 \oplus b, y_2 = x_2 \oplus b$$

$$\therefore y_1 \oplus y_2 = (x_1 \oplus b) \oplus (x_2 \oplus b) = x_1 \oplus x_2 \neq 0$$

That is, $y_1 \neq y_2$, so the relation between y_1 and y_2 is one to one correspondence.

$$\therefore P(y) = P(x) = \frac{1}{m}, that\ is\ ,\ y \sim \pi[0,m]$$

3.3 Passive Measurement Algorithm Analysis

Let network have n nodes, so sd pairs have $c = n \times (n-1)$ and $L \leq c$, and then it is $O(rc)$ under the worst instance, so algorithm complexity is $O(rn^2)$. In real network, the route probability, where $a_{ij} = 0$ is numerous, so algorithm complexity decline and space complexity is $O(rn^2)$.

4 Sampling Based on Packet Content

In order to reduce the sampling packet affecting the network running, we introduce sampling method. The sampling measurement on high-speed IP traffic is aimed at selecting partly traffic to estimate the total route information. A kind of sampling method is that the sampling event is generated by sampling model randomly, but Poisson random sampling randomly generates packet number or time. Before a packet arrives, we have known whether the packet will be sampling. The other sampling measurement method that has been studied in this paper is based on the sampling event generated by specific sampling model, yet the static attributes of sampling event are random statistically. After a packet arrives, according to its content, we can know whether the packet is sampled.

We regard the ordered bits of a packet x and of its invariant part $\phi(x)$ as binary integers and $\phi(m)$ as mask value. We use the sampling hash

$$h(\phi(x)) = \begin{cases} 1 & if \quad \phi(x) = \phi(m) \\ 0 & otherwise \end{cases} \tag{1}$$

When $h(\phi(x)) = 1$, the packet is sampled. In theory, sampling length decides sampling ratio. If mask length is m bits, there are $M = 2^m$ mask value, so in theory, packet sampling ratio $p = 1/M$. Considering the stability character, 16bits in identification field are chosen as random sampling matching bits.

Definition 1. Given network, the total traffic is n form all edge node to network, and total sampling packet traffic is n_c.

If the probe probability is p,

$$p = \frac{n_c}{n} \tag{2}$$

The measurement error includes two part: 1) sampling packet effect on router error shown σ_p; 2) route probing import error shown σ_m.

So, the total measurement error $\sigma = \sigma_p + \sigma_m$.

Given k_{th} period, s send sd node pairs i route probe packet $n_i^{(k)}$, where transiting link j probing packet is $n_{ij}^{(k)}$, that is

$$p_{ij}^{(k)} = \frac{n_{ij}^{(k)}}{n_i^{(k)}} \qquad (3)$$

Toward all route probe packet, p_{ij} follows Bernoulli distribution.

So the standard error of $p_{ij}^{(k)}$

$$p_{ij}^{(k)} = \sqrt{\frac{p_{ij}^{(k)}(1 - p_{ij}^{(k)})}{n_i^{(k)}}} \qquad (4)$$

From formula (2), (3) and (4), we have

$$\sigma_m = \frac{1}{\sqrt{np}} \sqrt{\frac{n_i}{n}(1 - \frac{n_i}{n})} \qquad (5)$$

If the background traffic of link e_i is n_i and the probe probability is p, the effect on router error σ_p is

$$\sigma_p = \frac{n \cdot p}{n_i} \qquad (6)$$

In test network, $n = 5,000$ (packet), $n_i = 1,000$ (packet), we may know relation between σ_m and σ_p shown in Figure 4. From Figure 4, we know the point of optimization is about 0.01, so the sampling probability p is 0.01.

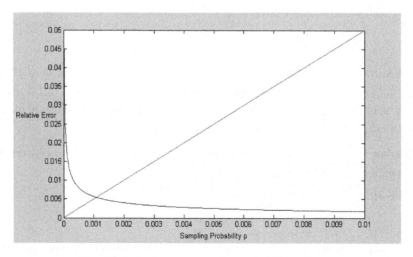

Fig. 4. Relation between sampling probability and error

5 Performance

In this paper, we use sampling measurement method to acquire *sd* pairs traffic. At the same times, we acquire route by non-intrusive measurement method, so we may acquire the traffic matrix. Owning to the length of paper, we only list the link measurement results from route1 to route2 in light load and the simulating results are shown in figure 5. In figure 5, the X axis is time and unit is minute, the Y axis is traffic and unit is packets/second, where RealTraffic is real traffic value and SampTraffic is traffic value using non-intrusive measurement method.

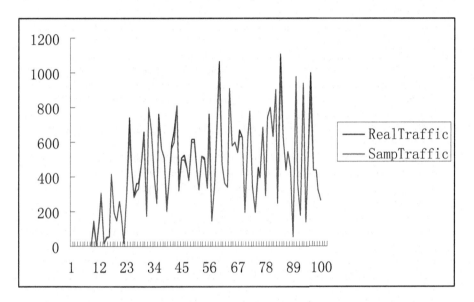

Fig. 5. Comparing traffic between real traffic and passive model measurement

6 Conclusions

Traffic matrix information is very important to networks. In this paper, a traffic matrix model is proposed based on passive measurement that can be used to high-speed IP network. Firstly, the traffic is measured at the edge node of network. We introduce passive measurement method to measure the node traffic. Because using the hardware measurement techniques outlined is expensive and complex, we use software measurement. Because our software is based on flow measurement, the flow matching, that is, packet classification is a key problem. In packet classification, we introduce the dual hash algorithm. We introduce the algorithm based on the absolute Non-Collision Hash and XOR hash. Secondly, the sampling method is introduced to acquire route probability and then the *sd* traffic is collected. Thirdly, the sampling probability is deduced so that the point of optimization is selected. Simulation results prove the effectiveness of this algorithm.

Acknowledgement

The author would like to thank the Doctoral Research Fund of Chongqing University of Posts and Telecommunications (A2006-08).

References

1. Gunnar, A., Johansson, M., Telkamp, T.: Traffic Matrix Estimation on a Large IP Backbone - A Comparison on Real Data. In: Internet Measurement Conference 2004, Taormina, Italy (October 2004)
2. Paxson, V., Almes, G., Mahdavi, J., Mathis, M.: Framework for IP performance metrics. IETF RFC 2330 (1998)
3. Cozzani, I., Giordano, S.: A passive test and measurement system: traffic sampling for QoS evaluation. In: IEEE Communications Society (ed.) GLOBECOM'98. Proceedings of the Global Telecommunications Conference, pp. 1236–1241. IEEE Press, Sydney (1998)
4. Awduche, D.O., et al.: MPLS and Traffic Engineering in IP Networks. In: IEEE Communications Magazine (December 1999)
5. Elwalid, A., et al.: MATE: MPLS Adaptive Traffic Engineering. In: Proceedings of INFOCOM'2001 (April 2001)
6. Tebaldi, C., West, M.: Bayesian inference on network traffic using link count data. Journal of the American Statistical Association 93(442), 557–576 (1998)
7. Vaton, S., Gravey, A.: Network tomography: an iterative bayesian analysis. In: Proc. ITC 18, Berlin, Germany (August 2003)
8. Medina, A., Taft, N., Salamatian, K., Bhattacharyya, S., Diot, C.: Traffic matrix estimation: Existing techniques and new directions. In: Proc. ACM SIGCOMM, Pittsburg (August 2002)
9. Zhang, Y., Roughan, M., Duffield, N., Greenberg, A.: Fast accurate computation of large-scale IP traffic matrices from link loads. In: Proc. ACM Sigmetrics, San Diego, CA (June 2003)
10. Nucci, A., Cruz, R., Taft, N., Diot, C.: Design of IGP link weight changes for estimation of traffic matrices. In: Proc. IEEE INFOCOM, Hong Kong, March 2004,
11. Feldmann, A., Greenberg, A., Lund, C., Reingold, N., Rexford, J., True, F.: Deriving traffic demands for operational IP networks: Methodology and experience. In: Proc. ACM SIGCOMM, Stockholm, Sweden (August 2000)
12. Trimintzios, P., et al.: A Management and Control Architecture for Providing IP Differentiated Services in MPLS-Based Networks. In: IEEE Communications Magazine (May 2001)
13. Vardi, Y.: Network tomography: Estimation source - destination traffic intensities from link data. Journal of the American Statistical Association (1996)
14. Duffield, N.G., et al.: Trajectory Sampling for Direct Traffic Observation. ACM Computer Communication Review 30(4) (2000)
15. Feldmann, A., et al.: NetScope: Traffic Engineering for IP Networks. IEEE Network (March/April 2000)
16. Dawn, X.S, Perrig, A.: Advanced and authenticated marking schemes for IP traceback [A]. In: [s.l.] Proceedings of Twentieth Annual Joint Conference of the IEEE Computer and Communications Societies [C], Alaska, pp. 878–886. IEEE Press, New Jersey (2001)

A Causal Analysis for the Expenditure Data of Business Travelers

Rob Law[1] and Gang Li[2]

[1] School of Hotel and Tourism Management, Hong Kong Polytechnic University,
Hong Kong
[2] School of Engineering and Information Technology, Deakin University,
221 Burwood Highway, Vic 3125, Australia
hmroblaw@polyu.edu.hk, gang.li@deakin.edu.au

Abstract. Determining the causal relation among attributes in a domain is a key task in the data mining and knowledge discovery. In this paper, we applied a causal discovery algorithm to the business traveler expenditure survey data [1]. A general class of causal models is adopted in this paper to discover the causal relationship among continuous and discrete variables. All those factors which have direct effect on the expense pattern of travelers could be detected. Our discovery results reinforced some conclusions of the rough set analysis and found some new conclusions which might significantly improve the understanding of expenditure behaviors of the business traveler.

1 Introduction

The process of information collection and comprehension from raw data is essential for a business to remain competitive. The useful information thus obtained, in whatever format, would assist decision makers to understand previous business performance. More importantly, decision makers can utilize the information to formulate future business strategy and predict future business performance. Such a process naturally applies to tourism, one of the fastest growing industries in the world. The existing literature has extensively documented the significant contributions of tourism receipts to a local economy [2,3]. Specifically, the expenditures from tourists can positively contribute to growth of local market businesses in accommodation, dining, retailing, transportation, and entertainment.

In view of the commonly agreed economic contributions that the tourism industry generates, researchers have long been advocating the need to examine tourism expenditures and to understand tourists' spending behavior [4,5,6]. Additionally, among different sectors in the tourism industry, business travel is particularly appealing not only in terms of volume. It also represents one of the highest spending groups among all categories of tourists [7,8]. Hence, the effort of applying innovative scientific techniques to raw tourism data would, undoubtedly, be beneficial to tourism practitioners and policy makers. Despite the emerging need for comprehending the increasing amount of raw tourism

R. Alhajj et al. (Eds.): ADMA 2007, LNAI 4632, pp. 545–552, 2007.

data, in general and particularly in business travel, the existing literature only has a limited number of prior studies that applied data mining technique to investigate the casual relationship between tourist expenditure and its associated variables. To fill in this gap in data mining, this research makes an attempt that incorporates a causal discovery algorithm into a data set of business travelers. Research findings would be of use to both practitioners and academics in the field of data mining applications in business.

In [1], Law *et al.* reported a survey of non-transit visitors from seven major tourist generating regions, and analyzed the survey data to build a prediction model for traveler expenses. In their paper, Law *et al.* used the rough set theory on the survey data, and drew a set of interesting decision rules. A major limitation of these rules is their complexity in application and the existence of 18% unclassified cases. As the advance of learning of graphical models [9], causal discovery can be carried out for this survey data and a set of causal results could be produced:

- A causal model on the domain which can represent the cause-effect relationships among the different attributes of the travelers. This can help us to understand the domain in a broad view.
- A small set of attributes which have direct or indirect effect on the expenditure patterns of business travelers. This result might help us to build more reliable prediction/classification models, and provide some support for tourism industry to better marketing.

2   Background

2.1   Business Travelers Data Set

Experimental data used in this research were collected in an Omnibus Survey, which was conducted at the Hong Kong International Airport [1]. The questionnaire asked trip profile and collected demographics of non-transit travelers to Hong Kong from its seven major tourist generating markets. At the end of the Survey and using a systematic sampling approach, $1, 282$ visitors were personally interviewed and 303 of them named themselves as business travelers. The demographic characteristics and trip profile of these 303 business travelers were identified by 27 variables, which in turn were grouped to three categories. These categories are broadly presented as follows.

Personal Information. Variables indicating traveler's personal information, like *Gender, Age*, etc.

Trip Information. Variables represents the characteristics of the trip, such as *the purpose of the trip, the length of stay in Hong Kong*, etc.

Expenditure Information. Variables represents the traveler's expense information, like *the total expenses*, etc.

Table 1 lists all those 27 variables. It is necessary to mention that the original design using *SPSS* allowed a maximum of eight characters for each variable. In this paper, we are going to analyze this survey data set using causal discovery algorithm.

Table 1. Variables in Business Travelers Expenditure Data

Personal Information		Trip Information		Expenditure Information	
Variable	Description	Variable	Description	Variable	Description
LANGUAGE	which language does the questionnaire use?	DESTINAT	Flight destination	EXPENSE	Total Expenses
COUNTRY	Country of Residence	RETURNHO	Does the respondent return home?	EXPENCUR	Total expenses (currency)
CHINA	Province in China	TOTALLOS	Total length of stay (whole trip)	EXPENHKD	Expenses in HK in HKD
WEUROPE	Western European Countries	HKLOS	Length of Stay in HK?		
GENDER	Gender	FIRSTVHK	First Visit to HK?	TOTALEXR	recoded total expenses in HKD
AGE	AGE	TRIPNO	Number of Trips made to HK including the current one		
EDUCATIO	Highest education level attained	ONLYDEST	Is H.K. the only destination you will visit during this trip?		
INCOME	annual household income	MAINDEST	H.K. as the main destination?		
		MAINPURP	Main purpose for visting H.K?		
		TRAVELMO	Mode of Travel		
		TTPARTY	Total travel party in your group		
		TOTALLOR	recoded Total length of stay (whole trip)		
		HKLOSR	recoded length of stay in HK		
		TTPARTYR	recoded total travel party in your group		
		RETURNR	recoded likelihood of return to H.K.		

2.2 Causal Model and Causal Discovery

Graphical Model is a succinct and efficient way to represent probabilistic (or causal) relations among a set of variables [10], it combines a representation for uncertain problems with methods for performing inference. When the structure of a graphical model is a directed acyclic graph (DAG), it is often referred to as a directed graphical model. If some additional assumptions are satisfied [10], a directed graphical model can also be viewed as a *Causal Model*, which is a representation of known or inferred causal relations among variables in domain. Accordingly, the automatic learning of a causal model is usually referred to as *Causal Discovery*.

According to the possible values of each variable, there exist two kinds of *Graphical Models*, i.e., *Linear Causal Model* and *Bayesian Network*. While *Linear Causal Model* can only deal with continuous data, *Bayesian Network* can only deal with discrete data. For the *business travelers* data set, if we learn the 'true' causal model structure from it, we could reveal many important aspects about factors which would affect the expense pattern of business travelers.

3 Discovery of Generalized Causal Models

Considering that the *business travelers* data set contains both discrete data and continuous data, we used the *Minimum Message Length*-based Discovery for *Generalized Causal Model* proposed in [9]. In this section, we recap the basic idea of this method.

3.1 Generalized Causal Models

The *Generalized Causal Model* extends the concepts of *Bayesian network* and *Linear Causal Model*, and allows both continuous and discrete variables, except that continuous variable cannot be the parent of discrete variables.

As a summary, the effect from discrete variables to a continuous variable is captured by a *Conditional Gaussian Distribution*. It is interesting that *Gaussian* distribution can be viewed as a special case of *Conditional Gaussian* distribution in which only one condition is allowed. For different local conditional distribution, the local parameters at each node can also include different elements. A summary of the local parameters is shown in Table 2.

Table 2. Summary of Local Parameters

Conditional Distribution	Local Parameter Θ_i
Gaussian	$\{\alpha_0, \alpha_1, \ldots, \alpha_{k_i}, \sigma_i\}^*$
Conditional Gaussian	$\{\alpha_0, \alpha_1, \ldots, \alpha_{k_i}, \sigma_i\}^*$
Multinomial	$\{\theta_{ij1}, \ldots, \theta_{ijr_i}\}^*$

* means for each configuration j of discrete parents

3.2   Discovery of Generalized Causal Model

According to the MML criterion [11], the shorter the encoding message length is, the better is the corresponding model. In the case of MML Discovery of *Generalized Causal Models*, the whole encoded message consists of 3 segments:

$$
\begin{aligned}
msgLen &= msgLen(S) + msgLen(\Theta_S) + msgLen(D|S, \Theta_S) \\
&= msgLen(S) + \sum_{i=1}^{n} (msgLen(\theta_i) + msgLen(D_i|\theta_i))
\end{aligned}
\tag{1}
$$

Where n is the number of nodes, θ_i is the local parameters at node V_i, and D_i is the data set confined to node V_i. $msgLen(S)$ is the encoding length of model structure S, while $msgLen(\theta_i)$ is the encoding length for the local parameters at variable V_i, and $msgLen(D_i|\theta_i)$ is the encoding length for the data set confined to variable V_i assuming the model. As the space is limited, we ignore the detailed encoding scheme, which can be found in [9].

3.3   Search Strategy

For a given sample data set, the number of possible model structures which may fit the data is exponential in the number of variables. To find out the best model structure from this huge space, an efficient search strategy is highly demanded.

In this paper we use the Message Length based Greedy Search (MLGS) algorithm: Starts with a directed acyclic graph provided by user or a null graph without any edge, *Message Length Based Greedy search* runs through each pair of nodes attempting to add an edge if there is none or to delete or to reverse it if there already is one. Such adding, deleting or reversing is done only if such changes result in a decrease of the total message length of the new structure. If the new structure is better, it is kept and then try another change. This process continues until no better structure is found within a given number of search steps, or the search from the whole structure space is completed.

4   Causal Analysis of *Business Traveler* Data Set

For the algorithm described in section 3, it is easy to find that the algorithm equals to the MDL/BIC discovery algorithm for *Bayesian network*, and equals to the MML discovery algorithm for *linear causal models*. In this section, we report the results of causal discovery from the *Business Travelers* data set using the algorithm described in section 3. The analysis was implemented in MATLAB with Bayes-Net Toolbox [12].

4.1   Prior Domain Knowledge

One advantage of causal discovery algorithm is that prior domain knowledge can be incorporated into the learning process, and when the domain knowledge is correct, the learning process can be speed up, and the better result can be constructed.

In our analysis, we provided the following domain knowledge to the causal discovery algorithm:

- "AGE" cannot be a child node of any other variable in this domain;
- "LANGUAGE" cannot be a child node of any other variable in this domain;
- "GENDER" cannot be a child node of any other variable in this domain;
- "EDUCATIO" cannot be a child node of any other variable in this domain;
- "TOTALEXR" cannot be a parent of any other variable in this domain.

4.2   The Discovered Causal Model

The generalized causal discovery algorithm was applied to induce the causal model from the Business Travelers Data set and the provided domain knowledge as mentioned above. Fig. 1 gives the best model structure found in the discovery process, and this model represents the causal relation among all *personal information variables*, *trip information variables* and *expense information variables*.

Validated Knowledge. From the discovered model, it is interesting to note that some processing in the survey data collection step was successfully discovered by the algorithm:

- "TOTALLOS" is a parent of "TOTALLOSR";
- "HKLOS" is a parent of "HKLOSR";
- "TTPARTY" is a parent of "TTPARTYR";
- "TOTALEXR" is the child of "EXPENSE" and "EXPENCUR";
- "MAINPURP" is an isolated node, this is not a surprise since in the survey data set, all instances are about Business travelers and their main purpose of trip are all "Business".

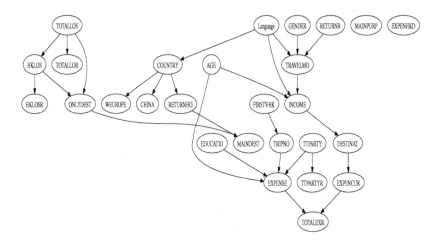

Fig. 1. The Casual Model Structure discovered from the Business Travelers Data

This result agree with the our common sense of the domain:

- "COUNTRY" is a parent of "WEUROPE" and "CHINA", since if the value of country is known, whether the traveler comes from Western Europe or China will be known.
- "FRISTVHK" is a parent of "TRIPNO", i.e., if it is the first visit to Hong Kong, then the trip number must be 1.

Predictors for Expenditure. For the tourism industry, two major interests are the expense and the return ratio of the travelers. Accordingly, we are interested in finding out which factors will affect these *Expenditure Information* variables.

In graphical models, an important characteristic is that: a variable $V - i$ is conditionally independent of all other variables given its *Markov blanket*, which consists of V_i's parents, children, and the parents of its children. Intuitively, the Markov blanket of V_i is the set of nodes that are directly correlated with V_i, at least in some circumstances.

For our business travelers data set, the Markov blanket of those *Expenditure Information* variables could be considered as the set of all factors which have directed effect on the expenditure of travelers. The following results can be induced from the discovered model:

1. "TOTALEXR" was used previously in Law et al's research work as the prediction variable. From the discovered causal model as shown in Fig. 1, it is interesting to find that: This variable was actually recoded from two other expenditure variables, "EXPENSE" and "EXPENCUR".
2. It is a surprise to find that the variable "EXPENHKD" has no connection with any other variables.

3. The Markov Blanket of variable "EXPENSE" includes the following variables: "AGE", "EDUCATIO", "TRIPNO", "TTPARTY", "TOTALEXR" and "EXPENCUR". Among which, the first four variables are either personal information or trip information, and they can be used as the predictor.
4. The Markov Blanket of variable "EXPENCUR" includes two variables: "DESTINAT" and "TOTALEXR". Among which the "DESTINAT" is the trip information, and it can be used as the predictor.

Therefore, in order to build a better prediction model for traveler's expenditure, our causal discovery results indicate that: five relevant variables will be adequate, and these 5 variables are "AGE", "EDUCATIO", "TRIPNO", "TTPARTY" and "DESTINAT".

5 Conclusion and Future Work

In the previous sections, this paper has discussed the importance of data mining in tourism, in general and in particular, in business travel. Also, the fundamental backgrounds of causal discovery algorithm and its induction details have been analyzed. The proposed approach was then applied to a data set of business travelers, and experimental outcomes indicated that five variables were identified as significant that could determine the expenses of business travelers. Non-technical tourism practitioners often find retrieval of useful information from raw data a difficult, slow, and mysterious process [13], and they have been criticized for making decisions based on guesswork instead of a scientific approach [14]. Applying these significant independent variables, industrial practitioners and official policy makers can confidently carry out more accurate planning activities for business travelers, as well as develop the related facilities for this important market segment. Moreover, this research sheds promising light on further work on data mining in tourism.

A natural extension of this research is to apply the proposed method to other tourism sectors. The general applicability of the method, once proved, would be of great value to the business world. Another future research possibility would be to do a qualitative study to examine the views of industrial practitioners on the induced independent variables. The effort of seeking industrial support would further enhance the credibility of the proposed method in the industry. Lastly, it would be beneficial to conduct a longitudinal study to check for the consistency of outcomes in different years. Such a attempt, however, requires a non-trivial amount of resources that needs careful planning.

Acknowledgement

This research was partly funded by the Hong Kong Polytechnic University under contract number: G-U314.

References

1. Law, R., Bauer, T., Weber, K., Tse, T.: Towards a rough classification of business travelers. In: Li, X., Zaïane, O.R., Li, Z. (eds.) ADMA 2006. LNCS (LNAI), vol. 4093, pp. 135–142. Springer, Heidelberg (2006)
2. Cai, L.A., Hong, G.S., Morrison, A.: Household expenditure patterns for tourism products and services. Journal of Travel and Tourism Marketing 4, 15–40 (1995)
3. Fesenmaier, D.R., Jones, L., Um, S., Ozuna, T.: Assessing the economic impacts of outdoor recreation travel to the texas gulf coast. Journal of Travel Research 28, 18–23 (1989)
4. Pyo, S., Uysal, M., McLellan, R.: A linear expenditure model for tourism demand. Annals of Tourism Research 18 (1991)
5. Sheldon, P.: Forecasting tourism: Expenditure versus arrivals. Journal of Travel Research 32, 41–57 (1994)
6. Smeral, E.: Tourism demand, economic theory and econometrics: An integrated approach. Journal of Travel Research, 38–43 (1988)
7. Lawson, F.R.: Trends in business tourism management. Tourism Management 3, 298–302 (1982)
8. Braun, B., Rungeling, B.: The relative economic impact of convention and tourist on a regional economy: a case study. International Journal Hospitality Management 11, 65–71 (1992)
9. Li, G., Dai, H.: The discovery of generalized causal models with mixed variables using MML criterion. In: Proceedings of The Fourth SIAM International Conference on Data Mining, Lake Buena Vista, Florida, USA, SIAM (2004)
10. Pearl, J.: Probabilistic Reasoning in Intelligent Systems (revised 2nd printing edn.) Morgan Kauffmann Publishers, San Mateo, California (1988)
11. Wallace, C., Freeman, P.: Estimation and inference by compact coding. Journal of the Royal Statistical Society B, 49, 240–252 (1987)
12. Murphy, K.: The Bayes Net Toolbox for Matlab. Computer Science and Statistics 33, 331–351 (2001)
13. Law, R.: Hospitality data mining myths. FIU Hospitality Review 16, 59–66 (1998)
14. Athiyaman, A., Robertson, R.: Time series forecasting techniques: Short-term planning in tourism. International Journal of Contemporary Hospitality Management 4, 8–11 (1992)

A Visual and Interactive Data Exploration Method for Large Data Sets and Clustering

David Da Costa[1,2] and Gilles Venturini[1]

[1] Laboratoire d'Informatique de l'Université de Tours, France
{david.dacosta,venturini}@univ-tours.fr
[2] Cohesium, France
ddacosta@cohesium.com

Abstract. We present in this paper a new method for the visual exploration of large data sets with up to one million of objects. We highlight some limitations of the existing visual methods in this context. Our approach is based on previous systems like Vibe, Sqwid or Radviz which have been used in information retrieval: several data called points of interest (POIs) are placed on a circle. The remaining large amount of data is displayed within the circle at locations which depend on the similarity between the data and the POIs. Several interactions with the user are possible and ease the exploration of the data. We highlight the visual and computational properties of this representation: it displays the similarities between data in a linear time, it allows the user to explore the data set and to obtain useful information. We show how it can be applied to standard 'small' databases, either benchmarks or real world data. Then we provide results on several large, real or artificial, data sets with up to one million data. We describe then both the successes and limits of our method.

1 Introduction

Visualization techniques have been studied since several years now and try to provide useful techniques in the data mining process [1]. Among the contributions of visual techniques to data mining (DM) and knowledge discovery in databases (KDD), one may mention the ability to easily analyze data or knowledge through a quickly understandable visualization, or the ability to allow the user formulate (or reformulate) queries in a graphical and interactive way.

One of the limitations and drawbacks often outlined for visual data mining (VDM) techniques is the handling of large data sets. When confronted to a large amount of data, many standard methods fail to produce high quality results in a reasonable amount of time. In the context of clustering for instance, it is known that ascending hierarchical clustering is an efficient method, but limited to small data sets, and therefore other work has been developed to overcome this limitation (Cure [2], Birch [3]). The same remark can be made with VDM techniques: the complexity of the visualization process is often too important to handle large data sets. So one of the most important aim and obtained result of this work is to show that VDM techniques based on points of interest may handle data sets with up to 1 million data in a reasonnable time and with useful interaction techniques.

R. Alhajj et al. (Eds.): ADMA 2007, LNAI 4632, pp. 553–561, 2007.

More precisely, we detail in this paper a new generic method for visualizing and exploring symbolic and numerical data (see the initial description in [4]). Our visual representation is an adaptation of the methods that use points of interest in the context of information retrieval in textual data [5]. Our objectives, in addition to those pursued by VDM techniques, are: 1) to be able to represent all types of data (many VDM techniques consider numerical data only), 2) to provide the domain expert with information and knowledge on the similarities between data and to allow him or her to perform interactive actions, 3) to display the data in linear time (a necessary condition in order to use dynamic display and to analyze very large data sets), 4) to require the shortest possible training time for potential users who are not considered as experts in data analysis.

The remaining of this paper is organized as follows: section 2 describes the state of the art in VDM with large data sets where we highlight the current limitations of traditional approaches. We also present in this section the existing methods based on points of interests. In section 3, we detail our approach: the selection of points of interest, the placement of the data, the interaction with the visualization (change of points of interest, zoom, queries). In section 4, we describe first the properties of our method. We provide typical results obtained on benchmark data as well as results on a real world application. Then we apply our method to large databases and provide computational results. Section 5 concludes on this work and proposes several perspectives.

2   State of the Art

2.1   Visual Data Mining Techniques and Large Databases

We review in this section the properties, possible behavior and limitations of existing VDM techniques when dealing with large databases.

There are several known methods to visualize and interact with data in a VDM process. One can mention "historical" methods such as Chernoff' faces [6]. This method represents each data in the form of a face, and it relies on the fact that the human brain easily analyzes the resemblances and differences between faces. However one may mention that occlusions and overlap between the displayed data is critical in this method when dealing with large data sets. The same comment can be made with parallel coordinates which is not well suited when the dimension of the data is high [7] [8]. One may mention also the "scatter plots" [9] which make it possible to obtain multiple views. Multiple views are however difficult to use when several thousands points must be considered.

Other more recent work is better suited for analyzing millions of data, like the pixel-oriented methods. A famous example is VisDB [10] which represents each data with a colored pixel which indicates the relevance of an article with respect to the user query for instance. Other examples have been studied like in the Tree-Map representation where hierarchical data can be displayed (with up to one million data) [11].

In general, it sounds difficult to try to visualize the dimensions of the data when dealing with large data sets (with many data and many dimensions). Therefore, we have concentrated our efforts on methods that are based on a similarity measure between the data. One possible method is to display directly the similarity matrix [12], but we argue that this method is not as intuitive as the one we propose in this paper.

2.2   Motivations for Using Methods Based on Points of Interest

One can notice that in general many VDM methods easily handle numerical data, whereas the representation of symbolic data seems to be much more difficult. The reason for this is the following one: defining a mapping between numerical attributes and visual attributes such as the position, length, orientation, seems rather "straightforward", while for symbolic values the standard visual representations are less adapted. Moreover, many VDM techniques try to visualize the dimensions of the data, and are thus limited to data sets with a small number of dimensions (or may require the use of attribute selection algorithms), like in the parallel coordinates for instance [7] [8].

Another point to consider is the learning time and adaptation required from the user which can be prohibitive. For instance, with the parallel coordinates, detecting a correlation between attributes is possible but requires considerable learning in order to detect the specific visual pattern.

We are thus interested in this paper with known techniques based on points of interest, but which have not really been used yet in symbolic/numerical data visualization. Among these methods, we can mention systems like VIBE [5], SQWID [13], Radviz [14] or Radial [15], which were used as methods for exploring a set of documents (in general, the results of a search engine).

3   Our Method

3.1   Basic Principles of Visualization

We consider a data set compound of n data $D_1, ..., D_n$ and a similarity matrix "Sim" between these data. $Sim(i, j)$ denotes the similarity between D_i and D_j. If $Sim(i, j) = 1$ then D_i and D_j data are identical, and if $Sim(i, j) = 0$ then they are completely different. This matrix is thus symmetrical and its diagonal is filled with 1s. It is the only assumption that we make about the data, and from this point of view, this allows us to be independent of data types (we may represent numerical/symbolic values, but also texts, images, etc), and of data dimensions. The similarities summarize all the information.

The basic principles of our method are the followings (more complex extensions will be explained in the next sections): initially, we will consider that the POIs are a subset of these data, and are denoted by $D_1, ..., D_k$. We display these k data on a circle, at equal distances.

If D_i is identically similar to all of the POIs, it will be displayed in the middle of the disc. On the opposite, if it is completely similar to one POI and completely different from the others, its position will be confounded with that POI. If its similarity is biased toward certain POIs, then it will tend to approach these POIs.

More generally, our method is such that two data close to one another in the initial representation will thus be also close with respect to POIs, and they will thus be close in 2D space. The visualized space thus becomes a space of distances between selected points (POIs) and the data. It is in this manner that this method can represent a set of data of any type. On the other hand, the reciprocity of this property is not true all the time : two data which are close in the 2D space are not necessarily close in the original space (all the points at equal distance of two POIs in the initial space form a mediating line and are thus displayed at the same 2D location). It will be necessary to use other methods to remove these ambiguities (see the next sections).

3.2 Cutting Down the Complexity

In general, when one wants to display the similarities between n data, the associated complexity is often in $\frac{n \times n+1}{2}$: if one wants to correlate the displayed distance with the similarity between data, it requires the computation of the similarity between all couples of data. In our method, the computation of similarities is limited to $(n - k) \times k$ where k is less than 10 in general, but the 2D display highlights the similarities between all data but without computing them: data which are similar to each others will be located at the same distances to POIs, and will thus be close to each others in the 2D representation. Once those similarities are computed, the display of the data is immediate (linear time). This property is fundamental in order to allow quick interactions and to display very large databases (see section 4).

3.3 Interaction

To be really efficient, the visualization must be interactive and must make it possible to dynamically refine the display and to answer to users graphic requests. In a visualization with POIs, the user can ask for the following information: what is this data (or this POI), how to enlarge this part of the visualization (zoom without loss of context), how to change POI (to remove some, to add some, to change their order, and possibly to define POIs which are not necessarily some data of the initial database).

When the mouse is on a point, we thus indicate what is this point. The user may also select a group of points in order to perform interactive clustering, i.e. labeling the data (see section 4.2).

As far as the POIs are concerned, we have represented the main possible interactions: first of all, it is possible to remove a POI. This is done very simply by dragging a POI inside the disc. This POI takes its place back within the data. The view is dynamically recomputed. A dynamic and progressive transition is performed so that the user can follow the change of representation. He then has the possibility to cancel its action, which causes to put the POI back on the circle. It is also possible to choose a data and to define it as a POI. For this purpose, the user drags the data on the circle. If the data is placed on a POI, it replaces this POI, and if it is placed between two POIs, it is inserted between them. The length of the arcs between POIs is kept constant. These functionalities are very important since they allow the user to redefine at will the representation.

Two zooming techniques have been implemented, one being based on a standard hyperbolic transformation of the display, and the second one being a distortion of the similarity function.

The hyperbolic zoom is triggered when the user clicks on a point: it places the selected data on the center of the disc, it increases the area centered on this data and pushes back the other data toward the edges of the visualization. The distortion is calculated using a hyperbolic function. This zoom makes it possible to enlarge the display while preserving the context of the data. It uses Cartesian coordinates but with we are currently studying a new version with polar coordinates.

The second zoom performs a kind of thresholding on the similarities, reducing the similarities below that threshold and increasing the similarities above this threshold. The resulting visual effect is that data are attracted by their most similar sets of POIs. One observes the creation of "straight lines" in the representation and the "crushing" of the data toward the POIs. The data that remain in the center are those which are the less attracted by the POIs. This zoom thus makes it possible to remove ambiguities.

3.4 Extensions

Several simple extensions can be added to this initial representation.

First of all, the initial choice of the POIs must be carried out. Initially, we consider that if the data are supervised (a class label is available), then we take the first representative of each class as initial POIs. There will thus be as many POIs as classes in the first visualization suggested to the user. If the data are not supervised (no data labels exist), we choose the k random data. Other automatic choices are possible (and certainly more judicious), and we try here to suggest initial choices that the user will be able to interactively and dynamically modify according to what is displayed.

A second extension is motivated by considering the order of the POIs: if a great number of data are attracted by two POIs, then it is desirable that these POIs are close to each others on the circle. A critical situation would consist in placing these POIs in a diametrically opposed way, which would generate unreadable visualizations (many data in the center). We propose an interactive solution to this problem: when the user adds a POI in the representation, he may add this POI at the location indicated by its mouse click (drag and drop), but our tool may also insert this POI at the best possible location in the circle. Given a new POI D_{k+1}, this POI is inserted in the ordered list $D_1, ..., D_k$ between its two most similar POIs. In this way, this should avoid the critical situation mentioned.

Then we have changed the constant distance between POIs to a variable one. The length of the arc between two POIs is a linear function of the similarity between them. In this way, POIs which are similar to each others are represented at close locations on the circle.

Lastly, it is possible to generalize POIs from specific data to any point in the space of data representation, and even more generally to knowledge which represent hypothesis to be checked by the user. Thus, one can represent "ideal" data, not really existing, and according to which the user would like to position the real data. Also, it would be possible to represent for example a decision rule as a POI, and to place the data according to their matching with this rule. This functionality offers many perspectives by visualizing not only data but also knowledge.

4 Results

4.1 Basic Tests and Illustrations

We illustrate the functionalities of our tool on standard databases. For this purpose, we have selected several databases with known characteristics from the Machine Learning Repository [16].

We present in figure 1 the effects of the zoom based on similarity distortion. Similarly, we have represented in figure 2 the use of the hyperbolic zoom.

We have represented in figure 2 the Iris data (150 data, 3 classes). In the Iris database, on may easily get correct information about the global shape of the classes (3 classes, 2 being non linearly separable). The same remark can be done for others databases. Our method is able to highlight outliers for instance, i.e. points which are far away from the others (and from more dense area).

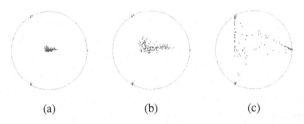

(a)   (b)   (c)

Fig. 1. Visualization of Wine data without zoom (a), with the zoom based on similarity distortion (b) and with increasing zoom factor (c)

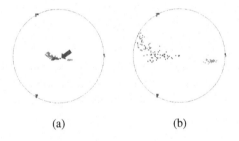

(a)   (b)

Fig. 2. Visualization of Iris data with data selection (a), and with hyperbolic zoom (b)

4.2   Interactive Clustering

In the case of an unsupervised database, our tool offers the possibility to create clusters from the 2D visualization. For this purpose, our method proposes to the user (or domain expert) a first visualization where the choice of POIs is not random (figure 3(a)). Then he has the possibility to select some displayed points and to assign them a label (figure 3(b) and (c)). We have represented this process in figure 3.

4.3   Dealing with Very Large Databases

We have applied our method to databases that contain up to one million data. The first database is the Forest CoverType data set as represented in figure 4. This data set contains a total of 581012 observations and each observation consists of 54 attributes, including 10 quantitative variables, 4 binary wilderness areas and 40 binary soil type variables. In our test, we have used all variables. There are seven forest cover type designations. We have represented first this database in figure 4.

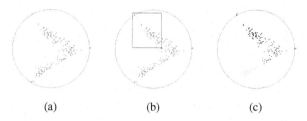

(a)   (b)   (c)

Fig. 3. Visual and interactive clustering of "Wine" data

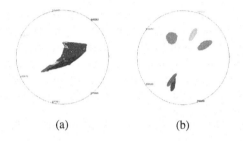

(a)   (b)

Fig. 4. Visualization of the Forest CoverType data set (a) and the 1 000 000 artificial data set (b)

The second series of databases have been generated using uniform laws. Each data consists in a vector in dimension 10. Several classes are defined, and a different set of parameters of uniform laws is defined for each class. We generate in this way databases with an increasing number of data, from 10000 to 1000000 data. We have then displayed the largest database in figure 4. For this database, the total reading/displaying time is about 1 minute on a standard computer (Pentium M at 2 GHz with 1Go RAM). An interactive action such as the change of a POI takes about 2 minutes. In table 1, we have shown how the computational cost evolves with an increasing number of data.

As mention in section 3.2, our method has a linear complexity while being able to let the user perceive the $n \times n$ similarities between n data. More precisely, our method performs the following operations: the first pass consists in reading the database for normalizing numerical attributes (detecting minimum and maximum values) and for choosing the POIs (first representative of each class). Then, the second pass consists in computing the $(n - k) \times k$ similarities between the $n - k$ data and the k POIs. If d denotes the dimension of the data (i.e. the number of variables/attributes), then the overall complexity of our method is in $O((n - k) \times k \times d)$.

Table 1. Performance results on the very large data set

Name	# of data	# of attributes	# of classes	Displaying time (sec.)
GenFile_10000	10 000	10	5	0.719
GenFile_100000	100 000	10	5	4.470
GenFile_1000000	1 000 000	10	5	49.505
Forest CoverType	581 012	54	7	91.069

5   Conclusion and Perspectives

We have described in this paper a new visualization method which is inspired from the work on points of interest in the context of information retrieval. This method consists in mapping a space of similarities into a 2D representation. This is performed by positioning the data according to a selected set of points of interest. These points can be a subset of the data or hypothesis to be tested. The main characteristics of our method are the following: it may represent the $n \times n$ similarities while needing only a linear computation time, it is easy to learn and to interact with, it visually represents the relationships between the data. We have shown how standard benchmark databases can be displayed with this method, and how it can help the user to perform interactive clustering, and how it can deal with real world databases. Finally, we have shown how it may handle large databases with up to one million data on a standard computer.

One limit of our visualization are the ambiguities which occur when points have similar display coordinates, especially when large data sets are displayed. For this purpose we intend to use forces and springs methods in order to separate points that are too close. Another extension consists in using a 3D representation such in Lyberworld [17]. We have already tested this approach with up to 60000 data.

References

1. Wong, P.C., Bergeron, R.D.: 30 years of multidimensional multivariate visualization. In: Scientific Visualization — Overviews, Methodologies and Techniques, pp. 3–33. IEEE Computer Society Press, Los Alamitos, CA (1997)
2. Sudipto, G., Rajeev, R., Kyuseok, S.: CURE: an efficient clustering algorithm for large databases. In: Haas, L.M., Tiwary, A. (eds.) Proceedings ACM SIGMOD International Conference on Management of Data, Seattle, Washington, USA, pp. 73–84. ACM Press, New York (1998)
3. Tian, Z., Raghu, R., Miron, L.: Birch: An efficient data clustering method for very large databases. In: Jagadish, H.V., Mumick, I.S. (eds.) Proceedings of the 1996 ACM SIGMOD International Conference on Management of Data, Montreal, Quebec, Canada, June 4-6, 1996, pp. 103–114. ACM Press, New York (1996)
4. Costa, D.D., Venturini, G.: An interactive visualization environment for data exploration using points of interest. In: Li, X., Zaïane, O.R., Li, Z. (eds.) ADMA 2006. LNCS (LNAI), vol. 4093, pp. 416–423. Springer, Heidelberg (2006)
5. Korfhage, R.: To see, or not to see: Is that the query? In: Bookstein, A., Chiaramella, Y., Salton, G., Raghavan, V.V. (eds.) Proceedings of the 14th Annual International ACM SIGIR Conference on Research and Development in Information Retrieval, Chicago, Illinois, USA, October 13-16, 1991, pp. 134–141. ACM, New York 1991(special Issue of the SIGIR Forum)
6. Chernoff, H.: Using faces to represent points in k-dimensional spae graphically. Journal of the American Statistical Association 68, 361–368 (1973)
7. Inselberg, A.: The plane with parallel coordinates. The Visual Computer 1, 69–91 (1985)
8. Fua, Y.H., Ward, M.O., Rundensteiner, E.A.: Hierarchical parallel coordinates for exploration of large datasets. In: VISUALIZATION '99: Proceedings of the 10th IEEE Visualization 1999 Conference (VIS '99). IEEE Computer Society, Washington, DC, USA (1999)
9. Becker, R.A., Cleveland, W.S.: Brushing Scatterplots. Technometrics 29, 127–142 (1987). In: Cleveland, W.S., McGill, M.E. (eds.) Dynamic Graphics for Data Analysis. Chapman and Hall, New York 1988 (reprint)

10. Keim, D.A., Kriegel, H.: VisDB: Database exploration using multidimensional visualization. In: Computer Graphics and Applications (1994)
11. Fekete, J., Plaisant, C.: Interactive information visualization of a million items proceedings of ieee symposium on information visualization (2002)
12. Jun Wang, B.Y., Gasser, L.: Classification visualization with shaded similarity matrices. Technical report, GSLIS University of Illinois at Urbana-Champaign (2002)
13. McCrickard, S., Kehoe, C.: Visualizing search results using sqwid. In: Proceedings of the Sixth International World Wide Web Conference (1997)
14. Hoffman, P., Grinstein, G., Pinkney, D.: Dimensional anchors: a graphic primitive for multidimensional multivariate information visualizations. In: NPIVM '99: Proceedings of the 1999 workshop on new paradigms in information visualization and manipulation in conjunction with the eighth ACM internation conference on Information and knowledge management, pp. 9–16. ACM Press, New York, NY, USA (1999)
15. Au, P., Carey, M., Sewraz, S., Guo, Y., Rüger, S.M.: New paradigms in information visualization. Research and Development in Information Retrieval, 307–309 (2000)
16. Blake, C., Merz, C.: UCI repository of machine learning databases (1998)
17. Hemmje, M., Kunkel, C., Willett, A.: Lyberworld visualization user interface supporting fulltext retrieval. In: SIGIR '94: Proceedings of the 17th annual international ACM SIGIR conference on Research and development in information retrieval, New York, NY, USA, pp. 249–259. Springer, New York (1994)

Explorative Data Mining on Stock Data –
Experimental Results and Findings

Lay-Ki Soon and Sang Ho Lee

Database Systems Laboratory, Department of Computing, School of Information Science,
Soongsil University, 1-1 Sangdo-dong, Dongjak-gu Seoul, 156-743 Korea
laykisoon@gmail.com, shlee@comp.ssu.ac.kr

Abstract. This paper presents our preliminary explorative data mining experiments on stock data. The experiments are performed using information gain evaluation, chi-square test and decision tree, which aim to explore the underlying patterns of the stock dataset. The explorative patterns obtained show some unexpected outliers in deriving the influential stocks with regard to Kuala Lumpur Composite Index.

1 Introduction

Time-series data mining is identified as one of the 10 challenging problems in data mining research [11]. It can be focusing on trend analysis, similarity search, segmentation, clustering and classification [1, 2, 3, 9].

To gain better understanding on mining time-series data, we have embarked by performing preliminary data mining experiments on a stock dataset. The main objective of our experiments at this stage is explorative, which aims to understand the underlying patterns of the stock data. We have performed three experiments using information gain, chi-square test and decision tree. The first two experiments will generate a list of stocks which are influential to Kuala Lumpur Composite Index (KLCI), while the third experiment will produce classification rules, which denotes the inter-relationships among the stocks in terms of their trading performance with respect to KLCI.

This paper is organized as follows. Section 2 briefly discusses some related works which motivate our experiments. Section 3 presents the stock dataset and the data pre-processing. Section 4 to Section 6 explain the experiments, the results as well as our findings. Lastly, we conclude this paper in Section 7.

2 Related Works

There are numerous issues and challenges in mining time-series data to be addressed [1, 3, 7, 8]. Active research works have been done on the representation of time series data for classification purpose, such as Fourier transformation and weight vector. At this stage, our data mining experiments are purely descriptive by using the symbolic

R. Alhajj et al. (Eds.): ADMA 2007, LNAI 4632, pp. 562–569, 2007.

representation of the stock dataset. In [9], sequential and non-sequential association rule mining (ARM) were used to perform intra and inter-stock pattern mining. Compared to [9], our exploratory mining experiments are similar to their inter-stock pattern mining, where associative relationships among different stocks were derived for inter-stock mining. In our work however, instead of deriving the expected association among the stocks, we are more interested in observing the performance of stocks with respect to a specific stock index, namely KLCI, which is always closely monitored in financial analyses.

3 Stock Dataset and Data Pre-processing

For the purpose of these experiments, we have used the stock trading dataset of Malaysia Exchange from December 31, 2001 till December 31, 2006, which amounts to 1233 days [4, 5]. Being one of the indices under Malaysia Exchange, KLCI is viewed as the local stock market barometer. It is a capitalization-weighted index, consisting 100 index components or stocks, where the list is updated every six months. KLCI's index components are selected based on several criteria, such as market capitalization, the company's performance, trading volume and so on [5].

The trading of the stock market within a day is recorded in a single text file. Figure 1 shows a snippet of a file. Each line represents the trading information of an index or a stock. Table 1 describes the information captured for the first line in Fig. 1.

```
COMPOSITE,D,12/29/06,1089.2800,1096.2400,1088.8000,1096.2400,9950,0
TRADING,D,12/29/06,151.3800,152.1300,151.1500,151.7400,1829,0
IND_PRODUCTS,D,12/29/06,91.5600,92.0700,91.3600,92.0700,886,0
PROPERTY,D,12/29/06,687.1400,693.4900,685.0600,693.4900,802,0
CONSTRTN,D,12/29/06,197.0300,200.0200,196.6100,200.0200,745,0
SECBOARD,D,12/29/06,91.5600,92.0300,91.3400,92.0300,713,0
FINANCE,D,12/29/06,8589.5100,8656.1200,8572.2400,8656.1200,363,0
PLANTATN,D,12/29/06,4215.6200,4315.9100,4215.6200,4315.9100,164,0
CONSUMER,D,12/29/06,266.5700,269.4600,265.5400,269.4600,140,0
MINING,D,12/29/06,526.8700,656.5700,526.8700,611.9900,124,0
INDUSTL,D,12/29/06,2206.8600,2225.4700,2204.4000,2225.4700,0,0
MESDAQ,D,12/29/06,118.8600,119.9200,118.4500,119.9200,238780100,0
TECHNOLOGY,D,12/29/06,28.3600,28.3600,27.9900,28.2800,7717700,0
7082WA,D,12/29/06,0.295,0.300,0.260,0.260,420,0
```

Fig. 1. Snippet of our stock dataset

Table 1. Information recorded at each line in the files

Information	Description
COMPOSITE	Kuala Lumpur Composite Index
D	Date
12/29/06	Dated December 29, 2006
1089.2800	Previous value (Last value of December 28, 2006).
1096.2400	Highest value of the day.
1088.8000	Lowest value of the day.
1096.2400	Closing / Last value of the day.
9950	Trading volume.
0	Constant.

Since noisy data is not an alien property of real-world datasets, we have spent almost 90% of our effort in these experiments on data pre-processing. Armed with the supporting information by Malaysia Exchange [5], we have developed a program to semi-automate the pre-processing activities.

The noise exists in our stock dataset can be generally classified into three main categories, which are duplicate records, inconsistent stock codes and incomplete files. Certain files contain duplicated records. For example, the transaction of stock 4863 is recorded twice in files dated December 9, 2002 and September 13, 2006. Thus, we have automated a thorough checking on all the files, ensuring only unique records on the same date.

In most of these files, stocks are represented by stock codes. For example, Telekom Malaysia Limited is recorded as 4863. However, in certain files, stocks are represented differently, such as TM or Telekom for Telekom Malaysia Limited. This inconsistency gets more complicated when the listed companies change their names or when companies merge. Having such inconsistencies, we have no choice but to rely on human intervention for the remedial action. Figure 2 depicts the steps in combating this problem. For the consistency of our experiments, we have chosen stock codes to represent the stocks, and COMPOSITE to represent KLCI. Note that T and I are provided by Malaysia Exchange for this purpose [5].

```
Input:
 • Raw dataset R (1232 files with approximately 1300 records.)
 • List of indices I, where each index i has its name nᵢ ∈ N.
 • List of stocks T, where each stock t has its code cₜ ∈ C and name sₜ ∈ S.
Output:
 • List L containing valid stock codes and indices' names.
 • Dataset D with stocks represented by stock codes and indices represented by names.
Automated Steps:
 (1)   for every file in R
 (2)         for every line in the file //refer Fig. 1
 (3)               split every record according to commas
 (4)               o = first item captured from step (3).
 (5)               if (o = sₜ, where sₜ ∈ S)
 (6)                     o = cₜ   // cₜ = stock code of sₜ
 (7)               end if
 (8)               if (o ∈ C) or (o ∈ N)
 (9)                     if (o ∉ L)
 (10)                          Lᵢ = o
 (11)                          i++
 (12)                    end if
 (13)         end for
 (14) end for
Manual Checking:
 • Given T, I, L and D, verify unique stocks and indices.
```

Fig. 2. Pre-processing steps for combating data inconsistency in the stock dataset

The files in the raw dataset contain all the market indices and stocks under Malaysia Exchange, where most of them contain approximately 1300 records. However, the file dated June 7, 2002 contains only stock indices without any record on stock trading. Up to the moment when this paper is prepared, the stock trading information on this date is still not available. Therefore, we have disregarded the empty file in order to maintain the integrity of our mining results. As a result, instead of the initial 1233 files, we used only 1232 files for our experiments.

Let *st* denotes a stock code together with its performance and *C* represents all the stock codes in our dataset, every record in these files is converted into the following format:

$$st: c = v \qquad (1)$$

where $c \in C$ and $v \in \{UP, DOWN, SAME\}$. *UP, DOWN* and *SAME* denote positive, negative and unchanged performance of a stock respectively. For example, the record shown in Table 1 is converted to COMPOSITE=UP. For every file in the dataset, after converting each record into format (1), we then combined them into one row, separated by commas. Every file was formatted as a single row composing the stock trading on that same day. These rows were then inserted into a file, which eventually contains 1232 rows of data, as illustrated in Fig. 3.

Fig. 3. Converted dataset, which is ready for the experiments

4 Information Gain Evaluation

Since KLCI serves as the benchmark for Malaysia equity market, we are particularly interested to identify influential stocks with respect to KLCI. Information gain is an attribute selection measure based on entropy, where it studies the value of certain piece of information [2]. This experiment uses information gain to derive a set of stocks which are sorted according to their collective worth with respect to KLCI within 31 December 2001 to 31 December 2006. The stocks with higher information gain imply greater worth or influence against KLCI. We used the *Select Attribute* feature provided by Weka [10], with 10-fold cross validation for this experiment.

Intuitively, the 100 stocks which constitute KLCI should appear as the top 100 stocks by information gain with respect to KLCI. However, some of the results generated by the experiment are not as expected. Out of the top 100 stocks by information gain, only 38 of them constitute the current KLCI. Table 2 shows the top 20 stocks by information gain with respect to KLCI. Since market capitalization is one of the criteria in selecting KLCI's constituents, it is listed for every company in the third column. For example, AMMB is the 23rd highest companies in terms of market capitalization in Malaysia Exchange, which is MYR6775.2 million [5]. TIME is not one of the top 100 companies by market capitalization, but it is included in current KLCI and has the market capitalization of MYR519.41 million.

Although some of these stocks are not KLCI's constituents, by the statistical analysis of information gain, their performance throughout this period of time are considered as worthy to KLCI's performance. Several reasons may lead to the unexpected results. One is because the dataset spread across 31 December 2001 to 31 December 2006. Hence, the experimental results represent the order of stocks by their collective information gain with respect to KLCI within this period of time.

Out of the top 20 stocks (Table 2), only 4 of them do not constitute the current KLCI. We consider LEADER and MRCB as exceptional cases as they belong neither to KLCI nor the top 100 companies by market capitalization [5]. Nevertheless, both of these companies have been performing steadily. Market analyses considered the performance of LEADER, which has MYR257.5 million of market capitalization is ahead of expectation. Although not included in KLCI, AMMB-WA and PBBANK-FOREI are sister companies of AMMB and PBBANK respectively.

Table 2. Top 20 Stocks by Information Gain Evaluation

No.	Stock	Status	Remarks
1	AMMB	KLCI	Market Cap.: 23rd
2	RHBCAP	KLCI	Market Cap.: 24th
3	TM	KLCI	Market Cap.: 3rd
4	MAYBANK	KLCI	Market Cap.: 2nd
5	GENTING	KLCI	Market Cap.: 8th
6	COMMERZ	KLCI	Market Cap.: 5th
7	TENAGA	KLCI	Market Cap.: 1st
8	LEADER	Non-KLCI	Exceptional
9	MRCB	Non-KLCI	Exceptional
10	PBBANK	KLCI	Market Cap.: 6th
11	AMMB-WA	Non-KLCI	Sister company of AMMB
12	YTL	KLCI	Market Cap.: 17th
13	MULPHA	KLCI	Market Cap.: 81st
14	MAS	KLCI	Market Cap.: 28th
15	BJTOTO	KLCI	Market Cap.: 24th
16	PBBANK-FOREI	Non-KLCI	Sister company of PBBANK
17	PLUS	KLCI	Market Cap.: 13th
18	DRBHCOM	KLCI	Market Cap.: 82nd
19	MAGNUM	KLCI	Market Cap.: 46th
20	TIME	KLCI	Market Cap.: MYR519.41 Mil.

In short, this experiment has sorted the stocks according to their worth or influence with regard to KLCI. We are surprised to observe that market capitalization of the stocks and their influence capabilities to KLCI may not be directly correlated.

5   Chi-Square Test

In this second experiment, we have applied χ^2 (chi-square) test to discover the statistical correlation between every stock and KLCI. However, note that correlation analysis does not imply causality [2]. Table 3 shows the contingency table between stock AMMB and KLCI. The columns represent the distinct values of AMMB. NULL value is possible when the trading information of AMMB is not available in certain trading days. The first and third rows represent the distinct values of KLCI. Out of 1232 days of stock trading, there are 372 days where both AMMB and KLCI performed positively; and 368 days where both of them performed negatively. The chi-square test for the correlation relationship between AMMB and KLCI can be computed as

$$\chi^2 = \frac{(372 - 254.16)^2}{254.16} + \frac{(171 - 284.81)^2}{284.81} + ... + \frac{(0 - 0.47)^2}{0.47} = 213.56$$

Table 3. Contingency Table Between KLCI and AMMB

KLCI \ AMMB	UP	DOWN	SAME	NULL	Total
UP	372	171	107	1	651
UP-Expected Frequency	*254.16*	*284.81*	*111.49*	*0.53*	
DOWN	109	368	104	0	581
DOWN-Expected Frequency	*226.84*	*254.19*	*99.51*	*0.47*	
Total	481	539	211	1	1232

By convention, we can reject the hypothesis that AMMB and KLCI are independent if its χ^2 value exceed a threshold (significance value). However, as we intend to explore the chi-square value of all the stocks, we do not apply any threshold in this experiment. We have used 10-fold cross validation in this experiment [10].

Table 4. Top 20 Stocks By Chi-Square Test

No.	Stock	Status	Remarks
1	AMMB	KLCI	Market Cap.: 23rd
2	RHBCAP	KLCI	Market Cap.: 24th
3	TM	KLCI	Market Cap.: 3rd
4	MAYBANK	KLCI	Market Cap.: 2nd
5	GENTING	KLCI	Market Cap.: 8th
6	COMMERZ	KLCI	Market Cap.: 5th
7	TENAGA	KLCI	Market Cap.: 1st
8	*MRCB*	*Non-KLCI*	*Exceptional*
9	*LEADER*	*Non-KLCI*	*Exceptional*
10	PBBANK	KLCI	Market Cap.: 6th
11	AMMB-WA	Non-KLCI	Sister company of AMMB
12	YTL	KLCI	Market Cap.: 17th
13	MULPHA	KLCI	Market Cap.: 81st
14	MAS	KLCI	Market Cap.: 28th
15	BJTOTO	KLCI	Market Cap.: 24th
16	PBBANK-FOREI	Non-KLCI	Sister company of PBBANK
17	*DRBHCOM*	*KLCI*	*Market Cap.: 82nd*
18	*PLUS*	*KLCI*	*Market Cap.: 13th*
19	*TIME*	*KLCI*	*Market Cap.: MYR519.41 Mil.*
20	*UEMBLDR*	*Non-KLCI*	*Market Cap.: 89th*

Table 4 shows the top 20 stocks sorted according to their chi-square values. Comparing the results from information gain and chi-square test, different stocks are presented at 8th, 9th, and 17th – 20th. Likewise, 8th and 9th stocks are considered as anomalies which may trigger further financial analysis. One significant difference between the lists generated by information gain and chi-square is on MAGNUM (19th in Table 2) and UEMBLDR (20th in Table 4). MAGNUM is placed at 22nd by chi-square test. Despite not being as one of KLCI's constituents, UEMBLDR is the 89th highest company in Malaysia Exchange by market capitalization.

Nevertheless, the overall results generated by both information gain and chi-square test are generally consistent, which can be considered as a positive verification of the results from both experiments. Again, we can conclude that market capitalization and the performance of KLCI are not directly correlated in most of the cases.

6 Decision Tree

Different from the previous two experiments, we only select the 100 index components of current KLCI in this experiment. Having KLCI as the class attribute, we expect to derive a set of classification rules. Our experiment on DT is performed using the implementation of C4.5 in Weka [6, 10]. The best accuracy – 74.51% is obtained at 14-fold cross validation.

Table 5. Classification Rules Derived from Decision Tree

No.	Classification Rules
1	*if* AMMB=UP *and* RHBCAP=UP *and* TIME=UP *then* COMPOSITE=UP **(191, 8)**
2	*if* AMMB=DOWN *and* MBB=DOWN *and* MAS=DOWN *then* COMPOSITE=DOWN **(118, 5)**
3	*if* AMMB=DOWN *and* MBB=SAME *and* COMMERZ=DOWN *then* COMPOSITE=DOWN **(80, 12)**
4	*if* AMMB=DOWN *and* MBB=UP *and* MMBCORP=DOWN *and* TA=DOWN *then* COMPOSITE=DOWN **(38,4)**
5	*if* AMMB=UP *and* RHBCAP=UP *and* TIME=SAME *then* COMPOSITE=UP **(36, 3)**
6	*if* AMMB=SAME *and* MAS=UP *and* YTL=UP *then* COMPOSITE=UP **(34, 1)**
7	*if* AMMB=UP *and* RHBCAP=DOWN *and* TENAGA=UP *and* GENTING=UP *then* COMPOSITE=UP **(28)**
8	*if* AMMB=UP *and* RHBCAP=UP *and* TIME=DOWN *and* BERNAS=UP *then* COMPOSITE=UP **(22)**
9	*if* AMMB=SAME *and* MAS=DOWN *and* GUTHRIE=DOWN *then* COMPOSITE=DOWN **(18)**
10	*if* AMMB=UP *and* RHBCAP=DOWN *and* TENAGA=DOWN *and* DRBHCOM=DOWN *and* COMMERZ=DOWN *then* COMPOSITE=DOWN **(18)**

With 220 leave nodes in the decision tree, it composes 293 classification rules. Based on the number of records covered by these rules, the top 10 rules are listed in Table 5. The figures followed after each classification rule indicate the number of records correctly and incorrectly classified by the rule respectively. Having the highest information gain, AMMB is selected as the root node. In the second level, the selected stocks are RHBCAP, MBB and MAS. All these 4 stocks are blue chip stocks which have high market capitalization. Using these classifications rules, we may observe the inter-relationships among these KLCI constituents as well.

7 Conclusion and Future Works

In this paper, we have presented our experience and findings on exploratory stock data mining using information gain, chi-square test and decision tree. Extensive data pre-processing steps has been performed. Within the generated underlying patterns, we obtained some unexpected results, which trigger further investigation by the financial analysts. With proper data pre-processing, we believe that these insightful patterns will be of valuable knowledge for stock market analysis.

For our future works, we plan to explore the temporal semantics of the dataset, such as the evolution of the correlation relationships among the stocks over a period of time. Having the patterns of stock data captured over time, we may perform similarity search among the stocks. Besides, the possibility of conceptualizing the relationships among stocks and indices ontologically will be investigated.

Acknowledgments. This work was supported by Seoul R&BD Program (10581cooperateOrg93112). We also thank Mr. Alex Tham Poh Seng and Mr. Abdul Rahim Hussein from Malaysia Exchange for providing us the information and advice throughout these experiments.

References

1. Antunes, C.M., Oliveira, A.L.: Temporal Data Mining: An Overview. In: Proceedings of KDD Workshop on Temporal Data Mining, San Francisco, pp. 1–13 (2001)
2. Han, J., Kamber, M.: Data Mining Concepts and Techniques. Morgan Kaufmann Publishers, Elsevier, San Francisco, CA (2006)
3. Keogh, E., Kasetty, S.: On the Need for Time Series Data Mining Benchmarks: A Survey and Empirical Demonstration. In: International Journal of Data Mining and Knowledge Discovery, vol. 7(4), pp. 349–371. Springer, Netherlands (October 2003)
4. Malaysia Exchange Dataset. (subscription) Available: http://www.klsedaily.com
5. Official Website of Malaysia Exchange. Available: http://www.bursamalaysia.com
6. Quinlan, J.R.: C4.5: Programs for Machine Learning. Morgan Kauffman, San Francisco (1993)
7. Roddick, J.F., Hornsby, K., Spiliopoulou, M.: An Updated Bibliography of Temporal, Spatial, and Spatio-Temporal Data Mining Research. In: Roddick, J.F., Hornsby, K. (eds.) TSDM 2000. LNCS (LNAI), vol. 2007, pp. 147–164. Springer, Heidelberg (2001)
8. Roddick, J.F., Spiliopoulou, M.: A Survey of Temporal Knowledge Discovery of Paradigms and Methods. IEEE Transactions on Knowledge and Data Engineering 14(41), 750–767 (2002)
9. Ting, J., Fu, T., Chung, F.: Mining of Stock Data: Intra- and Inter-Stock Pattern Associative Classification. In: Proceedings of 2006 International Conference on Data Mining, Las Vegas, USA, June 2006, pp. 30–36 (2006)
10. Witten, I.H., Frank, E.: Data Mining: Practical Machine Learning Tools and Techniques, 2nd edn. Morgan Kaufmann, San Francisco (2005)
11. Yang, Q., Wu, X.: 10 Challenging Problems in Data Mining Research. International Journal of Information Technology & Decision Making 5(4), 1–8 (2006)

Graph Structural Mining in Terrorist Networks

Muhammad Akram Shaikh, Jiaxin Wang, Zehong Yang, and Yixu Song

State Key Lab of Intelligent Technology and Systems
Department of Computer Science & Technology
Tsinghua University, Beijing, P.R. China
+86-10-87343432
alm04@mails.tsinghua.edu.cn

Abstract. Law enforcement agencies and intelligence analysts frequently face the problems of identifying the actor importance and possible roles among a specific group of entities in a terrorist network. However, such tasks can be fairly time consuming and labor-intensive without the help of some efficient methods. In this paper we will discuss how graph structural mining is applied in the context of terrorist networks using structural measures or properties from social network analysis (SNA) research. Structural properties are determined by the graph structure of the network. These properties are used to evaluate the relationship between entities and identifying different roles. The graph structural mining concept is also demonstrated by using publicly available data on terrorists network in two ways i.e., one for identifying different roles (leaders, brokers, and outliers) known as role structural mining and other for ranking important actors known as rank structural mining. In addition to this we also illustrate how terrorist network is disrupt by knowing the actor importance in a network using rank structural mining.

1   Introduction

Terrorists seldom operate in a vacuum but interact with one another to carry out various illegal activities. Organized crimes such as terrorism, drug trafficking, gang-related offenses, and frauds require collaboration among criminals. Relationships between terrorists form the basis for organized crimes [1]; which can be viewed as a network consisting of nodes (terrorists) and links (relationships). In terrorist networks, there may exist groups or teams. Individuals play different roles in their groups [2]. For example, some act as leaders to control activities of a group, others may serve as gatekeepers to ensure smooth flow of information and some act as outliers. Current practice of terrorist network analysis is a manual process because of the lack of advanced, automated techniques. Manual approaches may fail to generate knowledge in a timely manner. Fighting against terrorist networks requires a more nimble intelligence apparatus that operates more actively and makes use of advanced information technology. In this paper we will discuss how graph structural mining is applied in order to identify different roles and important actors according to their importance score. The rest of the paper is organized as follows: Section 2 discusses the background; Section 3

R. Alhajj et al. (Eds.): ADMA 2007, LNAI 4632, pp. 570–577, 2007.

gives an overview of the Graph structural mining using Role Structural and Rank Structural Mining; Section 4 describes how the concept of graph structural mining is helpful in network disruption; Section 5 shows experiments on terrorists network data set for identifying roles and ranking and discuss in detail how these methods are helpful to disrupt terrorist Networks by finding the most important actors with illustration; and section 6 concludes the paper.

2   Background

Law enforcement personnel have used social networks to analyze terrorist networks[3,4] and criminal networks [5]. The capture of Saddam Hussein was facilitated by SNA: military officials constructed a network containing Hussein's tribal and family links, allowing them to focus on individuals who had close ties to Hussein [6]. SNA, originating from social science research, is a set of analytical tools that can be used to map networks of relationships and provides an important means of assessing and promoting collaboration in strategically important groups [7]. SNA provides a set of measures and approaches for the investigation of terrorist networks. These techniques were originally designed to discover social structures in social networks [1] and are especially appropriate for studying criminal networks [8,9,10]. Graph structural Mining is used to identify criminals' roles and to uncover important actors. Specifically, in the literature the use of centrality and structural equivalence measures from SNA are used to measure the importance of each network member. Several centrality measures, such as degree, betweenness, closeness, and eigenvector, can suggest the importance of a node [1] and can identify the leaders, gatekeepers, and outliers in a network [11]. Baker and Faulkner [12] employed these measures, especially degree, to find the central individuals in a price-fixing conspiracy network in the electrical equipment industry.

3   Graph Structural Mining

Graph structural mining involves evaluating the different positions (roles) and rank important actors in a social network using structural (indices) measures or properties. Structural properties are determined by the graph structure of the network. In general, the network studied in this paper can be represented by an undirected and un-weighted graph $G = (V, E)$, where V is the set of vertices (or nodes) and E is the set of edges (or links). Each edge connects exactly one pair of vertices, and a vertex pair can be connected by (a maximum of) one edge, i.e., multi-connection is not allowed. A terrorist network consists of V set of actors (nodes) and E relations (ties or edges) between these actors. Mathematically, a network can be represented by an adjacency matrix A, which in the simplest case is an $N \times N$ symmetric matrix, where N is the number of vertices in the network. The adjacency matrix has elements.

$$A_{ij} = \begin{cases} 1 & \text{if } i \text{ and } j \text{ are connected} \\ 0 & \text{otherwise} \end{cases} \tag{1}$$

The matrix is symmetric since if there is an edge between i and j then clearly there is also an edge between j and i. Thus

3.1   Role Structural Mining

Role structural mining is used to identify different important roles specially the leader of the network using centrality measures. Freeman [13] provided definitions of the three most popular centrality measures: degree, betweenness, and closeness. The degree of a vertex in a network is the number of edges attached to it. In mathematical terms, the degree D_i of a vertex i is [14]:

$$D_i = \sum_{j=1}^{n} A_{ij} \tag{2}$$

A network member with a high degree could be the leader or "hub" in a network. Betweenness measures the extent to which a particular node lies between other nodes in a network. The betweenness B_a of a node a is defined as the number of geodesics (shortest paths between two nodes) passing through it:

$$B_a = \sum_{j}^{n} \sum_{a}^{n} g_{ij}(a) \tag{3}$$

Where $g_{ij}(a)$ indicates whether the shortest path between two other nodes i and j passes through node a. A member with high betweenness may act as a gatekeeper or "broker" in a network for smooth communication. Closeness C_a is the sum of the length of geodesics between a particular node a and all the other nodes in a network. It actually measures how far away one node is from other nodes and is sometimes called farness [12,13]:

$$C_a = \sum_{i=1}^{n} l(i, a) \tag{4}$$

Where $l(i, a)$ is the length of the shortest path connecting nodes i and a.

3.2   Rank Structural Mining

Rank structural mining is used to identify ranks among important actors according to their actor importance score in terrorist networks by applying following two methods.

Combine Centrality Actor Ranking. Both Closeness and Betweenness centralities are global measures, where as degree centrality is termed as local measure. Note that all these measures are relative ones. These three centrality measures may produce contrary results for the same graph. It can be a case in which an actor has a low degree centrality, with a high betweenness centrality. In the literature [13] it is shown that the betweenness centralities best "capture" the essence of important nodes in a graph, and generate the largest node variances, while degree centralities appear to produce the smallest node variances. In order to overcome the drawbacks of single centrality, here we used the concept of Combine Centrality Actor Ranking (CCR) as shown in equation 5:

$$CCR = D_i + B_a + C_a \tag{5}$$

Eigenvector Centrality Actor Ranking. This approach to order the vertices (nodes) of a graph was suggested by Bonacich [15]. His idea is based on the assumption that the value of a single vertex is determined by the values of the neighboring vertices. If we denote the centrality of vertex i by x_i, then we can allow for this effect by making x_i proportional to the average of the centralities of i's network neighbors [6]:

$$x_i = \frac{1}{\lambda} \sum_{j=1}^{n} A_{ij} x_j \tag{6}$$

Where λ is a constant. we can rewrite this equation in matrix form as:

$$\lambda x = Ax \tag{7}$$

Hence we see that x is an eigenvector of the adjacency matrix with eigenvalue λ. Assuming that we wish the centralities to be non-negative, it can be shown that λ must be the largest eigenvalue of the adjacency matrix and x the corresponding eigenvector. If A is an $N \times N$ matrix, it has N eigenvectors (for each node), and correspondingly many eigenvalues. The eigenvector of interest is the principal eigenvector, i.e. that with highest eigenvalue, since this is the only one that results from summing all of the possible pathways with a positive sign.

4   Network Disruption

Disruption techniques traditionally aim at neutralizing members of terrorist networks either through capture or death. These nodes are known as the 'critical' nodes within a network. The removal or isolation of these nodes ensures maximum damage to the network's ability to adapt, performance, and ability to communicate. In network analysis, node changes are the standard approach to network destabilization [5]. Kathleen Carley et al. proposed three indicators of network disruption [16]:

- The rate of information flow through the network has been minimized (perhaps to zero).
- The network, as a decision making body, cannot reach on a joint consensus.
- The ability of the network to accomplish tasks is totally impaired.

5   Experimental Results

In this section Role structural Mining and Rank Structural Mining are applied on September 11, 2001 data set of 62 hijackers and their affiliates, which was originally constructed by Valdis Krebs [3] and is reproduced here using NetMiner [17] as shown in figure 1.

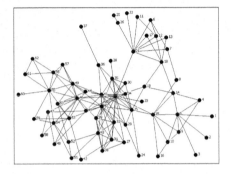

Fig. 1. The terrorist network of 9-11 hijackers and their affiliates

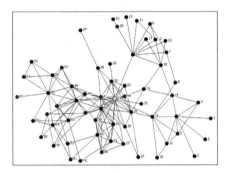

Fig. 2. The terrorist network showing yellow node (33) as most leader using role structural mining

5.1   Role Structural Mining

In terrorist network as shown in figure 1, three centrality measures were calculated: degree, betweenness, and closeness for each terrorist. Degree measure was used to identify the leaders in the network. Gatekeepers (brokers), members with high betweenness, hold special interest because gatekeepers are usually the contact person between several terrorists and play important roles in coordinating terrorist attacks. The closeness measure was used differently from the previous two centrality measures. Instead of terrorists with high closeness, we identified those with low closeness whom are usually called outliers in SNA literature. Outliers are of special interest because previous literature showed that, in illegal networks, outliers could be the true leaders [11]. They often direct the whole network from behind the scene, which prevents authorities from getting enough intelligence on them. Table 1 summarizes the different roles played by top 5 terrorists identified by the 3 centrality measures. Role structural mining

Table 1. Different roles suggested by role structural mining

S. No.	Leader	Gate Keeper (Broker)	Outlier
1	33	33	46
2	40	21	40
3	46	15	33

has identified 33 Mohamed Atta (the yellow node) as the leader, gatekeeper and outlier of the September 11 attack because he had the highest degree and betweenness and lowest closeness in the network as shown in figure 2.

5.2   Rank Structural Mining

Combine Centrality Actor Ranking. After Applying combine centrality actor ranking from equation 5 the top 6 terrorists are shown in table 2.

Table 2. Ranking of terrorists using combine centrality actor ranking

Ranking	Terrorist ID	Combine Centrality Actor Ranking Score
1	33	1.456
2	40	0.878
3	46	0.869
4	21	0.85
5	15	0.784
6	26	0.638

Eigenvector Centrality Actor Ranking. Applying Eigenvector Centrality Actor Ranking from equation 7, we calculate the ranking score of all terrorists. Among these actors the most prominent three terrorists in the 9-11 terrorist network are shown in table 3.

Table 3. The most important terrorists using Eigenvector Centrality Score

Ranking	Terrorist ID	Eigenvector Centrality Actor Ranking
1	33	0.407
2	40	0.403
3	41	0.286

This method identify that terrorist 33 (Mohamed Atta) is the top leader of the network, whereas actor 40 and 41 takes the position of second and third respectively in the network.

5.3   Network Disruption

Using the concept of graph structural mining as discussed, individuals who are key in the terrorist networks are identified and then removed. Their removal serves to weaken or break the network so that information flows slower and the network as a whole is no longer a single entity [18]. The centrality approach, consisting of measuring the centrality [13] of each node in the network, then selecting a small number of most central nodes as targets for removal, is an intuitive approach of network disruption. The six important actors (blue nodes) of the network are shown in figure 3 using higher Combine Centrality Actor

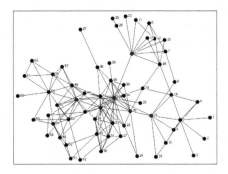

Fig. 3. Network showing important terrorists (blue nodes) as proposed by Combine Centrality Actor Ranking

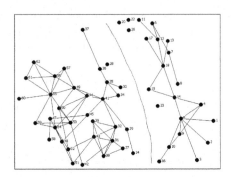

Fig. 4. Network's disruption by removing important actors as suggested by Combine Centrality Actor Ranking

Ranking score from Table 2. After removing them, the network is broken into two parts (separated by blue line) as shown in figure 4. The three most important actors (red nodes) are shown in figure 5 using Eigen vector centrality actor ranking score from table 3. After removing the top three actors (red nodes) from the network shown in figure, the network is broken into two parts (separated by red line) as shown in figure 6.

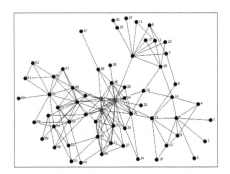

Fig. 5. Network showing important terrorists (red nodes) as proposed by Eigenvector Centrality Actor Ranking

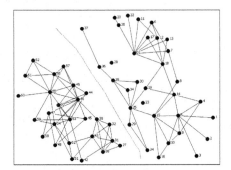

Fig. 6. Network's disruption by removing important actors as suggested by Eigenvector Centrality Actor Ranking

6   Conclusion

A clear understanding of network structures and individual roles can help law enforcement and intelligence agencies to develop effective strategies in order to prevent future terror attacks. The paper has discussed how different roles and important actors are identified and how network disruption strategy can be designed using the concept of graph structural mining. We have shown that not all the members play equal roles in a network. Instead, some members may have stronger social influences or higher social status than the others. Also we have shown that terrorist network members who occupy central positions should be targeted for removal or surveillance. Removal of these central members may effectively disrupt the network and put the operation of terrorist organizations out of action.

References

1. McIllwain, J.S.: Organized crime: A social network approach. Crime, Law & Social Change 32, 301–323 (1999)
2. Xu, J., Chen, H.: Criminal network analysis and visualization: A data mining perspective. Communications of the ACM 48(6), 101–107 (2005)
3. Krebs, V.E.: Mapping networks of terrorist cells. Connections 24(3), 43–52 (2001)
4. Stewart, T.: Six degrees of mohamed atta 2001 (accessed on March 10, 2006), http://money.cnn.com/magazines/business2
5. Borgatti, S.: The key player problem. In: Proceedings from National Academy of Sciences Workshop on Terrorism, Washington DC (2002)
6. Hougham, V.: Sociological skills used in the capture of saddam hussein 1991 (accessed on January 6, 2007), http://www.asanet.org/footnotes/julyaugust05/fn3.html
7. Chan, K., Liebowitz, J.: The synergy of social network analysis and knowledge mapping: a case study. Int. J. Management and Decision Making 7(1), 19–35 (2006)
8. Wasserman, S., Faust, K.: Social Network Analysis: Methods and Applications. Cambridge University Press, Cambridge (1994)
9. McAndrew, D.: The structural analysis of criminal networks. In: Canter, D., Alison, L. (eds.) The Social Psychology of Crime: Groups, Teams, and Networks, Offender Profiling Series, III, Aldershot, Dartmouth, pp. 53–94 (1999)
10. Sparrow, M.K.: The application of network analysis to criminal intelligence: An assessment of the prospects. Social Networks 13, 251–274 (1991)
11. Qin, J., et al.: Analyzing terrorist networks: A case study of the global salafi jihad network. In: Kantor, P., Muresan, G., Roberts, F., Zeng, D.D., Wang, F.-Y., Chen, H., Merkle, R.C. (eds.) ISI 2005. LNCS, vol. 3495, pp. 287–304. Springer, Heidelberg (2005)
12. Baker, W.E., Faulkner, R.R.: The social organization of conspiracy: Illegal networks in the heavy electrical equipment industry. American Sociological Review 58(12), 837–860 (1993)
13. Freeman, L.C.: Centrality in social networks: Conceptual clarification. Social Networks 1, 215–239 (1979)
14. Newman, M.E.J.: The structure and function of complex networks. SIAM Review 45, 167–256 (2003)
15. Bonacich, P.: Factoring and weighting approaches to status scores and clique identification. Journal of Mathematical Sociology 2, 113–120 (1972)
16. Carley, K.M., Lee, J.-S., Krackhardt, D.: Destabilizing networks. Connections 24(3), 79–92 (2002)
17. Accessed on January 9, 2007, http://www.netminer.com/NetMiner/home_01.jsp
18. Xu, J., Chen, H.: Crime net explorer: A framework for criminal network knowledge discovery. ACM Transactions on Information Systems 23(2), 201–226 (2005)

Characterizing Pseudobase and Predicting RNA Secondary Structure with Simple H-Type Pseudoknots Based on Dynamic Programming

Oyun-Erdene Namsrai and Keun Ho Ryu[*]

Database/BioInformatics lab,
School of Electrical & Computer Engineering,
Chungbuk National University
Cheongju, Chungbuk 361-763, Korea
Tel.: +82-43-261-2254, Fax: +82-43-275-2254
{oyunerdene, khryu}@dblab.chungbuk.ac.kr

Abstract. RNA is a unique biopolymer that has the ability to store genetic information, like DNA, but also can have a functional role in the cell, like protein. The function of an RNA is determined by its sequence and structure, and the RNA structure is to a large extent determined by RNA's ability to form base pairs with itself. Most work has been done to predict structures that do not contain pseudoknots. Pseudoknots are usually excluded due to the hardness of examining all possible structures efficiently and model the energy correctly. In this paper we will present characterization of Pseudobase and then we will introduce an improved version of dynamic programming solution to find the conformation with the maximum number of base pairs. After then we will introduce an implementation of predicting H-type pseudoknots based on dynamic programming. Our algorithm called "Iterated Dynamic Programming" has better space and time complexity than the previously known algorithms. The algorithm has a worst case complexity of $O(N^3)$ in time and $O(N^2)$ in storage. In addition, our approach can be easily extended and applied to other classes of more general pseudoknots. Availability: The algorithm has been implemented in C++ in a program called "IDP", which is available at http://dblab.cbu.ac.kr/idp.

Keywords: RNA secondary structure prediction, Pseudobase, H-type pseudoknots, dynamic programming.

1 Introduction

Determining the secondary structure of an RNA molecule is an integral part of understanding the function of the RNA molecules. Methods for determining the structure of an RNA molecule in a biology lab are expensive, and so a computational means for reliably predicting the secondary structure directly from the base sequence would be very valuable.

---

[*] Corresponding author.

R. Alhajj et al. (Eds.): ADMA 2007, LNAI 4632, pp. 578–585, 2007.

Thus, the knowledge of secondary and tertiary structures of an RNA molecule is highly desirable when investigating its role in a cell. Since experimental determination of RNA structure is time-consuming and expensive, its computational prediction is of great interest, and some efficient solutions are known.

Here we present an algorithm, IDP (Iterated Dynamic Programming), for prediction of RNA secondary structure, including H-type pseudoknots, which improve on the prediction quality of previous algorithms. [1],[2],[3].

This paper is organized as follows. First we introduce related works and then we will present some Preliminaries of RNA secondary structure. In Algorithms and Experimental review section, we give a detailed overview of our algorithm, summarize properties of sequences used in our evaluation of the algorithm, and describe our experimental settings. Finally we discuss results in the Discussion section, and conclude this section with an outlook on future work.

2 Preliminaries

An RNA molecule is a sequence of nucleotides which differs from other RNA molecules in its bases. The complementary Watson–Crick bases, C-G and A-U, form stable hydrogen bonds (*base pairs*) when they form a contact. The wobble pair G-U constitutes another strong base pair. These are the three most commonly considered base pairings and are also what we consider in our model.

2.1 RNA Secondary Structure

A secondary structure S is formally defined as the set of all base pairs (i, j) with $i < j$ such that for any two base pairs (i, j) and (k, l) with $i \leq k$ the two following conditions:

1. $i = k$ if and only if $j = l$.
2. There are no knots or pseudo knots allowed. For any two base pairs (i, j) and (k, l) the condition $i < k < l < j$ or $k < i < j < l$ must be satisfied.

2.2 RNA Secondary Structure with Pseudoknots

A folding of RNA secondary structure such as that shown in Fig. 2 is a pseudoknotted structure. To define a pseudoknot, let us consider an RNA sequence which consists of bases with index 0 to n. We also define a base-pair to be (i, j) if a base with index i form a pair with another base with index j in the same sequence. A pseudoknot is then defined as two base-pairs (i, j) and (i', j') such that $0 < i < i' < j < j' < n$ where 0 and n are the indices to the two ends of the RNA. In other words, if the base-pairs overlap each other, we have a pseudoknot as shown in Fig. 2.

Pseudoknots can be classified into 3 main types: I-type pseudoknot (interior loop); B-type pseudoknot (bulge loop); H-type pseudoknot (hairpin loop).

3 Characterizing Pseudobase and H-Type Pseudoknots

PseudoBase is a database containing structural, functional and sequence data related to RNA pseudoknots. [4],[5].

It can be reached at http://wwwbio.LeidenUniv.nl/~Batenburg/PKB.html.

As of May 2006, there are 245 unique pseudoknots in PseudoBase database. Of these, 189 or 77% are the simple "ABAB or ([ )] " H-type pseudoknots, 12 or 4.9% are the LL type pseudoknots (as shown Table1), 24 or 9.8% are the HLout type, 11 or 4.5% are the HLin type, 2 are double pseudoknotted structure where H stands for a hairpin loop; L stands for a bulge loop, internal loop or multiple loop.

H-type pseudoknots are the most abundant of all known pseudoknots, furthermore they are the only type of pseudoknot for which an energy model exists.

Hence, the prediction of H-type pseudoknots should improve our understanding of RNA structures and their associated functions. In principle, an H-type pseudoknots (H-stands for a hairpin loop), may contain 2 stems and three loops as shown in Fig 1.

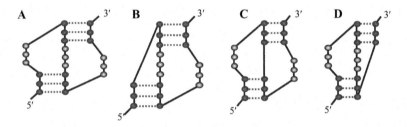

Fig. 1. Representations of simple H-type pseudoknots **A)** General presentation, B) Loop1 is eliminated, C) Loop2 is eliminated d) Loop3 is eliminated

Loop2 is often very short, in our observation 173 of the 189 unique simple H-type pseudoknots have Zero length (as shown in Fig2. B), 187 of the 245 or 80% have Loop2≤1.

After we had analyzed all the pseudoknots in Pseudobase, we classified it into six basic pseudoknot types as shown in Table 1 and Fig. 2.

Our algorithm is based on a dynamic programming solution to find the conformation with the maximum number of base pairs. The algorithm for loop matchings [6] is dynamic, short and exceedingly simple. For a molecule n nucleotides long, it finds the maximum number of base pairs using n^3 and n^2 memory units. One drawback of this algorithm is: all base pairs, i.e. AU, GC and GU are given equal weights and stacking interactions are ignored.

Based on our observation of total numbers of regular base pairs and wobble base pairings found in H-type pseudoknots (shown in Table 2), we give 10 and 6 for the regular base pairs, respectively, and 1 for GU non-regular base pair considering the oxygen –oxygen rebounding and 0 to other combinations.

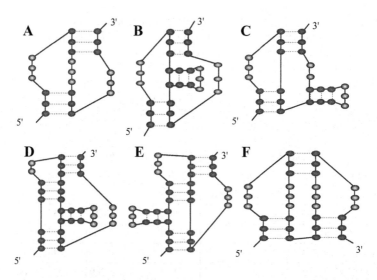

Fig. 2. Representations of Pseudoknot types: (A) Simple ABAB type, (B) LL type, (C) HLout type, (D) HLin type, (E) HH type, (F) HHH type. (ABACBC kissing hairpin structure). H: hairpin loop, L: internal loop, Bulge Loop or Multiple loop.

Table 1. Classification of 245 pseudoknots in Pseudobase

Type	PKB#	# of occ.	% of occ.
Loop in Loop	174, 140, 144, 145, 146, 143, 142, 141, 57 , 139, 150*, 76	12	4.9
Hairpin & Loop out	191, 35*, 134*,138* 136*,137*,168*,164* 173*,208 , 209,229,223,220,215,76*,212,232,226, 148,149,131, 132, 133	24	9.8
Hairpin & Loop in	135*,134*, 138*,136*, 137*,168*234, 164*, 205, 129 , 210	11	4.5%
HH type	181	1	0.4
HHH type	171, 169, 178, 173*, 150*, 163	6	2.4
Double pseudoknot PKB#71,#75		2	0.8
All others (ABAB or " ([ )] ")		**189**	**77**
Total		**245**	

4   Algorithms

As mentioned before, our algorithm is based on a dynamic programming solution to find the conformation with the maximum number of base pairs.

Firstly, we will give short introduction of this algorithm. After then we will show how an RNA secondary structure prediction with simple H-type pseudoknots which has the maximum number of base pairs can be computed by the dynamic programming algorithm in $O(N^3)$ time abd $O(N^2)$ space.

Table 2. Statistics of base pairings found in Pseudobase database. Columns are: Pseudobase categories, number of Pseudoknots, number of H-type pseudoknots, total number of base pairings found in H-type pseudoknots, total number of GC, AU and GU base pairings.

Pseudobase	#of PS	h-type	#BP	GC	AU	GU
I Viral ribosomal frameshifting signals	20	18	202	162	30	10
II Viral ribosomal readthroush signals	7	7	103	87	14	2
III Viral tRNA like structures	54	40	329	245	76	8
IV Other viral 5'–UTR	4	1	11	9	2	0
V Other viral 3'-UTR	81	80	792	391	354	47
VI Viral others	27	17	269	102	144	23
VII rRNA	6	3	41	19	15	7
VIII mRNA	10	9	165	91	64	10
IX tmRNA	19	8	123	69	39	15
X Ribozymes	6	0	0	0	0	0
XI Artamers	6	3	31	21	9	1
XII Artifical molecules	1	1	8	6	2	0
XIII Others	5	3	60	37	22	1
Total	246	190	2134	1239	771	124

4.1 RNA Folding by Dynamic Programming

One of the first attempts to predict RNA secondary structure was made by Ruth Nussinov and co-workers who used dynamic programming method for maximizing the number of base-pairs [6]. However, there are several issues with this approach, most importantly the solution is not unique. Nussinov et al published an adaptation of their approach to use a simple nearest-neighbour energy model in 1980 [7]. The algorithm 's search procedure is conducted in the following manner. Each nucleotide segment is examined for the structure with minimal energy:

$$E(i,j) = \min \begin{cases} E(i,k-1) + E(k+1, j-1) + E_{ki} \\ E(i,j-1) \quad i<=k<j-1 \min \end{cases} \tag{1}$$

That is, the energy (E) of the minimal energy structure formed between B_i and B_j is determined by examining each possible base pair $B_k B_j$ [4]. The basic approach is simular to the one described for the algorithm for maximal matching. Only one simple backtracking routine has been added to determine the location of each newly formed base pair with respect to its neighbors.

To improve prediction quality of dynamic programming solution to find the conformation with the maximum number of base pairs we introduce new scoring table. There are 2134 base pairs found in total 190 pseudoknots, of these 1239 or 58% are the GC pairs (as shown Table 2), 771 or 36% are the AU pairs, and 124 or 5.8% are the AU base pairs. According to the proportion of real world data and energy rules we give 10 and 6 for the regular base pairs, respectively, and 1 for GU non-regular base pair considering the oxygen –oxygen rebounding and 0 to other combinations. Based on these restrictions and the scoring table, we improved prediction quality of previous works.

One of the issues when predicting RNA secondary structure is that the standard recursions exclude pseudoknots. [6],[7] Now we will introduce a dynamic programming algorithm that could handle simple H-type pseudoknots.

4.2   Iterated Base Pair Maximizing

We now extend the basic dynamic programming algorithm to accommodate H-type pseudoknots. H-type pseudoknot can be thought as an interaction between two loop regions of a secondary structure, as illustrated in Fig. 4;

Therefore, we can run the previous improved base pair maximizing algorithm twice to identify simple H-type pseudoknots. In other words we could identify this kind of pseudoknot in two steps. Firstly we predict a secondary structure of simple H-type pseudoknot using the traditional algorithm. Secondly we remove all the paired bases and then predict remaining sequence. As mentioned before H-type pseudoknots the only type of pseudoknot for which an energy model exists.

A. ([ )] = ( )+[ ]   **B.**   ABAB =AA+ BB

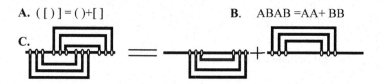

Fig. 4. A pseudoknot can be treated as two separate helices. First helices is identified by improved DP algorithm, second one is identified by IDP algorithm.

Our algorithm sketch as follows:

1) Prepare input sequence. It should contain RNA sequence only, without any comment or additional information

2) Run the improved dynamic programming method and retrieve the base-pairing score matrix

3) Write to output base pair list, which are found from the score matrix.

4) Eliminate those base pairs from input sequence and repeat steps 2-4 until no bases remain.

5) Merge output results and write it to final output. Output file contain only predicted H-type secondary structure of RNA.

Similarly, more complicated pseudoknots as shown in Fig. 2 (HHH, HH, HLin, Hlout type) may identified with more iterations.

We run the IDP algorithm multiple times, and each time we only accept the base-pairs that appear to be the most reliable, e.g. with the highest score. This modification attempts to avoid possible false predictions.

5 Experimental Review

We predicted the structure of 15 pseudoknotted RNA sequences taken from PseudoBase. Average lengths of our experimental sequences taken from PseudoBase were around 25 nucleotides.

Table 2. Some output of our IDP, PKB # is the Sequence ID on PeudoBase database, PBase is denotes the bracket view of structure on PseudoBase

	PKB #	PKB37
1.	Sequence	CCCCUUAACUUGAGGGAAAUCAAGC
	PBase	:(((:::::[[[[)))::::]]]]]:
	Prediction	:(((:::::[[[[)))::::]]]]:
	PKB #	PKB39
2.	Sequence	CCCCUAACUUGAGGGAAAUCAAGC
	PBase	:(((:::[[[[[)))::::]]]]]:
	Prediction	::(((::::[[[[)))::::::]]]]:
	PKB #	PKB20
3.	Sequence	CCCUUUUCCCCGGGUAGGGGC
	PBase	:(((::::[[[[)))::]]]]:
	Prediction	:(((::::[[[[)):::]]]]:

5.1 Experimental Settings and Results

As mentioned before PseudoBase is a database containing structural, functional and sequence data related to RNA pseudoknots. From here one can retrieve pseudoknot data as well as submit data for pseudoknots that are not yet in the database.

To evaluate our software we compared our result to that of the current best known software for globally optimal RNA pseudoknot prediction developed by Elena Rivas [8]. That algorithm has a worst case complexity of $O(N^6)$ in time and $O(N^4)$ in storage.

Our approach based on traditional base pair maximizing algorithm. The algorithm has worst case time and space complexities of $O(N^3)$ and $O(N^2)$, respectively.

IDP folds 14 pseudoknots out of 15 in the Pseudobase database. Our program predicts simple pseudoknots with correct or almost correct structure for 65% sequences. Table 2. Shows some output of our IDP program. From 120 pairs which are included in our total 15 pseudoknotted sequences, we predicted total 78 base pairs.

6 Discussion

In this paper, we analyzed whole pseudoknots stored in Pseudobase Database. We classified it into six main categories and represented graphical representations of them. Therefore we observed that H-type pseudoknots are the most abundant of all known pseudoknots.

We presented new improved algorithm for RNA secondary structure prediction. Our algorithm is based on dynamic programming solution to find the conformation with the maximum number of base pairs.

We also presented an algorithm for RNA secondary structure prediction with H-type pseudoknots, based on improved DP algorithm. Our experimental results indicate that our approach works well.

Our algorithm's advantage is, it has better space and time complexity than the previously known algorithms. But our approach is valid for only H-type pseudoknots.

In the future work we will check that possibility of IDP algorithm with more complicated pseudoknots and we will find a new solution for predicting RNA more general pseudoknots.

Acknowledgements. This work was supported by the Korea Research Foundation Grant funded by the Korean Government (MOEHRD) (The Regional Research Universities Program BITRC)

References

1. Namsrai, O.-E., Jung, K.S., Kim, S., Ryu, K.H.: RNA secondary structure prediction with simple pseudoknots based on dynamic programming. In: Huang, D.-S., Li, K., Irwin, G.W. (eds.) ICIC 2006. LNCS (LNBI), vol. 4115, pp. 303–311. Springer, Heidelberg (2006)
2. Namsrai, O.-E., Jung, K.S., Kim, S., Ryu, K.H.: RNA secondary structure prediction with simple H-type pseudoknots. In: Pacific symposium on Biocomputing, 2007.01, USA Stanford
3. Namsrai, O.-E., Jung, K.S., Kim, S., Ryu, K.H.: Characterizing Pseudobase and Predicting RNA secondary structure with simple H-type pseudoknots. In: Proceedings of the International Symposium on Bioinformatics Research and Applications, Atlanta, USA, May 2007, pp. 23–27 (2007)
4. van Batenburg, F.H.D., Gultyaev, A.P., Pleij, C.W.A.: PseudoBase: structural information on RNA Pseudoknots. Nucleic Acids Res. 29, 194–195 (2001)
5. van Batenburg, F.H.D., Gultyaev, A.P., Pleij, C.W.A, Ng, J., Oliehoek, J.: PseudoBase: a database with RNA pseudoknots. Nucleic Acids Res. 28, 201–204 (2000)
6. Nussinov, R.P., Eddy, S.R., Griggs, J.R., Kleitman, D.J.: Algorithms for loop matching. SIAM Journal of Applied Mathematics 35, 68–82 (1978)
7. Nussinov, R., Jacobson, A.: Fast algorithm for predicting the secondary structure of single-stranded RNA. Proc. Natl. Acad. Sci. 77, 6309–6313 (1980)
8. Rivas, E., Eddy, S.: A dynamic programming algorithm for RNA structure prediction including pseudoknots. Journal of Molecular Biology 285 (1999)

Locally Discriminant Projection with Kernels for Feature Extraction

Jun-Bao Li[1], Shu-Chuan Chu[2], and Jeng-Shyang Pan[3]

[1] Department of Automatic Test and Control, Harbin Institute of Technology, Harbin, China
junbaolihit@hotmail.com
[2] Department of Information Management, Cheng Shiu University, Kaohsiung, Taiwan
[3] Department of Electronic Engineering, National Kaohsiung University of Applied Sciences, Kaohsiung, Taiwan
jspan@cc.kuas.edu.tw

Abstract. Local Preserving Projection (LPP) is an unsupervised feature extraction method which considers the nearest neighbor information and has little to do with the class information, and it fails to perform well for the nonlinear problems due to its limitation of linearity. In this paper, we extend LPP to propose a novel feature extraction namely Kernel Locally Discriminant Projection (KLDP) by considering class label information and the nonlinear problems. The main work lies in: 1) the class label information is considered to create the similarity measure for the local structure graph; 2) the class-wise cosine similarity measure is applied to solve the selection of the free parameter of the similarity measure; 3) kernel method is applied to solve limitation of linearity of LPP. Besides some theory derivation, the experiments are implemented on ORL and Yale face database to evaluate the feasibility of the proposed algorithm.

1 Introduction

Recently locality preserving projection (LPP) [6] which shares some similar properties to LLE [1] discovers the submanifold structure of the data is widely applied to feature extraction as an effective dimensionality reduction, but LPP endures one constrained condition that the data can be modeled by a nearest neighbor graph which preserves the local structure of the input space. But the nearest neighbor method ignores the class label information, so it is not adapt to the supervised learning. So for the supervised learning, the class label information can enhance the performance of classification. Moreover, LPP only consider the data in the linear space with ignoring the nonlinear information, which limits its application.

In this paper, we consider the class label information enough to create the similarity measure to create the graph for LPP to propose Locally Discriminant Projection (LDP) algorithm for feature extraction. In order to avoid the problem that the selection of the free parameter influences the performance of the nearest neighbor similarity measure used in the traditional LPP algorithm, we apply the cosine similarity measure to create the graph, which dose not need to select the free parameter.

R. Alhajj et al. (Eds.): ADMA 2007, LNAI 4632, pp. 586–593, 2007.

Moreover we apply kernel method, which is widely used to solve the nonlinear problem successfully [4], [5], to enhance the LDP algorithm to propose the Kernel Locally Discriminant Projection (KLDP).

The rest of paper is organized as follows. In section 2, we introduce the Kernel Locally Discriminant Projection (KLDP) algorithm. Experimental results and conclusion are given in Section 3 and Section 4 respectively.

2 Kernel Locally Discriminant Projection (KLDP)

In this section, firstly we review locality preserving projection (LPP) algorithm, and secondly we introduce the class-wise similarity measures to propose the Kernel Locally Discriminant Projection (KLDP) and finally we apply kernel method to LDP to solve its limitation of linearity.

2.1 Locality Preserving Projection

Let us assume that the N-dimensional data set $\{x_1, x_2, ..., x_n\}$ are distributed on a low-dimensional submanifold. Our goal is to seek a set of data points $\{z_1, z_2, ..., z_n\}$ with the same local neighborhood structure as $\{x_1, x_2, ..., x_n\}$

$$\min \sum_{i,j} \|z_i - z_j\|^2 S_{ij} \tag{1}$$

where S is a similarity matrix with weights characterizing the likelihood of two points. To make the mapping not defined only for the original data set, LPP define a transformation:

$$z_i = w^T x_i \tag{2}$$

where $w = [w_1, w_2, ..., w_d] \in \mathbb{R}^{N \times d}$. The transformation matrix W that minimizes the objective function can be obtained by solving the generalized eigenvalue problem:

$$XLX^T w = \lambda XDX^T w \tag{3}$$

where $L = D - S$ and $D = diag\left[\sum_j S_{1j}, \sum_j S_{2j}, ..., \sum_j S_{nj}\right]$. Now assume N-dimensional

$$S_{ij} = \begin{cases} \exp\left(-\frac{1}{\delta}\|x_i - x_j\|^2\right) & \begin{array}{l} \text{if } x_i \text{ is among k nearest neighbors of } x_j \\ \text{or } x_j \text{ is among nearest neighbors of } x_i; \end{array} \\ 0 & \text{otherwise} \end{cases} \tag{4}$$

From the above theoretical derivation, we can easily find that LPP is an unsupervised feature extraction method. We propose Locally Discriminant Projection (LDP) algorithm for the class-wise feature extraction as follows.

2.2 Locally Discriminant Projection (LDP)

In class-wise feature extraction methods, the label information can be used to guide the procedure of feature extraction. In our work, our goal is to extend LPP to be a supervised feature extraction method, in which the similarity measure can be defined as follows.

Class-wise similarity measure 1

$$S_{ij} = \begin{cases} 1 & \text{if } x_i \text{ and } x_j \text{ belong to the same class;} \\ 0 & \text{otherwise} \end{cases} \tag{5}$$

Class-wise similarity measure 2

$$S_{ij} = \begin{cases} \exp\left(-\frac{1}{\delta}\|x_i - x_j\|^2\right) & \text{if } x_i \text{ and } x_j \text{ belong to the same class;} \\ 0 & \text{otherwise} \end{cases} \tag{6}$$

Class-wise similarity measure 3

$$S_{ij} = \begin{cases} \dfrac{x_i^T x_j}{\|x_i\|\|x_j\|} & \text{if } x_i \text{ and } x_j \text{ belong to the same class;} \\ 0 & \text{otherwise} \end{cases} \tag{7}$$

The LDP algorithm fails to perform well for the nonlinear problems, such as face recognition, due to its limitation of linearity. To overcome this weakness of LDP, the non-linear versions, Class-wise Kernel Local Structure Preserving Projection (KLDP) is proposed in the next section.

2.3 Kernel Locally Discriminant Projection (KLDP)

The nonlinear mapping Φ is used to map the input data space $\mathbb{R}^N$ into the feature space F, $\Phi : \mathbb{R}^N \to F$, $x \mapsto \Phi(x)$. Correspondingly, a pattern in the original input space $\mathbb{R}^N$ is mapped into a potentially much higher dimensional feature vector in the feature space F, that is

$$z^{\text{ker}} = \left(w^{\text{ker}}\right)^T \Phi(x) \tag{8}$$

Our goal is to seek a set of data points $\{z_1, z_2, ..., z_n\}$ with the same local neighborhood structure as $\{\Phi(x_1), \Phi(x_2), ..., \Phi(x_n)\}$ in the nonlinear mapping space.

$$\min \sum_{i,j}^{n} \left\| z_i^{\text{ker}} - z_j^{\text{ker}} \right\|^2 S_{ij}^{\text{ker}} \tag{9}$$

Similar to above linear algorithm, the transformation matrix w^{ker} is that minimizes the objective function can be obtained by solving a generalized eigenvalue problem. Since any of its eigenvector can be expressed by a linear combination of the observations in feature space, we have

$$w^{\text{ker}} = \sum_{p=1}^{n} \alpha_p \Phi(x_p) = Q\alpha \tag{10}$$

where $Q = \begin{bmatrix} \Phi(x_1) & \Phi(x_2) & \cdots & \Phi(x_n) \end{bmatrix}$ and $\alpha = \begin{bmatrix} \alpha_1 & \alpha_2 & \cdots & \alpha_n \end{bmatrix}$, since $k(x_i, x_j) = \Phi(x_i)^T \Phi(x_j)$, So it is easy to obtain the kernel matrix K, that is, $K = Q^T Q$.

Proposition 1. $\frac{1}{2}\sum_{i,j}^{n}\left\|z_i^{\text{ker}}-z_j^{\text{ker}}\right\|^2 S_{ij}^{\text{ker}}=\alpha^T K\left(D^{\text{ker}}-S^{\text{ker}}\right)K\alpha$, where S^{ker} is a similarity

matrix with samples in the nonlinear mapping feature space and

$$D^{\text{ker}}=diag\left[\sum_j S_{1j}^{\text{ker}},\sum_j S_{2j}^{\text{ker}},...,\sum_j S_{nj}^{\text{ker}}\right].$$

Matrix D^{ker} provides a natural measure on the data points. The bigger the value D_{ii} (corresponding to z_i^{ker}) is, the more "important" is z_i^{ker}. Therefore we impose a constraint as follows.

$$\left(Z^{\text{ker}}\right)^T D^{\text{ker}} Z^{\text{ker}}=1 \tag{11}$$

That is, $\alpha^T K D^{\text{ker}} K \alpha = 1$. Let $L^{\text{ker}}=D^{\text{ker}}-S^{\text{ker}}$, an optimization problem can be obtained as follows.

$$\min_{\alpha}\alpha^T K L^{\text{ker}} K\alpha$$

Subject to $\alpha^T K D^{\text{ker}} K\alpha=1$ \hfill (12)

Now let us consider QR decomposition of matrix K , $r_1,r_2,...,r_m$ are K 's orthonormal eigenvector corresponding to m largest nonzero eigenvalue $\lambda_1,\lambda_2,...,\lambda_m$.

$$K=P\Lambda P^T \tag{13}$$

where $P=\left[r_1,r_2,...,r_m\right]$ and $\Lambda=diag\left(\lambda_1,\lambda_2,...,\lambda_m\right)$. Let $\beta=\Lambda^{\frac{1}{2}}P^T\alpha$, i.e., $\alpha=P\Lambda^{-\frac{1}{2}}\beta$, $L_k=\Lambda^{\frac{1}{2}}P^T L^{\text{ker}}P\Lambda^{\frac{1}{2}}$, $D_k=\Lambda^{\frac{1}{2}}P^T D^{\text{ker}}P\Lambda^{\frac{1}{2}}$, then we can transform the above optimization problem (12) to the dual optimization problem as follows.

$$\min_{\beta}\beta^T L_k\beta$$

Subject to $\beta^T D_k\beta=I$ \hfill (14)

The solution of the above constrained optimization problem can often be found by using the so-called Lagrangian method. We define the Lagrangian as

$$L(\beta,\lambda)=\beta^T L_k\beta-\lambda\left(\beta^T D_k\beta-I\right) \tag{15}$$

with the parameter λ. The Lagrangian L must be minimized with respect to λ and β, and the derivatives of L with respect to β are must vanish, hence

$$L_k\beta=\lambda D_k\beta \tag{16}$$

Supposed λ is nonzero value, we can obtain, $D_k\beta=\frac{1}{\lambda}L_k\beta$. Let $\lambda^*=\frac{1}{\lambda}$, then we can solve the eigenvectors of the generalized equation problem shown as follows.

$$D_k\beta=\lambda^* L_k\beta \tag{17}$$

The solution of the above constrained optimization problem is equal to the eigenvector corresponding to the largest eigenvalue.

Proposition 2. Assume β^* is eigenvalue equation (17) corresponding to the largest nonzero eigenvalue λ^*, then $\beta^T L_k \beta$ has a minimum value at β^*.

Then, we consider the transformation matrix w^{ker} as follows.

$$w^{ker} = \left[w_1^{ker}, w_2^{ker}, ..., w_m^{ker} \right] = \left(QP\Lambda^{-\frac{1}{2}} \right)[\beta_1, \beta_2, ..., \beta_m] \qquad (18)$$

For a given sample x, the transformation can be divided into two items:

$$y = \left(QP\Lambda^{-\frac{1}{2}} \right)^T \Phi(x) \qquad (19)$$

and

$$z^{ker} = B^T y \qquad (20)$$

where $B = [\beta_1, \beta_2, ..., \beta_m]$. Now let us consider equation (19) as follows.

$$y = \left(QP\Lambda^{-\frac{1}{2}} \right)^T \Phi(x) = \Lambda^{-\frac{1}{2}} P^T Q^T \Phi(x)$$

$$= \left(\frac{r_1}{\sqrt{\lambda_1}} \quad \frac{r_2}{\sqrt{\lambda_2}} \quad \cdots \quad \frac{r_m}{\sqrt{\lambda_m}} \right)^T [k(x_1, x), k(x_2, x), ..., k(x_n, x)]^T \qquad (21)$$

Since $r_1, r_2, ..., r_m$ are K's orthonormal eigenvector corresponding to m largest non-zero eigenvalue $\lambda_1, \lambda_2, ..., \lambda_m$, the transformation in (21) is exactly the KPCA transformation, which transforms feature space F into Euclidean space $\mathbb{R}^m$.

Accordingly the class-wise similarity measure with kernel version can be defined as follows.

Class-wise similarity measure 1

$$S_{ij} = \begin{cases} 1 & \text{if } x_i \text{ and } x_j \text{ belong to the same class;} \\ 0 & \text{otherwise} \end{cases} \qquad (22)$$

Class-wise similarity measure 2

$$S_{ij} = \begin{cases} \exp\left(-\frac{1}{\delta}\left(k(x_i, x_i) + k(x_j, x_j) - 2k(x_i, x_j)\right)\right) & \text{if } x_i \text{ and } x_j \text{ belong to the same class;} \\ 0 & \text{otherwise} \end{cases} \qquad (23)$$

Class-wise similarity measure 3

$$S_{ij} = \begin{cases} \dfrac{k(x_i, x_j)}{\sqrt{k(x_i, x_i)}\sqrt{k(x_j, x_j)}} & \text{if } x_i \text{ and } x_j \text{ belong to the same class;} \\ 0 & \text{otherwise} \end{cases} \qquad (24)$$

Algorithm procedure of KLDP can be described as follows.

Step 1: Construct kernel matrix K, similarity matrix S^{ker} and D^{ker} with a given kernel function $k(x, y)$.

Step 2: Obtain the m eigenvectors $r_1, r_2, ..., r_m$ of kernel matrix K corresponding to m largest eigenvalue. $\lambda_1, \lambda_2, ..., \lambda_m$, where $m = rank(K)$.

Step 3: Construct matrix $L_k = \Lambda^{\frac{1}{2}} P^T L^{\text{ker}} P \Lambda^{\frac{1}{2}}$ and $D_k = \Lambda^{\frac{1}{2}} P^T D^{\text{ker}} P \Lambda^{\frac{1}{2}}$, where $L^{\text{ker}} = D^{\text{ker}} - S^{\text{ker}}$.

Step 4: Obtain $B = [\beta_1, \beta_2, ..., \beta_d]$ by solving the eigenvalue equation (17) corresponding to d largest nonzero eigenvalue.

Step 5: For a given sample x, obtain KPCA-transformed feature vector y with equation (21).

Step 6: Obtain the KLPP-transformed feature z^{ker} with equation (20).

3 Experimental Results

We implement experiments with two face databases, i.e., ORL face database [2] and Yale face database [3]. We carry out the experiments with three parts as follows. 1) Comparing the performance of k nearest neighbor measure and the proposed similarity measures, i.e., class-wise similarity measure 1, class-wise similarity measure 2, class-wise similarity measure 3; 2) Testing the feasibility of improving performance of the proposed algorithm using kernel method; 3) Testing the superiority of the proposed method compared with the existing methods.

In this section, our goal is to test the performance of k nearest neighbor measure and the proposed similarity measures on ORL face database. Firstly we find the optimal k value of k nearest neighbor measure, and secondly we compare the performance of the above similarity measures in the linear space, and finally we evaluate the performance of the above similarity measures with kernel version.

Firstly we find the optimal k value of k nearest neighbor measure, and we obtain the recognition accuracy of 0.6400, 0.6615, 0.6435 and 0.6870 when $k = 2$, $k = 3$, $k = 4$, $k = 5$ are selected for k nearest neighbor similarity measure in the experiments. So $k = 5$ is selected for k nearest neighbor similarity measure in the next part of experiments.

Secondly we evaluate the recognition performance of four similarity measures on ORL face database. As Table 1 shown, class-wise similarity measure 3 and 2 give a higher recognition performance compared with the other two similarity measures. But the class-wise similarity measure 3 gives a higher or equivalent recognition performance compared with the class-wise similarity measure 2, but the free parameter must be selected by experiments. Accordingly we also implement the four similarity measures with kernel version, and the polynomial and Gaussian kernels are used as the basic kernels as shown in Table 2. We select $\delta = 1 \times 10^{14}$ when the polynomial kernel $d = 2$ is selected; and $\delta = 1 \times 10^{22}$ for polynomial kernel $d = 3$, $\delta = 1 \times 10^{7}$ for Gaussian kernel with $\sigma^2 = 0.5 \times 10^{7}$ and $\delta = 0.5 \times 10^{8}$ for Gaussian kernel with $\sigma^2 = 0.5 \times 10^{8}$. And the dimension of feature vector is 160.

Table 1. Linear similarity measures

Similarity measure	ORL face database
k nearest neighbor measure	0.6870
Class-wise similarity measure-1	0.7580
Class-wise similarity measure-2	0.7680
Class-wise similarity measure-3	0.7635

Table 2. Similarity measures with kernel version

Similarity measures	Polynomial kernel ($d = 2$)	Polynomial kernel ($d=3$)	Gaussian kernel ($\sigma^2 = 0.5 \times 10^7$)	Gaussian kernel ($\sigma^2 = 0.5 \times 10^8$)
k nearest neighbor measure	0.7140	0.7085	0.6685	0.6885
Class-wise similarity measure-1	0.8185	0.8075	0.8660	0.7390
Class-wise similarity measure-2	0.8575	0.8845	0.9070	0.7575
Class-wise similarity measure-3	0.9110	0.9010	0.9035	0.7840

Table 3. Recognition accuracy performance on ORL face databases

Methods	Class-wise similarity measure-1	Class-wise similarity measure-2	Class-wise similarity measure-3
LDP	0.7580	0.7680	0.7435
KLDP	0.8075	0.8845	0.9010

Table 4. Recognition performance on ORL and Yale face databases

Methods	ORL face database	Yale face database
KLDP	0.9110	0.8525
LDP	0.7635	0.8306
LPP	0.6870	0.7911
KPCA	0.8450	0.8131
PCA	0.8000	0.8081

Thirdly we test the feasibility of improving the recognition performance with kernel method on the ORL face database. We compare the recognition performance of KLDP and LDP algorithm. In the experiments, the polynomial kernel with $d = 3$ is selected as the basic kernel. As shown in Table 3, kernel method gives the better performance than the linear method under the same parameters.

Finally we evaluate the recognition performance of KLDP comparing with LDP, LPP, PCA and KPCA on ORL face database and Yale face database. In the experiments, polynomial kernel with $d = 2$ is selected for KLDP and KPCA, and class-wise similarity measure-3 is selected for KLDP and LPP. As shown in Table 4, the proposed LDP algorithm can give a best recognition performance compared with other algorithms.

From the above experiments, we can obtain the following interesting cases: 1) when class label information is considered, we can obtain the higher recognition performance compared with the traditional Local Preserving Projection (LPP) algorithm, which applied the k nearest neighbor similarity measure. 2) Class-wise similarity measure 3 can give a higher or equivalent recognition performance compared with class-wise similarity measure 2, but the free parameter must be selected by experiments. If the free parameter is not adapted to be selected, the algorithm can not give a good recognition performance, which can be solved with the class-wise cosine similarity measure. 3) The face images of databases which are obtained under different lighting conditions, and it contains the nonlinear information. The nonlinear problems can be solved with kernel method, and kernel method enhances the performance of LDP algorithm.

4   Conclusion

A novel supervised feature extraction method, Kernel Locally Discriminant Projection (KLDP), is proposed in this paper. KLDP considers enough the class label information and solves the nonlinear mapping by kernel, which is to deal with the problem of LPP that LPP has little to do with the class information and fails to perform well for the nonlinear problems due to its limitation of linearity.

References

1. Roweis, S.T., Saul, L.K.: Nonlinear Dimensionality Reduction by Locally Linear Embedding. Science 290 (December 2000)
2. Samaria, F., Harter, A.: Parameterisation of a Stochastic Model for Human Face Identification. In: Proceedings of 2nd IEEE Workshop on Applications of Computer Vision, Sarasota FL, December 1994 (1994)
3. Belhumeur, P.N., Hespanha, J.P., Kriegman, D.J.: Eigenfaces vs. Fisherfaces: Recognition Using Class Specific Linear Projection. IEEE Trans. Pattern Analysis and Machine Intelligence 19(7), 711–720 (1997)
4. Lu, J.W., Plataniotis, K., Venetsanopoulos, A.N.: Face recognition using kernel direct discriminant analysis algorithms. IEEE Trans. Neural Network 14(1), 117–126 (2003)
5. Lu, J., Plataniotis, K.N., Venetsanopoulos, A.N.: A kernel machine based approach for multi-view face recognition. Proc. Int. Conf. Image Processing, I-265–I-268 (2002)
6. He, X., Niyogi, P.: Locality Preserving Projections. In: Proc. Conf. Advances in Neural Information Processing Systems (2003)

A GA-Based Feature Subset Selection and Parameter Optimization of Support Vector Machine for Content – Based Image Retrieval

Kwang-Kyu Seo

Department of Industrial Information and Systems Engineering, Sangmyung University,
San 98-20, Anso-Dong, Chonan, Chungnam 330-720, Korea
Tel.: +81-41-550-5371, Fax: +81-41-550-5185
kwangkyu@smu.ac.kr

Abstract. This paper presents the effectiveness of applying genetic algorithm (GA)-based feature subset selection and parameter optimization of support vector machine (SVM) for content-based image retrieval (CBIR). SVM, one of the new techniques for pattern classification, has been widely used in many application areas. The kernel parameters setting for SVM in the training process impacts on the classification accuracy. Feature subset selection is another factor that impacts classification accuracy. The objective of this study is to simultaneously optimize the parameters and feature subset without degrading the SVM classification accuracy using the GA-based approach for CBIR. In this study, we show that the proposed GA-based approach outperforms SVM to the problem of the image classification problem in CBIR. Compared with NN and SVM algorithm, the proposed GA-based approach significantly improves the classification accuracy and has fewer input features for SVM.

1 Introduction

As the need for effective multimedia information services is significant, the research on image classification and retrieval methods has become considerably important in various areas such as entertainment, education, digital libraries, and medical image retrieval. Content-based image retrieval (CBIR) techniques are becoming increasingly important in multimedia information systems in order to store, manage, and retrieve image data to perform assigned task and make intelligent decisions. CBIR uses an automatic indexing scheme where implicit properties of an image can be included in the query to reduce search time for retrieval from a large database [1].

Features like color, texture, shape, spatial relationship among entities of an image and also their combination are generally being used for the computation of multidimensional feature vector. The features such as color, texture and shape are known as primitive features. Images have always been an essential and effective medium for presenting visual data. With advances in today's computer technologies, it is not surprising that in many applications, much of the data is images. There have been considerable researches done on CBIR using artificial neural networks such as backpropagation, SOM, Fuzzy ART and SVMs and so on [2-5].

R. Alhajj et al. (Eds.): ADMA 2007, LNAI 4632, pp. 594–604, 2007.

Recently SVM which was developed by Vapnik [6] is one of the methods that is receiving increasing attention with remarkable results in pattern recognition. SVM classifies data with different class labels by determining a set of support vectors that are members of the set of training inputs that outline a hyperplane in the feature space. SVM provides a generic mechanism that fits the hyperplane surface to the training data using a kernel function. The user may select a kernel function for the SVM during the training process that selects support vectors along the surface of this function.

When using SVM, two problems are confronted such as how to choose the optimal input feature subset for SVM, and how to set the best kernel parameters. These two problems are crucial, because the feature subset choice influences the appropriate kernel parameters and vice versa [7]. Therefore, obtaining the optimal feature subset and SVM parameters must occur simultaneously.

Many practical pattern classification tasks require learning an appropriate classification function that assigns a given input pattern, typically represented by a vector of attribute values to a finite set of classes. Feature selection is used to identify a powerfully predictive subset of fields within a database and reduce the number of fields presented to the mining process. By extracting as much information as possible from a given data set while using the smallest number of features, we can save significant computation time and build models that generalize better for unseen data points. In addition to the feature selection, proper parameters setting can improve the SVM classification accuracy. The parameters that should be optimized include penalty parameter C and the kernel function parameters. To design a SVM, one must choose a kernel function, set the kernel parameters and determine a soft margin constant C.

Genetic algorithm (GA) has the potential to generate both the optimal feature subset and SVM parameters at the same time. We aim to optimize the parameters and feature subset simultaneously, without degrading the SVM classification accuracy. The proposed approach performs feature subset selection and parameters setting in an evolutionary way. In the literature, only a few algorithms have been proposed for SVM feature selection [8-10]. Some other GA-based feature selection methods were proposed [11-12]. However, these papers focused on feature selection and did not deal with parameters optimization for the SVM classifier.

This paper focuses on the improvement of the SVM-based model by means of the integration of GA and SVM in detecting the underlying data pattern for image detection and classification in CBIR using color features based joint HSV (Hue, Saturation and Value) histogram and texture features based on co-occurrence matrix.

The remainder of this paper is organized as follows. Section 2 presents the research background. The color and textural features of images are provided and the basic SVM and GA concepts are also described. Section 3 presents the proposed approach. Section 4 explains the experimental design and the results of the evaluation experiment. The final section contains some concluding remarks.

2   Research Background

This section presents how to extract image features. In this paper, color and texture information are used to represent image features. For color, joint HSV histogram extracted from local region is employed. For texture, entropies computed from local region are employed [5]. In addition, a brief introduction to the SVM and basic GA concepts are described.

2.1   Image Features

(1) Color

For representing color, we used HSV color model because this model is closely related to human visual perception. Hue is used to distinguish colors (e.g. red, yellow, blue) and to determine the redness or greenness etc. of the light. Saturation is the measure of percentage of white light that is added to a pure color. For example, red is a highly saturated color, whereas pink is less saturated. Value refers to the perceived light intensity.

Color quantization is useful for reducing the calculation cost. Furthermore, it provides better performance in image clustering because it can eliminate the detailed color components that can be considered noises. The human visual system is more sensitive to hue than saturation and value so that hue should be quantized finer than saturation and value. In the experiments, we uniformly quantized HSV space into 18 bins for hue (each bin consisting of a range of 20 degree), 3 bins for saturation and 3 bins for value for lower resolution.

In order to represent the local color histogram, we divided image into equal-sized 3×3 rectangular regions and extract HSV joint histogram that has quantized 162 bins for each region. Although these contain local color information, the resulting representation is not compact enough. To obtain compact representation, we extract from each joint histogram the bin that has the maximum peak. Take hue h, saturation s, and value v associated to the bin as representing features in that rectangular region and normalize to be within the same range of [0, 1]. Thus, each image has the $3 \times 3 \times 3 (= 27)$ dimensional color vector.

(2) Texture

Texture analysis is an important and useful area of study in computer vision. Most natural images include textures. Scenes containing pictures of wood, grass, etc. can be easily classified based on the texture rather than color or shape. Therefore, it may be useful to extract texture features for image clustering. Like as color feature, we include a texture feature extracted from localized image region.

The co-occurrence matrix is a two-dimensional histogram which estimates the pairwise statistics of gray level. The $(i, j)^{th}$ element of the co-occurrence matrix represents the estimated probability that gray level i co-occurs with gray level j at a specified displacement d and angle θ. By choosing the values of d and θ, a separate co-occurrence matrix is obtained. From each co-occurrence matrix a number of textural features can be extracted. For image clustering, we used entropy, which is mostly used in many applications. Feature extraction is performed by the following process:

(1) Conversion of color image to gray image
(2) Dividing image into 3×3 rectangular regions as in color case.
(3) Obtaining co-occurrence matrix for four (horizontal 0^0, vertical 90^0 and two diagonal 45^0 and 135^0) orientation in region and normalize entries of four matrices to [0, 1] by dividing each entry by total number of pixels.
(4) Extracting average entropy value from four matrices.

$$e = \frac{-\sum_k \sum_i \sum_j p(i, j) \log(p(i, j))}{4}, \qquad k = 1, 2, 3, 4, \qquad (1)$$

(5) Constructing texture feature vector by concatenating entropies over all rectangular regions.

Thus, each image has the $3 \times 3 (= 9)$ dimensional texture vector.

2.2 Support Vector Machine (SVM)

The goal of SVM is to find optimal hyperplane by minimizing an upper bound of the generalization error through maximizing the distance, margin, between the separating hyperplane and the data. SVM uses the preprocessing strategy in learning by mapping input space, X to a high-dimensional feature space, F. By this mapping, more flexible classifications are obtained. A separating hyperplane is found which maximizes the margin between itself and the nearest training points as shown in Fig. 1.

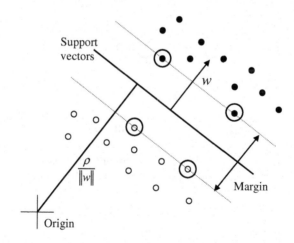

Fig. 1. Linear separating hyperplanes for two-class separation

The feature space is very high-dimensional space where linear separation becomes much easier than input space. This is equivalent to applying a fixed non-linear mapping of the data to a feature space, in which a linear function can be used.

A simple description of the SVM algorithm is provided as follows. Consider a pattern classifier, which uses a hyperplane to separate two classes of patterns based on given a training set $S = \{x_i, y_i\}_{i=1}^n$ with input vectors $x_i = (x_i^{(1)}, ..., x_i^{(n)})^T$ and target labels $y_i \in \{-1, +1\}$. Support vector machine (SVM) classifier, according to Vapnik's original formulation, satisfies the following conditions:

$$f(x) = y_i = \langle w \cdot x \rangle + b = \sum_{i=1}^n w_i x_i + b \qquad (2)$$

where w represents the weight vector and b the bias.

In the non-linear case, we first mapped the data some other Euclidean space F, using a mapping, $x = (x_1, ..., x_n) \mapsto \psi(x) = (\psi(x_1), ..., \psi(x_n))$. Then instead of the form of dot product, we use kernel function $K(x, y) = \phi(x)\phi(y)$.

The distance between two hyperplanes is the margin of the hyperplane w with respect to the sample S. The purpose of SVMs is to maximize this distance. Using a Lagrangian, this optimization problem can be converted into a dual form that is a quadratic programming (QP) problem where the objective function is solely dependent on a set of Lagrange multipliers α_i. The QP problem is as follows.

$$\min \frac{1}{2} \|w\|^2 \qquad (3)$$

subject to

$$y_i(w^T x_i + b) \geq 1, \quad i = 1, ..., n \qquad (4)$$

We can get the maximal margin hyperplane with geometric margin. And then the Lagrangian is as follows:

$$L(w, b, \alpha) = \sum_{i=1}^n \alpha_i - \frac{1}{2} \sum_{i,j=1}^n \alpha_i \alpha_j y_i y_j \langle x_i \cdot x_j \rangle \qquad 5)$$

subject to

$$0 \leq \alpha_i \leq C \qquad (6)$$

$$\sum_{i=1}^n \alpha_i y_i = 0 \qquad \text{for all } i = 1, 2, ..., n \qquad (7)$$

Some kernel functions include linear, polynomial, radial basis function (RBF) and sigmoid kernel, which are shown as functions (8), (9), (10) and (11). In order to improve classification accuracy, these kernel parameters in the kernel functions should be properly set.

- Linear kernel: $K(x_i, x_j) = x_i^T x_j$ $\qquad (8)$

- Polynomial kernel: $K(x_i, x_j) = (\gamma x_i^T x_j + r)^d$, $\gamma > 0$ $\qquad (9)$

- Radial basis function kernel: $K(x_i, x_j) = \exp(-\gamma \|x_i - x_j\|^2)$, $\gamma > 0$ $\qquad (10)$

- Sigmoid kernel: $K(x_i, x_j) = \tanh(\gamma x_i^T x_j + r)$ $\qquad (11)$

2.3 Genetic Algorithm (GA)

Genetic algorithm (GA) is an artificial intelligence procedure based on the theory of natural selection and evolution. GA uses the idea of survival of the fittest by progressively accepting better solutions to the problems. It is inspired by and named after biological processes of inheritance, mutation, natural selection, and the genetic crossover that occurs when parents mate to produce offspring [13]. GA differs from conventional non-linear optimization techniques in that it searches by maintaining a population of solutions from which better solutions are created rather than making incremental changes to a single solution to the problem. GA simultaneously possesses a large number of candidate solutions to a problem, called a population. The key feature of GA is the manipulation of a population whose individuals are characterized by possessing a chromosome. A fitness function assesses the quality of a solution in the evaluation step. The crossover and mutation functions are the main operators that randomly impact the fitness value. Crossover is performed between two selected individuals, called parents, by exchanging parts of their strings, starting from a randomly chosen crossover point. This operator tends to enable to the evolutionary process to move toward promising regions of the search space. Mutation is used to search for further problem space and to avoid local convergence of GA [14]. Associated with the characteristics of exploitation and exploration search, GA can deal with large search spaces efficiently, and hence has less chance to get local optimal solution than other algorithms.

3 The Proposed Approach

This study presents the GA-based approach to improve the performance of SVM for CBIR in two aspects such as feature subset selection and parameter optimization. GA is used to optimize both the feature subset and parameters of SVM simultaneously for CBIR. The chromosome design, fitness function, and system architecture for the proposed GA-based feature subset selection and parameter optimization are described as follows.

3.1 Chromosome Design and Fitness Function

To implement our proposed approach, this research used the RBF kernel function for the SVM classifier because the RBF kernel function can analysis higher-dimensional data and requires that only two parameters, C and γ be defined. We selected the RBF kernel and the parameters (C and γ) and features used as input attributes must be optimized using our proposed GA-based approach. Therefore, the chromosome comprises three parts, C, γ and the features mask. However, these chromosomes have different parameters when other types of kernel functions are selected. The binary coding system was used to represent the chromosome. Fig. 2 shows the binary chromosome representation of our design.

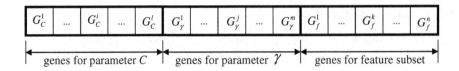

Fig. 2. The chromosome comprises three parts, C, γ and the features mask

In Fig. 2, $G_C^1 \sim G_C^l$ represents the value of parameter C, $G_\gamma^1 \sim G_\gamma^m$ represents the parameter value γ, and $G_f^1 \sim G_f^n$ represents the feature mask. l is the number of bits representing parameter C, m is the number of bits representing parameter γ, and n is the number of bits representing the features.

GA evolves a number of populations. Each population consists of sets of features of a given size and the values of parameters. The fitness of an individual of the population is based on the performance of SVM. The chromosome with high fitness value has high probability to be preserved to the next generation. We use SVM classification accuracy and the number of selected features to design a fitness function. The chromosome with high classification accuracy and a small number of features produce a high fitness value. The fitness function is designed as follows:

$$fitness \quad funtion = W_1 \times SVM\_Accuracy + W_2 \times Nonzeros \qquad (12)$$

where W_1 is the weight for SVM classification accuracy, $SVM\_Accuracy$ is SVM classification accuracy; W_2 is the weight for the number of features and $Nonzeros$ is the number of selected features.

3.2 System Architectures for the Proposed GA-Based Approach

To precisely establish a GA-based feature selection and parameter optimization system, the following main steps (as shown in Fig. 3) must be proceeded. The detailed explanation is as follows:

(1) Data preprocess (scaling): The main advantage of scaling is to avoid attributes in greater numeric ranges dominating those in smaller numeric ranges. Another advantage is to avoid numerical difficulties during the calculation. Feature value scaling can help to increase SVM accuracy according to our experimental results. Generally, each feature can be linearly scaled to the range [0, 1] by formula (13), where x is original value, x' is scaled value, x_{max} is upper bound of the feature value, and x_{min} is low bound of the feature value.

$$x' = \frac{x - x_{min}}{x_{max} - x_{min}} \qquad (13)$$

(2) Feature subset: After the genetic operation and converting each feature subset chromosome from the genotype into the phenotype, a feature subset can be determined.

(3) Fitness evaluation: For each chromosome representing C, γ and selected features, training dataset is used to train the SVM classifier, while the testing dataset is used to calculate classification accuracy. When the classification accuracy is obtained, each chromosome is evaluated by fitness function as the equation (12).

(4) Termination criteria: When the termination criteria are satisfied, the process ends; otherwise, we proceed with the next generation.

(5) Genetic operation: In this step, the system searches for better solutions by genetic operations, including selection, crossover, mutation, and replacement.

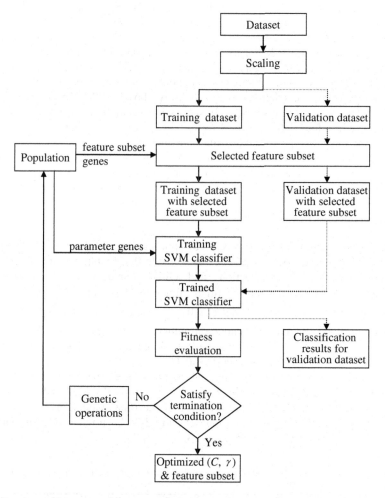

Fig. 3. System architectures of the proposed GA-based feature selection and parameters optimization of SVM for CBIR

4 Experiments

4.1 Experimental Design

To show the effective classification of the proposed method, we checked the classification accuracy. Classification results using color and texture features of real world images will be shown. All experiments were performed on a Pentium IV with 512 Mbytes of main memory and 100 Gbytes of storage. We experimented on 3,000 images where most of them have dimensions of 192×128 pixels. The 3,000 images can be divided into 10 categories with 300 images each such as such as airplane, eagle, horse, lion, polar bear, rose, zebra, tiger, valley and sunset.

In this paper, we use the term, 'hybrid GA & SVM' as the proposed approach which represents simultaneous optimization of SVM using GA. The dataset for GA-SVM is separated into two parts: the training dataset, and the validation dataset. The ratios are about 0.8 and 0.2. For example, the total dataset for each image category contains 600 images which consist of 300 images of each category and randomly chosen 300 images out of 2,700 images in the other image categories. Training for the proposed approach is carried out with randomly chosen 480 instances (i.e. training data) out of 600 dataset in each class. In succession, validation is carried out with 120 instances (i.e. holdout data) not used in training. These experiments for each image category are repeatedly performed by the same method.

Additionally, to evaluate the effectiveness of the proposed approach, we compare two different models with arbitrarily selected values of parameters and a given all feature subset. The first model uses neural network (NN) and the second model uses pure SVM.

4.2 Experimental Results

In order to evaluate the hybrid GA & SVM approach, we set the detail parameter for GA as follows: population size 200, crossover rate 0.9, mutation rate 0.1, two-point crossover, roulette wheel selection, and elitism replacement. We set $l = 20$, $m = 20$ and $n = 36$. According to the fitness function of Eq. (12), W_1 and W_2 can influence the experiment result. We defined $W_1 = 0.8$ and $W_2 = 0.2$ for experiments. The termination criteria are that the generation number reaches generation 500 or that the fitness value does not improve during the last 100 generations. The best chromosome is obtained when the termination criteria satisfy.

Table 1 shows both training and validation average success rates that were achieved under the different models. As can be seen, the hybrid GA & SVM approach has consistently given the best performance of the other models as shown 96.45% average success rate on the training dataset and 93.75% on the validation dataset in table 1 and average number of features is 13.

Table 1. Average image classification accuracy of NN, pure SVM and hybrid GA & SVM

Classifier	Type of kernel	Training (%)	Validation (%)
NN	–	88.33	86.15
Pure SVM	RBF	93.37	91.50
Hybrid GA & SVM	RBF	96.45	93.75

In order to test the superiority of the proposed approach, we perform the McNemar test which is used to examine whether the proposed approach significantly outperforms the other models. This test is a nonparametric test for two related samples using the chi-square distribution. This test may be used with nominal data and is particularly useful with 'before–and-after' measurement of the same subjects [15]. We performed McNemar test to compare the performance for the test data. As a result, the hybrid GA & SVM approach outperforms NN at the 1% statistical significance level, and pure SVM at the 10% statistical level.

5 Conclusions

In this paper, we presented the GA-based approach to improve the performance of SVM in CBIR. SVM parameters and feature subsets were optimized simultaneously in this work because the selected feature subset has an influence on the appropriate kernel parameters and vice versa. We proposed a GA-based approach to select the feature subset and set the parameters for SVM to improve image classification accuracy in CBIR. As far as we know, previous studies have not dealt with integration of GA and SVM although there is a great potential for useful applications in CBIR. This paper focuses on the improvement of the image classification accuracy by means of the hybrid GA and SVM approach in CBIR.

We conducted experiments to evaluate the classification accuracy of the proposed GA-based approach with RBF kernel, NN and pure SVM on 3,000 real-world image dataset with 10 image categories. Generally, compared with NN and pure SVM, the proposed GA-based approach has good accuracy performance with fewer features.

This study showed experimental results with the RBF kernel of SVM. In future work, we also intend to optimize the kernel function, parameters and feature subset simultaneously. We would also like to expand the proposed approach to apply to instance selection problems.

References

1. Smeulders, A.W.M., Worring, M., Santini, S., Gupta, A., Jain, R.: Content-based image retrieval at the end of the early years. IEEE Trans. Pattern Anal. Mach. Intelligence 22(12), 1349–1380 (2000)
2. Fournier, J., Cord, M., Philipp-Foliguet, S.: Back-propagation Algorithm for Relevance Feedback in Image retrieval. In: ICIP'01. IEEE International Conference in Image Processing, vol. 1, pp. 686–689 (2001)
3. Koskela, M., Laaksonen, J., Oja, E.: Use of Image Subset Features in Image Retrieval with Self-Organizing Maps. In: Enser, P.G.B., Kompatsiaris, Y., O'Connor, N.E., Smeaton, A.F., Smeulders, A.W.M. (eds.) CIVR 2004. LNCS, vol. 3115, pp. 508–516. Springer, Heidelberg (2004)
4. Pakkanen, J., Iivarinen, J., Oja, E.: The Evolving Tree - a Novel Self-Organizing Network for Data Analysis. Neural Processing Letters 20(3), 199–211 (2004)
5. Park, S.-S., Seo, K.-K., Jang, D.-S.: Expert system based on artificial neural networks for content-based image retrieval. Expert Systems with Applications 29(3), 589–597 (2005)
6. Vapnik, V.: Statistical learning theory. Springer, Heidelberg (1995)

7. Fröhlich, H., Chapelle, O.: Feature selection for support vector machines by means of genetic algorithms. In: Proceedings of the 15th IEEE international conference on tools with artificial intelligence, pp. 142–148. IEEE Computer Society Press, Los Alamitos (2003)

8. Bradley, P.S., Mangasarian, O.L.: Feature selection via concave minimization and support vector machines. In: Elomaa, T., Mannila, H., Toivonen, H. (eds.) ECML 2002. LNCS (LNAI), vol. 2430, pp. 82–90. Springer, Heidelberg (2002)

9. Weston, J., Mukherjee, S., Chapelle, O., Pontil, M., Poggio, T., Vapnik, V.: Feature selection for SVM. Advances in neural information processing systems 13, 668–674 (2001)

10. Mao, K.Z.: Feature subset selection for support vector machines through discriminative function pruning analysis. IEEE Transactions on Systems, Man, and Cybernetics 34(1), 60–67 (2004)

11. Raymer, M.L., Punch, W.F., Goodman, E.D., Kuhn, L.A., Jain, A.K.: Dimensionality reduction using genetic algorithms. IEEE Transactions on Evolutionary Computation 4(2), 164–171 (2000)

12. Yang, J., Honavar, V.: Feature subset selection using a genetic algorithm. IEEE Intelligent Systems 13(2), 44–49 (1998)

13. Goldberg, D.E.: Genetic algorithms in search, optimization and machine learning. Addison-Wesley, New York (1989)

14. Tang, K.S., Man, K.F., Kwong, S., He, Q.: Genetic algorithms and their applications. IEEE Signal Processing Magazine 13, 22–37 (1996)

15. Cooper, D.R., Emory, C.W.: Business research methods. Irwin, Chicago, IL (1995)

E-Stream: Evolution-Based Technique for Stream Clustering

Komkrit Udommanetanakit, Thanawin Rakthanmanon, and Kitsana Waiyamai

Department of Computer Engineering, Faculty of Engineering
Kasetsart University, Bangkok 10900, Thailand
{fengtwr, fengknw}@ku.ac.th

Abstract. Data streams have recently attracted attention for their applicability to numerous domains including credit fraud detection, network intrusion detection, and click streams. Stream clustering is a technique that performs cluster analysis of data streams that is able to monitor the results in real time. A data stream is continuously generated sequences of data for which the characteristics of the data evolve over time. A good stream clustering algorithm should recognize such evolution and yield a cluster model that conforms to the current data. In this paper, we propose a new technique for stream clustering which supports five evolutions that are appearance, disappearance, self-evolution, merge and split.

1 Introduction

Stream clustering is a technique that performs cluster analysis of data streams that is able to produce results in real time. The ability to process data in a single pass and summarize it, while using limited memory, is crucial to stream clustering.

Several efficient stream clustering techniques have been presented recently, such as STREAM [9], CluStream [2], and HPStream [1]. STREAM is a k-median based algorithm that can achieve a constant factor approximation. CluStream divides the clustering process into an online and offline process. Data summarization is performed online, while clustering of the summarized data is performed offline. Experiments show that CluStream yields better cluster quality than STREAM. HPStream is the most recent stream clustering technique, which utilizes a fading concept, data representation, and dimension projection. HPStream achieves better clustering quality than the above algorithms.

Since the characteristics of the data evolve over time, various types of evolution should be defined and supported by the algorithm. Almost of existing algorithms support few types of evolution. The objective of the research reported here is to improve existing stream clustering algorithms by supporting 5 evolutions with a new suitable cluster representation and a distance function. Experimental results show that this technique yields better cluster quality than HPStream.

The remaining of the paper is organized as follows. Section 2 introduces basic concepts and definitions. Section 3 presents our stream clustering algorithm called *E-Stream*. Section 4 compares the performance of E-Stream and HPStream with respect to the synthetic dataset. Conclusions are discussed in Section 5.

R. Alhajj et al. (Eds.): ADMA 2007, LNAI 4632, pp. 605–615, 2007.
© Springer-Verlag Berlin Heidelberg 2007

2 Basic Concepts and Definitions

In this section, we introduce some basic concepts and definitions that will be used subsequently.

The data stream consists of a set of multidimensional records $X_1...X_k...$ arriving at time stamps $T_1...T_k....$ Each data point X_i is a multidimensional record containing d dimensions, denote by $X_i = (x_i^1...x_i^d)$.

An isolated data point is a data point that is not a member of any clusters. Isolated data points remain in the system for cluster appearance computations.

An inactive cluster is a cluster that has a low weight. It can become an active cluster if its weight is increased.

An active Cluster is a cluster that can assemble incoming data if there is sufficient similarity score.

A cluster is a collection of data that has been memorized for processing in the system. It can be an isolated data point, an inactive cluster, or an active cluster.

Fading decreases weight of data over time. In a data stream that has evolving data; older data should have lesser weight. We decrease weight of every cluster over time to achieve a fast adaptive cluster model. Let λ be the decay rate and t be elapsed time, the fading function is

$$f(t) = 2^{-\lambda t} \tag{1}$$

Weight of a cluster is the number of data elements in a cluster. Weight is determined according to the fading function. Initially, each data element has a weight of 1. A cluster can be increased its weight by assembling incoming data points or merging with other clusters.

2.1 Fading Cluster Structure with Histogram: FCH

Each cluster is represented as a *Fading Cluster Structure (FCS)* [1] utilizing a α-bin histogram for each feature of the dataset. We called our cluster representation *Fading Cluster Structure with Histogram (FCH)*. Let T_i be the time when data point x_i is retrieved, and suppose t be the current time then $f(t-T_i)$ is the fading weight of data point x_i.

$FC1(t)$ is a vector of weighted sumation of data feature values at time t. The jth dimension is

$$FC1^j(t) = \sum_{i=1}^{N} f(t - T_i) \cdot (x_i^j) \tag{2}$$

$FC2(t)$ is the weighted sum of squares of each data feature at time t. The jth dimension is,

$$FC2^j(t) = \sum_{i=1}^{N} f(t - T_i) \cdot (x_i^j)^2 \tag{3}$$

$W(t)$ is a sum of all weights of data points in the cluster at time t,

$$W(t) = \sum_{i=1}^{N} f(t - T_i) \tag{4}$$

$H(t)$ is a α-bin histogram of data values. For the jth feature at time t, the elements of H^j are

$$H_i^j(t) = \sum_{i=1}^{N} f(t - T_i) \cdot (x_i^j) \cdot (y_{il}^j) \tag{5}$$

Where

$$y_{il} = \begin{cases} 1 & if & l \cdot b + left \leq x_i \leq (l+1) \cdot b + left \\ 0 & otherwise \end{cases} \tag{6}$$

$$left = \min(x_i^j) \tag{7}$$

$$right = \max(x_i^j) \tag{8}$$

$$b = \frac{left + right}{\alpha} \tag{9}$$

left is a minimum value, *right* is a maximum value in this cluster, b is a size of each bin, and y_{il} is a weigth of x_i in the lth bin.

2.2 Histogram Management

We utilize a histogram of cluster data values to identify cluster splits. A α-bin histogram summarizes the distribution of cluster data for each dimension of each cluster. The range of each bin is calculated as the difference between the maximum and minimum feature values divided by α. When the maximum or minimum value changes, we calculate a new range and update the values in each range from the intersection between the new and old ranges. Each cluster has a histogram of feature values, but the histogram is utilized only for the split of active clusters. Only an active cluster can assemble an incoming data point.

Cluster split is based on the distribution of feature values as summarized by the cluster histogram [1]. If a statistically significant valley is found between two peaks in histogram values along any dimensions, the cluster is split. If more than one split valley occurs in the histogram values, the value with the minimum height relative to the surrounding peaks is chosen. When a cluster is split, the histogram in that dimension is split and the other dimensions are weighted from the split dimension. FC1, FC2 and W are recalculated from the new cluster histograms.

2.3 Distance Functions

Cluster-Point distance is a distance from a data point to the center of a cluster, ormalized by the standard deviation (radius) of the cluster data in each dimension.

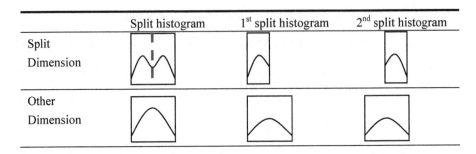

Fig. 1. Histogram management in a split dimensionop and other dimension

This function is used to find the closest active cluster for an incoming data point. A cluster with a larger radius yields a lower distance. Let C be an active cluster and x be a data point, the Cluster-Point distance is

$$dist(C, x) = \frac{1}{d} \cdot \sum_{j=1}^{d} \left| \frac{center_C^j - x^j}{radius_C^j} \right| \tag{10}$$

Cluster-Cluster distance is a difference between centers of two clusters. It is used to find the closest pair of cluster. If C_a and C_b are two clusters, the cluster-cluster distance is

$$dist(C_a, C_b) = \frac{1}{d} \sum_{j=1}^{d} \left| center_{C_a}^j - center_{C_b}^j \right| \tag{11}$$

3   The Algorithm

In this section, we first describe our idea. Then, we present the E-Stream algorithm composing of a set of sub-algorithms.

3.1   Our Idea

The behavior of data in a data stream can evolve over time. We can classify this evolution into five categories: appearance, disappearance, self evolution, merge, and split. A clustering model can start from empty. In the beginning, incoming data are considered as isolated clusters. A cluster is formed when a sufficiently dense region appears. The cluster assembles similar data and increases its existence. When a set of clusters has been identified, incoming data must be assigned to a cluster based on similarity score, or the datum may be classified as an isolated. In every change of a cluster we check if any of the following evolutions occur and handle it.

Appearance: A new cluster can appear if there is a sufficiently dense group of data points in one area. Initially, such elements appear as a group of outliers, but (as more data appears in a neighborhood) they are recognized as a cluster.

Disappearance: Existing clusters can disappear because the existence of data is diminished over time. Clusters that contain only old data are faded and eventually disappear because they do not represent the presence of data.

Self-evolution: Data can change their behaviors, which cause size or position of a cluster to evolve. Evolution can be done faster if the data can fade.

Merge: A pair of clusters can be merged if their characteristics are very similar. Merged clusters must cover the behavior of the pair.

Split: A cluster can be split into two smaller clusters if the behavior inside the cluster is obviously separated.

3.2 E-Stream Algorithm

This section describes E-Stream in details. Following is the list of notations used in our pseudo-code.

- $|FCH|$ = current number of clusters
- $FCH_i.W$ = weight of the ith cluster
- $FCH_i.sd$ = standard deviation of the ith cluster
- S = set of pair of the split cluster

E-Stream is the main algorithm. In line 1, the algorithm starts by retrieving a new data point. In line 2, it fades all clusters and deletes those having insufficient weight. Line 3 performs a histogram analysis and cluster split. Then line 4 checks for overlap clusters and merges them. Line 5 checks the number of clusters and merges the closest pairs if the cluster count exceeds the limit. Line 6 checks all clusters whether their status are active. Lines 7-10 find the closest cluster to the incoming data point. If the distance is less than *radius_factor* then the point is assigned to the cluster, otherwise it is an isolated data point. The flow of control then returns to the top of the algorithm and waits for a new data point.

Algorithm *E-Stream*
1   retrieve new data X_i
2   *FadingAll*
3   *CheckSplit*
4   *MergeOverlapCluster*
5   *LimitMaximumCluster*
6   *FlagActiveCluster*
7   (minDistance, index) ← *FindClosestCluster*
8   if minDistance < radius_factor
9   add x_i to FCH_{index}
10   else
11   create new FCH from X_i
12 waiting for new data

Fig. 2. E-Stream, stream clustering algorithm

FadingAll. The algorithm performs fading of all clusters and deletes the clusters whose weight is less than *remove_threshold*.

CheckSplit is used to verify the splitting criteria in each cluster using the histogram. If a splitting point is found in any cluster then it is split. And store the index pairs of split cluster in S.

CheckMerge is an algorithm for merging pairs of similar clusters. This algorithm checks every pair of clusters and computes the cluster-cluster distance. If the distance is less than *merge_threshold* and the merged pair is not in S then merge the pair.

LimitMaximumCluster is used to limit the number of clusters. This algorithm checks whether the number of clusters is not greater than *maximum_cluster* (an input parameter); if it exceeds then the closest pair of clusters is merged until the number of remaining clusters is less than or equal to the threshold.

FlagActiveCluster is used to check the current active cluster. If the weight of any cluster is greater or equal to *active_threshold* then it is flagged as an active cluster. Otherwise, the flag is cleared.

FindClosestCluster is used to find the distance and index of the closest active cluster for an incoming data point.

Algorithm *FadingAll*	**Algorithm** *CheckSplit*
for i ← 1 to |FCH|	for i ← 1 to |FCH|
fading FCH_i	for j ← 1 to d
if $FCH_i.W$ < *fade_threshold*	if FCH_{ij} have split point
delete FCH_i	split FCH_i
	S ← S U {(i, |FCH|)}

Fig. 3. FadingAll and CheckSplit algorithms

Algorithm MergeOverlapCluster	**Algorithm** *LimitMaximumCluster*
for i ← 1 to |FCH|	while |FCH| > maximum_cluster
for j ← i + 1 to |FCH|	for i ← 1 to |FCH|
overlap[i,j] ← dist(FCH_i,FCH_j)	for j ← i + 1 to |FCH|
m ← *merge_threshold*	dist[i,j] ← dist(FCH_i, FCH_j)
if overlap[i,j] > m*($FCH_i.sd$+$FCH_j.sd$)	(first, second) ← $argmin_{(i,j)}$(dist[i,j])
if (i, j) not in S	merge(FCH_{first}, FCH_{second})
merge(FCH_i, FCH_j)	

Fig. 4. MergeOverlapCluster and LimitMaximumCluster algorithms

Algorithm *FlagActiveCluster*	**Algorithm** *FindClosestCluster*
for i ← 1 to |FCH|	for i ← 1 to |FCH|
if $FCH_i.W$>= *active_threshold*	if FCH_i is active cluster
flag FCH_i as active cluster	dist[i] ←dist(FCH_i, x_i)
else	(minDistance, i) ← min(dist[i])
remove flag from FCH_i	return (minDistance, i)

Fig. 5. FlagActiveCluster and FindClosestClusterAlgorithms

4 Experimental Results

We tested the algorithm using a synthetic dataset consisting of two dimensions and 8,000 data points. This data changes the behavior of clusters over time. We can segment it into 8 intervals as follows

1. Initially, there are 4 clusters in a steady state. Data point from 1 to 1600.
2. The 5th cluster appears at position (15, 6). Data point from 1601 to 2600.
3. The 1st cluster disappears. Data point from 2601 to 3400.
4. The 4th cluster swells. Data point from 3401 to 4200
5. The 2nd and 5th cluster get closer. Data point from 4201 to 5000.
6. The 2nd and 5th are merged into a bigger cluster. Data point from 5001 to 5600.
7. A 6th cluster is split from the 3rd cluster. Data point from 5601 – 6400.
8. Every cluster is in a steady state again. Data point from 6401 – 8000.

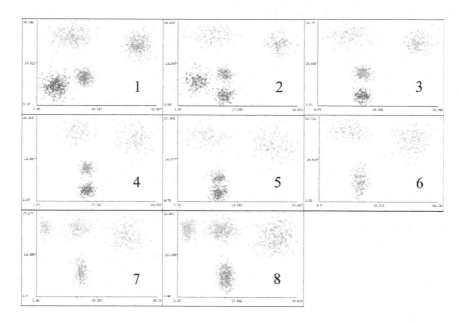

Fig. 6. The 8-Step evolution of the Synthetic Dataset

4.1 Efficiency Test

In this experiment, we set the parameters as in table 1. E-Stream allows the number of clusters to vary dynamically with the constraint of the maximum number of clusters, but requires a limit on the number of clusters. HPStream requires a fixed number of clusters. Since the synthetic dataset has at most 5 clusters in each interval, we used 5 as the cluster (group) count in HPStream, and 10 as the cluster limit in E-Stream. HPStream requires initial data for its initialization process before beginning stream clustering. We, therefore, set it to 100 points.

Table 1. Parameters of each algorithm

algorithm E-Stream	algorithm HPStream
maximum_cluster 10	num_cluster 5
stream_speed 100	stream_speed 100
decay_rate 0.1	decay_rate 0.1
radius_factor 3	radius_factor 3
remove_threshold 0.1	
merge_threshold 1.25	
active_threshold 5	

In the first interval, there are four clusters in a steady-state. Both algorithms yield clusters with little distortion, but E-Stream has a great number of clusters. Because at the beginning E-Stream does not have any active clusters, every incoming data point is considered as an isolated data point. As more data is accumulated, a cluster will appear. On the other hand, HPStream requires initial data (set to 100 points) for offline clustering, so HPStream exhibits better initial clustering than E-Stream.

In the second interval, a new cluster appears. HPStream still yields a better quality because it finds all clusters correctly, but E-Stream yields only little distortion.

In the third interval, an existing cluster disappears. E-Stream yields good clustering while HPStream tries to create a new cluster from existing cluster even though their do not have a significant difference, due to the fixed cluster-count constraint.

In the 4th interval, a cluster swells. In the 5th interval, two clusters are closer, an evolution that is supported by both algorithms. But, HPStream still tries to find five clusters.

In the 6th interval two close clusters are merged. Neither algorithms merge the two clusters in this interval even though their current behaviors are undistinguished.

In the 7th interval, a cluster splits. E-Stream can support this evolution when it receives enough data. But HPStream cannot detect this.

In the 8th interval there are four clusters in a steady state again. E-stream algorithm detects the previous merged case and identifies all clusters correctly within this interval. But HPStream still is confusing by the cluster behavior.

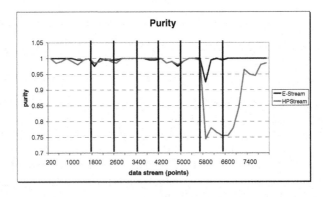

Fig. 7. Purity test between E-Stream and HPStream

In the purity test, E-Stream always has a purity greater than 0.9. But HPStream exhibits a big drop in the seventh interval (cluster split), because the algorithm cannot accommodate this evolution.

In the F-measure test, E-Stream yields an average value much better than HPStream although, there are two intervals where E-Stream has a lower F-measure. HPStream yields better results due to an initial offline process to find the initial clusters. The second instance is the second interval (1601-2600), where E-Stream merged two clusters incorrectly.

From the efficiency test, we can say that HPStream cannot support the evolution of number of clusters because the algorithm constrains it. E-Stream can support all the evolutions in Section 3, even though some evolutions such as merge require a lot of data to detect.

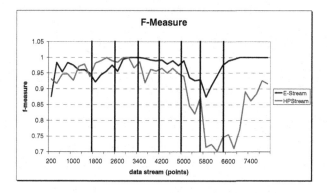

Fig. 8. F-Measure test between E-Stream and HPStream

4.2 Sensitivity with Number of Cluster (Input Parameters)

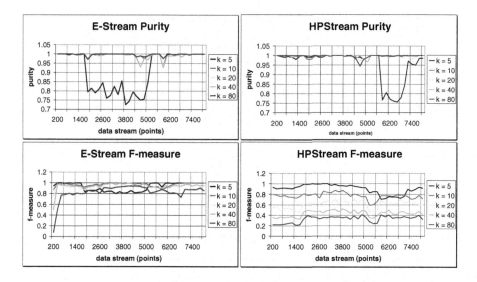

Fig. 9. Sensitivity with number of cluster (input parameter)

From the experiment, the Purity of both algorithms is not sensitive to this input parameter. But in F-measure terms, HPStream has a tendency to drop if the input number of clusters differs greatly from the actual number. E-Stream is not sensitive to this parameter since the number of clusters is not fixed. As long as the maximum number is not exceeded, E-Stream still yields good results.

4.3   Runtime with Number of Data

In this experiment we use a dataset consisting of 500,000 data points with two dimensions and five clusters.

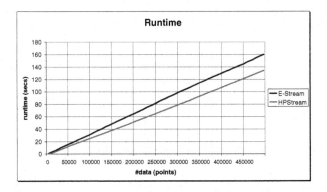

Fig. 10. Runtime with number of data (points)

Both algorithms exhibit linear runtime in number of data points, which is a constraint for stream clustering algorithms.

4.4   Runtime as a Function of Clusters and of Dimensions

To test the runtime as a function of the number of clusters, we use two dimensions, 100,000 data points, and vary the number of data clusters from 5 to 25 in increments of 5 clusters.

For runtime with the number of dimensions test, we use 5 clusters, 100,000 data points, and vary the number of dimensions from 2 to 20.

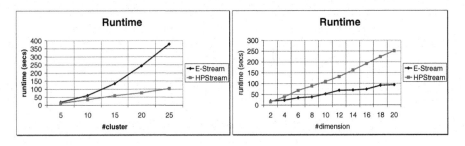

Fig. 11. Runtime with number of clusters and number of dimensions

The results of both experiments are summarized in Figure 15. HPStream exhibits linear runtime in both the number of clusters and the number of dimensions. E-Stream exhibits linear runtime in the number of dimensions but polynomial runtime in the number of clusters. This is due to the merging procedure, which requires $O(k^2)$ time in the number of clusters.

5 Conclusions

This paper proposed a new stream clustering technique called E-Stream which can support five cluster evolutions: appearance, disappearance, self-evolution, merge, and split. These evolutions can normally occur in an evolving data stream. This technique outperforms a well-known technique, HPStream. However, the runtime of the new approach is polynomial with respect to the number clusters.

Acknowledgment. Thanks to J. E. Brucker and P. Vateekul for their reading and comments of this paper.

References

1. Milenova, B.L., Campos, M.M.: Clustering Large Databases with Numeric and Nominial Values Using Orthogonal Projections. In: Proceedings of the 29th VLDB Conference (2003)
2. Aggarwal, C., Han, J., Wang, J., Yu, P.S.: A Framework for Projected Clustering of High Dimensional Data Streams. In: Proceeding of the 30th VLDB conference (2004)
3. Aggarwal, C., Han, J., Wang, J., Yu, P.S.: A Framework for Clustering Evolving Data Streams. In: Proceeding of the 29th VLDB conference (2003)
4. Barbara, D.: Requirements for Clustering Data Streams. In: SIGKDD Explorations (2002)
5. Gaber, M.M., Zaslavsky, A., Krishnaswmy, S.: Mining Data Streams: A Review. In: SIGMOD Record, vol. 34(2) (June 2005)
6. Oh, S., Kang, J., Byun, Y., Park, G., Byun, S.: Intrusion Detection based on Clustering a Data Stream. In: Proceedings of the 2005 Third ACIS International Conference on Software Engineering Research, Management and Applications (2005)
7. Guha, S., Meyerson, A., Mishra, N., Motwani, R., O'Callaghan, L.: Clustering Data Streams: Theory and Practice. TKDE special issue on clustering 15 (2003)
8. Song, M., Wang, H.: Highly Efficient Incremental Estimation of Gaussian Mixture Models for Online Data Stream Clustering. In: SPIE Conference on Intelligent Computing: Theory And Application III (2005)
9. Zhang, T., Ramakhrisnan, R., Livny, M.: BIRCH: An Efficient Data Clustering Method for Very Large Databases. In: Proc. ACM SIGMOD Int. Conf. Management of Data (1996)

H-BayesClust: A New Hierarchical Clustering Based on Bayesian Networks*

Morteza Haghir Chehreghani and Hassan Abolhassani

Department of Computer Engineering, Sharif University of Technology, Tehran, Iran
{haghir, abolhassani}@ce.sharif.edu

Abstract. Clustering is one of the most important approaches for mining and extracting knowledge from the web. In this paper a method for clustering the web data is presented which using a Bayesian network, finds appropriate representatives for each of the clusters. Having those representatives, we can create more accurate clusters. Also the contents of the web pages are converted into vectors which firstly, the number of dimensions is reduced, and secondly the orthogonality problem is solved. Experimental results show about the high quality of the resultant clusters.

Keywords: web clustering, hierarchy, Bayesian network, belief, cluster center.

1 Introduction

Clustering data is an important task in information mining which can be done as a pre-processing phase. On the web, this task has a special role; because it can be used for enhancing search engine results, improving web crawling task and etc. So far several methods have been developed for clustering the web data, which mostly include link analysis, content mining and combinations of them [2,14].

In an overall classification, we can divide the web clustering algorithms into two categories: Hierarchical and Partitioning algorithms [3,15]. The best known partitioning algorithm is *K-Means* [8] that in a simple form selects *K* points as cluster centers and assigns each data point to the nearest center. The reassigning process can be kept until a convergence criterion is met. The time complexity of this algorithm is $O(n)$. While using this algorithm, the most important challenge is the problem of selecting appropriate representative points. In an extension, first, some limited centers are selected and a threshold for assigning the web pages to the centers is defined. Then if a web page does not satisfy the threshold, it constructs a separate cluster by itself. In another extension a measure is used for finding cluster centers and updating them which contains links too [13].

Recently, with respect to the development of semantic web applications, some trends have been inclined toward using these concepts in data mining area [11,12]. One main layer of semantic web is *Ontology* which as is defined in [7] is "an explicit formalization of a shared understanding of a conceptualization". By constructing an

---

* This paper is supported by Iran Telecommunication Research Center (ITRC).

R. Alhajj et al. (Eds.): ADMA 2007, LNAI 4632, pp. 616–624, 2007.

ontology, we will have a limited number of concepts so that different data mining operations can be done with higher quality. Some examples of using ontologies are discussed in [1,5,6,10]. An alternative approach that can provide the inference among the vocabularies of a specific domain is using Bayesian Belief networks which is introduced in this paper.

The remaining of the paper is organized as follows: in section 2 the proposed method for inferring the cluster representatives is described. In section 3 the experimental results are discussed and finally required conclusion is given in section 4.

2   The Proposed Method for Inferring the Cluster Centers

As mentioned before, the proposed method uses the Bayesian network for finding the representative point of each cluster. The steps of the method are shown in Figure 1. At first, a specific domain is selected and a *vocabulary* (set of *concepts*) is extracted for it. Then this vocabulary is converted into a Bayesian network that provides the possibility of inference and extraction of the relationships between concepts. A subset of concepts with more general semantics is selected as *categories* which construct the main dimensions for creating page vectors. In the next step, these categories are grouped. Then, each web page of test collection is converted into a vector with dimensions equal to the number of categories and is assigned to an appropriate center. In the following we will explain each step in enough details.

1.	Find the **concepts** of the domain.
2.	Construct the Bayesian Belief Network and complete it by initial conditional probabilities.
3.	Identify the **categories** inside the network.
4.	Construct the vectors of the categories.
5.	Group the categories and consider them as cluster centers.
6.	Construct the vectors of the web documents (by categories).
7.	Assign the vectors of the test collection to the cluster centers.

Fig. 1. Steps of the proposed method for finding representative points

2.1   Constructing the Bayesian Belief Network

For constructing a vocabulary, in this paper, a statistical approach is used. After performing some pre-processing operations such as elimination of stop-words and stemming, remaining terms are ordered according to the *tf**idf measure and then among them, some more important terms are selected as candidates for constructing the *concepts* set.

Then the vocabulary is converted into a Bayesian network. For this we create a Bayesian network [9,14] that its nodes are concepts of the vocabulary. The network is organized in a way that concepts are placed from more specific concepts to more general concepts so that specific concepts become parents of general concepts.

After selection of nodes, we must create some base relations. A very popular algorithm is *K2* [14] that starts with a given ordering of the nodes and greedily adds edges from previously processed nodes to the current one. When there is no further improvement, the next node is selected. The result depends on the initial ordering, so several runnings are needed. With respect to the characteristics particular for web environment, we improve this algorithm for the web clustering. We create a relation if two associated nodes would have a certain degree of confidentiality. Also a relation always is created from more specific concept to more general concept and so the network will have no cycle finally. Using these enhancements there will not be a need for several repetitions. In particular, for each pair of concepts existing in network, the possibility of co-occurrence in different pages is calculated. If the frequency of the co-occurrences satisfies relation (1), then it is considered as a good occurrence. At the end, the relationship of two concepts is calculated by relation (2) and if it is more than a pre-specified threshold, a relation from parent to child (from more specific concept to more general concept) is created.

$$(n_1 \geq .3 n_2 \ Or \ n_2 \geq .3 n_1) \qquad , \qquad n_1, n_2 > 0 \tag{1}$$

$$\frac{\#of \ good \ occurences}{\#of \ all \ docs} \tag{2}$$

After creating initial relations, conditional probability tables are filled. While in reality we want to maximize the *conditional* probability, in most cases there is no closed-form solution for the maximum conditional-likelihood probability estimates [14]. However here we calculate the joint probability, instead of calculating the conditional probability. Thus for each node A, its joint probability with its parents is calculated and then the joint probability of its parents is calculated separately. By dividing the first value by second one, the probability $\frac{P(A, B_1, B_2, ..., B_n)}{P(B_1, B_2, ..., B_n)}$ is obtained. This is done for different values of binary variables B_1, B_2... In general we must obtain two classes of probabilities as shown in (3). After calculating the first probability, we can obtain the second probability by $1 - P(A = yes \mid B_1, B_2, ..., B_n)$. We say that a web page contains a concept if relation (4) holds.

$$\begin{aligned} &(1) \quad P(A = yes \mid B_1, B_2, ..., B_n) \\ &(2) \quad P(A = no \mid B_1, B_2, ..., B_n) \end{aligned} \tag{3}$$

Webpage contains the concept AND $W_{tf/idf}(concept, web \ page) >= .3 W_{max}(web \ page)$ (4)

After completing the Bayesian network, more general concepts are selected as *category* set as follows:

1. *Search the Bayesian network.*
2. *Select the nodes without any child.*
3. *Select the nodes that have more direct or indirect parents.*
4. *If necessary, limit the selection to a specific number (25 nodes).*

2.2 Grouping the Categories and Finding Cluster Centers

For each category, a vector with dimensions equal to the number of the concepts is created. Then the relationship of each category with each concept is calculated and this vector is fulfilled. So at the end, for each category we will have a vector. But how the relationship of each category with each concept is calculated? For this, for all pairs of *{"concept", "category"}* we obtain the directed paths from *"concept"* to *"category"* and eliminate all nodes that are not in the path. Then we set the existence probability of the root node (node containing *"concept"*) to 1 and obtain the belief of the last node (*"category"*). This process is repeated for all categories and is elaborated in algorithm of Figure 2.

For grouping, we can use several methods. Since the number of categories is limited, thus the main factor is the quality of the groups and the time complexity is not important. Therefore, we follow a process that is similar to hierarchical *average linkage* clustering [15] but produce the results in flat format. The method repeatedly combines smaller groups and generates bigger ones. At each iteration, the algorithm calculates the average distance of every two groups by averaging from sum of all pairwise distances inside the two groups and combines two groups which have the minimum average distance. This algorithm has the time complexity $O(n^3)$ but can create groups with high quality.

In the next step, we should assign the web pages to suitable groups (cluster centers). For comparing and finding similarity, we must have a representative vector for each cluster center. So for each cluster center, we construct a vector with dimensions equal to the number of categories. In this vector for categories existing in the cluster a constant value such as 20 is set and other values are set to 0.

1.	**for each** *category* in *network* **do:**
2.	**for each** *concept* in *network* **do:**
2.1.	Find all *paths* between *concept* and *category*;
2.2.	Eliminate the nodes that do not exist in *path*;
2.3.	Set the belief of the *path*'s root node (*concept*) to 1;
2.4.	Infer the belief of the *path*'s last node (*category*);

Fig. 2. Inferring the relationship of each category with the concepts of the network

2.3 Assigning the Web Pages to the Cluster Centers

Now we have the suitable cluster centers and only it remains to assign each web page to the nearest center. Before assignment, we must construct a vector for each web page so that we can calculate the similarities between the web pages and the cluster centers. For constructing this vector, in each web page, after elimination of stopwords and stemming, a vector with dimensions equal to the number of all concepts is created which each dimension shows the *tf/idf* weight of the associated concept. If there is a term in web page but there is not in the network, its importance is considered to be low and is neglected. Then, each vector must be reduced to a vector with dimensions equal to the number of categories. This is done as follows: for each concept, the belief of each category with having complete belief to the concept is inferred. As shown in Figure 2, the effect of the other nodes on the inference is removed and the result vector is updated by incremental adding for all concepts until it is completed.

Then, we should assign the vectors to the cluster centers. For assigning we use cosine similarity and define a threshold; if the similarity of a web page with a cluster center would be more than this threshold, the web page joins to that cluster.

2.4 H-BayesClust: Improving Toward Hierarchy

So far, we have proposed the *BayesClust* and now we extend the method for hierarchical clustering. The improvement involves following two steps:

1. *Constructing the cluster centers in a hierarchical manner.*
2. *Assigning each web page in test set to a suitable cluster in appropriate level.*

To do step 1, in grouping the categories, we define a threshold and if the similarity between two nearest clusters would be more than the threshold, we join them as a single cluster; otherwise they are joined in a higher level. In next stage, we must assign each web page of test collection to the cluster centers. For this, we define a measure and optimize it during the assignment process. This measure can be the cosine similarity. So, in the root node the cosine similarity between the root and the web page and as well as the cosine similarity between each child and the web page is calculated. Then if the root node would have the highest similarity, the web page is assigned to it; otherwise the web page is moved toward the child with highest similarity. The process is repeated in the child node. In this approach, each web page at most traverses a path with length equal to the depth of the hierarchy and in each level it is compared with the current node and its children. So the complexity of the assigning phase will be $O(dc)$, which d is the depth of the hierarchy and c is the average number of children of a hierarchy node. The total complexity will be $O(dcN)$ which is linearly dependent to the size of data.

3   Experimental Results

3.1 Examination of Proposed Method

In this paper we restrict the clustering to the web pages related to the Politics area. For constructing the vocabulary, 500 documents are selected as training set. After extracting the main concepts; they are converted to Bayesian network (with 124 nodes). For creating initial relations we obtain the probability of co-occurrence and if its value would be more than .65, a relation from the specific concept to the general concept is created. Then, the category set is created which is depicted in Table 1.

Table 1. Obtained categories from the Bayesian network

ADMINISTRATION	EAST	MUSLIM	ELECTION	THREAT
GOVERNMENT	OIL	POLITIC	PRESIDENT	PEOPLE
REVOLUTION	WAR	LEADER	RELIGIOUS	PARTY
DEMOCRACY	WEST	TERROR	ECONOMIC	COURT
PARLIAMENT	WIN	ENERGY	COUNTRY	WORLD

In the grouping step, for easier controlling, a variable is defined that shows the number of combinations of the base groups. The results of grouping the categories are depicted in Figure 3. According to the figure we can control the granularity of the clustering by choosing different repetition numbers.

| ☐ ADMINISTRATION
☐ COUNTRY + GOVERNMENT + POLITIC
☐ COURT
☐ DEMOCRACY
☐ EAST
☐ ECONOMIC
☐ ELECTION + PARLIAMENT
☐ ENERGY + OIL
☐ LEADER + REVOLUTION
☐ MUSLIM + RELIGIOUS
☐ PARTY
☐ PEOPLE
☐ PRESIDENT
☐ TERROR + THREAT + WAR
☐ WEST
☐ WIN
☐ WORLD

8 times repetition | ☐ ADMINISTRATION
☐ COUNTRY + GOVERNMENT + POLITIC
☐ COURT
☐ DEMOCRACY + ELECTION + PARLIAMENT
☐ EAST
☐ ECONOMIC + ENERGY + OIL
☐ LEADER + REVOLUTION
☐ MUSLIM + RELIGIOUS
☐ PARTY
☐ PEOPLE
☐ PRESIDENT
☐ TERROR + THREAT + WAR
☐ WEST
☐ WIN
☐ WORLD

10 times repetition | ☐ ADMINISTRATION + LEADER + REVOLUTION
☐ COUNTRY + GOVERNMENT + POLITIC + PRESIDENT + WORLD
☐ COURT
☐ DEMOCRACY + ELECTION + PARLIAMENT + PEOPLE
☐ EAST
☐ ECONOMIC + ENERGY + OIL
☐ MUSLIM + RELIGIOUS
☐ PARTY
☐ TERROR + THREAT + WAR
☐ WEST
☐ WIN

14 times repetition |

Fig. 3. The results of clustering the categories for different repetitions

In the test phase, we select 300 web pages from Politics domain by random. Then, for each web page, a 25-dimensional vector is created. Also for each cluster center a vector is created with the associated dimensions (to the cluster categories) get a constant value such as 20 and others becomes 0. Then the cosine similarity between each page vector and each cluster center vector is calculated and if the result would be more than *0.6*, it is added to the cluster members.

3.2 Evaluation of BayesClust Results

For evaluation of clustering results, we use the well known *F-measure* method [7]. If *P* and *R* show *Precision* and *Recall* respectively, this measure is defined by (8) and precision and recall are obtained by (9). n_j shows the size of cluster j, g_i shows the size of class i and $N(i,j)$ shows the number of pages of class i in cluster j.

$$F(i, j) = \frac{2(P(i, j) * R(i, j))}{(P(i, j) + R(i, j))}, \quad F = \sum_i \frac{g_i}{n} \max_j \{F(i, j)\} \tag{8}$$

$$P(i, j) = N(i, j) / n_j, \quad R(i, j) = N(i, j) / g_i \tag{9}$$

Table 2. Evaluation of clustering with 10 times repetitions

CLUSTER	F	R	P	N
ADMINISTRATION	0.550724638	.38	1	2
COUNTRY+GOVERNMENT+POLITIC	0.90989011	.92	.9	71
COURT	0.867126437	.82	.92	15
DEMOCRACY+ELECTION+PARLIAMENT	0.696503497	.6	.83	6
EAST	0.71630137	.63	.83	6
ECONOMIC+ENERGY+OIL	0.845764706	.79	.91	15
LEADER+REVOLUTION	0.717368421	.58	.94	4
MUSLIM+RELIGIOUS	0.760774194	.67	.88	8
PARTY	0.814969325	.81	.82	9
PEOPLE	0.648648649	.48	1	3
PRESIDENT	0.815421687	.72	.94	5
TERROR+THREAT+WAR	0.791194969	.74	.85	10
WEST	0.864971098	.87	.86	9
WIN	0.768152866	.67	.9	5
WORLD	0.919891304	.93	.91	12

Table 3. Evaluation of clustering with 14 times repetitions

CLUSTER	F	R	P	N
ADMINISTRATION+LEADER+REVOLUTION	0.7012987	.54	1	3
COUNTRY+GOVERNMENT+POLITIC+PRESIDENT+WORLD	0.9130434	1	.84	107
COURT	0.8671264	.82	.92	15
DEMOCRACY+ELECTION+PARLIAMENT+PEOPLE	0.793875	.73	.87	13
EAST	0.7163013	.63	.83	6
ECONOMIC+ENERGY+OIL	0.8457647	.79	.91	15
MUSLIM+RELIGIOUS	0.7607741	.67	.88	8
PARTY	0.8149693	.81	.82	9
TERROR+THREAT+WAR	0.7911949	.74	.85	10
WEST	0.8649710	.87	.86	9
WIN	0.7681528	.67	.9	5

Table 4. Evaluation of *K-Means* with 10 centers

CENTER ID	N	P	R	F
1	27	.65	.59	0.568421
2	11	.49	.53	0.484536
3	21	.71	.67	0.63496
4	26	.57	.71	0.579833
5	15	.48	.56	0.401590
6	12	.73	.71	0.719861
7	76	.68	.72	0.601652
8	19	.63	.71	0.600325
9	55	.59	.67	0.490909
10	38	.66	.74	0.617419

By calculating the value of total F-measure for clustering with 10 and 14 times repetitions the values 0.848685769 and 0.86534735 are obtained, respectively. Theses values and their associated precision and recall values shows the high quality of the proposed clustering method. For more comparison, we have implemented the *K-Means* and have applied it on the same data set. The results are shown in Table 4. The value of F-measure is 0.6549259 that has a considerable difference with the proposed method.

3.3 Examination of H-BayesClust

In this section we examine *H-BayeClust* and evaluate its efficiency. For this, we define the threshold as 0.65 and construct the hierarchy shown in Table 5. After constructing the hierarchy, we choose another data set containing 300 web pages of Politics area and assign each web page to the associated cluster center. Evaluation methods such as F-measure only are developed for evaluation of flat clusters (or lowest level of hierarchical clusters). So, we first apply F-measure on clusters without any child and then obtain the F-measure value of parent clusters from its children. Precision of a parent cluster can be calculated from the precision of its children using (10). P_R shows the precision of cluster members that do not belong to any other clusters. About recall a more simple way is calculating it in a straightforward manner without considering pre-calculated values for sub-clusters. Using this process the results for H-BayesClust are as shown in Table 5. The total F-measure for this case is *0.802054* that is acceptable for hierarchical clustering.

$$P_C = \sum_{\forall k \in C.children} \frac{n_k}{n} P_k + \frac{n - \sum_{\forall k \in C.children} n_k}{n} P_R \qquad (10)$$

Table 5. Results of assigning web pages to hierarchical cluster centers

ID	CENTER LABLES	N	P	R	F
1	WEST+COURT+CLUSTER2	293	.75	.84	0.792452
2	CLUSTER3+ CLUSTER4	274	.75	.82	0.783439
3	ADMINISTRATION+REVOLUTION+LEADER	18	.82	.83	0.824969
4	PARTY + CLUSTER5	251	.76	.81	0.784203
5	CLUSTER5+CLUSTER7	232	.77	.82	0.794213
6	CLUSTER8+CLUSTER9	107	.88	.86	0.869885
7	CLUSTER10+CLUSTER11	115	.75	.79	0.769480
8	EAST+CLUSTER12	39	.91	.90	0.904972
9	CLUSTER13+CLUSTER14	54	.86	.80	0.828915
10	WIN+CLUSTER15	46	.76	.79	0.774709
11	WORLD+CLUSTER16	57	.83	.81	0.819878
12	ECONOMIC + ENERGY + OIL	25	.93	.91	0.919891
13	MUSLIM + RELIGIOUS	19	.88	.83	0.854269
14	THERROR+THREAT+WAR	26	.86	.76	0.806913
15	PEOPLE+DEMOCRACY+ELECTION+PARLIAMENT	33	.79	.80	0.794968
16	PRESIDENT+GOVERNMENT+PLITIC+COUNTRY	45	.85	.81	0.829518

4 Conclusion

In this paper a new method is proposed for clustering the web pages which extracts the representatives by a knowledge extraction process. The main idea of the proposed method is that it constructs a Bayesian network for a domain and using this network and providing inference ability from it, infers the relations between concepts and categories of the domain. For learning the Bayesian network a method is proposed that is based on identifying initial relations with confidentiality avoiding the overfitting problem. Then the limited categories are grouped while providing a good opportunity for identification of cluster topics. The algorithm for grouping the categories is based on selecting and joining two groups with minimum average distance which can be done whether in flat or hierarchical mode. The experimental results of the both flat and hierarchical clustering methods show the improvements in the quality of resultant clusters.

References

1. Bloehdorn, S., Hotho, A.: Text classification by boosting weak learners based on terms and concepts. In: Fourth IEEE International Conference on Data Mining (2004)
2. Getoor, L.: Link Mining: A New Data Mining Challenge. ACM SIGKDD Explorations Newsletter 5(1), 84–89 (2003)
3. Grira, N., Crucianu, M., Boujemaa, N.: Unsupervised and Semi-supervised Clustering: a Brief Survey. In: ACM SIGMM workshop on Multimedia information retrieval, pp. 9–16 (2005)

4. Gruber, T.R: Towards Principles for the Design of Ontologies Used for Knowledge Sharing. In: Formal Ontology in Conceptual Analysis and Knowledge Representation, Netherlands (1993)
5. Hotho, A., Staab, S., Stumme, G.: Explaining text clustering results using semantic structures. In: Lavrač, N., Gamberger, D., Todorovski, L., Blockeel, H. (eds.) PKDD 2003. LNCS (LNAI), vol. 2838, pp. 217–228. Springer, Heidelberg (2003)
6. Koller, D., Sahami, M.: Hierarchically classifying documents using very few words. In: 14th International Conference on Machine Learning (ML), Tennessee, pp. 170–178 (1997)
7. Larsen, B., Aone, C.: Fast and effective text mining using linear-time document clustering. In: Proceedings of SIGKDD'99, CA, pp. 16–22 (1999)
8. McQueen, J.: Some methods for classification and analysis of multivariate observations. In: Fifth Berkeley Symposium on Mathematical Statistics and Probability, pp. 281–297 (1967)
9. Mitchell, T.M.: Machine Learning, ch. 6. McGraw-Hill, New York (1995)
10. Sebastiani, F.: Machine learning in automated text categorization. ACM Computing Surveys 34(1), 1–47 (2002)
11. Stumme, G., Hotho, A., Berendt, B.: Semantic Web Mining State of the art and future directions. Journal of Web Semantics: Science, Services and Agents on the World Wide Web 4(2), 124–143 (2006)
12. Flach, P.A., De Raedt, L. (eds.): ECML 2001/PKDD'01. LNCS (LNAI), vol. 2167/2168. Springer, Heidelberg (2001)
13. Wang, Y., Kitsuregawa, M.: Link Based Clustering of Web Search Results. In: Wang, X.S., Yu, G., Lu, H. (eds.) WAIM 2001. LNCS, vol. 2118, pp. 225–236. Springer, Heidelberg (2001)
14. Witten, I.H., Frank, E.: Data Mining, Practical Machine Learning Tools and Techniques, 2nd edn. ch. 6. Morgan Kaufmann, San Francisco, CA (2000)
15. Xu, R., Wunsch, D.: Survey of Clustering Algorithms. IEEE Trans. On Neural Networks 16(3) (2005)

An Improved AdaBoost Algorithm Based on Adaptive Weight Adjusting

Lili Cheng, Jianpei Zhang, Jing Yang, and Jun Ma

College of Computer Science and Technology, Harbin Engineering University, Harbin,
150001, China
{chenglili, zhangjianpei, yangjing, majun}@hrbeu.edu.cn

Abstract. The base classifier, which is trained by AdaBoost ensemble learning algorithm, has a constant weight for all test instances. From the view of iterative process of AdaBoost, every base classifier has good classification performance in a certain small area of input space, so the constant weight for different test samples is unreasonable. An improved AdaBoost algorithm based on adaptive weight adjusting is presented. The classifiers' selection and their weights are determined by full information behavior correlation which describes the correlation between test sample and base classifier. The method makes use of all scalars of base classifier's full information behavior, overcomes the problem of information losing. The results of simulated experiments show that the ensemble classification performance is improved greatly.

Keywords: Multiple classifiers combination, dynamic combination technology, full information behavior correlation.

1 Introduction

Multiple classifiers combination, which was proposed by Suen in 1990, has proven itself a powerful technology to overcome the limitations of individual classifier. It has been applied in many fields of pattern recognition, such as character recognition, handwritten and text recognition, etc[1],[2],[3]. It can obtain much better classification performance by effectively combine multiple classifiers which possess different performance identities.

Among many Boosting algorithms, the AdaBoost (Adaptive Boosting) which was proposed by Freuch and Schapire is most remarkable[4],[5].The base classifiers produced later depend on pre-produced classifiers during training phase, the samples that are mistakenly classified by pre-produced classifiers will be chosen into new training data set with high probability to train subsequent base classifiers. The voting is adopted to combine multiple classifiers, the weights of base classifiers are constants for all test instances depending on the classifiers' performance on their own training data [6]. From the training process of AdaBoost we can draw a conclusion that every base classifier mainly focuses on its own training data. The performance of base classifiers differ from each other in different areas which test samples belong to, so it

R. Alhajj et al. (Eds.): ADMA 2007, LNAI 4632, pp. 625–632, 2007.

is unreasonable that the constant weights of base classifiers for different test samples which possibly belong to different areas of input space. The classifiers' weights should change according to the area difference of test samples [7],[8].

A multiple classifiers combination algorithm based on adaptive weight adjusting is proposed. The definition of full information behavior correlation is given to describe the correlation between test sample and every base classifier. Classifiers' selection and their weights are determined according to base classifiers' local accuracy to combine every base classifier's result as the final decision.

The paper is organized as follows. In section 2, the drawback of AdaBoost is analyzed, weighted multiple classifiers combination based on full information behavior correlation is given in section 3. In section 4, we perform comparative experiments to demonstrate its effectiveness on UCI data set, some conclusions are outlined in section 5.

2   The Analysis of AdaBoost

About   AdaBoost,   the   weighted   classification   error   of   classifier

h_t is $\varepsilon_t = \sum_{i=1}^{m_k} D_t(i)[|h_t(x_i) \neq y_i|]$, $D_t(i)$ is the distributing of ith sample of tth time

iterative process, m_k is the number of samples, $D_{t+1}(i)$ is obtained from $D_t(i)$ which means increasing the weights of miss classified samples whereas decreasing the weights of correct classified samples. A conclusion can be drawn from the classification error that the produced single classifier mainly focuses on the samples that difficult to classify while ignoring other samples along with the iterative training process. The base classifiers have better classification performance in some small areas, not whole training space during iterative training process. The disadvantage of constant weights for different test samples is obvious.

To make use of base classifiers' difference in input space, the weights of base classifiers should be determined according to test sample. The full information behavior correlation is proposed to describe the correlation between test sample and base classifier, the estimated classification error of base classifier is obtained according to the correlation. The bigger of the error is, the smaller the weight will be, and vice versa.

3   Weighted Multiple Classifiers Combination

3.1   Efficient Neighborhood Set of Test Sample

Efficient neighborhood set of test sample includes the training data that close to test sample. There are some samples that the classification results given by base classifiers differ from test sample's class, so it has great probability that these samples belong to different classes with test sample. Such samples influence the classifiers' selection and combined weights of base classifiers and should be deleted to form efficient neighborhood set of test sample.

Suppose ω_j $(j=1,2,\cdots,M)$ is a classification problem having M classes, $H = \{h_i, i=1,2,\cdots,L\}$ is L classifiers, the output of classifier h_i to x is $h_i(x) = (c_{i1}, c_{i2}, \cdots, c_{iM})$ $0 \le c_{ij} \le 1$, c_{ij} indicates the probability that x belongs to ω_j given by classifier h_i.

Definition 1. $h_i(x) = (c_{i1}, c_{i2}, \cdots, c_{iM})$ is full information behavior of classifier h_i to x, $0 \le c_{ij} \le 1$, $i=1,2,\cdots,L$, $j=1,2,\cdots,M$.

The full information behavior is given to make full use of the output vector of base classifier, all scalars of sample x'output vector are arranged from big to small, the first scalar correspond to the first candidate class, the rest may be deduced by analogy.

Definition 2. $FIBM(x) = [h_1(x)', h_2(x)', \cdots, h_L(x)'] = \begin{bmatrix} c_{11} & \cdots & c_{1M} \\ \cdots & c_{ij} & \cdots \\ c_{L1} & \cdots & c_{LM} \end{bmatrix}$ is full

information behavior matrix of $H = \{h_i, i=1, 2, \cdots, L\}$ to x.

Some conclusions can be drawn from reference [9]: there exists certain relativity between the candidate class and the real class. High relativity between low ranked candidate classes(such as the first rank, the second rank) of classifier h_i to x and x's real class, on the contrary, low relativity between high ranked candidate classes of classifier h_i to x and x's real class .

Definition 3. The relativity between the tth scalar of full information behavior given by classifier h_i to training sample x_j, $j=1,2,...,n$ and test sample x is:

$$r_t = 1 - \left| \frac{c_{it}}{\sum\limits_{s=1}^{M} c_{is}} - \frac{c_{it}^j}{\sum\limits_{s=1}^{M} c_{is}^j} \right| \bigg/ \left| \frac{c_{it}}{\sum\limits_{s=1}^{M} c_{is}} + \frac{c_{it}^j}{\sum\limits_{s=1}^{M} c_{is}^j} \right| \tag{1}$$

c_{is}^j indicates the sth scalar of full information behavior given by h_i to x_j .If it correspond to different candidate class, then the relativity between them is nonsense, the relativity equal to 0, if it correspond to same candidate class and equal to each other, then the relativity is 1, so the relativity locates in [0,1]. The result of difference divided by sum demonstrates relative otherness, not absolute otherness [9].

Definition 4. Full information behavior correlation between training sample x_j and test sample x given by classifier h_i is:

$$R_i(x, x_j) = \sum_{t=1}^{M} \eta_t r_t \qquad (2)$$

η_t is support factor of the tth ranked candidate class and there are many forms, such as $\eta_t = e^{-\alpha(t-\delta)}$ or $\eta_t = 1.0 - \beta \times t$, α, δ and β are positive constants.

Definition 5. Full Information behavior correlation matrix (shortly FIBCM) of test sample x is:

$$FIBCM(x) = \begin{bmatrix} R_1(x, x_1), \cdots, R_i(x, x_1), \cdots, R_L(x, x_1) \\ \cdots \\ R_1(x, x_j), \cdots, R_i(x, x_j), \cdots, R_L(x, x_j) \\ \cdots \\ R_1(x, x_n), \cdots, R_i(x, x_n), \cdots, R_L(x, x_n) \end{bmatrix} \qquad (3)$$

Full information behavior correlation makes full use of the full information behavior of training samples, avoids losing classification information which produced by only using the maximum probability of the full information behavior. For example, supposing the outputs given by classifier h_i and h_j to x are [0.5,0.4,0.1] and [0.7,0.2,0.1], it is obvious that the latter reliability is higher than the former although these two classifiers give the same final classification result. If only the maximum probability is chosen to be the final classification result to take part in combination, the information of second class probability, third class probability will be lost, it will influence the final combination result.

According to the full information behavior correlation between test sample and training sample, the efficient neighborhood set of test sample can be obtained. We can draw a conclusion from formula(1),(2) and (3) that the correlation between the test sample x and training sample locates in (0,1), which means the smaller the correlation is, the bigger the divergence between base classifiers will be. Efficient neighborhood set of test sample includes the sample whose correlation with test sample is smaller than the threshold.

$$N(x) = \{x_j \mid \sum_{i=1}^{L} R_i(x, x_j) \Big/ L > threshold_1\} \qquad (4)$$

3.2 Adaptive Weight

Suppose classifier set $H = \{h_i, i = 1, 2, \cdots, L\}$, $X = \{(x_1, y_1), (x_2, y_2), \cdots, (x_n, y_n)\}$ is training sample set, x is test sample, local classification error vector is $e(x_j) = (e_1(x_j), e_2(x_j), \cdots, e_L(x_j))$.

Definition 6 Local estimated classification error of classifier h_i to x is:

$$E_i(x) = \frac{1}{|N(x)|} \sum_{j=1}^{k} \frac{e_i(x_j)}{R_i(x, x_j)} \tag{5}$$

$x_j \in N(x), k = |N(x)|$ is the number of training samples belonging to x's efficient neighborhood set, E_i describes the local classification error of test sample given by classifier h_i according to the correlation between x and x_j and the classification error of h_i to x_j, the weight is determined according to E_i:

$$\omega_i = \begin{cases} 0 & if \ E_i > threshold_2 \\ 1 - E_i & if \ E_i < threshold_2 \end{cases} \tag{6}$$

The classification effect of every base classifier to test sample is different according to its different characters, so the classifiers' selection and their weights can be determined according to local estimated classification error. The smaller of the classification error is, the bigger the weight will be. If the classification error is bigger than the threshold, the weight is 0, namely the classifier will not be chosen into combination system.

3.3 Multiple Classifiers Combination Based on Adaptive Weight Adjusting

The combination algorithm can be described as follows :

Step 1: Train and produce base classifiers according to AdaBoost;
Step 2: Input test sample x and compute full information behavior correlation, obtain efficient neighborhood set of x according to formula (4);
Step 3: Compute the local estimated classification error of x given by every base classifier according to formula (5);
Step4: obtain adaptive weights of base classifiers to x according to formula (6);
Step5: Obtain the combined output by weighted majority voting.

The proposed algorithm makes full use of the classifier's full information behavior correlation between test sample and training data to describe the performance difference of base classifiers in local area. Local estimated classification error is used to determine adaptive weight of base classifier to form dynamic combination scheme.

4 Simulated Experiments

4.1 Classifiers and Data Sets

In experiments, BP Neural network、 Support Vector Machine、 Bayes classifier are chosen to be learning algorithms of base classifiers. To validate the proposed method (named FIBC* here), 6 data sets of UCI machine learning test data sets are selected. To every data set, 2/3 samples as training data set, 1/3 as test data set. The data sets are shown in table 1:

Table 1. UCI data set

Name	Cases	Classes	Features	
			Numeric	Nominal
Anneal	898	6	9	29
Autos	205	6	15	10
Credit-a	690	2	6	9
Heart-c	303	2	8	5
Hepatitis	155	2	6	13
Colic	368	2	10	12

4.2 Analysis of Experiment Results

Table 2 shows the precision comparison among single classifier, from it we conclude that the classification performance is different for different algorithms to the same learning problem but no one gains satisfying performance to all problems. Multiple classifiers combination system is imperative under the situation.

Table 2. Precision comparison among single classifier

Data set	BPNN	SVM	Bayes
Anneal	0.832	0.869	0.845
Autos	0.848	0.871	0.857
Credit-a	0.852	0.882	0.849
Heart-c	0.829	0.869	0.831
Hepatitis	0.841	0.881	0.837
Colic	0.835	0.865	0.839

To multiple classifiers combination, the threshold of correlation is 0.5 to gain the efficient neighborhood of test sample, the threshold of classification error is 0.05. Table 3 shows the precision comparison among multiple classifiers combination, the former 3 columns adopt fix weighted multiple classifiers combination trained by AdaBoost, shortly FIX*, the latter 3 columns adopt adaptive weighted multiple classifiers combination, shortly FIBC*. From it we conclude that multiple classifiers combination with adaptive weights wholly improves performance.

Table 3. Precision comparison among multiple classifiers combination

Data set	FIX BPNN	FIX SVM	FIX Bayes	FIBC BPNN	FIBC SVM	FIBC Bayes
Anneal	0.829	0.843	0.839	0.882	0.881	0.885
Autos	0.844	0.856	0.852	0.880	0.892	0.887
Credit-a	0.861	0.875	0.853	0.880	0.894	0.881
Heart-c	0.827	0.854	0.825	0.881	0.885	0.892
Hepatitis	0.835	0.868	0.845	0.886	0.887	0.884
Colic	0.843	0.872	0.848	0.887	0.894	0.889

5 Conclusions

In pattern recognition, the multiple classifiers combination is a kind of development trend to improve the classification performance. The key point is to find a suitable combination rule to gain the optimal combination result.

The multiple classifiers combination algorithm based on full information behavior correlation makes full use of support factors of all ranked output vector to the real class, the adaptive weights are computed according to local classification accuracy of test sample to guide the combination. The experiments on UCI data sets show that weighted multiple classifiers combination based on full information behavior correlation improves classification performance greatly.

Acknowledgements. This paper is sponsored by the National Natural Science Foundation of China(General program) under Grant No. 60673131 and the Natural Science Foundation of Heilongjiang Province under Grant No. F2005-02 .

References

1. Lu, X., Wang, Y., Jain, A.K: Combining Classifiers for Face Recognition. In: IEEE International Conference on Multimedia &Expo. pp. 13–16 (2003)
2. Dietrich, C., Schwenker, F., Palm, G.: Classification of time series utilizing temporal and decision fusion. In: Proceedings of Multiple Classifier Systems (MCS), pp. 378–387. Cambridge University Press, Cambridge (2001)
3. Xu, L., Krzyzak, A., Suen, C Y.: Method for Combing Multiple Classifiers and Their Applications to Handwriting Recognition. IEEE Transactions on Systems, Man, and Cybernetics 22, 418–435 (1992)
4. Freund, Y., Schapire, R.E.: Yoav Freund and Robert E Schapire: A decision-theoretic generalization of on-line learning to boosting. Journal of Computer and System Science 55, 119–139 (1997)
5. Freund, Y., Schapire, R.E.: Experiments with a new boosting algorithm. In: Machine Learning: Proceedings of the Thirteenth International Conference, pp. 148–156 (1996)
6. Puuronen, S., Terziyan, V., Tsymbal, A.: A dynamic integration algorithm for an ensemble of classifiers. In: Raś, Z.W., Skowron, A. (eds.) ISMIS 1999. LNCS, vol. 1609, Springer, Heidelberg (1999)

7. Tang, C.-S., Jin, Y.-H.: A Multiple Classifiers Integration Based on Adaptive Weight Adjusting and it's Application on text Classification. Computer Science 20, 82–84 (2003)
8. Tang, C.-S., Jin, Y.-H.: A Multiple Classifiers Integration Based on Full Information Matrix. Journal of Software 14, 1103–1109 (2003)
9. Jing, X., Ynag, J.: Combining Classifiers Based on Analysis of Correlation and Effective Supplement. Acta Automatica Sinica 26, 741–747 (2000)

Author Index

Lecture Notes in Artificial Intelligence (LNAI)

Vol. 4384: T. Washio, K. Satoh, H. Takeda, A. Inokuchi (Eds.), New Frontiers in Artificial Intelligence. IX, 401 pages. 2007.

Vol. 4371: K. Inoue, K. Satoh, F. Toni (Eds.), Computational Logic in Multi-Agent Systems. X, 315 pages. 2007.

Vol. 4369: M. Umeda, A. Wolf, O. Bartenstein, U. Geske, D. Seipel, O. Takata (Eds.), Declarative Programming for Knowledge Management. X, 229 pages. 2006.

Vol. 4342: H. de Swart, E. Orłowska, G. Schmidt, M. Roubens (Eds.), Theory and Applications of Relational Structures as Knowledge Instruments II. X, 373 pages. 2006.

Vol. 4335: S.A. Brueckner, S. Hassas, M. Jelasity, D. Yamins (Eds.), Engineering Self-Organising Systems. XII, 212 pages. 2007.

Vol. 4334: B. Beckert, R. Hähnle, P.H. Schmitt (Eds.), Verification of Object-Oriented Software. XXIX, 658 pages. 2007.

Vol. 4333: U. Reimer, D. Karagiannis (Eds.), Practical Aspects of Knowledge Management. XII, 338 pages. 2006.

Vol. 4327: M. Baldoni, U. Endriss (Eds.), Declarative Agent Languages and Technologies IV. VIII, 257 pages. 2006.

Vol. 4314: C. Freksa, M. Kohlhase, K. Schill (Eds.), KI 2006: Advances in Artificial Intelligence. XII, 458 pages. 2007.

Vol. 4304: A. Sattar, B.-h. Kang (Eds.), AI 2006: Advances in Artificial Intelligence. XXVII, 1303 pages. 2006.

Vol. 4303: A. Hoffmann, B.-h. Kang, D. Richards, S. Tsumoto (Eds.), Advances in Knowledge Acquisition and Management. XI, 259 pages. 2006.

Vol. 4293: A. Gelbukh, C.A. Reyes-Garcia (Eds.), MICAI 2006: Advances in Artificial Intelligence. XXVIII, 1232 pages. 2006.

Vol. 4289: M. Ackermann, B. Berendt, M. Grobelnik, A. Hotho, D. Mladenič, G. Semeraro, M. Spiliopoulou, G. Stumme, V. Svátek, M. van Someren (Eds.), Semantics, Web and Mining. X, 197 pages. 2006.

Vol. 4285: Y. Matsumoto, R.W. Sproat, K.-F. Wong, M. Zhang (Eds.), Computer Processing of Oriental Languages. XVII, 544 pages. 2006.

Vol. 4274: Q. Huo, B. Ma, E.-S. Chng, H. Li (Eds.), Chinese Spoken Language Processing. XXIV, 805 pages. 2006.

Vol. 4265: L. Todorovski, N. Lavrač, K.P. Jantke (Eds.), Discovery Science. XIV, 384 pages. 2006.

Vol. 4264: J.L. Balcázar, P.M. Long, F. Stephan (Eds.), Algorithmic Learning Theory. XIII, 393 pages. 2006.

Vol. 4259: S. Greco, Y. Hata, S. Hirano, M. Inuiguchi, S. Miyamoto, H.S. Nguyen, R. Słowiński (Eds.), Rough Sets and Current Trends in Computing. XXII, 951 pages. 2006.

Vol. 4253: B. Gabrys, R.J. Howlett, L.C. Jain (Eds.), Knowledge-Based Intelligent Information and Engineering Systems, Part III. XXXII, 1301 pages. 2006.

Vol. 4252: B. Gabrys, R.J. Howlett, L.C. Jain (Eds.), Knowledge-Based Intelligent Information and Engineering Systems, Part II. XXXIII, 1335 pages. 2006.

Vol. 4251: B. Gabrys, R.J. Howlett, L.C. Jain (Eds.), Knowledge-Based Intelligent Information and Engineering Systems, Part I. LXVI, 1297 pages. 2006.

Vol. 4248: S. Staab, V. Svátek (Eds.), Managing Knowledge in a World of Networks. XIV, 400 pages. 2006.

Vol. 4246: M. Hermann, A. Voronkov (Eds.), Logic for Programming, Artificial Intelligence, and Reasoning. XIII, 588 pages. 2006.

Vol. 4223: L. Wang, L. Jiao, G. Shi, X. Li, J. Liu (Eds.), Fuzzy Systems and Knowledge Discovery. XXVIII, 1335 pages. 2006.

Vol. 4213: J. Fürnkranz, T. Scheffer, M. Spiliopoulou (Eds.), Knowledge Discovery in Databases: PKDD 2006. XXII, 660 pages. 2006.

Vol. 4212: J. Fürnkranz, T. Scheffer, M. Spiliopoulou (Eds.), Machine Learning: ECML 2006. XXIII, 851 pages. 2006.

Vol. 4211: P. Vogt, Y. Sugita, E. Tuci, C.L. Nehaniv (Eds.), Symbol Grounding and Beyond. VIII, 237 pages. 2006.

Vol. 4203: F. Esposito, Z.W. Raś, D. Malerba, G. Semeraro (Eds.), Foundations of Intelligent Systems. XVIII, 767 pages. 2006.

Vol. 4201: Y. Sakakibara, S. Kobayashi, K. Sato, T. Nishino, E. Tomita (Eds.), Grammatical Inference: Algorithms and Applications. XII, 359 pages. 2006.

Vol. 4200: I.F.C. Smith (Ed.), Intelligent Computing in Engineering and Architecture. XIII, 692 pages. 2006.

Vol. 4198: O. Nasraoui, O. Zaïane, M. Spiliopoulou, B. Mobasher, B. Masand, P.S. Yu (Eds.), Advances in Web Mining and Web Usage Analysis. IX, 177 pages. 2006.

Vol. 4196: K. Fischer, I.J. Timm, E. André, N. Zhong (Eds.), Multiagent System Technologies. X, 185 pages. 2006.

Vol. 4188: P. Sojka, I. Kopeček, K. Pala (Eds.), Text, Speech and Dialogue. XV, 721 pages. 2006.

Vol. 4183: J. Euzenat, J. Domingue (Eds.), Artificial Intelligence: Methodology, Systems, and Applications. XIII, 291 pages. 2006.

Vol. 4180: M. Kohlhase, OMDoc – An Open Markup Format for Mathematical Documents [version 1.2]. XIX, 428 pages. 2006.

Vol. 4177: R. Marín, E. Onaindía, A. Bugarín, J. Santos (Eds.), Current Topics in Artificial Intelligence. XV, 482 pages. 2006.

Vol. 4160: M. Fisher, W. van der Hoek, B. Konev, A. Lisitsa (Eds.), Logics in Artificial Intelligence. XII, 516 pages. 2006.

Vol. 4155: O. Stock, M. Schaerf (Eds.), Reasoning, Action and Interaction in AI Theories and Systems. XVIII, 343 pages. 2006.

Vol. 4149: M. Klusch, M. Rovatsos, T.R. Payne (Eds.), Cooperative Information Agents X. XII, 477 pages. 2006.